OXFORD MEDICAL PUBLICATIONS

The Genetics of Neurological Disorders

OXFORD MONOGRAPHS ON MEDICAL GENETICS

General Editors:

ARNO G. MOTULSKY   MARTIN BOBROW
PETER S. HARPER   CHARLES SCRIVER

Former Editors:

J. A. FRASER ROBERTS   C. O. CARTER

OXFORD MONOGRAPHS ON MEDICAL GENETICS · 18

The Genetics of Neurological Disorders

Second Edition

Michael Baraitser

B.Sc. M.B. Ch.B. F.R.C.P.

Consultant Clinical Geneticist, Hospital for Sick Children, Great Ormond Street, London, and Honorary Senior Lecturer in Clinical Genetics, National Hospital for Nervous Diseases, Queen Square, London

OXFORD NEW YORK TOKYO
OXFORD UNIVERSITY PRESS
1990

Oxford University Press, Walton Street, Oxford OX2 6DP

Oxford New York Toronto
Delhi Bombay Calcutta Madras Karachi
Petaling Jaya Singapore Hong Kong Tokyo
Nairobi Dar es Salaam Cape Town
Melbourne Auckland
and associated companies in
Berlin Ibadan

Oxford is a trade mark of Oxford University Press

Published in the United States
by Oxford University Press, New York

First edition 1982
First published in paperback (with revisions) 1985
Second edition 1990

British Library Cataloguing in Publication Data
Baraitser, Michael
The genetics of neurological disorders.
1. Man. Nervous system. Diseases. Genetic aspects
I. Title
616.8, 0442
ISBN 0-19-261814-8
ISBN 0-19-261813-X (pbk)

Library of Congress Cataloging in Publication Data
Baraitser, Michael.
The genetics of neurological disorders/Michael Baraitser.— 2nd ed.
p. cm. — (Oxford monographs on medical genetics; no. 18)
(Oxford medical publications)
Bibliography: p. Includes index.
1. Nervous system—Diseases—Genetic aspects. I. Title.
II. Series. III. Series: Oxford medical publications.
[DNLM: 1. Nervous Systems Diseases—genetics. WL 100 B224g]
RC346. B37 1989
616.8' 0442—dc 19 89-3073 CIP
ISBN 0-19-261814-8
ISBN 0-19-261813-X (pbk)

Typeset by Colset Private Limited, Singapore
Printed and bound in
Great Britain by Courier International
Tiptree, Essex

To Marion, Paula, Lisa, Vanessa, and
Alexandra

I have no faith in anything short of
actual Measurement and the Rule of three
Charles Darwin (*Life and Letters*, Vol. II)

Taking Three as the subject to reason about—
  A convenient number to state—
We add Seven, and Ten, and then multiply out
  By One Thousand Diminished by Eight.

The result we proceed to divide, as you see,
  By Nine Hundred and Ninety and Two:
Then subtract Seventeen, and the answer must be
  Exactly and perfectly true.
    from 'The Hunting of the Snark'
    by Lewis Carroll

Preface to the second edition

Over the past decade, there have been remarkable advances in DNA technology which have changed the practice of clinical neurology. A number of
neurological disorders have now been localized to specific regions on the
chromosomes and these include Huntington's chorea, Becker and Duchenne
muscular dystrophies, neurofibromatosis, tuberous sclerosis, and very
recently Friedreich's ataxia. This has led to prenatal and pre-symptomatic
diagnosis of some of these conditions which has allowed families to consider
prevention by appropriate tests. Each advance has brought with it new ethical
dilemmas and also a need for the neurologist to provide molecular biologists
with an accurate diagnosis so that the appropriate probes can be used. This
edition updates the clinical delineation of genetically determined neurological disorders and incorporates the new information about DNA.

London
January 1989

M.B.

Preface to the first edition

More than a decade has passed since the publication of Pratt's *The genetics of neurological disorders*. His was the first book devoted entirely to this subject and its publication sustained a core of interest in genetics at the National Hospital for Nervous Diseases, London, previously initiated by Eliot Slater, during the period 1946–1964.

Advances have been rapid and by 1980 more than 500 genetically determined conditions involving the nervous system have been described. Many of these disorders are rare, but taken together they contribute significantly to the diseases seen at teaching hospitals, specialized neurological units, and genetic clinics. Despite this, genetic counselling for those with chronic neurological diseases, or for the parents of children with nervous system malformations, is not freely available. Neurologists have successfully countered the accusation that they are more interested in the localization of defects than in treatment, by spectacular advances in therapy, but consideration of the patients' fears about the hereditary nature of disease has not been sufficiently attended to. It is a basic human need to know what went wrong, whether it will happen again, and whether it can be prevented, even if no action treatment is available. Advice on any of these subjects is hampered by the difficulty in identifying rare syndromes and by the different modes of inheritance after a descriptive medical diagnosis has been made.

The main purpose of this book is to help with the identification of these syndromes and to assist the clinician in the derivation of risks of recurrence. An awareness of the needs of the client, the timing of the interview, a perspective concerning the interpretation of risks, carrier detection, prenatal diagnosis, and the process of decision making, are all sensitive issues which, although only dealt with in the chapters on Huntington's chorea and muscle disease, apply throughout the book.

The author is indebted to Professor Cedric Carter for encouragement and help over a number of years, to Drs Anita Harding and Robin Winter for reading parts of the script, and to Mrs Mary Bravery who has patiently doubled as secretary and research assistant.

London
July 1981

M.B.

Contents

1
Cranial nerves

Cranial nerve I

Anosmia

A family in which anosmia was inherited as a dominant trait over three generations is described by Singh *et al.* (1970). In three of the five affected members anosmia was complete. There was no evidence of underlying disease.

Anosmia and hypogonadism (Kallmann syndrome)

At first thought to be limited to males (Kallmann *et al.* 1944), the syndrome has now been described in females (Tagatz *et al.* 1970), and transmission from father to son (Schroffner and Furth 1970; Merriam *et al.* 1977) has been reported. Cryptorchidism, midline craniofacial defects, deafness, and renal abnormalities can occur. Autosomal dominant inheritance with incomplete penetrance and expressivity is postulated (Santen and Paulsen 1973). A significant departure from the 1:1 sex ratio has been observed, presumably caused by sex limitation and not X-linkage. Hockaday (1966) described a man with the full syndrome, whereas his father and brother had anosmia alone. It is probable that Kallmann syndrome is not a single entity. It has been described in association with ichthyosis and epilepsy (Christian *et al.* 1971*b*) and by the same authors in association with Albright's hereditary osteo-dystrophy. Two brothers and their double first cousin have been reported to have Kallmann syndrome, with unilateral renal aplasia (Wegenke *et al.* 1975).

A four generation family in which six males were affected—transmission was through normal females—is reported by Hermanussen and Sippell (1985). Unilateral renal aplasia was seen in one male. The males had no pubertal development and no sense of smell. The condition as a whole is still likely to. be heterogeneous.

Anosmia and Parkinson's disease

Monozygotic twins with this combination are reported by Kissel and Andre (1976). The authors refer to two other families (Yamamoto *et al.* 1966; Constantinidis and De Ajuriaguerra 1970) and suggest that the primary defect is in the dopamine synthesis pathway.

Cranial nerve II

The hereditary optic neuropathies can be classified into three groups:

1. Those affecting only the eyes.
    (a) Dominant.
    (b) Recessive.
    (c) Leber's optic atrophy.
2. Those associated with disease elsewhere in the nervous system.
    (a) Cerebellum.
    (b) Peripheral nerve.
    (c) Basal ganglia.
    (d) Spastic paraplegia.
3. Those associated with craniosynostosis.
4. Others.

1a. Dominant optic atrophy

Large pedigrees have been reported. Brodrick (1974) described a family with affected members over six generations. The onset of symptoms was between 5–7 years and visual acuity ranged from counting fingers at a distance of one foot, to 6/36. Two other pedigrees are reported by Caldwell *et al.* (1971) and Werner and Benedikt (1971). Incomplete penetrance is reported by Shapiro *et al.* (1969). The onset of symptoms varies from early childhood to adulthood and the diagnosis is often made when applicants fail the visual test assessment for a driver's licence. Vision is seldom worse than 6/60. The peripheral fields are usually full, but centrocaecal scotomas are frequently found.

Genetic counselling is usually straightforward where a two-generation family is encountered in which the onset of symptoms is gradual. It is the reduced penetrance of the gene and its considerable variation in expression that makes counselling difficult. Both parents of a single case should be examined, but where they are found to be normal, risks to offspring of an the affected consultand are difficult to calculate. Ten per cent is a reasonable figure. A good review of the subject is that of Kline and Glaser (1979).

1b. Recessive optic atrophy

There have been no recent reports of this condition and looking back at the pedigrees published by Francois (1974), even those were not unequivocally recessive. Many infants thought to have an early onset optic atrophy have primary retinal disease, i.e. congenital retinal dystrophy with secondary pallor of the disc. Recessive optic atrophy should be diagnosed with caution.

1c. Leber's optic atrophy

Characterized clinically by an acute optic neuropathy affecting one eye and, within hours or days, the other, this is a disease of late childhood and adolescence affecting eight males for every female. After the initial abrupt decline in visual actuity, improvement occurs in some. In the acute phase tortuosity of the vessels around the disc and perifundal oedema are characteristic signs (Smith *et al.* 1973) and colour vision is defective for blue/green.

There is some evidence that Leber's disease is not restricted to the optic nerve. Pyramidal, cerebellar, and clinical evidence of peripheral nerve involvement is reported by de Weerdt and Went (1971). The occurence of mental disturbance (nervousness, restlessness, and irritability) is less easy to evaluate. The diagnostic criteria suggested by Lundsgaard (1944) include the presence of affected individuals in two generations, and examination of the patient in both the acute and chronic stage. This can seldom be achieved and most neurologists will be satisfied with a diagnosis based on an analysis of the pedigree in conjunction with the typical clinical picture.

The cases of Androp (1941), Wilson (1963), and Lees *et al.* (1964) all support the suggestion of involvement of structures other than the optic nerve.

Post-mortem studies (Adams *et al.* 1966; Wilson 1963; Kwittken and Barest 1958) add pathological evidence for more widespread central nervous system involvement. However, it should be noted that in most cases only vision is affected and those families with more widespread involvement might be genetically distinct. Leber's disease has never been shown to be transmitted by affected males to their daughters or to their daughters' offspring. The mode of inheritance in Leber's optic atrophy is still uncertain. Against simple X-linkage is the observation that 12 per cent of cases occur in females; this proportion is too large to be explained by either a single X chromosome (Turner's syndrome) or the Lyon hypothesis (the unequal random inactivation of those X-chromosomes that do not carry the mutant gene, leaving a preponderance of X chromosomes that do) in affected females. In addition, the frequency of the disorder in both the male offspring of carrier females and the frequency of their carrier daughters far exceed the expected 50 per cent. A carrier rate approaching 100 per cent has been suggested for the daughters of carrier females (Seedorf 1970). This high rate has been partially confirmed in a large four-generation family in which 86 per cent of daughters of carrier females were themselves carriers (Carroll and Mastaglia 1979).

In the pedigree reported by Wallace (1970) all the male offspring in the female line developed some interference with eyesight. In the Carroll and Mastaglia (1979) study in which more sophisticated methods of examination, including visual-evoked potential were used, 30 per cent of descendants of affected males showed minor changes. Nikoskelainen *et al*. (1982) have now confirmed these findings, especially the male-to-male transmission of a telangiectatic microangiopathy.

A mitchondrial enzyme rhodanese which is expressed in liver tissue has been found to be low but not absent (Cagianut *et al*. 1981), but there have been no recent studies confirming this finding.

Electrocardiological abnormalities have been reported to occur in Leber's disease—specifically Wolff–Parkinson–White (long P–R interval and an initial slow component of the QRS complex) and the Lown–Ganong–Levine syndromes (the PR interval is less than 0.12 seconds) called collectively pre-excitation syndromes (Nikoskelainen *et al*. 1987). There is some evidence from these authors that this might be the only manifestation. Males in this pedigree (Nikoskelainen *et al*. 1987) have transmitted the cardiac conduction defect alone.

Alternative mode of inheritance and counselling

Where pedigree analysis and the clinical presentation suggests Leber's disease, the risk for affected male offspring and carrier daughters of probable carrier females is high. These unusual non-Mendelian segregation ratios have given rise to speculation about cytoplasmic inheritance, as a likely explanation (Wallace 1970; Erickson 1972) for the mode of transmission. A carrier is defined as the mother of two affected males or a female with an affected brother and affected son. Her daughter is likely to be a carrier and will have a greater than 1 in 10 probability of having an affected male child. The female offspring of affected males can on present evidence be reassured. Recently, deletions have been found in mitochondrial DNA. (Wallace *et al*. 1988)

Leber's disease and HLA linkage

A single family with four members (including two females) affected over three generations is reported in which there was an association between the disease and HLA-A$_2$, B$_8$ antigens (Stendahl-Brodin *et al*. 1978).

X-linked optic atrophy with spastic paraparesis

A large pedigree was reported by Bruyn and Went (1964)—18 males and one female were affected over seven generations. The optic atrophy developed between the fifth and tenth year, and clinically was indistinguishable from Leber's optic atrophy. Immediately thereafter, or within a period of 2 years, a picture of spasticity, athetosis, marked dysarthria, and scoliosis developed. The relationship of this disorder to Leber's optic atrophy is uncertain.

Another family similar to the one reported by Bruyn and Went has been described by Novotny *et al.* (1986). Eight subjects with optic atrophy occurring simultaneously in both eyes had a median age of onset of about 17 years. All were left with scotoma. Fourteen had rigidity or dystonia beginning in the lower extremities which then became generalized. Clinically the condition looked like a striatal degeneration. One subject had both eyes and the central nervous system involved. The transmission was always maternal and the onset of neurological symptoms was between 1.5 and 9 years. All had low density areas on CT scan in the region of the putamen and later the caudate became involved. One had the lesions in the thalami.

X-linked optic atrophy

A family with at least eight males affected over three generations had optic atrophy with an early onset—possibly from birth—making Leber's disease unlikely (Went *et al.* 1975). The visual fields were essentially normal. Some members had an ataxic gait, intention tremor, absent ankle jerks, and extensor plantar reflexes.

X-linked optic atrophy and G6PD Worcester deficiency

A pedigree was reported by Snyder *et al.* (1970) in which there appeared to be linkage of an acute optic neuritis followed by optic atrophy with the allele for G6PD Worcester. Obligate female carriers could be detected by ophthalmoscopy.

2a. Optic atrophy and cerebellar ataxia

Behr syndrome—see page 186.
Friedreich's ataxia—see page 196.
Dominant cerebellar atrophy—see page 198.

2b. Optic atrophy with Charcot-Marie-Tooth disease

See page 237.

2c. Optic atrophy in diseases of the basal ganglia

Three brothers developed progressive rigidity starting in the legs and then affecting the arms. Visual symptoms began towards the end of the first decade (Wolpert 1916). A post mortem on one of the brothers (Kalinowsky 1927) who died at the age of 26 years revealed extensive degeneration of globus pallidus and substantia nigra. This is probably not a separate entity (see Hallervorden–Spatz disease—p. 484).

Optic atrophy, congenital cataracts, ataxia, and extrapyramidal features

In a family described by Garcin *et al.* (1961), this combination occurred in two sisters. A third sister had only the ataxia (described as Friedreich's ataxia). One of the affected sisters had two daughters, both of whom had congenital cataract, optic atrophy, and a cerebellar ataxia. The second sister had a child with cataracts and optic atrophy alone.

2d. Spastic ataxia and optic atrophy of early onset

A pedigree in which two children, the product of a second-cousin marriage, had a spastic ataxia and optic atrophy is reported by Hogan and Bauman (1977). No evidence of a leukodystrophy was found. Behr's disease was excluded as it is essentially non-progressive and is associated with mental retardation.

3. Optic atrophy in the craniosynostoses

See page 37.

4. Others

Dominant optic atrophy and tritanopia

This combination in two families is reported by Krill *et al.* (1971). It is uncertain whether these two conditions are separate genetic entities.

Familial bilateral optic nerve hypoplasia

This is rarely genetically determined. A dominant pedigree with five affected individuals is reported by Hackenbruch *et al.* (1975). Other reports have been suggestive of recessive inheritance. Kytilä and Miettinen (1961) described the condition in two brothers, but the vast majority of cases have been single and recurrence risks are usually small.

Leber's amaurosis (congenital retinal blindness)

Most patients have congenital retinal blindness without neurological problems. In a retrospective study of the condition (Vaizey *et al.* 1977), some patients with severe visual deficit had disequilibrium, hypotonia, and severe subnormality suggesting that neurodevelopmental abnormalities can occur, but do not manifest in all patients. Deafness and a cystic renal dysplasia have also been described. The characteristic signs are those of blindness with an isoelectric electroretinogram and minimal macular pigmentation. Inheritance is recessive.

It is still uncertain whether those with or without neurological problems breed true within families. Note too the association with Joubert syndrome. See page 103. Congenital retinal dystrophy (Leber's) with neurological problems similar to Zellweger syndrome is known to occur. For instance the boy reported by Ek *et al.* (1986), had an early onset of psychomotor retardation, and no visual contact. The ERG was absent, he was deaf, and had a big liver. The biochemistry was similar to that found in Zellweger syndrome. He could have had infantile Refsum syndrome.

Optic atrophy and deafness

Six persons in four generations with severe congenital deafness developed optic atrophy in adulthood (Konigsmark *et al.* 1974). Previously another dominant pedigree had been reported by Michal *et al.* (1968).

Optic atrophy, deafness, and a sensory ataxia

Jéquier and Deonna (1973) reported two sisters with this combination. Deafness began at 8 years, the optic atrophy at 11, and the sensory ataxia with loss of reflexes at 15–20 years.

Optic atrophy, diabetes mellitus, and diabetes insipidus (Wolfram's syndrome or Didmoad)

The association of diabetes mellitus, diabetes insipidus, and optic atrophy has been known since 1938 (Wolfram). The population frequency is about 1 in 100 000 and 1 in 150 patients with juvenile diabetes could have this syndrome (Gunn *et al.* 1976). Nerve deafness is a feature in about one-third of cases (Fraser and Gunn 1977). In a review of 21 families selected from the literature because the probands had the full triad, Fraser and Gunn (1977) concluded that the data were consistant with a homozygous autosomal gene causing juvenile diabetes mellitus and one or more of diabetes insipidus, optic atrophy, and deafness.

A description of some reported sibships follow. A sister and brother had the full triad of signs (Carson *et al.* 1977), but incomplete expression of the gene is commonly found.

Pilley and Thompson (1976) reported two sibs with this syndrome—both were deaf. The manifestations differed in severity in the two children. Page *et al.* (1976) described the disorder in four sibs in which the parents were first cousins. None of the patients was deaf, but audiograms showed each to have bilateral nerve deafness which was more marked at higher frequencies. Parental consanguinity has also been reported by Ikkos *et al.* (1970) and Najjar and Mahmud (1968).

Other sibships are those of Sunder *et al.* (1972)—two sibs and two second cousins; Bretz *et al.* (1970), Ikkos *et al.* (1970)—three sibs; Damaske *et al.* (1975)—three sibs.

In some of the reported cases (Marguardt and Loriaux 1974) there has been urinary tract dilatation and amnioaciduria. The syndrome is reviewed by Cremers *et al.* (1977).

It remains uncertain whether families with diabetes mellitus and diabetes insipidus alone or other combinations are part of the syndrome, but the considerable variability in expression raises this possibility.

Septo-optic-dysplasia with hypopituitarism

Four single cases were reported by Patel *et al.* (1975). Hypotonia, seizures, defective vision, and developmental delay occur and some patients have agenesis of the septum pellucidium. Recurrence risks are less than 1 per cent. Recently, the subject has been reviewed by Blethen and Weldon 1985.

Unfortunately, the condition is variable in its manifestations making a confident diagnosis difficult. Indeed, it might even be unusual for a patient to have the full spectrum. In addition to midline defects of the septum pellucidum, corpus callosum, and basal ganglia, other abnormalities include hydrocephalus, cerebral atrophy, encephalocoele, porencephaly, cerebellar hypoplasia, and hydranencephaly (Morishima and Aranoff 1986). In the review by these authors, only a third of patients with bilateral optic nerve hypoplasia had the classical spectrum and 62 per cent had evidence of pituitary dysfunction. In those patients in whom a scan had been performed, about 60 per cent had a structural abnormality.

Cranial nerve III

Ptosis

Congenital

Six pedigrees with dominant transmission, two with incomplete penetrance, are reported by Rank and Thomson (1959). In an early description of a mother and two of her children, ptosis was part of a more complex external ophthalmoplegia (Rodin and Barkan 1935).

Late onset

Large pedigrees were reported in the early literature (Spencer 1917). Onset in one proband was at 16 years and six other members had a similar onset. Inheritance was as a simple dominant. Because these pedigrees and those of Faulkner (1939) were published before sophisticated muscle tests were available, the relationship of this condition to those discussed on page 380 (progressive external ophthalmoplegia) is uncertain.

Ptosis with strabismus and ectopic pupils

A family in whom four individuals were affected, the mother and three of her children, is discussed by McPherson *et al.* (1976).

Ptosis blepharophimosis and epicanthus inversus

The association of ptosis and blepharophimosis with epicanthus is also known as the blepharophimosis syndrome. Dominant inheritance is reported by Fontaine *et al.* (1974)—six individuals were affected over four generations and other pedigrees are reported by Owens *et al.* (1960) and Smith (1976b). MacIlroy (1930) reported affected family members in four generations, as did Ross (1932). The subject is reviewed by Oley and Baraitser (1988).

Blepharophimosis

It is uncertain whether the difference between ptosis and blepharophimosis has always been correctly reported. A remarkably large pedigree extending over five generations is reported by Stoll *et al.* (1974).

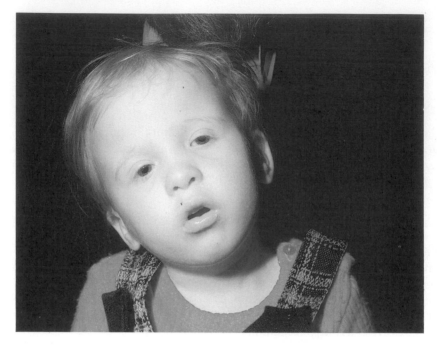

Fig. 1.1   Smith–Lemli–Opitz syndrome.

Ptosis in other conditions

(a) Smith–Lemli–Opitz syndrome (Fig 1.1)

Ptosis in this syndrome is accompanied by mental retardation, microcephaly, short stature, syndactyly of toes 2 and 3, hypospadias, cryptorchidism, and anteverted nostrils (see Smith 1976*a*). Inheritance is recessive.

(b) Congenital myasthenia

Ptosis with weakness of the external ocular muscles, bulb, and limbs are all features, although ptosis might not be as prominent as in adult-onset myasthenia. For inheritance see page 411.

(c) Congenital myotonic dystrophy

Ptosis, in addition to bilateral facial muscle weakness and respiratory problems in a neonate whose mother has myotonic dystrophy is the typical presentation. See page 481.

(d) Congenital myopathies

Ptosis is a feature in the myotubular (centronuclear) and nemaline myopathies.

(e) Other conditions with ptosis

Aarskog syndrome; Dubowitz syndrome; Freeman–Sheldon syndrome; Möbius syndrome; Schwartz–Jampel syndrome; Noonan syndrome; Coffin–Lowry syndrome; Rubinstein–Taybi syndrome; Saethre–Chotzen syndrome; fetal alcohol syndrome; fetal hydantoin syndrome (see Smith 1976*a*).

Familial paralysis of horizontal gaze with kyphoscoliosis

Four sibs, the offspring of unrelated Chinese parents, had a syndrome of paralysis of horizontal gaze, pendular nystagmus, and progressive scoliosis. The eldest sib had bilateral facial myokymia (Sharpe *et al.* 1975). Two further families (Dretakis and Kondoyannis 1974) with three sibs in one and two in the other had essentially the same syndrome, although the three sibs in one of the families showed mild mental retardation. Riley and Swift (1979) reported two brothers with a congenital horizontal gaze palsy and kyphoscoliosis whose parents were first cousins.

Horner syndrome

There are only rare references in the literature to familial Horner syndrome. A report describing a large pedigree in which five members were affected refers to two other reports since 1900 (Durham 1958).

Adie syndrome

Adie (1932) described a combination of dilated, unresponsive pupils with absent knee and ankle reflexes. The mother and daughter reported by De Rudolf (1936), and a father and son (McKinney and Frocht 1940) suggest that the syndrome is occasionaly dominantly inherited.

Duane syndrome (see also Wildervanck syndrome)

This consists of a restriction of abduction and bulbar retraction with narrowing of the palpebral fissures on attempted adduction. In approximately 20 per cent of patients the syndrome is bilateral. Most cases are single; but transmission through many generations has been recorded (Kiricham

1970). Of interest to the neurologist is the association with deafness and a Klippel–Feil anomaly known then as Wildervanck syndrome. There is also an association with radial defects (Okihiro *et al.* 1977).

Nystagmus

In an investigation of 40 patients who had nystagmus and were registered as blind, Pearce (1978) found that a single gene was responsible for the condition in the majority of cases. In 15 patients there was X-linkage, in 15 autosomal recessive inheritance, and three were either sex-linked or autosomal recessive. The main causes of blindness underlying the nystagmus were congenital cone dystrophy, albinism, Leber's congenital amaurosis, optic atrophy, and stationary night blindness.

Genetic counselling is difficult when there are no obvious signs of an underlying disease. Two patients in the above series had nystagmus alone, segregating as an X-linked trait. A third male patient had an affected brother. Visual acuity in all of the idiopathic cases was decreased. Four patients were isolated males. Taking the group as a whole, 75 per cent of the various forms of congenital nystagmus are genetically determined (Pearce 1978).

Congenital nystagmus

This is often divided into a jerk nystagmus and a pendular nystagmus. In the former there are no structural abnormalities of the eye whereas, in the latter, cataract, optic atrophy, and macular dystrophy might be present (Yee *et al.* 1976). The division is not exact, and in the X-linked pedigree of Yee *et al.* (1976) both pendulum and jerk nystagmus were present in a single sibship. In this family, the unaffected mother married twice and had affected males by different fathers. Forssman (1971) found that the disorder is sex-linked with reduced penetrance not only for females, but also for males. Expression is variable. No father-to-son transmission was found. Females were affected at a population frequency of 1 in 2800 as compated to 1 in 1000 males.

Familial spasmus nutans

This is a transient disorder of childhood with a triad of symptoms, namely nystagmus, involuntary head nodding, and a tilt of the head. In a paper reporting 14 children with the disease, concordance in twins (probably monozygotic) is reported (Hoefnagel and Biery 1968).

Monocular nystagmus

Environmental aetiology is more important than genetic factors, although this condition has been reported in families (Yawger 1917). Monozygotic twins concordant for monocular nystagmus are reported by Hoyt and Aicardi (1979).

Hereditary vertical nystagmus

Members of an Italian family had congenital vertical nystagmus in three generations (Forsythe 1955). A rotary movement of the head occurred in four of the five patients. The head movements were not in the same plane as the nystagmus and the rate was much slower. The movement disappeared during sleep. In three families in which members had a predominantly vertical nystagmus some also had mild ataxia (Marmor 1973). Most of those affected showed absent optokinetic nystagmus and hyperactive vestibulocular responses. Inheritance in all three was possibly dominant with incomplete penetrance. Sogg and Hoyt (1962) reported intermittent vertical nystagmus in a father and son.

Hereditary voluntary nystagmus

This is not a pathological condition. In two families with this facility, the pattern of transmission suggested dominant inheritance (Aschoff *et al.* 1976; Keyes 1973).

Three sibs (parents unaffected) are reported by Goldberg and Jampal (1962).

Nystagmus, essential tremor, and duodenal ulceration

A large dominant pedigree has been described in which many members in a Swedish–Finnish family were affected (Neuhauser *et al.* 1976*a*). Those members most severely affected had a cerebellar ataxia. The nystagmus was present from birth or early infancy.

Marcus–Gunn phenomenon

Recessive

Unilateral ptosis increased by moving the jaw to that side, with lid retraction on jaw movement to the opposite side, has been described in sibs (Kirkham 1969*a*). A brother and a sister had this phenomenon affecting the left upper lid. Both parents were normal. Inheritance in the family is uncertain as a maternal aunt had photographic evidence of a unilateral ptosis.

Dominant

A mother, her brother, and her son had the syndrome (Falls *et al.* 1949), and the authors quote an earlier report by Leri and Weill (1929).

Ocularmotor apraxia (Cogan's syndrome)

Cogan's (1952) ocularmotor apraxia is a defect of voluntary and optically induced horizontal eye movement with a retention of random eye movement and of voluntary vertical gaze. Vassella *et al.* (1972) described the condition in a father and daughter, sibs are reported by Sachs (1967) and Redding (1970) and monozygotic concordant twins by Robles (1966). Most cases are single.

Cranial nerve V

Congenital trigeminal anaesthesia

Isolated corneal anaesthesia was reported by Purcell and Krachmer (1979) in 13 members of a family over four generations. For a review of this condition, see Rosenberg (1984).

Tic douloureux

In his extensive investigation of trigeminal neuralgia, Harris (1936) found 30 instances of familial tic douloureux in 2500 personally examined cases. He described one three-generation family with nine out of 19 members affected. In the non-familial cases, 4.3 per cent were bilateral, whereas 20 per cent were bilaterally affected in the familial cases. Allan (1938) reported a three-generation family, and Castaner-Vendrell and Barraquer-Bordas (1949) a four-generation family. More recently, Daly and Sajor (1973) described tic douloureux in a mother and three of her six offspring, and Herzberg (1980) described four members of a family over three generations (the onset of two of the cases was in their thirties and all were unilateral).

There seems to be a dominantly inherited trigeminal neuralgia in a small group of families. The most useful clinical criterion for possible genetic determination is the bilaterality of the condition.

A single pedigree (Auld and Bauermann 1965) with six affected in a sibship of seven could be an example of recessive inheritance. Parents were unaffected and unrelated.

Tic douloureux, facial weakness, and peripheral neuropathy

A man (and possibly also his brother) had an early onset hypertrophic neuropathy, bilateral facial weakness, unilateral facial spasms, and trigeminal neuralgia (Kalyanaraman *et al.* 1974). The condition is similar to that of Hellsing's syndrome—see page 23.

Cranial nerve VII

Familial aggregation of Bell's palsy is unusual, although Alter (1963) found that more than one-quarter of those with Bell's palsy reported an additional affected family member. A small stylomastoid foramen or a narrow canal might be genetically determined. In Alter's series, 4.3 per cent of parent, 2.7 per cent of sibs, and 2.3 per cent of children were affected.

Reported families are those of Cawthorn and Haynes (1956) where one patient had five attacks and a brother had three. Both had extremely cellular mastoids. De Jong (1950) described a family in which a brother, sister, and a son of the brother had facial palsy. Stone (1950) reported three brothers and their father in one family who, in addition to episodes of Bell's palsy, had ophthalmic migraine with an external ophthalmoplegia. Danforth (1964) described a father and son with Bell's palsy, and Auerbach *et al.* (1981) a parent and two children. There is a suggestion that the familiar type tends to relapse more frequently. A large family with Bell's palsy reported by Amit (1987), was unusual in that the age of onset was juvenile—around the age of 12 years. No recurrent episodes were recorded in the older generations.

Familial oculomotor paresis with Bell's palsy

A Jewish sibship is reported in whom the parents were first cousins (Currie 1970). Between the four affected sibs there were a total of seven episodes of Bell's palsy and four episodes of external ophthalmoplegia. One brother had proven diabetes and another latent diabetes. The seventh nerve palsy and the ophthalmoplegia did not occur together.

Progressive hemifacial atrophy (Parry–Romberg syndrome)

Franceschetti *et al.* (1953) reported two unrelated women, both products of a consanguineous marriage, with progressive wasting of one side of the face. The onset in one was in the first decade and the other in middle life. Inheritance in that family was thought to be recessive, but most cases are sporadic. Localized scleroderma, cortical hemiatrophy with contralateral Jacksonian epilepsy, and a progressive hemiparesis are well recognized manifestations of the disorder.

The suggestion that this condition is a localized form of scleroderma is supported by the report of Lewkonia and Lowry (1983) of a boy with mild hemifacial atrophy and localized scleroderma on legs and trunk. No recent familial cases have been reported.

Facial nerve palsies associated with bony lesions of the cranium

Cranial nerve palsies, specially involving the VII, VIII and optic nerves, occur in those primary bone dysplasias associated with abnormal modelling of the skeleton and increased density of bone.

They are divided by Beighton (1978) into the following classes.

A. Craniotubular dysplasias (abnormal modelling of long bones with sclerosis of the cranium)
   1. Metaphyseal dysplasia (Pyles disease)
   2. Craniometaphyseal dysplasia
   3. Craniodiaphyseal dysplasia
   4. Frontometaphyseal dysplasia
   5. Dysosteosclerosis
   6. Osteodysplasty

B. Osteoscleroses (sclerosis predominates and changes in configuration are minor)
   1. Osteopetrosis—autosomal dominant
   2. Osteopetrosis—autosomal recessive
   3. Pycnodysostosis
   4. Osteosclerosis with abnormalities of nervous system

C. Craniotubular hyperostoses (bony overgrowth rather than defective modelling)
   1. Endosteal hyperostosis (van Buchem)
   2. Sclerosteosis
   3. Diaphyseal dysplasia (Camurati–Engelmann disease)
   4. Infantile cortical hyperostosis (Caffey syndrome)
   5. Osteoectasia with hyperphosphatasia

A. Coaniotubular dysplasias

A1. Metaphyseal dysplasia (Pyles disease)

(a) There is little cranial involvement and neurological complications are unusual.

(b) Metaphyseal dysplasia, anectoderma, and optic atrophy   Two Egyptian sisters whose parents were first cousins had this combination (Temtamy *et al.* 1974*a*). The metaphyseal dysplasia resembled Pyles disease.

A2. Craniometaphyseal dysplasia

(a) Dominant type   Fifteen individuals in five generations were affected in the family reported by Beighton *et al.* (1979*a*). Facial palsy, either unilateral

or bilateral, was present in 30 per cent of cases. Another large pedigree (Rimoin *et al*. 1969) had members with compression at the upper cord level, optic foramina, and facial nerve canals. Clinically, the bony prominence initially affects the nasal bridge causing obstruction.

(b) Recessive type This type is more severe and cranial nerve palsies are probably more common than in the dominant type. Sibs are reported by Lehmann (1957) and Millard *et al*. (1967). Two further sibships are described by Penchaszadch (1980), and in one member the optic and auditory nerves were affected.

A3. Craniodiaphyseal dysplasia

Overgrowth of the bones of skull and face can lead to cranial nerve entrapment. Deafness and optic atrophy have been recorded (Joseph *et al*. 1958). Those patients described by MacPherson (1974) had a progressively bizarre facial deformity leading to nasal obstruction and severe immobility of the face. Most are dominantly inherited.

A4. Frontometaphyseal dysplasia

This is characterized by marked overgrowth of the supra-orbital ridge, absence of the frontal sinuses, defective modelling of the metaphyses of the long bones, arachnodactyly, deafness, and hirsutism (Gorlin and Cohen 1969).

A dominant pedigree is reported by Weiss *et al*. (1976) and a family with only the frontal component of the syndrome by Levine *et al*.1975). In a recent review of the condition Gorlin and Winter (1980) suggest that the condition might be inherited as an X-linked recessive, although this is not certain. A pedigree reported by Fitzsimmons *et al*. (1982) seemed dominant.

A5. Dysosteosclerosis

The clinical features are those of short stature, fragile bones, enamel hypoplasia, optic atrophy, and sometimes mental retardation. Sclerosis of the skull, and other bones, platyspondyly, and metaphyseal splaying of the long bones are the main radiological features. Sibs are reported by Field (1939) and consanguinity by Spranger *et al*. (1968), but a sex-linked type is suggested by Pascual-Castroviejo *et al*. (1977), who reported eight males affected in more than one generation.

A6. Osteodysplasty

(a) Dominant type Members of two families with this condition had a prominent forehead, small mandibles, kyphoscoliosis, genu valgum, short

distal phalanges, and bowing of limbs (Melnick and Needles 1966). Deafness may be a feature. Bones at the base of the skull were thickened, and the ribs and long bones show constrictions.

(b) Recessive type   Four sibs, the product of a consanguineous mating, had osteodysplasia (Anderson *et al.* 1972), and a severe early onset type is reported by Danks *et al.* (1974).

B. Osteosclerosis

B1. Albers–Schönberg disease (osteopetrosis)

(a) Dominant type   A mother and son have been reported with this type of osteopetrosis (Thomson 1949). Both had striking parietal and frontal bossing. Obstruction of the foramina led to optic atrophy in the child. Two further dominant pedigrees with variable penetrance have been reported (Johnston *et al.* 1968). These authors found 161 (16 per cent) of reported cases in the literature to have cranial nerve palsies, especially involving the I I, I I I and V I I cranial nerves.

A remarkable pedigree in which 13 cases of facial palsy were seen in five generations is documented by Welford (1959). The proband had unilateral palsy and his father bilateral palsy. All those affected had osteopetrosis.

(b) Recessive type.   The recessive form is often more malignant than the dominant (Enell and Pehrson 1958). These authors describe two sibs; both had facial palsies, optic atrophy, nystagmus, and hydrocephalus. Seven further patients with severe neurological complications, including cerebral atrophy, developmental delay, and optic atrophy, were reported by Loria-Cortes *et al.* (1977). Two of the children had facial nerve palsies. A mild recessive type also exists (Kahler *et al.* 1984).

B2. Osteopetrosis with renal tubular acidosis and cerebral calcification

Four Saudi children with this condition were reported by Ohlson *et al.* (1980). The report included one pair of sibs. Stunted growth, mental retardation, and abnormal teeth are features of the condition. Previously, Sly *et al.* (1972) described this combination in three sibs in which optic atrophy was a feature; two brothers were reported by Guibaud *et al.* (1972). Inheritance is recessive.

B3. Pycnodysostosis

The features of this syndrome are osteosclerosis with bone fragility, deformity of the skull (including wide sutures), and short stature. The facial features are characteristic: patients have a small face, parrot-like nose, small chin, and abnormal teeth. The cranium is prominent as are the eyes. The fingers are short and the nails dysplastic (Beighton *et al.* 1977; Waziri *et al.* 1976). Inheritance is recessive (Sedano *et al.* 1968).

B4. Osteosclerosis with abnormalities of nervous system and meninges

A mother and daughter with this condition were described by Lehman *et al.* (1977). The daughter had generalized osteosclerosis and pyknodysostosis was thought of, but dismissed because of the absence of the other features of the disease. The patients presented by Lehman *et al.* (1977) had multiple meningoceles, and hypoplasia of the brain and spinal cord. Mandibular hypoplasia, a large sella turcica, platybasia, basilar impression, and a spacious foramen magnum were also noted.

(a) Acro-osteolysis with osteoporosis (Hajdu-Cheney syndrome) This is a short-stature syndrome of slow progression, affecting the skull, spine, and long bones with facial features which include synophrys, thick lashes, dental maleruption, short neck, and low-set ears (Weleber and Beals 1976). Hearing loss and optic atrophy occur. The skull is characterized by a prominent occiput, thickening of the posterior fossa, and wormian bones. Basilar impression is frequent. Most cases are sporadic except the one reported by Cheney (1965) where inheritance was probably dominant.

(b) Mandibuloacral dysplasia Dominant inheritance of a progeroid-like syndrome with marked hypoplasia of the subcantaneous tissue, short stature, partial alopecia, a bird-like facial appearance, small jaw, total or partial absence of the clavicle, bellshaped thorax, and severe acro-osteolysis was noted by Welsh (1975). Two single cases had previously been reported by Young *et al.* (1971).

C. *Craniotubular hyperostoses*

C1. Hyperostosis corticalis generalisata or hyperphosphatasemia tarda (van Buchem disease)

Characterized by increased bone formation resulting in thickening of the base of the skull, mandible, and long and short tubular bones, this condition can be present clinically with facial diplegia and focal neurological signs (hemiplegia). The markedly thickened lower jaw is typical, but prognathism is not a feature and the teeth, tongue, and nose are normal. An increase in alkaline phophatase might occur.

First described by Van Buchem *et al.* in 1955, with eight new cases reported in 1971—all eight appearing to be related to each other. Inheritance is as a recessive.

C2. Sclerosteosis

This rare syndrome leads to severe bony overgrowth and facial distortion. It is particularly prevalent in South Africa where it has a minimum prevalence of 1 in 75 000 (Beighton *et al.* 1976). In a nation-wide survey in that country, Beighton and co-authors found 25 affected members in 15 kindred. Deafness and facial nerve palsies developed in the second decade. Gigantism and syndactyly of fingers 2 and 3 are associated features. Inheritance is as a recessive.

Beighton (1988) have now examined 50 persons with this condition and 15 individuals in Holland with Van Buchem disease. The only difference they found was that the South African disease was more severe and syndactyly did not seem to occur in Van Buchem's disease. They concluded that both conditions could still be the result of homozygosity of the same mutant genes.

C3. Diaphyseal dysplasia (Camurati–Engelmann's disease)

The skull is frequently involved in this condition which is characterized by progressive cortical thickening of bone. The tibia, femur, fibula, humerus, ulnar, and radius are also commonly affected.

The age of onset varies from 3 months to 57 years with a mean at 19 years (Hundley and Wilson 1973). Muscle atrophy and a waddling gait occur in a third of patients and the children tend to be tall and thin. Inheritance is dominant (Girdany 1959; Hundley and Wilson 1973). The musculature seems underdeveloped and in a report on five children, the EMG showed evidence of a myopathy (Naveh *et al.* 1985). The electron microscopic appearance of muscle revealed fibre atrophy, accumulation of endomysial collagen, and thickening of the perivascular basement membrane.

C4. Infantile cortical hyperostosis (Caffey syndrome)

Neurological complications have not been described.

C5. Osteoectasia with hyperphosphatasia

This condition presents with the same clinical picture as Paget's disease. The onset is, however, in early childhood with bowing of the limbs and an increasing head circumference. Optic atrophy and deafness are late complications. Radiologically there is generalized demineralization and the skull shows patchy areas of calcification. Inheritance is recessive (Temtamy *et al.* 1974*b*).

Two sibs with this condition (the parents were first cousins) had angioid streaks with pigmentary changes (Iancu *et al*. 1980). Alkaline phosphatase is raised.

Other conditions involving the skull with occasional involvement of cranial nerve

Dominant osteosclerosis This disorder is distinct from van Buchem disease. Exomphalos, hypertelorism, increased head circumference, nasal obstruction, and cranial nerve involvement are all part of the syndrome (Gorlin and Glass 1977).

Radiology of the skull shows endosteal sclerosis of the calvarium with loss of diplöe, and osteosclerosis and hyperostosis of the mandible. The pelvis, metacarpals, and metatarsals are also involved.

Inheritance is dominant (Gorlin and Glass 1977; Gelman 1977).

Hyperostosis frontalis interna (Morgagni–Stewart–Morel syndrome) This consists of thickening of the inner table of the frontal bone, with obesity and hypertrichosis. Females are affected many times more than males and although familial cases are rare, dominant pedigrees are known. Knies and Le Fever (1941) reported a mother and three of her children to be affected, and a family with the transmission of this syndrome through four generations is reported by Rosatti (1972). Twelve members were affected of whom 10 were females. There was no case of father-to-son transmission so that the inheritance pattern is still uncertain. In the family reported by Rosatti (1972), the proposita was the offspring of an affected female who had married her first cousin.

Benign hyperostosis corticalis generalisata (Worth) An autosomal dominant type of hyperostosis is reported by Worth and Wollin (1966), and Maroteaux *et al*. (1971). There is increased bone density especially involving the skull and tubular bones, but without any changes of bone morphology. The condition is benign.

Oculodento-osseous dysplasia This rare condition can lead to serious neurological deficit owing to nerve and cord compression. Clinically there is microphthalmia, microcornea, a narrow nose, and hypoplasia of dental enamel (Beighton *et al*. 1979*b*). An important clue to the diagnosis is syndactyly between fingers 4 and 5. These authors describe two patients who were the product of marriages between a pair of brothers and a pair of sisters. The cranial hyperostosis and mandibular overgrowth in these recessive pedigrees is of a greater degree than previously noted and might constitute a separate genetic entity. Previous reports have suggested dominant inheritance (Rajic and De Veber 1966), but the group as a whole is probably heterogeneous.

Osteopathia striata with cranial sclerosis   Deafness is not infrequent in this disorder (Horan and Beighton 1978). These authors describe an alteration in the shape of the skull (frontal bossing) in four families with the disorder. Inheritance is as a dominant. The characteristic multiple parallel lines running lengthwise in the long bones can be a feature in focal dermal hypoplasia (Goltz syndrome) and in osteopetrosis.

Cleidocranial dysostosis   The association of a broad head (frontal and parietal bossing), hypertelorism, clavicular hypoplasia, and excessive mobility of the shoulder is called cleidocranial dysostosis. The other cranio-facial features are incomplete closure of the anterior fontanelle, persisting metopic sutures, and the presence of wormian bones. Occasionally, kyphosis, scoliosis, or short stature are problems.

Autosomal dominant inheritance is well established (Jarvis and Keats 1974; Jackson 1951), but a report of a rare recessive type in which sibs were affected (consanguineous, unaffected parents) suggests genetic heterogeneity of the condition (Goodman *et al.* 1975).

Melkersson syndrome   The combination of facial palsy, facial oedema, and lingua plicata (scrotal tongue) is known as Melkersson syndrome (Carr 1966). Sibs were reported by Rosenthal (1931), and by New and Kirch (1933). If the scrotal tongue is taken as a minimal manifestation then some of the pedigrees might be dominant (Carr 1966; Scott 1964). Kunstandter (1965) reported a dominant pedigree with incomplete penetrance, and Lygidakis *et al.* (1979) a family with seven affected members in four generations.

Hemifacial spasms   This has been reported as a dominant by Stocks (1923) in a family in which members in four generations were affected. The majority of cases are not thought to be genetically determined.

Congenital hypoplasia of anguli oris muscle   Asymmetrical crying facies is considered to be due to hypoplasia of the anguli oris muscle on one side of the mouth. Thirty-seven cases were found in 4500 consecutive live births (Papadatos *et al.* 1974). In no case was there an association with other severe congenital malformations. A family study showed first- and second-degree relatives to be affected to a much greater degree than in the general population, but the pattern favoured multifactorial inheritance rather than a single gene defect.

An association between this condition and other congenital defects—especially heart disease—has been suggested by Caylor (1969), and Pape and

Pickering (1972). This was not a feature of the nine cases reported by Nelson and Eng (1972). The condition as a whole is probably heterogeneous.

Hemifacial spasm plus (Hellsing's disease) In 1930, Hellsing described a familial condition characterized by facial spasms, decreased tendon reflexes, and Argyll Robertson pupils. Members of this Uppsala family were affected over five generations and the trait was inherited as a dominant. Members of the family emigrated to Sweden where a completely new generation with a different clinical picture now exists (Lundberg and Westerberg 1969). The onset of the facial spasms occurred in the middle years (30s), first on one side (lasting about 1 minute) and then the other. The facial muscles were in addition weak. Marked weakness of the extremities has now occurred in generation seven. Sensation (except for vibration loss in one female) was normal. Two sibs had a coarse tremor of the arms and trophic ulcers in the feet which were slow to heal. In five cases pes cavus was found. A coarse nystagmus, optic atrophy, and reduced motor conduction velocity are part of this slowly progressive disease. Note, too, the family of Cruse *et al.* (1977).

Congenital facial paralysis (Möbius syndrome) Facial diplegia in the presence of bilateral abducens palsies is known as Möbius syndrome. Since the original description by Möbius in 1888, many other congenital anomalies have been included in the syndrome designation. These are bony abnormalities (syndactyly, polydactyly, agenesis of digits, brachydactyly), epicanthic folds, absent muscles (especially pectorals and trapezii), and mental retardation. Families in which more than one member is affected have not included those with bony deformities. The dominantly inherited family of Van der Wiel (1957) had only the facial diplegia. One family reported by Hicks (1943) had a mother and son affected. Only the son had an extra thumb and webbing of the fingers. Both had bilateral facial weakness and abducens palsies.

In a study of 15 cases Baraitser (1977) found sibs to be affected in two instances. In both of these, primary muscles disease was the underlying disorder.

Möbius syndrome might occur in association with Poland's anomaly (absent pectoralis muscle and symbrachydactyly mostly on the same side). In addition, Hanhart syndrome (a small jaw and tongue in association with a transverse defect of an arm or occasionally of the lower limb) has been described in association with Möbius syndrome. Dextrocardia can also occur (Bosch–Banyeras *et al.* 1984). A pedigree widely quoted and suggestive evidence that Möbius syndrome can be inherited as a dominant is that of Ziter *et al.* (1977), but in that pedigree an ophthalmoplegia did not occur. A child with Möbius syndrome is reported whose father only had a transverse unilateral limb defects (Collins and Schimke 1982), but the connection between the two must at present be interpreted with caution. Counselling— Recurrence risks are small—approximately one per cent to sibs.

Lower cranial nerves

Laryngeal-abductor paralysis

Congenital laryngeal-abductor paralysis due to nucleus ambiguus dysgenesis in three brothers was reported by Plott (1964). The brothers had mild to moderate mental retardation. One had a sixth-nerve palsy and was deaf. A fourth had stridor and died.

Watters and Fitch (1973) described three further males (siblings and nephew). All the patients had other neurological involvement. Opitz *et al*. (1978) in their review of severe mental retardation found two large pedigrees in which many males were affected in at least two generations. Inheritance is as an X-linked recessive (see p. 471).

Laryngeal-adductor paralysis

A family with dominant transmission of this disorder had members affected over five generations (Mace *et al*. 1971). All affected members were hoarse. This mode of inheritance is rare.

Hereditary gustatory sweating

Sweating when eating spices or sour foods was described in five members of a black American family over three generations (Mailander 1967).

2

Cerebral and vertebral malformations

Dandy–Walker syndrome

Hydrocephalus associated with a posterior fossa cyst and hypoplasia of the cerebellar vermis is known as the Dandy–Walker syndrome. In a series of 28 autopsied cases, Hart *et al.* (1972) found 68 per cent to have other anomalies. These included cerebral gyral anomalies (agyria, polymicrogyria, heterotopia), agenesis of the corpus callosum, and other cerebellar anomalies (including heterotopia). Two cases had cleft palate, three had polydactyly/syndactyly, and one each the Klippel–Feil and Cornelia de Lange syndromes. All the cases were sporadic.

In another series of 12 cases of (Tal *et al.* 1980) agenesis of the corpus callosum occured in four, an occipital meningocele in two, and an aqueduct stenosis in one. Two had a cleft palate and one polycystic renal disease. All were single cases.

The autopsy findings in ten cases were presented by D'Agostino *et al.* (1963*a*). This included two male sibs.

Two sisters with the Dandy–Walker syndrome (Benda 1954) had another sib who was similarly affected.

The majority of reported cases are sporadic with only a few families known in which sibs are affected. The cases in which the risk is probably much higher (possibly 1 in 4) are those with polydactyly (could be Joubert syndrome) and those with polycystic renal disease (could be Meckel–Gruber syndrome) if these are excluded the recurrence risks should be no greater than 2 per cent. Murray *et al.* (1985) suggest 1–5 per cent.

Dandy–Walker/A–V septal defects

Two sisters had in addition to the above combination, macrocephaly, a prominent forehead and occiput, parietal foramina, hypertelorism with downslanting palpebral fissures, and a depressed nasal bridge (Ritscher *et al.* 1987). Both sibs were moderately mentally retarded.

Microcephaly

This is defined as a head circumference of more than two standard deviations below the mean for age.

Böök *et al.* (1953) and Komai *et al.* (1955) estimate the incidence of genetic microcephaly to be between 1 in 25 000 to 1 in 50 000 live births. Data on 113 cases of microcephaly in the Netherlands (van der Bosch 1959) suggest that about one-third are of genetic origin. In the genetic group, consanguinity was found in 54 per cent of cases. In a Japanese study, parents were first cousins in 44.8 per cent of the families (Komai *et al.* 1955). In a study of microcephaly in a population isolate (Kloepfer *et al.* 1964), 22 cases were distributed in 13 sibships in a very inbred single kinship.

Microcephaly can be divided as follows.

A. Simple microcephaly

    1. Autosomal recessive microcephaly.
    2. Autosomal dominant microcephaly.
    3. X-linked microcephaly.

B. Complicated microcephaly

    1. Microcephaly with retinopathy.
    2. Microcephaly with intracranial calcification.
    3. Microcephaly with severe spasticity.

Note—this subdivision concerns those cases of microcephaly which are not part of a malformation syndrome. Angelman syndrome must be excluded—see p. 178.

A.1. Autosomal recessive microcephaly

The importance for the genetic counsellor is to distinguish between genetic, or primary true microcephaly, and the many causes of secondary or environmental microcephaly (radiation, rubella, drugs—methyldopa). The following clinical criteria are suggestive of the genetic variety—microcephaly uncomplicated by other deformities, a well preserved personality despite poor intellect, a facial profile with a receding forehead often parallel with the slant of the nose, receding chin, and normal-sized face. The ears in particular appear large (Qazi and Reed 1973).

Other useful criteria suggestive of true microcephaly are consanguinity, microcephaly from birth, normal delivery and post-natal period, relatively normal early developmental milestones, only mild spasticity, and few seizures.

However, it may be extremely difficult to predict genetic risks. There have been cases where one of a sib pair has had a clinical picture compatible with true recessive microcephaly while the other had considerable neurological deficit. A CT brain scan is useful in that it shows a normal, but small brain. It will also exclude those cases of microcephaly associated with porencephaly.

A.2. Autosomal dominant microcephaly

Haslam and Smith (1979) have indicated the existence of dominant microcephaly. The degree of intellectual retardation is not as severe as the autosomal recessive type and the reduction in skull circumference is not as marked. Four families are reported with at least two generations affected.

A.3. X-linked microcephaly

This rare mode of inheritance has been reported in a family by Deshaies *et al.* (1979). At least eight males were affected over three generations. All showed, in addition, growth retardation and obesity, and some had cryptorchidism, tapering fingers, contractures, proximally implanted thumbs, and club feet. Distal displacement of the palmar axial triradius was constant in all five patients examined dermatoglyphically.

Genetic counselling in microcephaly (see also p. 29)

The clinical differentiation between the genetic and non-genetic type is not absolutely reliable. One of discordant monozygotic twins reported by Brandon *et al.* (1959) had the facial features of true recessive microcephaly. At present it is realistic to assume that about half of the single cases of microcephaly who conform to the clinical appearance of so-called 'true' microcephaly, as described above, are homozygous for a recessive gene causing the condition. Recurrence risks are therefore between 1 in 8 and 1 in 10. If the parents are first cousins the risk approaches 1 in 4.

In the series of Optiz *et al.* (1978) a recurrence risk for pure microcephaly was 20 per cent (1 in 5). In another series reported by Bartley and Hall (1978) the recurrence risk was 11 per cent. When microcephaly is accompanied by multiple congenital malformations the latter authors found a recurrence risk of 6 per cent. Antenatal diagnosis, using an ultrasound measurement of head circumference and comparing this with the abdominal circumference is a useful procedure in those who choose to face the high recurrence risk, but no patient should be assured that all cases will be picked up in this way at present.

Successful prenatal diagnosis was achieved by Pescia *et al*. (1983) for true microcephaly.

B.1. Microcephaly with chorioretinopathy

Five members in one branch and two of their second cousins had microcephaly and a chorioretinal dysplasia (Cantu *et al*. 1977). These members were from an inbred Pennsylvanian family, members of a 'horsebuggy' Mennonite sect. Inheritance is recessive.

There has been a small increase in the number of recessive pedigrees in which there is an association of severe microcephaly and a pigmentary retinopathy. Most patients do not have the facial features of the 'true recessive' type of microcephaly.

In the patient described by Cantu *et al*. (1977) intelligence was low, whereas in the Parke *et al*. (1984) report, intelligence was average or low-average.

Microcephaly, microphthalmia and falciform retinal folds

Two brothers were reported with this combination (Jarmas *et al*. 1981). The mother did not have the eye problem, but had severe microcephaly and mild mental retardation. Inheritance is probably dominant.

Microcephaly with pigmentary retinal changes—autosomal dominant

In the family reported by Tenconi *et al*. (1981) three generations were affected, but grandmother had only the retinopathy and not microcephaly. Intelligence was average or low-average.

B.2. Microcephaly/intracranial calcification

This can be subdivided as follows.

(a) Those sibs with abnormal gyral patterns which include lissencephaly and polymicrogyria (Burn *et al*. 1986*b*). The sibs in this report had corneal opacities, bile duct hypoplasia and hepatomegaly.

(b) Those sibs with (on CT) a normal looking brain, but with intracranial calcification (Baraitser *et al.* 1983*a*).

(c) Tuberous sclerosis (see p. 142) must be excluded.

(d) The syndrome of Aicardi and Goutieres (see p. 336).

Genetic counselling in single cases is extremely difficult, but recessive inheritance must be considered if tests for intrauterine infection are negative.

B.3. Microcephaly with severe spasticity

The first of the sib pair reported by Schinzel and Litschgi (1984) had severe neurological problems with serious spasticity causing flexion contractures. Seizures were also a feature. The second affected sib (a fetus) was terminated at 24 weeks because of microcephaly. In support of the postulate that these sibs did not have pure or true microcephaly is that the CT on the first-born showed structural abnormalities of the brain.

Bundey and Griffiths (1977) report three sibs with microcephaly and severe spasticity. The father's maternal cousin had three children with microcephaly without spasticity. In a study by Tolmie *et al.* (1987) of a series of cases sent for genetic counselling there were in all, nine sibships with recurrences (29 isolated cases). Therefore just over half (five out of nine) were in this category. Clinically, only one sib pair fitted the usual uncomplicated 'pure' recessive microcephaly pattern.

Five of the nine kindreds had sibs with seizures and/or spasticity (with variation between sibs). Seven of the eleven sibs died before 12 years, mostly from respiratory infections. One further sib pair had a neonatal onset of myoclonic seizures and profound retardation, and in the final two sibships there was spasticity and seizures, but these were not prominent features.

It must be concluded that these patients constitute a second group of recessive microcephaly, i.e. with spasticity and seizures, but it is not yet possible to differentiate the genetic, high risk cases, from the majority who seem to be sporadic.

Summary of counselling in microcephaly

'Pure' microcephaly with 'characteristic' craniofacial features:
    non-consanguineous mating   (12 per cent);
    with cousin parents   (25 per cent).

Microcephaly with symmetrical spasticity   (13 per cent).

Microcephaly with severe retardation:
    spasticity and seizures   (3–5 per cent).

Microcephaly with retardation but
    without craniofacial dysmorphism   (3 per cent).

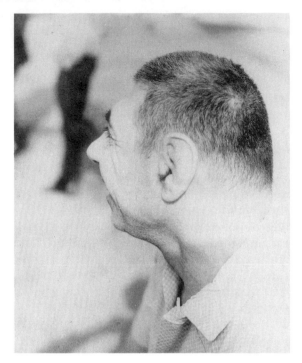

Fig. 2.1   Microcephaly.

Microcephaly with normal intelligence (A.R.)

Autosomal recessive microcephaly with normal intelligence has also been reported (Teebi *et al*. 1987). These authors report a large inbred Arab kindred in which eight members had small heads, bird-like faces, and short stature. Some have an immune deficiency (Seemanova *et al*. 1985).

Microcephaly with normal intelligence (A.D.)

This certainly occurs and parental head size should always be measured. See page 27.

Microcephaly in other conditions

Microcephaly is also a feature of the following conditions (only the additional features are listed).

Bloom syndrome—short stature, facial telangiectatic erythema made worse by exposure to sunlight, malar hypoplasia, immunoglobulin deficiency, increased sister chromatid exchange.

COFS syndrome—microcephaly, microphthalmia, cataract, and blepharophimosis. These features plus a prominent root of the nose, small jaw, and flexion contractures of the elbows and knees, are known as COFS syndrome (cerebro-oculo-facial-skeletal syndrome). Sibs are described by Pena and Shokeir (1974), and Lurie *et al.* (1976).

Cockayne syndrome—see page 472.

Coffin–Siris syndrome—mild microcephaly with coarse facial features and thick lips in the presence of hypoplastic or absent fifth finger and toe nails is of unknown aetiology.

de Lange syndrome—mental and growth retardation, synophrys, long philtrum, thin upper lip, anteverted nostrils, low hairline, skeletal abnormalities (Fig. 2.2).

Dubowitz syndrome—microcephaly, low birthweight, dwarfism, and an eczematous skin eruption. Flat nasal bridge in line with the forehead, large low-set ears, and small jaw were features described in a girl by Dubowitz (1965). A sib was said to be similarly affected.

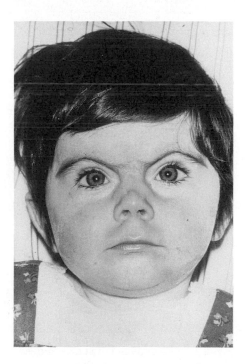

Fig. 2.2 Cornelia de Lange syndrome.

Langer–Giedion syndrome—also called tricho-rhino-phalangeal syndrome type II, the syndrome comprises a combination of a bulbous nose, sparse hair, large, poorly lobulated and everted ears, multiple exostoses, and cone-shaped epiphyses of the bones of the hand. Mental retardation is a significant feature. The genetics are uncertain. Chromosome 8 deletions occur in some.

Meckel syndrome—exencephalocele, microphthalmus, polydactyly, polycystic kidney, cleft lip/palate. Sibs are reported by Hsia *et al.* (1971) and Crawfurd *et al.* (1978).

Rubinstein–Taybi syndrome—mental retardation, broad thumbs and great toes, beaked nose, antimongoloid slant to eyes, small maxilla (Figs 2.3 and 2.4).

Seckel syndrome—severe short stature, bird-like facies with prominent nose.

Smith–Lemli–Opitz syndrome—see page 10.

Williams syndrome—microcephaly is often mild. Supravalvular aortic stenosis, transient hypercalcaemia, prominent lips, and a elfin-like face

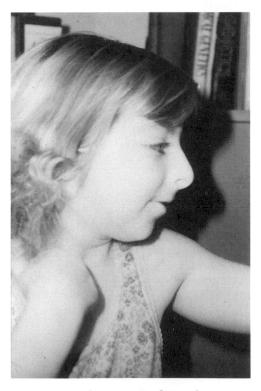

Fig. 2.3   Rubinstein–Taybi syndrome.

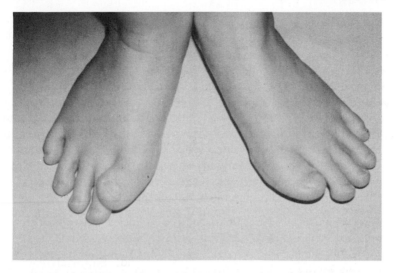

Fig. 2.4 Feet in Rubinstein–Taybi syndrome.
Courtesy of the Department of Medical Illustration, Great Ormond Street Hospital for Sick Children, London.

(short palbebral fissures, a prominent epicanthic fold, stellate pattern in the iris, anteverted nostrils and a long philtrum are the facial features). Most cases are sporadic.

Microcephaly with cervical spine anomaly

Two brothers, the product of a consanguineous union, were short, had fusion of cervical vertebrae, a funnel-shaped chest, kyphosis, and lordosis (Zackai *et al.* 1972). Both had the typical features of the genetic type of microcephaly and were retarded.

Microcephaly, nystagmus, short stature, and epiphyseal dysplasia

Three brothers had this syndrome (Lowry and Wood 1975). The parents were unrelated and unaffected.

Agenesis of the corpus callosum

Agenesis of the corpus callosum is frequently associated with gyral anomalies including heterotopia, hypoplasia and microcephaly. Hydrocephalus was found in 16 cases of which three had the Dandy–Walker syndrome (the subject is reviewed by Parrish *et al.* 1979). The most frequent facial abnormalities were hypertelorism, malformed ears, and a hypoplastic mandible.

Of the approximately 210 cases reported in the literature up to the mid-1960s there were only five families with more than one affected (Shapira and Cohen 1973). About 13 per cent of cases are asymptomatic (Slager *et al.* 1957). Both recessive and dominant X-linked forms are known.

Recessive—Von Ziegler (1958) described the condition in two brothers, and Naiman and Fraser (1955) in two sisters. Shapira and Cohen (1973) described two sisters who were the product of first-cousin parents. Both sisters were mildly retarded. Three sibs (unrelated parents) had a clinical syndrome including infantile spasms with hypsarrythmia, microcephaly, severe mental retardation, and a spastic quadriparesis (Cao *et al.* 1977). It is uncertain whether the brothers of Von Ziegler (1958) or those described by Wilson *et al.* (1983) fall into this or the following category. The sibs described by Wilson were both retarded.

X-lined—Menkes *et al.* (1964) reported five males in two generations who had intractable seizures and mental retardation. A post-mortem on one showed partial agenesis of the corpus callosum, polymicrogyria, and heterotopia of grey matter. Another X-linked pedigree with partial agenesis had in addition macrocephaly and an imperforate anus (Opitz and Kaveggia 1974). (See page 42 the FG syndrome.) A third family of X-linked recessive agenesis of the corpus callosum was reported by Kaplan (1983).

Dominant—This is a rare mode of inheritance, but a father and son were reported by Lynn *et al.*(1980). Both had enlarged heads and the boy was in the low–normal intellectual range.

Counselling

Recurrence risks in the absence of a family history are small.

Agenesis of the corpus callosum with hypothermia

The condition has been described in sibs (Pineda *et al.* 1984). Both children died in the first months of life. A post-mortem revealed, in addition to the abnormalities of the corpus callosum, extensive spongiosis of the white matter.

The condition is not confined to children. A single case in an adult was associated with hypothermia (episodic) and diaphoresis. Spongiosis was not a feature (Shapiro *et al.* 1969, Lewitt *et al.* 1983).

Lipomas of the corpus callosum

More than 80 cases have been reported, but all are sporadic.

Other associations with agenesis of the corpus callosum

(a) Interhemispheric cysts (Diebler and Dulac 1987).
(b) With agyria (Josephy 1944).
(c) Frontonasal dysplasia.
(d) FG syndrome.
(e) Optic nerve hypoplasia.
(f) Vermis hypoplasia.
(g) Numerous chromosomal abnormalities (see Diebler and Dulac 1987).
(h) Acrocallosal syndrome—with pre-axial polydactyly (Schinzel 1988).

(i) Agenesis of the corpus callosum with macrocephaly

A brother and sister had this combination with mental retardation (Young *et al.* 1985). The brother also had unilateral cerebellar hypoplasia. It should be noted that agenesis of the corpus callosum and macrocephaly can occur in the FG syndrome—see page 42.

(j) Agenesis of the corpus callosum with anterior horn cell disease (Charlevoix County disease)

Two brother presented with mental retardation in two separate unrelated French-Canadian families (Andermann *et al.* 1972). The two brothers came from the same region as the two sisters reported by Naiman and Fraser (1955) (see above). Recently, Andermann *et al.* (1977) have found 45 patients in 24 sibships with this condition in an inbred community in Charlevoix County, Quebec.

(k) Aicardi syndrome

In 1965, Aicardi *et al.* described a syndrome characterized by agenesis of the corpus callosum, infantile spasms, and punched out retinal lesions. Since then, 70 cases have been reported. A chorioretinopathy, mental retardation, microcephaly, vertebral anomalies (fusion or multiple hemivertebra), and poor life expectation are now considered as part of the syndrome (De Jong *et al.* 1976). The electroencephalogram shows hypsarrhythmia and independent electrical activity of the two hemispheres (Dennis and Bower 1972).

Only females have been described and the pattern fits that of an X-linked dominant with lethality in the affected males. No two-generation families have been reported. Recently (Curatolo *et al.* 1980*a, b*), a male with Aicardi syndrome was reported. He was not unduely severely affected.

Four females with the syndrome are reported by Denslow and Robb (1979). These authors reviewed the ocular manifestation of 53 patients reported in the literature. Increased retinal pigmentation, epipapillary glial tissue, and colobomas were the most frequent findings. Microphthalmia and retinal detachment were other complications.

The other CNS abnormalities include porencephaly, cortical heterotopia, papillomas of the choroid plexus, lissencephaly, poly and microgyria and even holoprosencephaly. (Sato *et al.* 1987). In a review of six Japanese cases (Yamamoto *et al.* 1985), all were female and all were sporadic. Most parents can be reassured after the birth of a child with Aicardi syndrome that recurrence risks are small.

Cavum septum pellucidum and cavum vergae

It should be noted that 97 per cent of premature infants have a cava of the above structures and that by 1 month they disappear. Only occasionally will a cyst of the septum pellucidum cause hydrocephalus.

Absent septum pellucidum and porencephaly

This association has been reported on three occasions (Aicardi and Goutieres 1984; Shimozawa *et al.* 1986). Clinically, there is often a congenital hemiplegia or double hemiplegia with mental retardation. All cases have been single.

Schizencephaly

It should be noted that schizencephaly (clefts in the region of the primary cerebral fissures, usually bilateral and symmetrical) may have the abnormalities listed in the previous section. Seven of the eleven cases reviewed by Miller *et al*. (1984) had midline lesions.

Cerebral gigantism (Sotos syndrome)

The diagnosis is suggested by a large baby with accelerated growth during childhood, advanced bone age, but normal eventual adult stature, early eruption of teeth, a large head, prominent jaw, dolichocephaly, high arched palate, and large hands and feet. Mental retardation is variable and not always present. Most cases are sporadic, but three families are reported by Zonana *et al*. (1977) with a dominant pattern of inheritance.

Evidence is accumulating that the condition is autosomal dominant. A father and four offspring were reported by Winship (1985) and a further family by Bale *et al*. 1987 included a mother and two children.

A comprehensive study of the phenotype based on 22 cases was published by Wit *et al*. (1985). Two subjects have reached an exceptionally tall stature. It should be noted that all were not large at birth. CT scan abnormalities include mild enlargement of the ventricles, although occasionally other abnormalities including agenesis of the corpus callosum are found.

Craniofacial abnormalities

Only those conditions of interest to neurologists will be mentioned.

A. Craniosynostosis syndromes.
B. Midfacial syndromes:
   (1) holoprosencephaly;
   (2) frontonasal dysplasia.

A. Craniosynostosis

Craniosynostosis has an incidence of 4 per 10 000 live births (Hunter and Rudd 1976). Sutures are normally open at birth, are interdigitated by 7½ months, and closed by the fourth decade. Normal growth of the skull takes place at right angles to the sutures. Thus, coronal growth adds length to the forehead, the lambdoids length to the posterior part of the head, and the sagittal to the breadth (Gordon 1959). The shape of the skull bears some relationship to the order in which the sutures are closed (Armendares 1970).

Within a family where more than one member is affected there is not always a close correlation between the suture involved nor a relationship between the prematurely stenosed suture and the shape of the head. Patients with the same genetic disorder might have premature fusion of different sutures. A skull which is small but of normal shape is unlikely to be the result of craniosynostosis unless all the vault sutures are equally inactive (Till 1975). In this series, the sagittals were affected by premature closure in 40 per cent, both coronals in 20 per cent, one coronal in 16 per cent, all in 9 per cent, sagittal and coronal in 9 per cent, metopic in 4.5 per cent, and sagittal and lambdoid in 1 per cent of cases.

Clinical terminology

A head that is too long is dolichocephalic, whereas a head that is too short is brachycephalic. In scaphocaphaly, the head is in the form of an inverted boat, whereas in oxycephaly it is keel shaped or sharp. The acrocephalic skull has a short anterior posterior diameter with a flat occiput and high forehead. Plagiocephaly is an asymmetrical skull shape—often in the form of a rhomboid when viewed from above.

Relationship between the clinical picture and the suture

Sagittal—premature fusion results in a dolichocephalic or scaphocephalic head shape.

Coronals—when both are prematurely fused an acrocephalic or brachyce-phalic skull results.

One coronal—results in a flat forehead and prominent eye on that side—plagiocephaly.

All sutures—presents a striking appearance, often resulting in a small head. However, oxycephaly or acrocephaly can also occur.

Metopic—the patient presents with a prominent keel-shaped mid-forehead.

Sagittal-coronal—when retardation of bony growth begins before the closure of the anterior fontanelles, bone surrounding that fontanelle is elevated to a turret-like deformity (Till 1975). When the retarded growth and early closure of the sutures occur after the fontanelle closes the skull is broad anteriorly and elongated in the anterior—posterior dimension.

Genetics of 'simple' craniosynostosis without an identifiable syndrome

In 'simple' craniosynostosis, neither prominent craniofacial dysmorphism, nor digital anomalies should be present. The condition most frequently missed and therefore erroneously included in the 'simple' or non-syndrome group is the Saethre–Chotzen syndrome. The presence of ptosis, plagiocephaly, abnormal ear crus, a sloping forehead, and skin syndactyly should suggest Saethre–Chotzen syndrome.

The other condition easily mistaken for simple craniosynostosis is Crouzon syndrome. The facial features, including the head shape, can change dramatically with age without operative intervention and only by looking at early photographs is it possible to exclude the condition.

Family studies of 'simple' coronal craniosynostosis

A sib had premature closure of the coronal suture in three out of 104 families reported by Hunter and Rudd (1977). There were approximately 146 unaffected sibs. Twenty-two patients with uni- or bilateral coronal synostosis had 26 sibs of whom two were similarly affected (Carter *et al.*, 1982).

Family studies of simple sagittal craniosynostosis

In a recent study (Carter *et al.* 1982) the 109 sibs of probands with premature closure of the sagittal suture were all unaffected.

In a study of sagittal synostosis (Hunter and Rudd 1976), out of 218 sibs of the index patients, three were affected.

Counselling

Unaffected parents after the birth of an affected child with premature fusion of the sagittal suture have a 1 per cent recurrence risk. Risks to offspring of those affected are in the vicinity of 2 per cent for sagittal synostosis (these are approximate figures). Where the coronal sutures (either bilateral or unilateral) are involved, risks to sibs are higher—about 6 per cent (Carter *et al.* 1982), but risks to offspring are not yet known.

The author has now seen, on a number of occasions, patients (especially infants) who appear to have 'simple craniosynostosis' where a parent has suggestive features and in early pictures definite features of the same condition. Simple craniosynostosis should not be diagnosed and a smallish recurrence risk given without due consideration.

Trigonocephaly

Trigonocephaly results from premature closure of the metopic sutures. It occurs in 4–16 per cent of craniostenosis (Anderson and Geiger 1965; Shillito and Matson 1968). Very few cases are described in which trigonocephaly is familial. A mother and son with neonatal trigonocephaly and multiple suture synostosis is described by Hunter *et al.* (1976). Both patients were of normal intelligence and had no major malformations. A similar family also with normal intelligence and showing transmission as a dominant is reported by Frydman *et al.* (1984). Trigonocephaly in association with mental retardation occurs with the deletion of the short arm of a number 9 chromosome and the long arm of chromosome 11 or 3.

Trigonocephaly C (Opitz C) should be excluded. This recessive syndrome is characterized by a prominent metopic suture, upslanting palpebral fissures, and severe mental retardation. Other features include contractures, polysyndactyly, aberrant oral frenulae, and dysplastic ears. Recent experience with this syndrome emphasizes the difficulty in clinically recognizing it. It might be that a wide variety of chromosomal abnormalities cause these clinical features and the syndrome as a recessive could be extremely rare (Sargent *et al.* 1985).

Syndromes involving premature closure of sutures (based on Cohen 1979 and Cohen 1986)

1. Craniosynostosis and radial aplasia (Baller–Gerold syndrome)

An infant with ocular hypotelorism and a prominent metopic ridge, continuous with a prominent nasal bridge, was reported by Greitzer *et al.* (1974). At post-mortem there was polymicrogyria. The radial defects were bilateral. In a previous report (Baller 1950) the parents were consanguineous. In a report of a single case (Anyane-Yeboa *et al.* 1980) the child had in addition an imperforate anus and ectopic kidneys.

2. Craniostenosis with radio-ulnar synostosis (Berant syndrome)

Sagittal suture synostosis in four sibs (of whom two were identical twins) was reported by Berant and Berant (1973). The radio-ulnar lesions were present in the mother and daughter (unilaterally) and bilaterally in her twin sons. The mother had no evidence of craniosynostosis. Radio-ulnar synostosis is occasionally present in Saethre–Chotzen syndrome.

3. Craniosynostosis and the absence of the fibula (Lowry syndrome)

Two sibs, the offspring of second-cousin Greek parents, had this combination. Both brothers had cryptorchidism. The coronals were the sutures involved although in one the sagittal suture was in addition prematurely closed (Lowry 1972).

4. Craniosynostosis, brachydactyly, symphalangism, strabismus, hip osteochrondritis, and carpotarsal fusion

An Italian pedigree with five affected members in three generations is reported by Ventruto *et al.* (1976). No other pedigrees are known.

5. Craniosynostosis and acrocephalopolydactylous dysplasia

Two sibs (first-cousin parents) had hexadactyly of the upper limbs, premature fusion of all the sutures, acrocephaly, multiple anomalies of nose, face, auricles, and greatly increased thickness of the skin (Elejalde *et al.* 1977). Both sibs were of giant size due to the excessive amount of connective tissue deposited everywhere except in the cerebrum.

6. Craniosynostosis with additional congenital abnormalities (Gorlin-Chaudhry-Moss syndrome)

Two sibs had translucent grey teeth, patent ductus arteriosus, hypoplasia of the labia majora, and hypertrichosis (Gorlin 1960). The facial features included defective upper eyelids, an antimongoloid slant to the eyes, severe dysostosis of the maxilla, zygoma, and nasal bones.

7. Sensenbrenner syndrome

Five children including sibs had in common dolichocephaly (with sagittal synostosis in three), epicanthic folds, sparse fine hair, hypodontia, brachypodia, and brachydactyly (Levin *et al.* 1977). The children were of small stature.

8. Christian syndrome: craniosynostosis, microcephaly, laryngoma lacia, cleft palate, and preaxial polysyndactyly

Ophthalmoplegia, antimongoloid stant to the eyes, muscular fasciculation, and eventual death was described in a sibship (Christian *et al.* 1971a, b). Neuropathological studies showed dysmyelination. Three of the four children were from the Amish community and inheritance is recessive. All those affected had adducted thumbs and the authors suggest that 'The adducted thumbs syndrome' is an appropriate name.

Other craniofacial syndromes (based on Cohen 1979)

9. Craniofacial dyssynostosis

Two sisters of Spanish ancestry with craniosynostosis, craniofacial dysostosis (broad forehead, hypertelorism), and short stature, were reported by Neuhauser *et al.* (1976*b*). Hypoplasia of the maxilla, a high narrow palate, and nasal obstruction were part of the clinical picture.

10. Summitt syndrome

The association of craniosynostosis, obesity, strabismus, syndactyly of digits in hands and feet with normal sized thumbs and toes has been described in sibs (Summitt 1969*a*; Sells *et al.* 1979).

11. Herrmann syndrome I

The combination of craniosynostosis, mental retardation, hypoplastic supraorbital ridges, bitemporal flattening, and ocular hypertelorism, was reported by Herrmann and Opitz (1969) in a single case. This may constitute a separate syndrome but resembles the aminopterin syndrome.

12. FG syndrome

A large head, frontal bossing, mental retardation, high narrow palate, imperforate anus, sacral dimple, and partial syndactyly of toes 2–3 constitute a distinct syndrome (Optiz and Kaveggia 1974). One patient had a severe craniosynostosis. Inheritance is X-linked recessive. Three brothers are reported by Riccardi *et al.* (1977).

Many more cases have now been reported (Thompson and Baraitser 1987). All cases should have severe neonatal hypotonia and in the absence of an imperforate anus, severe constipation. Some affected males have been subjected to a rectal biopsy to exclude Hirschprung's disease. A CT brain scan might reveal an agenesis of the corpus callosum.

13. Herrmann syndrome II

Craniosynostosis with both radial and fibular aplasia, in addition to cleft lip and nasal speech is reported as a single case by Ladda *et al.* (1978).

A similar syndrome was reported by Herrmann *et al.* (1969). The patient had an absent thumb, unusual ears, and severe varus deformities of the feet. The other features were the same as the case reported by Ladda *et al.* (1978).

Acrocephaly polysyndactyly

Type I: Noack syndrome

First described by Noack in 1959 in a father and daughter, type I is characterized by mild acrocephaly, broad big toes, syndactyly and brachydactyly in the feet with or without syndactyly of the fingers. The polydactyly in Noack's original case involved a duplication of the whole of the first toe including the metatarsal bone. This condition is probably the same as Pfeiffer syndrome.

Type II: Carpenter syndrome

Acrocephaly, soft tissue syndactyly of digits 3 and 4 with brachymesophalangy, preaxial polydactyly, short stature, coxa valga, obesity, mental retardation, and congenital heart defect was reported by Carpenter (1901) who described two sisters with this combination. Inheritance is recessive.

The syndrome has also been reported in two children with normal intelligence (Frias *et al*. 1978).

Recently, Robinson *et al*. (1985) have reported three further sibs and reviewed the literature. They again emphasize that intelligence might be normal. Obesity with an onset in infancy persisted into adolescence. Early motor delay can resolve.

Up to 1979 33 per cent of the reported cases had congenital heart defects (Cohen 1979).

Acrocephalopolysyndactyly III

The combination of acrocephalopolysyndactyly, cardiac disease, polydactyly of the fingers, and polydactyly and syndactyly of the toes, bowed, femurs, hypoplastic tibiae, and deformities of the ears was reported by Sakati *et al*. (1971). This was a single case.

Acrocephalopolysyndactyly IV

Three sibs in a Jewish–Iranian family had acrocephalopolysyndactyly, brachydactyly, clinodactyly, camptodactyly, and marked ulnar deviation (Goodman *et al*. 1979). The parents were unaffected. Intelligence was normal and intrafamilial variation was considerable. Differentiation from Carpenter's syndrome is not easy as a patient with the latter syndrome and normal intelligence has been reported (Frias *et al*. 1978). The clinical picture of type IV is similar to the Summitt syndrome (see p. 42), but polydactyly has not been reported in the Summitt syndrome.

Greig's cephalopolysyndactyly syndrome

In 1926 Greig described a mother and daughter with symmetrical webbing of the fingers of both hands, polydactyly (in daughter), and a broad, prominent forehead. Since then families have been reported by Temtamy and McKusick (1969), Marshall and Smith (1970), Duncan *et al.* (1979), Baraitser *et al.* (1983*b*), and Gollop and Fontes (1985). Craniosynostosis is not a feature. Tommerup and Nielsen (1982) reported a balanced 3:7 translocation in six affected members of a family.

Hootnick–Holmes syndrome

The clinical features are similar to Greig's cephalopolysyndactyly syndrome. Post- and preaxial polysyndactyly occur and craniosynostosis is a feature (see above). In the original description, Hootnick and Holmes (1972) reported the association in father and son.

Acrocephalosyndactyly

The classification of the acrocephalies is uncertain as entities thought previously to be separate are now thought to be variable expressions of a single condition.

Acrocephalosyndactyly	type I	typical Apert syndrome	autosomal dominant
Acrocephalosyndactyly	type II	Vogt, pseudo-Crouzon. Probably not a separate entity	autosomal dominant
Acrocephalosyndactyly	type III	Saethre–Chotzen	autosomal dominant
Acrocephalosyndactyly	type IV	Waardenberg	?
Acrocephalosyndactyly	type V	Pfeiffer	autosomal dominant

Cohen (1979) regards Vogt, pseudo-Crouzon, and Waardenberg acrocephalosyndactyly syndromes to be spurious entities.

Type I: Apert syndrome (Fig 2.5)

Apert syndrome has as its main features, a tower-shaped skull, a flat occiput, maxillary hypoplasia, downslanting palpebral fissures, strabismus, hypertelorism, flattened midface, high arched palate, mitten hands (fusion of digits 2 to 4 with a mid-hand digital mass) (Fig. 2.6), and fusion of toes 2 to 4. In more than 150 reported cases of Apert all but four have been sporadic (Cohen 1979)—see reports of Blank (1960), Bergstrom *et al.* (1972), and

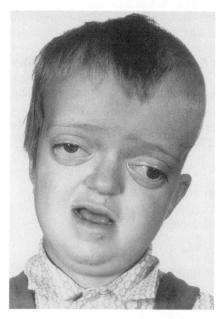

Fig. 2.5   Apert syndrome.

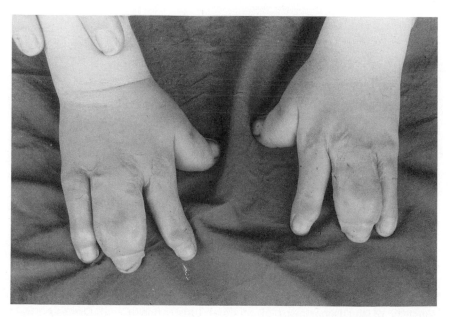

Fig. 2.6   Apert syndrome showing hand after surgery.

Weech (1927) for dominant pedigrees. Recurrence risks when both parents are normal are low. Blank (1960) reported an incidence of 1 in 100 000. Mental retardation is said to occur in a one-third of cases (Cohen 1979) and in some instances there can be considerable hydrocephalus. The co-existence of Apert and Pfeiffer syndromes has been reported in a single kindred (Escobar and Bixler 1977), but this might be a different, possibly allelic genetic condition (Cohen 1979). It should be noted that the survivors are not all seriously mentally handicapped. In a recent report about half had meaningful employment (Patton *et al.* 1988).

Type II: Vogt or Apert–Crouzon syndrome

Probably not a separate entity and will be omitted.

Type III: Saethre–Chotzen syndrome (Fig. 2.7)

This is characterized by a combination of a low-set frontal hairline, asymmetry of cranium and face, ptosis, telecanthus, heavy lateral, but sparse medial eyebrows, low-set posteriorly rotated ears, a prominent metopic

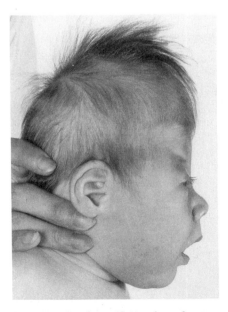
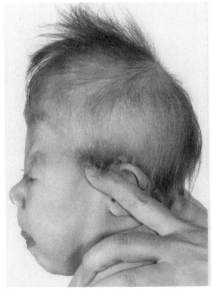

Fig. 2.7   Saethre–Chotzen syndrome.
Courtesy of the Department of Medical Illustration, Great Ormond Street Hospital for Sick Children, London.

ridge, cleft palate (occasionally), brachydactyly, and cutaneous syndactyly of fingers 2–3. The auditory crus forms a prominent horizontal bar. The toes and thumbs are normal. The pedigrees of both Saethre (1931) and Chotzen (1932) showed two generations affected, and other dominant pedigrees are reported by Kreiborg *et al.* (1972), and Aase and Smith (1970).

The condition was reviewed by Bianchi *et al.* (1985). The most constant features are acrocephaly, low frontal hairline, ptosis, facial asymmetry, a prominent ear crus and other than soft tissue syndactyly, relatively normal hands. Parietal foramina have been reported (Thompson and Baraitser 1984).

Type IV: Waardenberg syndrome

A single case of a male infant with acrocephaly, a beaked and pointed nose, hypertelorism, cleft palate, congenital cataracts, ambiguous genitalia, and limb anomalies was reported by Waardenberg (1934). The limb anomalies were a partial soft tissue syndactyly in the hands, duplication of the distal phalanx of the index and middle fingers which led to a bulbous termination of the phalanx with two nails. There were four toes on each foot. No other cases are known (Temtamy and McKusick 1969).

Type V: Pfeiffer syndrome (Figs 2.8 and 2.9)

This condition is characterized by acrocephaly, normal intelligence, hypertelorism, antimongoloid slant to the eyes, prominent eyes, bossed forehead, flat nasal bridge, soft tissue syndactyly, broad thumbs and toes, brachymesophalangia, ulnar deviation of the proximal phalanx of the thumb, radially deviated phalanges of the fingers, and a medially displaced big toe. Inheritance is dominant (Martsolf *et al.* 1971; Saldino *et al.* 1972; Jackson *et al.* 1976; Naveh and Friedman 1976). Variable expression can make the diagnosis difficult (Baraitser *et al.* 1980*a*). These authors describe a three-generation family in which the proband had the full Pfeiffer syndrome whereas at least four other members had only partial syndactyly and large big toes. Recently, Cohen (1986) has suggested that this family is more likely to have the same condition as that described by Jackson *et al.* (1976) (see below). The difference is in the normality of the thumbs in this latter condition.

Robinow–Sorauf syndrome

A large kindred was reported by Robinow and Sorauf (1975). Intra-familial variation was so great that some members had a clinical picture suggestive of Crouzon syndrome, whereas others could be classified as

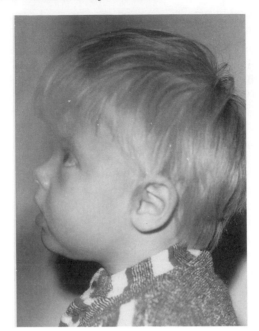

Fig. 2.8   Pfeiffer syndrome.

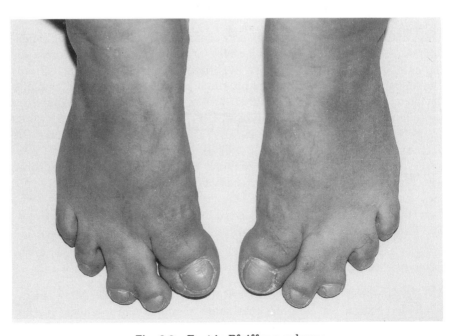

Fig. 2.9   Feet in Pfeiffer syndrome.

acrocephalosyndactyly type V—Pfeiffer syndrome. Although called by the authors Noack syndrome the duplication involved only the terminal phalanx with lateral displacement of the big toe. It is still disputed whether this is separate from Saethre–Chotzen syndrome. Facially, they are alike, but differ in that patients with so-called Robinow–Sorauf have a broad, big toe and to a lesser extent, a broad thumb with duplication of the distal digits of the big toe on radiography. The inheritance pattern is dominant.

Jackson–Weiss syndrome

A very large pedigree with many affected members was reported by Jackson *et al.* 1976. There was considerable variation within the pedigree. Some had skin syndactyly of toes 2 and 3 and medialy deviated broad great toes. [Cohen (1986) thinks that the family reported by Baraitser *et al.* (1980) had this syndrome and not Pfeiffer Syndrome.] One member had pre-axial polydactyly in the feet and one had fusion of the phalanges of the thumb. Facially, there was ocular hypertelorism, proptosis, a broad nasal root, and a high forehead.

B. Mid-facial syndromes

Holoprosencephaly (Fig. 2.10)

The terms holoprosencephaly, holotelencephaly, and arhinencephaly have been used interchangeably.

Arhinencephaly refers to a malformation of the forebrain characterized by an absence of the olfactory tracts and bulbs. Because the rhinencephalon is not completely or invariably absent and the initial defect is of the prosencephalon, De Myer *et al.* (1964*b*) prefer the term holoprosencephaly. Holotelencephaly excludes malformation of the diencephalon. Embryologically, the prosencephalon divides into the diencephalon and the telencephalon. From the diencephalon develops the thalamus, epithalamus, and hypothalamus. Where the diencephalon is affected by congenital malformation, pituitary agenesis and secondary endocrine abnormalities (adrenal aplasia) occur. From the telencephalon the cerebral hemispheres develop. The olfactory tracts and bulbs arise from the anterior part of the telencephalon. All the conditions have a variable degree of fusion of the cerebral hemispheres, hence the preference for the term holoprosencephaly for the whole group.

More recently, Probst (1979) has challenged the older classification of De Myer and Zeman (1963) in which the three pathological categories of alobarprosencephaly, semilobarprosencephaly, and lobar holoprosencephaly, are used to describe a spectrum of malformation with lobar holoprosencephaly the most organized.

In the older classification cyclopia, ethmocephaly, cebocephaly, and premaxillary agenesis were associated with alobar holoprosencephaly. Lobar holoprosencephaly was clinically associated with microcephaly or trigono-cephaly, hypotelorism, or normally spaced eyes.

It is, however, clear that the clinical relationship between the cerebral malformation and the facial defect is imperfect, and the division into lobar, alobar, and semilobar types inadequate. Corpus callosal defects and arhin-encephaly should not be a part of the classification of the prosencephalies.

The genetic contribution is small in all groups. An incidence of holoprosencephaly has been assessed at between 1 in 16 000–54 000, but a recent study from Bristol (Saunders *et al.* 1984) suggests an incidence of 1 in 5200 (two of these were familial).

The clinical conditions of the mid-face are associated with holopros-encephaly, are listed.

(a) Cyclopia—a single orbital cavity associated with a probosis projecting above the orbit.

(b) Ethmocephaly—where two orbits form, but are situated close together. The probosis or nose when present is positioned between them.

(c) Cebocephaly—an association of orbital hypotelorism with a single nasal cavity correctly placed, but with no septum or columella.

(d) Premaxillary agenesis—orbital hypotelorism with median cleft lip and absent philtrum.

(a) Cyclopia and clefts

Most cyclopian fetuses are sporadic and two-thirds are thought to have a chromosomal abnormality (Mollica *et al.* 1979). Nishimura (1970) found the frequency of cyclopia and cebocephaly to be 5 in 1000 in induced abortions and 0.06 per 1000 in the newborn. Sibs with cyclopia and cebocephaly (Klopf-stock 1921) and sibs with ethmocephaly and cebocephaly are known (Welter 1968). Parental consanguinity was reported by Klopfstock (1921) and Grebe (1954). In a pedigree reported by Cohen and Gorlin (1969) a cyclops was the product of a consanguineous mating.

A remarkable pedigree was published by Pfeitzer and Müntefering in 1968. Four cyclopian stillborn infants appeared over two generations in a single family. They were the offspring of three sisters and the daughter of one of their normal daughters' children. An unbalanced translocation involving a C-group chromosome was found.

(b and c) Ethmocephaly and cebocephaly

Trisomy 13 (Lazjuk *et al.* 1976) and deletion of the short arm of chromosome 18 (Gorlin *et al.* 1968) have both been found in cebocephaly (Holmes *et al.* 1974; Uchida *et al.* 1965), but there might be a bias toward reporting those cases with chromosomal abnormalities. It is suggested that chromosomal abnormalities are more likely to be found when cebocephaly is accompanied by visceral malformation (Lazjuk *et al.* 1976).

Under the title 'familial cebocephaly', James and Van Leeuwen (1970) described two sibs and report hearsay evidence of a cousin similarly affected. There was no history of consanguinity. The infant boy had a single nostril, orbital hypotelorism, but no clefting. Post-mortem revealed fused hemispheres, fused thalami, and lateral ventricles with no corpus callosum. There was agenesis of the pituitary and adrenals, and thyroid hypoplasia.

Three sibs, one with cyclopia and two with holoprosencephaly, are known (Dominok and Kirchmair 1961).

(d) Recessive holoprosencephaly associated with a midline cleft philtrum (Fig. 2.10)

Two sibs with the typical midline philtrum and lip defect are described by (Hintz *et al.* 1968). A post-mortem on one showed holoprosencephaly, an absence of the pituitary gland, and hypoplasia of the adrenals. A further sibship (De Myer *et al.* 1963) is known, with normal karyotypes.

Two sisters with median cleft lip had an alobar holoprosencephaly (Godeano *et al.* 1973). The parents were healthy, unrelated Ashkenazi Jews. Karyotypes were normal. Sibs in two families were reported in the same paper by Godeano *et al.* (1973) as a personal communication from Roach *et al.* (1975), and two sibs were reported by Nivelon-Chevallier and Nivelon (1975).

Dominant holoprosencephaly

The families of Dallaire *et al.* (1971) and Patel, *et al.* (1972*a*) were evaluated by Lowry (1974). Patel, *et al.* (1972) described a single case, but Lowry points out that the mother had a single centrally placed medial incisor tooth, suggesting that the inheritance could be autosomal dominant with variable expression.

Another dominant pedigree in which mother and seven of her children were affected is reported by Cantu *et al.* (1978). The mother was minimally affected. She was mildly retarded, microcephalic, and retained both deciduous canines. No X-ray or pathological evidence of a cerebral disorder was presented.

A family (Dallaire *et al.* 1971) in which the proband had at least four first cousins affected by holoprosencephaly is complicated by the presence of other members with cleft lip and palate, and congenital abnormalities. The authors propose that all the malformations are produced by a single abnormal gene with variation in expression and penetrance.

A large family with six affected members over three generations was

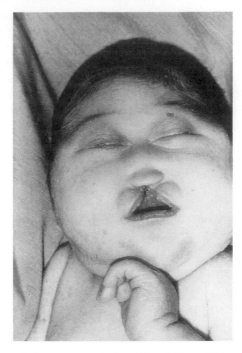

Fig. 2.10   Holoprosencephaly with midline philtrum defect.
Courtesy of Dr Martin Crawford.

reported by Benke and Cohen (1983). Transmission was through unaffected members.

Another autosomal dominant pedigree is reported by Berry *et al*. 1984. Two sibs had classical holoprosencephaly whereas the father and his sister were found to have single central maxillary incisors and hypotelorism. The following chromosomal abnormalities have been found in holoprosencephaly: in addition to those of chromosome 13 and 18. 3p duplication and trisomy 21 (Gillerot *et al*. 1987).

Genetic counselling in holoprosencephaly irrespective of clinical type

All patients should have a karyotype. If this is not possible the parents should be checked. The recurrence risk for sibs of a single case is 4–5 per cent. This refers to the holoprosencephaly group as a whole. Prenatal diagnosis is possible using ultrasound scanning of the brain and should be used where possible.

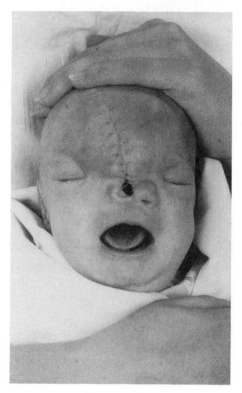

Fig. 2.11 Frontonasal dysplasia with encephalocele.
Courtesy of the Department of Medical Illustration, Great Ormond Street Hospital for Sick Children, London.

Frontonasal dysplasia (median facial cleft syndrome) (Fig. 2.11)

The combination of ocular hypertelorism, broad nasal root, medial clefting of the nose, a widow's peak hairline, cranium bifidum occultum with cleft lip and/or palate, constitutes a distinct entity. An anterior encephalocele might be a part of the syndrome. The primary defect is a failure of the nasal capsule to develop, allowing the primitive brain vesicle to occupy the space normally taken by the nasal capsule (Stewart 1978). The brain is displaced inferiorly.

Sedano *et al.* (1970) suggests a subdivision into four types.

(a) Hypertelorism, broad nasal root, medial nasal groove, and a bifid nasal tip. True clefting in the midline is absent.

(b) Hypertelorism, a broad nasal root, deep median facial groove, and true cleft affecting the nose or nose and upper lip and sometimes the palate.

(c) Hypertelorism, broad nasal root, and unilateral or bilateral notching of the ala nasi.

(d) Combination of (b) and (c).

Frontonasal dysplasia is usually sporadic, but Cohen *et al.* (1971) have reported a three-generation family. The documentation of sibs (Fox *et al.* 1976) and half sibs (Warkany *et al.* 1973) suggest that this is not a homogeneous syndrome. Three affected members in an inbred pedigree are reported by Moreno-Fuenmayor (1980). Several other members had hypertelorism and the authors stress that relatives of those with the full spectrum of frontonasal dysplasia may have minor manifestations suggesting that in some cases (the minority) dominant inheritance might be operating. A mother and daughter with considerable hypertelorism, midline nasal flattening, and notching reported by Montford in 1929 probably have the same condition.

Greig's 1924 paper on ocular hypertelorism with mental deficiency describes patients who probably fall within this category.

Counselling

Risks to sibs are one or two in 100. Risks to offspring are less certain, but probably small.

Frontonasal dysplasia and craniosynostosis

This combination is described in three sibs by Slover and Sujansky (1979). Father and maternal grandmother had a mild form of the same syndrome suggesting dominant inheritance. Another sibship is reported by Pendl and Zimprich (1971).

The condition was recently reviewed by Young (1987). There is a preponderance of females and it is either dominantly inherited or X-linked dominant. A cleft lip and palate is an occasional feature.

Holoprosencephaly and polydactyly

There have now been two case reports of this combination (Young and Madders 1987; Moerman and Fryns 1988). Both were single events.

Atelencephaly

This is characterized by an absence of cerebral hemispheres and ventricular system, but not involving the skull nor the brainstem (Danner *et al.* 1985). All cases to date have been single.

Holoprosencephaly and endocrine dysgenesis

Two brothers were reported by Begleiter and Harris (1980) with holoprosencephaly and a micropenis. One had marked hypoglycemia. At postmortem on one, no pituitary could be seen.

A midline cleft syndrome resembling familial holoprosencephaly

Five members in a single family (three generations affected) had mental retardation, microcephaly, cleft lip and anterior cleft palate, hypotelorism, and anomalies of feet and spine (Martin *et al*. 1977). Those affected had an appearance indistinguishable from familial holoprosencephaly.

Hypertelorism (Fig. 2.12)

Hypertelorism, dysphagia, and hypospadias (G syndrome)

Opitz *et al*. (1969) described four brothers with this combination. The mother had minimal stigmata. A four-generation family (Pedersen *et al*. 1976) and

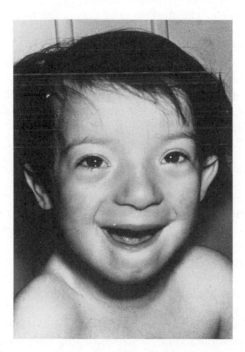

Fig. 2.12 Hypertelorism and mental retardation.

three brothers whose mother was mildly affected (Van Biervliet and Van Hemel 1975) suggest that the syndrome is either dominantly inherited with male sex limitation or X-linked. The major problems in these patients are due to aspiration and this was found in a two-generation family reported by Greenberg and Schraufnager (1979). The swallowing difficulties are related to laryngotracheoesophageal defects. Mental retardation occurs in about half of the cases.

Recent experience has confirmed dominant inheritance although expression does seem to be more severe in males. A cleft lip and palate and hydrops might be part of the clinical spectrum (Tolmie *et al*. 1987).

Telecanthus

A family with pure telecanthus had many affected members over five generations (Juberg and Hirsch 1971).

Telecanthus and/or hypertelorism with hypospadias (G/BBB syndrome)

This association was first noted by Optiz *et al*. (1965). A mother with hypertelorism had a son with hypertelorism and glandular hypospadias. Reed *et al*. (1975) reported six more families. Seven of those affected were males, only one female. The mothers in all the six families had hypertelorism and/or telecanthus. Inheritance might be either autosomal dominant with male sex limitation or X-linked.

Christian *et al*. (1969) described a five-generation pedigree in which ten individuals had telecanthus. Two of the four affected males had hypospadias. Mental retardation occurred in about a half of the cases. The differentiation between the G and BBB syndromes is uncertain. [See Cordero and Holmes (1978).]

Hypertelorism and persistent metopic fontanelle or suture

The metopic fontanelle is the anterior extension of the anterior fontanelle. Its persistence is noted in craniofacial dysostosis, cleidocranial dysostosis, parietal foramina, in association with an encephalocoele, and as a sporadic event (Tan 1972).

Familial occurrence of hypertelorism with a persistent metopic fontanelle was noted by the above author. The skull was brachycephalic and the condition occurred in mother and three of her offspring. In this family there was no association with mental retardation.

In Greig's (1924) original case of hypertelorism (Greig's syndrome) there was a persistent metopic fissure, a structure which is probably related to a metopic fontanelle.

Hypertelorism and mid-facial clefting (HMC syndrome)

Two sisters with hypertelorism, microtia, and clefting of the lip, palate, and nose are reported. Both were retarded and microcephalic. In addition, atretic auditory canals, ectopic kidneys, and congenital heart disease were features (Bixler *et al.* 1969; Baraitser 1982)

Macrocephaly, mental retardation, hypertelorism, kyphoscoliosis

Two brothers with this combination are reported by Jammes *et al.* (1973). The disorder was non-progressive, and either autosomal or X-linked recessive inheritance is likely.

Telecanthus with deafness (facio-oculo-acoustic-renal syndrome)

Sibs with epiphyseal dysplasia and deafness, true hypertelorism, and telecanthus are reported by Holmes and Schepens (1972).

The sib pair had several features in common with Waardenburg's syndrome, but the inheritance is different and neither child had a white forelock.

Other causes of hypertelorism and nervous system involvement

Williams syndrome

See page 32.

Multiple lentiginosis

See page 157.

Noonan's syndrome

Short stature, hypertelorism, antimongoloid eye slant, mental retardation (not a constant feature), ptosis, webbing of the neck, low-set ears, low hairline, cardiac lesions—especially pulmonary stenosis—and *café-au-lait* patches are the main features. Autosomal dominant inheritance is established (Baird and De Jong 1972; Allanson *et al.* 1987).

Leprechaunism (Donohue syndrome)

In the original report (Donohue and Uchida 1954) two sisters were described (consanguineous parents) with small grotesque facies, hypertelorism, saddleshaped noses, upturned nostrils, low-set ears, thick lips, large clitoris, and mental retardation. Males have a large phallus and all the children fail to thrive, with a poor prognosis. Hyperinsulinism is the rule.

Facio-digital-genital syndrome (Aarskog syndrome)

Short stature, hypertelorism, antimongoloid eye slant, ptosis, a broad nasal bridge, short nose, long philtrum, a shawl scrotum which surrounds the penis, and short fingers with clinodactyly are the main features. Inheritance is X-linked with occasional manifestation in females of hypertelorism or short stature.

Fetal aminopterin syndrome

Hypertelorism is a prominent feature in a syndrome of short stature, prominent eyes, triangular face with small chin, upswept hair, and sometimes mild mental retardation.

Wormian bones and delayed closure of the fontanelle are the main radiological features.

Hydrocephalus

The incidence of congenital hydrocephalus not associated with spina bifida is about 0.4 per 1000 live births and of this group one-third have an aqueduct stenosis. It is uncommon for the condition to present in more than one member of a family but when it does the inheritance is usually X-linked recessive. It is estimated that about 2 per cent of all cases are X-linked and in these cases the hydrocephalus is usually associated with stenosis of the aqueduct of Sylvius. Edwards *et al.* (1961) described 15 affected males in a single family. Since the first paper of Bickers and Adams (1949) who reported eight hydrocephalic males in three generations, there have been at least eight families with more than one affected child. A description of some of these follows.

Two families were reported by Needleman and Root (1963). In one, four males were affected in a single generation (parents unrelated), and in the other two half-brothers (same mother) were similarly affected. Abnormalities of the thumb (hypoplastic and flexed) have been reported by Warren *et al.* (1963) and previously by Edwards (1961). A large X-linked pedigree had members with a flexion–adduction deformity of the thumb in just under half of those males with hydrocephalus (Gilly *et al.* 1971). One had a post-mortem which confirmed the aqueduct stenosis.

Unfortunately, there is at present no way of recognizing the X-linked cases, although abnormalities of the thumbs are suggestive. Shannon and Nadler (1968) reported four males in two generations and Jansen (1975) reported males in successive generations. Six males were affected in two generations in the report by Faivre *et al.* (1976). All had adduction–flexion deformities of the thumb. In some instances the abnormal position of the thumb is due to corticospinal tract involvement, but as an absent thumb can be part of the syndrome the abnormality is not always secondary.

Holmes *et al.* (1973) reported a Wisconsin/Madison family in whom eight hydrocephalic males have occurred, and an extraordinary family in which three sisters had between them eight affected males with hydrocephalus and aqueduct stenosis was described by Søvik *et al.* (1977).

There is no anatomical characteristic of the stenosed aqueduct that will help differentiate between the sporadic and familial cases. Other families were reported by Cassie and Boon (1977), and Fried and Sanger (1973).

Three males with hydrocephalus were born in a sibship of seven (the four other sibs were normal males)(Abdul-Karim *et al.* 1964). The parents were second cousins.

Genetic counselling

1. The genetic contribution to simple hydrocephalus is relatively small— about 2 per cent.
   Empiric risks in males with non-communicating hydrocephalus:
   4 per cent risk to male sib;
   2 per cent risk to female sib.
2. The relationship to neural tube defects is unclear. Some authors have suggested a higher frequency of spina bifida in subsequent pregnancies— but not above 2 per cent.
3. Autosomal recessive inheritance is rare, but Raja Abdul-Karim *et al.* (1964) reported a family with two girls and a boy affected. The parents were related (first cousins once removed). A brother and sister with aqueduct stenosis were reported by Howard *et al.* (1981).
4. Twenty-five per cent of aqueduct obstruction in males could be X-linked. Other studies (Halliday *et al.* 1986) suggest a fequency of X-linkage among those with aqueduct stenosis as being between 13 and 23 per cent.
5. The recurrence risk after the birth of a single male with aqueduct stenosis is 12 per cent for another affected male (Burton 1979).

X-linked hydrocephalus without aqueductal stenosis

A family in which males had progressive head enlargement had three members with hydrocephalus, but without signs of obstruction (Willems *et al.* 1987). However, it was found that five other males in this family had been considered to have non-specific mental retardation. Three of these had biggish heads and were found to have moderate to severe hydrocephalus and two had normal ventricles, but were retarded. No fragile sites were found.

The authors suggest that the aqueduct stenosis might be the consequence of hydrocephalus in the syndrome of X-linked hydrocephalus (see above) and that all males with non-specific retardation and biggish heads need not only a karyotype looking for fragile sites, but also a CT scan (see also Varadi *et al.* 1987).

Familial aqueduct stenosis and basilar impression

Two brothers with this combination are reported by Sajid and Copple (1968).

Hydranencephaly

Potter (1961) defined hydranencephaly as an abnormality in which there is a complete or almost complete absence of the cerebral hemispheres although the meninges are in the correct position. Male twins with this anomaly are reported by Haque and Glasauer (1969), but vascular disturbances, irradiation, rubella, and toxoplasmosis (Muir 1959) have all been causally associated, and the genetic contribution is small.

Congenital hydrocephalus-hydranencephaly

Fowler *et al.* (1972) reported hydrocephalus occurring in four successive pregnancies. All those affected were females. A fifth pregnancy was a normal female, and the sixth a hydrocephalic female. All the pregnancies were complicated by hydramnios and all the infants had severe arthrogryposis. Autopsies were performed on two and showed a diffuse destructive process of the brain and spinal cord resulting in gliosis which blocked the aqueduct of Sylvius. There seemed to be a defect in the vascularity of the central nervous system with microscopic calcification.

A similar sibship was reported by Harper and Hockey (1983). Two sisters had an underlying proliferative vasculopathy throughout the central nervous system causing local ischaemia and destruction.

Hydrolethalus

This is predominantly a Finnish disorder characterized by hydrocephalus and early death. The jaw is usually small and post-axial polydactyly of the hands and pre-axial polydactyly of the feet are usually present. Occipito schisis extends to the foramen magnum (Salonen *et al.* 1981). A non-Finnish patient who has survived (till 5 months) is reported by Aughton and Cassidy (1987).

Multicystic encephalmalacia or polyporencephaly

In this condition there are multiple porencephalic cysts in both hemispheres. The genetic contribution is very small. Claireaux (1972) described a sibship with the condition.

There have been other families with familial porencephaly. Sibs are reported with microcephaly, a motor deficit and seizures, by Berg *et al.* (1983) and sibs are reported by Airaksinen (1984). A very unusual family is reported by Smit *et al.* (1984) in which mother and her two children have cavities in the white matter. No other autosomal dominant pedigrees are known.

In general recurrence risks are less than 1 in 100.

Lissencephaly (Miller–Diecker syndrome)

The brain in lissencephaly is small and without convolutional pattern. Dobyns *et al.* (1984) divides them as follows:

Type I—Classical Associated with specific craniofacial abnormalities in (a) and (b).
(a) Miller–Diecker syndrome
(b) Norman and Roberts syndrome
(c) Lissencephaly sequence

Type II Lissencephaly with hydrocephalus and other severe malformations of eye or brain, e.g. Walker–Warburg or H A R D ± E syndrome.

There are pathological differences between types I and II (Dobyns *et al.* 1984). In type I there is microcephaly and a thickened cortex with four rather than six layers. In type I I there is agyria, obstructive hydrocephalus, heterotopia, and other malformations.

This division is not conclusive—see under Lissencephaly sequence.

Type I (a)—Miller–Diecker syndrome Two sibs, born to unrelated parents, had lissencephaly (Miller 1963). Both infants had small heads, with low-set ears and small mandibles beneath prominent maxillae. There was failure to thrive and death occurred at 3 and 4 months, respectively. Two further sibs (Daube and Chou 1968) had an early onset of convulsions, severe motor retardation, an absent response to external stimuli, and death in infancy. A maternal first cousin was also affected (Diecker *et al.* 1969). Parental consanguinity was present in the sibship of Norman *et al.* (1976), providing further evidence that some cases are recessive. The sibs described by Reznik and Alberca-Serrano (1964) had marked hypertelorism and survived, despite intractable epilepsy and progressive spastic paraplegia, till the age of nine and 19 years. More recently Garcia *et al.* (1978) have reported a sib pair in which the diagnosis in the second child was highly suggestive on the computer tomography (CT) scan. Chromosomal analysis is essential because an abnormality of chromosomes has now been consistently found.

Indeed, the term Miller–Diecker syndrome is now reserved for those with lissencephaly who have vertical furrows on the forehead and who have a deletion of band 17p 13.3. All the families with more than one affected child have had a defect of 17p.

Type I (b)—Norman and Roberts syndrome   These authors (Norman *et al.* 1976) described a single patient with severe microcephaly, short sloping forehead, a prominent nasal root, and a small jaw. There have been no recent reports although Dobyns *et al.* (1984) suggest that the sibs described by Barth *et al.* (1982) may have had this condition.

Type I (c)—Lissencephaly sequence   These patients have severe neurological problems which include severe mental retardation, early hypotonia, subsequent spasticity, poor feeding, and seizures. A CT scan might show, in addition to the lissencephaly, enlarged ventricles and heterotopias.

Facially, there is microcephaly, bitemporal narrowing, a prominent occiput, and a small jaw (data from Dobyns *et al.* 1984). Most cases are single. This is probably the commonest type and probably results from unidentifiable intrauterine insults which interfere with neuronal migration. All patients need a karyotype performed.

Counselling — Recurrence risks are small. High risks only when parents carry a chromosomal rearrangement.

Type II   HARD ± E (Walker Warburg) see below

Other causes of lissencephaly (mostly partial)

COFS (cerebro-oculo-facial-skeletal) (see p. 31)
Neu-Laxova (lethal)
Cerebro-ocular dysplasia–muscular dystrophy (COD-MD) (see p. 398)
Congenital muscular dystrophy (Fukuyama syndrome) (see p. 397)
XK aprosencephaly syndrome (aprosencephaly and radial defects)

Neuronal migration defects (heterotopia and microgyria)

In microgyria there are small wandering gyri without intervening sulci or sulci which are bridged by fusion of the overlying molecular layers (Barth 1987). The term is used synonymously with polymicrogyria and micropolymicrogyria.

There are many causes which include infection (CMV) (Friede and Milcolaser 1978) and carbon monoxide poisoning *in utero*.

Genetic contribution

Meckel–Gruber syndrome
Fukuyama congenital muscular dystrophy
Zellweger syndrome
Neonatal adrenoleukodystrophy
Glutaric-aciduria Type II
HARD ± E
Menkes disease
Chromosomal disorders
Neurocutaneous syndromes
Joubert syndrome

Congenital hydrocephalus and cerebellar agenesis

In a sibship of three brothers, two died shortly after birth with congenital hydrocephalus and cerebellar agenesis. A maternal great-uncle might also have been affected. Post-mortem showed an absence of the foramina of Luschka and Magendie (Riccardi and Marcus 1978). The authors suggest that cerebellar agenesis may sometimes be a cause of familial congenital hydrocephalus. Despite similarities, the authors suggest that this is not the Dandy–Walker syndrome.

Hydrocephalus, agyria, retinal dysplasia, HARD ± E (encephalocele syndrome), or Warburg syndrome

There are now many families described with this condition. The syndrome is variable in expression. Agyria and hydrocephalus are the most constant features. Cerebellar dysplasia, agenesis of the corpus callosum, Arnold–Chiari malformation, an aqueduct stenosis, and a Dandy–Walker malformation, have all been reported as part of the condition (Whitley *et al*. 1983).

Sibs were reported by Pagon *et al*. (1978) and Chemke *et al*. (1975) where the sibs (three out of seven) were the offspring of a consanguineous marriage (third cousin Jewish Yemenite parents). The cerebral lesion consisted of lissencephaly and the Dandy–Walker malformation. Congenital cataracts, retinal dysgenesis, and coloboma of the choroid were found.

The eye abnormality now include anterior and/or posterior chamber defects with secondary cataracts and microphthalmia. Small genitalia are not infrequent. The infants are nearly always floppy and until recently it has been suggested that the hypotonia was of central origin. One report has now questioned this as the CPK was very raised and the differentiation from COMS or COD-MD (see page 398) is in dispute. They could be the same condition [see Donai and Farnden (1986) for a review]. Inheritance—autosomal recessive.

Hydrocephalus and polydactyly (Biemond syndrome II)

(Biemond 1934*b*) described a condition which has iris colobomata, hypogenitalism, hydrocephalus, retardation, polydactyly, and obesity as the main features. Three brothers were affected. It is uncertain whether this is a separate disorder. The sibship of Blumel and Kniker (1959), thought to be another example of this syndrome, did include in one sib deafness, and in another, blindness and deafness.

Benign familial macrocephaly

A head circumference which exceeds two standard deviations above the mean for age, is macrocephalic. A family with five affected males over two generations is known (Asch and Myers 1976). All had normal intelligence, but three had an enlarged ventricular system and in two the ventricular size was at the upper end of normal.

There have been two previous reports: De Myer (1972) described three families, and Schreier *et al.* (1974) a family with 12 affected members over three generations. Two children were hydrocephalic and intellectually retarded and two normocephalic, but also retarded.

Normal children with large heads—benign familial macrocephaly (circumference above the 98th centile)—are reported by Day and Schutt (1979). One parent was affected in 11 out of their 13 families. A normal CT scan excluded hydrocephalus.

Male predominance was found by Asch and Myers (1976) in their dominant pedigree.

De Myer (1986) has reviewed the whole subject of megalencephaly and suggests the following criteria for a diagnosis of the benign familial type.

1. OFC more than 2 SD > mean.
2. No increased intracranial pressure.
3. Normal neurological examination.
4. No neurocutaneous stigmata.
5. A family history of megalencephaly.
6. Normal or only slightly enlarged ventricles.
7. Negative screen for metabolic disorders.

Hemimegencephaly

This is a rare anomaly which might present clinically with seizures and developmental delay. The hypertrophied hemisphere shows abnormalities of neuronal migration and micropolygyria (Dambska *et al.* 1984). Many giant neurones have been observed. There are clearly some similarities to the pathology described in tuberose sclerosis and this condition should be excluded clinically. If no skin lesions are present in the infant and parents, recurrence risks are small.

Macrocephaly, pseudopapilloedema, and multiple haemangiomata

See page 169.

Other causes of macrocephaly (based on table in De Myer 1972)

(a) *With generalized gigantism*
Sotos syndrome (see p. 37).

(b) *With dwarfism*
Achondroplasia.

(c) *Neurocutaneous syndromes*
Neurofibromatosis.

(d) *Metabolic*
Spongy degeneration.
Alexander disease.
GM_1 gangliosidosis.
Tay–Sachs disease.
Metachromatic leucodystrophy.

(e) *Chromosomal abnormalities*
Must be excluded—especially the fragile X syndrome and Klinefelter syndrome.

Familial communicating syringomyelia

Communicating syringomyelia is considered to be a non-inherited developmental disorder of the spinal cord most frequently found in association with a Chiari-type I malformation (Gimenez-Rolden *et al.* 1978).

Most previous families described as hereditary lumbosacral syringomyelia probably had hereditary sensory neuropathy. Two sibs, their father, and maternal aunt (Gimenez-Roldan *et al.* 1978) had syringomyelia. Three had, on air-myelography, ectopic cerebellar tonsils, and a postural collapse of the cord.

Van Bogaert (1929) reported two sisters with syringomyelia, and Wild and Behnert (1964) monozygotic twins.

The occurrence of syringomyelia in two sisters in one family and a brother and sister in another is described by Bentley *et al.* (1975).

Syringomyelia is often associated with skeletal abnormalities, especially at the base of the skull. Spillane *et al.* (1957) found basilar impression in one-third of patients.

Foramina parietalis permagna (Catlin mark)

A family in which six members were affected over three generations is known (Lipinski and Stenzel 1974). Penetrance was incomplete. This condition has been described previously by Pepper and Pendengrass (1936) and Lother (1959). It is benign and occurs without neurological defect.

A father and son with parietal foramina were reported by Thompson *et al.* (1984), both had the Saethre–Chotzen type of craniosynostosis.

Familial occurrence of basilar impression (occipital dysplasia/Chiari Type I)

Using the angle between the plane of the hard palate and the plane of the atlas as a measure, Bull *et al.* (1955) found 20 patients in whom this angle was three or more times the standard deviation of the mean. Seven affected individuals were found to have a relative with the same condition. Inheritance seemed to be autosomal dominant with occasional failure of penetrance. A similar conclusion was reached by Morariu and Taranu (1968) in an analysis of two families.

Erickson *et al.* (1950) reviewed 22 patients with basilar impression. One had an affected brother.

Another large dominant pedigree with variable expression and penetrance (Paradis and Sax 1972) showed the propositus to have syringomyelia and syringobulbia. Another member, a 6-year-old boy, had nystagmus and extensor plantar reflexes.

Metzger *et al.* (1962) described the same combination in a mother and child.

Another three generation family with variable expression has been described by Coria *et al.* (1983). The main clinical features were a short neck, dysarthria, other cerebellar signs, and brisk reflexes.

The occipital dysplasia (an underdevelopment of the enchondral part of the occiput bone) showed itself in various ways. In some there was basilar invagination, but in the milder cases there was a short, flat occipital squama which reduced the size of the posterior fossa.

However, not all cases are clearly dominant. Some, as pointed out by Coria *et al.* 1983, might show incomplete penetrance (see above) especially if basilar invagination is used as an exclusive criterion for gene carriers. Some patients will have minor degrees of occipital dysplasia without basilar invagination.

Two sisters, one with syringobulbia and syringomyelia and one with syringomyelia are reported by Busis and Hochberg (1985). The onset of symptoms in both was in the teens. Other families are reported by Gimenez-Rolden *et al.* (1978), Bentley *et al.* (1975), Caraceni and Giovannini (1977), and Barre and Reys (1924). Of these only the Gimenez-Rolden family was dominant whereas the others were sibs.

Counselling

Most cases are single. When inherited it is, at present, not certain whether the pattern is autosomal dominant with variable expression or whether a separate recessive type exists.

Arnold Chiari Type II and III

These malformations occur with meningomyeloceles and occipital encephaloceles and counselling is as for neural tube defects.

Vertebral column malformations

Spina bifida

Birth frequency

The frequency of anencephaly and spina bifida in London is 0.3 per cent (Carter and Evans 1973*a*), in Belfast 0.87 per cent (Elwood and Nevin 1973), and in South Wales 0.76 per cent (Carter *et al*. 1968).

Risks after one affected child

Family studies have shown that in South Wales the percentage of sibs affected is 5.2 per cent and in London 4.4 per cent, hence the recurrence risk for genetic counselling purposes lies between 1 in 20 and 1 in 25. No reliable figures are available for second- or third-degree relatives, but in both South Wales (Carter *et al*. 1968) and London (Carter and Evans 1973*a*) there was a two-fold increase in mother's sisters' children.

There is some evidence that there has been a natural fall in the frequency of neural tube defects and that recurrence risks in south-east England might be closer to 1 in 30.

Risks to sibs after two affected children

After the birth of two children with spina bifida, the risk for a subsequent child is approximately 10 per cent (Carter and Evans 1973*a*). In Northern Ireland, an area of high prevalence, the risk after two children was 1 in 5 and after three, 1 in 4 (Nevin and Johnston 1980).

Risks to children of adult survivors with spina bifida cystica

The results of a study by Carter and Evans (1973*b*), combined with that of Tünte (1971), show the risk to the offspring of parents with spina bifida cystica was 3 per cent irrespective of the sex of the parent. In a smaller study, Lorber (1971) found three out of 81 children were affected. For genetic counselling purposes, a 3 per cent risk is the best figure to date.

Spinal dysraphism

This term is used for conditions in which the spinal cord is split. The cord is abnormally tethered by means of fibrous bands and the condition is often associated with intraspinal lipomata, dermoid cysts, hamartomata, angiomata, hairy patches, skin dimples or sinuses, and cutaneous naevi. On X-ray of the lower dorsal and lumbar region, spina bifida, fused vertebral bodies abnormal, lamina, and a widened canal might be found.

In a study by Carter *et al.* (1976) the proportion of affected sibs with anencephaly and spina bifida cystica approximated the proportion found if the proband had had an open spina bifida, indicating an association between spinal dysraphism and spina bifida cystica.

The evidence is that multiple vertebral anomalies belong to the same group as open spina bifida and anencephaly (Wynne-Davies 1975). Those patients with a minor vertebral defect, single hemivertebra, or with a kyphoscoliosis due to a missing vertebral body do not belong to the spina bifida cystica group.

Prevention of neural tube defects (practical considerations only)

(a) General population screening using materal alpha-fetoprotein

This investigation is now being offered in many maternity units at 16–18 weeks. Accurate assessment of gestational age is critical. The majority of single raised levels will subsequently be found to be normal, but twinning, fetal death, exomphalos, low birth weight, congenital nephrosis, and fetomaternal haemorrhage can all cause raised levels.

About three-quarters of mothers with open spina bifida fetuses will have serum values greater than the 97th percentile at 16–18 weeks.

Many centres employ two times the median as a cut-off level to determine who should have the blood test repeated. If the second specimen shows a raised level and the dates and other possible causes of a raised level are checked by an ultrasound examination, the patient proceeds to amniocentesis.

A raised level of acetylcholine esterase in amniotic fluid has been found to be associated with open spina bifida, and may be a better test than alpha-fetoprotein, especially when amniotic fluid is contaminated with blood.

(b) Vitamin supplementation

Many clinicians throughout the world have now been faced with persuasive evidence that pre-conceptual supplementation with multivitamins and folic-acid reduces the recurrence risks of spina-bifida back to general population levels. The exact component of the 'cocktail' responsible for the reduction has not been determined and there is some controversy in genetic circles about whether it is ethical to participate in clinical trials to pinpoint it.

(c) After the birth of a child with a neural tube malformation

Amniocentesis is offered and counselling will include the following information:

(a) There is a risk of 0.3 per cent that amniocentesis might induce a miscarriage.

(b) In addition, there is a small risk of remediable congenital deformities such as talipes and congenital dislocation of the hip.

(c) Most of the open neural tube defects will be detected (95 per cent). The false positive rate is less than 0.5 per cent.

(d) The closed spina bifida which will be missed do not usually have serious neurological complications.

(e) Raised level of alpha-fetoprotein have been found in congenital nephrosis, Turner's syndrome, exomphalos, duodenal atresia, and bladder outflow obstruction.

(f) In the event of borderline alpha-fetoprotein values amniocentesis might have to be repeated. Only about 0.7 per cent of woman with borderline values have abnormal pregnancies.

(g) Ultrasonography will detect all the anencephalic fetuses and in experienced hands is useful in helping to resolve borderline cases of spina bifida. Some practitioners think that good ultrasonography is all that is needed.

All women who are offered maternal serum testing or amniocentesis should have the procedure fully explained to them—risks and benefit—and they should be offered the opportunity to opt into the programme.

Spondylolisthesis

Four cases in one family are described by Shahriaree and Harkess (1970). In three the condition was asymptomatic.

Five members in three generations had spondylolisthesis of L5 on S1 and in addition had a defect in the posterior spinous processes of these vertebrae (Amuso and Mankin 1967).

The familial type was thought to be secondary to a defect in the pars interarticularis and has been called 'true spondylolisthesis'. Neither spondylolisthesis nor spondylolysis is rare in the general population with an approximate prevalence of between 2 and 5 per cent. However, in a Finnish family reported by Haukipuro *et al.* (1978), autosomal dominant inheritance with about 75 per cent penetrance could account for the familial aggregation.

Caudal regression anomalad (sacral agenesis)

The association of this condition with maternal diabetes has frequently been noted (Passarge and Lenz 1966). Most of the cases are sporadic, but there have been four familial reports unassociated with maternal diabetes (Stewart and Stoll 1979). Two sibships are reported by Robert *et al.* (1974) and another by Finer *et al.* (1978).

One sibship in which mother was diabetic is reported by Stewart and Stoll (1979).

Taken as a whole the recurrence risks are small if the mother is diabetic, and probably small if non-diabetic.

Cervical spondylosis

Cervical spondylosis increases with age, but genetic factors might play a small role (Bull *et al.* 1969).

In a twin study these authors found concordance for severity in both their monozygous and dizygous twin pairs, but although the ratio was high in the identical twin boys, inheritance is probably not monogenic (77.5 and 42.5 per cent were the concordance rates).

Partial sacral agenesis with anterior sacral meningocele

A family with seven affected females over four generations with the above combination is reported by Bay-Nielsen and Cohn (1969). Inheritance is possibly X-linked dominant. Three other females in two generations are reported by Say and Coldwell (1975).

Vertebral anomalies (Jarcho–Levin syndrome)

The major manifestations are block vertebrae, hemi- and butterfly vertebrae, and spina bifida.

Clinically, the child has a short neck and dwarfism with a prominent thorax (Gellis and Feingold 1976). The facial features include a broad forehead, mongoloid eye slant, and wide nasal bridge. Cantu *et al.* (1971) report five cases in an inbred kindred. Inheritance is autosomal recessive and most families have been Puerto Rican. Pérez-Comas and Garcia-Castro (1974) reported sibs with the condition as did Lavy *et al.* (1966) and Castroviego *et al.* (1973).

Costovertebral dysplasia—autosomal recessive

Affected persons with this condition have short stature, a short trunk, and extensive vertebral anomalies such as fusion and scoliosis. The ribs are fused and might differ in thickness or are bifid (Silengo *et al.* 1978). The differentiation from Jarcho–Levin is not clear, the latter condition being mostly lethal and predominantly involving the thoracic cage.

Costovertebral dysplasia—autosomal dominant

Dominant inheritance of gross vertebral anomalies including butterfly and hemivertebrae was reported by Rimoin *et al.* (1968). Four were affected in four generations. Looking at the severity of the two types it appears as if the recessive form is more likely to cause early death.

Cerebrocostomandibular syndrome

Three sibs had a cleft palate, micrognathia, and extensive vertebral anomalies and rib gaps (McNicholl *et al.* 1970). The commonest presenting feature is neonatal respiratory distress. Microcephaly has occurred in a small proportion of cases (Silverman *et al.* 1980).

Congenital intraspinal extradural cysts

Three cases are reported in one family (Bergland 1968). The cyst wall is composed of arachnoid and the cyst enlarges to cause cord compression. In all three sibs congenital lymphoedema of the legs and a double row of eyelashes were present. A similar condition in a mother and two children was reported by Schwartz *et al.* (1979). A double row of eyelashes was again one of the manifestations. A progressive myelopathy was the reason for investigation.

Klippel–Feil syndrome

Anomalies of the cervical vertebrae—most often fusion—might be genetically determined. The clinical picture of the Klippel–Feil syndrome is that of short neck, low hairline, and restriction of neck movement. About one-third of cases have some degree of deafness (Palant and Carter 1972) and renal abnormalities are common (Moore *et al.* 1975). Three types are recognized: in type I several adjacent vertebrae are fused, whereas in type II, there is local fusion of two vertebrae with other minor malformations. In type III fusion of the vertebrae of the lower dorsal or lumbar spine occurs in conjunction with type I or II. It is particularly type II where monogenic inheritance has been suggested (Gunderson *et al.* 1967). They found three families where inheritance was as an autosomal dominant. In one family a C5–C6 fusion was possibly recessively inherited. In two families with variable expression and penetrance, inheritance was possibly dominant. No familial aggregation occurred in type I.

A single case with consanguineous parents (Juberg and Gershanik 1976) suggests that some cases might be recessive.

Klippel–Feil, abducens paralysis, retracted bulbi, and deafness (Wildervanck syndrome)

This combination is genetically distinct from the above and affects predominantly girls. The mode of inheritance is uncertain, but might be X-linked dominant.

3

Disorders of higher cerebral function

Dyslexia (specific reading disability)

Dyslexia tends to aggregate within families. However, an assessment of the mode of inheritance is difficult as dyslexia is unlikely to be a single condition and there is little accord between workers on the precise definition of the term or even the relationship between minimal brain dysfunction and dyslexia.

The World Federation of Neurology in 1968 defined dyslexia as a distur-bance in children who do not achieve, in spite of conventional instruction, the reading, spelling, and writing abilities appropriate to their intellectual qualities and sociocultural opportunities. The cognitive disabilities usually associated are disorders in speech and language, incoordination, disordered temporal orientation, naming colours, and recognizing the meaning of pictures. Left/right confusion has not been found to be a feature.

Incidence figures vary considerably. Estimates are 10 per cent in the United Kingdom (Critchley 1970), 1.7 per cent in Czechoslovakia (Zahalkova *et al.* 1972), 10 per cent in Australia (Walsh and Morris 1973), and 0.6–0.9 per cent in Japan (Makita 1968).

Most workers find a difference in incidence between boys and girls. In Australia an average of 8 per cent of boys and 2 per cent of girls were affected. The sex difference could be explained on the polygenic model with a lower threshold for males, or a single gene whose expression is modified by sex, or occasional X-linked inheritance.

Of the family studies the proportion of probands with other affected close family members is as follows:

	Other affected family members (%)
Klasen (1968)	39
Naidoo (1972)	36
Zahalkova *et al.* (1972)	42
Mattis *et al.* (1975)	79

In the study of Finucci *et al.* (1976), 45 per cent of 75 first-degree relatives were affected and male relatives were more commonly disabled than females. A unifying explanation of the pedigrees was not possible. Unaffected parents had two out of two affected children and in three families where both parents were affected there were both affected and unaffected offspring.

The same authors (Finucci and Childs 1983) have re-examined their data in order to analyse the relationship between the severity of the reading disability in the parents and their children. The findings were interesting in that severely disabled index cases tend not have severely affected parents, but have severely disabled sibs and normal sibs suggesting recessive inheritance in a subgroup. Where the index cases are borderline or mildly affected a parent was either severely affected or borderline. This would suggest that those dyslectics on the borderline or lower end of the normal spectrum possibly arise because of multifactoral inheritance.

Twin studies

Twin studies show monozygotic twins with concordance rates far in excess of dizygotic twins:

	Monozygous twins	Concordant	Discordant
Norris (1939)	9	9	0
Hermann (1956)	10	10	0
Bakwin (1973)	31	26	5
Weinschenk (1962)	14	14	0
	Dizygous	Concordant	Discordant
Norris (1939)	30	10	20
Hermann (1956)	33	11	22
Bakwin (1973)	31	9	22
Weinschenk (1962)		(40%)	(60%)

Where males and females are looked at separately, 83 per cent of females and 84 per cent of male monozygotic twins were concordant (Bakwin 1973). For dizygotic twins, 8 per cent of females and 42 per cent males were concordant.

Counselling

Risks on an empiric basis are higher when a parent and sib are affected (> 10 per cent), and smaller when normal parents have a single affected child and want recurrent risks for further sibs (between 5–10 per cent). High-risk situations should alert parents and teachers to the possibility that a child might need special tuition.

A linkage analysis in a number of families showing apparent autosomal dominant inheritance was undertaken by Smith *et al.* (1983). There was a suggestion in that study that perhaps one form of reading disability might be localized to chromosome 15.

Hereditary bilateral obligatory associated movement—mirror movement

Fourteen families with this condition were reported up to 1977 (Schott and Wyke 1977). Typical examples are the families of Regli *et al.* (1967), the family of Cohn and Kurland (1958), and more recently the family of Somers *et al.* (1976) (four affected over two generations).

Inheritance is dominant. Other pedigrees are those of Haerer and Currier (1966) and Crawford (1960).

A single case (Rasmussen and Waldenström 1978) was the offspring of first-cousin parents and suggests the possibility of a recessive form.

Cerebral palsy

It is helpful to divide cerebral palsy into the following groups.
(a) Spastic syndromes:
   hemiparesis
   paraparesis
   tetraparesis
(b) Dyskinetic syndromes:
   athetosis
   dystonia
(c) Ataxic syndromes:
   congenital ataxia
   ataxic diplegia

Taking cerebral palsy as a whole, approximately 2 per cent of all cases have a significant genetic component. Asher and Schonell (1950) looked at 400 cases and found low figures, as did Bundey and Griffiths (1977).

It is within the groups of symmetrical spastic paraparesis, ataxia, and athetoid cerebral palsy that most of the high-risk situations lie. This is evident in the survey by Gustavson *et al.* (1969). Familial cases of cerebral palsy were traced in Sweden. Sixteen families were identified in which those affected had a similar syndrome. In three additional families the syndromes were identical but there was also an abnormal perinatal period. In a further 24 families the syndromes were non-identical. These were the only familial cases out of a total of 3150 families suggesting that familial aggregation is rare. Of the familial cases the most frequent clinical picture was that of a non-progressive ataxia and mental retardation (10 families). The mode of inheritance in most was autosomal recessive. In two families, sibs had an ataxic diplegia and in one family this syndrome was present in three generations.

It is emphasized by Fisher and Russman (1974) in a review of genetic syndromes associated with cerebral palsy that, when a positive family history is obtained, a specific diagnosis should be sought. They found nine familial cases in their series, which were regarded as cerebral palsy, to have heredity spastic paraplegia, hereditary ataxia, Behr syndrome, Lesch–Nyhan disease, Sjögren–Larsson disease, or Marinesco–Sjögren disease.

(a) Spastic cerebral palsy

Patients with spastic diplegia or quadriplegia from birth were studied by Bundey and Griffiths (1977), and 24 were selected because they had symmetrical spasticity and normal birth histories. Of their 55 sibs, six had a similar disorder. The recurrence risk for this form of cerebral palsy in the absence of an abnormal perinatal history is about 1 in 9. Some single cases might be fresh dominant mutations. Later, Bundey *et al.* (1978) reported that an index patient from the series above married and had an affected daughter, emphasizing the need for caution in counselling this group of patients. Additional evidence for monogenic inheritance in a proportion of patients comes from the study of Penrose (1938) who found nine pairs of sibs with spastic diplegia out of 65 persons with this type of cerebral palsy. The parents were related in six instances. In Adler's (1961) series of cases, there were three siblings with mental retardation and a spastic quadriparesis and two non-identical twins with spasticity in the lower limbs, whose parents were related. Allport (1971) reported four out of eight siblings with mental retardation and a spastic paraparesis. In the series of Bundey and Carter (1974), a higher risk for a sib was found in index patients with a normal head circumference and symmetric spastic quadriparesis. These authors confirm the empiric risk of about 10 per cent. The recurrence risk for asymmetric cerebral palsy is 1–2 per cent.

(b) Dyskinetic cerebral palsy/athetosis/dystonia

Environmental causes have been emphasized (Griffiths and Barrett 1967), but Bundey and Carter (1974) described athetoid cerebral palsy in brothers, as did Adler (1961).

An X-linked athetoid spastic cerebral palsy with myoclonus was found in the Bundey and Griffiths (1977) study. This might be the same condition as that of Baar and Gabriel (1966, see p. 288). Recurrence risks for sibs are in the region of 1 in 10. It might be that genetic causes are becoming relatively more common, hence the recurrence risk for sibs of 1 in 10. Lesch–Nyhan must be excluded.

(c) Ataxia

It is estimated that one-third to one-half of all cases of congenital ataxia with mental retardation are genetically determined (Gustavson *et al.* 1969). These authors found mental retardation with a non-progressive cerebellar ataxia to be present in more than one sib on ten occasions. A pair of brothers with ataxia were included in a series of 33 patients with either congenital ataxia or congenital ataxic diplegia (Ingram 1964). Some of these cases now fit into the classification of congenital cerebellar agenesis (see p. 175). Recurrence risks are unknown but might be in the vicinity of 10 per cent.

Dysequilibrium syndrome

Within the group of ataxic cerebral palsy there are some children in whom the predominant sign is a disturbance of posture and equilibrium (Mackeith *et al.* 1959). This is called the dysequilibrium syndrome. Sibs have been reported by Hagberg *et al.* (1972) and by Alajouanine *et al.* (1943). In a survey by Sanner (1973) consanguinity was found in four of the families of 23 patients and two pairs of sibs were ascertained (see p. 175). The dysequilibrium syndrome is prevalent in the Hutterite population in North America (Pallister and Opitz 1985).

4

Disorders associated with retinitis pigmentosa

Those neurological conditions associated with retinitis pigmentosa are listed below (adapted from Shibasaki *et al.* 1979).

(A) With mental retardation

Laurence–Moon–Biedl (see p. 79)
Biemond syndrome (see p. 64)
Sjögren–Larsson (see p. 286)
Microcephaly and chorio-retinal
 dysplasia (see p. 28)

(B) With deafness

Alström syndrome (see p. 82)
Usher syndrome (see p. 82)
Hallgren syndrome (see p. 84)
Edwards syndrome (see p. 82)
Hersh syndrome (see p. 84)
Infantile Refsum disease
 (see p. 470)

(C) With ataxia

Bassen-Kornzweig (see p. 194)
Olivopontocerebellar
 degeneration (see p. 198)
Refsum disease (see p. 236)
Vitamin E deficiency (see p. 194)

(D) With spasticity

Kjellin syndrome (see p. 281)
Homocarnosinosis (see p. 282)
Gordon's pedigree (spasticity,
 deafness, retinitis pigmentosa
 with mental retardation)
 (see p. 82)

(E) With extrapyramidal features

Winkleman pedigree (see p. 308)

(F) With degenerative disease

Batten's disease
(with liver fibrosis, see p. 178)

	Cockayne syndrome (see p. 472)
	Hallervorden–Spatz disease
	(see p. 484)
(G) With myopathy	Kearns–Sayre (see p. 380)
	Mitochondrial myopathy-
	cytochrome oxidase deficiency
	(see p. 384)

(A) Retinitis pigmentosa and mental retardation

Laurence–Moon–Biedl syndrome

An association of a pigmentary retinopathy, spastic paraplegia, hypogonadism, and mental retardation was reported in 1866 by Laurence and Moon. Polydactyly and obesity (Figs 4.1 and 4.2) were not commented on until Bardet (1920) and Biedl (1922) described these features in addition to mental retardation, pigmentary retinopathy, and hypogonadism. This led to confusion and it was suggested in 1925 (Solis-Cohen and Weiss) that all four names should be used. In most English speaking countries the syndrome has remained Laurence–Moon and on the continent Bardet–Biedl syndrome.

McKusick (1988) lists Laurence–Moon in which mental retardation, spastic paraparesis and hypogonadism are the features, separately from Bardet–Biedl syndrome in which a pigmented retinopathy, polydactyly, obesity, hypogenitalism, and mental retardation are the main features. They might be separate conditions although both are recessive.

As pointed out by Cantoni *et al.* (1985), nystagmus is more frequent in Laurence–Moon whereas polydactyly more commonly occurs in Bardet–Biedl.

Spasticity occurs in 33 per cent and spino cerebellar ataxia in 90 per cent but only in Laurence–Moon syndrome

Variability of the clinical picture

If the syndrome is taken as a whole, then only about one-quarter of those affected have the full picture (Warkany and Weaver 1940). A typical pedigree illustrating the variability of the clinical picture within a single family is that of Ciccarelli and Vessell (1961). One sib had the complete clinical picture and two had obesity, mental retardation, and polydactyly. In addition to the five

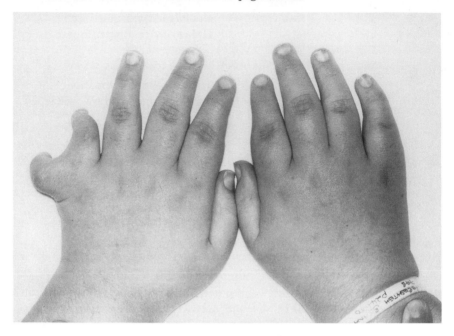

Fig. 4.1   Polydactly in Laurence–Moon–Biedl syndrome.
Courtesy of the Department of Medical Illustration, Great Ormond Street
Hospital for Sick Children, London.

main features, namely retinitis pigmentosa, polydactyly, obesity, hypoplastic genitalia, and mental retardation, many other signs have been described.
These are syndactyly, congenital heart disease, dwarfism, cataracts, ataxia,
deafness, and macrocephaly. Renal lesions were present in 71 per cent of the
14 cases autopsied up to 1969 (Nadjmi *et al*. 1969). These include renal
hypoplasia and hydronephrosis, but no systematic study was made at post-
mortem of the whole urinary tract.

The incidence of heart disease in Laurence–Moon–Biedl syndrome might
be higher than thought (McLaughlin *et al*. 1964). It has been suggested (Klein
and Ammann 1969) that it is useful to divide the syndrome into five types.

1. Complete—all five cardinal signs are present.

2. Incomplete—one or more features are missing.

3. Abortive form—only one or two signs are present.

4. Atypical—optic atrophy instead of retinitis pigmentosa.

5. Extensive—additional features such as epilepsy or extrapyramidal
    manifestations.

In Bell (1958) series of cases the author found 91 per cent to have retinal or
optic nerve changes, 93 per cent to have the obesity, 87 per cent to be mentally

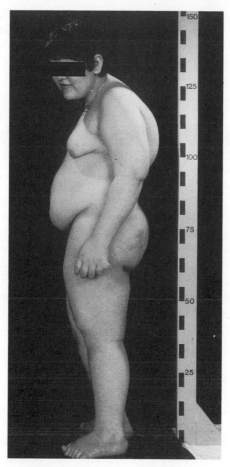

Fig. 4.2   Obesity in Laurence–Moon–Biedl syndrome.
Courtesy of the Department of Medical Illustration, Great Ormond Street Hospital for Sick Children, London.

retarded, 71 per cent to have polydactyly, and just over half the cases to have hypogenitalism. The obesity is of the truncal type.

It should therefore be noted that patients might present with only the typical renal lesion, polydactyly, and obesity, but with normal intelligence. The diagnosis in this situation is extremely difficult. Very occasionally (Roussel *et al.* 1985) periportal liver fibrosis has been described.

There have been more than 400 cases reported in the literature (Bauman and Hogan 1973). The frequency of the syndrome in Switzerland was 1 in 160 000 with a heterozygote frequency of 1 in 200 (Klein and Ammann 1969). In the monograph of Bell (1958) the author found a parental consanguinity rate of 23.4 per cent based on over 300 families.

In the genetic study (57 probands) carried out by Klein and Ammann (1969), consanguinity occurred in 52.6 per cent of 38 sibships. A segregation analysis came close to the 25 per cent expected for an autosomal recessive disease. Prenatal diagnosis using fetoscopy to detect polydactyly is possible, but this feature is only present in three-quarters of cases.

(B) Retinitis pigmentosa and deafness

Alström diseases

Alström et al (1959) described a syndrome consisting of obesity, diabetes mellitus, pigmentary retinopathy with nystagmus, and nerve deafness with on onset of visual problems before 1 year. They thought the combination was distinct from Laurence–Moon–Biedl syndrome. Polydactyly and mental retardation were not features. Hypogondism was not clinically obvious, but a testicular biopsy showed germinal cell aplasia. Associated findings are a progressive chronic nephropathy, acanthosis nigricans, and hypertriglyceridemia in some (Cantoni *et al.* 1985). Alström's patients were third cousins. Goldstein and Fialkow (1973) reported the syndrome in three sisters.

Gordon syndrome

This is very similar. Retinitis pigmentosa, deafness, and mental retardation occur.

Two brothers with this combination were reported by Gordon *et al.* (1976). The parents were fourth cousins. The pigmentary changes were diffuse and not central. The optic discs were pale. Both sibs had myoclonic jerks and brachydactyly.

Edwards syndrome

The four sibs reported by Edwards *et al.* (1976) were severely mentally retarded but had the deafness and diabetes as described by Alström *et al.* (1959). (The Alström patients are not retarded.)

Usher syndrome

The combination childhood onset of retinitis pigmentosa and congenital deafness has been frequently reported and was first described by Usher

Table 4.1 (Based on Cantoni et al.)

Features	Laurence–Moon	Bardet–Biedl	Alström	Edwards et al.	Gordon et al.	Usher	Hersh et al.
Retinopathy	+	+	+	+	+	+	+
Deafness	–	–	+	+	+	+	+
Mental retardation	+	+	–	+	+	+	+
Hypogenitalism	+	+	+	+	–	–	+
Abnormal glucose metabolism	–	–	+	+	–	–	–
Spasticity	+	–	+	+	+	–	–
Obesity	+	+	+	–	+	–	–
Polydactyly	–	+	–	–	–	–	–
Acanthosis nigricans	–	–	+	+	–	–	–

(1935). A thorough genetical-statistical analysis was carried out by Lindenov (1945) in seven Danish counties. His results suggest the syndrome is recessive, although variable expression among sibs did occur. The condition is reviewed by Nuutila (1970) who emphasizes the possible relationship with Hallgren syndrome.

Hallgren syndrome, retinitis pigmentosa, deafness with vestibular disturbance, cerebellar ataxia, and mental retardation in a proportion of cases

A survey of 177 affected members belonging to 102 families with this combination of signs showed recessive inheritance. Consanguineous marriage rate was twice the frequency of the general population. Deafness was congenital and the retinitis pigmentosa had a comparatively good prognosis (42 per cent were blind in the 41–50 years of age group). Cataract was a frequent complication (35 per cent of cases). The otoneurological investigations suggested that there was a labyrinthine lesion. Mental retardation was a feature in 24 per cent of cases (Hallgren 1959).

Familial tapetoretinal degeneration and epilepsy

Two Ethiopian sibs had this combination. The seizures were generalized and began in childhood or adolescence (Cohen *et al.* 1979).

Retinitis pigmentosa, deafness, mental retardation, and dysmorphic features

Two sibs were described with these problems (Hersh *et al.* 1982). The facial features included an open anterior fontanel, frontal bossing, downslanting palpebral fissures, and a normal to slightly small head circumference. The midface was flat. For a comparison with other conditions featuring retinitis pigmentosa see Table 4.1.

Note: All patients with a pigmentary retinopathy, deafness, and mental retardation should be investigated to exclude infantile Refsum disease.

5

Dementia

Alzheimer's disease

In 1910, Alzheimer described a woman aged 51 with a memory deficit, disorientation, and delusions of persecution. The characteristic autopsy findings are senile plaques and neurofibrillary degeneration (Landy and Bain 1970). Alzheimer's disease cannot be clinically nor pathologically differentiated from senile dementia, and family studies do not support the division between presenile and senile dementia as both early and late onset cases have been reported in the same family (Sjögren *et al.* 1952; Heston and Mastri 1977). Shields (1975) pointed out that in the Larsson *et al.* report (1963) 40 presenile cases were found among relatives where the probands had senile dementia.

Prevalence

In people over the age of 65 years, senile dementia occurs in 4.4 per cent and two-thirds of these have Alzheimer's disease (Terry 1976). A more recent figure is in line with this estimate and gives a prevalence of dementia in people over 65 years of age to be 2–7 per cent (Kay 1986). Mild dementia occurs in 10.8 per cent of that age group. Katzman (1976) estimated that the senile form of Alzheimer's disease is the fourth or fifth commonest cause of death in the United States. Another estimate (Heston and Mastri 1977) suggests that 70/100 000 Minnesota deaths in persons over the age of 40 are due to Alzheimer's disease. If criteria are stringent, then presenile dementia of the Alzheimer's type has an onset between the ages of 40 and 65 years and is characterized by an 'insidiously progressive loss of memory for recent events' (Heston and Mastri 1977). The course is relentlessly progressive with gradual decline in mentation over a period of 5–10 years. Parkinsonism with rigidity and poor facial mobility might develop late in the disease. Atrophy affects the frontal and temporal lobes, and there are changes in the basal ganglia.

The frequency of familial Alzheimer's disease

Between 15 and 35 per cent of probands with Alzheimer's disease have affected first degree relatives. Breitner and Folstein (1984*a*) have suggested an estimate of about 40 per cent. In their study of nursing home residents with Alzheimer's disease a lifetime risk for dementia in relatives was about 50 per cent compared with a 0 per cent risk for those with a language disorder (used as controls). Unfortunately, no pathological difference between sporadic and familial Alzheimer's can be detected.

A recent twin study of 22 pairs (zygosity was determined by childhood photographs and in some by blood tests) was undertaken by Nee *et al.* (1987). Seven monozygotic pairs were concordant whereas ten were discordant. Two dizygotic pairs were concordant and three were discordant. This study confirms the genetic contribution, but there is still no adequate explanation for the ten monozygotic twin pairs that were discordant. The possibilities are that Alzheimer's disease is heterogeneous or alternatively the variable age of onset will underestimate those who might still have become affected. There is in addition the possibility that an environmental trigger is needed for expression.

Older family studies showed a smaller frequency of affected first degree relatives. For instance, studies based on a consecutive series of patients have shown the following: Sjögren *et al.* (1952) found 10 per cent of parents and 4 per cent of sibs to be similarly affected, and Constantinidis *et al.* (1962) found a 3 per cent recurrence risk for first-degree relatives. Neither study is without bias. In the Heston and Mastri (1977) report the risk to a sib was not higher than 3 per cent. When one parent and one sib were affected the risk for another sib was about 16 per cent (Sjögren *et al.* 1952).

A pedigree in which a mother and four of her eight offspring had presenile dementia is reported by Beighton and Lindenberg (1971). Histology was available on one and the changes were compatible with Alzheimer disease. Three sibs with an onset in their late thirties are reported by Landy and Bain (1970) with cerebral biopsy confirmation of the diagnosis on the index patient. The mother of the three sibs had a dementing illness with onset at 52 years. Wheelan (1959) reported a mother and five of her 10 offspring with this disease (in two there was post-mortem confirmation), and in the family reported by Feldman *et al.* (1963) 11 members had the disease in two recent generations and two additional members in the previous two generations.

Other families: *two-generation families* (Friede and Magee 1962; Essen-Möller 1946; English 1942; Van Bogaert *et al.* 1940; McMenemy *et al.* 1939; Lowenberg and Waggoner 1934); *three-generation families* (Van Bogaert *et al.* 1940; Zawuski 1960); *five-generation families* (Lauter 1961); *more than one sib* (Neumann and Cohn 1953; Lüers 1947). A pedigree with multiple sibs (three out of four) and first-cousin parents (McMenemy *et al.*

1939) is suggestive of recessive inheritance. Two sets of discordant mono-zygotic twins are known (Hunter *et al*. 1972; Davidson and Robertson 1955).

A Canadian family with 51 affected members in eight generations was reported by Nee *et al*. (1983). Autopsies in four confirmed the diagnosis. Whereas the inheritance pattern seemed clearly dominant, the family came from a relatively isolated area and intermarriage in the family had taken place in previous generations. However, inbred matings did not produce more affected people than those in which outbreeding occurred.

Genetic counselling

Where family aggregation does occur autosomal dominant inheritance is the most likely mechanism. High recurrence risks should only be counselled in those families in which an analysis of the pedigree suggests that at least two generations are affected. Risks to the offspring of single cases are small. They range from 2–3 per cent (Constantinidis 1978) to 10 per cent (Sjögren *et al*. 1952). However, it is possible (Breitner and Folstein 1984*a*) that the risk could be higher if those at risk lived longer. This should not, for practical purposes, alter the 3–10 per cent risk given above.

Alzheimer's disease/Down syndrome and the localization of the gene for Alzheimer's

Two important clues to the localization of the gene for Alzheimer's disease emerged over the past decade. Firstly, it was noted that a very high propor-tion of patients with Down syndrome who survive into their late thirties and come to post-mortem, have clear evidence of Alzheimer's disease. However, three-quarters of these persons do not show clinical evidence of dementia, so that the relationship is still puzzling.

Secondly, in the Heston and Mastri (1977) study of Alzheimer's disease, there were six cases of Down syndrome in 777 near relatives of probands with Alzheimer's disease, a figure that was higher than would be expected by chance.

The recent breakthrough in Alzheimer's disease has evolved from these findings. Complementary DNA clones coding for the brain amyloid found in Alzheimer's disease, was isolated and mapped to chromosome 21 (Goldgaber *et al*. 1987). It was postulated that an over-expression of this gene or even an over production at a post-translational level due perhaps to an environmental stress was at the basis of Alzheimer's disease. However, linkage between familial Alzheimer's and the β-amyloid locus is not very close (Van Broeckhoven *et al*. 1987). Data in some familial Alzheimer's disease (St. George-Hyslop *et al*. 1987) have confirmed linkage to the 21q

11.2–q21 region and there seems little doubt about the localization, but the conclusion at present is that the amyloid β-protein locus is not the site of the inherited defect (Tanzi *et al.* 1987).

Alzheimer's disease and myoclonus

In a review of this association Jacob (1970) cites six examples of this subgroup. He added a family of his own in which six members were affected over three generations with dominant transmission. The myoclonic jerks developed with the initial amnesia. It is likely that this syndrome is distinct from Alzheimer's disease (Pratt 1970).

Pick's disease

Pick's disease is clinically difficult to differentiate from Alzheimer presenile dementia. Early language involvement, relative retention of learned material and of personal topographic orientation appear to be peculiar to Pick's disease and there is no obvious apraxia nor agnosia (Constantinidis *et al.* 1969/1978). Frontal and temporal atrophy are the most characteristic gross morphological features. Pathologically, Pick's disease is characterized by an absence of senile plaques and neurofibrillary tangles, and the presence of neuronal swelling and agyrophilic inclusions. Of the presenile dementias Pick's disease is the rarest.

The frequency of Pick's disease in Minnesota in persons dying after the age of 40 years was 24 per 100 000 deaths compared with 70 per 100 000 for Alzheimer's, and 19 per per 100 000 for Huntington's chorea (Heston and Mastri 1977). Up to 1974 there have been 16 pedigrees in the literature in which more than one histologically proven case has been found (Constantinidis *et al.* 1969/1978). Fourteen of the 16 showed more than one generation to be affected and in two pedigrees there were only affected sibs. In these latter two families, those affected had cortical gliosis without neuronal swelling (Grunthal 1930; Hascovec 1935—quoted by Constantinidis *et al.* 1969/1978). A similar pedigree is presented by Constantinidis (1969) in which two cousins were affected. They were the offspring of two brothers who married two unrelated sisters. The parents were unaffected.

Dominant pedigrees are those of Malamud and Waggoner (1943) and Keddie (1967). Three pedigrees showing dominant inheritance were reported by Sim and Bale (1973).

Heston (1978) examined 11 families which were ascertained because of a proband who was found to have Pick's disease on histopathological examination. The same disease was found in seven first-degree relatives (one sister, one brother, five parents). In one family three generations were affected.

Five new cases in a six-generation pedigree, previously described by Verhaart (1930) and Sanders *et al.* (1939) have now been reported (Groen and Endtz 1982). In four out of twelve patients at risk CT scan abnormalities were found. One man subsequently developed the disease at 43 and one at 28. It might in the future be possible to reduce risks in gene carriers on the basis of normal scans at various ages.

Genetic counselling

In the study of Heston (1978) the frequency of the disease in parents was 25 per cent. This is a disturbingly high figure, but the numbers are small. The risk to sibs is 1 in 13 and is therefore higher than for Alzheimer disease. Neither risk is applicable for offspring. The risks to sibs in the older study of Sjögren was 1 in 14 and 1 in 5 parents were affected. Risks to the offspring of single cases are unavailable.

Creutzfeldt–Jakob disease

Creutzfeldt–Jakob disease is characterized by a progressive presenile dementia, myoclonus, pyramidal, and extrapyramidal signs. In most cases progression is rapid with death within 18 months of onset. Several different clinical types have been postulated, namely (i) corticospinal (spastic pseudosclerosis) with both upper and lower motor neuron involvement; (ii) dyskinetic type; (iii) those with disorders of vision (occipitoparietal type); (iv) with ataxia of a cerebellar type. Most cases are single and recurrence risks are small. In a survey of 124 consecutive cases of Creutzfeldt–Jakob disease in France, five patients had a family history, and six others a possible history (Brown *et al.* 1979). In this group the age of onset was earlier—46 ± 6 years —than in most cases (60–64 years).

Familial cases have been reported by Friede and de Jong (1964) in the 'B' family (a father and his two daughters). A third affected daughter in the family was subsequently reported by May *et al.* (1968). In the 'Backer family' (Kirschbaum 1924; Jacob *et al.* 1950), an affected brother and sister in one generation and three out of seven of the brothers' offspring were similarly affected. The age of onset was between 30–40 years.

In the patients discussed by Roos *et al.* (1973) two had other affected family members. In one, the patient's mother, maternal grandmother, and two maternal aunts died of a similar illness. In the other, a sister of the proband was affected. Other dominant pedigrees (proband, mother, maternal grandmother, and two maternal aunts affected) were reported by Bonduelle *et al.* (1971) and Haltia *et al.* (1979).

Despite this familial aggregation the emphasis has changed from genetic to environmental factors—in particular, a slow virus. The transmission of the disease has been reported after a corneal graft from an affected donor (Duffy *et al.* 1974), to subhuman primates (Roos *et al.* 1973; Gibbs *et al.* 1968), and also serially from primate to primate (Gajdusek and Gibbs 1975). Intra-cerebral injection of brain tissue from 12 Creutzfeldt–Jakob patients caused a similar disease in chimpanzees 10–14 months later.

A family (Buge *et al.* 1978) with three pathologically proven cases of Creutzfeldt–Jakob disease, including transmission in one case to a squirrel monkey, is remarkable in that there was a total of eight members affected in three generations. All lived together in the north of the Ardennes. In the third generation, five out of six sibs were affected. Five familial cases are reported by Gajdusek and Gibbs (1975).

A further pointer to the viral origin of the disease was the geographical clustering and possible contact between cases noted by Matthews (1975) and there is a pocket of high incidence in Israel where an ethnic group of Libyan origin had 26 per 1 000 000 verified cases compared with 0.38 per 1 000 000 in non-Libyan Israelis (Kahana *et al.* 1974). This finding has been confirmed by Neugut *et al.* (1979).

A genetic susceptibility to the disease is not ruled out. In an analysis of more than 1000 patients, Masters *et al.* (1981) found that 15 per cent of cases were familial. Transmission of Creutzfeldt–Jakob disease to primates (non-human) was successful in 1 out of 40 familial cases.

Masters *et al.* (1981) again pointed out that familial Creutzfeldt–Jakob disease occurred in 15 per cent of 2000 cases. Eighty families were included— the largest was the Backer family—see above. An English based study by Will and Mathews (1984) found a positive family history in 6 per cent.

The frequency in Japan is in line with Western countries, but differs in the diffuse nature of the white matter involvement (Yamamato 1985).

Counselling

If there is a family history the mode of inheritance is usually dominant and the risks are high (single figures). Risks to sibs are likely to be small if both parents are normal. Risks to offspring of single cases, i.e. where there is no family history, are also likely to be small on the basis that 10 per cent are familial (between 6 and 15 per cent). Some cases could be fresh dominant mutations. The probability, therefore, of the 'at risk' person developing the condition should be no more than a couple of per cent, and to his/her offspring a half of that small risk.

Provided that genetic counselling is expertly given and that means that the patient should understand how these small risks are derived, then this type of reasoning should be acceptable.

Other clinical problems

(i) Mitochondrial disease

Ragged red fibres in muscle have been reported in a family with Creutzfeldt Jakob disease. Rosenthal *et al*. (1976) reported members of a family affected by a spongiform encephalopathy with a clinical course typical of Creutzfeldt–Jakob disease in the propositus. A first cousin had a chronic dementia, but no spongiform changes were present. Muscle biopsy in the propositus revealed ragged red fibres.

(ii) Ataxic form of Creutzfeldt–Jakob disease

The ataxic form of Creutzfeldt–Jakob disease with Hirano bodies has been reviewed by Cartier *et al*. (1985) who, in addition, reported three sibs in a single generation. Previously, Brownell and Oppenheimer (1965) had reported four cases—all sporadic—with a progressive cerebellar ataxia, dementia, infrequent or absent myoclonus, and no periodic EEG changes. Pyramidal and extra-pyramidal features were absent. Pathologically, there were marked hippocampal changes (rare in Creutzfeldt–Jakob) and Hirano bodies in the Ammons horn, without specific Alzheimer's senile changes. (Note too the Gerstmann–Straussler syndrome in which ataxia and dementia occur, but without Hirano bodies; p. 93.)

(iii) Conditions masquerading clinically as Creutzfeldt–Jakob disease

A family included on clinical grounds in the survey of familial Creutzfeldt–Jakob disease by Masters *et al*. (1981), was subsequently found to have thalamic degeneration as the main pathological features (Little *et al*. 1986). There were nine affected members who were demented and had myoclonus. The characteristic EEG changes were not present.

(iv) Clinical differentiation between Creutzfeldt–Jakob disease and Alzheimer disease

The clinical differentiation between Creutzfeldt–Jakob disease and that of Alzheimer can be difficult. A family in which members were affected by a presenile dementia over two generations is reported by Ball (1980). An autopsy on one (with a prolonged course) showed her to have a spongiform encephalopathy of the Creutzfeldt–Jakob type. Two other members had a dementing illness lasting from 5 to 10 years and had at autopsy a post-viral temporal lobe encephalopathy in one, and a spongiform encephalpathy in the other. In the study of Masters *et al*. (1981) there were two families with Alzheimer's disease in which another member died from Creutzfeldt–Jakob disease.

Hereditary multi-infarct dementia

This condition, first delineated by Haschinski *et al.* (1974), has been described in a family (Sourander and Walinder 1977). Three generations without skips were affected. The age of onset was between 29 and 38 years (five members) of an episodic neurological disorder with eventual dementia. No amyloid deposits were found.

A Finnish family (16 affected) is reported by Sonninen and Savontaus (1987). The onset was between 30 and 55 years of relapsing strokes and neuro-psychiatric problems.

It is probable that the same disorder was described by Stevens *et al.* (1977) under the name 'chronic familial vascular encephalopathy'. In this report, inheritance was also as a dominant, but the age of onset was slightly later (39–57) than in the family of Sourander and Walinder.

Non-specific familial presenile dementia (Kraepelin type)

A family in which six members were affected by a dementia beginning in the thirties is described by Schaumburg and Suzuki (1968). A post-mortem showed no evidence of infarctions, neurofibrillary tangles, senile plaques, Pick cells, or any other lesions. Clinically, other than the dementia, the neurological examination was unremarkable.

Hereditary polycystic osteodysplasia with progressive dementia

Often incorrectly diagnosed as Alzheimer's disease, this disease has been reported in Finland, Sweden, and Japan. It has an onset at 20–30 years of age and the initial symptoms are skeletal pain and tenderness and swelling of the ankles and wrists; on X-ray symmetrically located cystic lesions of bone are found (Adolfsson *et al.* 1978). Mental deterioration with involvement of upper motor neurons, epilepsy, and myoclonus follows. Two families in which seven out of 18 children were affected were reported by Adolfsson *et al.* (1978). This, and the cases of Hakola *et al.* (1970) and Hokola (1972), strongly suggest recessive inheritance.

Gerstmann-Sträussler syndrome

Recently a number of families, including that of Worster-Drought *et al.* (1940) (see pp. 93 and 284) and that of Adams *et al.* (1979) in which a relentless dementia, ataxia, and plaque-like deposits occur, have been grouped together and called the Gerstmann–Sträussler syndrome. The relationship of these apparently dominant conditions to Creutzfeldt–Jakob is discussed by Masters *et al.* (1981).

There is also evidence that Gerstmann–Sträussler syndrome can be transmitted to animals. For a full discussion see Prusiner (1987) which includes a discussion about prions, the possible causitive agent and Owen *et al.* (1989) for an insertion in the prion protein gene as a possible cause for the condition.

Counselling

When there is a family history the pattern is usually dominant.

Familial cerebral amyloid angiopathy with plaques

Atypical Gerstmann–Sträussler syndrome/dyshoric cerebral angiopathy

The proband (Courten-Myers and Mandybur 1987) had intermittent cerebellar signs starting at 46 years which then became more permanent. Those affected started to dement at 54, and at post-mortem showed a remarkable number of neurofibrillary tangles and widespread amyloid deposits. The patient's mother and brother had a similar gait disturbance for 40 and 20 years, respectively, but no dementia. This condition is probably the same as that described by Griffiths *et al.* (1982) and others (see p. 129). Another family was reported by Love and Duchen (1982).

Familial transient global amnesia

A family in which four brothers had this condition was reported by Corston and Godwin-Austen (1982). The onset was late in life. During an attack the brothers were unable to recall current events, but insight was retained during these episodes. Subsequently, a brother and sister pair (Munro and Loizou

1982) were reported and twin sisters who had in addition migraine (Dupuis *et al*. 1987).

Thalamic dementia

Little *et al*. (1986) reported nine cases in a family who had a clinical picture which looked like familial Creutzfeldt–Jakob disease. (Indeed they had been included in the Masters *et al*. paper 1981.) At post-mortem there was marked gliosis with neuronal loss in the thalami and olivary hypertrophy. Spongiform changes were minimal or absent. The authors refer to the twelve Japanese families described by Oda (1976) in which the post-mortem lesion was predominantly thalamic. There was a rapidly progressive illness with dementia, ataxia, rigidity, and late in the illness, pyramidal signs. The condition might be dominantly inherited with variable age of onset (teens to late middle life). At post-mortem there could, in addition, be involvement of the cerebellum and the inferior olives.

Some of the cases have been described as a system degeneration with involvement of the thalamus.

For instance, Katz *et al*. (1984) describe a post-mortem performed on a member of the Rothner *et al*. family (see page 281) who had a slowly progressive intellectual decline (onset at 7 years). At 21 he was severely demented and had optic atrophy and a spastic paraparesis. At the age of 30 his mother developed a progressive disturbance of gait, was hyper-reflexia, and had optic atrophy, and by the age of 45 years was demented and spastic. The proband had four affected sibs, with an age of onset in childhood. At post-mortem the striking changes were degeneration of the thalami and marked attenuation of hemispheric white matter.

Martin *et al*. (1983) described a condition which is very similar, but the course in Katz' patient was much more protracted.

The authors also indicate that the cases previously called the Stern–Garcin syndrome could have this condition.

X-linked early onset ataxia and dementia of adult onset

This combination was reported by Farlow *et al*. (1987). There was initially delayed walking and tremor, and in the teens a progressive ataxia and pyramidal involvement. This family is unique.

Syndromes involving speech and sleep

Worster–Drought syndrome

Also called congenital supra-bulbar palsy, this condition is characterized by dysarthria and troublesome dribbling due to selective weakness and impairment of movement of the orbicularis oris muscle, the tongue, and the soft palate. A genetic contribution was hinted at by Worster–Drought in 1977 when he noted a familial occurrence in 12 of the patients (out of 200) and a three generation family with skips was reported by Patton *et al.* (1986). Recurrence risks in single cases to both sibs and offspring is somewhere in the order of 5 per cent.

Note: this condition should not be confused with the other Worster-Drought syndrome (see p. 284).

Narcoleptic syndrome

The four characteristic symptoms are narcolepsy, cataplexy, sleep paralysis, and hypnagogic hallucinations. Familial aggregation has long been known and one of the first reports (Westphal 1877) described a mother and child with narcolepsy. Yoss and Daly (1957) found that 30 per cent of index patients with the narcoleptic syndrome had an affected relative. Variation within families is the rule and Daly and Yoss (1959) described a family in which 12 were affected over four generations, but three subjects had only cataplexy, whereas Gelardi and Brown (1967) describe a family in which 15 members had cataplexy, three had sleep paralysis, and three narcolepsy. In a review of the syndrome (Baraitser and Parkes 1978), 52 per cent of the patients had an affected first-degree relative. Of the group as a whole, 16 per cent of parents, 14 per cent of sibs, and 41 per cent of the offspring (age corrected) were affected. Probands had at least two manifestations of the syndrome narcolepsy being the most frequent. For genetic counselling it is also helpful to know that after the age of 25 years nearly 85 per cent of this risk has gone.

An interesting HLA association has been reported in those with the narcoleptic syndrome by Langdon *et al.* (1984) who found that nearly 100 per cent of those with the syndrome are HLA-DR$_2$ (or DQ$_1$).

Only 1–2 per cent of patients with classical narcolepsy do not have this haplotype (Parkes *et al.* 1986) whereas in Britain, 20 per cent of unaffected people have this haplotype. Kramer *et al.* (1987) looked at six North American black people with narcolepsy and all had the appropriate HLA-DR$_2$ subtype.

6

Epilepsy

Hippocrates called epilepsy a familial disease; this and the unfortunate association of epilepsy with mental retardation and degeneracy has led in the past to epileptics being prohibited from marrying and to sterilization.

In a study of social aspects of epilepsy in childhood, Ward and Bower (1978) found that one-quarter of parents expressed fears about the familial nature of the disease—and yet only a minority of families (two out of 81) had more than one sib affected.

Prevalence

In a study of Pond *et al.* (1960), who reported the findings of a survey of 14 general practices in Great Britain, the prevalence of epilepsy was between 420 and 620 per 100 000 of the population. In the Kurland study (1959) the prevalence was 365 per 100 000. More recently Ross *et al.* (1980) found that by the age of 11 years 6.7 per cent of 1043 unrelated children had a history of seizures and a clear-cut diagnosis of non-febrile convulsion by the age of 11 years was obtained in 4.1/1000 children. The general population risk against which individual risk will need to be judged is therefore 1 in 250.

Family studies

The many large family studies of epilepsy are difficult to interpret because there is no consensus about the differentiation of seizures from epilepsy, nor one type of seizure from another. Some series have studied epileptics in institutions, others concern those seen at neurological out-patient departments. There is also a significant difference in the results according to whether children or adults are used as index patients. Older children are less inclined to convulse. Maturation raises the seizure threshhold. In 1951, Ounsted found that 45 per cent of children under four years with purulent meningitis convulsed, whereas in those over 4 years only 8 per cent did so.

There have been a number of recent books on the genetics of epilepsy, but for the practising clinician the summary by Blandfort *et al.* (1987) is by far the most helpful. The classification and some of the figures quoted below are derived from her paper.

(1) Epilepsy with focal origin

(A) Simple Jacksonian
(B) Complex partial seizures (temporal lobe epilepsy)
(C) Benign partial:
    (a) with centro-temporal sharp waves
    (b) with occipital sharp waves
(D) Generalized seizures of focal or multifocal origin:
    (a) West syndrome
    (b) Lennox–Gastaut syndrome

(A) Simple Jacksonian or epilepsy associated with focal sharp waves

This category is often due to brain damage. Seizures of different types are associated with focal sharp waves and in the study by Gerken *et al.* (1977), febrile convulsions, *grand mal* (18 per cent), focal seizures (71 per cent), and *petit mal* (3 per cent) were found.

In the study of Gerken *et al.* (1977), 23 per cent had sibs, parents, aunts, uncles, and grandparents with a seizure disorder. About 3 per cent of sibs and a similar number of parents were affected.

When E E Gs are performed on the sibs (Doose *et al.* 1977), then 18 per cent of the sibs showed at least one centrencephalic episode. Abnormal sensitivity to light was a frequent finding; genetic suceptibility might play an important role in epilepsy with focal waves.

Counselling

Risks to sibs and offspring are small.

(B) Complex partial seizures temporal lobe epilepsy

The lesions most frequently associated with temporal lobe epilepsy are Ammon horn sclerosis and mesial temporal sclerosis (Falconer 1971). About half the cases of controlled seizures who came to lobectomy had mesial temporal selerosis (Davidson and Falconer 1975). In the study of Cavanagh and Meyer (1956) 64 per cent of patients had been in status epilepticus during the first few years of life. As Ammon horn sclerosis is thought to be caused by an asphyxial insult in infancy (Norman 1964), it is suggested that of all the possible aetiological factors in temporal lobe epilepsy, severe febrile convulsions with asphyxia appear to be the most important. The inherited predisposition of temporal lobe epilepsy could be related to the inheritance of febrile convulsions, in that uncontrolled febrile convulsions, could result in lesions in the medial temporal lobe.

Falconer (1971) found a significant family history in 13 per cent of those with mesial temporal sclerosis and in a Danish study (Jensen 1975) a positive family history was found in 30 per cent of those with resistant temporal lobe epilepsy who underwent temporal lobectomy (most were not first-degree relatives). In sibs the frequency was 2.9 per cent.

One-hundred children suffering from temporal lobe epilepsy were reviewed by Ounsted *et al.* (1966), and the pedigrees re-analysed by Lindsay (1971). A third of the patients had a history of a severe cerebral insult, a third had been in status epilepticus, and a third neither of the two. The percentages of affected sibs for the three groups were 1.9 per cent for the cerebral insult group, 30 per cent for those who had been in status epilepticus, and 9 per cent for neither. When the type of seizure recorded in the sib was examined, most had febrile convulsions (70 per cent) and only 15 per cent were epileptic. This suggests a relationship between temporal lobe epilepsy, early status epilepticus, and febrile convulsions, but not all studies confirm this.

Counselling

The overall risk to a sib of those patients with temporal lobe epilepsy, where there is evidence of a focal temporal lobe lesion or a previous history of febrile convulsions and status epilepticus is between 10 and 20 per cent. Most of this risk is for a sib with febrile convulsions. Only 3 per cent will have temporal lobe epilepsy.

It should be noted that few patients with febrile convulsions proceed to have temporal lobe epilepsy and usually only those with complicated febrile convulsions (prolonged and frequent seizures) will have this complication.

(C) Benign partial epilepsy with centrotemporal spikes

Mid-temporal spikes or rolandic discharges

Patients with benign epilepsy of childhood and centrotemporal electroence-phalographic foci (rolandic discharges) have been studied (Heijbel *et al.* 1975). Seizures started between the ages of 1 and 13 years, and 95 per cent of patients became seizure-free at 15 years. Rolandic discharges were not observed after 16 years. Fifteen per cent of the sibs had seizures and rolandic discharges, and 19 per cent had rolandic discharges alone. Eleven per cent of parents had seizures in childhood, but none in adult life. Only one parent (3 per cent) had rolandic discharges. In an earlier study (Bray and Wiser 1964), 36 per cent of sibs and offspring had focal discharges.

Counselling

The 15 per cent risk to sibs is high, but the condition is generally benign.

(D) Infantile spasms

(a) West syndrome

Infantile spasms with hypsarrhythmia occur between the age of 3 and 8 months and are characterized by flexion spasms, mental retardation, and a grossly abnormal electro-encephalogram (Jeavons and Bower 1964). Prognosis for mental development is poor. Family data were assessed by Fleiszar *et al*. (1977). Two-thirds of probands were mentally retarded. No cases of infantile spasms were reported among relatives of control children. Three secondary cases (two sibs and one aunt) were found in the affected group. The authors suggest a risk to a sib of about 15 per 1000. Tuberose sclerosis and metabolic conditions especially the aminoacidurias must be excluded.

In a large follow-up study of 200 patients with infantile spasms (Matsumoto *et al*. 1981) 40 per cent had a positive family history if epilepsy in 1st, 2nd, and 3rd degree relatives is counted. They had one instance of three affected sibs.

Counselling

It is not clear how to interpret the high frequency of a positive family history of generalised epilepsy as found by Matsumoto *et al*. (1981) or even the figure of 9 per cent by Lacy and Pendry (1976). Most geneticists will be asked about a sib risk for infantile spasms and this should not be higher than 2 per cent.

(b) Lennox–Gastaut syndrome

The existence of the condition as an independent entity is still not clear. The age of onset is between 2 and 7 years and is characterized by absence seizures without the typical 3 c.p.s. spike and wave pattern. Instead, there is an EEG pattern of diffuse slow spike waves (*petit mal* variant) and clinically, head dropping, brief tonic fits (especially in sleep) or absences might occur.

Counselling

The vast majority of cases are single and recurrence risks are small.

(2) Epilepsy with primary generalized seizures

(A) Absences
(B) Myoclonic-astatic *petit mal*
(C) Juvenile myoclonic epilepsy (impulsive *petit mal*)
(D) *Grand mal*
(E) Photogenic epilepsy
(F) Ohtahara syndrome (with burst-suppression on EEG)—mostly sporadic.

(A) Absences

Genetic counselling where the EEG abnormality is known
centrencephalic epilepsy or 3 c.p.s. spike and wave activity

This was defined by Metrakos and Metrakos (1960, 1961) as *petit mal* or generalized epilepsy accompanied by an electroencephalogram showing 3 c.p.s. spike and wave activity. The frequency of any type of seizure in sibs, parents, and offspring was 12 per cent in this study. In the control group the frequency of a convulsion, irrespective of cause, was 3 per cent.

When only electroencephalographic criteria were considered, about 30–40 per cent of offspring and sibs had a centrencephalic dysrhythmia if tested between the ages of 4 and 17 years, but this was found in only 7 per cent of parents. The figure approached 50 per cent if other electroencephalographic abnormalities usually associated with epilepsy were included. In the 1966 study, the same authors looked at 82 offspring of those with 3 c.p.s. spike and wave activity and found the same pattern in 35 per cent. Most of these genetically predisposed near relatives do not proceed to clinical epilepsy and Metrakos and Metrakos (1961) estimate the prevalence of epilepsy of centrencephalic origin in sibs and offspring to be between 8 and 10 per cent. They suggest that if the proband has centrencephalic epilepsy and one of the parents is epileptic (although not necessarily proven to be centrencephalic) the risk should be elevated to 13 per cent.

In the Gerken and Doose (1973) study 9.5 per cent of sibs of probands with 3 c.p.s. spike and wave activity had an abnormal EEG, and 7 per cent had seizures. The same authors found that 4.8 per cent of sibs of probands with irregular spike and wave activity had an EEG abnormality.

Risks to offspring of those who have 3 c.p.s. spike and wave activity, irrespective of the type of clinical seizure, lie somewhere between 4 and 8 per cent. The risk needs to be altered if more than one family member is affected. It should be noted that whereas the risks to the offspring of having an EEG abnormality is high, the risk of epilepsy is only small.

Clinical petit mal

This was defined by Penry *et al.* (1975) as a brief blank stare acompanied by unawareness and amnesia. In their series, only 9–10 per cent had typical clinical seizures although all had 3 c.p.s. spike and wave activity.

Four per cent of parents in a large series (Dalby 1969) had suffered from epilepsy; the majority had *grand mal* seizures. No parent had *petit mal*. Seven per cent of sibs had repeated seizures and of this group only 15 per cent had *petit mal* alone. The 139 patients of Gibberd (1966) all had clinical *petit mal*. Twenty-five had a first-, second-, or third-degree relative with *grand mal*, and six had a family history of *petit mal*. Matthes and Weber (1968) found that 3.1 per cent of parents and 3.7 per cent of sibs had epilepsy, and Doose *et al.* (1973) found 6.7 per cent of sibs to be epileptic.

Counselling

For counselling purposes, risks to offspring of those with clinical *petit mal* of those who had *petit mal* as a child, are relatively small and lie between 3 and 10 per cent. Four per cent for the offspring of single cases, and 10 per cent if in addition a parent is affected [see also Blandfort *et al.* (1987) who quote 8–10 per cent for offspring risks].

(B) Akinetic seizures or centrencephalic myoclonic-astatic *petit mal*

Akinetic seizures and myoclonic astatic convulsions were first described by Hunt (1922) as seizures characterized by a sudden loss of tone. They are often associated with severe mental retardation and are resistant to therapy. Many patients show a combination of seizures including *grand mal*. In a study by Schneider *et al.* (1970), nine of 40 children had another affected relative (22 per cent), but the relationship is not stated. A similar figure (27 per cent) was obtained by Lennox and Davis (1950). Age of onset is from 1 to 6 years.

Fifty children with myoclonic and/or astatic seizures with bilateral synchronous irregular or regular 2–3 c.p.s. spike, and wave and/or polyspike activity were seen and their relatives were assessed (Doose *et al.* 1970). Seizures occurred in 12.6 per cent of sibs and 7.1 per cent of parents. Pathological electroencephalograms were obtained in 46 per cent of the sibs and 14 per cent of the parents examined. Taking the electroencephalograms and clinical data together, clear evidence of a familial predisposition to convulsive disease was found in almost 50 per cent of first-degree relatives, but the risk of actual convulsions is between 7 and 12 per cent.

The study was further extended and reviewed (Doose and Baier 1987). All probands manifested absences and or myoclonic/astatic seizures. The Lennox syndrome was not included—it is similar, but the E E G is multifocal. In the sample as a whole, incidence in sibs was 10 per cent.

Counselling

The risk figure of 7–12 per cent as stated above is appropriate.

(C) Juvenile myoclonic epilepsy (impulsive *petit mal*)

In this condition myoclonic jerks occur mostly in the arms. Seizures are triggered by sleep withdrawal and occur shortly after wakening. Ninety-five per cent of those affected also have *grand mal* seizures.

Counselling

Offspring risk—7.5 per cent (from Blandfort *et al*. 1987); data from Tsuboi (1977).

(D) Generalized epilepsy (idiopathic *grand mal* epilepsy)

Most studies show the frequency of epilepsy in the close relative of patients to be between 3 and 6 per cent compared with 0.5 per cent in the general population. Alström's (1950) patients came from the out-patient department of a neurological clinic. When all the patients were considered together 1.3 per cent of parents, 1.5 per cent of sibs, and 3 per cent of children were similarly affected. When the idiopathic group was taken separately then 1.7 per cent of parents, 1.9 per cent of sibs, and 4 per cent of children were affected. These are small increased risks. Another early study was that of Conrad (1937) who found that 6 per cent of offspring of idiopathic epileptics suffered the same disorder. Signorato (1963) analysed 1500 patients with epilepsy. The frequency of affected first-degree relatives when the proband had so-called idiopathic epilepsy, was 2.7 per cent in parents, 1.7 per cent in sibs, and 4.1 per cent in children. Harvald (1954) chose institutionalized epileptics as index patients; 4 per cent of their children and sibs, and 3 per cent of their parents were similarly affected. Lennox (1947) analysed 1237 offspring of patients with idiopathic and symptomatic epilepsy. Thirty-four (2.7 per cent) were affected—3.2 per cent in the idiopathic group and 0.9 per cent of the symptomatic group. In the study of Beaussart and Loiseau (1969)

5200 cases who attended an out-patient clinic were analysed. In the category of generalized epilepsy, 3.2 per cent had another affected member. There have been other studies (Ortiz de Zarate and Rodriguez 1958) with similar results.

A study by Annegers *et al.* (1976) looked at the frequency of seizures in the offspring of parents who themselves had a history of seizures. Where the mother had epilepsy the proportion of her offspring with epilepsy was 3 per cent (excluding febrile convulsions and isolated seizures), but when the father was affected the numbers were no different from the general population levels.

Family studies where children are probands

In studies where children were chosen as index patients, the figures are often higher. Metrakos and Metrakos (1960, 1961) chose children with a history of any type of convulsion. Approximately 10 per cent of sibs had convulsions compared with 3.4 per cent among control children. In Ounsted's (1955) study the risk to sibs for epilepsy was 7.5 per cent.

Genetic counselling in generalized epilepsy

Taken as a whole, the studies of idiopathic epilepsy show that where a specific electroencephalographic or clinical diagnosis is not available, recurrence risks for the offspring and sibs are in the vicinity of four in 100. There are pedigrees with affected members in three or more generations but these are rare. In Alström's (1950) study, there were seven pedigrees which could be interpreted as examples of dominant transmission. The 3–6 per cent recurrence risk for the group as a whole should be manipulated on an empiric basis to reach 10 per cent or higher in those families with many affected members.

(E) Photoconvulsive response

The following situations need to be defined (Doose *et al.* 1969):

1. Photosensitivity—patients who show a photoconvulsive pattern on electroencephalogram. Seizures also occur without a light stimulus.
2. Photogenic epilepsy—seizures occur only on exposure to a flickering light, such as television, and self-induced epilepsy.

·Organic lesions play an important role in this type of epilepsy. Birth injuries were more common in the probands than in the control group (Doose *et al.* 1969).

Familial association of light-sensitive epilepsy has been noted by Daly and

Bickford (1951), Daly *et al.* (1959), Davidson and Watson (1956), Schwartz (1962), and Watson and Davidson (1957). Doose *et al.* (1969) found that about 23 per cent of sibs and 6.8 per cent of control children had electroencephalographic evidence of a photoconvulsive response. In the 16 families described by Davidson and Watson (1956) there were 39 affected individuals (both two-generation generation families and brothers and sisters pairs are reported), and in a later study (Watson and Davidson 1957) 62 per cent of patients had one or more light-sensitive relatives. Penetrance within a family was high and in some three generations were affected. Abnormal discharges evoked by flickering light only were found in 36 per cent of 112 relatives.

The transmission could be accounted for by either as an incomplete dominant or multifactorial inheritance. As with centrecephalic epilepsy there is a pronounced age dependency of photosensitivity and the condition occurs more commonly in females (Rabending and Klepel 1970; Doose and Gerken 1973). These latter authors looked at the photic response in the sibs of a large group of epileptic (not photosensitive) and in a control group of children. The frequency of abnormal responses (photoconvulsive) in the first group was only slightly higher than the control group and significantly different from the frequency in the sibs of probands with epilepsy and positive photoconvulsive responses. In the study a total of 15.9 per cent of the sibs of probands positive for the photoconvulsive responses had a similar response and 5.5 per cent of controls.

Counselling

Risks to sibs—6–10 per cent; risks to offspring—6–10 per cent.

Febrile convulsions

Taylor and Ounsted (1971) suggest that there is a general propensity of infants to respond to an insult (fever) by convulsing. This tendency subsides with age, but it improves more rapidly in females than males. The tendency begins at six months, reaches a peak at 18 months, and then recedes.

Febrile convulsions were defined by Fishman (1979) as those which follow a temperature elevation of 38 °C in a child less than 6 years, with no evidence of central nervous system infection and no acute systemic metabolic disorder. Benign febrile seizures last less than 15 minutes and do not have focal features. When the first seizure occurs before the age of 13 months (Frantzen *et al.* 1968) there is a greater chance of developing further febrile seizures.

The incidence of febrile convulsions in the general population has been reported to range between 1 and 15 per cent. A figure of about 3 per cent (Schumann and Miller 1966; Tsuboi and Endo 1977) is the one most often quoted.

Genetics

Lennox (1953) suggested that a family history of seizures was more common in febrile convulsions than in non-febrile convulsions, *petit mal*, or any other type of epilepsy. There is also agreement that children with a family history of febrile convulsions are more likely to convulse than those without a family history (Herlitz 1941; Lennox 1953; Lennox 1949). Three risk situations will be discussed.

Frequency of epilepsy in relatives of probands with febrile convulsions

In Ounsted's (1955) study the risk of epilepsy for sibs of patients with febrile convulsions was 9.7 per cent. In the study of Frantzen *et al.* (1970) the rate of epilepsy in near relatives of children with febrile convulsions was not different from the population and significantly below the rate in parents and sibs of patients with generalized epilepsy. Conversely, when the patient had chronic epilepsy the risks to sibs for febrile convulsions is 8 per cent and for epilepsy 5 per cent (Ounsted 1955). In two studies of febrile convulsions in isolated populations the incidence of febrile convulsions was much higher than elsewhere, whereas the prevalence of *grand mal* epilepsy was no different from other non-isolated populations, suggesting that the two conditions are genetically distinct (Lessell *et al.* 1962; Mathai *et al.* 1968).

Frequency of febrile convulsions in relatives

A positive family history is found in between 25–35 per cent of patients (Doose *et al.* 1966, 1968; Hrbek 1957). Ounsted (1955) found that 18 per cent of sibs had febrile convulsions.

In the study of Frantzen *et al.* (1970), the incidence of febrile convulsions in sibs was 20 per cent. If one parent was affected, the incidence in sibs was 36.5 per cent. The more severe the febrile convulsion the greater the incidence among sibs.

In the Schiøttz-Christensen (1972) series out of 114 sibs of probands with febrile convulsions, 16 were similarly affected giving a frequency of 14 per cent.

Van der Berg (1974) ascertained children with febrile convulsions and found 11.5 per cent of their sibs to have had febrile convulsions. The frequency in a control group was 2.4 per cent. He concluded that the incidence of febrile convulsions among siblings of index children is 3–5 times higher than among sibs of control children.

Tsuboi and Endo's (1977) figures were 14.6 per cent for parents, 20.7 per cent for sibs, 5.2 per cent for second-degree relatives, and 4.7 per cent for third-degree relatives.

Ounsted (1955) concluded that his data suggested inheritance of febrile convulsions as a single gene disorder with dominant transmission. Frantzen *et al.* (1970) came to the same conclusion. They favour dominant inheritance with reduced penetrance. For counselling purposes, whatever the mode of inheritance, the risk for another sib having a febrile convulsion is between 1 in 10 and 1 in 5.

In a more recent study by Tsuboi (1986), the current situation is summarized.

1. There is a 56 per cent concordance for monzygotic twins and 14 per cent for dizygotic twin pairs.

2. Familial cases have an age of onset between 8 and 9 months. They are characterized by many recurrences even after the age of 3 years.

3. The risk to sibs is 24 per cent. It should be noted that this is not a recessive risk as a parent was affected in 12 per cent of cases.

Tsuboi has concluded that the condition is multifactorial with an heritability of 75 per cent.

Frequency of any sort of convulsion in the families of those with febrile convulsions

In various series of patients from 2 to 60 per cent of the probands have had a relative who had had at least one convulsion (Millichap 1968). The study of Ounsted (1955) on 1000 children with febrile convulsions showed that the risks to sibs of these children with single remitting febrile convulsions was 25 per cent for any sort of convulsion, and when Frantzen *et al.* (1970) examined the family history of children with febrile convulsions, a history of convulsions was obtained for 40 per cent of the probands and one-third of them had more than one affected relative. Five to seven years later the figures reached 50 per cent and one-half.

Prognosis of febrile convulsions

Death related to febrile convulsions occurs in less than 1 per 1000 children. Recurrent febrile convulsions occur in 25–40 per cent and in 2 per cent last longer than 15 minutes. Two to four per cent develop epilepsy [quoted by Fishman (1979) who refers to publications by Nelson and Ellenberg (1978) to substantiate this].

The development of non-febrile convulsions after repeated febrile convulsions are reported as follows: Frantzen *et al.* (1970) 1 per cent, Van der Berg and Yerushalmy (1969) 3 per cent, Taylor and Ounsted (1971) 16 per cent, and Tsuboi and Endo (1977) 17 per cent.

Counselling

Risks of febrile convulsions to sibs might be high, but few (< 5 per cent) procede to epilepsy.

Reading epilepsy

Jaw jerks triggered only by reading and accompanied by abnormal discharges on the electroencephalogram have been described in families. Two sisters were reported by Lassater (1962), and a mother and daughter by Matthews and Wright (1967). Rowan *et al.* (1970) described a mother with reading epilepsy whereas her daughter had attacks on watching television and on reading.

Three sibs and a first cousin in a family where inbreeding was not present had epilepsy (Daly and Forster 1975). The proband had seizures related only to reading whereas the two sibs had discharges in their resting electroencephalograms but did not have seizures. The parents were normal but they were beyond an age when the electrical discharges might have been apparent. A pair of identical twins concordant for reading epilepsy has also been recorded (Daly and Forster 1975). The mode of inheritance of reading epilepsy is uncertain and the data could fit either incomplete dominant or multifactorial inheritance.

Seizures triggered by eye closure

Two brothers and their half-sisters (same father) showed identical generalized polyspike and wave and 3–5 c.p.s. second spike and wave activity on eye closure (Tieber 1972). All were mentally retarded.

Others

Familial epilepsy and yellow teeth

A remarkable family in which five male children suffered from severe seizures with an onset between 11 months and four years leading to death before the age of 10 years, was described by Kohlschütter *et al.* (1974). All the affected children had conspicuously yellow teeth. Cerebral deterioration and spasticity were the main features. The parents were unrelated and this autosomal recessive or X-linked syndrome is thought, by the authors, to be a neuro-ectodermal condition.

Pyridoxine-dependent epilepsy

Hunt *et al.* (1954) reported two sibs with this syndrome. The seizures could not be controlled on phenobarbitone alone. Three sibs of Italian descent (Waldinger 1964) were similarly affected.

In a remarkable pedigree (Bejsovec *et al.* 1967) the mother felt intrauterine convulsions during all three pregnancies. Two of the children died after only days or weeks, but the third survived. Another pedigree is reported by Robins (1966) where sibs were affected.

All infants who have seizures for no obvious reason should be given a trial of pyridoxin, but there is a worry that despite prompt treatment the eventual outcome for intelligence might not always be good. There is also some evidence that patients might present later in childhood.

Epilepsy with infantile baldness and mental retardation

Moynahan (1962) described two male sibs with generalized convulsions, congenital alopecia, and oligophrenia. The hair started to grow in one sib at the age of two years. Congenital baldness occurred without epilepsy or mental retardation on both sides of the family and, in addition, the mother had late-onset epilepsy. Universal permanent alopecia, mental retardation, psychomotor epilepsy, and pyorrhoea were reported to occur with variable expression over four generations (Shokeir 1977). Six members over the last two generations had the full syndrome. Two sibs (boy and girl), the offspring of a consanguineous marriage, with alopecia, convulsions, and mental retardation are reported by (Perniola *et al.* 1980). The boy had a sensorineural deafness.

X-linked epilepsy

The reported cases refer to syndromes of X-linked mental retardation with epilepsy as one of the features of the syndrome.

A family reported by Feinberg and Leahy (1977) had five affected males, four of whom died before the age of 6 years with infantile spasms. Three generations were affected.

Benign familial neonatal convulsions

In 1968, Bjerfe and Cofnelius reported a Swedish family with 14 members in five generations who had convulsions during the first week of life. Some had, in addition, febrile convulsions and sporadic seizured occurred up to the age of 10 years. Subsequent development was normal. A similar family (eight members in three generations) was reported by Rett and Teubel (1964), others by Carton (1978) and Quattlebaum (1979). A confident diagnosis is difficult in the early stages but inheritance seems to be as a simple dominant.

A more recent report by Kaplan and Lacey (1983) suggest two types. In one the members do not go on to have generalized epilepsy and in the other type, there is a 20 per cent risk of this happening if it has already happened once in the family.

Familial Q–T prolongation syndrome

An abnormally prolonged Q–T interval on the electrocardiogram can lead to syncope and convulsions. Both recessive and dominant syndromes exist.

Dominant (Romano–Ward syndrome)

The dominant syndrome is not associated with deafness. A family in this category was described by Roy *et al.* (1976). Five members were affected in three generations. Another family with dominant inheritance and incomplete penetrance was reported by Singer *et al.* (1974). Reviewing about 75 cases in the literature, Singer *et al.* reported that syncope occurred as the initial symptom in 17 per cent of patients.

Recessive (syndrome of Jervell and Lange-Nielson)

Originally described by Jervell and Lange-Nielson (1957), congenital deaf mutism is the distinguishing feature not found in the dominant pedigrees.

Epilepsy-telangiectasia

Six out of seven siblings with mental retardation, epilepsy, palpebral telangiectasia, and reduced IgA were reported by Aguilar *et al.* (1978*a*).

Epilepsy is not infrequent in mental retardation as a whole, but in this family not all of those with epilepsy were severely retarded.

Convulsions in sibs associated with familial hypomagnesemia

Two brothers developed tetanic convulsions at the ages of 15 and 23 days (Strømme *et al.* 1969). The main biochemical findings were hypocalcemia and hypophosphatemia. The younger had severe hypomagnesemia (elder not determined). Calcium administered had no effect on the elder who died. His younger brother responded dramatically to magnesium supplementation. There has also been a single case reported by Friedman *et al.* (1967) in which the patients were first cousins, and another sibship by Salet *et al.* (1970).

Fetal hydantoin syndrome

Over the past decade reports have suggested that hydantoin is potentially teratogenic. The association between hydantoin usage during the early months of pregnancy and cleft lip with or without cleft palate and a distinctive dysmorphic facial appearance was suggested by Meadow in 1968. These features include a broad nasal bridge, epicanthic folds, short upturned nose, hypertelorism, ptosis, prominent, slightly malformed low-set ears, wide mouth, and prominent lips (Hanson and Smith 1975). Microcephaly (not present in all) is the most serious defect and the syndrome has been associated with mild to moderate mental retardation.

Growth retardation, hypoplasia of the distal phalanges and nails, and a low hair line are all features of the syndrome.

The risk of malformations in offspring of those mothers taking hydantoin during early pregnancy is about 2–3 times the overall risk of a single major malformation.

In a study from Cardiff, Lowe (1973) found the malformation rate of 6.7 per cent in children born to epileptic mothers on anticonvulsants against a 2.7 per cent general population risk.

Cleft lip and palate occurred in 1.8 per cent as against 0.22 per cent for infants in the general population.

Kelly (1984) has critically reviewed the literature and concludes:

(1) there is a two- to three-fold increase in congenital malformation (but this is retrospective data) mostly affecting the lip and palate, growth, and cognitive function;

(2) minor digital and minor facial dysmorphism occur in 30 per cent of women on Epanutin (in most instances this is not a major problem).

Prospective studies of women on anticonvulsants are still needed. Probably, the best study is that of Kelly (1984). There were 171 infants of whom only seven (4.1 per cent) had a major malformation and an evaluation of these seven infants did *not* suggest that there was necessarily a relationship between the malformation and the anticonvulsant. Thirty per cent had distal digital hypoplasia and 3.3 per cent had a head circumference below the 5th percentile. Seven per cent had an interpupillary distance of greater than the 90th centile. No patient had a cleft lip or palate.

Fetal trimethadione syndrome

According to Feldman *et al.* (1977) there have been over 50 reported pregnancies in which a malformed fetus has been exposed to this drug. Most of these pregnancies resulted in either fetal loss or a congenital malformation, with either malformed ears, cleft lip, cardiac, urogenital, or skeletal defect. In none of these pregnancies was trimethadione the only drug involved. In the additional family reported by Feldman *et al.* (1977) there were seven pregnancies resulting in four infants who died from multiple congenital malformations and three spontaneous abortions. Three children born to separate mothers, but with similar features (developmental and speech delay, V-shaped eyebrows, epicanthus, low-set ears and folded helix, and high arched palate) were reported by Zackai *et al.* (1975).

Fetal valproate syndrome

This is now recognizeable as a specific entity (Diliberti *et al.* 1984). The facial features include a prominent forehead an inferior epicanthus with lateral extension which continues to form a crease below the orbit, a flat nasal bridge, a small upturned nose, shallow philtrim, thin upper lip and prominent lower lip. Most worrying is the association with spina bifida with a 2–10 per cent risk for those who are exposed. Psychomotor delay and an increased perinatal mortality might be features (Jager-Roman *et al.* 1986).

The management of epileptic parents who desire children

Despite the increased risk—two to three times the general population risk—for any malformation to those women who are epileptic and on medication, the management is still controversial. Any attempt to reduce the dose of hydantoin has to be weighed up against the risks of provoking an increased number of seizures during pregnancy.

It is estimated that between one- and two-thirds of epileptic women have an increased number of fits during pregnancy, but *status epilepticus* is rare. The effect of seizures on the fetus is uncertain and fetal death as well as a normal outcome has been reported after severe status. One patient reported by Goodwin and Lawson (1947), had severe status towards the end of the third month of pregnancy and produced a normal infant, but there is little information in the literature about correlating the time of status to the outcome of the pregnancy.

The geneticist in conjunction with the neurologist should seek to maintain pregnant epileptic women on minimum medication (preferably a single drug) and the evidence suggests that phenobarbitone is less teratogenic than hydantoin. Valproate is the most worrying drug during pregnancy and Tegretol is, on present evidence, the least toxic. If a change to Tegretol is not possible then it might still be argued that the risk of microcephaly and mental retardation using hydantoin is small and probably acceptable to many women.

Hyperexplexia—a hereditary startle syndrome

An inherited congenital disorder in which attacks of stiffness provoked by surprise occurred in 10 individuals over three generations (eight females and two males; Klein *et al*. 1972). A large family had several members with severe startle reaction—hyperexplexia (Suhren *et al*. 1966). At least four generations were affected. The reaction could be so severe that the patient would feel frozen and fall to the ground, unable to execute protective movement. The EEG showed subcortical discharges and the condition is related to those described on page 339. The condition is characterized by hypertonia, flexed postures, and little spontaneous movement. The excessive startle is provoked by a variety of stimuli (visual, tactile, auditory). Later in life the hypertonia improves, but the response to startle remains (Kurczynski 1983). Sleep is interrupted by frequent severe myoclonic jerks. A large family, Morley *et al*. (1982) confirms dominant inheritance. Many of those affected in this family had hip dislocation.

The two conditions mentioned above (stiff-man syndrome and hyperexplexia) are clinically similar—the most distinguishing feature is the presence of continuous electrical activity of motor units, at rest, in the stiffman syndrome. The jumping Frenchmen of Maine had a similar condition, but differed in that the startle response did not improve on anti-epileptic medication. The inheritance appears dominant, but pseudodominance cannot be excluded. 'The Jumping Frenchmen of Maine responded to a loud command by repeating it. They showed both echolalia and echopraxia.'

Recently, Saint-Hilaire *et al*. (1986) have studied eight members of the 'Jumping Frenchmen of Maine'. They are of the opinion that this is not a neurological disorder, but can be explained in psychological terms.

Other syndromes related to the above

Described under the heading 'an unidentified hereditary disease', Kok and Bruyn (1962) reported 29 affected individuals in six generations with generalized hypertonia in flexion, disappearing in sleep. Prognosis was excellent. They also describe in these patients an exaggerated startle response to a sudden stimulus. Epilepsy occurred in some.

A family with 11 members affected in three generations had periodic ataxia (Van Dyke *et al.* 1975*b*). Attacks were provoked by shock, caloric-vestibular stimulation, or sudden movement, and improved with age and anticonvulsants. Electroencephalographic changes were found in some of those affected. Attacks lasted from 1 to 2 minutes and consisted of jerking movements of the head, arms, and legs. Additional features were constant myokymia in the face and extremities with onset in the second decade, and a hand posture resembling carpopedal spasm.

Familial stiff-man syndrome (neuromyotonia, continuous muscle activity, Isaac's syndrome)

In this condition the EMG shows continuous activity. A large dominant pedigree is reported by Sander *et al.* (1980). The infants were hypertonic at birth, improved by three years, and then worsened during adolescence (the stiffness could be precipitated by cold or movement; see p. 241).

7

Myoclonic epilepsy

Genetically determined myoclonic syndromes

A. Recessive

(a) Unverricht and Lundborg—myoclonus, epilepsy, without dementia or with only slowly progressive dementia.

(b) Lafora body disease—myoclonus with rapid dementia.

(c) Lafora body negative disease—myoclonus, with rapid dementia.

(d) Dyssynergia myoclonica cerebellaris (Ramsay Hunt syndrome—recessive type). It is uncertain whether this is a different entity from the Unverricht–Lundborg type.

(e) Myoclonic epilepsy as part of lipidosis.
    (i) Gaucher's disease—juvenile (Tripp *et al*. 1977)
    (ii) Sialidosis (Rapin *et al*. 1978)
    (iii) Late infantile, juvenile and adult ceroid lipofuscinosis
    (iv) Krabbe's disease

(f) With cardiomyopathy and retinopathy.

B. Dominant

(a) Dominantly inherited epilepsy, myoclonus, and ataxia (also called Ramsay Hunt syndrome—dominant type).*

    Note about the classification of Ramsay Hunt—autosomal dominant. This could be the same as D R P L A (dentatorubro-pallido-luysian atrophy) described by Naito in five patients and quoted by Iizuka *et al*. (1984). Some of Naito's cases had extra-pyramidal features. D R P L A has been subdivided by Iizuka into three, but all three types have been found in the same family and they must be the result of the same mutant gene.) It might be that D R P L A is a single entity with variable expression and an evolving pathology. For the geneticist an autosomal dominant inheritance pattern is the most likely so that the distinction between this and Ramsay Hunt A D is not crucial (see p. 298).

(b) Cerebellar ataxia, myoclonus, peripheral neuropathy.
(c) Essential myoclonus or Friedreich's paramyoclonus multiplex.
(d) Dominantly inherited cerebellar ataxia and myoclonus.
    (i) with chorea (p. 124);
    (ii) with hearing loss;
    (iii) with photomyoclonus.

C. X-linked

A pedigree with five affected males over three generations was reported by Wienker *et al.* (1979). Affected males developed the seizures and myoclonus towards the end of the first decade and demented over the next. Carrier females were mildly affected, and developed a cerebellar type of ataxia in middle and later life. A similar pedigree was published by Vogel *et al.* (1965).

D. Other conditions associated with myoclonus

Mitochondrial disease with ragged red fibres (Fukuhara *et al.* 1980).

E. Focal/palatal

F. Others

(a) Peroneal muscular atrophy

(b) Dancing eyes, dancing feet syndrome

(c) Creutzfeldt/Jakob disease

(d) Alzheimer's disease

(e) Dementia/unnamed/but see also Ramsay Hunt syndrome p. 123

(f) Restless legs syndrome

(g) Dystonia

(a) Progressive myoclonic epilepsy or Unverricht Lundborg (Baltic epilepsy)

In 1891, Unverricht reported one family with myoclonic epilepsy occurring in five out of 11 children born to unaffected parents. The onset of symptoms was between six and 15 years, and the myoclonus was complicated by the presence of generalized seizures (often the first symptom). Dementia was not commented on.

In 1903 and 1912, Lundborg reviewed 'Unverrricht's myoclonie' and added personal material of this own. He reported 17 patients from nine families. His cases occurred in sibs and, although progression was variable, dementia occurred. He re-examined two of Unverricht's cases and found signs of dementia and Parkinsonism. Some of Lundborg's cases were probably of the Unverricht type, whereas others with rapidly progressive dementia were not. Progressive myoclonic epilepsy has come to be known as Unverricht–Lundborg disease. Lundborg's families came from Sweden and the Unverricht–Lundborg disease has been considered specific to the Nordic race.

A study of Finnish cases (Harenko and Toivakka 1961) showed two families with three affected sibs, nine families with two affected sibs, and 21 singletons. No two-generation families were found. The disease began with myoclonus in 14 cases and epilepsy in 12.

In a large series (93 cases) of Finnish descent thought to resemble those described by Unverricht (1891, 1895) myoclonus occurred as often as *grand mal* epilepsy as a first symptom (Koskiniemi *et al*. 1974). The myoclonus was stimulus sensitive and increased in severity until the patient was no longer able to walk (late teens). The average age of death was 25 years. The earlier the myoclonus appeared, the worse the prognosis, but in some the disease appeared to become stationary.

Histopathological examinations in 31 patients revealed no Lafora bodies. The incidence of the disorder in Finland is in the order of 1 in 27 000 live births (Koskiniemi *et al*. 1974).

The two families of Unverricht were Estonians—a neighbouring Swedish province of Finland, and Lundborg's 10 sibships were all descendants of one pair of ancestors in Sweden (Norio and Koskiniemi 1979). Because of the close contact between these three countries, these authors suggest that the cases could be related. Fifteen families in the USA without Scandinavian connections are now known. Affected members show on the EEG 3 c.p.s. spike and wave activity enhanced by photic or tactile stimulation (Eldridge *et al*. 1983).

The main clinical difference between this group of patients and the Lafora body positive group is the relative sparing of intelligence. Mental deterioration could be seen in almost all patients, but it was not as rapid as in Lafora body disease. The average age of onset was 10 years. Neither optic atrophy nor tapetoretinal degeneration occurred. The disease was familial in 25 out of 67 instances and transmission was compatible with autosomal recessive inheritance. In two families the parents were first cousins.

Other case reports are those of Halliday (1967) in which three out of seven sibs were affected by epilepsy and myoclonic jerks. *Pes cavus* was also present. No intellectual deterioration occurred, but the affected sibs were of low intelligence.

Two families with sibs affected with myoclonus and epilepsy were reported by Harriman and Millar (1955), and Wohlfart and Höök (1951). These latter authors do not think there is any difference between Unverricht–Lungborg disease and dyssynergia cerebellaris myoclonica (see p. 118). Action tremor of a cerebellar type and action myoclonus is difficult to distinguish clinically.

(b) Lafora body disease

The typical clinical picture is that of an onset in childhood or early adolescence of progressive dementia, myoclonus, ataxia, dysarthria, and seizures. Progressive visual loss in the presence of normal fundi occurs. The disease is fatal within a decade of onset. The exact enzyme defect is unknown but a diagnosis can be made prior to death on brain biopsy or skeletal muscle biopsies (Neville *et al*. 1974). Post-mortem studies have shown pathological changes to be present in spinal cord, peripheral nerve, skin, heart, muscle, liver, and retina.

Consanguinity is frequent and inheritance is recessive. Sibs were described by Schwarz and Yanoff (1965) and Van Heycop ten Ham and de Jager (1963). The parents were first cousins in the sibship of Lope *et al*. (1974). The occurrence of the disease in an inbred family is reported by Janeway *et al*. (1967). Three sibs were described by Scelsi *et al*. (1976), and many sibs in three families (no histology) by Rustam *et al*. (1975).

(c) Lafora body negative progressive myoclonic epilepsy

A family in which two children showed progressive myoclonus, recurrent major fits, and progressive neurological disability was described by Matthews *et al*. (1969). The onset was between nine and 18 years. No Lafora bodies were found at post-mortem. The clinical features are indistinguishable from the Lafora body positive group.

Similar families (all with a strong suggestion of recessive inheritance) were described by Ammermann (1940), Yokoi *et al*. (1966), and Vignaendra and Loh (1978). The three affected sibs in the last named paper were the offspring of first cousins. Haltia *et al*. (1969) described the post-mortem findings in sibs with a clinical picture of Lafora body disease but where no inclusions were found. They found instead non-specific degeneration of neurons—especially the Purkinje cells.

Clinical differentiation of Lafora body negative disease from the Unverricht–Lundborg type can be difficult. In most published reports, those affected with Lafora body disease have died before their mid-thirties. In contrast, those classified under Unverricht–Lundborg disease can survive until middle life.

Atypical cases

A late-onset form of Lafora body disease was reported by Kraus-Ruppert
et al. (1970) in two brothers. Towards the end of the second decade generalized epilepsy occurred, followed by an increasing frequency of myoclonus
over periods of 28 and 43 years. Lafora bodies were found in the brain and
spinal cord. A single male with late onset (in his thirties) was reported by
Dastur *et al.* (1966). Progressive mental deterioration and myoclonus eventually resulted in death.

Another atypical case with late onset was reported by Yerby *et al.* (1986).
This was a single case with an unusual microscopic appearance of the Lafora
bodies.

*(d) Dyssynergia cerebellaris myoclonica (Ramsay–Hunt syndrome
recessive type)*

It is uncertain whether this category differs from Unverricht–Lundborg
disease. Hunt (1921) described six cases in which there was a combination of
myoclonus, intention tremor, and epilepsy. Two of those affected were twins
with an onset in the second decade. They had, in addition, a sensory deficit
recorded as characteristic of Friedreich's ataxia. Dementia did not occur.
One died at 37 years of exhaustion and a post-mortem revealed a spinocerebellar degeneration, with dentate nucleus atrophy and rarefaction of myelin
in the superior cerebellar peduncle. The twins differed clinically from the
other cases in that they did not have generalized epilepsy.

The Ramsay Hunt syndrome is best accepted as a disorder with an onset
between five and 20 years, and characterized by cerebellar ataxia, dysarthria,
tremor, nystagmus, epilepsy, myoclonus, and occasionally *pes cavus*.
Rapidly progressive dementia, pyramidal tract sings, marked sensory loss,
and optic nerve involvement are usually absent.

Other sibships that may fall into this category are the following.

Noad and Lance (1960) described four affected sibs who were the product
of a first-cousin marriage. *Grand mal* epilepsy was a feature in two, and there
was evidence of mental retardation and minimal intellectual deterioration.

Jacobs (1965) described four sibs (unrelated normal parents) with a
syndrome of ataxia and myoclonus without intellectual deterioration, as did
Kreindler *et al.* (1959).

Sibs with progressive myoclonus and epilepsy are reported by Bird and
Shaw (1978). The onset was in the second decade with a slow but relentless
progression till death. Dentatorubral, spinal cord, and focal cortical degeneration were found at autopsy. The clinical picture consisted of myoclonus,
epilepsy, ataxia, absent tendon reflexes, *pes cavus*, tremulousness, decreased

appreciation of vibration and position sense, and reduced motor conduction velocities. Dementia was part of the syndrome.

Friedreich's ataxia and myoclonus

The association of myoclonic epilepsy with Friedreich's ataxia was noted in two families (Skre 1975) but the myoclonus occurred late in the disease and is seldom a prominent early sign in Friedreich's ataxia.

(B) Dominantly inherited syndromes

It might be that the syndromes featured below (see also myoclonus and dementia page 123) are the same condition. Only paramyoclonus multiplex is definitely different.

(a) Cerebellar ataxia and myoclonus (Dominantly inherited Ramsay Hunt syndrome)

Dominant pedigrees have been described by Bradshaw (1954), Gilbert *et al.* (1963), and Otsuki *et al.* (1965). The Gilbert *et al.* dominant pedigree showed incomplete penetrance. There were eight affected members with a syndrome comprising a cerebellar ataxia, tremor, and myoclonus. Progression was slow. Diebold *et al.* (1974) reported five cases in several generations of what he called dyssynergia myoclonica (Ramsay Hunt syndrome).

A family with chorea, ataxia, myoclonus, convulsions, mental retardation, and dementia is described by Takahata *et al.* (1978). Many members were affected over five generations. In the first, second, and third generations, dementia and ataxia occurred. In the fourth and fifth generations, mental retardation, convulsions, and chorea were features. Transmission was as a dominant. Pathologically, there were changes in the dentate nucleus and globus pallidus but this could be the same condition as the Ramsay Hunt dominant syndrome.

Naito and Oyanagi (1982) reported a family with this combination in which transmission was dominant with incomplete penetrance. They suggest that the choreoathetosis differentiated this condition from the Ramsey Hunt syndrome. At autopsy dentatorubral and pallidoluysian degeneration were found (see note, page 114).

(b) Cerebellar ataxia, myoclonus, peripheral neuropathy or Ramsay Hunt syndrome with a peripheral neuropathy

This refers to those patients with myoclonus, epilepsy dementia, ataxia, i.e. Ramsay Hunt with a peripheral neuropathy. It includes the family of Smith *et al.* (1978).

Another dominant syndrome in this category was reported by Ziegler *et al.* (1974). These authors described a patient with 'myoclonic epilepsia partialis continua and Friedreich's ataxia'. The proband presented at nine years of age with generalized convulsions, *pes cavus*, weakness, generalized areflexia, and a mild kyphoscoliosis. One week prior to admission repetitive myoclonic seizures developed. The patient's 4½-year-old daughter had seizures and impaired co-ordination. A similar family in which a mother and three of her offspring were affected by myoclonic epilepsy, ataxia, and a neuropathy resembling the neuronal form of peroneal muscular atrophy was reported by Smith *et al.* (1978). One sib was of low intelligence, but dementia was not a feature. Manifestations were variable—one of the sibs had optic atrophy and two had *pes cavus*. They had, in addition to the myoclonic epilepsy, generalized convulsions.

A pedigree in which a woman of 26 years was first considered to have a progressive cerebellar degeneration and then dyssynergia cerebellaris myoclonica because she developed action myoclonus, was described by Bonduelle *et al.* (1976). Her mother and two sibs were similarly affected. At post mortem she had olivopontocerebellar atrophy, but the dentate system was unaffected. The authors felt that dyssynergia cerebellaris myoclonica is a syndrome complex and not a single disease.

The family reported by Rondot *et al.* (1983) (dominant inheritance) had three affected members with an onset in their twenties. An ataxia (cerebellar) probably occured first. Horizontal eye movements were slow and upward gaze was defective. Myoclonus was prominent. All demented and lost their deep tendon reflexes (but with a Babinski sign). The pathology was of OPCA (olivo-ponto-cerebellar-atrophy) and the dentate was spared.

Note: It can again be seen that the clinical picture was in the category described in this section whereas at post-mortem OPCA was diagnosed. This combination of signs has been reported in a family by Skre and Löken (1970). In one sib unsteadiness of gait occurred from the seventeenth year. No progression took place until he was 32 years. Within four years he was a comple invalid. Another sib was similarly affected and other members over three generations had an abortive form of the disease (ataxia, amyotrophy, and spasticity). The same authors describe another single case with this clinical picture whose parents were first cousins.

Progressive myoclonic epilepsy, seizures, scoliosis pes cavus, retinopathy, cardiac involvement, ophthalmoplegia, and dementia

Two brothers had an onset at 2 and 3 years (Logigian *et al.* 1986). No ragged red fibres were found, but at post-mortem there was pan-cerebellar cortical atrophy, cell loss in the inferior olivary nuclei and an old middle cerebral artery territory infarction.

Note: There has been one previous case report of retinopathy associated with progressive myoclonic epilepsy (Klein *et al.* 1968), but the rest of the clinical description was very different.

(c) Paramyoclonus multiplex or hereditary essential myoclonus

This condition is associated with the name of Friedreich who, in 1881, described a man who at the age of 45 years had 'shakes and twitches during waking hours'. No relevant family history was obtained.

Since then, many families have been documented. Becker and Wieser (1964), and Mahloudji and Pikielny (1967) suggest the following diagnostic criteria for this entity.

(a) Onset of paramyoclonus in the first or second decades. Paramyoclonus is characterized by bilaterally symmetrical jerks often inhibited by movement (Wohlfart and Höök 1951), whereas myoclonic jerks are often provoked by action, but this division is not perfect.

(b) Benign course—compatible with normal life span.

(c) Dominant inheritance with variable in expression.

(d) Absence of generalized seizures.

(e) No gross dementia nor ataxia.

(f) Normal electroencephalogram.

The myoclonus has a predeliction for the face, trunk, and proximal musculature. Some families described as 'familial essential myoclonus' have had members with myoclonus and an essential tremor.

A large family in which seven members had only essential tremor and four tremor with myoclonus was described by Korten *et al.* (1974). A relationship between essential tremor and esential myoclonus is suggested by their coexistence in this family and, indeed, many characteristics of the two conditions are similar. Both are inherited as autosomal dominants and in both the dyskinesia is dependent on or intensified by change of posture and movement. In both onset is early but variable, and the movement disorder is influenced by stress. The myoclonus and the tremor can be abated by alcohol. In this family, myoclonus was predominent in the younger generation and tremor in the older. Nearly all patients with tremors developed the symptoms in adult life.

Under the title 'heredity essential myoclonus', Daube and Peters (1966) described two families in which myoclonus occurred between the years of one and 20. The myoclonic jerks involved arms, legs, trunk, and neck. Progress was slow and no intellectual impairment occurred. The older and more severely affected patients developed an unsteadiness of gait and speech impairment, but these were mild.

Biemond (1963) described a similar pedigree with autosomal dominant inheritance and variable expression. He emphasized that facial muscles are often involved and that the myoclonus can be made worse by emotional stress. A recent family is reported by Lundemo and Persson (1985)

(d) Dominantly inherited cerebellar ataxia and myoclonus

Myoclonus, ataxia, and deafness

A dominant pedigree with myoclonus, ataxia, and, in addition, hearing loss was reported by May and White (1968). The analysis of the pedigree is made difficult because of consanguinity in the parents of the affected mother. However, she married an unrelated male and had an affected son. Dementia was not a feature. An additional family is reported by Baraitser *et al.* (1984). The deafness in this family was of cochleal origin and one of the three patients probably had a sensory neuropathy. Post-mortem showed loss of neurons in the dentate nucleus, gliosis in the inferior olives, mild loss of Purkinje cells and white matter in the cerebellar hemispheres and loss of myelin in the fascicular gracilis in the spinal cord.

Myoclonic epilepsy and deafness (without ataxia)

The association of myoclonic epilepsy and congenital deaf mutism occurred in five out of eight sibs (Latham and Munro 1937–38). The myoclonic jerks began in the second decade and major fits began at about the same time.

Cerebellar ataxia and photomyoclonus

A dominantly inherited syndrome (at least five affected over three generations) was described by Ekbom (1975). The onset of the disease was between 35–40 years and the initial symptom was extreme sensitivity to photic stimulation resulting in symmetrical myoclonic jerks. The myoclonus could be violent enough to cause the patient to be thrown to the ground. A cerebellar gait and dysarthria were additional features, but neither sensory nor pyramidal tracts were clinically affected. All patients had kyphosis and *pes cavus*. Four were demented; lipomata of the neck, shoulders, and back occurred in three. The author suggested that the absence of epilepsy and the presence of dementia, lipomata, and skeletal deformities made this family different from those described by Ramsay Hunt (1921).

In at least one family (Grinker *et al.* 1938) dominance with variable expression occurred. Some members only had epilepsy whereas others had myoclonus and epilepsy.

(D) Other conditions associated with myoclonus

Infantile polymyoclonus

Also known as 'myoclonic encephalopathy' (Kinsbourne 1962) and 'dancing eyes, dancing feet syndrome' (Dyken and Kolăr 1968), the clinical features are those of chaotic rapid ocular movement (opsoclonus), myoclonus, and ataxia. The onset is in infancy and the course is non-progressive with fluctuation. An association with neuroblastomas is known (Solomon and Chutorian 1968; Bray *et al.* 1969; Brandt *et al.* 1974). The occurrence of this rare syndrome in second cousins (Robinson *et al.* 1977*a*) raises the possibility of occasional genetic transmission. Recurrence risks are small.

Focal myoclonus

Under the title 'a partial form of familial myoclonus' Van Leeuwen and Lauwers (1947) described sibs (unaffected unrelated parents) with an onset towards the end of the first decade of progressive myoclonic epilepsy and generalized epilepsy. Mental deterioration was not a feature. One of the affected sibs only had myoclonus of the tongue and palate.

Myoclonus as part of Creutzfeldt–Jakob syndrome

The association of dementia and myoclonus in middle life is suggestive of Creutzfeldt–Jakob disease. Differentiation from Alzheimer's disease can be difficult, but 90 per cent of cases survive fewer than 2 years and the periodic high-amplitude electroencephalographic complexes characteristic of Creutzfeldt–Jakob disease are seldom seen in Alzheimer's disease.

Myoclonus in Alzheimer's disease

Myoclonus is known to occur in Alzheimer's disease (Gimenez-Roldan *et al.* 1971) and when it does it follows the dementia by a couple of years. Falden and Townsend (1976) reported a pathologically proven case, and noted at least eight previous cases with pathology in the literature.

Myoclonus and dementia

Adult onset myoclonic epilepsy (without generalized epilepsy) and dementia occurred in a dominantly inherited pedigree (Sandyk 1982). The EEG showed spike and wave activity. The authors commented that the clinical picture resembled paramyoclonus multiplex but the dementia (and EEG findings) suggested that it was separate.

Photomyoclonus, diabetes mellitus, nephropathy, and deafness

Some of the affected members in a family described by Herrmann *et al.* (1964) had progressive neurological deterioration including dementia. All of those affected over five generations had photomyoclonus. Cerebral pathology suggested a lipid storage disorder.

Familial restless-leg syndrome (includes periodic movement in sleep)

An irresistible urge to move the legs in association with a deep, ill-defined unpleasant sensation within the limbs has a familial incidence in about one-third of cases (Ekbom 1960; Bornstein 1961). In a large dominant pedigree with 18 affected members over five generations, 10 of those affected had evidence of myoclonic jerking (Boghen and Peyronnard 1976). Huizinga (1957) reported a pedigree in which members had a 'crawling sensation' deep in the limbs reminiscent of Tinel's 'nocturnal acromelalgia' (1939). Patients would walk on a cool stone floor or hang their legs out of bed to find relief.

A family in whom nine members were affected over five generations was reported by Montplaisir *et al.* (1985). They suggest that periodic movement in sleep and the restless-leg syndrome are identical conditions (see p. 341).

Palatal myoclonus—progressive ataxia

There have been a total of six cases with this combination in which an acute lesion of brainstem or cerebellum does not seem to be the etiological factor (Sperling and Herrmann 1985). The onset is in middle life with slurred speech and jerking of the lips and soft palate and a progressive cerebellar ataxia affecting the gait. Pyramidal tract signs and dementia are variable features. All cases have been single.

Myoclonic dystonia

This condition is similar to benign essential myoclonus (see page 121), but has dystonia as an additional feature. Intellect is normal and the arms, trunk, and face are predominantly affected with sparing of the lower limbs. A family with dominant inheritance is reported by Quinn and Marsden (1984) and by Obesco *et al.* (1983). There might be a dramatic response to alcohol. The age of onset is between 5 and 47 years, and there is truncal and neck dystonia with superimposition of rest and action myoclonous.

Myoclonus and chorea

This condition resembles both hereditary benign chorea (see page 329) and hereditary essential myoclonus. Onset is in childhood with multifocal myoclonus, chorea and a mild ataxia. A mother and two children are described by Kurlan *et al.* (1987). Inheritance is dominant.

8

Vascular lesions

Migraine

Patients with classical migraine (episodic headaches, nausea, and vomiting preceded by a visual aura) frequently have affected first-degree relatives. The problems of definition, as well as the high prevalence, and reports of hearsay evidence of other affected members, have all contributed to estimations of the condition ranging from 14–90 per cent in close relatives (Allan 1928; Dalsgaard-Nielson 1965; Friedman 1972).

Prevalence

A Finnish study (Sillapää 1976) of migraine in children aged 7 years showed the prevalence of headaches to be 37.7 per cent and of migraine 3.2 per cent. A similar figure for migraine (3.9 per cent) was found by Bille (1962) in children between 7 and 15 years. In another study (Waters 1974) it was estimated that 15–19 per cent of men and 25–29 per cent of women experience a migraine attack during their lifetime.

Frequency in relatives

Figures vary considerably. Taking three studies as examples, Allan (1928) found 60 per cent of patients had an affected parent, whereas Brewis *et al*. (1966) found 1 per cent, and Waters and O'Connor (1970) 19 per cent. In the Allan (1928) study, when both parents were affected 83 per cent of children had migraine. Waters and O'Connor (1970) classified relatives as having no headache, headaches alone, or headaches with one, two, or three features of migraine (unilateral headache, nausea/vomiting, visual or sensory symptoms). If headaches plus two or three features were accepted for a diagnosis of migraine, then 19 per cent of close relatives were affected, but this prevalence figure in immediate relatives is not significantly greater than that of the general population. In a survey of eight reports in the literature of migraine in childhood (Prensky 1976) a close relative was affected by either classical or common migraine in 44–87 per cent of children.

Twin studies confirm that single gene inheritance is unlikely to operate. Harvald and Hauge (1965) investigated 1900 unselected twins in Denmark and found migraine was present in 84 (2.2 per cent). Twelve out of 24 monozygotic twin pairs were concordant, and of the dizygotic pairs three were concordant and 54 discordant. Two more recent twin studies by Lucas (1977) and Ziegler *et al.* (1975) found a low concordance rate in both monozygotic as well as dizygotic twin pairs. The ratios were 26 per cent for monozygotic and 13 per cent for dizygotic twins (Lucas) and 29 per cent for monozygotic and 17 per cent for dizygotic (Ziegler).

Counselling

Risks for first-degree relatives are probably in the order of at least 1 in 5. This high risk might not be significantly different from the prevalence in the population, but it is not known whether the risk is for more frequent and serious migraine as opposed to an occasional attack.

The best recent figures of the frequency of migraine in sibs comes from Baier and Doose (1985). In summary, if one child is affected, but neither parent is affected the risk to a sib is 12.8 per cent. If both parents are affected it is 66.7 per cent. If father is affected (and one child) the risk to the sib is 20 per cent. If mother is affected (and one child) the risk is 27 per cent. It should be noted that in half of the cases mother had migraine and in 1 in 5 it was father.

Basilar migraine

First defined by Bickerstaff in 1961, the symptoms are visual disturbance (usually bilateral loss or distortion of vision), vertigo, ataxia, dysarthria, and tinnitus. These might be followed by headache, or headaches might alternate with the above symptoms.

In a review of 30 cases of basilar artery migraine, Lapkin and Golden (1978) found a family history of migraine in 26 instances (86 per cent). Three out of the four patients without a family history were adopted. Mothers and maternal grandmothers were more often affected than fathers.

In a study of eight children (Golden and French 1975) seven had another affected first- or second-degree relative. The mother was affected in five and the father in one. All the affected relatives had classical migraine and not basilar migraine. There is clearly a relationship between the two types and counselling of risks should be for the disease as a whole.

Recurrence risks are greater than 10 per cent, but the condition is not usually serious.

Familial hemiplegic migraine

In familial hemiplegic migraine the visual aura is replaced by a focal neurological deficit, which includes hemiplegia, hemianesthesia, and a speech deficit (Rosenbaum 1960). There is possibly a separate type in which headache precedes the aura, followed by a neurological deficit with a more prolonged course before recovery. Whitty (1953) separated patients with hemiplegic migraine into two types: those with hemiplegic migraine in which there is a family history of classical migraine and those families in which all the affected had hemiplegic migraine.

Glista *et al.* (1975) reported a family in which 10 of the 26 members were affected by stereotyped attacks. In three of the 10 the attacks were associated with minor head trauma and in one a permanent deficit resulted. Clarke (1910) described a large pedigree (11 affected members) with migrainous headaches and transient neurological signs. The weakness (in five) and numbness (in one) persisted for hours or days. One patient, aged 54 years, died during an attack. In the five families described by Whitty (1953), there was another affected member with hemiplegic migraine in three. Out of the five probands reported by Rosenbaum (1960), although a history of migraine was found in a close relative in all cases, only the father of one had migraine associated with an alternating numbness in both arms. Further families were described by Blau and Whitty (1955), Parrish and Stevens (1977), and Bradshaw and Parsons (1965). These latter authors state that about one-half of the 36 patients in the literature had another affected relative and in their own series familial incidence of hemiplegic migraine was 18 per cent (75 patients). They do, however, point out that about one-third of migrainous patients have transient unilateral limb symptoms.

Inheritance is as a dominant.

The prognosis is usually good—some patients are known to have had more than 100 attacks without permanent ill effect (Bradshaw and Parsons 1965). Progressive dementia (Symonds 1951) and residual hemiplegia have been noted, and respiratory arrest was reported in one member of a three-generation family with seven affected members (Neligan *et al.* 1977).

Familial migraine triggered by mild head trauma

In three families five members in four generation had typical migraine related to trauma (Haas and Sovner 1969). Transient paresthesia, hemiparesis, and visual disturbances were features of the attacks.

Familial hemiplegic migraine with cerebellar manifestation ('cerebellar migraine')

At least four members with this combination of signs occurred in three generations (Ohta *et al*. 1967). The cerebellar signs developed during an attack of severe migraine and were persistent in two out of the four patients. This family could be classified under basilar migraine. A similar family (two generations were affected) was reported by Zifkin *et al*. (1980). The affected members had hemiplegic migraine, nystagmus, and tremor.

Families with recurrent coma inherited as a dominant condition have been reported (Zifkin *et al*. 1980). During the attack there is hyperpyrexia and CSF pleocytosis with the result that the condition is often diagnosed as 'meningitis' following trivial head injury. These patients also have a cerebellar ataxia precipitated by migraine attacks and then subsequently develop persistent ataxia. Radiologically, they show cerebellar atrophy.

Familial hemiplegic migraine with retinal degeneration, deafness, and nystagmus

Four cases of hemiplegic migraine and seven of classical migraine occurred in four generations (Young, *et al*. 1970*b*). The authors conclude that the retinal degeneration and deafness was probably Usher's syndrome (recessively inherited) which was segregating separately.

Cerebrovascular disease

The prevalence of cerebral haemorrhage, thrombosis, or embolism in the relatives of index patients with non-embolic cerebral infarction was slightly in excess in the mothers of male index patients (Marshall 1971). However, taken as a whole, the genetic contribution of cerebrovascular disease is small and risks to sibs and offspring are not greater than those in the general population.

In a later study on the familial incidence of cerebral haemorrhage (based on the clinical picture, blood in the cerebrospinal fluid, operative or post-mortem findings) Marshall (1973) did not find a statistically significant increase in the frequency in first-degree relatives, except among the brothers of female index patients, but this increase does not constitute an appreciable risk.

Internal carotid artery hypoplasia

In a pedigree of 11 sibs, three brothers had hypoplasia of the internal carotid arteries (Austin and Stears 1971).

There was no known consanguinity. The first symptoms of vascular disease occurred before the age of 40 years.

Fibromuscular dysplasia

This is a recognized cause of strokes in adults and children (Shields *et al*. 1977). Autosomal dominant inheritance with variable expression has been postulated.

Familial stroke syndrome associated with a mitral valve prolapse

In a family, eight individuals over three generations had a prolapsed mitral valve. Four of the eight had strokes within the first four decades of life—probably on an embolic basis (Rice *et al*. 1980).

Cerebral angiopathy (dyshoric)—congophilic or amyloid angiopathy

This is a degenerative condition of small arteries or capillaries occurring in old age. Histologically, there is a deposit within the walls of blood vessels and in the perivascular spaces there was a substance with the histochemical properties of senile plaques. In a review of 78 cases Surbek (1961) found all patients to be demented and of advanced age. In two families, four sibs were affected in one and two sisters in another.

Richard *et al*. (1965) studied eight families with dyshoric angiopathy of the cerebral arteries. In six families sibs were affected. The relationship of this condition to senile dementia associated with amyloid deposition is unclear. They could be the same disorder.

A similar condition (dyshoric cerebral angiopathy) was described by Kinney *et al*. (1980) and in two sibs by Griffiths *et al*. (1982). These latter authors state clearly that their patients have the same condition described by Worster-Drought *et al*. (1940; see pages 93 and 284). Inheritance was dominant (previous generations were affected).

Familial cerebrovascular disease in association with a midline nevus flammeus

See page 167.

Familial cerebrovascular disease in association with livedo reticularis—Sneddon's syndrome

This dominantly inherited condition has an onset between 10–33 years. The middle calibre are affected and the skin mottling is the clue to the diagnosis (Rebello *et al.* 1983). Most have peripheral involvement as well.

Moya-moya disease

Twenty families in which first-degree relatives were similarly affected are reported in Japan (Kitahara *et al.* 1979). In 16 families, sibs only were affected, and in four, two generations. Two non-Japanese sibs were reported by Praud *et al.* (1972) and by Søgaard and Jorgensen (1975). Familial aggregation, however, occurs only in the minority of cases—about 7 per cent of 600 cases from Japan. In the families of Søgaard and Jorgensen (1975) three sibs were affected and likewise in an inbred family reported by Narumi *et al.* (1976). Neurofibromatosis must be excluded.

Moya-moya type phenomenon is know to occur in:

(i) Down's syndrome; (ii) Periarteritis nodosa; (iii) collagen disorders; (iv) Sickle cell disease; (v) infection/trauma.

Cerebral aneurysms

The familial occurrence of aneurysms is rare. They might occur as part of another genetic syndrome, and have been reported in the Ehlers–Danlos syndrome (Graf 1965) and in polycystic kidney disease.

Population frequency

For the calculation of risks an assessment of the frequency of affected sibs in a consecutive series of patients needs to be evaluated. In the 276 patients of Kragenbühl and Yasargil (1958) another sib was affected in two instances. Logue *et al.* (1968) studied 90 patients with an anterior cerebral artery aneurysm; two parents had died of a subarachnoid haemorrhage, but no post-mortem evidence of a ruptured aneurysm was available. In a report by

Chakravorty and Gleadhill (1966) three affected relatives were found in a survey of 337 proven cases.

If the population frequency of subarachnoid haemorrhage is taken as 6–16 per 100 000 (Crawford and Sarner 1965), the increased risk to sibs or offspring is negligible. The population frequency of aneurysms at autopsy was 2.1 per cent in the study of Housepian and Pool (1956) and 1.1 per cent in Walton's (1956) series. A figure as high as 7.2 per cent (McCormick, personal communication reported by Bannerman *et al.* (1970)) has been ascertained. It has been suggested (Kak *et al.* 1970) that the familial occurrence of aneurysms might be higher if more radiographic or necropsy evidence were available.

The frequency of aneurysms in the 0–20-year group is lower—1.9 per cent (Patel and Richardson 1971).

Reports of familial occurrences

Hashimoto (1977) pointed out in a review of 46 families with more than one affected member, that rupture occurred at a significantly younger age than most aneurysms (peak in the 30–40-year-old range as compared with 50–60 in non-familial cases).

In familial cases both sib pairs and parent–offspring pairs are reported.

Sibs

Sibs in two families were reported by Bannerman *et al.* (1970), and Brisman and Abbassioun (1971) described two sisters who had in addition 'Friedreich's ataxia'. Two pairs were reported by Kak *et al.* (1970). The aneurysms were situated on the same site in each of the pairs, and the two sib pairs of Graf (1966) had aneurysms with strikingly similar anatomical locations and angiographic appearances. Other sib pairs were reported by Ullrich and Sugar (1960) and Phillips (1963)—two sisters with aneurysms at the same site. Three out of eight sibs were found to have berry aneurysms— not on the same vessel (Beaumont 1968) and in another study (Acosta-Rua 1978) 12 cases of familial intracranial aneurysms (ruptured) came from six families. The aneurysms were in the same location in members of three families. In 10 of the cases sib pairs were affected, and in two a mother and daughter.

Identical twins concordant for aneurysms have been reported by Jokl and Wolffe (1954), Koch (1957), Fairburn (1973), and by O'Brien (1942) in twin brothers (only one proven).

Some of the relationships are more remote. In the report of Ros (1959), as well as two pairs of brothers and one sister-brother pair, there was also one uncle and nephew in a separate family.

Two-generation families

A remarkable family history in which a father and four of his 10 children had symptomatic saccular cerebral aneurysms was reported by Edelsohn *et al.* (1972). Four further sibs had infundibular widening. The aneurysms and infundibular widening were all situated at the junction of the internal carotid and posterior communicative arteries. A mother and two of her daughters (Bentzen 1972) died from ruptured aneurysms. In other families a father and son pair (Chambers *et al.* 1954) and mother and daughter pair (Ross 1959) are reported. The daughter had, in addition to her aneurysm, two angiomas. Aneurysms in a mother and son, probably on the basis of a thinning of the elastic lamina, were described by Nagae *et al.* (1972). Brisman and Abbassioun (1971) described three separate families with mother and daughter affected. Hashimoto (1977) reported a family in which four members had an intracranial aneurysms. Elective angiography on five other members showed an aneurysm in two. Many other members died from cerebrovascular accidents. A family with incomplete penetrance is reported by Ter Berg *et al.* (1987).

Counselling

From the study by Halal *et al.* (1983) the familial cases of aneurysms are not always congenitally present but develop with age. Marked stenosis can occur which might be related to previous haemorrage. When familial cases do occur it is usually as a dominant with variable expression or less likely (author's view) polygenic/multifactorial inheritance.

For counselling purposes Halal *et al.* (1983) suggest that where at least two first degree relatives are affected the risks are as follows (for offspring):

Parents and one child   40–46%
Two sibs alone   18.6%

This could be a low estimate and it might be necessary where two close family members are affected to scan those who are at risk.

Cerebrovascular diseases associated with other genetic disorders

(a) Ehlers–Danlos disease

Familial aggregation of cerebrovascular disease can sometimes be accounted for by an underlying genetic disease. Two sibs with a caroticocavernous fistula (Bannerman *et al.* 1967) and two sibs, one with a caroticocavernous fistula and the other the same lesions, and in addition numerous aneurysms were recorded by Imahori *et al.* (1969). All had the Ehlers–Danlos syndrome. A young Indian had Ehlers–Danlos disease and a vertebral artery aneurysm (Brodribb 1970). Bilateral caroticocavernous fistulae in a young woman with the Ehlers–Danlos syndrome was reported by Schoolman and Kepes (1967). Her mother died from a cerebrovascular accident at the age of 33 years. Tridon *et al.* (1969) reported a six-generation family with Ehlers–Danlos disease in whom one patient had multiple intracranial aneurisms, and five had caroticocavernous fistulae. Kissel *et al.* (1972) point out that cerebro-meningeal haemorrhages as well as caroticocavernous fistulae can occur in the arterial form of Ehlers–Danlos disease (Sack syndrome).

(b) Central nervous system involvement in hereditary haemorrhagic telangiectasia

Only rarely have neurological lesions been described in Rendu–Osler–Weber disease. Recent reviews are those of Waller *et al.* (1976), Adams *et al.* (1977), and King *et al.* (1977). A family (King *et al.* 1977) in which eight members over four generations suffered from the disease, contained four affected with central nervous system vascular disease. Two of these had proven arterio-venous malformations. The authors list over 20 other single cases with both skin and central nervous system manifestations of the disease. Most of the lesions were arteriovenous malformations but fistulae, angiomas, aneurysms, and telangiectasia were also noted. Another possible familial form was described by Boynton and Morgan (1973). Many members had the skin lesions alone, but a child developed heart failure and her father had surgical removal of a vascular spinal cord tumour.

(c) Aneurysms associated with polycystic renal disease

See Ditlefsen and Tonjum (1960).

(d) Familial intracranial aneurysms in Marfan's syndrome

Seven members over three generations had aneurysms. Two others in this family had Marfan's syndrome and other congenital malformations, but not aneurysms (Ter Berg *et al.* 1986).

Cerebral blood vessel hamartomas

There are three groups.

1. FCMCR (familial cavernous malformation with cavernous haem-angioma of the retina).

2. Hereditary neurocutaneous vascular malformation (also called hereditary neurocutaneous angioma syndrome).

3. Familial arterio-venous-malformation.

1. FCMCR/Gass syndrome or familial cavernous angioma/cavernous haemangioma of the retina

This condition was reviewed by Dobyns *et al.* (1987) who described a three-generation family. Two out of the four affected had a retinal cavernous malformation. In the father and daughter pair reported by Gass (1971) the daughter had a raised red nodule on her inner thigh which on histology looked like a capillary haemangioma. In the family reported by Goldberg *et al.* (1979) many were affected over four generations—some had only the retinal lesions, others only the cutaneous lesions (one is shown in the lumbar region). Some had both retinal lesions, seizures and a stroke-related death.

Gass commented that the vascular tumour of the retina in Von Hippel disease is sufficiently characteristic to be easily distinguished from the cavernous lesions in this syndrome.

2. Hereditary neurocutaneous vascular malformation

There have been a small number of families with cutaneous haemangiomata and intracranial arteriovenous malformations without retinal lesions. The first was reported by Zaremba *et al.* (1979). Four members in three generations showed variable expression and penetrance. Two similar families have been reported by Hurst and Baraitser (1988). The cutaneous lesions appeared on the face, neck, trunk, and elsewhere, and were small and diffuse. They often need to be surgically removed because of haemorrhage. Some patients had only one or two lesions. A similar family reported by Foo *et al.* (1980) had members affected over three generations.

The intracranial lesions are arteriovenous malformations in the cerebrum or upper cervical spine. In all of these families some members had skin and intracranial lesions, some had only the skin lesions and some had only the arteriovenous malformation.

3. Arteriovenous malformations and cavernous haemangiomas (without skin or retinal lesions)

Two reports (Kufs 1928; Michael and Levin 1936) suggest occasional dominant inheritance. In each of these families only one patient was examined pathologically. In the family reported by Clark (1970), father and daughter were affected. Both had cavernous angiomas and died in their twenties. For two further families (dominant transmission) see Bicknell *et al.* (1978). Another large pedigree—five with cerebro-vascular malformation, three with cavernous angiomata, was reported by Hayman *et al.* (1982). Forty-three relatives were examined by C T scan and lesions were found in 15. Inheritance was as a dominant with variable expression.

A three-generation family with members (at least four) having arterio-venous malformations is reported by Boyd *et al.* (1985). No significant skin lesions were found.

Tonnis and Lange-Cosack (1953) reported the occurrence of arteriovenous malformations in two sibs and two sisters were reported by Laing and Smith (1974). Both sisters presented with subarachnoid haemorrhages. A brother and a sister had arteriovenous vascular malformations (Barre *et al.* 1978). Both died from subarachnoid haemorrhages.

Aberfeld and Rao (1981) reported sibs (parents were well and unrelated) and Snead *et al.* (1979) described arterio-venous malformations in three half-sibs (same mother). The onset in all was in early childhood and there were no skin lesions.

Bannayan–Zonana syndrome

In this condition macrocephaly and subcutaneous lipomas are the main features and intracranial arteriovenous malformations are unusual, but they have been reported (Miles *et al.* 1984).

Counselling in whole group

Most cases are single. It is unfortunate that high risks can only be predicted where a positive family history already exists.

Familial cirsoid aneurysm of the scalp

Cirsoid aneurysms are rare. Definitely two and possibly a third member of a Persian family out of a sibship of seven were affected. No mention of consanguinity was made (Khodadad 1971).

Familial capillary-venous leptomeningeal angiomatosis (Divry-van Bogaert syndrome)

Two sibs with an onset during the second decade of migraine, with focal paresthesia, Jacksonian epilepsy, post-ictal paresis, visual-field defects, aphasia, apraxia, and progressive dementia were reported by Martin *et al.* 1973*a*. At post mortem the authors found diffuse capillary-venous angiomatosis (non-calcifying) of the leptomeninges and, in addition, fibrosis which selectively involved some tracts in the brain stem. The authors thought that the condition was similar to that described by Divry and Van Bogaert

(1946) in three brothers. These three brothers had cutis marmorata. Another three brothers with cutis marmorata and a dementing illness (Guazzi and Martin 1967) might have had a sudanophilic leucodystrophy.

Four sibs in a Dutch family of 10 children had leucodystrophy and angiomatosis (Bruens *et al.* 1968). Three boys and a girl were affected and they showed early convulsive episodes leading to a decerebrate state. Two further sibs with the infantile form of diffuse sclerosis and meningeal angiomatosis were reported by Arseni *et al.* (1973).

An angiographic study of a single case (Bussone *et al.* 1984) showed a diffuse angiomatous circulation mainly in the distal branches of the middle and posterior cerebral arteries. The differential diagnosis is Moya-moya disease (see page 130) and Binswanger dementia (in this condition there is always sclerosis and atheroma—it is not genetic).

9

Cerebral tumours

Familial occurrence of cerebral tumours is rare and the families reported below are exceptions. In general, recurrence risks for sibs and offsprings are small.

Meningiomas

Meningiomas in a sister and brother in whom there was no evidence of neurofibromatosis was reported by Gaist and Piazza (1959). Identical twins concordant for spinal and cranial meningiomas have also been reported (Sedzimir *et al.* 1973).

Familial occurrence of meningiomas in brothers and their first cousin once removed was reported by Sahar (1965) and Joynt and Perret (1961) described meningiomas in a mother and daughter, as did Wagman *et al.* 1960).

Most of the familiar cases will be found to have dominantly inherited BANF (see p. 147). There is also an interesting finding that a chromosome 22 (see Bolger *et al.* 1985) abnormality identical to those seen in BANF can be detected in tumour cells of isolated meningiomas.

Gliomas

Up to 1977 there were 31 proven brain tumours (unspecified gliomas, astrocytomas, spongioblastomas, medulloblastomas, and ependymonas) reported in sibs (Von Motz *et al.* 1977). There have, however, been few recent studies of a consecutive series of patients with gliomas.

In one large study by Harvald and Hauge (1956), 1344 relatives of 169 patients with glioblastomas were studied. The conclusion reached was that genetic factors did not play a major role whereas van der Wiel (1960) in a similar investigation of more than 5000 relatives of 100 patients with a proven glioma found seven relatives with a verified glioma compared with none in the control group.

Single case reports include those of Fairburn and Urich (1971) who report a pair of identical twins with a part astrocytic, part oligodendroglial tumour in both members, and Von Motz *et al.* (1977), who described a sibship in which three sisters had astrocytomas. The parents were unrelated. Armstrong and Hanson (1969) reported three sibs with gliomas, and Kjellin *et al.* (1960) noted seven families in which six had members with the same type of tumour. A father and his two children, a mother and son, a mother and daughter, twin brothers, and two sets of half-sibs were the pairs involved. Inheritance in the six cases of cerebral gliomas reported by Thuwe *et al.* (1979) was suggestive of dominant inheritance with incomplete penetrance and there was a strong suggestion of a common origin of those affected in the eighteenth century.

Two male sibs out of a sibship of five died at 38 and 50 from the complication of a glioblastoma multiforme (Kaufman and Brisman 1972). The same authors report two male first cousins, one with a benign cystic cerebellar astrocytoma and the other with a malignant cerebellar astrocytoma.

Three cases of presumed malignant gliomas occurred in a family (Munslow and Hill 1955); only two were verified histologically. Two brothers had an onset of symptoms in middle life and an affected cousin died at 46 years of age.

Isamat *et al.* (1974) reported six families in which another close relative was affected. In 10 of the 12 affected there was concordance for the type of tumour (astrocytoma of various grades in all) and in one mother–son pair the mother had a glioma and her son an ependymoblastoma. There were two pairs of sisters, three parent–offspring, and one aunt–nephew combinations.

The simultaneous appearance of frontal lobe oligodendrogliomas in brothers were reported by Parkinson and Hall (1962).

An unusual inbred family in which five males were affected with a childhood onset of an anaplastic glioblastoma multiforme (Chemke *et al.* 1985) has been described. Four of the males were sibs and one was a first cousin. Inheritance seemed recessive.

There have also been identical twins, concordant for oligodendrogliomas and for uterine leiomyomas (Roelvink *et al.* 1986).

In a review of the literature, Challa *et al.* (1983) who also reported two families of their own with multiple generations affected with skips, and members with other types of cancer conclude:

(1) it is unusual to find familial aggregation;

(2) when it occurs it is usually in twins, sibs or parent/offspring (first degree relatives);

(3) many generational families are exceptional but note those mentioned above. Some of those families are so called 'cancer families' with many types of neoplasm, including astrocytomas, segregating in the same family.

Gliomas and polyposis coli

The association of cerebral glioma and polyposis coli has been noted twice. Turcott *et al*. (1959) reported a brother and sister, one with a medulloblastoma and the other with a frontal lobe glioblastoma (both had polyposis). The family of Braughman *et al*. (1969) was remarkable. Out of a sibship of six, three had gliomas and polyposis, and one only a glioma. A post-mortem was refused so that the presence of polyposis could not be confirmed. The onset of the cerebral tumour was in childhood in all the members.

More recently, Todd *et al*. (1981) and Itoh and Ohsato (1985) have reported further families.

Familial cerebral sarcoma

Gainer *et al*. (1975) reported four cases of cerebral fibrosarcomas in two families. The affected members were a father and daughter in one family, and two sisters in another. Onset in both families was late in life. In one of the cases, there seemed to be a multicentric origin of the tumour. From the histology in all four cases it could be argued that the tumours were meningiomas that had undergone malignant change.

Medulloblastoma

The rare occurrence of medulloblastomas in Chinese newborn sisters was noted by Belmaric and Chau (1969). These authors reported another instance of familial medulloblastoma in identical twin girls aged 8 and 11 weeks originally described by Griepentrog and Pauly (1957). The same histopathology was found in tumours developing in a half-brother and half-sister (the common parent was not stated) reported by Kjellin *et al*. (1960).

Medulloblastoma in twins (probably identical twins) was reported by Leavitt in 1928. Both children died before their tenth year.

Other cerebellar tumours

Two sisters, one aged 5 and the other 2 years, had brain tumours. One had a ganglioneuroma with malignant changes and the other had a desmoplastic medulloblastoma (Thomas *et al*. 1977).

Familial frontonasal cysts

A father and all three of his children had frontonasal dermoid cysts (Plewes and Jacobson 1971).

Familial benign intracranial hypertension

Familial aggregation occurs in only a small percentage of cases. Three sisters were reported by Trawiesa *et al.* (1976). Obesity was a striking feature in all three. In one of the sisters progressive optic atrophy over a period of nine years ended in blindness. Descriptions of two sisters (Bucheit *et al.* 1969), and two further sisters (Howe *et al.* 1973) confirm occasional recessive inheritance. However Rothner and Brust (1974) reported a mother and son with this condition.

10

Neurocutaneous malformations

A. Tuberose sclerosis.

B. (a) Neurofibromatosis.
    (b) Conditions confused with neurofibromatosis.

C. von Hippel–Lindau disease.

D. (a) Sturge–Weber syndrome.
    (b) Conditions confused with Sturge–Weber syndrome.

E. Klippel–Trenaunay–Weber syndrome.

F. Riley–Smith syndrome.

G. Xeroderma pigmentosa.

H. Incontinentia pigmenti.

I. Hypopigmentation syndromes.

J. Cerebrotendinous xanthomatosis.

A. Tuberose sclerosis (Figs 10.1 and 10.2)

Tuberose sclerosis is characterized clinically in adults by mental retardation, epileptic seizures, and adenoma sebaceum. Gunter and Penrose (1935) found the frequency in a hospital for the mentally defective to be 1 in 300, Zaremba (1968) 1 in 100, and Ross and Dickenson (1943) 1 in 175.

All previous studies have underestimated the frequency of the condition. If minimal manifestations such as hypomelanotic macules are attended to, the frequency is way above the original estimates. The figures of (Hunt and Lindenbaum 1984) are 1 in 30 000. For those under 30 it is 1 in 21 000 and those under 5, 1 in 15 400. Recently, R. H. Lindenbaum (1987, personal communication) has reviewed these figures and suggests an even higher

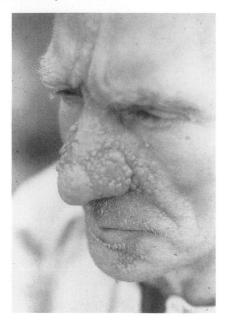

Fig. 10.1   Adenoma Sebaceum in tuberose sclerosis.
Courtesy of Dr M. Ridler.

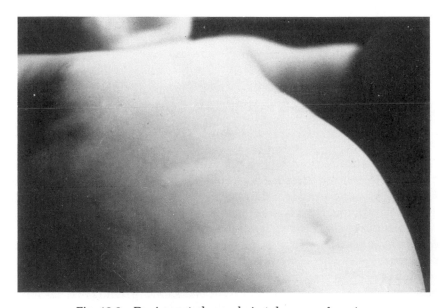

Fig. 10.2   Depigmented macule in tuberose sclerosis.

prevalence rate, i.e., 1 in 18 000 for those under 30 years, and 1 in 7 500 for under 5 years.

Skin lesions

The full clinical picture evolves gradually and differences are to be expected in the presentation in children and adults. In 100 children (up to the age of 12 years), Pampiglione and Moynahan (1976) found the main clinical features in the first two years of life to be seizures (98 per cent of cases), particularly infantile spasms (69 per cent). The adenoma sebaceum and intracranial calcification became evident after the age of 4 years, but experienced observers (Pampiglione and Moynahan 1976) could recognize the skin lesions in nearly one-third of cases by the age of two years. In adult life the percentage of patients with intracranial calcification reached 51–60 per cent (Lagos and Gomez 1967; Zaremba 1968). The skin manifestations of tuberose sclerosis consist of an initial telangiectasis around which fibrous tissue develops. These have been inadvertently called adenomata sebaceum. Hypopigmented macules (leucoderma), pigmented naevi, sublingual or subungual papillomas, and shagreen patches are the other cutaneous signs.

Hurwitz and Braverman (1970) found that only 13 per cent of children with tuberose sclerosis developed the facial lesions of adenoma sebaceum before the first year of life. These dome-shaped papules are symmetrically distributed on the nasolabial folds, cheeks, and chin, but the upper lip and philtrum are spared. The same authors describe typical shagreen patches as flat or slightly elevated lesions flesh-coloured and wrinkled, like pigskin. They develop at the same time as the adenoma sebaceum and occur in 20–50 per cent of patients. In adults they occur more frequently—83 per cent (Nevin and Pearce 1968). The periungual fibromata develop at puberty. If, by puberty, no skin lesions have appeared they are unlikely to develop in adolescent or adult life.

Hypopigmented macules are the earliest signs of tuberose sclerosis although they are often looked for later in the course of the disease after the other more recognizable manifestations have been found. These macules vary in shape and are only occasionally in the form of a 'mountain ash leaf' as described by Fitzpatrick *et al.* (1968).

They are best seen using a Wood's lamp and all children and adults in whom the diagnosis of tuberose sclerosis is suspected should be looked at under an ultraviolet lamp in a dark room. Histologically, there are adequate numbers of melanocytes present, but the melanosomes are small and partially or totally deficient in melanin, whereas in vitiligo the melanocytes are few in number. However, there is nothing diagnostic about the hypomelanotic maculae and a biopsy is not useful for diagnostic purposes.

In a study of the hypopigmented macules by Jimbow *et al.* (1975) the authors compared the histology with that of two other congenital hypo-melanoses (piebaldism and nevus depigmentosus). It is suggested that in the lesion of tuberose sclerosis there is a decrease in the synthesis, melanization, and size of the melanosomes which is quite distinct from the other two conditions. Zaremba (1968) found white naevi in 3.3 per cent of children in a mental institution. In light-skinned children a Wood's lamp is helpful in the detection of the white patches. With the passage of time pigmentation does occur within the macules. The macules are found in 78 per cent of patients with the highest concentration on the trunk and legs (Hurwitz and Braverman 1970). *Café-au-lait* occur in 5 (Vinken and Bruyn 1972) to 28 per cent of cases (Nevin and Pearce 1968).

Smooth, slightly raised fibrous plaques with a red/yellow colour are typically found on the forehead and scalp.

CNS manifestations

Seizures occur in 70 per cent of patients. It is a useful rule of thumb that seizures might occur without mental retardation, but the diagnosis is unlikely in those with mental retardation without seizures.

Mental retardation occurred in 61 per cent of patients (Nevin and Pearce 1968) and in 56 per cent in the Mayo Clinic series (Gomez 1979). Its pathogenesis is obscure, although Gomez (1979) suggests that there is a relationship between the severity of the seizures and mental retardation. Nineteen of their patients who had never suffered seizures had average intelligence, but the relationship might not be simple.

Four cases of tuberose sclerosis (typical skin lesions) with normal skull X-rays were shown to have periventricular opacities in CT scans (Gomez *et al.* 1975). The usefulness of a CT scan is illustrated by Martin *et al.* (1976) who found lesions on the scan in eight patients in whom an ordinary skull X-ray showed no abnormality. The children were aged from 8 months to 12 years. All had skin lesions. Similarly, when 20 children with tuberose sclerosis were examined by CT scan (all had seizures and other suspicious manifestations of tuberose scleosis) all had areas of increased density on the scan whereas the standard skull X-ray showed calcifications in only three out of nine cases (Kuhlendahl *et al.* 1977).

Two infants were investigated soon after birth because of seizures. A CT scan within the first week of life showed intracranial paraventricular calcification and both were subsequently found to have tuberose sclerosis (Sugita *et al.* 1985). (One had depigmented macules and the other had evidence of the condition on brain biopsy.)

Central nervous system lesions are of two types (Chalhub 1976). Pale, hard gliotic areas in the convolutions known as tubers, and multiple tumour-like nodules which have a predeliction for the subependymal region and project into the ventricles giving the radiological picture of candle guttering.

A giant cell astrocytoma is the main intracranial tumour associated with tuberose sclerosis. It is not particularly malignant, but growth might cause obstruction to the flow of CSF and results in hydrocephalus.

MRI scan in tuberose sclerosis

A comparison with CT was made by Roach *et al.* (1987). Multiple high-signal MRI lesions involving the cerebral cortex were seen, probably corresponding to hamartomas and gliotic areas. These were only occasionally seen on CT. The periventricular calcification is better seen on CT.

Eye

Involvement of other organs is frequent. Retinal phakomatoses are present in about half of the gene carriers at birth.

Kidney

Occasionally, the presenting sign in tuberose sclerosis in infancy or childhood is an intra-abdominal mass. These are mainly angiomyolipomas of the kidney and they are usually bilateral and multiple. About 50 per cent of gene carriers have renal lesions (Fryer and Osborne 1987). Whether renal cysts apart from angiomyolipomas occur is not clear although there is evidence that this is a separate lesion associated with tuberose sclerosis (Stapletone *et al.* 1980).

Heart

Single or multiple rhabdomyomata occur in 30 per cent of tuberose sclerosis patients coming to post mortem (see Fryer and Osborn's review 1987). Gibbs (1985) found a 50 per cent incidence in a neurology clinic population, but most patients seem to be asymptomatic. The relevance of rhabdomyomata in the absence of other lesions is not yet certain.

Lungs

This is a rare organ to be involved but lymphangiomata (multiple hamarto-matas of the lymphatic ducts) have been reported.

Other features

Hamartomatous lesions have been found in the pancreas, thyroid, and adrenals. Poliosis (patches of grey hair) and depigmented white hair have also been observed (McWilliam and Stephenson 1978).

Malignant tumours are not rare and include giant cell astrocytomas (Fowler and Williams 1973), leiomyosarcomas, fibroblastic sarcomas, and angiosarcomas. Multiple huge rhabdomyomata were found in a 31-week-old fetus in whom the cerebral lesions were typical of tuberose sclerosis (Probst and Ohnacker 1977).

There are few reported cases of tuberose sclerosis in infancy. One such infant who was born at 33 weeks gestational age died after 36 hours (Thibault and Manuelidis 1970). Multiple nodules were found in the cortex beneath the ventricular ependyma and several rhabdomyomata of the heart were seen.

Thirteen other cases reported within the first week of life are known and summarized by Thibault and Manuelidis (1970). Skin lesions were only occasionally mentioned, but 11 out of the 13 reported cases had rhabdomyomata of the heart. There is a suggestion from the authors that a single rhabdomyoma of the heart may represent a forme fruste of the disease (see above).

Genetics

Inheritance is as a dominant and many families have been described. In 1913, Berg reported a three-generation pedigree. Two generations were affected in four families (Nevin and Pearce 1968), and Marshall *et al.* (1959) found 16 cases in two families. Bundey and Evans (1969) calculated that 86 per cent of tuberose sclerosis are fresh mutations. In their own series, only 10 out of 71 index cases had an affected parent, and these were only mildly affected. Four had adenoma sabaceum as the only clinical manifestation, whereas four had in addition fits which subsequently did not recur. In four others, the parent was so mildly affected that the diagnosis was not made until the child developed obvious signs and symptoms.

The frequency of fresh mutations in other studies is slightly lower: in the Nevin and Pearce (1968) series 75 per cent, Zaremba (1968) 50 per cent, and Stevenson and Fisher (1956) 78 per cent.

No skips occurred in the published series of Nevin and Pearce (1968), Zaremba (1968), Francois (1975), and Ponsot and Lyon (1977). Variability in expression and some variability in penetrance can make genetic counselling difficult.

Dizygotic twins with epiloia are reported (Primrose 1975). The mother was normal and the father had a single lightly pigmented birthmark measuring 7.3 × 8 cm and other small pigmented lesions. He had no obvious evidence of tuberose sclerosis. Two sibs with unaffected parents and normal CT scans were reported by Wilson and Carter (1978) and two first cousins with normal parents are known to the author. In the last two instances, none of the parents had dermatological evidence of tuberose sclerosis. These cases are exceptions —most family studies show a regular pattern of transmission without skips.

A unique pedigree with tuberose sclerosis in three generations had two gene carriers who were asymptomatic heterozygotes (Rushton and Shaywitz 1979). A complete physical examination of the asymptomatic carriers included a skin examination by Wood's lamp, but no depigmented macules were found.

A further family with skips—half sibs with the same unaffected mother —is reported by Connor *et al*. (1986). Mother had normal skin and normal renal and brain scans. They refer to one other family where two so called normal parents who refused detailed investigation, had two affected children.

A full examination of both parents for minimal manifestations in possible carriers should include an examination of the skin, including the use of a Wood's lamp, if any suspicious lesions are seen, then proceed to a CT scan of the brain (possibly including the kidneys). In general, however, those without skin manifestations are unlikely to have intracranial signs so that a brain scan is not mandatory. Notwithstanding these rare families with skips, recurrence risks are small when parents are normal. A 1 per cent risk seems appropriate.

Gene localization

The gene for tuberose sclerosis has now been localized to chromosome 9q (Fryer *et al*. 1987). If sufficiently close polymorphic markers are found it does seem possible that pre-natal diagnosis will shortly become available. The problem will initially be that many cases are fresh mutations and these cannot be prevented. There might be heterogeneity with another locus on the chromosome (the rarer locus).

B. Neurofibromatosis

There are two well described entities and three other possibly distinct conditions.

1. Bilateral acoustic neurofibromatosis (BANF) or central neuro-fibromatosis.

2. Peripheral neurofibromatosis or von Recklinghausen disease.

3. Segmental neurofibromatosis.

4. Late onset—non genetic neurofibromatosis (not discussed further).

5. *Café-au-lait* patches alone (not discussed further).

(1) BANF

Central neurofibromatosis is characterized by bilateral acoustic neuromas with an onset of symptoms at about 20 years, with only mild skin changes. Three large families were reported by Kanter *et al.* (1980). No patient had more than six *café-au-lait* patches although half of the patients had between one and five.

The condition might solely present with multiple intracranial neoplasms, whereas the only common intracranial tumour in peripheral neurofibromatosis is the optic nerve glioma. It is highly likely that many pedigrees previously reported as having familial meningiomas are examples of central neurofibromatosis.

The following families are examples.

(a) *Those with bilateral acoustic neuromas.* Gardner and Turner (1940) reported a follow-up of a remarkable family described previously by Gardner and Frazier (1930). At least six generations were affected and 38 individuals were regarded as definitely or probably affected. The pedigree was re-examined by Young *et al.* (1970a; Gardner was again a co-author). Peripheral signs of neurofibromatosis were few, suggesting that the family had a central form of the disease that might be genetically distinct from the peripheral form. In two other families (Feiling and Ward 1920; Moyes 1968) bilateral acoustic neuromas were the chief clinical manifestations, but in the Moyes pedigree subcutaneous nodules (no histology) were reported.

(b) *Those with meningiomas.* A remarkable family with multiple meningiomas in grandmother, mother, and daughter was reported by Girard *et al.* (1977). The mother had two tiny neurofibromas on her hand and had bilateral deafness. In a series of meningiomas of childhood, Merten *et al.* (1974) found that almost a quarter of their patients had manifestations of neurofibromatosis. A family with minimal features of neurofibromatosis had members over three generation with multiple meningiomas (Battersby *et al.* 1986).

The confirmation that central neurofibromatosis or BANF is distinct from peripheral neurofibromatosis has now emerged by the localization of the central condition to chromosome 22 (Seizinger *et al.* 1987) and peripheral neurofibromatosis to chromosome 17 (Barker *et al.* 1987).

(2) Peripheral neurofibromatosis

This is characterized by tumours of both the peripheral and central nervous system, *café-au-lait* patches, and involvement of many organs. The disease varies considerably in its manifestation (Figs 10.3 and 10.4). The birth

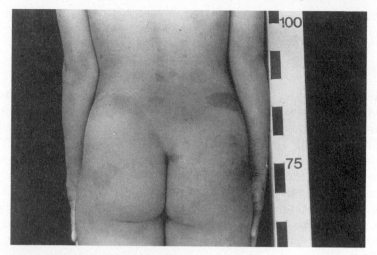

Fig. 10.3 Neurofibromatosis showing multiple. *café-au-lait* patches.

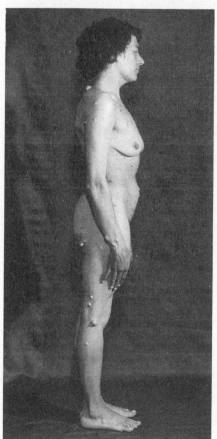

Fig. 10.4 Neurofibromatosis showing multiple neural tumours.

incidence is approximately 1 in 3000 and it is suggested that between 50–70 per cent of cases are fresh mutations.

Skin lesions

Café-au-lait patches are essential for the diagnosis of neurofibromatosis.

Whitehouse (1966) found that more than two spots occur in normal children under the age of 5 years in only 0.75 per cent of the population, and that five spots with a diameter of more than 0.5 cm is suggestive of the diagnosis. In adults, Crowe (1964) considers six *café-au-lait* spots exceeding 1.5 cm in the broadest diameter to be significantly abnormal. Axillary freckles when present are helpful indicators of the disease (Crowe 1964).

In a series of childhood cases (Fienman and Yakovac 1970) 43 per cent of the children showed clinical signs of the disease at birth, and 63 per cent by one year of age. The signs were multiple *café-au-lait* spots and neurofibromas.

The *café-au-lait* spots tend to increase in size and number during the first and second decades. Histologically, they contain giant granules and an increased number of melanocytes. They can undergo melanomatous degeneration (Knight *et al.* 1973; Perkinson 1957). Other skin tumours, including basal cell carcinoma, have been described (Rodriguez and Berthrong 1966).

Although *café-au-lait* patches tend to increase in early life, especially during adolescence, they tend to disappear in middle and old age especially in those with multiple subcutaneous tumours. There might be a separate entity (see page 147) in which the sole manifestation is the appearance of multiple *café-au-lait* patches, but this is extremely rare and no patient should be promised this benign course.

Plexiform neuromas are large, raised subcutaneous masses whose margins are difficult to palpate. Many develop during the first few years of life, and nearly a half of those in the Huson *et al.* (1988) series developed before 14 years of age. In their study a third of the total number of patients had this lesion. They most often appear on the trunk, and least commonly on the head and neck, but it is in this latter position that they are cosmetically the most troublesome.

Severe bony lesions

The bone lesions are erosions, cyst overgrowths, pseudoarthrosis tibiae, hemihypertrophy, bowing, scoliosis, and skull and facial deformities (Fienman and Yakovac 1970). Anterolateral bowing of the tibia tends to occur before the age of 2 years and can precede the onset of the skin lesions (Crawford 1978). The same author states that 55 per cent of patients with congenital pseudoarthrosis of the tibia have neurofibromatosis although looked at the other way around, pseudoarthrosis occurs in 3 per cent of neurofibromatosis.

Scoliosis has been said to occur in 20 per cent of patients, but less frequently—4 per cent—in the Huson *et al.* (1988) paper. Lateral meningoceles are common, but they seldom cause clinical problems. Anterior vertebral beaking in early childhood is a sign which needs frequent evaluation for this might be an indication that rapid acute angulation will occur.

Malignancy

Sarcomas

Sarcomatous changes can occur in a plexiform neuroma. These are ill defined soft tissue masses of early onset occuring in 30 per cent of patients often in the distribution of the V cranial nerve, but they occur on the limbs and elsewhere. The detection of malignant change in a plexiform neuroma is difficult and the only useful criteria, are rapid growth and pain.

Sarcomatous degeneration of neurofibromas occurs as follows: Knight *et al.* (1973), 4.4 per cent of patients; D'Agostino *et al.* (1963*b*), 3.1 per cent of patients; Preston *et al.* (1952), 16.5 per cent of patients.

Peripheral nerve malignancy

In the Brasfield and Das Gupta (1972) series, the overall incidence of malignant schwannomas was 29 per cent, but it should be noted that malignant change in a single well defined peripheral neurofibroma is rare. Only the plexiform lesions tend to become malignant.

Malignancy as a whole

There is still dispute about the frequency of malignancy in neurofibromatosis. In the Huson *et al.* (1988) study the incidence of malignancy including CNS tumours related to neurofibromatosis was approximately 6 per cent.

Cerebral tumours

In the series of Brasfield and Das Gupta (1972), nine patients (8 per cent) had cerebral tumours. Seven of the nine died early and in four in whom an autopsy was performed all had optic nerve gliomas and mid- and forebrain tumours. The cerebral tumours occur more frequently in childhood. The spinal cord involvement, because of extension of spinal root tumours in four patients, occurred later (20–25 years of age). A disseminated astrocytic

glioma in the brain and brainstem (Barnard and Lang 1964), and gliosar-
comas (Roberts 1967) are typical tumours but are rare.

Optic nerve gliomata

Ten to 40 per cent of patients with optic nerve gliomas show evidence of
neurofibromatosis (Marshall 1954). In their series of cases, Crowe *et al.*
(1956) found this complication occurred in only one of 223 patients with
neurofibromatosis, but there were cases of blindness of unknown cause. In a
survey of 64 optic nerve gliomas the incidence of neurofibromatosis was 9.4
per cent (Font and Ferry 1972).

Tumours of the orbital floor (Gurland *et al.* 1976) and tumours of the eye
lid (Brownstein *et al.* 1974) have been described. Melanomas of the choroid
(Wiznia *et al.* 1978) and choroidal neurofibromas also occur as well as glial
tumours of the retina (Martyn and Knox 1972).

Other cancers

The incidence of a second primary tumour is difficut to assess. Melanomas,
cancer of thyroid, breast, and lung have all been described in association with
the disease. There have been single cases of cancer of the colon (Jenkins and
Gill 1972) and of leukaemia (Reich and Wiernik 1976; Bader and Miller
1978). Non-lymphocytic leukaemia, a rarity in childhood, predominated.

Vascular disease

Vascular disease most commonly affects the renal arteries, but also affects
the aortic, celiac, and mesenteric arteries (Tomsick *et al.* 1976). Three
patients with occlusive cerebrovascular disease were reported by Hilat *et al.*

Table 10.1

	Number	%
Bone lesions	52	47.2
Non-tumerous central nervous system lesions	13	12
Gastrointestinal tract involvement	12	11
Vascular lesions	4	3.6
Benign schwannomas	65	59
Malignant schwannomas	32	29
Other type of cancer	16	14.5
Miscellaneous cancer	5	4.5

More than one system was involved in many patients.

(1971), and a picture indistinguishable from Moya moya disease can occur. Multiple aneurysms may be a manifestation (Fye *et al.* 1975).

Mental retardation with von Recklinghausen's disease

This well-known association is difficult to quantify. The figures of 8 per cent (Allen 1964) and 10 per cent (Canale *et al.* 1964) are useful for counselling purposes. The pathology is that of a disordered cerebral architecture with neuronal heterotopia (Rosman and Pearce 1967).

In the series of cases collected by Fienman and Yakovac (1970), 24 per cent of patients were retarded or slow, and in that of Chalhub (1976) 10–20 per cent. All of these figures may be biased and a true figure is probably somewhere between 1.4 and 9.9 per cent (Carey *et al.* 1979).

It is now clearly established that speech delay accounts for a proportion of those who were previously considered to be retarded. Extra speech therapy at school is indicated and improvement is often encouraging, enabling children to attend normal school. In the study reported by Huson *et al.* (1988) 10 per cent needed special school education and a further 10 per cent attended remedial classes. Only one patient had severe retardation.

(3) Segmental neurofibromatosis

Two instances where the distribution of the *café-au-lait* spots and neurofibroma was segmental were reported by Miller and Sparkes (1977). In the Crowe *et al.* (1956) series, four similar instances were noted. These authors suggest that a somatic mutation might explain the phenomenon. All the case reports have been of isolated cases.

Genetics and counselling

The disease has a frequency (State of Michigan—Crowe *et al.* 1956) of 1 in 2000–3000 live births. Presier and Davenport (1918) suggested that inheritance is dominant and many observers have confirmed this. In the series of Crowe *et al.* (1956) and Fienman and Yakovac (1970) about 50 per cent of patients had an affected relative, and in the Brasfield and Das Gupta (1972) series, 48.2 per cent. In the latter study, five families were affected through four generations.

In a large dominant pedigree (Norman 1972) seven sibs out of eight were affected. Their mother had the disease. Four of her children were retarded and two had optic nerve gliomas. All had *café-au-lait* spots. Six had macrocrania.

Table 10.2   Based on Carey *et al.* 1979

	%
IQ > 70 with learning difficulty	11.4
Mental retardation < 70	9.9
Seizures	6.1
Scoliosis	5.3
Precocious puberty	2.2
Optic glioma	1.5
Other intracranial tumours	5.3
Macrocephaly	35.5
Short stature	3.8

Over two-thirds have relatively mild involvement and one-third develop serious health problems.

Consultands take risks according to the magnitude thereof and the burden of the disease. Parents might accept a 50 per cent risk for *café-au-lait* patches and subcutaneous tumours but not for the severe manifestation. Table 10.2 shows the frequency of serious complication.

Another useful way of viewing the problem is provided by Huson *et al.* (1988). About 19 per cent have problems in childhood which cause lifelong problems e.g. orthopedic problems and plexiform neuromas. Six per cent have additional problems which can if detected be easily managed e.g. renal artery stenosis, and there is a 6 per cent occurrence of malignancy. These are the medical problems. However, the personal anguish of having numerous subcutaneous tumours should never be underestimated and is often the most debilitating complication.

A frequent counselling problem concerns advice to those who have an affected parent, but who themselves at marriageable age have no skin manifestations. In general, those who inherit the gene will show clinical evidence of the disease before the end of the second decade. No absolute assurance can be given, and it would not be unreasonable to suggest a longer wait, but by the early thirties the risk would have fallen to acceptable levels. The penetrance of the gene is high. In one report a father had an osteosarcoma (without skin lesions), and his daughter had a rhabdomyoma and the skin lesions of neurofibromatosis. Reports of non-penetrance are rare.

Given that the gene has now been localized to chromosome 17 (Seizinger *et al.* 1987), pre-natal diagnosis, where appropriate, should soon be on offer. Until a gene specific probe is available, the test cannot be used for diagnostic purposes.

Maternal effect on severity of neurofibromatosis

In a study by Miller and Hall (1978) it was found that members of six sib pairs with neurofibromatosis, whose mothers were affected, had severe complications, whereas no such association was found when father was the affected parent. This has not been confirmed by other studies.

Neurofibromatosis and Sturge–Weber syndrome

This rare combination is reported in one family (Kissel and Arnould 1954). The mother had neurofibromatosis and her daughter had Sturge–Weber syndrome.

Neurofibromatosis and von Hippel–Lindau disease

Meredith and Hennigar (1954) reported a lady with a cystic cerebellar haemangioma, von Recklinghausen's disease (proven by skin biopsy), and a pheochromocytoma. No retinal lesions were seen and there was no family history.

Conditions which might be confused with neurofibromatosis

Neurocutaneous melanosis

Abnormal skin pigmentation—usually a single large nevus or multiple smaller areas of dense pigmentation in association with an infiltration of melanocytes and melanoblasts in the meninges—has been described many times (Fox 1972). The skin lesions are present from birth and are not usually malignant although a patient with malignant leptomeningeal melanoma was reported by Kaplan *et al.* (1975*a*). Raised intracranial pressure from defective absorption of c.s.f. or obstruction is the usual presenting sign to the neurologist. Familial occurrence has not been reported for the full syndrome.

Nevus of Ota

Pigmentation of the face and sclera is also known as oculocutaneous melanosis. The lesion is mainly benign although malignant melanomas have been described (Jay 1955). In a series of 240 patients, Hidano *et al.* (1967) found no neurological involvement. Most of the described cases are sporadic but in his series there was a sib pair (unrelated parents), and a father and son similarly affected. There has been a report of an association with sensori-neural deafness (Reed and Sugarman 1974), with Sturge–Weber syndrome, with Klippel–Trenaunay–Weber syndrome (Furukawa *et al.* 1970), and with an intracranial arteriovenous malformation (Massey *et al*. 1979). These latter authors report that taken as a whole 80 per cent of those affected are female and most cases are Japanese.

Basal cell nevus syndrome

The common manifestations are multiple nevoid basal cell tumours, palmar and plantar dyskeratosis, cysts of the mandible or maxilla, calcification of the dura mater, and bifid or splayed ribs (Codish *et al*. 1973). More than 200 case reports have been published and inheritance is as a dominant. This mode of inheritance was first suggested by Gorlin and Goltz (1960), and other pedigrees are those of Herzberg and Wiskemann (1963) and Swift and Horowitz (1969).

Blue rubber bleb nevus (Bean's syndrome)

Single or multiple, blue and raised, the blue rubber bleb nevi vary in size from 0.1 to 5 cm and are frequently situated on the trunk, extremities, or face (Morris *et al*. 1978). Of diagnostic importance is that blood can be expressed from the lesions. Most are single cases, but a few reports of dominant inheritance are known (Fine *et al*. 1961). Vascular visceral lesions have been described on many occasions (Morris *et al*. 1978) and the association with Maffucci's syndrome noted (Sakurane *et al*. 1967). Inheritance through five generations was reported by Berlyne and Berlyne (1960; the bowel was affected in four of the 14 patients), and two pedigrees (both dominant) were reported by Walshe *et al*. (1966).

Giant pigmented nevi

These giant hairy nevi, present at birth, vary in size and are frequently situated on the back or neck (Goodman *et al*. 1971). The surface is irregular and deeply pigmented. These authors suggest that, at least in some cases, the lesions are genetically determined. They found that close relatives of three probands had multiple small raised pigmented nevi often associated with hairy patches. Inheritance in these families could be interpreted as a dominant. Giant pigmented hairy nevi have also been reported in sibs (Voigtländer and Jung 1974). Malignant changes occur in between 2 and 20 per cent of cases. Although 5 per cent of patients with neurofibromatosis have giant pigmented nevi the two conditions are distinct (Solomon *et al*. 1980).

Counselling

Most are sporadic and recurrence risks are small. The risk of malignancy is between 5 and 10 per cent.

LEOPARD syndrome (lentiginosis profuso)

The mnemonic LEOPARD stands for lentigines, electrocardiogram abnormalities, ocular hypertelorism, pulmonary stenosis, abnormalities of genitalia, retardation of growth, and deafness (Gorlin *et al.* 1969). Mental retardation occurs only occasionally (Pickering *et al.* 1971). Inheritance is as a dominant and families are reported by Lassonde *et al.* (1970), Seuanez *et al.* (1976), and Smith *et al.* (1970).

Other conditions which are probably the same as the LEOPARD syndrome (or the Noonan/neurofibromatosis syndrome)

Pulmonary artery stenosis, abnormal genitalia, retardation of growth, and deafness

Inheritance is as a dominant (Watson 1976; Matthews 1968). The skin lesions are dark brown and they become obvious from 2 to 5 years with an increase in number at puberty.

Lentiginosis and cardiomyopathy

Eight cases and their families were reviewed by Polani and Moynahan (1972), and the inheritance was thought to be dominant with variable expression and penetrance.

A patient with lentiginosis described by Selmanowitz *et al.* (1971) had Crowe's sign (multiple axillary freckles), but in general, the skin lesions of von Recklinghausen's disease are larger, paler, and usually asymetrical (Polani and Moynahan 1972). Heart disease is not a feature of neurofibromatosis.

Naegli syndrome

This hyperpigmentation syndrome has, in addition, hypohidrosis, nail dystrophy, poor teeth, moderate hyperkeratosis of palms and soles, and hypoplasia of the print patterns as features. A report by Sparrow *et al.* (1976) described a family in which seven persons were affected in seven generations. A similar family is described by Franceschetti and Jadassohn (1954).

C. von Hippel–Lindau disease

von Hippel (1904) established as a clinical entity the condition of retinal angiomatosis. Lindau (1926) subsequently noted that 20 per cent of the cases of von Hippel's disease had haemangioblastomas of the cerebellum or lower brain stem. In addition, hypernephromas, polycystic disease of kidney and pancreas were suggested by him as part of the disease now known as von Hippel–Lindau's disease. The manifestations of this single-gene disorder are variable. The first symptoms are usually related to the cerebellum or retina and the age of onset of symptoms is between 18 and 50 years (Melmon and Rosen 1964).

It was Lindau in his original paper (1926) who described the association between cerebellar haemangioblastomas and benign tumours of various organs including the kidney. Since then, the association of the cerebellar tumour with renal carcinoma has been shown in families by Olivecrona (1952) and Tonning *et al.* (1952), and with liver tumours (Craig *et al.* 1941). In a series of 22 cases, 13.5 per cent had renal carcinoma (Malek and Greene 1971). These lesions occurred in relatively young persons and were multifocal in origin in one case. A prospective study of a family with von Hippel–Lindau's disease was undertaken looking for renal lesions. Two brothers were already affected (Richards *et al.* 1973). Polycythemia was suggestive of the syndrome in this family and probably occurs in 10–20 per cent of all cerebellar haemangioblastomas.

Polycystic disease of the pancreas occurred in just over half the members in a pedigree in which 11 members were affected (Christoferson *et al.* 1961). The age of onset in this family ranged from 21 to 45 years.

Spinal haemangiomas were reported in Lindau's disease in six members of a single family (Otenasek and Silver 1961).

Phoeochromocytomas occur in only 10 per cent of patients and it might be that some families are more pronc to this complication than others.

A recent paper by Huson *et al.* (1986) emphasizes the variability of the clinical picture and the need to follow up those at risk and those who have minimal manifestations. In their study, 20 patients with cerebellar haemangioblastomas were followed and a diagnosis of von Hippel–Lindau was subsequently established in eight. In retrospect, seven of eight were known to be at risk. These authors calculated that 40 per cent of those with cerebellar haemangioblastomas had von Hippel–Lindau syndrome. The age of onset does not help to differentiate between isolated cerebellar haemangioblastomas and von Hippel–Lindau disease. The mean age of presentation of the cerebellar tumour was 32 years, the renal carcinoma was 43 (31–69) and the retinal lesions 21 years (8–67).

Genetics

Lindau, in his 1926 paper, showed that dominant inheritance occurred in 20 per cent of his initial cases, but going back even further, von Hippel's original patient's father came to autopsy and was found to have tumours of the cerebellum, cauda equina, kidneys, and spleen. Cushing and Bailey (1928), Møller (1952), Levin (1936), and Macrae and Newbigin (1968) have added dominant pedigrees. Nicol (1957) described affected members in five generations, Rho (1969) 12 cases in five generations, Horton *et al.* (1976*b*) nine families with 50 affected members.

Counselling is difficult because of partial expression and the late onset of the disease. Gene carriers can be asymptomatic and can only be identified by special investigations or post-mortem. In addition, there might be a lapse of 20 years between the onset of the eye symptoms and those symptoms related to the cerebellum in the same patient. Unlike some of the other neuro-cutaneous malformations there is no age after which the disease no longer manifests. In the study by Horton *et al.* (1976*b*) one patient developed a renal tumour at 61 and one had visual loss from retinal angiomatosis at 67.

Variability in expression of the disease was shown in the family described by Såebø (1952). Five were affected in three generations. One had only the angioma of the retina, another the retinal lesion plus a cerebellar haemangioblastoma and two had clinical evidence of cerebellar tumours. Similarly, Bird and Krynauw (1953) reported five patients with cerebellar cysts but only two had retinal angiomata. One patient had the retinal lesions alone. In Møller's (1952) pedigree, there was inheritance through unaffected members.

Nicol (1957) described a remarkable family in which there were no fewer than 12 deaths from intracranial tumours. In two there were four pathologically proven cerebellar cysts and two of these had, in addition, an associated haemangioblastoma. In this pedigree, there were three suspected cases of pancreatic cysts. The disease occurred over five generations and in none was there a history of ocular involvement. Cerebellar haemangioblastomas were found in five and the same lesion in the brain stem in three. Six had hypernephromas.

Other pedigrees are those of Melmon and Rosen (1964) who reported an enormous family with 10 proven and two clinically unequivocal cases in three generations, and in a large pedigree Goodbody and Gamlen (1974) reported six members over three generations who had cerebellar haemangioblastomas, but only the propositus had, in addition, a renal carcinoma.

Shokeir (1970*a*) published three pedigrees—one a simple dominant, the second a dominant with incomplete penetrance, and the third two sibs and their second cousin. The sibs were the offspring of a second-cousin marriage. This recessive pedigree is probably unique, but dominant inheritance with reduced expression cannot be excluded. However, the ratio of two affected out of six sibs and the five intervening unaffected members between the two branches of the pedigree make recessive inheritance more likely.

Supratentorial haemangioblastomas as part of von Hippel–Lindau's disease are rare. In a family reported by Lee *et al.* (1978) a supratentorial leptomeningeal haemangioblastoma occurred in association with other features of the syndrome.

Screening is advisable for those at risk using CT scanning (of the brain and cord) routine ophthalmoscopic examination and renal ultrasonography as the main investigations.

Go *et al.* (1984) calculated that 95 per cent of gene carriers developed some manifestation by 50 years.

Suggested routine (Huson *et al.*)

1. Ophthalmological lesions should be sought from 5 years of age.

2. Urinary VMA and metadrenaline from 10 years.

3. CT brain scan every 2 years from 15 years.

4. CT kidney scan every 2 years from 20 years.

Recently there has been a report of linkage to chromosome 3 (Seizinger *et al.* 1988).

Note: The following entity is probably the same as von Hippel–Lindau disease.

Phaeochromocytoma and cerebellar haemangioblastoma

This combination was first recorded by Mandeville and Sahyoun (1949) in a 55-year-old woman; subsequently a mother and son with this association were described by Chapman and Diaz-Perez (1962) and in another family the father of a young woman with a phaeochromocytoma and cerebellar haemangioblastoma (Nibbelink *et al.* 1969) had a cystic cerebellar haemangioblastoma. His brother died from a cerebellar haemorrhage and was found to have a phaeochromocytoma. The subject was reviewed by Lowden and Harris (1976) who added two families of their own. In one, the two conditions co-existed in three members. Seven further families were reported by Hagler *et al.* (1971).

The association of phaeochromocytoma and von Recklinghausen's disease is better known. Five to 25 per cent of patients with these adrenal tumours have neurofibromatosis.

Isolated haemangioblastoma of the spinal cord

In a review of 80 cases of patients who presented with haemangioblastomas of the spinal cord, a family history was present in 11 (Browne *et al.* 1976). Retinal and cerebellar lesions usually became symptomatic before the spinal cord lesion so that a family history of the retinal and cerebellar components should be sought.

Association of adenomas of the epididymis with von Hippel–Lindau disease

Papillary cystadenomas were described in three cases of Lindau's disease by Melmon and Rosen (1964). One of three sibs with this tumour (Tsuda *et al.* 1976) had evidence of Lindau's disease. In a review of the subject, Price (1971) pointed out the most patients with bilateral papillary cystadenomata of the epididymis have von Hippel–Lindau's disease.

von Hippel–Lindau disease and Sturge–Weber syndromes

This association has been noted by Bentsen *et al.* (1938), and by Delay and Pichot (1946). The combination is rare and might be fortuitous.

von Hippel–Lindau and von Recklinghausen's disease

Tishler (1975) described a large kindred where the two syndromes co-existed in the same family. Some members had angiomatous lesions of the retina, one in association with a renal carcinoma and pancreatic cysts. When ten members of this family were re-examined by Thomas *et al.* (1978*b*) three members were found to have von Recklinghausen's disease in association with retinal angiomas.

Two other reports (Chapman *et al.* 1959; Fracassi and Parachu 1935) are known, but these concern the association in single cases.

D. Sturge–Weber syndrome (Fig. 10.5)

The classical clinical picture is that of a vascular port-wine facial nevus in the distribution of the trigeminal nerve, mental retardation, buphthalmos,

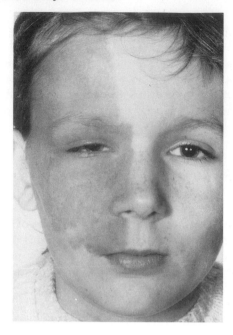

Fig. 10.5   Sturge–Weber syndrome.
Courtesy of the Department of Medical Illustration, Great Ormond Street
Hospital for Sick Children, London.

and contralateral hemiparesis and/or focal epilepsy. Solitary choroidal
hemangiomata are commonly found (Font and Ferry 1972).

No convincing familial cases have been reported. The mother of a patient
had a haemangioma of the occipital region, and mother's brother and sister
had a haemangioma of the arm, but in the absence of the lesion over the first
division of the trigeminal nerve a diagnosis of Sturge–Weber syndrome is
untenable (Chao 1959*a*, *b*).

Discordance in monozygotic twins has been reported (Koch 1966), as well
as concordance in monozygotic twins by Teller *et al.* (1953) and Tiedemann
(1951).

Many unusual single cases are known. The typical intracranial calcification has been reported without a facial nevus (Andriola and Stolfi 1972). A
large series reviewed by Lund (1949) showed seven without a facial nevus,
two proven at operation, and a similar finding (five out of 35 without the
facial nevus) was reported from the Mayo clinic (Peterman *et al.* 1958).

The reported pedigrees in Koch's series (1972) are inconclusive. Two sibs
are quoted by Koch (1972) from a pedigree published by Yakovlev and
Guthrie (1931). The proband in the family had Sturge–Weber syndrome, but

the sib who was epileptic had two facial nevi which disappeared before his first birthday. In other reported pedigrees (Koch 1940; see others in Koch 1972) only the proband had the skin lesion whereas other affected members had epilepsy and cerebral palsy. Recurrence risks are less than 1 per cent.

There has been one report (I am indebted to Dr Koch for drawing my attention to it) in which father and son seemed to have the full spectrum of Sturge–Weber syndrome (Debicka and Adamczak 1979). This is an exceptional case and should not detract from small risks to parents after one affected child. Offspring risks might not be so well documented as many do not reproduce, but it remains likely on present evidence that the risks are small.

Association of Sturge–Weber with other conditions

Neurofibromatosis and Sturge–Weber syndrome

A single case (Streiff 1947) had both lesions. There were biopsy-proven neurofibromas, and photographic evidence of the typical skin lesions of Sturge–Weber (see also p. 155).

Adenoma sebaceum and Sturge–Weber syndrome

A male with this combination had no other affected family members (Solomon and Bolat 1972). He had a unilateral angiomatous lesion involving all three divisions of the trigeminal with intracranial calcification, epilepsy, and glaucoma. In addition, he had typical adenoma sebaceum. This is a rare occurrence, but it has previously been recorded that solitary angiofibromas may occur in normal adults. A similar case was reported by Tramer (1951). Greig (1922) reported a case of a meningeal nevus associated with adenoma sebaceum.

Sturge–Weber syndrome and von Hippel–Lindau disease

This association has been noted by Van Bogaert (1950) and by Bentsen *et al.* (1938). Both were single cases.

E. Klippel–Trenaunay–Weber syndrome (Fig. 10.6)

The relationship of the classic triad of cutaneous haemangiomata, varicose veins, and hypertrophy of the underlying bone to Sturge–Weber syndrome

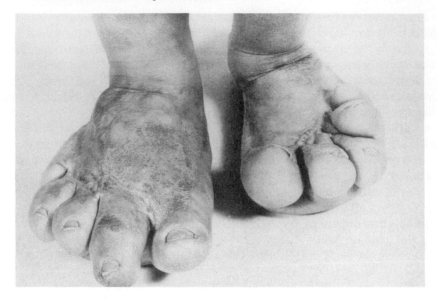

Fig. 10.6   Asymmetrical hypertrophy in Klippel–Trenaunay–Weber syndrome.

is unclear. The two have been associated (Kramer 1963; Harper 1971; Furukawa *et al.* 1970; Meyer 1979). Two sibs with Klippel–Trenaunay–Weber syndrome were reported in a series of 18 cases (Lindenauer 1965). However, most cases are single and recurrence risks are small.

Intracranial lesions in the Klippel–Trenaunay–Weber syndrome are rare, but macrocephaly is not infrequent.

However, an ischaemic infarct of the brain stem is reported (Albert 1976) in a man with hypertrophy of the right side of his body and moderate varicosities of the right foot. The brain stem lesion was an aneurysmal dilatation of the terminal part of the basilar artery and ectasia at the junction of the right vertebral with the basilar artery. A similar patient is reported by Benhaiem-Sigaux *et al.* (1985).

Klippel–Trenaunay–Weber syndrome and neurofibromatosis

This combination is recorded in six Dutch pilots undergoing a medical examination (Van Wulfften-Patthe 1957). Most of the *café-au-lait* patches were single although some were multiple and extensive, and the association seems valid in some cases.

Klippel–Trenaunay–Weber syndrome and cord angiomas

The association of the Klippel–Trenaunay–Weber syndrome with an underlying spinal cord angioma was found in five cases (Djindjian *et al*. 1977) All were singletons.

Klippel–Trenaunay–Weber and Tuberose sclerosis

This association in a single patient is reported by Troost *et al*. (1975). A 41-year-old lady had adenoma sebaceum, mental retardation, and seizures. Her left limb was hypertrophied and displayed prominent varicosities. The author (M.B.) has personal experience of a similar case.

Conditions confused with Sturge–Weber

Jadassohn's nevus phakomatosis (epidermal nevus syndrome) or Linear sebaceous nevus syndrome (Fig. 10.7)

This condition consists of a linear sebaceous nevus, mostly midline, and in its full expression the lesion extends in the midline from the forehead through the nose, upper and lower lip to the chin (Zaremba 1978). The association of the skin lesion with central nervous system lesions is known (Feuerstein and Mims 1962). Epilepsy occurs in more than 70 per cent of cases and mental retardation in just over half. Cortical atrophy, hydrocephalus, and cerebrovascular malformations have all been described (Solomon and Esterly 1975). The location of the nevus on the face does not necessarily mean that the central nervous system is involved. When the nevus occurs on the trunk it is often in the distribution of a peripheral nerve. When the scalp is involved, alopecia is frequent. Congenital malformations of the eye, bone, and vascular system occur.

The genetics are uncertain. No family is known where other members have the same nevus but epilepsy occurs more frequently within families than within the general population (Zaremba 1978). Single cases were reported by McAuley *et al*. (1978) and Solomon and Esterly (1975). Recurrence risks are small.

Four cases with neurological complications were investigated by Baker *et al*. (1987). In addition, they reviewed 60 reported cases. These authors

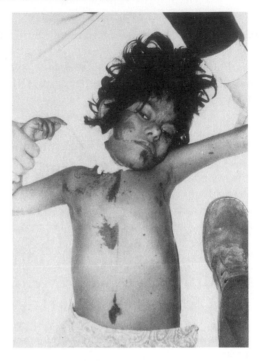

Fig. 10.7   Jadassohn's nevus phakomatosis.
Courtesy of Dr R. M. Winter.

suggest that central nervous system complications are more likely to occur when the nevus is situated on the head.

Encephalocraniocutaneous lipomatosis

Typically there is a large area of scalp alopecia, soft subcutaneous craniofacial lipomas, yellow papules on the eyelids and connective tissue nevi of the oculoconjunctiva. The neurological complications are mental retardation and seizures (Alfonso *et al.* 1986). These authors review the literature and report the third case with spinal cord lipomatosis. At least eight patients have had porencephalic cysts and five have had intracranial calcification. The main differential diagnosis is linear sebaceous nevus syndrome and differentiation is made difficult in that a linear sebaceous nevus might be part of the syndrome.

Nevus flammeus familial

Multiple 'birthmarks' over four generations affected 12 members (Shelley and Livingood 1949). Biopsy showed these to be nevi flammei. Most of the multiple lesions were not on the face and this rare syndrome should not be confused with Sturge–Weber syndrome.

Nevus flammeus of the forehead

Sturge–Weber syndrome must be differentiated from other facial haemangiomata. The nevus flammeus of the forehead, glabellar region, and upper eyelid is common in infancy and tends to fade in childhood. The familial nature of the lesion was noted by Selmanowitz (1968) where at least five were affected in two generations.

Midline nevus flammeus with intracranial vascular disease

Three children, sibs and a first cousin, developed arterial occlusive disease before the end of the first decade. Both sibs had the midline skin lesion, but their first cousin did not. Five other maternal relatives had the skin lesion but not the arterial disease. The authors (O'Tauma *et al.* 1972) suggested that this was a new neurocutaneous syndrome, but chance co-existence cannot be excluded. A family was reported by Kaplan *et al.* (1975*b*) in which a nevus flammeus was inherited over three generations with incomplete penetrance. In a 16-month-old girl the skin lesions were associated with a spinal arteriovenous malformation.

Angioma of the skin and nervous system

In those families in which this combination occurs the nevi are on the trunk and therefore unlike Sturge–Weber disease. The histology of the nervous system lesion differs from the cerebellar haemangioblastomas of von Hippel–Lindau disease, but resembles that found in Rendu–Osler–Weber syndrome (Zaremba *et al.* 1979; Kidd and Cumings 1947; see p. 133).

Coat's disease

This disorder may mimic Sturge–Weber syndrome. It is characterized by telangiectatic retinal vessels with thickening of the walls and occasional retinal detatchment. It is sometimes associated with a port-wine skin stain (Allen and Parlette 1973). Most cases are single.

Note the association with muscular dystrophy (see page 368).

Cobb syndrome or cutaneomeningospinal angiomatosis

Cobb syndrome is an association of a vascular skin nevus and an angioma of the spinal cord. The skin lesions are flat angiomata situated in relationship to the cord lesion. Most cases to date have been sporadic (Jessen *et al*. 1977), but a 16-month-old girl with the syndrome had evidence in the family history of members in three successive generations with the skin haemangiomas.

A patient with a cervical cutaneous angioma (nevus planus) and with progressive hoarseness over 15 years is reported by Vaquero *et al*. (1981). Lesions of the V, VIII, IX, X, XI, and XII cranial nerves occured. An otologist saw a vascularized polyp of the middle ear and CT showed a tumour—probably a glomus tumour. The cutaneous angioma was present on the left side of the neck.

Bonnet–Dechaume–Blanc syndrome

This is characterized by a telangiectatic facial nevus, and retinal and intracranial angiomatosis, especially involving the thalamus and midbrain. In the original report by Bonnet *et al*. (1938) a mother and grandmother were similarly affected.

Wyburn Mason syndrome

The arteriovenous aneurysms involve the retina and midbrain and only rarely the face, however, two of the 30 patients with this combination had facial nevi (Paillas *et al*. 1951; Wyburn Mason 1943). Most cases are single.

Erythrokeratoderma with ataxia

Twenty-five members in five generations had a papulosquamous skin lesion which appeared soon after birth but disappeared after 25 years of age (Barbeau and Giroux 1972). All those with the skin lesion developed in middle life a cerebellar-like ataxia, with reduced reflexes. Some were spastic. Muscular pain was prominent. The skin lesions on initial inspection looked typically ichthyosiform.

Vascular cutaneous lesions associated with intracranial arterio-venous malformations

There have been a small number of families with this condition (see p. 134).

F. Riley–Smith syndrome

The classic triad of macrocephaly, pseudopapilloedema, and haemangiomata was first described by Riley and Smith in 1960. Four out of seven sibs were affected and mother had the full triad of the syndrome, but mildly. The dermatological features were those of subcutaneous and cutaneous nodules, initially colourless, but later typically haemangiomatous in appearance. One sib had the lesions over the forehead.

G. Xeroderma pigmentosa

Characterized by abnormal sensitivity to sunlight, this results in telangiectasia, freckle-like lesions, depigmentation, skin atrophy, and malignancy (melanomas, basal cell, or squamous cell carcinomas). Xeroderma pigmentosa is known to be associated with mental retardation and microcephaly (Thrush *et al.* 1974). The authors describe sibs in whom a peripheral neuropathy and pyramidal tract signs were present and, in addition, a progressive dementia, chorea, and deafness. When severe neurological signs and symptoms are present, the disorder has been called the De Sanctis–Cacchione syndrome and evidence from cell-hybridization suggests that it might be a separate genetic entity from xeroderma pigmentosa (Afifi *et al.* 1972). In both syndromes, repair of DNA is deficient. Sibs are reported by Reed *et al.* (1977), in one of whom an olivopontocerebellar atrophy occurred as a major central nervous system abnormality.

H. Incontinentia pigmenti (Bloch–Sulzberger syndrome)

Neurological involvement occurs in about a quarter of all cases of incontinentia pigmenti. Mental retardation, spasticity, microcephaly, hydrocephalus, and cortical atrophy have all been reported (Carney 1976). The skin lesions appear during the first two years of life. Linear vesiculation occurs within the first 4 months and the verrucous stage within the first year. These lesions resolve spontaneously leaving atrophy or depigmentation. The pigmentary stage, often with lesions on the torso, occurs by the age of two years (Morgan 1971) and need not occur in relation to the bullous lesions.

The first cases were described by Bardach (1925) in identical twins. Reviewing the world literature, Carney (1976) found that 97 per cent of cases were females. Analysing the 74 reported sibships, the abortion rate was 23 per cent. Of the 16 reported males, only one was severely mentally retarded. No male patient has reproduced.

Inheritance is as an X-linked dominant. Female carriers should be counselled that one pregnancy in four could end in a miscarriage, but that half her female offspring could be affected. Most single cases with unaffected mothers should be regarded as fresh mutations, but possible female carriers should be carefully examined for depigmented or hypopigmented skin lesions and for abnormalities of the teeth (conical deciduous teeth or agenesis of permanent teeth). Other pedigrees are those of Iancu *et al.* (1975, mother and two daughters), Curth and Warburton (1965), and Gordon and Gordon (1970). Recently, an affected father and daughter were reported (Sommer and Liu 1984), but this is exceptional.

The gene for incontinentia pigmentosa could be situated at Xpll based on finding two unrelated girls with an X-autosomal translocation (Hodgson *et al.* 1985) the break point on the X involving that region.

I. Incontinenti pigmenti achromians (Hypomelanosis of Ito)

In 1952 Ito described an entity characterized by hypopigmented whorls and streaks either bilateral or unilateral affecting any part of the body. Multiple systems might be affected and delayed development has occurred in about half of the patients; in a series of eight patients reported by Schwartz *et al.* (1977) four had seizures and six had developmental delay. Eye malformations occur frequently and microphthalmia, esotropia, strabismus, corneal opacities, and choroidal atrophy have all been reported. Rubin (1972) described an affected girl whose two brothers had hypopigmented areas and whose father and his brother had darkening of the skin lesions with age.

Grosshans *et al.* (1971) reported this syndrome in a mother and three of her daughters but most cases are single.

The neurological complications have been reviewed by Golden and Kaplan (1986). Seizures, mental retardation, and focal and generalized cortical atrophy (frontal atrophy, porencephalic cysts, atrophy involving one hemisphere, asymmetry of the ventricular system with cerebellar hypoplasia) can all occur. About 60 per cent of patients have non-cutaneous abnormalities, including facial hemiatrophy extremely dysplastic teeth and alopecia. Occasionally, anterior horn cell degeneration might be a feature (Larsen *et al.* 1987).

Cutaneous albinism, ataxia, and peripheral neuropathy

Cutaneous hypopigmentation in three Lebanese sibs has been reported in association with a progressive spastic paraplegia and a peripheral neuropathy (Abdallat *et al.* 1980). Two possible heterozygotes had white patches of hair but were otherwise normal. A similar sibship (Lison *et al.* 1981) had as the main dermatological features diffuse vitiligo, hyperpigmentation of exposed areas, diffuse lentigines, and premature greying of scalp and body hair. Unlike the Abdallat family there was a peripheral neuropathy in addition to progressive spastic paraparesis.

Oculo-cerebral syndrome with hypopigmentation (Cross syndrome)

Four sibs had cutaneous hypopigmentation eye anomalies including spastic ectropion, small opaque corneas, and coarse nystagmus. Spasticity and athetoid movements occurred (Cross *et al.* 1967). Inheritance is as a recessive.

Oculo-osteo-cutaneous syndrome

Three sibs with small upper and prominent lower jaw, short stature, short toes except the big toe, partial odontia, sparse hair growth and hypopigmentation of the skin were reported by Tuomaala and Haapanen (1968). Lenticular opacities, strabismus, nystagmus, antimongoloid eye slant, and distichiasis were the ocular features.

J. Cerebrotendinous xanthomatosis

This rare disorder, originally described by Van Bogaert *et al.* (1937), is characterized by a spastic ataxic syndrome. Onset is in childhood. Xanthomatosis, especially of the Achilles tendon, dementia, and early cataracts are the main signs (Schimschock *et al.* 1968). In the final stages of the disease, progressive bulbar paralysis and distal muscular wasting are present. Serum cholestanol is markedly raised.

Sibs were reported by Schimschock *et al.* (1968) and Schreiner *et al.* (1975). A sibship of three affected out of 12 (De Jong *et al.* 1977) and parental consanguinity in Van Bogaert's original report suggest that inheritance is most likely to be recessive. Other families were reported by Farpour and Mahloudji (1975).

Six Sephardic Jews of Moroccan origin were reported by Berginer and Abeliovich, (1981). They belonged to three unrelated families, and in two there was consanguinity. These authors calculate the heterozygote frequency in the Moroccan Jews to be 1:54.

11

Cerebellar syndromes

No satisfactory classification of the cerebellar ataxias has emerged. Only Friedreich's ataxia is easily recognizable on genetic, clinical, and pathological grounds. It has also become clear that there is a poor correlation between the clinical presentation and the post-mortem findings. The clinician is faced with an array of symptoms that he needs to formulate as a clinical entity in order to draw on other workers' experiences and to form an opinion on prognosis both clinical and genetic. It is, moreover, unlikely that he will get pathological verification for his clinical diagnosis.

Classification

A. Congenital cerebellar ataxia.
B. Early childhood onset (excluding Friedreich's ataxia). This group is predominantly recessive ataxia which can be divided as follows based on the additional features:
   (a) eye involvement
   (b) skin lesions
   (c) hypogonadism
   (d) hearing loss plus optic atrophy (Behr's syndrome)
   (e) hearing loss
   (f) other hereditary cerebellar ataxias
   (g) known biochemical abnormalities
C. Friedreich's ataxia
D. Adult onset cerebellar ataxia or cerebellar ataxia 'plus'. This includes those with one or more of the following manifestations:
   (a) i. extra-pyramidal signs
       ii. ophthalmoplegia
      iii. optic atrophy
       iv. dementia
These are not distinct entities and there is considerable overlap and variation within families.

    (b)   i. Machado's disease

         ii. Parynchymatous degeneration

         iii. Spinopontine degeneration

The designation of olivo-ponto-cerebellar atrophy (OPCA) is frequently used for the 'adult onset' group. (see below). Adult onset cerebellar ataxia with retinitis pigmentosa is discussed separately because it seems to breed true and might be a separate genetic entity.

E. X-linked cerebellar degeneration

Additional notes

(a) *Marie's* (1893) *cerebellar ataxia* is excluded from this classification. Marie's data was based on a collection of cases taken from reports by Fraser (1880), Nonne (1891), Brown (1892), and Klippel and Durante (1892). The group was heterogenous. The cases included members with retinal, optic nerve, and eye movement disorders. They were also genetically different. Nonne's cases were sibs with unaffected parents and Sanger-Brown reported a large dominant pedigree. Later, Marie's 'hérédoataxië cérébelleuse' was changed to 'spastic ataxia' by Holmes (1907), but neither description has any precise meaning.

(b) *Dejerine and Thomas'* cases had no family history, nor did those of:

(c) *Marie, Foix, and Alajouanine* who reported single cases of a predominantly truncal ataxia with minimal involvement of the upper limbs and minimal dysarthria.

(d) For cerebello-olivary degeneration (Holmes 1970*b*) see p. 184.

(e) The family (dominant inheritance) reported by Menzel had ataxia, brisk jerks, and choreiform movements, and this is included in 'adult onset or cerebellar ataxia plus.'

There is confusion about the relationship between OPCA and Dentato-Rubro-Pallido-Luysian atrophy (DRPLA). These terms need an explanation because they recur in the literature and are difficult for geneticists to comprehend. OPCA, when used by pathologists, usually means atrophy of the cerebellar cortex, white matter, the pons, and often the striatum and substantia nigra. It does not affect the dentate, red nucleus, luysian body, or the pallidum.

Extrapyramidal features might occur, but they manifest after the cerebellar signs. DRPLA implies lesions, in the dentate, red nucleus, pallidum, and luysian body. Clinically, the patients present with chorea, athetosis, and ballismus.

Guide for the geneticist

1. Where Friedreich's ataxia has been excluded (a rare cause after the age of 20 years) the majority of adult-onset hereditary cerebellar ataxias are dominantly inherited.
2. The rare adult onset recessive and X-linked pedigrees are noted on page 207. These should be diagnosed with caution.
3. The proportion of adult onset cerebellar degenerations that are genetically determined, is unknown. Even if half are genetically determined the risk to offspring would still be high.
4. The cerebellar degenerations with an eye movement disorder or extrapyramidal signs are usually dominantly inherited.
5. The cerebellar degeneration with hypogonadism or ichthyosis is often recessively inherited.
6. Reflex changes are of little help in the classification with the exception of Friedreich's ataxia, where tendon reflexes in the lower limbs are reduced or absent. In the adult-onset cerebellar ataxias patients with absent or brisk reflexes occur in the same family.
7. The cerebellar degenerations with retinitis pigmentosa can be either dominant or recessive. The adult-onset combination is more often dominant.
8. The pathological reports of primary parenchymatous cerebellar degeneration often refer to families with dominant inheritance, but the patients can usually be incorporated into the clinical entity of 'cerebellar ataxia plus' as they commonly have additional features. Single cases are probably under reported. Members of the pedigree originally described by Holmes (1907*a*) and often used as a prototype for parychymatous degeneration had hypogonadism and were homozygous for a recessive gene.
9. Dementia can occur in any of the categories, but especially the late-onset dominant types.
10. A cerebellar ataxia might be a prominent feature of hereditary motor and sensory neuropathy.
11. Hexosaminidase A deficiency might present in late childhood or adulthood in this way.
12. A cerebellar ataxia might be a prominent sign in mitochondrial disease.

A. Congenital cerebellar ataxia

Cerebellar hemisphere hypoplasia

Wichman *et al.* (1985) described three sib pairs and noted clinical similarity within families. In one sib pair there was severe ataxia and moderate retardation, whereas in another pair there was only subtle motor clumsiness and intelligence was in the low normal range. In the third pair, both of those affected had mild retardation.

It could be that most of the cases described as ataxic cerebral palsy have this condition. Schult (1963) reviewed 32 cases and found two pairs of affected sibs—one pair had normal intelligence and the other sib pair were moderately retarded. Gustavson (1969) thought that 50 per cent of cases were autosomal recessive.

This category probably also includes the previously designated congenital granular cell hypoplasia. Jervis (1950) described three sibs of Italian extraction with this type of anomaly. No consanguinity was recorded. One of the sibs was severely mentally retarded and had unco-ordinated movements of both arms. At post-mortem there was extensive degeneration of the granular cell layer and, to a lesser extent, of the Purkinje cells.

Two sibs whose parents were unrelated had delay in walking (Norman 1940). One sibling had an increase in inco-ordination during the last years of life, whereas in the other the clinical condition remained stationary. Microcephaly, mental retardation, and seizures (in one) were part of the pattern. Monozygotic twins (Jervis 1954) had a similar pathology.

Counselling

It is still not possible to distingish between the genetic and the sporadic. A recurrence risk of 1 in 8 is appropriate.

Congenital cerebellar atrophy—autosomal dominant

This is a rare mode of inheritance and there have been only two reports (Furman *et al*. 1985; Tomiwa *et al*. 1987). In both, the ataxia was preceded by hypotonia and delayed motor milestones. Mental retardation was either mild or did not occur. A brain scan in both families showed atrophy of the superior vermis of the cerebellum.

Recessive vermis agenesis (Joubert syndrome)

Familial agenesis of the cerebellar vermis was described in four children in a sibship of six (remote consanguinity) (Joubert *et al*. 1969). The clinical picture was that of episodic hyperpnoea with periods of apnoea, abnormal eye movements, ataxia, and mental retardation. Two sibs with the same syndrome were observed by Boltshauser and Isler (1977). A post-mortem on a patient whose parents had a common ancestor four generations back (Friede and Boltshauser 1978) showed atresia of the lateral foramina and the authors discuss the similarity of this syndrome with that of the Dandy–Walker syndrome. A CT scan is recommended as a useful procedure when this condition is suspected (Curatolo *et al.* 1980*b*). Polydactyly does occur in Joubert syndrome. Polydactyly and a picture similar to the OFD syndrome (oro-facio-digital), i.e. with hamartoma of the tongue was reported by Egger *et al.* (1982*a*) in sibs. Vermis aplasia was present.

De Haene (1955) described two brothers with vermis agenesis. Gross hypotonia, cerebellar ataxia, and early death were the clinical features. The phenotype of the Joubert syndrome has been extended to include an occipital meningocoele. (Joubert *et al.* 1969), chorioretinal coloboma (Lindhout *et al.* 1980), Lebers congenital amaurosis (King *et al.* 1984), and unilateral ptosis (Houdou *et al.* 1986). The outcome, although usually poor, is variable. Two children are reported by Casaer *et al.* (1985). One died at 4 months, whereas the other (unrelated) seemed normal at 18 months. It should be stressed that the diagnosis of Joubert syndrome should not be made on radiological grounds alone—at least the abnormal breathing pattern and/or the eye movement disorder should be present.

Pre-natal diagnosis has been achieved using ultrasound scanning, but it is not certain whether the condition can always be successfully diagnosed in this way (Campbell *et al.* 1984).

Congenital ataxia with aniridia (Gillespie 1965; Sarsfield 1971)

See p. 180.

Congenital ataxia with choroidal colomba (Pfeiffer *et al.* 1974)

This might not be a separate condition. The author has seen a pair of sibs, one with cerebellar hypoplasia and a coloboma, and the other with cerebellar ataxia alone.

Granular cell hypertrophy (Lhermitte–Duclos disease)

Of the nearly 40 cases described up to 1969, only one had another close relative affected (Ambler *et al.* 1969). These authors describe this pathology in a boy who had, at the age of 2½ years, a large head. At the age of 22 years, he presented with papilloedema. His mother died at the age of 51 years and was similarly affected.

Pontoneocerebellar hypoplasia

Hypoplasia of the lateral lobes of the cerebellum in association with underdevelopment of the griseum ponti and middle peduncle is called ponto-neocerebellar hypoplasia. This was described by Norman and Urich (1958) in two unrelated young infants with cerebellar hypoplasia. The parents of one of the infants were second cousins. A sister of one of the patients died at 2½ years and was mentally handicapped and microcephalic. The clinical picture was that of failure to thrive, spastic paraparesis, and no intellectual progress. Both children died during the first year of life.

Hypoplasia of the cerebellar olives and pons (Paine syndrome)

A syndrome in a French Canadian family of mental retardation, myoclonic jerks, microcephaly, epilepsy, mild optic atrophy, and spastic diplegia was recorded by Paine (1960). An autopsy on one showed hypoplasia of the cerebellum, pons, and olives. Many males were affected over two generations and the inheritance was probably X-linked. Seemanova *et al.* (1973) reported a similar family.

Neonatal cerebellar atrophy with retinal dystrophy

Two sibs are reported by Harding *et al.* (1988). The onset was at or soon after birth. Both had a retinal dystrophy and limitation of joint movements. A systemic illness was prominent in both, and this included vomiting, diaorrhoea, edema, and percardial effusion, finally culminating in death. At post-mortem there was liver fibrosis and a cerebellar atrophy also involving the pons and olives. The pathological picture was identical to that seen in adult olivo-ponto-cerebellar degeneration and is probably the same condition as on page 196.

Angelman syndrome

Angelman's syndrome is characterised by mental retardation with a profound speech delay, jerky voluntary movement, a happy disposition with paroxysms of laughter, tongue thrusting and a characteristic facial appearance (Robb *et al.* 1989). The facial dysmorphism is subtle and consists of a prominent lower jaw, a wide mouth and midfacial hypoplasia. The electroencephalogram is unusual and helps to confirm the diagnosis.

The abnormalities consist of generalised, rather rhythmic intermediate slow waves at 4–6 cycles/second (c/s) and runs of 2–3 c/s activity anteriorly often with discharges. On closure of the eyes spikes with 2–4 c/s activity are seen posteriorly (Boyd *et al.* 1988).

Earlier reports were all of single cases (Angelman 1965), and, although Williams and Frias (1982) in their review noted a pair of Japanese sibs (Kuroki *et al.* 1980) with the condition, they concluded that there was little evidence of Mendelian inheritance in the group as a whole. There have been two further reports of affected sibs – two brothers (Pashayan *et al.* 1982) and one further sib pair (Fisher and Burn, 1987). Seven patients from three sibships were reported by Baraitser *et al.* 1987. Recently a deletion in the long arm of chromosome 15 has been reported in 40 per cent of cases. (Pembrey *et al.* 1988).

B. Specific recessive syndromes of early onset

(a) With eye involvement

1. *The association of ataxia and optic atrophy with mental retardation*

See Behr's syndrome (p. 186).

2. *Progressive visual and hearing loss, progressive spastic paraplegia, cerebellar ataxia, and dementia*

Defects in the cochlear vestibular and optic systems can co-exist. They might appear early in the course of the disease or, as in the family of Van Bogaert and Martin (1974), late. Both dominant and recessive pedigrees have been reported.

(i) Dominant

Sylvester (1958), under the title of 'Some unusual findings in a family with Friedreich's ataxia', described a father with progressive impairment of vision (bilateral optic atrophy) and hearing, but without ataxia, who had six of his nine children affected by an onset within the first decade of life of failing vision, hearing disturbance, and unsteadiness of gait. The face was expressionless. Wasting of shoulder girdle muscles and hands was noted in four, and in three of these fasciculations were seen.

(ii) Recessive

Van Bogaert and Martin (1974) described two branches of a family in which the parents were first cousins. In one branch two brothers developed an ataxia and deafness at 6 years of age, with blindness by puberty. A third brother had only the ataxia. In another branch of the family the clinical picture was that of ataxia and a mental disturbance. Progression was slow. All the affected individuals needed surgical correction for *pes cavus*.

There are other families with an infantile onset of progressive visual loss.

Examples of these sibships are those of Muller and Zeman (1965, two brothers) and Levy (1951, two brothers). In the family reported by Meyer (1949), four male sibs had the disease. Nyssen and Van Bogaert (1934) described a brother and sister with this condition in whom the motor disturbance had an onset towards the end of the first decade. Zeman (1975) reviewed the subject, and commented on its rarity and the difficulty of making a diagnosis. Onset is usually in infancy with rapid deterioration to idiocy within months or a year.

3. Cerebellar ataxia and pupillary paralysis

A mother and her 10 children had this condition with bilateral *pes cavus*, kyphoscoliosis, and paralysis of the sixth cranial nerve (Sutherland *et al.* 1963). The mother had bilateral cataracts, but not her children. The condition was not progressive.

4. Ataxia, aniridia, and mental retardation (Gillespie syndrome)

Gillespie (1965) described sibs with this combination whose parents were unrelated. Chazot *et al.* (1975) described the same combination in monozygotic twins who had, in addition, myoclonic epilepsy. One of the twins had a scoliosis. A third interesting pedigree (Crawfurd *et al.* 1979) was a South Yorkshire family with two affected sibs. The affected female married a local inhabitant (not thought to be related) and had an affected son. It was presumed that the unaffected husband was a heterozygote. The single case reported by Sarsfield (1971) also came from South Yorkshire and a further single patient is reported by Lechtenberg and Ferretti (1981).

Inheritance is probably as an autosomal recessive.

5. Marinesco–Sjögren syndrome

Superneau *et al.* (1985) has pointed out that Moravscik *et al.* gave an excellent description of the condition in the Hungarian literature of 1904, that is well before Marinesco or Sjögren. Superneau *et al.* (1985) described three sibs, the offspring of a consanguineous mating. Soon after birth strabismus, nystagmus and cataracts were noted. Later, dysarthria, ataxia, an intention tremor, and pyramidal signs developed. One female at 23 was short and had never menstruated. Involvement of skeletal muscle with raised CPK has been noted.

The triad of mental deficiency, ataxia, and bilateral congenital cataracts is still known by the eponym Marinesco–Sjögren syndrome. Pathologically, there is atrophy of the cerebellar cortex, but this excludes the nodulus, flocculus, and the paraflocculus (Todorov 1965). The report of Marinesco *et al.* (1931) was of a Rumanian sibship with four affected siblings. Sjögren (1950) pointed out that five out of six of his sibships were the product of first-cousin marriages.

A very inbred family with 11 affected in four generations (Gayral and Gayral 1966) showed variation in the clinical manifestations. Some patients had the full syndrome, whereas others only had the cataracts.

Two sibs, descendants from the same great-great-grandparents, had a congenital non-progressive cerebellar ataxia, with evidence of pyramidal tract involvement, moderate mental retardation, and cataracts (Garland and Moorhouse 1953). In one-third of cases, there is a progressive motor weakness (Alter *et al.* 1962).

The condition has been reported throughout the world. Iranian sibs, whose parents were second cousins, are reported by Mahloudji *et al.* (1972). Andersen (1965) reported four sibs, the product of a consanguineous mating, and two other sib pairs, All the patients lived in the north of Norway relatively near to one another.

6. Variants of Marinesco–Sjögren syndrome

A man with the classic signs of Marinesco–Sjögren syndrome (mental retardation, cerebellar ataxia, and cataracts) had, in addition, epilepsy, microcephaly, and corticospinal involvement (Ron and Pearce 1971*a*). This was a single case, but four sibs in an isolated south Norwegian family had the syndrome, including short stature and more spasticity than is usually found, including bladder paralysis (Nyberg-Hansen *et al.* 1972). An association of the Marinesco–Sjögren syndrome with a peripheral neurogenic lesion (Serratrice *et al.* 1973), and with a myopathy (Alter and Kennedy 1968) in a singleton, and in two sibs whose parents are first cousins (Chaco 1969) is documented.

7. Marinesco–Sjögren syndrome with hypogonadism

See page 185.

8. Late onset cataracts, choreo-athetosis, ataxia, progressive hemiballismus, and dementia

Sibs developed this disorder (Titica and Van Bogaert 1946). The parents were unaffected and unrelated.

9. Ataxia, cataracts, psychosis, and/or dementia with deafness

Nine members with this combination of signs and symptoms were affected in five generations (Strömgren *et al.* 1970). The onset of the ataxia occurred in adult life.

10. Cerebellar ataxia with downbeat nystagmus

Two brothers with this condition had an onset during the fourth decade; the limb reflexes were markedly exaggerated. The parents were unrelated and unaffected (Schott 1980). The subject is reviewed by Halmagyi *et al.* 1983.

11. Ataxia, macular corneal dystrophy, spasticity and cataracts

An inbred Beduin family was reported (Mousa *et al.* 1986) with an onset of the corneal dystrophy at 10 years. By 40 years vision was severely affected. Mental retardation was not a prominant feature and occured in only one patient. The gait ataxia had an onset in early childhood and there was evidence of cerebellar atrophy on the CT scan in all the adult ataxic patients. Neither upper limb ataxia, nystagmus, nor dysarthria ever became prominent. Spasticity always occurred in the ataxic patient but was not severe.

12. Recessive cerebellar degeneration with corneal dystrophy

Two sisters (possibly a third was affected) were born to cousin parents (Der Kaloustian *et al.* 1985). They were normal for the first year, but then at 1–2 years developed corneal clouding. At 6 years in one and 15 years in the other they developed a cerebellar syndrome which included a gait ataxia, head tremor, and an abnormal finger-nose test. There was some evidence of pyramidal tract involvment. The third sister had only mild cerebellar disease. The inheritance pattern is recessive and there are similarities with the previous entity.

(b) Ataxia with skin or hair change

1. Erythrokeratodermia with ataxia

A pedigree with 25 members affected over five generations had this combination (Giroux and Barbeau 1972). No patient without the skin lesion had the neurological syndrome which consisted of an adult-onset gait ataxia, some spasticity, and muscle pain. The skin lesion which appeared soon after birth disappeared after the age of 25 years. A skin biopsy showed the lesion to be an ichthyosiform erythroderma.

2. Ataxia and albinism

In an inbred kindred, four persons had a cerebellar ataxia, nystagmus, and dysarthria in middle life (Skre and Berg 1974). One had a peripheral neuropathy and posterior column loss. All those neurologically affected had tyrosinase negative albinism.

Three sibs are described by Bamezai *et al.* (1987), but only one had both traits whereas one had ataxia alone and one had only the albinism.

3. Ataxia, deafness, white forelock, and mental retardation

This combination occurred in two Pennsylvanian families (Telfer *et al.* 1971). Those affected had piebaldism and inheritance was as a dominant.

4. Ataxia, ichthyosis, and hepatosplenomegaly

Two girls had the condition (Harper *et al.* 1980). A third had only the skin lesion and a fourth brother was possibly affected. A dementing illness in two and the presence of supranuclear ophthalmoplegia in one suggested that this might be a storage disease.

5. Chediak-Higashi syndrome

This is characterized by partial albinism, nystagmus, decreased lacrimation, hyperhidrosis, and enlarged liver, spleen, and lymph glands. The association of this syndrome with a spinocerebellar ataxia is reported by Sheramata *et al.* (1971) who described three male sibs with the disease. Onset was in their thirties (later than usual).

A peripheral neuropathy in a single case was reported by Lockman *et al.* (1967) and one of two brothers with this disease (Weary and Bender 1967) had foot drop. Inheritance is probably as an autosomal recessive.

6. Ataxia and skin hyperpigmentation

Three sibs developed weakness (pyramidal), dysarthria and ataxia with an onset in middle life (Daras *et al.* 1983). The condition progressed in one to cause the patient to enter a wheelchair by 36 years of age. On examination there was severe weakness with brisk reflexes, extensor plantar responses, *pes cavus* and nystagmus. The skin lesions were small. There were multiple hyperpigmented naevi on the back, trunk, and limbs (see Abdallat, page 171 for a similar condition involving the peripheral nerves).

(c) Cerebellar ataxia and hypogonadism

The association of cerebellar ataxia with hypogonadism, either alone or with deafness or mental retardation, can occur as a recessively inherited condition.

The families of Holmes (1907*a*), Matthews and Rundle (1964), and others will be considered under this heading.

In the Holmes (1907*a*) family neither parent was affected and both lived to a good age. The four affected children had a cerebellar ataxia with an onset in the fourth decade. This was followed by dysarthria. Reflexes were normal or slightly exaggerated and there was no obvious mental deterioration. The three males had small external genitalia and little body hair. Their sister was married, but died at the age of 48 years, childless.

Matthews, and Rundle (1964) described two sibs, the product of unrelated parents, with ataxia and hypogonadism. One of the two was deaf and in both the ataxia was progressive. A third sib developed the neurological picture, but her secondary sexual characteristics were normal. The two males had low gonadotrophin excretion.

Neuhauser and Opitz (1975) reported four sibs (four affected out of fifteen children in the sibship), two boys and two girls with an almost identical picture of a cerebellar ataxia, nystagmus, and dysarthria with hypogonadotrophic hypogonadism. The onset of the cerebellar disturbance was between 12 and 20 years in three, and 33 and 38 years in one. The parents were second cousins and autosomal recessive inheritance is likely.

Some patients with hypogonadism with cerebellar ataxia could have the Laurence–Moon–Biedl syndrome.

Boucher and Gibberd (1969) described sibs with early onset of visual disturbance due to choroidoretinal degeneration. *Pes cavus*, cerebellar ataxia, depressed jerks, and extensor plantar reflexes were additional features. Both had hypogonadism due to inadequate gonadotrophin production.

Note: some of the following conditions combine hypogonadism with deafness.

Ataxia, deafness, mental retardation, and hypogonadism

Richards and Rundle (1959) described five sibs with this combination who were the offspring of second-cousin parents. Deaf-mutism, hypogonadism, mental retardation, a cerebellar ataxia starting in infancy, and progressive limb weakness and wasting were the main signs. In 1972, Sylvester performed a post-mortem on a male and female sib of family. He found primary gonadal dysgenesis and degeneration of the spiral ganglia cells in the inner ear.

Cerebellar ataxia and hypergonadotrophic hypogonadism

Skre *et al.* (1976) described two kindreds with Marinesco–Sjögren syndrome. In both kindreds inbreeding was likely those affected. Another family with this condition is reported by Lowenthal *et al.* (1979) had anosmia and were retarded.

Early onset cataract, myopathy, oligophrenia, hypergonadotrophic hypogonadism, and premature menopause

This syndrome is similar to Marinesco–Sjögren syndrome (Skre *et al.* 1976), but the presence of hypergonadotropic hypogonadism and the virtual absence of ataxia suggest a separate entity. Patients with this association were reported by Skre and Berg (1974) in two inbred kindred.

(d) Behr's syndrome

Optic atrophy, ataxia, spasticity, and posterior column loss were features described by Behr (1909) in mentally retarded children with an onset in infancy. The disease was not progressive. The genetics of the syndrome are unclear.

Landrigan *et al.* (1973) reported two sibs with this syndrome who had, in addition, a peripheral neuropathy.

Francheschetti (1966) reported the syndrome with recessive inheritance, as did Francois (1976) and Horoupian *et al.* (1979). The four sibs with Behr's syndrome reported by Van Leeuwen and Van Bogaert (1942) had a mother and two maternal males with optic atrophy and their grandfather was said to have been affected. A more recent report (Monaco *et al.* 1979) also indicates that a careful examination of first-degree relatives is necessary to exclude dominant inheritance with incomplete penetrance. The proband's brother, sister, and possibly a niece had neurological signs which the authors interpreted as minimal expressivity of the gene, but the evidence is not conclusive. A further family was reported by Thomas *et al.* (1984) with five affected in two generations. This was an inbred Asian family and the authors suggested that the inheritance pattern was recessive despite vertical transmission (pseudodominance). The index patient had a peripheral neuropathy (axonal degeneration) and the condition seemed to progress in adult life.

Ataxia, juvenile cataracts, myopathy, and mental retardation

Four sibs had this combination. They were hypotonic from birth and developed cataracts in early childhood (Herva *et al.* 1987). The sibs were subsequently found to have a myopathy characterised by vacuolar degeneration and increased adipose tissue. A CT scan showed cerebellar hypoplasia.

(e) Cerebellar ataxia and deafness (see also under (c))

Ataxia and deafness

Schimke (1974) reported sibs with an adult-onset cerebellar ataxia. A perceptive deafness preceded the onset of the ataxia and over a 10-year period, neither spasticity or extrapyramidal signs were noted.

Cerebellar ataxia, deafness, and amyotrophy

Four brothers had this combination, of whom two were deaf from birth (Matthews 1950). The author suggested that the inheritance is an autosomal recessive with reduced penetrance for deafness. The onset of the cerebellar ataxia occurred in their teens. The other unusual feature was the wasting and weakness of hand muscles and quadriceps. Pyramidal tract involvement was a feature in one.

Ataxia/amyotrophy of the hands/deafness and spastic paraparesis

This unique combination is reported in two brothers by Gemignani (1986). Onset was in early adult life with unsteadiness of gait, but by this time hand weakness had probably begun. One brother was, in addition to the above, mentally handicapped (mild) and his onset was with deafness and only later did the ataxia develop. Hypogonadism was a feature. A sural nerve biopsy showed demyelination.

Ataxia, deafness, and myoclonus

See May and White p. 122.

Ataxia, deafness, hyperuricaemia, and renal insufficiency

A large dominant pedigree with this combination was reported by Rosenberg *et al.* (1970). Expression was variable.

Ataxia, deafness, and mental retardation

Lichtenstein and Knorr (1930) described three sibs with a progressive ataxia, hypotonia, depressed reflexes, and deafness. Inheritance is probably as a recessive.

Ataxia, oligophrenia, myocardial sclerosis, and deafness

A brother and sister with this combination are described by Jeune *et al.* (1963). The disorder was progressive in both sibs. The parents were first cousins. Subsequently, Konigsmark and Gorlin reported (1976) that a third sib was affected.

Flynn–Aird syndrome

A large dominant pedigree (10 members over five generations) with a clinical picture of bilateral nerve deafness, severe myopia, ataxia, peripheral neuritic pain, joint stiffness, and muscle wasting, was reported by Flynn and Aird (1965). Convulsions and mental retardation were features in some. Three cases had retinitis pigmentosa and eight had bilateral cataracts. Kyphoscoliosis, dental caries, atrophy of the skin, poor skin healing, and blindness were also present.

Cerebellar ataxia, deafness, amyotrophy, and dementia

Sibs had an onset in the first decade of ataxia, intellectual deterioration, and neural hearing loss (Berman *et al.* 1973). The ataxia began at walking age and up to that time milestones were normal. Then followed atrophy of leg muscles, with shortening of Achilles tendons and facial weakness. Electromyography suggested neurogenic atrophy. All had upper motor neuron signs (exaggerated tendon reflexes and extensor plantars).

Three sisters from a consanguineous Turkish union had this condition (Koletzko *et al.* 1987). Two had delayed motor development whereas in one (who walked at 1 year) only the speech was delayed. When seen later all seemed to be moderately mentally retarded and all had severe ataxia. Tendon reflexes were normal, but the plantar reflexes were extensor. The hearing loss was progressive and there was some electrical evidence of peripheral nerve involvement. All three had a myopathic facial appearance.

(f) Other hereditary cerebellar ataxias

Ataxia and fasciculations

An Indian family with a dominantly inherited cerebellar ataxia in conjunction with widespread fasciculation but no wasting was reported by Singh and Shain (1964).

Cerebellar ataxia, brachydactyly, and nystagmus

Biemond (1934*a*) described a dominant pedigree with this combination. There were, however, some members with the brachydactyly alone or in conjunction with strabismus, but without ataxia.

Progressive cerebellar ataxia and myoclonus

This combination is discussed on page 122.

Ataxia telangiectasia

Patients with oculocutaneous telangiectasia and progressive cerebellar ataxia were found to have an immunological deficiency, an increased tendency to develop malignancy (mostly lymphoreticular), increased sensitivity to mutagenic agents, an increased chromosome breakage, and an elevated serum alpha-fetoprotein (Waldmann and McIntire 1972). The thymus is abnormal in all patients in whom it has been examined (Hansen *et al.* 1977). The onset occurs soon after birth and the neurological manifestations are progressive.

The condition was first described in 1941 by Louis-Bar who reported a nine-year-old child with ocular telangiectasia and similar lesions over the ears, arms, and legs. The child presented with a cerebellar syndrome accompanied by involuntary movements, often choreiform. The children are prone to develop respiratory and other infections and this is the commonest cause of death. Most of those affected learn to walk and talk, but soon after these milestones are reached deterioration begins. Mental changes (dementia) occur later (second decade) in about one-third of cases and are not usually severe.

A review of 101 cases is presented by McFarlin *et al.* (1972). Genetic studies were carried out on 24 patients from five families (Cohen *et al.* 1975). In all the families there were affected siblings. Parental consanguinity was found in three families. Inheritance is as a recessive. Cytogenetic investigations on patients with the disease show a high frequency of abberation, especially involving chromosome 14 and chromosome 7. More specifically the break points are 7p14, 7q35, and 14q11-12 (Hollis *et al.* 1987). Human T-cell malignancies often show breaks at 14q11, hence the higher frequency of T-cell tumours in ataxia telangiectasia.

Antenatal diagnosis has been attempted (Giannelli *et al.* 1981) using the increased sensitivity of cultured fibroblast to X-radiation as a diagnostic test. In their case the cultured amniotic fibroblasts responded normally and this was confirmed at birth.

Atypical cases of ataxia telangiectasia

An atypical case is reported by Fiorilli *et al.* (1985). The proband had a progressive cerebellar degeneration with an onset at 18 months. Recurrent episodes of lung and gut infections occured and the other features included ocular dyspraxia and mental retardation. Another child in the same paper had a progressive cerebellar ataxia starting at the age of 2 years. Ocular telangiectasia developed at 6 years when he was also noted to have auricular telangiectasia, progeroid appearance of the skin, ocular dyspraxia, dysarthria, and chorioathetosis.

Immunological changes were minimal and the levels of induced chromosomal aberation were intermediate between ataxia telangiectasia and normal controls.

Two other variant forms are discussed by Taylor *et al.* (1987). Cultured fibroblasts from two patients in whom the clinical features were very suggestive of ataxia telangiectasia showed a smaller increase of radiosensitivity than is usual although the response of DNA synthesis after irradiation was no different from that found in other patients with ataxia telangiectasia.

Cerebellar ataxia and deficient cellular immunity without telangiectasia

Sibs with a progressive ataxia and hypotonia, but without telangiectasia are reported by Hagberg *et al.* (1970). Both sibs died from overwhelming infection. One of the sibs had low IgG. Sibs with a similar condition were described by Graham-Pole *et al.* (1975). It is uncertain whether this is separate from ataxia telangiectasia.

Three sibs had an onset in early infancy or childhood of a progressive neurological disorder predominantly a gait ataxia (Byrne *et al.* 1984). In their teens they developed abnormal movements—mostly torticollis, athetosis, and dystonia followed by dementia and an internuclear ophthalmoplegia. They were IgE deficient. One sib developed leukaemia. Cytogenetically there was increased chromosomal breakage, but without a specific pattern.

(g) Progressive or intermittent cerebellar ataxia with specific biochemical abnormalities

The following is a list of known metabolic disorders which may present with a cerebellar ataxia and are discussed elsewhere.

Hexosaminidase deficiency
Cerebro-tendinous xanthomotosis

Metachromatic leucodystrophy
Globoid-cell leucodystrophy
Adrenoleucodystrophy
Mitochondrial disease

Other conditions in this group

Amino-acids were measured in the autopsied brains of two patients in a family with dominantly inherited olivopontocerebellar atrophy (or cerebellar ataxia plus; Perry *et al*. 1977). Five generations were affected. The family had been previously described by Currier *et al*. (1972; see p. 199). Aspartic acid, gamma-aminobutyric acid, and homocarnosine were reduced in the cerebellar cortex and dentate nucleus. Taurine was elevated in the same region.

(a) Ataxia and disturbance of pyruvate and lactate metabolism

There are many causes of a chronic metabolic acidosis with raised blood levels of lactate and pyruvate. Some of these are glycogen storage disease type I, fructose diphosphate deficiency, methylmalonic acidemia, pyruvate carboxylase and decarboxylase deficiencies, and Leigh's disease.

It should be noted that a severe metabolic acidosis and ketonuria is suggestive of an organic acidemia rather than a urea cycle defect. Also, an absence of fasting hypoglycemia suggests that the raised lactate is not due to a primary enzyme defect in gluconeogenesis, but to a deficiency of pyruvate carboxylase, fructose 1–6 disphosphatase, phosphoenolpyruvate carboxylase or glucose -6- phosphatase.

In this group is pyruvate dehydrogenase deficiency.

As summarized by McCormick *et al*. (1985) the pyruvate dehydrogenase complex has three enzyme components:

E_1—decarboxylase
E_2—dihydrolipoyl transacetylase
E_3—dihydrolipoyl dehydrogenase

The following section refers to reports of these three types.

E_1

The clinical picture of E_1 deficiency varies considerably.

There is a suggestion that the degree of deficiency of the enzyme might relate to the severity of the disease (Cavanagh 1978). A 15 per cent level could be associated with a severe infantile ataxia, whereas those with a 50 per cent level could manifest as a childhood ataxia.

A boy with intermittent cerebellar ataxia and a choreo-athetoid movement disorder was found to have elevated levels of pyruvic acid in the blood and an elevated urinary alanine content (Blass *et al.* 1971). Onset of the ataxia was during the second year of life and a brother was said to be similarly affected. A similar patient was described by Lonsdale *et al.* (1969).

A boy with lactic acidosis and severe neurological signs died at 6 months, and had no activity of the pyruvate decarboxylase nor dehydrogenase enzymes (Farrell *et al.* 1975).

Note The true designation of these cases to E_1 must at present remain uncertain and few recent cases have been reported. Counselling is equally uncertain.

E_2

Cederbaum *et al.* (1976) reported a boy who had hypotonia at birth and who made little psychomotor or growth progress. He was microcephalic and developed flexion deformities of his hands and feet. He had evidence of pyramidal tract involvement despite his hypotonia. Seizures did not occur and he had minimal elevation of pyruvate and lactate on a normal diet, but developed life threatening lactic acidosis on 65 per cent carbohydrate and 15 per cent fat in his diet. Two sisters were similarly affected. Data are presented to support a defect in E_2 of pyruvate dehydrogenase.

E_3

The patient described by Robinson *et al.* (1981) was floppy at one hour and had an odd cry. At 10 weeks following diarrhoea and vomiting she had a respiratory arrest, but was resuscitated. She had a significant acidosis with a high lactate and thereafter had frequent episodes of acidosis and deteriorated neurologically. Activities of alpha-keto glutarate dehydrogenase and branched chain keto acid dehydrogenase were deficient. E_3 in skin fibroblasts was 5 per cent of that in controls. The same assay in liver and muscle showed E_3 to be undetectable.

(b) Spinocerebellar degeneration with hexosaminidase A and B deficiency

See page 440.

(c) Familial spinocerebellar degeneration, haemolytic anaemia, and glutathione deficiency

Sibs with this combination had an onset of symptoms in their twenties. Besides the ataxia, there was evidence of a peripheral neuropathy and a cardiomyopathy (Richards *et al.* 1974).

(d) Disorders of glycine metabolism (non-ketotic hyperglycinaemia)

Classification:

1. Infantile—severe
2. Infantile—mild
3. Late onset

Infantile—severe

The onset of this often lethal condition occurs in most instances during the first few days of life. Of the 19 children from 13 families reported by Wendt *et al.* (1979) only seven survived the first year of life. The incidence in Finland is about 1 in 12 000. Hypotonia, lethargy followed by myoclonic convulsion, and progressive depression of special function were the main signs. All the survivors were mentally retarded.

   A review of the Finnish patients (Wendt *et al.* 1979) looking for hetero-zygote manifestations was undertaken, but the results were inconclusive.

Infantile—mild

Two mentally retarded young adults were reported by Flannery *et al.* (1983). Looking back there was poor feeding and hypotonia in infancy. Later, seizures, athetosis, hyperreflexia and poor speech were noted. Both had optic atrophy and were thought to have athetoid cerebral palsy. Sisters with the mild variant were reported by Haan *et al.* (1985). One sister aged 15 was profoundly retarded with epilepsy whereas her younger sister was treated with strychnine and is only mildly handicapped.

Late infantile

The infant reported by Trauner *et al.* (1981) developed normally for the first 6 months and then became irritable, began to regress and develop seizures. She died at 15 months and at post-mortem there was cystic degeneration of the white matter.

Late onset

Three brothers with non-ketotic hyperglycinaemia (parents unrelated) developed progressive spasticity between the ages of 8 and 20 years (Bank and Morrow 1972). Both upper and lower motor neuron signs developed in the legs.

   A 15-year-old boy with progressive optic atrophy and a spinocerebellar degeneration had non-ketotic hyperglycinaemia (Steiman *et al.* 1979). The children in this report and in that of Bank's came from the Lebanon. Cole and Meek (1985) described non-ketotic hyperglycinaemia in three sibs. One was of near normal intelligence and the others had an expressive speech disturbance. Although the authors call it 'juvenile', this was because of long survival. Onset was possibly in infancy.

A raised level of glycine in the cerebrospinal fluid helps to distinguish this condition from the following disorder.

(e) Ketotic hyperglycinaemia

First described by Childs *et al.* (1961), all the patients have a primary organic acidaemia with a secondary deficit in glycine metabolism (Shuman *et al.* 1978). These latter authors describe the condition secondary to a defective propionic acid metabolism and the same phenomenon occurs in methylmalonic and isovaleric acidaemia.

Other biochemical disorders associated with intermittent ataxia are listed by Blass *et al.* (1971). These are:

(f) Intermittent branch-chain ketoaciduria

(g) Hereditary hyperammonaemia

  (i) Congenital hyperammonaemia type II. see p. 498
 (ii) Citrullinaemia.
(iii) Argininosuccinic aciduria.
(iv) Hyperornithinaemia.

(h) Hartnup disease

This is characterized by an association of a cerebellar ataxia, a pellagra-like skin rash, and aminoaciduria. Progressive dementia and spasticity are prominent and the clinical signs can be intermittent. Mental retardation is sometimes the only clinical manifestation of the disease (Shih *et al.* 1971*b*; Pomeroy *et al.* 1968). About one-third of patients with the amino-acid abnormality have no clinical abnormalities. Four sibs were the offspring of a consanguineous mating (Lopez *et al.* 1969).

Ataxia with pellagra-like skin rash

A 14-year-old boy (a child of first cousin parents) had a Hartnup-like illness with intermittent ataxia. Biochemically, there was a suggestive block in tryptophan degradation (Freundlich *et al.* 1981).

(i) Abeta-lipoproteinaemia

The Bassen–Kornzweig syndrome was first described in 1950. The early symptoms are those of malabsorption and a failure to gain weight. Coeliac disease is often suspected.

A progressive ataxia begins in the middle of the first decade and by 10–15 years of age walking becomes progressively difficult (Kornzweig 1970). The deep reflexes are absent and there is evidence of posterior column loss with diminished position and vibration sense. The plantar responses are extensor. At this stage, the ocular involvement begins. Partial ptosis and nystagmus are present, but the poor vision that develops is due to a pigmentary retinopathy with or without macular degeneration. Mild mental retardation is part of the clinical picture and death is due to cardiac decompensation. The diagnosis is suggested by the low plasma cholesterol and crenated red blood cells (acanthocytes).

The disorder is recessively inherited and no abnormalities are found in carrier parents (Kayden 1972) although Salt *et al.* (1960) found both parents of an affected child to have half the normal quantity of lipoprotein. Parental consanguinity is often reported and, although Jewish and Italian parentage is common, cases have been described in Blacks and Maoris.

The peripheral nerve lesions were studied by Wichman *et al.* (1985). In the severest cases there was paranodal demyelination.

(j) Vitamin E deficiency

The neurological problems in abetalipoproteinemia are probably related to a deficiency of vitamin E. A very similar clinical picture has been reported in vitamin E deficiency due to malabsorption. In general, the following neurological signs are characteristic: a progressive ataxia, areflexia, peripheral sensory loss—especially position sense loss, dysarthria, ophthalmoplegia, retinitis pigmentosa, and upgoing plantar responses.

It should be noted that the pigmentary retinopathy is much more frequent in vitamin E deficiency itself than in vitamin E deficiency due to malabsorption.

It the two adult patients reported by Harding *et al.* (1985) the neurological features (ataxia and poor balance) became evident 20 years after the symptoms of malabsorption.

Stumpf *et al.* (1987) reported a lady with an onset at 4 years of an ataxia. She lost her reflexes and began to walk on her toes. She developed dysarthria at 15 years followed by head titubation and posterior column involvement. Her ataxia progressed and she went into a wheelchair at 20 years of age. On examination there was constant dystonia of her face, but no weakness. She had neither liver disease nor malabsorption, but had a low vitamin E.

There is some resemblance to the patient of Burck *et al*. (1981) who described a male who at 3 years fell frequently and was found to have a cerebellar ataxia with absent deep tendon reflexes. At 9 years he had fasciculations of the tongue and generalized muscle hypotonia, but little weakness. There were no pyramidal tract signs. He came from an inbred family. Vitamin E levels were low and two relatives were similarly affected.

(k) Hypobeta-lipoproteinaemia

There is uncertainty about the clinical correlation between this biochemical entity and neurological disease. Mars *et al*. (1969) examined 30 members of a family in which the propositus had a chronic progressive demyelinating disease with frequent remissions. The clinical picture was that of an intermittent but progressive weakness, clumsiness, and dysesthesia in both lower limbs. Co-ordination was only minimally impaired. Tendon jerks were hyperactive and extensor plantar reflexes were present. Sensation was normal. The propositus had low serum beta-lipoprotein and cholesterol levels. Acanthocytes were demonstrated in eight persons.

Korula *et al*. (1976) described a family with a spinocerebellar degeneration and hypobeta-lipoproteinaemia. The propositus had a dysarthria (onset in childhood) of an explosive type, and an impairment of proprioceptive sensation (absence of vibration and joint position below the ankle). Touch and pain were only minimally diminished, and the tendon reflexes were normal. His mother was similarly affected.

Aggerbeck *et al*. (1974) reported four sibs with hypobeta-lipoproteinaemia of whom three had a progressive neurological disorder. In the proband a gait disturbance was noted at two years of age. A cerebellar ataxia, extensor plantar reflexes, absent reflexes in the lower limbs, and *pes cavus* were features. Unlike Friedreich's ataxia, there was a glove and stocking sensory deficit for all sensory modalities, and an electrocardiogram was normal. No acanthocytes were found. All patients do not have a cerebellar ataxia.

Mawatari *et al*. (1972) studied a 19-year-old Japanese man with a clinical picture of a mixed motor and sensory neuropathy with insidious onset and chronic evolution. No acanthocytosis was present. Seven members in two generations had the biochemical defect. Brown *et al*. (1974) reported a large pedigree with dominantly inherited hypobeta-lipoproteinaemia. One infant was mentally retarded.

There has been a suggestion of a relationship between abeta-lipoproteinaemia and hypobeta-lipoproteinaemia, as both conditions have been described in a single family (Biemer and McCammon 1975). The proband had absent beta-lipoproteins, whereas her two sibs, two nieces, and her own daughter had reduced beta-lipoproteins. The proband's parents were first cousins. All those with hypobeta-lipoproteinaemia were neurologically intact, but the proband with abeta-lipoproteinaemia had minimal neurological signs, especially balance impairment and depressed reflexes. It is suggested that abeta-lipoproteinaemia can result from the homozygous state of the gene, which in single dose results in hypobeta-lipoproteinaemia and that this condition should be called familial homozygous hypobeta-lipoproteinaemia—a milder disease than abeta-lipoproteinaemia (Cottrill *et al*. 1974). The classical Bassen–Kornzweig abeta-lipoproteinaemia is quite distinct from homozygous hypobeta-lipoproteinaemia.

(l) Cerebellar disease and acanthocytosis with normal lipoproteins

Most patients have extrapyramidal and not cerebellar disease, but in a paper entitled 'Acanthocytosis and neurological disorders without abetalipo-proteinemia', Critchley *et al*. (1968) described ataxia in the youngest of 10 children (see p. 323). Three sibs were thought to be similarly affected. Two further sibs had acanthocytosis but were without neurological disease. A female child of one of the sibs with only acanthocytosis had a disorder resembling Friedreich's ataxia (with myocardial involvement) and had a few acanthocytes.

Kuo and Bassett (1962) presented in a brief report a male proband with ataxia, a peripheral neuropathy, retinopathy, and steatorrhoea. His brother had neuromuscular disease. Two sisters were symptom-free and had acanthocytes. Subsequent investigation of this family showed normal lipoprotein levels and intestinal biopsies were histologically normal (Critchley *et al*. 1970).

(m) Liver and ataxia

Two sibs with failure to thrive and with poor development had ascites and splenomegaly (Agamanolis *et al*. 1986). Both were very hypotonic. The CT scan suggested cerebellar atrophy. Both had a dyslipoproteinaemia with lipid in the liver (see p. 178).

C. Friedreich's ataxia

The inheritance of Friedreich's ataxia has recently been reviewed and, if strict diagnostic criteria are adhered to, Friedreich's ataxia is without exception inherited as a recessive disorder (Andermann *et al*. 1976). The diagnostic criteria suggested in the Quebec study by Barbeau (1978) are an onset in childhood or adolescene of a progressive cerebellar ataxia without remission, dysarthria, depressed or absent knee and ankle jerks, and extensor plantar reflexes. Posterior column sensory loss, especially vibration and/or position sense loss, occurred in 100 per cent of their typical group. Kyphoscoliosis and *pes cavus* often develop within 2 years of the onset. Small hand muscle wasting and optic atrophy are less frequently observed.

The most useful paraclinical finding is the absence of sensory action potentials and reduced motor conduction velocity (mild). Cardiomyopathy is found in the majority of patients, but is sometimes only evident if special tests are done. Using only an electrocardiogram, 90 per cent of patients have changes (Thoren 1962). Cardiac catheterization, angiography, and echo-cardiography suggest a hypertrophic obstructive cardiomyopathy. The post-mortem changes are those of a dilatation of the chambers and thickened endocardium in about one-third of the hearts examined.

Diabetes occurs clinically in 19 per cent of cases, but 40–50 per cent have an abnormal glucose tolerance test. That inheritance is as a recessive is clearly shown in the Quebec study as well as in a personal series (Baraitser *et al.* 1980).

One family is known (Harding and Zilka 1981) in which members in two generations were affected by a condition indistinguishable from Friedreich's ataxia. The age of onset was slightly later than usual, but all the other diagnostic criteria were present. It is presumed that an affected member married an (unrelated) heterozygote, although the chance of this happening is less than 1 per cent. What appears to be dominant inheritance, because two generations are affected, is pseudodominance. A few other families exist in which the relationship is more remote (A.E. Harding, unpublished data 1980; M. Baraitser, unpublished data). Marriage between heterozygotes would account for this.

Risks to the offspring of those who had Friedreich's ataxia are small and is equal to half the carrier rate in the general population, i.e. half of 1/110 (Harding 1981c). This is a small risk and the patient and spouse can be reassured.

Where inbreeding occurs and an affected member marries a first cousin there is a 1 in 4 chance that the cousin carries the same genetic factor and hence the risk to offspring is 1 in 8. The main problem for the genetic counsellor is that the family is usually complete before the diagnosis is made so that prevention is not easy to achieve. The biochemical abnormalities suggested by Blass *et al.* (1971) (a defect in the pyruvate dehydrogenase enzyme complex) have not been substantiated by other workers. There have been a number of other biochemical hypotheses to explain the metabolic abnormalities in Friedreich's ataxia but none has yet been confirmed. A deficiency in malic acid oxidation (Stumpf *et al.* 1982) looked persuasive, but this has not been substantiated. Recently, Friedreich's ataxia has been linked to probes on chromosome 9 (Chamberlain *et al.* 1988).

Early-onset recessive cerebellar ataxia with brisk reflexes

There is an early-onset progressive cerebellar ataxia indistinguishable from Friedreich's ataxia in age of onset and clinical presentation except that the upper limb and knee tendon reflexes are brisk in contradistinction to the diminished or absent reflexes found in Friedreich's ataxia. A cardiomyopathy is not thought to be present, and there was no increase in the frequency of diabetes. An analysis of the family history of 20 cases suggests autosomal recessive inheritance (Harding 1981*b*), and the prognosis was better than in Friedreich ataxia. Ankle jerks are absent in half of the cases (Harding 1981*b*), but despite this the sensory action potentials are normal.

Other recessive syndromes related to Friedreich's ataxia are reported by Barbeau (1980). The families were ascertained in isolated Canadian communities.

(i) Beauce type: the clinical picture is compatible with Friedreich's ataxia, but the course is more rapid.

(ii) Rimouski type: a basal ganglia component (dystonic posturing) was an added feature.

(iii) Acadian type: a typical Friedreich ataxia constellation of signs, but with late onset (16–24 years), slow progression, and no scoliosis. Movements were slow.

(iv) Matane type: reflexes were normal and the age of onset at 20 is later than most cases of Friedreich's ataxia.

All the above are recessively inherited.

The subject was reviewed by Barbeau *et al.* (1984) who described another branch of the original family with Acadian ataxia.

D. Cerebellar ataxia plus or adult onset cerebellar ataxia

Cerebellar ataxia ophthalmological signs/extrapyramidal/pyramidal

This is a group of dominantly inherited cerebellar degenerations with onset in middle life. It is characterized by ataxia of gait, dysarthria, tremor, involuntary movements (usually chorea), and occasionally ophthalmoplegia and pyramidal tract involvement. In the previous edition an attempt was made to separate those with ataxia and extrapyramidal signs from those with eye signs, but the variability is such that the division is no longer tenable.

Variability within families is considerable. The families described by Menzel (1891), Keiller (1926), Waggoner *et al.* (1938), Gallemaerts *et al.* (1939), Rosenhagen (1943), Gray and Oliver (1941), and Destunis (1944) all fall into this category. All the pedigrees are dominant—that of Waggoner's had 27 members affected in five generations, and Gray and Oliver 24 members in four generations. The onset of the disease is in early adulthood. It is now established that the pedigree of Gray and Oliver (1941) is the same family as that of Schut and Haymaker (1951), emphasizing the difficulty in delineating one family to the types according to the classification of Konigsmark and Weiner (1970).

The ocular signs most often found in association with cerebellar ataxia are optic atrophy and ophthalmoplegia. Unfortunately, the signs do not breed true, so that the separate designation of cerebellar ataxia plus ophthalmoplegia or cerebellar ataxia plus optic atrophy is not tenable.

The huge family of Schut and Haymaker (1951) had members of six generations affected, but the pedigree shows extensive variation in severity and manifestation of the disease. First described by Schut (in 1950), the family was re-examined in 1974 by Landis *et al*. The onset is usually before the age of 30 years. So much intrafamilial variation was observed that four subtypes were distinguished.

(a) Hereditary spinal ataxia (one case), that resembled Friedreich's ataxia.

(b) Cerebellar ataxia (seven cases). A cerebellar type of inco-ordination with normal or depressed reflexes.

(c) Cerebellar ataxia with brisk reflexes (12 cases).

(d) Hereditary spastic paraplegia with only minimal ataxia (two cases).

When Landis *et al*. (1974) re-examined some of the family, not only did 100 per cent of those seen have a cerebellar ataxia, but also optic atrophy and increased tendon jerks. Muscular atrophy, posterior column sensory loss, dementia, facial muscle atrophy, dysphagia, and slow apraxic eye movements were also found. Involvement of cranial nerves IX, X, and XII occurred in some members.

A similar pedigree was described by Currier *et al*. (1972). In a huge family that could be traced back to the 1700s, 65 were examined. There was an onset of symptoms between 20 and 40 years. Ataxia of gait was followed by speech disturbance. Variation in expression of the disease was remarkable. One branch had extrapyramidal signs with masking of facial expression and staring eyes. Random movements of the hand (misdiagnosed as Huntington's chorea) were seen in some. Neither optic atrophy, retinal degeneration, nor dementia were noted, but ophthalmoplegia occurred. As the disease progressed, hyperactive reflexes became hypoactive.

Konigsmark and Weiner (1970) used two kindreds as protocols for olivopontocerebellar atrophy with progressive dementia and ophthalmoplegia. Extrapyramidal features were part of the clinical picture. Both pedigrees are dominant. The pedigree of Carter and Sukavajana (1956) showed a father and a six of his children to be affected. Ataxia and speech difficulty occurred between seven and 43 years. Reflexes were increased and the plantar responses were extensor. When re-examined a severe supranuclear ophthal-moplegia had developed. The face was expressionless and dementia occurred within a decade of the onset.

In the second pedigree (Konigsmark and Lipton 1971), the onset was in the second or third decade. Ataxia, tremor, rigidity, paucity of movement, and mental deterioration were prominent. Ocular movement was affected by a paresis of upward gaze. Another dominant pedigree (Chandler and Bebin 1956) had a similar combination of signs.

In a family with three (possibly four) affected over three generations an affected mother had affected offspring by two separate marriages, confirming dominant inheritance (Whyte and Debakan 1976). The onset of ataxia varied from adolescence to the early thirties. Eye movements were slow. Scanning was absent and tracking (pursuit) was slow. Mental deteriora-tion occurred.

In a brief report, Starkman *et al.* (1972) describe a family of Italian descent in whom 10 members had abnormal eye movements as well as spinocerebellar degeneration. One had the eye signs alone and six only the ataxia. Abnormal saccadic movements, but normal pursuit were the chief ocular signs.

Variability within a family is the rule. A disorder of ocular movement occurred in only one member of the family described by Gerstenbrand and Weingarten (1962).

Koeppen and Hans (1976) described a Scottish family where nine were affected over four generations. The onset was at 40 years and a post-mortem in one confirmed the diagnosis of an olivopontocerebellar atrophy. The clinical picture was that of a gait disturbance, speech defect, disordered eye movement, and an extrapyramidal movement disorder (athetosis). All tendon reflexes were absent. Clinically, the disorder of eye movement consisted of an intact but slow pursuit. Saccadic eye movements on command were abnormal.

Zee *et al.* (1976) described members of a family with dominant inheritance in middle life of a cerebellar ataxia without involvement of any other systems. Progression was slow. The affected proband showed a slowness of horizontal eye movement.

Wadia and Swami (1971) described members of nine different Indian families in which 30 members were affected. Cerebellar ataxia and slow eye movements were the main clinical features. The age of onset was between eight and 30 years. The ataxia progressed to involve upper limbs and trunk. Diminished reflexes, facial weakness, clawing of the feet, kyphoscoliosis, chorea, and extensor plantar reflexes were features in some. Electro-myography on 11 showed evidence of chronic denervation in the limbs in seven patients. Those affected were slow to move their eyes spontaneously, or on command. Following movements were full. Fifteen per cent were described as mentally backward.

Philcox *et al.* (1975) described a Cape coloured family with members affected over four generations. Onset of ataxia was in middle life and 'staring eyes' were noted. Gaze was full in all directions, but rapid shifts were not always possible. Optokinetic responses were absent or altered in all those tested.

A dominant pedigree in which at least three members had slow saccadic and smooth pursuit movement in the hoizontal was noted by Singh *et al.* (1973*a*). Vertical eye movement, convergence, and vestibular-ocular movements were intact. Cogwheel rigidity was found in addition to the cerebellar ataxia.

In another dominant pedigree (Euziere *et al.* 1952) members were affected over five generations. In addition to the cerebellar ataxia and ophthalmoplegia, pyramidal tracts were affected.

A pedigree with a cerebellar ataxia, spasticity, and a progressive ophthalmoplegia (Hariga 1959) showed considerable variation over four generations, as did the dominant pedigree of Devos (1957).

An isolated family in West Virginia, possibly inbred although no actual instance of cousin marriage is reported (Heck 1964), has had many members since the middle of the nineteenth century affected by a progressive ataxia, scoliosis, nystagmus, mirror movements, reduced reflexes, equivocal plantar responses, high arches, and extra-ocular palsies. Onset occurred before the age of 20 years.

The Schut (1950) family can be placed in this category. Similarly, the four-generation family reported by Brown (1892) which contained 21 affected members is difficult to categorize. At least four had ptosis and ophthalmoplegia. In addition, some members had hyper-reflexia, choreiform movements, optic atrophy, atrophic retinal changes, and dementia. Sjögren (1943) reported six families of whom 40 per cent of those affected with a hereditary ataxia (he called it Marie's ataxia) were thought to have optic nerve involvement. An associated ophthalmoplegia occurred in one-third of cases.

In the family of Piton and Tiffeneau (1940), mother and daughter had ataxia, optic atrophy, and an ophthalmoplegia.

In 1942 Van Leeuwen and Van Bogaert reported a boy with stiffness of the lower limbs from early childhood. Poor vision was noted from 3 years of age and at 6 years he had bilateral optic atrophy. At 8 years, bilateral Babinski reflexes, increased jerks, and an ataxic gait were noted. He died at 11 years.

Seven years later (1949) the same authors reported that his mother, grandfather, and several others had optic atrophy. Three of the proband's four brothers and sisters showed a severe optic atrophy and a neurological syndrome which was variable, but consisted of ataxia and spasticity.

Counselling

The mode of transmission of this combination of neurological signs in the majority of families is dominant. Offspring risks to single cases should be in single figures. (There might be phenocopies and rare recessives, hence the risk might not be 1 in 2, but perhaps 1 in 4 to 1 in 8.) Three entities probably fall within this group.

(a) Machado/Joseph disease

A disorder called Machado disease (Nakano *et al.* 1972) or Azorean or Joseph's disease probably falls into this category. The onset is in early adulthood of progressive ataxia, nystagmus, reduced reflexes, a variable degree of amyotrophy, and peripheral sensory loss, mostly of posterior column type. All affected members stem from William Machado, who lived on the island of Sao Miguel in the Portugese Azores. The family described by Romanul *et al.* (1977) came originally from the region of San Miguel and had predominantly a cerebellar ataxia and polyneuropathy. The disease has now been subdivided into Types I, II, and III, but might be a single autosomal dominant disorder affecting inhabitants (who could be related) originally from the island of Flores (Coutinho and Andrade 1978). The clinical subdivision of:

Type I.   The features are an early onset (about 25 years), extrapyramidal (dystonia, athetosis, rigidity), and pyramidal signs.

Type II.   Intermediate onset (early 30s) of cerebellar ataxia and spasticity.

Type III. Late onset (fifth decade)—called specifically Machado disease.

The features are a motor polyneuropathy and a cerebellar deficit, usually beginning after the age of 40. The family reported by Romanul *et al.* (1977) can be assigned to type III.

In a study (Rosenberg *et al.* 1978) of the families still resident in the Azores (Souza family), two clinical variants referred to as type I and type II were evident within the same families. These authors report that Antone Joseph emigrated to the United States from the Azores Islands of Flores, where a huge pedigree with affected members over seven generations can be found. An unsteady gait, rigidity, and spasticity are the main features. Later the speech becomes slow and indistinct because of pharyngeal weakness. An ophthalmoplegia develops early and is characterized by a failure of upward gaze and horizontal and vertical nystagmus. The intellect is unaffected. Dystonia and athetoid movements are a feature in some of the patients. A post-mortem on one showed striatonigral degeneration (Rosenberg *et al.* 1976) with, in addition, lesions of the dentate nucleus and cerebellum, but the pathological description in this case is uncertain (Nielsen 1977) and the condition should not be confused with striatonigral degeneration discussed on page 297. Inheritance is clearly dominant. A large Japanese family had a clinical and pathological picture identical to that of Joseph's disease (Sakai *et al.* 1983). Members showed two types of neurological abnormality. One was predominantly cerebellar and pyramidal with or without extrapyramidal signs, and the other had cerebellar signs and a peripheral neuropathy as the main features.

A Japanese family (Yuasa *et al*. 1986) had involvement of the dentato-rubro-pallidal system but pathologically (not clinically) it differs from DRPLA (see page 298) in that in this latter condition neither Clarkes column, nor the anterior horn cells are involved.

An Indian family had a combination of ophthalmoplegia, ataxia, pyramidal signs, amyotrophy and facial fasciculations (Bharucha *et al*. 1986*a*). The onset was in the 3rd or 4th decades. Mental state was normal.

In the report by Goto *et al*. (1982) the family clinically resembled Joseph's disease, but the pathology was dentatorubro pallidoluysian atrophy.

The marriage of two affected members with this disease is known (Rosenberg *et al*. 1978). They produced a male child with an onset (of incoordination) at 8 years. At 12 years, he was dystonic with marked spasticity. Athetoid movements and difficulty with swallowing made spontaneous walking or eating impossible. The boy had no obvious cerebellar signs and yet both parents had type II with cerebellar ataxia and spasticity.

(b) Parenchymatous degeneration

This is unlikely to be a distinct entity, and most reports are consistent with its inclusion under the category of cerebellar ataxia 'plus'. It cannot be diagnosed clinically and patients with pure cerebellar ataxia cannot be assumed to have this pathology.

Pedigrees previously assigned to parenchymatous degeneration, but with more widespread clinical involvement, are those of Weber and Greenfield (1942), whose patient had an onset in middle life of ataxia and progressive dementia (the disease was observed to occur over three generations and one patient had a Parkinsonian-like facial immobility), and Hoffman *et al*. (1970) who described a pedigree in which four generations were affected with an onset in the fifties.

Hall *et al*. (1941) described a pedigree in which members were affected over several generations by a gait ataxia, and speech and swallowing difficulties. One member had athetosis. In one of the other affected persons a 'Parkinsonian-like' immobility of the face was noted at 45 years of age.

Five members were affected over three generations in the family reported by Akelaitis (1938). The age of onset was between 40 and 50 years, and the course of the disease, which included dementia, was rapid. Diplopia occurred in one patient and all had difficulty with swallowing early on in the illness.

A sibship (there is no information about the parents) in which three members had a middle-life onset of ataxia is reported by Richter (1940). The pathology at post-mortem was of a parenchymatous cortical cerebellar atrophy, but one of the affected sibs demented and had facial contortions and another had optic atrophy and sphincter disturbance, emphasizing that the pathology does not correlate well with the clinical features.

Richter (1950) reported 15 cases over three generations with onset of ataxia in the fifth or sixth decade. Voluntary conjugate eye movements were impaired in all directions. Dementia occurred towards the end of the disease. A cousin had a similar clinical course, but only upward gaze was impaired and there were those affected without an eye movement disorder. Although the pathology was suggestive of primary parachymatous degeneration, the clinical picture cannot be distinguished from the cerebellar atrophy plus category.

(c) Spinopontine degeneration

Using the pathological designation of spinopontine degeneration or striatonigral-dentatal degeneration, Boller and Segarra (1969), Woods and Schaumburg (1972), and Taniguchi and Konigsmark (1971) have described dominant pedigrees that are clinically indistinguishable from the families described above.

Clinically, there occurred an adult-onset gait ataxia, nystagmus, dysarthria, and an extrapyramidal type of rigidity. Muscle atrophy, brisk reflexes, and, in some cases extensor plantar reflexes were evidence of the motor involvement. Woods and Schaumberg's patient had a severe ophthalmoplegia. Boller and Segarra's case had, in addition, a failure in upward gaze and diplopia on looking to the right. In 1978, Pogacar *et al.* studied two additional members of the 'W' family reported by Boller and Segarra (1969). In their branch of the family reflexes were at first brisk, but were later lost; flexion contractures occurred in the lower limbs and dementia was a feature. In one case there was, in addition, optic atrophy and palatal myoclonus. A post-mortem showed olivopontocerebellar atrophy.

Neither on clinical nor pathological grounds alone, is it possible to differentiate between dominant spinopontine atrophy and cerebellar ataxia plus. The Boller–Segarra family could be fitted into the category above and the Woods–Schaumberg pedigree is now known to come from the Azores and the condition is discussed on page 202. The same argument applies to the three generation family reported by Bale *et al.* (1987). In this family most patients had a limitation of upward gaze.

Cerebellar ataxia plus retinitis pigmentosa

Families in which retinal degeneration has been prominent are described by Björk *et al.* (1956), Bergstedt *et al.* (1962), Carpenter and Schumacher (1966), and Weiner *et al.* (1967). Both dominant and recessive pedigrees are known.

There is so much variability in this condition that it is often difficult to categorize families. In the report of Colan *et al.* (1981), a mother and three of her children (from two different marriages) were affected. The age of onset varied between 5 months and 10 years. One child died at 2½ years because of anterior horn cell involvement whereas mother and one of her daughters had a milder illness with macular degeneration. An impairment of lateral and upward gaze and hyporeflexia were part of the syndrome.

(i) Dominant

In Bjork's (1956) families, inheritance was as a dominant. A gait disturbance occurred before the end of the second decade and the reflexes were noted to be 'lively'. Franceschetti and Klein (1948) described 22 cases in 12 families. The fundal changes in some were reported as retinitis pigmentosa and in others as macular degeneration. In the pedigree of Bergstedt *et al.* (1962) (dominant with four members affected over three generations), the age of onset varied between six and 55 years and the visual symptoms occurred several years after the cerebellar gait disturbance. The disorder was progressive, but affected neither sensation nor intelligence.

Havener (1951) described a father and one of his five children, both of whom developed the cerebellar ataxia, but only the father had the macular degeneration and under the title 'Hereditary cerebelloretinal degeneration', Halsey *et al.* (1967) described 11 members (in a Negro family) affected over three generations.

In the family described by Weiner *et al.* (1967), 27 members were affected over five generations—a writhing athetosis was part of the clinical picture which was that of progressive visual disturbance, ataxia, and slurred speech. Pigmentary changes were originally in the region of the macula but later involved the whole retina and were followed by optic atrophy. The clinical picture included extrapyramidal, pyramidal, and cerebellar signs. A pedigree which might be remotely related to Weiner's had, with some intrafamilial variability, the same ocular signs, ataxia, pyramidal, and possibly extrapyramidal features (Murdoch and Nissim 1971).

In the family described by Carpenter and Schumacher (1966), a father and his three children are described. The authors stress the pathological and clinical differences that occur between generations. Whereas the three children had an early onset, the father became ataxic in his twenties. The clinical picture was that of a cerebellar ataxia, spasticity, visual loss, and involuntary movements (athetoid). He was found to be mentally dull, to have early optic atrophy, and sensory loss in hands and feet. All of his three children were dead before the third year of life. All three had retinal degeneration and ataxia. A post-mortem on the father showed features of olivopontocerebellar atrophy whereas the children had only cerebellar changes.

Woodworth *et al.* (1959) described, under the title 'A familial disorder with features of olivopontocerebellar atrophy, Leber's optic atrophy and Friedreich's ataxia', a dominant pedigree in which the affected member had brisk reflexes and optic atrophy with or without obvious retinal pigmentation. Extra-ocular movements were sluggish. The age of onset varied from childhood to adulthood.

Stadlin and Van Bogaert (1949) reported two sisters, the offspring of first cousins, with Friedreich's ataxia. Much later, cochlear disabilities and a tapetoretinal dystrophy developed. The mother is thought to have had an abortive form of Friedreich's ataxia (Van Bogaert and Martin 1974). In a Negro family eight members showed a progressive external ophthalmoplegia, retinal degeneration, and a spinocerebellar ataxia (Jampel *et al.* 1961). Paralysis of upward gaze and defective convergence were the first signs of the ophthalmoplegia. The age of onset varied from early childhood to adulthood.

(ii) Recessive

The following pedigrees are recessively inherited and include retinitis pigmentosa and ataxia as features.

Sjögren (1943), in his monograph on the hereditary ataxias, found only one family in which an ataxia was associated with a pigmentary retinopathy. Both sibs (unaffected parents) had the ataxia, but only one a salt and pepper pigmentary retinopathy.

Other families with ataxia and retinitis pigmentosa and evidence of more widespread disease are those of Ledic and Van Bogaert (1960) who reported two families. In the one, two sibs had retinitis pigmentosa, epilepsy, and mental retardation. Their first cousin had macular retinitis pigmentosa and a spastic ataxia. Two third cousins were affected, but one had the retinitis pigmentosa with a spastic paraparesis and the other had the paraplegia without retinal involvement (see also p. 281).

Counselling

In general most of the adult onset cerebellar ataxias with retinitis pigmentosa are dominantly inherited. Recessive inheritance is rare. For other recessive syndromes with retinitis pigmentosa see page 78.

Cerebellar ataxia with a peripheral neuropathy

The classification of these families is difficult. It is a matter of emphasis whether they are classified under the peripheral neuropathies or the cerebellar ataxias. If predominantly ataxic they are best classified here. In the family described by Carenini *et al.* (1984), eleven patients from nine families had a neuropathy and cerebellar ataxia. In two patients with early onset there was severe axonal degeneration and in six there was evidence of demyelination (see also p. 231).

There is no doubt that patients with HMSN Types I and II can be considerably ataxic, but the neuropathy predominates.

E. X-linked cerebellar ataxia

This is an unusual mode of inheritance for cerebellar disorders. Three separate families were reported by Shokeir (1970*b*). In one pedigree, six males in two generations were affected, and in two families there were affected males in three generations. The clinical picture was that of an onset between 16 and 21 years of an ataxia which progressed until the thirties when it became stationary. Marked pyramidal signs were present in some members, but neither atrophy nor sensory deficit occurred. An unusual finding was a female with this disorder who had Turner's syndrome.

Malamud and Cohen (1958) reported an unusual X-linked pedigree with progressive cerebellar ataxia in two male first cousins. There was hearsay evidence that six other males were affected. The onset was in the latter half of the first year of life. Pyramidal tract signs developed, together with progressive deterioration of motor and mental abilities. Rigidity replaced the cerebellar signs, and an external ophthalmoplegia and optic atrophy were added features. One of the two boys had myoclonic epilepsy. A post-mortem showed a combination a cerebello-olivary and dentatorubral degeneration.

An X-linked pedigree of Turner and Roberts (1938) has been cited as an example of X-linked Friedreich's ataxia. Nine males were affected in two generations. A description of one patient is given and, although the age of onset and progressive nature of the disease resembles Friedreich's ataxia, the boy developed immobility of facial muscles—a most unusual feature for Friedreich's disease—and a muscular dystrophy cannot be excluded.

An extensive kindred is reported by Spira *et al.* (1979). Ten males were affected in four generations. Delayed walking (18 months to 2 years) was the earliest feature, followed by an ataxia in the first or second decade. Affected members were confined to a wheelchair in their thirties. *Pes cavus*, scoliosis, and slowly progressive wasting of the lower limbs and pronounced pyramidal tract involvement occurred.

Three males—two full brothers and their half brother—through their mother had a non-progressive cerebellar ataxia with an onset in infancy and moderate retardation (Young *et al.* 1987). One of the boys had vermis atrophy and the other more generalized involvement of the cerebellum. (See also p. 94.)

Familial posterior column ataxia (Biemond syndrome)

A rare familial form of ataxia due to posterior column involvement was first described in 1929 by Mollaret. The familial nature of the disease was shown by Biemond (1951). In his family a father, his four children and his brother were affected, but the mode of inheritance is uncertain as the father married his cousin. Loss of vibration and position sense were the main features. Pain and temperature were intact.

Familial posterior column ataxia with scoliosis

This is probably not a separate genetic entity. Three brothers with this combination are reported by Singh *et al.* (1973*b*). A fourth brother had a positive Romberg sign and, as in his three affected brothers, all tendon reflexes were absent. Pyramidal tract signs and cerebellar signs were not features. The condition is likely to be the same as hereditary motor and sensory neuropathy (see p. 227).

F. Late-onset cerebellar atrophy without associated features

(a) Recessive

There are no clinical grounds for diagnosing recessive cerebellar atrophy (recessive pedigrees are much rarer than dominants). Only pedigree analysis is helpful. Fickler (1911) and Winkler (1923) reported siblings who may be examples of this category, but even here the onset was before the end of the first decade in one of the sibs reported by Fickler (1911).

More recent reports of a late-onset recessive type of cerebellar ataxia without associated features are difficult to find.

Recessive late onset cerebellar syndromes with associated features

Downbeat nystagmus (see p. 182).
Hypogonadism (see p. 184).
Hearing loss (see p. 186).
Retinitis pigmentosa (see p. 194, 204).
Optic atrophy—rare (see p. 184).

(b) Dominant late-onset cerebellar ataxia without associated features

These families are surprisingly rare and there have been few reports in the literature (Harding 1981*d*). Most cases without associated features have affected relatives with cerebellar ataxia plus and a pure disorder should be diagnosed with caution. In addition, it should be noted that the associated features, especially the extrapyramidal and eye movement disorders, may be minimal in individual members and can easily be overlooked.

Counselling of single late-onset cases of cerebellar ataxia

It is uncertain what proportion of single adult-onset cases are fresh dominant mutants as opposed to those caused by as yet obscure—non-genetic environmental factors. In general, the onset of the cerebellar ataxia is sufficiently late not to seriously influence the reproductive fitness so that a large number of fresh dominant mutants would not be expected. It has been observed that no single cases of late onset have had affected offspring (Harding 1981*a*), but a longer period of observation is necessary. The clinical differentiation between the genetic and non-genetic causes of adult-onset cerebellar ataxia is difficult, but optic atrophy, ophthalmoplegia, and retinal pigmentation are less common in singleton cases, and the absence of these signs might indicate a non-genetic aetiology (Harding 1981*a*). At present the counsellor can only indicate that there might be a risk to offspring somewhere between 5 and 10 per cent.

Cerebellar disorders and HLA linkage

Yakura *et al.* (1974) report a family with so-called Marie's ataxia in which HLA-A typing suggested that the gene locus responsible is on the sixth chromosome near the HLA-A locus. The father in this family was deduced to be HLA-A 9,5/HLA-A 11, W 10. The three affected offspring, out of a total of five children, received the HLA-A 9,5 haplotype.

Jackson *et al.* (1977) reported an olivopontocerebellar atrophy (cerebellar ataxia plus) in which HLA-A typing predicted with 90 per cent accuracy the gene carrier in that family.

More recently, Nino *et al.* (1980) reported a similar study in a family with 'cerebellar ataxia plus' and found that if the lod score values from the pedigree of Yakura *et al.* (1974) and Jackson *et al.* (1977) were pooled with their own (a dubious manipulation) that odds for linkage with the HLA-A loci was approximately 48 000:1. It was also found that the recombination fraction was considerably higher than previously reported and the suggestion is that the gene is situated about 20 centimorgans from the HLA-A locus.

Not all studies have shown linkage but as the clinical information has been scanty it is difficult to assess many of the reports. Five dominant pedigrees were studied by Koeppen *et al.* (1980) and no linkage was found. No definite conclusions can be drawn about the relationship between HLA-A markers and cerebellar ataxia plus. Although it is at present appropriate to consider the group as a single entity, the presence of linkage is some families and not in others might, if confirmed, suggest that the group is heterogeneous.

Kuru

This subacute cerebellar degeneration is characterized by a tremor and a progressive dementia leading to death within a year.

Harper (1977) used kuru as an example of a disease thought at first to have a strong hereditary basis, but found subsequently (Gajdusek *et al.* 1966; Gajdusek 1973, 1977) to be transmitted by an infective agent. The disease is localized to an area of the New Guinea Highlands, occupied by the Fore people, a linguistic group from which 80 per cent of kuru cases originate. The first two villages where kuru is said to have started were at Keiagana and a Fore village on the Fore–Keiagana boundary. The only factor which appeared to distinguish kuru-free from kuru-affected clans was intermarriage with the Fore.

A genetic model was suggested on the basis of familial aggregation of the disease. It was thought by Bennett *et al.* (1959) that a single gene was responsible with expression in the heterozygous state in females, but in the homozygous state in the males. However, although the sex ratio in adults showed a female preponderance (3:1 in some villages), this was not so in children. Affected children were therefore considered to be homozygous. Other pecularities noted were the fairly recent occurrence of the disease in this region and that the people in that area were not of a single tribe.

By 1962 Gadjusek was considering the existence of a slow virus infection, and by 1966 he and others had successfully transmitted the illness to primates by intracerebral injection of brain tissue. The ritual consumption of cerebral tissue, especially by mourning women and children then became clear as the mechanism of spreading of the disease. It is possible that an inherited immune mechanism might play a role in susceptability, but by and large, kuru is not a genetic disease. It has been disappearing from the area during the past 15 years.

Hypertrophia musculorum vera in familial ataxia

O'Donnell *et al.* (1986) reported a family with an onset in the 30's of gait disturbance, dysarthria, and calf hypertrophy. Cramps, fasciculation, ataxia, and hyporeflexia, were additional features. A raised CPK occurred in at least one, and a biopsy showed normal or hypertrophic fibres. There were widespread fasciculations and it is uncertain whether the underlying pathology was myopathic or neurogenic.

Periodic ataxia

(i) Autosomal dominant

Familial periodic ataxia

Patients with this disorder experience episodes of vertigo, ataxia, and nystagmus. Only six families up to 1979 have been recorded (Donat and Auger 1979). All the families have shown dominant transmission of the disease. The original families were described by Parker (1946) and others are those of White (1969), Hill and Sherman (1968), and La France *et al.* (1977). Donat and Auger (1979) added a family of their own. The attacks were often precipitated by fatigue and emotion and lasted from seconds to weeks. Only in the families of White (1969) and Farmer and Mustian (1963) was there marked and persistent ataxia between attacks.

Hereditary paroxysmal ataxia with response to acetazolamide

Eight patients in a family had paroxysmal ataxia, dysarthria, and nystygmus (Griggs *et al.* 1978). Inheritance was dominant. Attacks lasted from 1 to 6 hours, with a mild cerebellar ataxia between attacks. Acetazolamide completely abolished all attacks.

Note: These two conditions are probably the same—see review by Gancher and Nutt (1986) who add cases of their own.

(ii) X-linked periodic ataxia

A pedigree suggestive of X-linked recessive inheritance was reported by Livingstone *et al.* (1984). The cerebral pathology in one case had some features of Leigh's disease and in two patients there was an abnormality of pyruvate metabolism similar to that found in pyruvate dehydrogenase deficiency. Response to a acetazolamide was partial.

12

Hereditary neuropathies

Sensory neuropathies

The following classification of the hereditary sensory neuropathies is based on that of Dyck and Ohta (1975):

HNS Type I —a dominant hereditary sensory neuropathy based on the cases of Denny-Brown (1951).

HNS Type II —a recessive hereditary sensory neuropathy. Also called congenital sensory neuropathy or Morvan's disease.

HNS Type III—dysautonomia (Riley–Day syndrome).

HNS Type IV—congenital insensitivity to pain with anhidrosis (tactile sensation is usually normal).

HNS Type V —congenital pansensory neuropathy with morphometric features typical of Type II (therefore not only peripheral).

It might be helpful for geneticists to know that in Types I, II, and III there are absent sensory action potentials, absent or reduced reflexes, and unequivocal pathological abnormalities of the nerve. Types IV and V are neuropathic disorders (congenital insensitivity to pain) despite normal reflexes and normal nerve conduction velocities.

To the above can be added:

The above classification will be followed.

A more genetically orientated classification (Donaghy *et al.* 1987) is as follows.

A. *Autosomal dominant*
1. Denny-Brown or HSN I (see p. 213).
2. With paraplegia (Cavanagh, see p. 289).

B. *Autosomal recessive*
3. Type II (p. 216).
4. Type III (p. 217).
5. Type IV (p. 218).
6. With keratitis (Donaghy *et al.* 1987).
7. With paraplegia (Cavanagh *et al.* 1979*a*).

C. *X-linked*
8. See p. 226 (Jestico *et al.* 1985).

D. *Uncertain*
9. Nordborg *et al.* (1981; p. 222).
10. HSN V—small myelinated fibre loss (Low *et al.* 1978; Dyck *et al.* 1983; p. 219).

Note: Congenital indifference to pain does not appear in the classification (see p. 219).

Type I dominant hereditary sensory neuropathy

Nelaton (1852) observed foot ulcers in several members of a family. In 1922, Hicks examined 34 members over four generations, 10 of whom had 'perforating ulcers of the feet, shooting pains about the body and deafness'. Sensation in both lower limbs was severely affected, especially the modalities of pain and temperature. Hicks considered that his cases resembled syringomyelia. Denny-Brown (1951) examined a member of the third generation of Hick's family and described the post-mortem findings as a primary degeneration of the dorsal root ganglia. He called the condition hereditary sensory radicular neuropathy. Large dominant pedigrees were described in 1939 (Tocantis and Reimann) under the title 'an instance of probable myelodysplasia'.

It is, however, Denny-Brown's hereditary sensory radicular neuropathy that has been used as the prototype for type I. Up until 1951, when Denny-Brown showed that the primary lesion was in the cells of the posterior root ganglia, various other names such as acropathic ulcers, mutilating lumbo-sacral syringomyelia, and myelodysplasia had been in use (Cambell and Hoffman 1964). Type I is a slowly progressive sensory neuropathy with an onset in the second, third, or later decades. Pain and temperature are predominantly affected and the distribution of the sensory deficit is distal in the legs. Ulcers develop on the soles of the feet and in the lower limbs, often in the third or fourth decades. Reflexes are decreased or absent, especially in the lower limbs, but both upper and lower limbs can be affected. Deafness is occasionally part of the disease (Fitzpatrick *et al.* 1976).

Other type I pedigrees are those of Mandell and Smith (1960), Jackson (1949), and Nair (1976).

Histologically, there is a complete loss of myelinated fibres especially those of small diameter, but also including those with large diameter. There is also an absence (loss) of unmyelinated fibres (Danon and Carpenter 1985).

Type I with severe muscle wasting (Fig. 12.1)

Muscle wasting and weakness is seldom prominent but, when it does occur (Cambell and Hoffman 1964) differentiation from the motor neuropathies can be difficult. England and Denny-Brown (1952) described a huge pedigree with what they suggested was Charcot–Marie–Tooth disease (57 were known to be affected). Inheritance was as a dominant. Particularly remarkably in the family was the degree of sensory loss. In two the loss of sensation led to perforating ulcers, osteomyelitis, and amputation. In others, the sensory

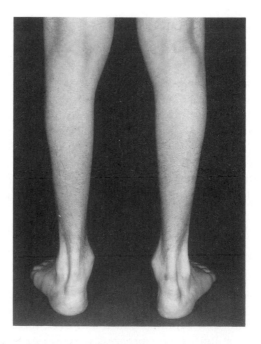

Fig. 12.1 Distal wasting in hereditary motor and sensory neuropathy.
Courtesy of the Department of Medical Illustration, National Hospital for Nervous Diseases, Queen Square, London.

changes were minimal. The nerves were not thickened and ataxia was not a feature.

A Virginian kinship had such prominent wasting that Dyck *et al.* (1965) thought the disease was related to Charcot–Marie–Tooth disease. Twenty-two persons were affected in four generations.

The families described by Riley (1930), Cambell and Hoffman (1964), and Barraquer-Forré and Barraquer-Bordas (1953), in which the muscle wasting was marked, fall to some extent halfway between the type I hereditary sensory neuropathies and the motor neuropathies, as does the report of Van Bogaert (1957) which described possible monzygotic twins, who, at the age of 10–11 years developed painless lesions on the toes. X-ray examinations revealed rarefaction of bone. Neither child had ever walked normally and their psychomotor development was retarded. Their legs became thinner and they developed a clinical picture of amyotrophy with severe involvement of the posterior columns. Later, small ulcers developed on the tips of the fingers. A large dominant pedigree (Reimann *et al.* 1958) had additional, but unrelated features (cleft lip and palate).

Recessive pedigrees should not be included in type I. In some sibships incomplete information about the parents has made the interpretation of the genetic data difficult. Three sibs (Iwabuchi *et al.* 1976) had the disease, but their father, who died at 55 years, had hearing loss and had to have a toe amputated. In these three sibs, IgA and IgG levels rose as the disease progressed.

Another family with type I sensory neuropathy and increased synthesis of IgA is reported by Whitaker *et al.* (1974) in many members over five generations. Unaffected family members did not show the immunological change.

Sibs (two out of five) were reported by Schoene *et al.* (1970), but there is some uncertainty about the inheritance in this family as the father, who had no evidence of a peripheral nerve disorder, had bilateral nerve deafness. His two unaffected children had a neuropathy with an onset after the age of ten years. Neither child was deaf and the motor system was virtually spared.

In some reported pedigrees (Kuroiwa and Murai 1964) sibs only had the disease, but the mother was suspected of being a gene carrier on the basis of electrophysiological studies. No trophic or sensory changes were observed. The occasional families in which inheritance seems recessive might be examples of variable expression and penetrance of a dominant condition. However, careful examination of the 80-year-old parents of two sibs (Pallis and Schneeweiss 1962) revealed no abnormal findings.

Sporadic cases occur, but it is uncertain if they represent fresh dominant mutants or are non-genetic.

Counselling

There is a 50 per cent offspring risk to those who are affected.

Type II congenital sensory neuropathy (recessive)

This is characterized by an onset in infancy or childhood of a multilating neuropathy with ulcers and fractures developing in the peripheral parts of all four limbs. There is a marked loss of peripheral sensation, especially touch and pressure. Tendon reflexes are lost or diminished. Sweating is impaired distally and nerve biopsy shows an almost complete absence of myelinated fibres and a decrease in the number of unmyelinated fibres.

Most of the patients to date have had a non-progressive or only slowly progressive course. It is as yet not certain whether type II is a single genetic entity and in their book *Acrodystrophic neuropathy*, Spillane and Wells (1969) suggest that congenital sensory neuropathy be divided into a progressive and non-progressive type.

Up to 1973 there were 20 familial cases and 13 isolated instances (Murray 1973). The author described sibs in whom the lack of pain caused recurrent ulceration, osteomyelitis, and neuropathic joints. No progression took place over a period of 26 years.

Consanguinity was present in the parents of two unrelated sib pairs reported by Johnson and Spalding (1964). In a single case reported by Ogden *et al.* (1959) the parents were related. Fedrizzi *et al.* (1972) reported a rapidly progressive neuropathy in which the parents were second cousins. Like Johnson and Spalding's cases, tactile sensitivity was more severely affected than temperature. Most other cases show a syringomyelic type of dissociated sensory loss.

Sibs (unrelated parents) with a congenital sensory neuropathy were described by Haddow *et al.* (1970).

A Quebec kinship (Ohta *et al.* 1973) had four affected sibs with an onset before the end of the first decade. The parents were unrelated and unaffected. Tendon reflexes were absent in the lower limbs and distal sensory loss of touch/pressure was more marked than the loss of pain/temperature.

Sibs with an onset in the first decade of life but not at birth (Jedrzejowska and Milczarek 1976) had histological evidence that the process was probably progressive.

Counselling

As for autosomal recessive conditions.

Variants of congenital sensory neuropathy

A pair of sibs (Miller *et al.* 1976) had, in addition, tonic pupils and scoliosis and in an unusual pedigree reported by Heller and Robb (1955) there was evidence of dominant inheritance with variable expression and reduced penetration. Onset was at or soon after birth.

It is not certain that all the mutilating neuropathies of childhood described under various headings such as 'l'acropathia ulcero-mutilante familialis' (Jusic *et al*. 1973; Thevenard 1953) or familial lumbosacral syringomyelia (Van Epps and Kerr 1940) are all type II. They are excluded if inheritance is dominant. The two sibs (normal parents) described by Jusic *et al*. (1973) had excessive proliferation of Schwann sheaths and endonurium suggesting a different pathology, but at present they are best classified as part of the congenital sensory group with recessive inheritance.

Some families are known with a clinical picture conforming to type II, but with spastic paraparesis as the first significant clinical finding. These are discussed on page 289.

Type III Riley–Day syndrome

Riley *et al*. in 1949 described five Jewish New York children with episodic hypotension, excessive sweating, skin blotching, acrocyanosis, cyclical vomiting, and absent lacrimation. Since the original description, there have been more than 200 published cases (Brunt and McKusick 1970). The condition predominantly affects Ashkenazi Jews. The birth frequency in this population group is between 1 in 10 000 to 1 in 20 000 with a gene frequency of 0.01 per cent. In the study of Brunt and McKusick (1970) 172 families were studied, and 210 affected patients had the following signs and symptoms, in addition to the ones mentioned above—absent fungiform papillae of the tongue, inco-ordination and unsteadiness of gait, swallowing difficulty in infancy, and episodic fever.

In a review by Axelrod *et al*. (1974), the authors state that they do not recognize a well documented case in a non-Jew. However, Levine *et al*. (1977) described a male Brazilian child with all the major criteria, and the Riley–Day syndrome in a non-Jewish family has been reported by Klebanoff and Neff (1980). Other claims by Linde (1955) and Hutchinson and Hamilton (1962) are not accepted by most authorities as fulfilling the Riley and Moore criteria (1966). In the Brunt and McKusick (1970) paper, most of the Ashkenazi Jews could trace their ancestry back to the Pale region in Eastern Europe.

The criteria, as summarized by Levine *et al*. (1977) from Riley *et al*. (1949), Riley (1952), Riley and Moore (1966), and Axelrod *et al*. (1974) are: feeding problems in infancy, delay in motor milestones, skin blotching with excitement, breath-holding spells, unexplained fever, frequent pneumonia, short stature, reduced or absent overflow tears, impaired taste perception, absent lingual papillae, hyporeflexia, impaired pain sensation, postural hypotension, hyperhidrosis, absent flare after intradermal histamine, miosis after methacholine, and Jewish ancestry.

No instance of autonomic dysfunction has been reported in a parent of a child with familial dysautonomia. Rubenstein and Yahr (1977) reported a parent of a child with familial dysautonomia who developed in adult life progressive impotence, urinary incontinence, postural hypotension, and cerebellar ataxia. His parents were first cousins and, he, in turn, married a first cousin. The clinical picture is that associated with Shy–Drager (type II), where the pathology (Bannister and Oppenheimer 1972) is that of a multi-system disorder.

For the parents of children with this recessively inherited disorder who want to know the prognosis and the extent of the serious manifestations, early mortality is high, but according to Axelrod *et al.* (1974) the Dysautonomic Foundation in the USA has records of 35 individuals with the disease who are alive at ages between 20 and 35 years. Pulmonary infection, cor pulmonale, extreme hyperpyrexia, dehydration, and vomiting crises are the main causes of death. Febrile seizures are common but chronic epilepsy less so. A reduction in IQ occurs in a significant proportion of children (Sak *et al.* 1967).

Axelrod *et al.* (1938) described a 'congenital sensory neuropathy with skeletal dysplasia'. All five cardinal signs of familial dysautonomia were present, i.e.

(a) There was no histamine flare.

(b) Fungiform papillae were absent.

(c) Miosis occurred after instillation of 2.5 per cent methacholine chloride.

(d) Deep tendon reflexes were reduced.

(e) Diminished tear flow.

Unlike the Riley–Day syndrome the myelinated fibres bore the brunt of the pathology. Those affected had proximal shortening of the limbs. X-rays showed flared metaphyses, short humeri, femurs and tibial bones, and thick cortices.

In HSN III it is the unmyelinated axons which are predominantly affected.

Counselling

As for autosomal recessive disorders.

Type IV congenital insensitivity to pain with anhidrosis. Also called familial dysautonomia type II

Two male sibs—the offspring of unrelated normal parents—had a congenital syndrome consisting of a universal insensitivity to pain accompanied by an inability to sweat (Swanson 1963). Pain and temperature were affected but other sensory modalities were intact. Their intelligence quotients were 62 and 77. Tear secretion was normal. Self-mutilation is often a feature including tongue biting.

Two non-Jewish sibs with hypo- or anhidrosis, insensitivity to pain and temperature, normal lacrimation an absent pupillary response to metha-choline, unexplained fever, an abnormal Valsalva, and abnormal oeso-phygeal motility were described by Pinsky and DiGeorge (1966). The children were hyperactive and repeatedly self-mutilating. They could feel a tickle sensation, but did not respond to pain or temperature. Reflexes were absent. Wolfe and Henkin (1970) re-examined these sibs. An inability to taste, unaltered by parenteral methacholine, with normal lingual and palatal papillae were additional finding.

These authors stress the points of difference between type I and type II dysautonomia. In type II there is no response to parenteral methacholine, in that tear secretion, the triple response, and miosis do not occur. In type I there is a positive response to methacholine.

Two further sibs were reported by Vassela *et al.* (1968). The sibs were products of a consanguineous marriage—the father had married his niece.

Itoh *et al.* (1986) reported a child who was the offspring of consanguineous parents. Peripheral nerves seemed to be normal on light microscopy, but electron microscopic studies showed extreme paucity of unmyelinated fibres and to a lesser extent of small diameter myelinated fibres. After many fractures the child died at 17 years from recurrent osteomyelitis.

Similar pathology was found by Matsuo *et al.* (1981), i.e. a marked decrease in unmyelinated fibres and to a lesser extent small myelinated fibres.

Counselling

As for autosomal recessive disorders.

Type V congenital insensitivity to pain or congenital pansensory neuropathy without anhidrosis

There is now doubt whether the condition previously called 'insensitivity to pain' is a separate entity (Donaghy *et al.* 1987). Most patients would, using modern day techniques be shown to have a sensory neuropathy. Retention of reflexes is the rule. Even the case of Thrush (1973*a, b*) might have had a small fibre neuropathy and it is now considered that morphometry is essential in view of the fact that only a histogram might show small myelinated fibre loss.

In a paper entitled 'Not indifference to pain but varieties of hereditary sensory and autonomic neuropathy' Dyck *et al.* (1983) dispute the existence of this as a separate entity. They include also those cases previously described as having congenital insensitivity to pain. Three children are described (consanguinity was suspected, but not confirmed in one) with initially mild developmental delay but in whom intelligence was probably normal. They showed no response to pain and often bit their tongue and lips. Reflexes were normal and whereas they seemed to feel touch, pain was not felt. One did not sweat, another did intermittently, but probably inappropriately, and a third had a neurogenic arthropathy and a pigmentary retinopathy. One had corneal scars. All had normal motor and sensory nerve conduction and the sural nerve showed small and large myelinated and unmyelinated fibres to be present. Only on more detailed analysis was an absence of small myelinated fibres—selectively involving the A delta afferent fibres, detected. This could not be detected on routine examination.

Despite similar pathology in all three the authors thought that the designation HSN V was appropriate to the one child with the congenital insensitivity to pain alone. Congenital insensitivity to pain is characterized by defective pain sensation distributed over the whole of the body surface. The other sensory modalities are normal or only minimally impaired and tendon reflexes are preserved (Thrush 1973b). Intelligence of those affected is dull/normal, and unlike type III, complexion is often fair. The sensations of temperature and touch are normal. The histamine skin test produces an appropriate response.

There are plentiful normal unmyelinated fibres, which distinguishes the group from congenital insensitivity with anhidrosis (Type IV) and the Riley–Day (type III) syndrome. Inheritance is as a recessive. A family described by Fanconi and Ferrazzini (1957) had two affected children who were the products of a first-cousin marriage, and consanguinity was reported by Ogden *et al.* (1959), Ramos and Schmidt (1964), Gwinn *et al.* (1966), and Durand and Belotti (1957). Sibs were also reported by Silverman and Gilden (1959), Chandra (1963), Khan and Peterkin (1970), and Becak *et al.* (1964).

In a sibship of seven, three boys and one girl were affected (Thrush 1973a). In this latter sibship self-mutilation and possibly a neurogenic bladder were added manifestations. Some patients are difficult to classify.

A girl with distal impairment of pain and temperature and a sural nerve biopsy which showed a selective loss of the small myelinated fibres is reported by Low *et al.* (1978). Pathologically and clinically this is unlike the congenital sensory neuropathies in which there is seldom selective pain or temperature loss and congenital insensitivity to pain was unlikely as sensation over the trunk was normal.

Ervin and Sternbach (1960) noted a family with six affected in two generations. Neither headache nor labour pain were felt and the author thought the lesion was at the thalamic level.

Another dominant pedigree is reported by Comings and Amromin (1974), but the classification of that family is not certain and there is doubt about the interpretation of the histology (Donaghy *et al.* 1987).

Counselling

It is likely that this group is recessively inherited and single cases should be counselled accordingly.

Congenital asymbolia

This implies a congenital indifference to all sensory input including sound and sometimes vision. Two Yoruba sibs and their half-sister (same mother) were said to turn to noise, but not to understand spoken words (Osuntokun *et al.* 1968). They knew the difference between sharp and blunt, but were indifferent to pain. Tendon reflexes were normal, as was touch. Temperature was impossible to test. This too might not be a separate condition and further reports with sural nerve histology are needed.

HSN with keratitis

Three members of a Kashmiri family had a congenital insensitivity to pain (Donaghy *et al.* 1987). They had a selective loss of pain and temperature, maximum distally, and those affected presented with either a mutilating arthropathy or facial/tongue injury. They had widespread anhidrosis and corneal opacification. Sural nerve biopsy showed a selective reduction in small myelinated fibres.

Histologically, the cases resembled those of Low *et al.* (1978) and Dyck *et al.* (1983; see type V), but are clinically different.

Four Navajo Indian children (two sibs in one family) had a mutilating neuropathy with severe anaesthesia for all sensation (Appenzeller *et al.* 1976). All had delay in walking, and had corneal ulceration, muscle weakness, atrophy, and absent reflexes. The patients superficially resembled type II sensory neuropathy, but the severe motor disorder, corneal ulceration, high c.s.f. protein make this designation unlikely. As in type II, there was selective involvement of the myelinated fibres.

Donaghy *et al.* (1987) discuss this case report and note the similarities to their cases. However, they suggest that it differs in that the keratitis was thought to be just traumatic and as stated above the nerve biopsy was similar to type II, i.e. there was not a selective dimunition of the small myelinated fibres.

Inheritance is as an autosomal recessive.

Other unclassified motor and sensory neuropathies

HSN—similar to Riley–Day.

Nordborg *et al.* (1981) reported 3 single cases of a neuropathy characterised by hypotonia, apnoea, poor swallowing and delayed motor milestones. No reflexes could be obtained. There was insensitivity to pain and temperature and absent deep sensation, corneal reflexes and taste. All had poor temperature control and none sweated. They cried tears but the volume was reduced. Other problems included deafness and poor co-ordination. They had normal IQ's. A sural nerve biopsy showed total lack of myelinated fibres and a histamine skin test produced no flare. The authors noted the similarities to Riley–Day syndrome, but thought this was a separate condition.

Hereditary sensory neuropathy with ataxia, deafness, and scoliosis

A dominant pedigree (with onset of the disease during the first or second decades) had affected members over four generations. The authors (Robinson *et al.* 1977*b*) consider this to be a new variety of hereditary sensory neuropathy. Touch and pressure were affected to a greater degree than pain and temperature. The diminished touch and pressure is clinically unlike type I, and the inheritance is unlike type II.

Sensory neuropathy, gastro-intestinal abnormalities, and deafness

Hirschowitz *et al.* (1972) described three sisters with progressive nerve deafness and a sensory neuropathy. Two of the sisters had multiple diverticulae and jejunoileal ulceration.

Two sisters are reported by Potasman *et al.* (1985). One was well until her early twenties, but then began to lose weight associated with vomiting and nausea. Neurologically she had slight dysarthria, bilateral ptosis, complete external ophthalmoplegia, deafness, and absent deep tendon reflexes. On neurophysiological examination she had a neuropathy with moderately reduced conduction velocities. Her sister had an age of onset at 13 and died at 20 years. The parents were first cousins.

Fabry disease

The deficient enzyme in angiokeratoma corporis diffusum (Anderson–Fabry disease) has been identified as alpha-galactosidase. The deficiency is not absolute, since 10–20 per cent of normal activity remains. Most children experience excruciating pain of a causalgic nature. The other clinical features are the angiokeratomous skin lesions on the trunk and scrotum, corneal opacities, fever, oedema, proteinuria, and cardiovascular disease. Obligate female carriers frequently manifest corneal opacities, skin lesions, and inter-mittent pain. Peripheral oedema, vascular lesions of the conjunctiva, and signs of renal involvement have all been reported (Desnick *et al.* 1978).

All reported cases have been males—see pedigrees of Kocen and Thomas (1970) and Pierides *et al.* (1976). Inheritance is X-linked recessive.

A peripheral neuropathy involving small myelinated and unmyelinated fibres occur as the disease progresses (Cable 1982).

Hereditary aspects of accessory deep peroneal nerve

Anomalous innervation of extensor digitorum brevis by the accessory deep peroneal nerve occurred in 22 per cent of 100 healthy unrelated individuals (Crutchfield and Gutmann 1973). In five subjects the anomaly was dominantly inherited.

Tangier disease

Also called hereditary high-density lipoprotein deficiency, this condition was described by Fredrickson *et al.* in 1961. Since then Kocen *et al.* (1967, 1973), Engel *et al.* (1967), and Haas *et al.* (1974) have all described patients who presented with neurological abnormalities in addition to splenomegaly, hepatomegaly, lymphadenopathy, and corneal ulcerations.

Kocen's *et al.* (1973) case was unusual in that this patient presented with disassociated pain and temperature loss and focal motor weakness. The patient of Haas *et al.* (1974) had an onset at 22 years of progressive weakness and wasting of hands and shoulders. At 35 years there was weakness of eye closure, wasting and weakness of the upper limbs, and impaired pain, touch, and temperature sensations. At 47 years, stabbing pains, developed in the hands. Inheritance is as a recessive. Sibs are reported by Assmann *et al.* (1977) and in this family many heterozygotes were identified over three generations. Other pedigrees are those of Fredrickson *et al.* (1961) and Hoffman and Fredrickson (1965).

Pollock *et al.* (1983) described four patients of whom three had a relapsing and remitting multiple mononeuropathy with de- and re-myelination. Two of their patients were brothers.

These authors suggest that there are two clinical forms of Tangier disease.

(a) A symmetrical neuropathy with pseudo-syringomyelic features, i.e. proximal dissociated sensory loss, and weakness and wasting of proxi-mal upper limb muscles.

(b) A mononeuropathy-multiplex. They also emphasize that normal serum cholesterol levels do not exclude the diagnosis, and that both high density lipoproteins and serum cholesterol should be measured.

Amyloidosis

Type I (dominant Portuguese type)

This dominantly inherited familial polyneuropathy was first described by Andrade in 1952. The skin, muscle, and digestive system, as well as the peripheral nerves, are affected and, although the original description was in Portuguese families, emigration has resulted in the appearance of this neuropathy elsewhere (Dyck and Lambert 1969).

The clinical picture (Andrade 1963; Thomas 1975) is that of a slowly progressive neuropathy initially affecting the lower limbs and later the hands (with an age of onset which varies considerably but is mostly in the third and fourth decades). Perception of pain and temperature is lost early and spontaneous pain might occur. In a review of the subject Thomas (1975) mentions the possibility that the disease in all the Portuguese families could stem from a single mutant gene. Andrade *et al.* (1969), in a study of 148 families, found that many came from a fishing town in the Oporto district of Portugal and cases of Portuguese origin have been reported in the USA (Munsat and Poussaint 1962), Brazil (Juliao *et al.* 1974), and elsewhere.

Cases from Japan in which no Portuguese ancestry could be traced are reported by Araki *et al.* (1968), Nakao (1966), and Izawa *et al.* (1969), but as pointed out by Thomas (1975) some of these patients came from an area of Japan where the Portuguese had a trading station in the sixteenth century. Many families are reported from Sweden by Andersson and Hofer (1974). Other dominant pedigrees are those of Coimbra and Andrade (1971), and Carvalho *et al.* (1976).

A genealogical survey of familial amyloid polyneuropathy in the Arao district of Japan was undertaken by Sakoda *et al.* (1983). There were 92 patients in nine families, the inheritance pattern being autosomal dominant with 83 per cent penetrance. An Irish family from the north-west of the country was reported by Staunton *et al.* (1987). There were six other single cases in the report all from the same region.

A single amino acid substitute of methionine for valine at position 30 in the prealbumin molecule has been shown in the Japanese (Tawara *et al.* 1983) and Swedish families (Dwulet and Benson 1984).

This gene has been mapped to the long arm of chromosome 18 (Sparkes *et al.* 1987)

There have been other families in which the neuropathy is associated with disease elsewhere—kidney, eye, heart, and bowel. One such Swedish family

(Benson and Cohen 1977) had members over seven generations with a progressive motor/sensory neuropathy beginning in the lower extremities with subsequent involvement of the kidney, heart, gastro-intestinal tract, skin, and the eye. A similar family was reported by Andersson (1970).

Recessive type I

Three English brothers with a late-onset (50–70 years) neuropathy were thought by the authors to belong to this type (Zalin *et al*. 1974). Parents were unaffected, and died at 78 and 84 years. Dissociated sensory loss was not a feature.

Type II (Rukavina or Indiana type)

This was first described in a group of Swiss families living in Indiana (Rukavina *et al*. 1956) and later by Mahloudji *et al*. (1969) in a number of unrelated families living in Maryland. Involvement of the hands (including a carpal tunnel syndrome) occurs at an early stage, distinguishing this group from the type I families. Onset is usually in the fifth decade and progression is slow. Pain, parasthesia, and numbness are common early complaints (Mahloudji *et al*. 1969). This is followed by weakness and wasting, and the picture develops of a mixed motor and sensory polyneuropathy. Signs and symptoms spread to the legs from between 5 and 20 years after the onset.

Inheritance is dominant with males and females equally affected, but males seem to be more severely affected.

The amino acid substitution in this condition is serine for isoleucine at position 84 (Dwulet and Benson 1986).

Type IV (Van Allen or Iowa type)

Van Allen *et al*. (1969) described a family of British origin with onset in the third or fourth decade of a symmetrical mixed sensory and motor polyneuropathy. The kidneys were involved and renal failure was a common cause of death.

Type V (cranial neuropathy)

Meretoja (1969), and Meretoja and Teppo (1971) published clinical and histopathological data on three families with hereditary amyloidosis beginning with a lattice dystrophy of the cornea in the third decade and cranial nerve palsies in the fifth decade. A mild peripheral neuropathy might occur. A genetic analysis showed autosomal dominant transmission (Meretoja 1973). In 46 families the disease occurred in two consecutive generations. No skips were observed. As many as 16 families came from the same parish and the author feels that all the cases are the consequence of a single mutation that occurred at least 500 years ago. However, Boysen *et al.* (1979) have reported a Danish family with no Finnish connection. Two American cases have been reported (Darras *et al.* 1986).

Type VI (hereditary cerebral haemorrhage with amyloidosis

A family is reported in which members are known to have a high incidence of cerebral haemorrhage (Gudmundsson *et al.* 1972). Eighteen members were affected in three generations and some of them at an early age. A post-mortem in five cases substantiated the cerebral haemorrhages and microscopy revealed thickening of the arterial walls with amyloid. Inheritance is as a dominant with high penetrance.

However, only a small number of elderly persons who die of cerebral haemorrhage or who develop dementia late in life will have a genetically determined condition. The mean age of onset in the Cosgrove *et al.* (1985) series was 73 years. All were single cases.

Furthermore, the relationship of this condition to 'angiopathie dyshorique' or to congophilic angiopathy (see p. 129) is uncertain.

Type VII Familial amyloidosis with vitreous opacities (Oculoleptomeningeal type)

A Japanese family with many members affected with vitreous opacities and extensive meningovascular deposition of amyloid resulting in cerebral infarction is reported by Ogata *et al.* (1978). The vitreous opacities were deposits of amyloid.

Another large family (dominantly inherited) is reported by Goren *et al.* (1980). This was a German family living in Ohio (USA). They developed dementia, seizures, strokes, coma, and visual deterioration due to vitrious opacities. The age of onset was between 20–57 years. At post-mortem there was diffuse amyloidosis of the leptomeninges and subarachnoid vessels associated with patchy fibrosis and obliteration of the subarachnoid spaces.

A Jewish family was described by Gafni *et al.* (1985). Vitreous opacities developed in a father and his son in the third decade of life. This was followed by severe autonomic dysfunction and a progressive peripheral neuropathy in the lower limbs. Analysis of the amyloid showed that glycine replaced threonine at position 49. The father died from a bleeding gastric erosion at 36 and amyloid was present in the blood vessels of every organ examined.

X-linked HSN

Five members in a family had this mode of transmission (Jestico *et al.* 1985). Four developed deformities of the feet within the first 10 years (2 within the first 3 years). The sensory changes were restricted to the feet and ulcerations developed. Sural nerve action potentials were reduced or absent and a biopsy showed loss of myelinated fibres especially those of small diameter. The changes were similar to HSN I, but the inheritance pattern suggested a separate disorder.

Hereditary motor and sensory neuropathy (HMSN)

Several subdivisions have been proposed (Dyck 1975).

(I)   Charcot–Marie–Tooth disease. Dominantly inherited hypertrophic neuropathy, peroneal muscular atrophy, type I hereditary motor, and sensory neuropathy including Roussy-Lévy syndrome (reduced motor conduction velocity).

(II)   Dominantly inherited neuronal or axonal variety of Charcot–Marie–Tooth disease or type II hereditary motor sensory neuropathy (normal or near normal motor conduction velocities).

(III) Dejerine and Sottas disease.

(IV)   Refsum disease.

(V)   HMSN associated with spastic paraplegia.

(VI)   HMSN associated with optic atrophy.

(VII) HMSN associated with retinitis pigmentosa.

Distal progressive spinal muscular atrophy is classified with the anterior horn cell diseases.

Especially difficult to classify are those disorders which have, in addition to the sensory/motor neuropathy, signs affecting other systems. This has led to the suggestion that the spinocerebellar degenerations, spastic, paraplegia, Friedreich's ataxia, Charcot–Marie–Tooth, and amyotrophic lateral sclerosis might all be seen as a continuous disorder with similar aetiology. In most instances, it is possible to differentiate between these conditions.

As can be seen from the above classification, Charcot–Marie–Tooth disease has been subdivided on the basis of both the pathology and nerve conduction studies into types I and II. A segmental demyelination and proliferation of Schwann cells—so-called hypertrophic neuropathy—is typical of type I, whereas an axonal neuropathy occurs in type II. The motor conduction velocity is reduced in type I to less than 38 m/s, but not in type II. This division is not accepted by all. Bradley *et al.* (1977) suggest the following categories: hypertrophic neuropathy with median nerve conduction velocity of less than 25 m/s, an intermediate category with velocities of between 24 and 45 m/s and then the axonal type with a velocity above 45 m/s.

Other mixed neuropathies

HMSN   X-linked
HMSN   with deafness and mental retardation
HMSN   with optic atrophy and deafness
HMSN   with deafness
HMSN   with basal ganglia signs
HMSN   with cerebellar ataxia
HMSN   with joint contractures
HMSN   with heart block
HMSN   scapulo peroneal distribution (Davidenkow)
HMSN   with continuous motor activity (stiff man/neurogenic)
HMSN   with paraproteinaemia
HMSN   with chronic inflammatory changes
HMSN   with hyperkeratosis

HMSN type I

Peroneal muscular atrophy (dominantly inherited hypertrophic neuropathy)

This condition was well documented by Charcot and Marie (1886) and independently by Tooth (1886). Clinically, there is an onset in childhood or adolescence of weakness and wasting of the lower limb muscles resulting in foot drop. *Pes cavus*, clawing of the toes, and an equinovarus deformity are early manifestations. The atrophic process spreads proximally to involve the calf muscles but usually stops at mid-thigh level to give the characteristic inverted champagne bottle appearance. The hands are affected some years after the onset in the legs when clawing, similar to that which occurs, in the toes, involves the fingers. Sensory loss is nearly always found, but there is considerable variation in the extent of the loss, even within families. Thomas *et al.* (1974) quote two families in which the sensory change was so severe that the diagnosis of lumbosacral syringomyelia had been made previously in one member. In the one family, the severe sensory loss seemed to breed true. In the other family, sensory changes in the younger members were mild, but in a previous generation a member sustained severe injuries without intense appreciation of pain.

   In many pedigrees, the clinical picture is similar in all members, but it is the intrafamilial variability in some pedigrees that has made the nosology difficult. Thomas *et al.* (1974) suggest that the variation within families of probands selected for having Charcot–Marie–Tooth disease is so great as to contain cases indistinguishable from the Roussy–Lévy syndrome. They suggest that the term Roussy–Lévy is dropped, and that the entity be included in hereditary motor and sensory polyneuropathy, type I. The clinical differentiation between type I and type II can be difficult, although Thomas *et al.* (1974) suggest that ataxia, foot deformity, and sensory loss, are more pronounced in type I.

Ataxia, tremor, and scoliosis are accepted as part of the syndrome. Dyck and Lambert (1968*a,b*) recorded the presence of ataxia in only the severe examples of Charcot–Marie–Tooth disease. In the study by Thomas *et al.* (1974) ataxia occurred in 11 out of the 17 families and, unlike the Dyck and Lambert study, there was not a good correlation between the severity of the motor and sensory changes and the ataxia. However, the ataxia was only severe in members of three families.

It was suggested by Dyck and Lambert (1968*a,b*) that reduced conduction velocity known to occur in some families with CMT ran true within families and this is the main criterion for differentiating type I (slow conduction) from type II (normal conduction). This was supported by Thomas *et al.* (1974), but Davis *et al.* (1978) did not find the bimodal distribution. They found, as did Humberstone (1972) and Salisachs (1974, 1975) a spectrum of conduction velocities from very low to normal with a unimodal distribution. Salisachs *et al.* (1979) also found significant differences within families. Davis *et al.* (1978) suggested that families could be divided into three groups, normal, slow, and intermediate on the basis of conduction velocity. A study by Harding and Thomas (1980*b*) undertaken to clarify the position, again suggested that the classification of type I from type II on the basis of conduction velocity is valid and electrical studies remain a useful criterion for subdivision.

The gene for hereditary motor and sensory neuropathy, type I, has now been localized to chromosome I where it is linked to the locus for Duffy blood group (Guiloff *et al.* 1982), but this has not been found in all families. Bird *et al.* (1983) suggested that HMSN Type I might be divided into two types.

Ia—Intermediate nerve conduction velocities. Not linked to Duffy on chromosomal I. There are less prominent onion bulbs on nerve biopsy.

Ib—Linked to Duffy on chromosome I. There are very slow conduction velocities and more onion bulbs on sural nerve biopsy.

This subdivision was not confirmed by the large family reported by Leblhuber *et al.* (1986) which showed slow conductive velocity, but no linkage to Duffy. The matter is not yet resolved. Chance *et al.* (1987) again showed some evidence of linkage of Ib to Duffy.

Important for the geneticist who will often be asked by young patients about the eventual prognosis, is that those who are affected remain ambulant for many decades (especially type II) and the disease is compatible with a normal life span.

Roussy-Lévy syndrome or Charcot-Marie-Tooth disease with essential tremor (see type I)

Much controversy surrounds the existence of this syndrome as a separate entity. It has been accepted as such by Symonds and Shaw (1926), Gordon (1933), Spillane (1940), Yudell *et al.* (1965), and Oelschlager *et al.* (1971). Alternatively, it has been regarded as a forme fruste of Friedreich's ataxia or included as part of Charcot-Marie-Tooth disease. Because they found a clinical picture indistinguishable from the Roussy-Lévy syndrome in first-degree relatives of patients with Charcot-Marie-Tooth disease. Thomas *et al.* (1974) included both diseases under the term hereditary motor and sensory neuropathy. Before Roussy and Levy's (1926) paper, Raymond (1900-01), Marie (1906), and Boveri (1910) had drawn attention to the association between peroneal muscular atrophy and essential tremor (from Salisachs 1976). Boverie, in his paper, noted at post-mortem the presence of a hypertrophic neuropathy with plentiful 'onion bulbs'. Four out of the seven original patients reported by Roussy and Lévy (1926) had an essential tremor and at least two members of that family had a clinical picture indistinguishable from Charcot-Marie-Tooth disease (Salisachs 1976).

The alternative viewpoint is that there are clinical differences between Roussy-Lévy syndrome and Charcot-Marie-Tooth disease in that the former has a slightly earlier age of onset, slower progression, and has a tremor with all the features of a benign essential tremor. It is suggested by Oelschlager *et al.* (1971) that the combination of weakness of the small hand muscles plus the tremor is responsible for the clumsiness of the hands, so often reported in Roussy-Lévy syndrome. Spillane (1940) described 20 members of a single family with an onset in early childhood of difficulty in walking. However, so great was the intrafamilial variability that Spillane believed that 16 of the patients had Roussy-Lévy, three had Charcot-Marie-Tooth disease, and others showed features of Friedreich's ataxia. Other families in which some members had what looked clinically to be Roussy-Lévy syndrome and others Charcot-Marie-Tooth disease are those of Salisachs (1976) and Hierons (1956).

The third alternative, that essential tremor and the Roussy-Lévy syndrome are separate dominant conditions occurring together, is unlikely, for they occur too frequently within families to suggest separate loci with chance occurrence. The least satisfactory designation is that Roussy-Lévy syndrome is a 'forme fruste' of Friedreich's ataxia. Lapresle (1956) found that the original cases described by Roussy and Lévy eventually developed a clinical picture intermediate between Friedreich's ataxia and Charcot- Marie-Tooth disease. The gait deteriorated, the hand muscles become atrophic, kyphoscoliosis developed, and all four had extensor plantar reflexes. In 1973, Lapresle and Salisachs looked both clinically and pathologically (nerve biopsy) at one of Roussy-Lévy's original patients 29 years after she was first seen. When originally seen she had *pes cavus*, bilateral club feet and a normal gait. No atrophy of the hands nor in the lower half of her legs was detected—but reflexes were absent. Plantar responses were flexor and sensation was normal. At the age of 61, the deformity of the feet was more marked. She walked on a slightly wide base and there was wasting of the distal third of her legs. The plantar reflexes were now extensor. Vibration and position sense were impaired in both lower limbs and a kyphoscoliosis was noted. At the age of 79 she was seen again. All the signs were more marked and she had a tremor which had the characteristics of a benign essential tremor.

The majority of cases of Roussy-Lévy syndrome have reduced motor conduction velocities compatible with that found in type I. Those patients with normal velocities who clinically have Charcot–Marie–Tooth disease and an essential tremor (Salisachs *et al.* 1979) should be classified under type II HMSN.

In the family studied by Delwaide and Schoenen (1976), the age of onset was in middle life and the clinical picture that of tendon areflexia, intension tremor, and impairment of deep sensation. In some of the affected members, *pes cavus* and amyotrophy were present. Only in the amyotrophic patients was motor nerve conduction moderately slowed and nerve biopsy showed no hypertrophic changes.

Although the first case description (Marie 1906) and Boverie (1910) reported sibs only to be affected, inheritance is now established to be as a dominant.

Summary of the current position regarding Roussy-Lévy syndrome

The evidence is strongly suggestive that the Roussy-Lévy syndrome is but a part of the expression of type I HMSN and the genetic counsellor should assess the risks on that basis.

Ataxia plus peripheral nerve lesions

The families of Bing (1905), Van Bogaert and Moreau (1939), Monaco (1964), Hopf and Port (1968) have been thought to belong to a separate group.

The Hopf and Port (1968) cases were sibs and those of Van Bogaert and Moreau a family in which three cases occurred in two generations. In 1948 Van Bogaert re-examined a member of this family and found the clinical picture to be that of classical peroneal muscular atrophy. It is possible that many of the families will ultimately be found to belong the groups of HMSN type I, II, or III.

Although called by Hopf and Volles (1972) 'spinocerebellar ataxia with neural myatrophy' the early-onset severe kyphoscoliosis, distal atrophy, areflexia, raised c.f.s. protein, and reduced motor conduction velocities with hypertrophic changes on biopsy are suggestive of Dejerine and Sottas disease.

HMSN type II

The patients with type II in the Davis *et al.* (1978) and in the Bradley *et al.* (1977) studies tended to be less severely affected and to progress more slowly. Buchthal and Behse (1977) and Behse and Buchthal (1977) found that more than two-thirds of their type I patients had an onset before 10 years and that two-thirds of type II began symptoms after the age of 10 years.

A more recent study (Brust *et al.* 1978) could not differentiate clinically between the two groups. They agree that it is even possible that there is on neurophysiological grounds a third intermediate group, as suggested by Davis *et al.* (1978).

Sensory changes are less frequently found in type II (60 per cent of Thomas *et al.*'s (1974) patients had a normal peripheral sensory examination). When these changes are not found, it is often only the absence of the sensory action potentials that differentiates this group from the distal spinal muscular atrophies (see p. 264).

Genetic counselling in type I and II motor and sensory neuropathy

Because of the variable age of onset in Charcot–Marie–Tooth disease of all types, genetic counselling can be difficult. The apparently healthy offspring of an affected parent might, at the age of 20 years, want to marry and have children. The following guidelines are suggested.

(a) Dominant   This is by far the commonest mode of inheritance. Dominant pedigrees are ten times more common than recessive (Harding and Thomas 1980*b*). Analysing 109 individuals from 15 unrelated families, Bird, and Kraft (1978) found that using physical examination and motor nerve conduction the penetrance of the gene was 28 per cent in the first decade and was virtually complete by the middle of the third decade. They suggest that those over the age of 27 years who have no clinical manifestation whatsoever have a less than 3 per cent risk of having inherited Charcot–Marie–Tooth disease. (This applies especially to type I.) High arched feet or obvious *pes cavus* were taken as a clinical expression of the disorder. The usefulness of motor nerve conduction as one of the criteria for counselling families is not applicable to type II where conduction velocity is less severely affected. In type I there is evidence (P. K. Thomas, personal communication) that the slow conduction preceeds the clinical abnormality and might well be present in early childhood.

(b) Recessive Except for the 1935 paper by Bell, the occurrence of unequivocally recessive pedigrees has only rarely been reported. Dyck and Lambert (1968a, b) reported sibs with unaffected parents and Beighton (1971) reported an inbred family with the recessive type. More than one generation was affected, but the considerable inbreeding in this family suggested that dominant inheritance was less likely ('pseudodominance').

Out of six families with autosomal recessive inheritance, four were type I and two were type II (Harding and Thomas 1980c). The recessive types tended to show more weakness, ataxia, and scoliosis than the dominant form. Consanguinity was found in three out of the six families. The age of onset was not found to be useful in distinguishing the type I recessive from type I dominant (both around 9.5 years), but in type II the mean age of onset for the recessive groups was 12 years and for the dominant 25 years (with considerable variation). The most difficult differentiation is between type I recessive and type III (also recessive).

Three sibs with an early-onset mixed polyneuropathy were the offspring of first-cousin parents (Mahloudji 1969). The cerebrospinal fluid protein was raised. Progression was slow. Weakness of dorsiflexion of the feet and absent tendon jerks were the main findings. In the Rossi *et al.* (1983) report there were two sets of siblings with unaffected parents.

Genetic counselling in the absence of family history for HMSN types I and II

Segregation analysis suggests that approximately 70 per cent of single type I cases are autosomal recessive and 30 per cent are possibly new dominant mutations (Harding and Thomas 1980c). These authors suggest a recurrence risk to the offspring of an isolated type I patient to be 1 in 8. Risks to offspring of an isolated type II patient are 1 in 5. Harding and Thomas (1980c) data suggest a higher proportion of new dominant mutations in type II. Risks to sibs will be less frequently needed as the age of onset is such that the diagnosis will seldom be made in the first sib before the family is complete. However, 1 in 6 for type I and 1 in 16 for type II are suggested as empiric risks to sibs of single cases by Harding and Thomas (1980c).

The homozygous expression of the gene was reported by Killian and Kloepfer (1979). Two affected members married and had a child with an early onset of a severe neuropathy and skeletal deformity.

Risks of still developing the disease after the age of 30 years, in type II, are not yet known, but on present evidence—see age of onset data (Harding and Thomas 1980c)—the risks are small.

It should also be noted that in a study of 14 children with HMSN I Vanasse and Dubowitz (1981) found that five of eleven affected parents were asymptomatic and they recommend electrodiagnostic studies on both parents.

HMSN type III

DeJerine and Sottas

In 1893, these authors described two sibs with a neurological disorder with an onset in one in infancy and in the other at 14 years. As reported by Dyck and Lambert (1968*a,b*), the DeJerine and Sottas syndrome includes clubfoot, kyphoscoliosis, distal weakness and wasting initially in the legs and, later, in the arms, areflexia, marked sensory loss in all four limbs, incordination in the arms, and Romberg's sign.

DeJerine and Sottas has now been subdivided:

(A1) with hypomyelination and onion bulb formation;

(A2) without myelin formation at all (severe);

(B)   neuronal form; in this form the onset is congenital, but with conduction velocities that are normal or only slightly reduced.

(A1) Hypomyelination with onion bulbs

In Dyck and Lambert's (1968*a,b*) series of patients, there were two sibs with the DeJerine and Sottas type of hypertrophic neuropathy. Motor development was delayed. Weakness and wasting was more severe distally, but also involved proximal muscles, reflexes were lost, and all modalities of sensation were affected (especially light touch, vibration, and joint position). Cerebrospinal fluid protein was raised. Conduction velocities are markedly reduced. Dyck (1975) suggests that the fundamental defect is one or hypomyelination and demyelination. Reported cases which satisfy these criteria are rare.

Two sibs were reported by Kalyanaraman *et al.* (1970). The two children had an early onset, rapid progression and both were in a wheelchair by 10–11 years of age. The authors felt that these cases were different from those described by DeJerine and Sottas, in which total paralysis by the end of the first decade was unusual. The disease then appeared to become stationary. The cerebrospinal fluid was normal, and, although the nerves were not thickened clinically, a typical hypertrophic appearance was found on biopsy.

Fundamental for a diagnosis of type III is the onset of the symptoms at or soon after birth (despite the original description).

Inheritance is as a recessive. Sibs are reported by Dyck and Gomez (1968). The sibs (brother and sister) described by Weller (1967) are difficult to place. The proband had a gross kyphoscoliosis, weakness in all four limbs, an onset at two years, ataxia, and loss of vibration and joint-position sense. An unusual feature was bilateral seventh nerve palsies. A hypertrophic neuropathy was present on biopsy. A single case is reported by Kennedy *et al.* (1977).

Note. It is difficult in infants to test sensation; therefore, the clinical presentation is similar to acute spinal muscular atrophy. Two sibs are reported by Koto *et al.* (1978).

(A2) Severe amyelination

There are no myelin sheaths and no onion bulbs. Peripheral neuropathies have only rarely been reported as being the underlying pathology in arthrogryposis multiplex congenita but a report by Seitz *et al.* (1986) records this. Central myelination was normal, but cranial nerves and nerve roots showed no myelin although the axons were well preserved.

The patient reported by Hakamada *et al.* (1983) also had extreme hypotonia with contractures whereas the infants reported by Karch and Urich (1975) were floppy, but had no contractures. In the patient of Schröder and Bohl (1978) there was a similar myelin deficiency in the central nervous system. In the majority, this condition is clinically severe with extreme wasting, respiratory distress, and often death.

Another six cases were described by Lutschg *et al.* (1985). Onset was between birth and 12 months. Two patients (onset 3 months and birth) were brothers. All had weakness, hypotonia, absent reflexes, and most had atrophy. None had palpable nerves. Two had fasciculations of the tongue including one of the brothers (the other brother did not). Motor nerve conduction velocities were slow (markedly).

Severe but not complete hypomyelination

Ono *et al.* (1982) reported a floppy child who could not walk at 3 years. Speech was normal. Guzzetta *et al.* (1982) included an infant with severe hypotonia and swallowing difficulties. (Most survive in spite of marked but not progressive muscle wasting). Atypical onion bulbs consisting of double-layered basement membranes and a marked amount of endoneural interstitium were found.

(B) Neuronal form

One of the six children with severe hypotonia and arexflexia reported by Lutschg *et al.* (1985) had only moderately slow motor conduction velocities. A biopsy showed complete lack of large myelinated fibres but no hypo- or demyelination. Similar cases are reported by Ouvrier *et al.* (1987). Differentiation from Type II is difficult, but the condition is more severe and has an earlier onset.

Note: Differentiation between HMSN Type I (AD) presenting in childhood and Type III is not always easy (Ouvrier *et al.* 1987). Some authors even doubt the existence of the DeJerine and Sottas syndrome (Hagberg and Westerberg 1983).

However, no case is on record of a type I (AD) parent having a severely affected child fitting into Type III. Even in the most severely affected childhood case of HMSN I the pathological findings are of I and not III. Whether HMSN I (AR) and III are genetically distinct is not certain.

Pointers of difference are the greater frequency of ataxia, areflexia, and clinical nerve enlargement in III compared with type I (AR).

HMSN type IV. Refsum's disease

In 1945, Refsum described a syndrome that he called heredoataxia hemeralopica polyneuritiformis. He described five cases from two families; all the children were the product of consanguineous unions (Quinlan and Martin 1970). Later, Refsum changed the name to heredopathia atactica polyneuritiformis. The clinical features are those of a chronic polyneuropathy, cerebellar ataxia, an atypical pigmentary retinopathy and an increased cerebrospinal fluid protein. Less frequently, deafness, anosmia, ichthyosis, cardiomyopathy, naevi, lens opacities, *pes cavus*, and ephiphyseal dysplasia have been noted. Onset is in the first or second decade.

The underlying defect was first shown by Klenke and Kahlke (1963) to be an accumulation of phytanic acid in the blood and tissue, and the nature of the defect relates to a failure of the alpha oxidation of phytanic acid.

The disease is rare (only 75 cases reported to 1979).

Inheritance is as a recessive. Sibs have been reported many times—see the family of Quinlan and Martin (1970) and Barolin *et al.* (1979). Most cases have occurred in Norway and it has been suggested that the disease was spread by the Vikings to England, France, and Ireland (Richterich—quoted by Barolin *et al.* 1979). The American family reported by Alexander (1966) was originally from Norway.

Refsum's syndrome with normal phytanic acid excretion

A single case (unrelated parents) with a typical clinical picture of Refsum's syndrome is reported by Ron and Pearce (1971b), but without the biochemical defect. The authors suggest genetic heterogeneity for the syndrome as a whole.

HMSN type V

Hereditary motor and sensory neuropathy with spastic paraplegia

The cases described by Gee (1889), Holmes (1905), Garland and Astley (1950), Dick and Stevenson (1953), and Silver (1966) are discussed under hereditary spastic paraplegia with amyotrophy.

The eight patients described by Dyck and Lambert (1968*a,b*) from two kindreds (dominant inheritance) probably fall into group V. The onset was in childhood and the cases differed from the more usual type of Charcot–Marie–Tooth disease in that knee jerks were brisk and the plantar reflexes in some were extensor. The pyramidal tract features bred true within families and unlike the hereditary spastic paraplegia with amyotrophy (see p. 283), sensory conduction was abnormal. Motor conduction was unaffected.

Twenty-five cases from 15 families were reported by Harding and Thomas (1984). The onset was within the first two decades of life. Reflexes in the upper limbs were normal, but often increased at the knees. The ankle jerks were usually absent. The plantar responses were extensor in 22 of the patients. The condition is slowly progressive, but did not usually lead to severe disability.

Note: Silver syndrome especially the Van Gent *et al.* (1985) pedigree could be this condition (see p. 283).

Type VI

Hereditary motor and sensory neuropathy or progressive muscular atrophy of peroneal type with optic atrophy

This is a rare category and no recent papers have reported this association. Two brothers with this combination are reported by Milhorat (1943). The parents were first cousins and the inheritance is recessive. The onset of the disease was in early childhood and the visual disturbance and peripheral weakness occurred together. The disease was slowly progressive. A sensory deficit for pain, touch, and vibration (and also position sense in the feet) was found. Nystagmus was present in both brothers. A sister was affected by a neurological disorder, which included optic atrophy, but she was thought to have multiple sclerosis. This case is often used to illustrated that optic atrophy can occur in families with Charcot–Marie–Tooth disease. However, it would seem from the occurrence of this disorder in sibs with unaffected first-cousin parents that the inheritance is recessive and unlike Charcot–Marie–Tooth disease which is usually inherited as a dominant.

Taylor (1912) reported a man with wasting below the knees, optic atrophy, and deafness. His sister was said to be similarly affected (it is uncertain whether the sister had optic atrophy; see also Iwashita *et al.* 1970; see below).

Other cases are reported by Ballet and Rose (1904), and Hoyt (1960), both single case. Two cases occurring in brothers are reported by Schneider and Abeles (1937). Both parents were unaffected, but were first cousins. These are the same cases subsequently reported by Milhorat in 1943. McLeod *et al.* (1978) described a pedigree with Leber's disease and Charcot–Marie–Tooth disease, but suggest that the two conditions were inherited separately.

In most of the families described above, inheritance is as a recessive.

HMSN VII

Hereditary motor and sensory neuropathy with retinitis pigmentosa

Retinitis pigmentosa plus distal muscle weakness and atrophy were described in sibs by Massion-Verniory *et al.* (1946). Several sibships were seen, but not described by Dyck (1975).

Four patients were described by Tuck and McLeod (1983). The onset in two was at 6 years and in two others (all unrelated) in their 40's. All had a pigmentary retinopathy, an ataxia with an intention tremor, and a mixed motor and sensory neuropathy. Three had mild deafness and one had a dry skin. Electrical studies revealed abnormalities of sensory conduction and normal or mild slowing of motor conduction. The biochemistry was normal. A sural nerve biopsy showed reduction in the density of myelinated fibres. This condition is similar to HMSN type VII, but the ataxia is more prominent and the atrophy less pronounced. The authors suggest that the family described by Furukawa *et al.* (1968) is similar, but there was no evidence in that report of demyelination. Inheritance is uncertain. The condition is probably heterogeneous.

X-linked Charcot–Marie–Tooth disease

The existence of an X-linked type of Charcot–Marie–Tooth disease was suggested by Bell (1935). In a review of Charcot–Marie–Tooth disease in Norway, Skre (1974*b*) could distinguish three genetically distinct types. The dominantly inherited groups had a frequency of 36 in 100 000, X-linked 3.1 per 100 000, and recessive group 1.4 per 100 000. However, in the series of more than 100 patients reported by Harding and Thomas (1980*d*), no X-linked pedigrees were found. In most series X-linked pedigrees are much rarer than 3.1 per 100 000 and this mode of inheritance should be diagnosed with caution.

Recently, there has been increased interest in X-linked conditions as laboratories using DNA probes are localizing many of the X-linked conditions using gene tracking. For this reason X-linked Charcot–Maric–Tooth disease has become disproportionly of interest. It should be noted that authors have used both X-linked dominant and X-linked recessive desgnations to describe the observed genetic phenomena depending on whether females are affected albeit mildly (X-linked dominant) or not at all (X-linked recessive). These are semantic differences and the two conditions are the same.

A large pedigree was described by Phillips *et al.* (1985) with no male to male transmission. Females were affected, but never as severely as males.

Conduction velocities were slow in the males but not consistently slow in females.

The original family reported by Allan in 1939, was recently reviewed by Rozear *et al.* (1987). Phenotypically it resembled the dominant form. Affected males showed slow nerve conduction velocities and on sural nerve biopsy, there was a demyelinating hypertrophic neuropathy. Some carrier females had mild manifestations with slow velocities. Linkage to X_q13 region has been confirmed.

An X-linked dominant form of CMT

This is now sited in the McKusick catalogue based on a pedigree described by Gordon in 1933. A five-generation kindred and family was initially reported by Woratz (1964) and re-reported in 1985 (Gal *et al.*) with a suggestive linkage to the proximal long arm of the X chromosome (see previous section).

For other reports see family of Sladky and Brown (1984) and the report by (Fischbeck *et al.* 1986).

Combination of motor and sensory neuropathy with disease in other parts of the nervous system

(a) Progressive muscular atrophy of peroneal type with optic atrophy and deafness

Rosenberg and Chutorian (1967) reported this combination in three males in two generations (two male sibs and a nephew). The nephew was only mildly affected.

In a letter to the editor, Pauli (1984) reports as a personal communication that in the original Rosenberg–Chutorian family, mother, grandmother, and great grandmother of the affected nephew had slowly progressive hearing loss.

Iwashita *et al.* (1970) reported a similar clinical syndrome in male and female sibs. Both parents were unaffected and unrelated. Sensory loss was present below the knees with moderate diminution of position sense and vibration, and slight loss of touch/pain and temperature. Rombergism was present and the gait was ataxic.

(b) Charcot–Marie–Tooth disease with cerebral atrophy and Parkinson's disease

See pp. 242 and 272.

(c) Charcot–Marie–Tooth disease with ataxia

Spillane (1940), Roth (1948), Shepherd (1955), Greenfield (1912), Plowright (1928), and Hierons (1956) all reported pedigrees of this type. As discussed above, ataxia is well known to occur in some families with Charcot–Marie–Tooth disease and in those families classified as having Roussy–Lévy syndrome, now type I HMSN. Most are dominant and the evidence suggests that they are at present best classified under Charcot–Marie–Tooth disease. For additional families see p. 206.

(d) Peroneal muscular atrophy plus arthrogryposis multiplex congenita

This is a rare occurence. When *in utero* neurogenic atrophy or hypokinesis leads to arthrogryposis present from birth the underlying pathology nearly always involves the anterior horn cells. Yuill and Lynch (1974) described a dominant pedigree with an onset at or soon after birth of foot deformities. Weakness and wasting of the feet, calves, peronei, quadriceps, and hamstring muscles were found. The plantar reflexes were extensor. Neither ataxia nor tremor were features and there was little evidence of progression. Reflexes were depressed or absent in the lower limbs and arthrogryposis multiplex congenita was present (not in the arms). An electromyogram was only performed on one patient—the father—and this suggested a neuropathic process. See also p. 233.

(e) Charcot–Marie–Tooth disease with heart block

A family in which a motor and sensory peripheral neuropathy was segregating is described by Lascelles *et al.* (1970). The proband had a cardiomyopathy compatible with that found in Friedreich's ataxia.

A three-generation family with three members affected by both the cardiac defect and peroneal muscular atrophy had six members with the cardiac defect alone and one with only the neurological disorder. The relationship of the cardiac to the neurological manifestation is uncertain (Littler 1970). A single case is reported by Leak (1961).

A different entity in which sibs had a childhood onset of a mixed motor and sensory neuropathy (glove and stocking peripheral sensory loss) with deafness and scoliosis is reported by Bouldin *et al.* (1980). The ECG showed evidence of a conduction defect in both sibs. The peripheral nerves were not thickened.

(f) Scapulo-peroneal syndrome (neurogenic type)

The huge pedigree described by Davidenkow (1939) is now thought to have the scapulo-peroneal syndrome on a neurogenic basis (see also p. 260).

A further dominant pedigree is that of Schwartz and Swash (1975).

The review of the subject by Serratrice *et al.* (1979) contains a family with four members affected over three generations. *Pes cavus*, areflexia, and distal diminution of sensation were features. Motor conduction was considerably reduced and the histology revealed a hypertrophic neuropathy.

An interesting family reported by Harding and Thomas (1980*a*) had members with type I HMSN and a single member with a clinical picture identical with Davidenkow's scapulo peroneal syndrome. It is suggested that Davidenkow's syndrome might not be genetically distinct from HMSNI.

A family (dominant inheritance) with a scapulo peroneal syndrome (neurogenic) with sensory signs is described by Sand and Hestnes (1985). The age of onset was between 15 and 35 years, and the sensory signs were in one, unilateral proximal hypesthesia and dysesthesia, whereas defective position sense with a sensory ataxia occured in another.

Counselling is as for dominant conditions.

(g) Charcot–Marie–Tooth disease and deafness

This combination without optic atrophy was reported in three brothers who were the offspring of consanguineous unaffected parents (Cornell *et al.* 1984). All had foot drop and the peripheral wasting included the hands. Slow velocities were found on nerve conduction studies. The inheritance pattern is probably autosomal recessive.

Following the report of Cornell *et al.* 1984, Pauli (1984) reported moderately severe sensorineural hearing loss, pigmentary retinal deterioration, and mixed motor and sensory neuropathy (conduction velocities not stated) in a three-generation family with male to male transmission. This could be the same as HMSN VII (see p. 238).

Motor and sensory neuropathy/deafness/mental retardation

An X-linked pedigree in which the neuropathy began in infancy is reported by Cowchock *et al.* (1985). The weakness and wasting was distal, and the hearing loss was sensorineural and not congenital. On further studies the nerve conduction velocity was slow and nerve biopsy showed a complete lack of onion bulb formation, but with evidence of re- and de-myelination.

(h) With facial and trigeminal involvement—see p. 23.

Charcot–Marie–Tooth disease with continuous motor unit activity

Patients are reported (Vasilescu *et al.* 1984) with type I neuropathy. They had fasciculation, muscle stiffness, myotonia, percussion myotonia, myokymia, cramps, and impaired relaxation of muscle groups. One patient had muscle hypertrophy. The EMG showed continuous firing of motor unit potential with variable periods of electrical silence. Charcot–Marie–Tooth can there-fore cause this clinical picture (see p. 113). The families of Lance *et al.* (1979) and that of Welch (1972) are similar. Percussion myotonia was, however, not present in the Lance family. Inheritance is dominant.

Peripheral neuropathy, high IgM, and paraproteinaemia

This is usually sporadic, but a mother and son have been reported (Busis 1988). The age of onset was in late/middle life and started with sensory symptoms in hands and feet. IgM was raised.

Chronic inflammatory demyelinating neuropathy

A chronic inflammatory demyelinating neuropathy was reported in sibs by Gabreels-Festen and Hageman (1986). The onset was about 1 year of age with the development of *pes cavus* and scoliosis. The sibs had a raised CSF protein and slow conduction velocities. A sural nerve biopsy showed re-myelination of the axon, 'onion bulbs' and an inflammatory infiltrate.

Peripheral neuropathy and palmoplantar hyperkeratosis

Three generations were affected with this combination (Tolmie *et al.* 1987). The nail dystrophy was present from birth or developed in early childhood whereas the palmoplantar keratosis became apparent in late childhood. Motor conduction velocities were only slightly reduced.

Peripheral neuropathy and basal ganglia disease

Byrne *et al.* (1982) have described a large family (10 affected over three generations) with the above combination. They suggest that the family of Biemond and Beck (1955), and Biemond and Sinnege (1955) have the same condition and that the lesion is not at anterior horn cell level.

The onset was between 40 and 50 years. The extrapyramidal signs were a coarse rest tremor and cogwheel rigidity. At least two patients had peripheral neuropathy (axonal rather than demyelination) and were therefore similar to type II hereditary motor and sensory neuropathy. Biemond and Sinnege (1955) described a family with six affected over three generations (dominant inheritance). The onset was between 20 and 25 years. A cerebellar ataxia was a feature in two.

Biemond and Beck (1955) described two brothers and a sister with onset of a neurological disorder in middle life (40–50). The proband had weakness of the extensors of her feet and toes on both sides, absent Achilles reflexes, and later an extensor plantar reflex. Her brother had distal atrophy, absent reflexes, bilateral Babinski reflexes, monotonous speech, a mask-like facies, and vibration loss in both feet. A post-mortem showed neural atrophy and degeneration of the substantia nigra. The third member had Parkinson's disease from the outset. Later, he developed areflexia in the legs and a right extensor plantar response. No atrophy or sensory changes were observed (see p. 272 for other comments).

It is uncertain how to place this family as these were sibs with no evidence of an affected parent.

Peroneal muscular atrophy, epilepsy, chorea

A family (with variable expression of chorea and *pes cavus*) is reported by Routsonis and Georgiadis (1984). Seven members were affected over three generations, five had epilepsy. The onset was in early childhood of a progressive peripheral neuropathy. In addition, one had a cerebellar ataxia, two had choreoathetosis, and three were retarded.

Counselling

The most likely explanation at present is that this combination of signs does not constitute a single entity. Both autosomal dominant and recessive patterns of inheritance have been observed.

Neuralgic amyotrophy

Acute brachial plexus neuropathy typically presents with pain in the arm, followed by a rapid paralysis of proximal muscles (Bradley *et al.* 1975). The distribution of weakness follows that of nerve root, or plexus and sensory signs are not prominent. Paralysis lasts for weeks or years and recovery is usually full. Parsonage and Turner (1948) have called the same disorder neuralgic amyotrophy, but the familial form of neuralgic amyotrophy is unlikely to be the same condition as brachial neuritis, although clinical differentiation can be difficult.

Arts *et al.* (1983*a*) suggested that there are two types.

(i) Classic type—pain and weakness lasting from weeks to months leaving atrophy and residual weakness.

(ii) Recurrent pain—weakness lasting a few days and recurring about once a week for months; also causing eventual atrophy and weakness. Both can, however, occur in a single family (see Arts *et al.* 1983*a*)

Other parts of the peripheral nervous system might be affected subclinically. Sural nerve biopsy shows a decreased number of myelinated fibres.

Genetically determined neuralgic amyotrophy

Inheritance is usually dominant although in 1886 Dreschfeld described a 43-year-old woman and her sister with recurrent pain, weakness, and atrophy of the proximal and distal arm muscles. When Ross and Bury (1893) subsequently re-examined the proband there was some suggestion that minor injuries provoked the episodes (see following section). The first manifestation is usually in childhood (unlike the sporadic case where the onset is usually after the age of 20 years).

Dunn *et al.* (1978) described a 28-year-old father, his 5-year-old son, and two first cousins, who were similarly affected. The pedigree showed dominant transmission with reduced penetrance, as did the pedigree of Guillozet and Mercer (1973). Inheritance in the majority of families is as a regular dominant (Ungley 1933; Poffenbarger 1968; Geiger *et al.* 1974) three families (Wiederholt 1973) two families.

Occasionally, lower cranial nerves are affected (Jacob *et al.* 1961). Vocal cord paralysis and swallowing difficulties occurred in some members of three separate families, each with dominant transmission.

In the family described by Gardner and Maloney (1968), the brachial neuropathy included a cranial neuropathy, and dysphagia and hoarseness were symptoms. The affected members had webbed toes, hypotelorism, and a small mouth. Nine persons were affected over four generations.

Geiger et al. (1974) suggest that the following differences exist between sporadic and familial cases. In the familial cases hypotelorism and syndactyly occur, but this is not found in the sporadic type. Hypotelorism has been noted in the families of Jacob *et al.* (1961), Gardiner and Maloney (1968), Guillozet and Mercer (1973), and Geiger *et al.* (1974). The onset is earlier in the familial cases (childhood or young adulthood) and symptoms can be precipitated by pregnancy (Jacob *et al.* 1961; Ungley 1933). Furthermore, the distribution in the familial type encompasses the whole shoulder girdle, whereas in sporadic cases isolated parts of the shoulder girdle are affected. The sensory loss is predominantly radial forearm and occasionally in the distribution of the axillary nerve in the familial type, whereas the axillary nerve is more commonly affected than the radial in the sporadic cases. Lesions outside the brachial plexus are most often found in the familial form.

Familial recurrent pressure palsy

This is probably different from the brachial plexus neuropathies as distal nerves, ulnar, and lateral popliteal, are more commonly affected. It is a motor and sensory neuropathy, acute and intermittent, and provoked by pressure. Pain is usually not a feature.

Davies (1954) described a father and his two children with a tendency to develop peripheral nerve palsies. The father had three episodes of a lateral popliteal nerve palsy—once after kneeling, and a radial nerve palsy after pushing a lawn mower. His son had a similar illness, as did his daughter, but not all the episodes could be positively related to pressure.

In the four families described by Earl *et al.* (1964), two of the pedigrees were clearly dominant, and in the two other families where sibs only were affected, the parents were not available for study.

Roos and Thygesen (1972) followed a huge Danish family and found 19 affected over five generations. The attacks of motor and/or sensory symptoms occurred between four and 24 years of age and the symptoms were either spontaneous or provoked by pressure upon nerves. Behse *et al.* (1972) described the same phenomenon in three parent–child pairs.

Lhermitte *et al.* (1973) described a father and son with recurrent neuropathies, mostly occurring after local minor compression, and a similar pedigree is reported by Dubi *et al.* (1979). Behse *et al.* (1972) and Madrid and Bradley (1975) emphasize the sausage-shaped thickening of the myelin sheath and suggested the term 'tomaculous neuropathy' for this condition. However, the same histological picture occurs in chronic motor polyneuropathy affecting the upper limbs (Madrid and Bradley 1975; see next section). In half of Madrid and Bradley's four cases there was a clear dominant pattern of inheritance.

A median nerve palsy in a mother aged 30 years after prolonged use of a screwdriver probably belongs to this group (Attal *et al.* 1975). Her two children were similarly affected. Other families with dominant inheritance were reported by Staal *et al.* (1965) and Cruz-Martinez *et al.* (1977).

Counselling

As for autosomal dominant inheritance.

Multiple excercise related mononeuropathy with abdominal colic

This condition differs from those in the previous section in that painful palsies which developed after strenuous work occured concurrently with episodes of abdominal colic (Trockel *et al.* 1983). Porphyria was excluded. Sausage-like swellings of the myelin sheath were demonstrated. Inheritance was autosomal dominant.

Hereditary motor peripheral neuropathy predominantly affecting the arms

Many affected members over four generations had an exclusively motor neuropathy (with slow motor conduction velocity; Lander *et al.* 1976).

Progression was slow and only the distal part of upper limbs was affected. Brisk reflexes in some patients suggested the possible involvement of upper motor neurons.

The relationship of the condition to Silver's (1966) family and what Dyck and Lambert (1968*a*,*b*) called 'familial spastic paraplegia with amyotrophy of the hands' is uncertain (see p. 283).

Multiple peripheral nerve entrapment

Three adult brothers with normal intelligence, but short stature, had multiple entrapments involving median and ulnar nerves at elbow and wrist (Karpati *et al.* 1974). It is suggested that these patients had an unusual variant of Hurler's syndrome.

Familial carpal tunnel syndrome

The familial occurrence of this entrapment neuropathy of the median nerve is rare. Danta (1975) reported four affected members over three generations. In two, symptoms began in the first decade of life and one patient had at operation a thickening of the transverse carpal ligaments. It is suggested that this thickening might be constitutional.

There have been other familial cases described. McArthur *et al.* (1969) reported a bilateral carpal tunnel syndrome in siblings and another sib pair had mucolipidosis type II as the primary reason for the compression (Starrenveld and Ashenhurst 1975). A similar sibship (Bundey *et al.* 1974) also had an underlying mucolipidosis.

Mochizuki *et al.* (1981) report, a huge family (eight affected) with autosomal dominant inheritance and, in another family, Fowler *et al.* (1986) report a mother and four children affected. Two of the affected children also had tarsal tunnel syndrome.

Familial meralgia paresthetica

Only a few families have been reported. The family of Malin (1979) with five affected over the generation was clearly dominant and Massey (1978) and Goldstein (1921) reported a similar family. A consecutive series of 158 cases (Bollinger 1961) found no familial cases.

Hypertrophic neuropathy with cataract

Three Hindu sisters developed a peripheral neuropathy of early onset (Gold and Hogenhuis 1968). The nerves were thickened.

Giant axonal neuropathy

A chronic slowly progressive polyneuropathy characterized pathologically by giant axons packed with neurofilament, de- and re-myelination has been reported in a number of children. Those affected have been noted to have kinky hair and some have signs (epilepsy and long tract signs) related to lesions in the central nervous system.

The histology of the hair is normal under light microscopy although Fois *et al.* (1985) gave found opposed longitudinal grooves on electron microscopy—so called pili canaliculis.

The onset of symptoms is between 5 and 8 years usually with weakness, wasting, hypotonia of the lower limbs a Babinski sign, and optic atrophy. Scoliosis and progressive intellectual deterioration occur. The diagnosis can be made on nerve biopsy where discreet masses of cytoplasmic filaments are seen in endoneurial fibroblasts, Schwann cells and perineural cells. Recent case reports are those of Ionasescu *et al.* (1983) and Takebe *et al.* (1981) who report sibs. The two sibs were sisters and were normal for the first 18 months. Serious deterioration occured at 5–7 years. Both had a mixture of central and peripheral nervous system involvement with a Babinski sign. One had optic atrophy. Although most cases have been single the condition is probably autosomal recessive.

Sibs have recently been reported by Maia *et al.* (1988).

13

Anterior horn cell disease

Spinal muscular atrophy

The subdivision of spinal muscular atrophy into separate genetic and clinical entities is still controversial. The criteria used are age of onset, age of death, rapidity of progression and different modes of inheritance. These differences do not adequately allow simple classification and biochemical criteria are not available. Only the acute type I spinal muscular atrophy (Werdnig–Hoffmann's disease) has emerged as a distinct entity. There are those authors who believe that the childhood and juvenile onset spinal muscular atrophies (excluding acute spinal muscular atrophy) are a single disease ('lumpers') and those who believe that a subdivision of the childhood group is possible ('splitters'). Emery (1971) proposed a classification to cover most of the described cases and until there is a biochemical means of differentiating one from the other, an amended subdivision as he proposes is useful for counselling purposes.

A. Proximal spinal muscular atrophy

Type I.   *Infantile* spinal muscular atrophy has an age of onset before 6 months, and death before 3 years.

Type II.   *Intermediate group*:
  (i) with an onset usually after 6 months and survival into adult life;
  (ii) early onset (before 6 months) and slow progression;
  (iii) with early onset (before 6 months) and total arrest.

Type III.   Juvenile—onset from 4 to 24 years, described by Kugelberg and Welander (1954, 1956).

Type IV.   *Adult* onset, autosomal recessive, autosomal dominant, X-linked.

Table 13.1 Summary of risks in chronic proximal spinal muscular atrophy (based on Pearn—see text)

Age of onset	Sibs	Offspring
1–3	1 in 5	1 in 50
3–5	1 in 10	1 in 10 to 1 in 20
5–15	<1 in 10	1 in 10
15–20	1 in 5	1 in 20

B. Distal spinal muscular atrophy

C. Anterior horn cell disease with particular distribution

(i) Scapuloperoneal
    (a) autosomal dominant
    (b) autosomal recessive
    (c) X-linked recessive

(ii) Quadriceps

(iii) Bulbar

(iv) Ophthalmoplegia

(v) Oculopharyngeal

(vi) Facio-scapulo-humeral

(vii) Segmental

(viii) Others

A. Type I—Werdnig–Hoffmann disease

The clinical signs of type I spinal muscular atrophy may be present at birth or soon after. The atrophy and weakness is proximal and more severe in the legs than the arms. Hypotonia is profound, and reflexes are absent. Weakness affects both bulbar and respiratory muscles and tongue fasciculations are common. Jerky, non-rhythmical movements of the head and limbs, so-called mini-polymyoclonus, are sometimes seen (Spiro 1970). The disease ends in death before the third year.

The disease frequency is between 1 in 5000 and 1 in 50 000.

The gene frequency is approximately 1 in 160 in Great Britain and the carrier frequency 1 in 80 (Pearn 1974).

Spinal muscular atrophy type I is possibly the second or third most common fatal recessively inherited disease of childhood (Pearn 1974). In the Karaite community in Israel, the disease frequency was 1 in 400 with a gene frequency of 1 in 20 (Fried and Mundell 1977).

Numerous studies have established that type I is recessively inherited. Where the disease begins before the age of six months, and often by 3½ months (95 per cent), is fatal before the third year, and has on examination normal nerve conduction velocities and cerebrospinal fluid protein, the diagnosis is established. The recurrence risk is 1 in 4. No antenatal diagnosis is possible. It is sometimes acceptable to parents who face high risks (risks greater than 1 in 10) to endure the consequences of being unlucky where the duration of the illness is short (less than two years), in contrast to their reluctance to accept a risk where a child is clinically incapacitated for many years. Thus, in acute spinal muscular atrophy where the disease is invariably fatal (mostly in the first two years), parents might be inclined to take the risk. There is evidence (Pearn *et al.* 1973; Pearn 1978*b*) that type I is a homogeneous condition. Parental consanguinity occurred in six of the 28 families in which more than one sib was affected (Zellweger and Hanhart 1972). Electromyographically, regular spontaneous motor units occur frequently in type I spinal muscular atrophy and not in other conditions involving anterior horn cells (Buchthal and Olsen 1970).

Unfortunately, prenatal diagnosis is not available.

Type I with contractures

The female infant in the report by Mitsumoto *et al.* (1982) had marked flexion contractures of fingers, elbows, and knees. She was severely hypotonic and could only move her eyelids. At post-mortem there was extensive neuronal loss and gliosis throughout the cord. The cranial nerves were involved.

Type I with sensory changes

Clinically detectible sensory changes do not occur in SMA, but extensive sensory changes have been described at post-mortem (Probst *et al.* 1981). They were found in the dorsal root ganglia and there was evidence of primary axonal damage in a sural nerve biopsy. One case, clinically, had pale optic discs.

Type II—Intermediate type (Fig. 13.1)

There is an intermediate type of proximal spinal muscular atrophy milder than Werdnig–Hoffmann disease, but more severe than Kugelberg–Welander disease. Type II is suggested when infants learn to sit unaided, but do not stand or walk. Progression is slow and survival into adulthood is not unusual. The age of onset is between three to 15 months, but the main differentiating feature from acute Werdnig–Hoffmann disease is survival beyond the age of three years (Guinter *et al.* 1977).

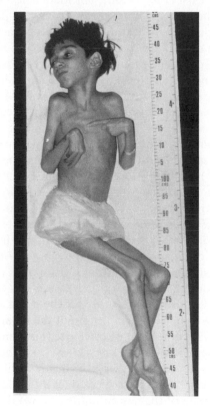

Fig. 13.1 Spinal muscular atrophy. Intermediate group: early onset, slow progression.
Courtesy of the Department of Medical Illustration, Great Ormond Street Hospital for Sick Children, London.

A subdivision of the group into three types is suggested by Lugaresi *et al.* (1966):

(a) early onset, but more chronic than acute spinal muscular atrophy;

(b) slow evolution with later onset;

(c) early onset with total arrest.

The occurrence of the early onset long surviving type of neurogenic muscular atrophy accounts for more than 50 per cent of patients in the reviews of Munsat *et al.* (1969*b*), and Pearn and Wilson (1973). In a review of over 500 cases of spinal muscular atrophy of all types, Emery *et al.* (1976) found that 80 per cent of patients had sat unsupported suggesting that the less severe category is common. The classification into three types is imperfect as within families there is sufficient variability for individuals to be fitted into more than one group. Differentiation from type I (acute spinal muscular atrophy) is well established (Pearn *et al.* 1978*a*), but there is less certainty about the differentiation of the intermediate group from the juvenile Kugelberg–Welander group (type I I I). In 10 families in the Newcastle series (Gardner-Medwin *et al.* 1967) there were significant differences in the ages of onset within families. Two extreme instances concerned two brothers, one with an onset at 18 months and the other at 15 years, and another sib pair—one with onset at birth and the other at four years. In a sib pair (Martin-Sneessens and Radermecker 1965) one had an onset in infancy and died at 27 months. The second is now 6 years old. The onset was at 15 months and the progression was slow. Kessler (1968) described two sibs, one of whom was normal until 15 months. The other never learnt to sit. Data from these authors support Munsat *et al.* (1969*b*) who suggested that the clinical differentiation, based on age of onset or rapidity of progression, is entirely artificial. They described families with an onset from birth to 18 months (all recessively inherited) with prolonged survival in some, and the acute classical Werdnig–Hoffman disease with short survival within the same family. In a survey of childhood onset spinal muscular atrophy (Winsor *et al.* 1971) 91 cases were ascertained, including 15 sibships with two or more affected. In 14 out of the 15 there was a good correlation between sibs for age of onset and clinical course, but the authors conclude that there was not a clear bimodal distribution for there to be any certainty about the separation into infantile and juvenile forms. Emery (1976), who reported six families in which death occurred in the proband in infancy, but with sibs who had the non-infantile form, states that this discrepancy in age of death within families need not contradict the hypothesis that there is a distinct infantile type as it would be expected that a few cases of the non-infantile type would die in infancy. His data suggest a continuous variation in expression in the non-infantile cases which could be consistent either with a single entity or with several genetic entities with considerable overlap in expression.

For genetic counselling purposes, Pearn and Wilson (1973) suggest that a subsequent sib is likely to have the same prolonged course as the first. The decision about taking the high recurrence risk is less easily reached than in type I because of the chronicity of the illness. Both dominant and recessive inheritance have been reported within this group.

Dominant inheritance

Dominant pedigrees are rare. This type of inheritance has been reported in families in which the condition in those affected progressed over a few years, and then remained static. This was reported by Lugaresi *et al.* (1966) in a pedigree with infantile onset and in a similar pedigree by Magee and de Jong (1960) with an onset at 18 months to 3 years. In the dominant pedigree of Garvie and Woolf (1966) the affected children had an onset between 12 and 18 months. The family reported by Zellweger *et al.* (1972) had 21 affected members over seven generations. The onset was in early infancy, but the disease was only slowly progressive.

Some of the other reported pedigrees with dominant inheritance have a later onset (Armstrong *et al.* 1966) and are described under Kugelberg–Welander disease (type III). In the collaborative study reported by Emery (1976) the family histories of 403 individuals were assessed. In 308, one was affected ('sporadic'), in 81 two or more sibs (total 186 affected), and in only three was there parent-to-child transmission (dominant). Eleven were possible examples of dominance with reduced penetrance (cousins were affected). No clinical difference between the sporadic group and the sibship group was found.

Recessive inheritance

It can be seen from the Emery *et al.* (1976) study that where more than one family member is affected recessive inheritance is much more common than dominant inheritance. Sibs were reported by Fried and Emery (1971).

Pearn *et al.* (1978*a*) looked at the segregation analysis of chronic childhood spinal muscular atrophy. He included in his study all those cases with an onset before the age of 15 years (including type III, but excluding type I). He interpreted his data as compatible with recessive inheritance in the majority of families, but suggested that some of the small proportion of sporadic cases were either new dominant mutations or environmentally caused. His data also suggest that fresh dominant mutations do not occur in cases where the onset is before 12 months.

Very similar conclusions can be drawn from the study of Bundey and

Lovelace (1975) and possibly from the data of Winsor *et al.* (1971), and the earlier study of Brandt (1949).

Sex ratio

Males are affected more commonly than females (Emery 1976; Namba *et al.* 1970; Furukawa *et al.* 1968).

Genetic counselling

If the proband has an age of onset before 3 years and has the chronic form of spinal muscular atrophy (type II or type III) then the risk for another affected sib is 1 in 5 (Bundey and Lovelace 1975; Pearn *et al.* 1978*a,b*). If the onset is after 3 years, then the risk for a sib is 1 in 10. When the onset is after 5 years the risk for sibs is smaller than this, but in this group there may be fresh dominant mutations with a risk to offspring of about 1 in 10.

Type II—with contractures

A family with congenital, non-progressive anterior horn cell involvement and contractures is reported by Fleury and Hageman (1985). In twenty-one instances the signs were non-progressive. The inheritance pattern was autosomal dominant. A remarkable pedigree in which a total of seven infants were affected in a sibship of 10 was reported by Rosenmann and Arad (1974). The parents were second cousins. Severe muscle wasting was present at birth and all the affected were either stillborn or died within the first year of life. Muscle biopsies supported the diagnosis of neurogenic atrophy.

A family described by Bargeton *et al.* (1961) in which the parents were first cousins is thought to be similar. Two children were affected by a severe neurogenic atrophy with arthrogryposis multiplex congenita.

It could be that the autosomal recessive type has an onset in infancy and the dominant type occurs later.

Type III—Kugelberg–Welander juvenile spinal muscular atrophy (Fig. 13.2)

This might not be distinct from type II, but the names of Wohlfart (1942), and Kugelberg and Welander (1954, 1956) are linked to a form of spinal muscular atrophy with an onset of the disease between three and 17 years

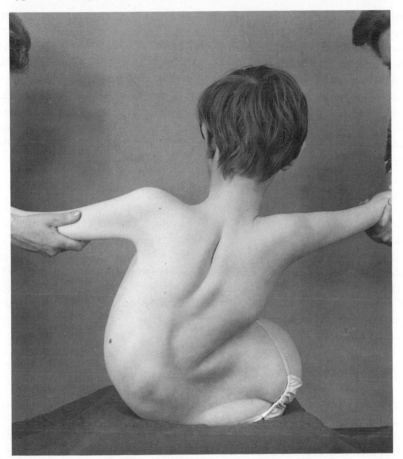

Fig. 13.2 Juvenile spinal muscular atrophy (Kugelberg and Welander syndrome).
Courtesy of Dr A. Harding.

(Wohlfart *et al.* 1955). Proximal muscles are more affected than distal and the course of the disease is protracted. Deformities of the chest and spine, and severe restriction of joint mobility are frequent complications. The incidence of chronic spinal muscular atrophy (groups II and III) is about 1 in 24 000 live births with a prevalence of 1.2 per 100 000 (Pearn 1978).

According to Namba *et al.*'s (1970) review of 382 patients, 48.8 per cent with chronic spinal muscular atrophy, had an onset between 3 and 18 years, and 15.2 per cent over 18 years (36 per cent had an onset before 2 years, but after 6 months and are considered in the intermediate group—type II). It should, however, be noted that the age of onset is difficult to assess. In a study of Kugelberg–Welander disease, Hausmanowa-Petrusewicz (1978) found that the assessment of the age of onset by the parents was often inaccurate and early symptoms went unrecognized for a long time. This makes the precise classification on the basis of age of onset, especially those with onset of symptoms between 2 and 3 years, difficult (see also Van Wijngaarden and Bethlem 1973).

Within this group both dominant and recessive, families have been described and clinically, there is as yet no way of separating them. There is some indication (not conclusive) that when the strictly proximal distribution of weakness and wasting is not absolute, then dominant inheritance is more likely (Pearn 1974; Bundey and Lovelace 1975).

(i) Recessive

Recessive pedigrees are those of Kugelberg and Welander (1956), Byers and Banker (1961), Levy and Wittig (1962), McLeod and Williams (1971), Dubowitz (1964), Bundey and Lovelace (1975), and Garvie and Woolf (1966). Consanguinity has been noted by Hanhart (1962), Spira (1963), Bundey and Lovelace (1975), and Kugelberg and Welander (1956). Radu *et al.* (1966) suggest that the recessive forms appear to be 10 times more common than the autosomal dominant types. However, this applies to those with an onset before the age of 3 years.

(ii) Dominant inheritance

Dominant pedigrees with an onset before 10 years are less frequent. None was found in the 32 patients followed by Van Wijngaarden and Bethlem (1973). In one study (Emery *et al.* 1976) parent-to-child transmission accounted for 0.75 per cent of cases and a previous estimate (Emery *et al.* 1973) was 2.4 per cent. Dominant pedigrees are those of Gamstorp (1967), Garvie and Woolf (1966, mother and two children), Cao *et al.* (1976), and Bundey and Lovelace (1975, mother and son).

A dominant pedigree with reduced penetrance (Amolg and Tal 1968) might represent a quasidominant due to inbreeding (Emery 1971).

Intrafamilial variability in the severity of the illness occurs not uncommonly when the age of onset is between 4 and 24 years. In the case of Cao *et al.*'s (1976) dominant pedigree, two members were chair-bound at 13 years, whereas in the previous two generations, affected members were still ambulant in the fourth and fifth decades. Variability was also pronounced in the families of Nelson and Amick (1966) and Paunier *et al.* (1973).

Armstrong *et al.* (1966) described a Negro family with 19 members affected over three generations with slowly progressive neurogenic weakness. Onset was at about 10 years. Tsukagoshi *et al.* (1966) described a Japanese family with affected members over four generations. Some had bulbar palsy. Incomplete penetrance was a feature.

Genetic counselling

See p. 249.
The figures from Hausmanowa-Petruvewicz *et al.* (1984) if used instead of those of Pearn (see above) are as follows for sibs:

Age of onset (years)	*Risk* (sibs)
1–3	1 in 8
3–8	1 in 5
9–18	1 in 5

Both studies confirm the major recessive contribution. A sib risk of between 1 in 4–1 in 10 is useful for the category as a whole for practical purposes this is a fairly high recurrence risk.

Type III—severe

A form of spinal muscular atrophy with an age of onset in the Kugelberg–Welander range (SMA III) was reported by Scrimgeour and Mastaglia (1984). However, it differs from SMA III in the unusual rapid progression and the early involvement of both proximal and distal muscle.

All the family members were Melanesians and lived in two small fishing villages on the Gulf of Papua. Cousins were affected, but not their parents and the mode of inheritance is uncertain.

X-linked bulbo-spinal spinal muscular atrophy

X-linked pedigrees are mostly of later onset than 18 years (the arbitrary upper limit for the Kugelberg–Welander syndrome).

Kennedy *et al.* (1968) described two families in which a progressive muscular atrophy occurred in 11 male members. In one family, three generations were affected, transmission being through unaffected females. Progression was slow. Extraocular muscles were not affected, but bulbar palsy was a feature. Seven males in two generations were described by Tsukagoshi *et al.* (1970) without father-to-son transmission. Three unaffected females had affected sons. As with Kennedy's cases, bulbar palsy was a feature of the illness. The age of onset in Tsukagoshi's patients varied between 15 and 38 years, and a tremor resembling an essential tremor was present in all the patients and was often the presenting manifestation. Kennedy *et al.*'s (1968) patients also had an essential tremor.

Magee (1960) described three male sibs with an onset in their 40s and 50s. A similar family (Stefanis *et al.* 1975) had three males over two generations with an onset in the third to fifth decades. Bulbar muscular atrophy was a feature of the disease.

Gynaecomastia was described in the affected males of Kurland (1957), Nishigaki *et al.* 1966), Kennedy *et al.* (1968), Tsukagoshi *et al.* (1970), and Stefanis *et al.* (1975) and might be a pointer to the X-linked category. Like-wise, an essential tremor and facial fasciculation are unusually common in the X-linked spinal muscular atrophy. In the family described by Arbizu *et al.* (1983), three had testicular atrophy and two of these males had oligo-spermia. A biopsy in one showed germinal layer atrophy.

Ross *et al.* (1974) described a family in which three brothers were affected. They had five unaffected sisters. Symptoms started in the third decade. A family with remarkable variation in both the distribution of weakness and the severity thereof is reported by Skre *et al.* (1978). Inheritance was an X-linked recessive. The index case had facio-scapulo-humeral involvement and a cousin had an onset in infancy with proximal weakness. In two of the males the plantar reflexes were extensor (see also page 263).

Recently, Wilde *et al.* (1987) described two brothers and a nephew with this condition. At post-mortem, the sensory pathways were involved and the authors interpreted their findings as showing additional pathology of the dorsal root ganglia.

Genetic counselling. As for X-linked disease

(iv) Pedigrees in which the relationship is more remote

There have been pedigrees where simple, recessive, dominant, or X-linkage is thought to be less likely. Zellweger *et al.* (1969) described second cousins, and White and Blaw (1971) described sibs and a first cousin with spinal muscular atrophy.

Becker (1964*b*) proposed an irregular dominant mode of inheritance. The gene a^+ is the mutant gene. Another allele a' is needed for manifestations of mild clinical signs and a third allele a'' for severe signs, thus $a^+ a'$ = mild form of the disease, $a^+ a$ = normal, and $a^+ a''$ = severe.

Genetic counselling

See previous section for those index patients with an onset before the age of three years. In the series of Pearn *et al.* (1978*c*) all those index cases with an onset after the age of 5 years were single cases. Some of the late-onset cases might be fresh mutants and the risk to their children may be significant (possibly about 10 per cent). Risks to sibs in the 3–17-year group are 1 in 10 or less, but see p. 257.

Type IV adult-onset proximal spinal muscular atrophy

In a review of nine cases, Pearn *et al.* (1978*c*) found a range of onset from 15 to 50 years with a mean of 35 years as categorizing this group. All the patients had normal strength in childhood and this is unlike the Kugelberg–Welander patients where a below average strength is often noted by parents and children (Pearn *et al.* 1978*b*). The course of the disease is benign with slow progression over many years, but some patients, especially those in dominant pedigrees, could not walk completely unaided 5 years after the onset. In the nine kindred with adult onset spinal muscular atrophy three were dominant, three recessive and three either new dominant mutations or recessives (Pearn 1974). The author felt that new dominant mutations cannot be totally excluded in single cases.

Three genetic types of adult onset occur.

(a) Autosomal recessive

Of the nine cases ascertained by Pearn *et al.* (1978*c*) there were three sib pairs and three single cases. These authors state that in the adult-onset proximal type of spinal muscular atrophy if the parents are unaffected and the patient is female, autosomal recessive inheritance is the most likely. If the patient is male, then only pedigree analysis is helpful. Sibs are reported by Bundey and Lovelace (1975). One had an onset at 22 and the other at 30 years.

Another late-onset recessive pedigree is reported by Tsukagoshi *et al.* (1965).

(b) Dominant

One case reported by Bundey and Lovelace (1975) had an onset at 45 years. His father and paternal half-brother were said to be similarly affected. Three dominant pedigrees (onset in middle life) are reported by Pearn *et al.* (1978*c*) and one by Peters *et al.* (1968).

Two very large pedigrees with this condition (one with 86 affected persons) were reported from Brazil (Richieri-Costa *et al.* 1981). The age of onset was 48.8 years (mean). Cramps occurred early on and 13 out of 25 experienced spontaneous feelings of 'suffocation'. The condition always first appeared in the legs and progressed slowly to involve the upper limbs, but not the bulbar muscles. The condition is also called spinal muscular atrophy (SMA) type Finkel.

In the Huang *et al*. (1983) report, one of the four cases had a three generation family history with an age of onset between 36–59 years. The other cases were single.

(c) X-linked

See page 257.

Genetic counselling in the adult types

Pearn *et al*. (1978c) suggest a 1 in 5 risk for sibs of sporadic cases and a 1 in 4 risk for familial cases. The risk to children of affected individuals might be 1 in 20 but there are insufficient data to be sure. These authors also suggest that the risk to a sib should fall to 1 in 10 if they are unaffected at 35 years of age.

Recently (see under dominant above), there have been further dominantly inherited families. Offspring risks in single cases could be in the order of 10 per cent.

C. Anterior horn cell disease with particular distribution

(i) Scapulo-peroneal atrophy (Stark–Kaeser)

Dominant

Kaeser (1964) reported 12 members in five generations (same family as Stark 1958) who were affected by a slowly progressive spinal muscular atrophy. The foot and toe extensors were affected first and the process then spread to the leg, thigh, and pelvic girdle muscles. The onset of the disease was between the ages of 30 and 50 years. Clinical manifestations were variable and, in the subsequent generations, some patients had atrophy of the shoulder girdle, upper arms, neck, face, soft palate, pharynx, and external eye muscles. Fasciculation was present in one, and in the three members of the family that were examined no sensory changes occurred. An autopsy on one patient gave convincing evidence of neurogenic atrophy. Pathologically, the findings closely resemble those found in Kugelberg–Welander disease, but the age of onset is much later and it is the distal group of muscles that are first affected in scapulo-peroneal muscular atrophy. The cases reported by Seitz (1957) and Hausmanowa-Petrusewicz and Zielinska (1962) might be muscular dystrophies, but dominant inheritance was likely in four families (Serratrice *et al.* 1976). Single cases are reported by Ricker *et al.* (1968) and Emery *et al.* (1968).

Recessive

A group of patients presenting with a clinical picture of atrophy in the scapulo-peroneal distribution was studied by Feigenbaum and Munsat (1970). Some seemed to be myopathic, others neurogenic, and in some cases it was impossible to decide to which group the cases belonged. Three cases, which were thought to be neurogenic, included two male sibs. Four of the indeterminant cases were dominantly inherited. Recently, two brothers (Mercelis *et al.* 1980) were reported with an onset in childhood.

X-Linked

This mode of inheritance is extremely rare. A scapulo-peroneal muscular atrophy with cardiopathy (Mawatari and Katayama 1973) occurred in five male first cousins with inheritance through unaffected sisters. The onset was between 7 and 10 years with rapid progression. No pyramidal, bulbar, nor sensory signs were observed. Fasciculation occurred in one patient. A disorder of cardiac conduction which resulted in a bradycardia suggests that the primary pathology could include muscle or be solely muscle. An X-linked pedigree with wide variation in the age of onset was reported by Skre *et al.* (1978). The distribution of weakness was facio-scapulo-humero-peroneal.

Uncertain mode of inheritance

Takahashi *et al.* (1974) described a single male, the child of a consanguineous mating, with scapulo-peroneal distribution of weakness and wasting, and two further males (brothers) with the same disorder who had heart muscle involvement. The authors suggest that both neurogenic and myopathic lesions were present. Inheritance was probably recessive in the single male and either X-linked or autosomal recessive in the brothers.

Single cases are reported by Schuchmann (1970).

(ii) Chronic neurogenic quadriceps amyotrophy

Two patients with an onset in childhood of predominantly quadriceps involvement were a brother and sister (Furukawa *et al.* 1977). Their parents were cousins. The disease in the brother was more widespread and severe.

(iii) Progressive childhood-onset bulbar paresis (Fazio–Londe disease)

Fazio (1892) described a mother and her 4½-year-old child with a progressive bulbar palsy, whereas in Londe's (1894) report, two brothers were affected whose parents were first cousins. For a condition which is clearly heterogeneous, the old term Fazio–Londe disease is not meaningful. A juvenile form was described by Albers *et al.* (1983) with an age of onset between 11 and 20 years. There was marked facial weakness with an inability to move the forehead or to close the eyes. Fasciculations of the tongue were noted.

In the report by Benjamins (1980) the onset was similar in sibs and was within the first year of life. These were very similar to the cases of Alexander *et al.* (1976) and Gomez *et al.* (1962).

In a review of the condition, Alexander *et al.* (1976) found 14 cases, five of whom were familial. Marinesco (1915) reported the condition in a pair of sibs, as did Paulian (1922). Single cases have been reported by Gordon (1968) and Gregoriou *et al.* (1969).

Counselling

Most familial cases are in sibs and the sib risk is likely to be high. Offspring risks might be in the vicinity of 5 per cent.

Progressive adult-onset bulbar paresis

A pedigree in which four members died in their 60s from a progressive bulbar palsy without pyramidal tract signs or amyotrophy elsewhere is reported by Lovell (1932). Inheritance was as a dominant.

Bulbopontine paralysis and deafness (Vialetto–Van Laere syndrome)

A woman of 22 years with weakness of the muscles of mastication, temporal, and facial muscles but without sensory loss, is reported by Boudin *et al.* (1971). Bulbar and scapular muscles were also affected. Three sibs with this combination are reported by Van Laere (1966). Bilateral nerve deafness with motor weakness of cranial nerves VII, IX, and XII are the cardinal features of this rare recessive syndrome.

Further sibs are reported by Brucher *et al.* (1981). Onset was in the middle teens and death occured in the 20's from respiratory complications. Alberca *et al.* (1980) reported sibs which an age of onset of deafness at 2–5 years whereas the cranial palsies occured at 14 and 6 years.

Gallai *et al.* (1981) reported two sibs. One died at 20 months from respiratory problems, whereas the other was alive but ill at 14 years. In most instances deafness was followed by the lower cranial nerve palsies within a couple of years, but occasionally the interval was several years.

Spinal muscular atrophy/optic atrophy/deafness

A single case developed distal weakness and wasting in her early twenties (Chalmers and Mitchell 1987). At the age of 10, she developed visual problems and a sensorineural deafness at 22 years. Progression of the illness was very slow. Note similarly with HMSN type VI.

(iv) Spinal muscular atrophy with ophthalmoplegia

Cases are described by Aberfeld and Namba (1969) with onset in childhood. No other family members were affected. Pachter *et al.* (1976) described another single case with onset in infancy.

Oculopharyngeal muscular atrophy

Seven members in two generations were described by Matsunaga *et al.* (1973). Only four were examined. Two of these had a distal muscle atrophy in the limbs in conjunction with the oculopharyngeal atrophy. One had the oculopharyngeal atrophy alone and in the other the proximal part of the upper limbs was also affected.

(v) Facio-scapulo-humeral atrophy

The family described by Cao *et al.* (1976) had proximal spinal muscular atrophy with facial involvement but with rapid progression to involve the legs. A mother and daughter with facial and shoulder girdle weakness with an onset in adolescence and only very slow progression was reported by Fenichel *et al.* (1967). A similar family is that of Furukawa and Toyokura (1976) in which a mother, son, and daughter had a facio-scapulo-humeral type of chronic spinal muscular atrophy. The mother had profuse fasciculations in the shoulder girdle muscles. The authors point out that one of the original cases described by Landouzy and Dejerine (1885) was a 25-year-old man with 'remarkable fasciculations'.

(vi) Juvenile type of distal and segmental muscular atrophy of upper extremities

This disorder has been described many times in the Japanese literature (Hirajama 1972), but familial cases have only rarely been reported. In a series of 71 cases (Sobue *et al.* 1978) there was only one case of familial incidence (father–son).

A group of patients from India was reported by Gourie-Devi *et al*. (1984). Onset was in the second and third decades. Wasting and weakness were confined to one upper limb and one lower limb, and after slow progression there followed a long stationary period. All were sporadic.

(vii) Distal spinal muscular atrophy and GM$_2$ gangliosidosis

The patient, a 24-year-old Ashkenazi Jewish man, had an onset at 15 years. A paternal relative had classic Tay–Sachs disease (Navon *et al*. 1980).

B. Distal spinal muscular atrophy

Pedigrees with predominantly distal involvement are rare, but both dominant and recessive inheritance have been noted. The condition can be differentiated from the hereditary motor and sensory neuropathies by the absence of sensory loss, the relative preservation of tendon reflexes, and the finding of normal motor nerve conduction velocities and normal sensory action potentials (Harding and Thomas 1980*a*).

Recessive

Martin-Sneesens (1962) reported a pedigree of three affected males with spinal muscular atrophy of slow evolution (intermediate type or type II) in whom wasting was predominantly distal. The parents were cousins.

Sibs are reported by Meadows and Marsden (1969, although father had a chronic neurological disorder), Harding and Thomas (1980*a*), and Pearn and Hudson (1978). Consanguinity was present in three families ascertained by Pearn and Hudgson (1979). Two were single cases and three sibs were affected in the third family.

Dominant

See pages 253 and 256 for the pedigrees of Nelson and Amick (1966) and Lugaresi *et al*. (1966). These and the family of Biemond (1955), all had predominantly distal weakness and wasting, although the Lugaresi *et al*. (1966) family had in addition shoulder girdle wasting. Some of the patients reported by Meadows *et al*. (1969) had marked upper limb weakness.

Sporadic cases are reported by Gardner-Medwin *et al*. (1967) and Ricker *et al*. (1968).

The group of patients described by Dyck and Lambert (1968*a*) under 'progressive spinal muscular atrophy type of Charcot–Marie–Tooth disease' could be included in this category.

Four patients with autosomal dominant transmission had an onset in the third and fourth decades, and the authors (McLeod and Prineas 1971) suggest that the age of onset in the dominant cases is later than in the recessive.

In a series of 34 patients with this distribution of muscle wasting (Harding and Thomas 1980*a-d*) there were four families with autosomal dominant inheritance.

Counselling

Pearn (1979*a, b*) suggests that a risk of 1 in 4 should be given for sibs of sporadic cases and 1 in 8 for their offspring. In the Harding and Thomas (1980*a*) study a recurrence risk for sibs of single cases was approximately 1 in 16 and for offspring 1 in 4.

It is likely that the group of distal spinal muscular atrophies is heterogeneous to account for the preponderance of recessive families in the Pearn study and the dominant pedigrees in the Harding and Thomas (1980*a*) study. There is some suggestion that the rate of progression in the recessive pedigrees reported by Pearn was more marked than in the dominant families.

The genetic counsellor's main problem will be to calculate risks to the offspring of single cases, and this will lie between 1 in 4 and 1 in 8.

Distal SMA with vocal cord paralysis

A large family with hereditary distal spinal muscular atrophy with vocal cord paralysis was reported by Young and Harper 1980.

Most members had an onset in their teens of distal muscular wasting, beginning in the upper limbs then proceeding to involve the lower limbs. When seen in middle life only one had a normal voice. All the others showed a unilateral abductor paralysis causing an immobilized, fixed cord.

A similar family, also from Wales, but unrelated to the above, is known to the author. The voice changes preceeded the distal upper limb weakness.

Others

Microcephaly, mental retardation, and spinal muscular atrophy

Three sibs (unrelated parents) had this combination (Spiro *et al.* 1967). The onset of weakness was at 18 months and progression was slow in two and rapid in one.

Hereditary myoclonus and progressive spinal muscular atrophy

Three members of a Louisiana–Texas family had an adult-onset stimulus-sensitive myoclonus and slowly progressive distal weakness and wasting. Seizures, dementia, and cerebellar dysfunction were 'infrequent or absent' (Jankovic and Rivera 1979).

Spinal muscular atrophy and 'ragged red fibres'

In a family (Dobkin and Verity 1976) six members were affected over three generations (dominant inheritance). The propositus had had a nasal intonation to the voice since childhood, but the weakness of the hands only began at the age of 25 years. A muscle biopsy showed ragged red fibres.

Note clinical similarity to the condition described on page 265.

Ryukyuan muscular atrophy

Studies by Kondo *et al*. (1970) on the Ryukyuan island located between Kyushu, Japan, and China revealed 32 cases of bilateral flaccid weakness of the lower extremities with onset soon after birth. Walking was delayed, and later proximal weakness of the arms occurred. Progression was slow and most patients reached middle life. Fasciculations were noted in eight. The head and neck were normal, but in some cases *pes cavus* and a kyphoscoliosis were observed. A high rate of parental consanguinity and the occurrence of the disorder in sibs make recessive inheritance likely.

Spinal muscular atrophy with calf hypertrophy

In a study of 100 patients from 79 families with an age of onset between birth and 32 years, Bouwsma and Van Wijngaarden (1980) found 23 with hypertrophy. All were males and all showed a raised C P K.

There was some evidence of concordance within families (except in two families), but the authors state that hypertrophy might disappear with increasing progression. Only two males were in a wheelchair before 10 years. Confusion with Duchenne muscular dystrophy is possible and electrophysiological studies and muscle biopsy are important in differentiating the two. Similar patients were described by Pearn and Hudgson (1978).

14

Motor neuron disease

The following classification is adapted from Morariu (1977).

 (i) Sporadic

(ii) Familial
   A. Dominant
      1. Pure
      2. Complicated
         (a) Hirano–Lewy body disease
         (b) Dementia (including amyotrophic lateral sclerosis with Parkinsonism/dementia complex in the Western Pacific Islands.)
         (c) extrapyramidal.
   B. Recessive
      (a) Pure
      (b) With extrapyramidal signs
      (c) Amytrophic lateral sclerosis with dementia (Staal and Went 1968)

Introduction

The genetic types tend to differ from the common sporadic type in the following ways:

(a) the course is sometimes protracted; (but mostly not—mean survival 2.5 years)

(b) in some families all do not have pyramidal tract signs;

(c) the age of onset is often earlier; (mean age of onset 45 years)

(d) post-mortem studies show involvement of the posterior columns and spinocerebellar tracts.

Family studies on consecutive series of patients with motor neuron disease of adulthood show that recurrence risks are small. Gross prevalence of motor neuron disease is 4 per 100 000 (Kurland 1957). In Israel it is 3 per 100 000 (Kahana and Feldman 1976). In the Chamorro population on the island of Guam, the prevalence is 420 per 100 000. Kurland (1957) found five out of 58 had another affected member in an American population, and MacKay (1963) studied 67 patients with familial occurrence in four, giving a 6 per cent familial incidence. In a study of 255 Finnish patients (Jokelainen 1977) the frequency of familial cases was 0.8 per cent, and therefore much lower than the 5–10 per cent of all cases found by Kurland (1977) and Ting-Ming *et al.* (1988).

In the review of Haberland (1961) 12.0 per cent of 251 patients gave a positive family history, but other studies from the USA (Wechsler *et al.* 1944) and Sweden (Müller 1952) found no additional family member to be affected. Intermediate figures come from the study from Rumania of Ionasesco and Drinca-Ionesco (1964).

A. Dominant

1. Pure

Takahashi *et al.* (1972) pointed out that the dominant type—and there are few when compared with the common sporadic cases—is characterized pathologically by the presence of not only pyramidal and anterior horn cell changes, but also by degeneration of the posterior columns and the spino-cerebellar tracts. The family these authors describe had five affected over three generations. Onset was after the mid-fifties and in one patient the reflexes in the legs were hyperactive. Bulbar muscules were affected termi-nally. Horton *et al.* (1976*a*) reported dominant inheritance in seven out of 14 families (seven were recessive).

Within families there is usually concordance for the pathological findings. In two families in the Horton study, as well as five others in the literature, where more than one member has been examined at post-mortem, the patho-logical changes have been identical. One exception is the family of Farmer and Allen (1969). Of the 31 families (Horton *et al.* 1976*a*) in the literature, 27 showed a short course (1–5 years) and two a long course (mean survival 14 years). The remaining two could not be classified. This division would be useful if families 'bred true' for the duration of the illness. However, in a large dominant pedigree (Gimenez-Roldan and Estaban 1977) death occurred from 26 months to 12 years after the onset of the illness. There was also considerable variation in age of onset (33–60 years), and not all cases had pyramidal tract involvement. In the family described by Alajouanine and Nick (1959) the course ranged from less than a year to 20 years. The presence of bulbar palsy might have been the main factor determining the course of the illness but this has not been confirmed by Horton *et al.* (1976*a*).

The family described by Green (1960) had an early onset. Only one of the patients had pyramidal tract involvement and that was late in the disease. The Farr family of Vermont, described by Osler (1880), had familial amyotrophic lateral sclerosis. The mean age of onset was 32 years with a span between 19 and 50 years. In all except one (seven were affected in three generations) the distal musculature of the legs was affected first and not all had increased reflexes. Death occurred after 3–5 years. Post-mortem (Hirano *et al*. 1967) has confirmed the diagnosis. The Massachusetts branch of the same family is known as the Wetherbee Ail family (Brown 1951).

Espinosa *et al*. (1962) described amyotrophic lateral sclerosis in a mother and seven of her 11 offspring. The duration of the illness was longer than 10 years in six and less than 3 years in one. In this family, the duration of the disease in women was significantly greater than in males. Both upper and lower motor neuron signs were present.

Poser *et al*. (1965) reported a dominant pedigree with onset in middle life (both pyramidal tract and anterior horn cell disease were clinically present), and in a paper entitled 'Hereditary amyotrophic lateral sclerosis', Thompson and Alvarez (1969) describe a pedigree in which five (all males) were affected over three generations. There was male-to-male transmission. The clinical picture was that of bulbar and proximal limb-girdle wasting with fasciculation, and death occurred at the age of 37, 2 years after the onset. Most of the other affected members died in their early 30s. In another two families with rapid progression to death (but without mental deterioration), inheritance was as a dominant (Roe 1964).

The mode of inheritance (Gardner and Feldmahn 1966) is sometimes dominant with incomplete penetrance. Three members had affected offspring without themselves being affected. However, as one man developed the illness at 71 years, 15 years after his son had died at the age of 30, the wide variation in age of onset might account for the 'incomplete penetrance'. In this family, 20 persons in seven generations were affected. Onset varied between 25 and 75 years.

A dominant pedigree is reported by Swerts and Van den Bergh (1976) with onset in middle age, and similar ones by Wolfenden *et al*. (1973) and Husquinet and Franck (1980). In a large study by Ting-Ming *et al*. (1988), 15 per cent of familial cases had sensory features.

Twin studies in amyotrophic lateral sclerosis

Monozygotic twins brought up separately were discordant for the rapidly progressive type of amyotrophic lateral sclerosis (Jokelainen *et al*. 1978). Onset in the affected sister was in her mid-sixties.

Dizygotic twins who both developed amyotrophic lateral sclerosis in middle life (Estrin 1977) and dizygotic male twins concordant for the disease (Dumon *et al*. 1971) are known.

Counselling

A summary of the clinical/familial data was provided by Mulder *et al.* (1986). An analysis of 72 families with 329 known to be affected indicated that most were compatible with autosomal dominant transmission. There was a family with only one generation affected but an incomplete history or incomplete penetrance were the most likely explanations.

The familial cases could be characterized as follows:

Median survival of 2.4 years. Mean age of onset 45 years.
Hyperreflexia or spasticity — 82%
Weakness and atrophy   — all
Minor sensory symptoms   — 20%
Dementia   — 7%

Most clinicians would advise a low recurrence risk for sibs and offspring of those patients with the common, late onset type of motor neuron disease. Where a positive family history is obtained it is mostly dominant. In single cases with an onset below 50 years, its risk to offspring could be 5%

It should also be noted that there is some evidence for intrafamilial concordances for age of onset and duration (Chio *et al.* 1987) but there are many exceptions (Ting-Ming *et al.* 1988).

2(a). Amyotrophic lateral sclerosis with Hirano–Lewy bodies

This is probably not different from familial ALS (see page 268).

Hirano *et al.* (1967) described a familial type of amyotrophic lateral sclerosis in which four were affected over two generations with inheritance through an unaffected female. Other members of the family have been reported by Kurland and Mulder (1955), and by Engel *et al.* (1959). The age of onset in the various members differed from 36 to 55 years. At post-mortem in one patient who had no evidence of sensory or cerebellar involvement clinically, there was severe demyelination of the spinocerebellar tract and involvement of Clarke's columns. Intracytoplasmic and axonal accumulation of hyalinized material resembling Lewy bodies was noted in the anterior horn cells.

Few neurologists would accept a diagnosis of amyotrophic lateral sclerosis in the presence of sensory signs or even persistent sensory symptoms; familial amyotrophic lateral sclerosis is, however, associated with post-mortem changes in the long sensory tracts. The family reported by Metcalf and Hirano (1971) had 21 members affected over four generations in which two of the five patients who were examined showed evidence of glove and stocking sensory impairment. Other members in the family complained of numbness or tingling. The course of the illness in this family was slow (some were alive 20 years after onset). The age of onset varied, but most were affected in their thirties when both upper and lower motor neuron signs occurred. At post-mortem, intraneuronal inclusion bodies were found in anterior horn cells [see the family of Takahashi *et al.* (1972); page 268]. An additional pedigree is reported by Alter *et al.* (1974). A mild dementia was part of the clinical picture.

2(b) Motor neuron disease and dementia (including amyotrophic lateral sclerosis/Parkinsonism dementia complex in the Western Pacific Islands and Japan)

Dementia is not a usual feature of motor neuron disease, but it occurs frequently in those on the Island of Guam and in Japan. All the evidence points to this being a similar condition, but distinct from motor neuron disease (or A L S) described above.

The Guamian village of Umatac has the highest known incidence of the disorder and an extensive family survey has been undertaken by Plato *et al.* (1967) and Reed *et al.* (1966).

Inbreeding in the village is such that almost all the persons born in Umatac since 1830 could be represented in one large pedigree. However, the mean inbreeding coefficient of the sibships with one or more affected was close to that of sibships with none affected and the data fits best with dominant transmission or with a communicable factor. The disease is completely penetrant in the male, but only 50 per cent in females.

Not all patients with this combination of signs originate from the Pacific Islands.

Moffie (1961) called a disease in two brothers (information about the parents was scanty) 'Familial occurrence of neural muscle atrophy (Charcot–Marie–Tooth) combined with cerebral atrophy and Parkinsonism'. The head and hands showed a tremor of 'the Parkinsonism type'. At *post mortem,* lesions were found in the cord, cerebellar cortex, dentate nucleus, inferior olives, red nucleus, cortex, and substantia nigra.

A family with late-onset dominantly inherited bulbar paresis, progressive muscular atrophy, and dementia was described by Robertson (1953). At least four members were affected, but in three instances there was transmission through an unaffected parent. Pinsky *et al.* (1975) reported a father with motor neuron disease (onset at 71 years) whose daughter died of a similar disorder with, in addition, dementia. A brother had predominantly bulbar signs. A second Canadian family is reported by Finlayson *et al.* (1973).

It has been known for the past 50 years that there is a high frequency of amyotrophic lateral sclerosis in the Kii peninsula of Japan. A prevalence of

152 per 100 000 was noted for an area called Hobara (Shiraki and Yase 1975). The clinical picture is similar to the classical type of amyotrophic lateral sclerosis in that the average age of onset is between 50 and 60 years with a short duration of the illness until death. Neuropathologically, the findings resemble the Guam cases. The genetics are uncertain, but in one area another family member with the illness was found in 70 per cent of cases (Hobara). It was 0 per cent in another area (Kozagawa). Autosomal dominant inheritance with reduced penetrance might fit the data.

Three unrelated females with ALS/dementia are reported by Wikström *et al.* (1982). The age of onset was in late middle life and the signs of ALS were mainly bulbar. In two, dementia was the initial symptom, but in general the picture was similar to that seen in the Japanese patients.

(c) Motor neuron disease, Parkinsonian signs, sensory signs, and dementia

Alter and Schaumann (1976) described two Scandinavian families who might have had a common ancestor, in which 14 members were affected over five generations. Members of the one family had involvement of both upper and lower motor neurons. The mean age of onset was 46 years and the illness was quickly progressive. In the second family, in addition to the upper and lower motor neuron signs, rigidity occurred in three members and in two there were glove and stocking sensory changes. Mental deterioration occurred in several. A diagnosis of familial Creutzfeldt–Jakob disease was contemplated, but the classical picture of amyotrophic lateral sclerosis was present at post-mortem.

Motor neuron disease and Parkinsonism

The combination of Parkinsonism and motor neuron disease is seen in Creutzfeldt–Jakob disease, the amyotrophic lateral sclerosis complex of Guam, and occasionally in the Shy–Drager syndrome. Familial cases not belonging to any of the above disease have been described. Van Bogaert and Radermecker (1954) first reported a family with both of these diseases. It occurred in three generations, in five males, but there was male-to-male transmission making dominant inheritance likely. Not all of the five males had both disorders. Two had Parkinson's disease, two had amyotrophic lateral sclerosis, and only one had both.

Motor neuron disease, Parkinsonism, and ataxia

It is uncertain how to classify this group. Brait *et al.* (1973) described a dominantly inherited condition characterized predominantly by a disturbance of balance. The patient walked by 'flopping his feet on the ground' and

2 years later he had features of Parkinsonism. He became progressively weaker and muscle twitchings were noted. His sister and his one son were ataxic. Both were examined and found to have a cerebellar ataxia without Parkinsonism or amyotrophic lateral sclerosis. His mother and aunt had a progressive gait disturbance. The entity could be the same as that of Byrne *et al.* (1982; see p. 242).

A disorder reported by Ziegler *et al.* (1972), which is probably different in that a peripheral neuropathy was part of the clinical picture, occurred in 31 members over four generations. Onset was in the third decade. Basal ganglia features were bradykinesia, rigidity, and rest tremor. An intention tremor and cerebellar type of ataxia were features, and there were signs of both upper and lower motor neuron involvement (atrophy, fasciculation, and spasticity). On the sensory side, vibration sense was lost.

The condition began with ataxia and there was some evidence of segmental demyelination. It could be classified as 'ataxia-plus' (see p. 198).

Motor neuron disease with peripheral neuropathy

Kurent *et al.* (1975) report a family with many affected over two generations.

B. Recessive

(a) Pure

This mode of inheritance is rare. Fleck and Zurrow (1967) described three sisters and their brother with what clinically appeared to be amyotrophic lateral sclerosis. However, their mother died of an ill-defined condition involving her extremities, especially on her left side, although this was said to have been a stroke.

Two female sibs with amyotrophic lateral sclerosis and a brother and his sister are reported by Jokelainen (1977). Both sib pairs had the rapidly deteriorating type. Three separate sib pairs (MacKay 1963) are reported to have had amyotrophic lateral sclerosis.

(b) With extrapyramidal signs

See p. 242. Most of these families have now been shown to have neuropathies and not anterior horn cell disease.

(c) Amyotrophic lateral sclerosis with dementia/early onset

First-cousin Dutch parents had 15 children of whom seven had this disorder (Staal and Went 1968). Onset of the illness occurred toward the end of the first decade. Only two were of normal intelligence and one of these demented. The picture was one of a gait disturbance, and both upper and lower motor neuron signs developed. Sensation was normal. A single case, the product of first-cousin marriage, was reported by Staal and Bots (1969).

(d) Motor neuron disease and GM$_2$ gangliosidosis

A single male Ashkenazi Jewish patient with onset at 20 years was described by Yaffe *et al.* (1979). (See also page 264.)

15

Spastic paraplegia/HSP

A. Hereditary spastic paraplegia

In 1880, Strümpell described two brothers with an onset in middle life of progressive weakness and spasticity in the lower limbs. Clinically, other parts of the nervous system were not affected although a post-mortem on one patient showed that the lateral spinocerebellar tracts in the cervical and thoracic regions were involved. Many of the cases subsequently described have not been 'pure'—a term which means that on clinical examination spasticity and increased reflexes are the only signs. Holmes and Shaywitz (1977) have found only 100 cases of the pure type in the literature.

Biological fitness did not seem to be impaired and fresh mutations might be rare. Harding (1981g) divided the dominant group into two types. Type I (much more common) with an onset before 35 years (on examination and not according to patients' symptoms) and a smaller group Type II, with an onset after 35. The usefulness of this division is that patients at risk in a family of Type I could have that risk reduced if at a certain age no signs were present (see Table 15.1).

A. Type I   i. Spasticity more than weakness.
             ii. Onset < 35 years
           iii. Progression slow

Table 15.1 Risk to offspring at various ages if proband is clinically unaffected

Years	Risk
25	1 in 4.6
30	1 in 5.4
35	1 in 7.7
40	1 in 9
45	1 in 11

All start with a risk of 1 in 2.

Type II   i. Weakness and spasticity.
         ii. Distal sensory loss (mild)
        iii. Urinary problems
         iv. Onset > 35 years
          v. More rapid progression.

Both types are usually dominant in inheritance—they are discussed together, but there is some evidence that the types breed true within families.

B. Complicated HSP
    Additional features might be:
          i. Optic nerve involvement
         ii. A pigmentary retinopathy
        iii. Extra-pyramidal signs
         iv. Distal atrophy
          v. Dementia
         vi. Ataxia
        vii. Skin lesions
       viii. Oligophrenia
         ix. Neuropathy
          x. Deafness
         xi. Troyer syndrome
        xii. MASA syndrome
       xiii. MAST syndrome
        xiv. With cone-shaped epiphyses

A. Type 1

(a) Dominant pedigrees

This is the usual mode of inheritance. Progression is slow and spontaneous arrest is not uncommon. The legs are usually affected first and, where the onset is early, spread to the arms can likewise be early (by the age of 20 in two patients of Behan and Maia 1974). *Pes cavus* is a frequent finding. Posterior column sensory loss (especially vibration) occurs in a considerable proportion of the patients in which the disease is present for some years (Rischbieth 1973). Cartlidge and Bone (1973) published a dominant pedigree in which three members all had an early onset of bladder disturbance as part of the pyramidal tract syndrome and members of the pedigree reported by Rischbieth (1973) had urinary frequency and urgency. In the series of familial cases collected by Holmes and Shaywitz (1977), 70 per cent were thought to be dominantly transmitted. Behan and Maia (1974) recorded six families with uncomplicated, pure familial spastic paraparesis. Dominant inheritance was the mode of transmission in all of the families. In a survey of hereditary spastic paraplegia in Norway (Aagenaes 1959) one family is described in which 10 members in four generations had a pure spastic paraplegia. Weakness occurred after many years and the course was usually benign.

Reduced penetrance occurs in some pedigrees. One of monozygotic twins, the offspring of an affected father, was affected (Bone *et al*. 1976), suggesting that environmental factors can be important.

Other dominant pedigrees are reported by Schwarz and Liu (1956) including one pedigree with 22 members affected in six generations Bickerstaff (1950), Philipp (1949), Roe (1963), Rhein (1916), and Holmes and Shaywitz (1977).

Static form

This might be a separate entity but is also dominantly inherited. The onset dates from the time when the individual starts to walk. Affected members in a large pedigree (Thurmon and Walker 1971) walked on their toes in a bizarre fashion, but ataxia is not a part of the syndrome. A family with 12 members affected over four generations had little progression of the illness (McLeod *et al*. 1977). Onset was early in childhood and there were no sensory changes nor muscle wasting.

(b) X-linked spastic paraplegia

There have been a few well documented pedigrees in which spastic paraparesis has been inherited as an X-linked recessive condition. The pedigrees in which the paraparesis is 'pure' are probably different from those in which the clinical manifestations are more widespread.

Zatz *et al*. (1976*b*) described a large pedigree in which 24 males were affected over five generations. The disease began in late childhood or adolescence, the course was slow, and only the lower limbs were involved. The pedigree of Thurmon *et al*. (1971) is X-linked, but that of Raggio *et al*. (1973), although reported as X-linked, showed male-to-male transmission.

The X-linked paraparesis described by Johnston and McKusick (1962) (17 males in five generations) began as a pure spastic paraparesis. However, over the years, patients developed signs in the upper limbs, and evidence of brain stem and optic nerve dysfunction. Posterior column sensory loss, dysarthria, nystagmus, and progressive dementia occurred in some. The onset in the family was early and most of the patients never learnt to walk normally. A post-mortem on one patient (Ginter *et al*. 1974) showed corticospinal and spinocerebellar tract involvement and the family could be classified as cerebellar ataxia plus (see p. 198).

Blumel *et al*. (1957) described four males over two generations with what

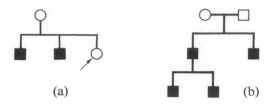

Fig. 15.1

he called 'Hereditary cerebral palsy'. The males were thought to be normal at birth, but developed paraplegia and a horizontal nystagmus. Sensation was normal and cerebellar signs were not present. In one family, two members had wasting of the lower third of the calves without wasting of the hands. In no case did the disorder progress to total disability.

This entity remains a problem in that the pure form continues to be exceedingly rare. X-linked recessive spastic paraplegia families have been reported with very suggestive linkage to specific areas on the X-chromosome, but they are mostly complicated by other features. In the report of an X-linked family by Kenwrick *et al.* (1986) the males were mentally handicapped. Two of the male sibs had at birth bilateral absence of extensor pollicis longus and one, in addition, had an absence of the extensor indices. Both developed multiple flexion contractures. Four uncles were similarly handicapped and at least two showed optic atrophy. It should be noted, however, that this family probably has the MASA syndrome and is likely to be distinct from pure X-linked hereditary spastic paraparesis (see p. 290).

As stated above, the proportion of families with X-linked hereditary spastic paraplegia is very small. This mode of inheritance should be considered for the situation shown in Fig. 15.1a. However, the likelihood of recessive inheritance is much greater with considerably smaller risks to the males of the unaffected sister than would be the case for X-linked inheritance. If the category of X-linked inheritance constitutes only 1 per cent of the whole—and this might well be an overestimation—then the risks to the arrowed patient's offspring would be much smaller than 0.5 per cent.

Warning—One of the males in pedigree has had two affected sons, suggesting dominance with incomplete penetrance (Fig. 15.1(a)). The parents were examined in their 70's (see Fig. 15.1(b) and p. 280).

(c) Recessive

Although frequently cited as a possible mode of inheritance for spastic paraparesis, reports in the recent literature are few. Bell and Carmichael (1939) suggest that 49 out of 74 pedigrees were possibly inherited in this way, but no other published series contains this high a proportion of recessive pedigrees.

In a review of the literature up to 1977 (Holmes and Shaywitz) 30 per cent of reported families showed recessive inheritance. In the Harding study (1981*g*) there were three families (out of 22) in which autosomal recessive inheritance was considered to be likely.

There is a suggestion that the age of onset is earlier (before the age of 10 years) in the recessive type (74 per cent of cases), whereas an onset in the first decade in the dominant form occurred in 59 per cent of cases. However, the mildness of the disease in the dominant type makes the exact age of onset difficult to assess, and these data are of limited use in genetic counselling. The severity of the illness is similar in both groups. The clinical difficulty in deciding the difference between the recessive and the dominant form is illustrated by the two pedigrees of Vernea and Symington (1977). The onset was in middle or late life in both families. In the one family only sibs were affected and they were the product of a first-cousin marriage, whereas in the other sibship three were affected; both parents had died in their 60s and were said to be unaffected.

Many of the reported recessive pedigrees are not of the so-called 'pure' type. Skre (1974*a, b*) in a survey of hereditary spastic paraplegia in Norway, found five sibships in whom those affected were the offspring of consanguineous parents. There appeared to be two types—an infantile and an adult form. The infantile type included patients with dementia and the late-onset type patients with retinitis pigmentosa.

A sibship (Rothschild *et al.* 1979) living in rural Louisiana, had six affected. Ten deceased had the same disease. All those affected were the offspring of consanguineous marriages. An abnormal visual evoked response to an optic stimulus was suggestive of demyelination of the optic nerves.

Genetic counselling

Where pedigree evidence of either dominant or recessive inheritance is available counselling is straightforward. Parents of single cases or of sibs should be personally examined as cases are known (A. E. Harding, personal communication) where only minimal evidence has been found of pyramidal involvement in a pedigree where only simple dominant transmission could explain the large number of individuals affected. More difficult is the calculation of risks in single cases. No data is available to help the geneticist.

It should also be noted that the estimate by Holmes and Shaywitz (1977) that, of the familial cases of hereditary spastic paraplegia (pure) 30 per cent are recessive, is based on published reports and it is not known how many so called unaffected parents were personally examined. It is possible that future studies will show that the adolescent onset of slowly progressive spastic paraplegia in sibs is less frequently recessively inherited than the quoted 30 per cent.

The practical problem often put to the geneticist is whether a patient in the mid-twenties is still likely to become affected and have a substantial risk to offspring. If the age of onset within that particular family is early then the geneticist can be fairly encouraging. It should be noted, however, that the age of onset bears only a tenuous relationship to the onset of symptoms, in that many of those with an early onset do not develop substantial mobility problems until late in life. Normal tendon reflexes in a relaxed person at the age of 30 in the early onset variety would be encouraging—and a risk to offspring of possibly less than 10 per cent appropriate. Males are often more severely affected than females.

There is some indication that there is an excess of males in both the autosomal dominant and the recessive forms. There is also a tendency for males to have an earlier age of onset and to be more disabled (Harding 1981g).

Non-penetrance as a problem in genetic counselling

The author has had personal experience of a pedigree in which both grandparents were examined in their 80's and found to be normal, Fig. 15.1(b).

In a paper by Burdick and others (1981) non-penetrance occurred to the extent that penetrance was calculated to be 0.71.

B. Complicated HSP

(i) Spastic paraparesis, optic atrophy, dementia, and cardiac involvement

Three brothers presented with spastic paraparesis—their parents were distant cousins (Sutherland 1957). In two there were mild cerebellar signs and in one vibration sense was decreased in the legs (aged 30 years). All had *pes cavus* and all demented. The age of onset in the three brothers was in the 20s and optic atrophy was a feature. There was electrocardiographic evidence of heart disease in the affected brothers.

Spastic paraparesis and optic atrophy

The association of a hereditary spastic paraparesis and optic atrophy alone is rare. Monozygotic twins are reported with this combination (Nyberg–Hansen and Refsum 1972). The parents were unrelated. Spasticity developed during early childhood and visual symptoms at about 10 years of age. One boy was of normal intelligence, the other slightly dull. A sister developed optic atrophy and an intention tremor with increased reflexes.

The pedigree of Bruyn and Went (1964), in which six males were affected, is discussed under Leber's optic atrophy (p. 5).

Familial spastic paraparesis, optic atrophy, and dementia

A mother and five of her six children developed a syndrome with an onset in the children at the end of the first decade, and in the mother in her 30s. Progression to dementia was slow (Rothner *et al.* 1976). The inheritance is dominant.

(ii) Spastic paraparesis with retinitis pigmentosa

In a personal communication by Sylekus (1973) to Francois (1974) the author found that 10 per cent of 48 cases of spastic paraplegia had a peripheral retinal degeneration.

There is no certainty about the subdivision of retinal degeneration into macular or peripheral types. Francois (1974) suggests that the cases of spastic paraparesis and pigmentary retinopathy of the peripheral type are those of (only familial cases mentioned): Stewart (1937), Froment *et al.* (1937), Jequier *et al.* (1945), Jequier and Streiff (1947), Evans (1950), Walsh (1957), Mahloudji and Chuke (1968), and Sylekus (personal communication to Francois 1974).

The association of spastic paraparesis with macular degeneration is found in the families of Louis-Bar and Pirot (1945), Kjellin (1959), Landau and Gitt (1951; two patients had a chorioretinitis; see p. 116), and Ledic and Van Bogaert (1960). The family of Landau and Gitt (1951) also had clinical manifestations of cerebellar ataxia plus.

Most of the families cited by Francois (1974) had other involvement of the nervous system and they will be grouped as follows.

Macular pigmentary retinopathy

Familial spastic paraplegia, with amyotrophy, oligophrenia, and central retinal degeneration (with or without deafness)—Kjellin syndrome

Two families in which pairs of brothers with mental retardation, a spastic paraplegia beginning in their 20s with slow progression, and small hand muscle wasting (onset in 30s) were described by Kjellin (1959). In the one family, the parents were first cousins. The mother's sisters (two) were deaf mutes and had bilateral cataracts and pigmentary changes in both eyes.

Ledic and Van Bogaert (1960) described a syndrome of spastic paraplegia and retinal deterioration in sibs. The optic discs were pale and there was macular degeneration, and mild hearing loss. Intelligence was normal.

Peripheral pigmentary retinopathy

Recessive.

Familial spastic paraplegia with peripheral retinal degeneration

Mahloudji and Chuke (1968) described a progressive spastic paraplegia with an onset in the 30s. In addition, there was reduction in vision due to a pigmentary retinopathy (macular and peripheral). Three sibs were affected. The parents were alive and well, and inheritance was probably as a recessive. Progressive dementia was not a feature but the affected sibs were duller than those unaffected. The third sister had only the spastic condition. Her fundi were normal. Later, Macrae *et al.* (1974) found that the sister had developed the retinopathy and that two other sibs had become affected.

In one sibship reported by Skre (1974*a*,*b*) of the adult type all three affected had retinitis pigmentosa and the same clinical picture was found in another single patient whose parents were first cousins.

Spastic paraparesis, retinal degeneration, and sensory loss

Jequier *et al.* (1945) described a pedigree in which only sibs were affected. A severe peripheral sensory neuropathy leading to ulceration was preceded by a spastic paraparesis.

A progressive spastic paraparesis, mental retardation, and retinitis pigmentosa (Sjaastad *et al.* 1976) occurred in sibs who had a markedly elevated homocarnosine in the cerebrospinal fluid.

Dominant   An unusual family with six affected over two generations (regular dominant inheritance) had a severe progressive spastic paraparesis with a retinal pigmentary degeneration (Evans 1954). The discs were pale.

(iii) Hereditary spastic paraplegia with associated extrapyramidal signs

A family with seven members affected over three generations without skips was described by Dick and Stevenson (1953). The onset was in childhood, and the progressive extrapyramidal features were athetosis and/or dystonia with extreme rigidity. Grimacing movements of the face occurred in some. Reflexes were brisk. Dementia was not a feature.

Gilman and Horenstein (1964) described a familial spastic paraparesis with extrapyramidal signs in which dystonia affected the bulbar, limb, and girdle musculature. One other member was spastic, but a dystonic posture became evident on walking and another nine had only spasticity. Inheritance of this variable syndrome was as a dominant.

Some of the cases classified elsewhere on pathological grounds had this combination (see Davison family under Pallidopyramidal degeneration, p. 299) and Miyoshi *et al.* (1969) (see p. 301).

Not all the pedigrees are dominant. A sibship (parents well and unrelated) of three members with a spastic paraparesis and a fourth with in addition mental deterioration and an extrapyramidal syndrome at the age of 21, was reported by Jackson (1934).

(iv) Familial spastic paraplegia with amyotrophy

Families are known where wasting of the small muscles of the hands has been the first and often a marked manifestation of the disease. This is not a single condition and both dominant and recessive pedigrees have been described. Classification is made difficult because spastic paraplegia and amyotrophy are not always easy to differentiate from hereditary amyotrophic lateral sclerosis.

Dominant

Garland and Astley (1950) described a dominant pedigree with nine members affected over three generations. The onset was around the end of the first decade of life and transmission did occur once through an unaffected member. The clinical picture was that of a spastic gait without ataxia, with in addition a marked degree of wasting below the knee, including *pes cavus*. Extensor plantars were present.

Silver (1966) described a dominant pedigree with seven affected over five generations in one family and many over three generations in another. The presenting complaint was that of weakness in the hands, and severe wasting was noted. The wasting appeared between the ages of 15 and 30 years and progression, if at all, was slow. Occasionally, weakness spread to the shoulder girdle. In the lower limbs there was spasticity and weakness. Similar families have been described by Gee (1889) in a father and his two children, and in a family of Holmes (1905) two sisters and a first cousin were affected —both sets of parents were unaffected. The proband's hands were wasted and clawed. Skre (1974*a,b*) in his Norwegian series of patients also reported a dominant pedigree.

A large Dutch family with features similar to those reported by Silver were described by Van Gent *et al.* (1985). In older patients sensory changes occurred, whereas in others it presented like motor neuron disease. Some seemed to have distal spinal muscular atrophy alone and others a pure pyramidal tract disorder. There was histological and electrophysiological evidence of a neuronal/axonal neuropathy. (Silver's patients were not tested.)

This condition is similar to HMSN type V, but there is too much spasticity.

Possible recessives

In a Jewish family (Ormerod 1904) there were three affected brothers. The onset was in childhood. In only two of the three, weakness and wasting of the hands were noted and the eldest son had nystagmus on lateral movement of the eyes.

In another family (Maas 1904) sibs were affected. One demented slightly. Three brothers have been described by Refsum and Skillicorn (1954). The onset was within the first few years of life and progression was slow. Initially there was a spastic paraparesis in the lower limbs extending to the upper and then involving the bulbar muscles. Wasting was generalized. Sensation and intelligence were normal and a severe kyphoscoliosis was part of the clinical picture. A similar pedigree was observed in an Amish isolate by Gragg *et al.* (1971). The patients were two brothers who developed the illness within the first decade of life, but it is difficult to know whether the Gragg patients had this condition as they might have had amyotrophic lateral sclerosis (ALS).

Counselling

There are no convincing recently described recessive cases. Most are dominantly inherited.

(v) Familial spastic paraparesis and dementia

A huge pedigree with affected members over three generations was reported by Worster-Drought *et al.* (1940). In most of those affected the spastic paraparesis preceded the mental disturbances and dysarthria was an early symptom. The onset occurred in the fifth decade of life and the average duration of the disease was 8 years. Eight of the 10 cases were females. Pathologically, widespread demyelination and 'peculiar perivascular structures' around small blood vessels were seen. The exact placement of this familial illness is uncertain. Inheritance is as a dominant (see p. 129).

(vi) HSP—and cerebellar ataxia

This combination occured in five members of a family (over 3 generations including male to male transmission) (Scholtz and Swash 1985). Onset was in the fifth decade and clinically there was spasticity in the lower limbs with a cerebellar ataxia in lower and upper limbs. In some there was distal sensory loss and absent ankle jerks. Necropsy showed an almost complete loss of cerebellar Purkinje cells, whereas in the cord there was degeneration of corticospinal, spino-cerebellar and posterior collumn tracts.

• *Note*: clinically there is a similarity with dominantly inherited cerebellar ataxia plus (see p. 198).

Spastic paraplegia, cerebellar ataxia, and extrapyramidal features (see also under cerebellar ataxia plus, p. 198)

Families described by Ferguson and Critchley (1929), Mahloudji (1963) and Brown and Coleman (1966) could be included in the cerebellar ataxia plus classification and there is a marked resemblance between the patients described in these reports and those in the pedigree of Schut (1950). A similar family was described by Landau and Gitt (1951) in which the patients had a combination of a spastic ataxia affecting 21 members in seven generations with the onset in middle life. Many patients had a fixed expression and one had remarkable muscular atrophy confined to the legs. Two had a chorio-retinitis. A slurring speech and a wide-based gait suggested cerebellar involvement. One member in the family had a picture indistinguishable from that reported by Brown and Coleman (1966), who described spastic ataxia, dysarthria, and emotional lability with relative preservation of mental function. A pseudo-ophthalmoplegia of upward and lateral gaze, lid retraction, and nystagmus were the ocular features. On the extrapyramidal side, rigidity, poor facial expression, peri-oral tremor, dystonic smiling, brady-kinesia, and tremor were the main features. Inheritance was as a dominant.

The family of Ferguson and Critchley (1929) was described as a 'heredo-familial disease resembling disseminated sclerosis'. The clinical picture was that of weakness and unsteadiness in middle life. On examination there was a cerebellar ataxia, spasticity, distal sensory loss, and diplopia (a supranuclear ophthalmoplegia). Most of the cases also had extrapyramidal signs (15 affected over three generations). Recently, a branch of this family was examined and the clinical picture is now that of a typical progressive cere-bellar ataxia 'plus' ophthalmoplegia (see p. 198) (Harding 1981*f*). Mahloudji's (1963) family was similar. The onset was between 30 and 40 years and the disease degenerative and ultimately fatal.

Four sibs (out of six) and a first cousin (parents unrelated in both instances) developed a gait disturbance at the age of two years (Martin *et al.* 1974). This was probably due to a bilateral pyramidal syndrome and was followed within the first decade by cerebellar signs. Seizures, dementia, extrapyramidal signs (choreo-athetosis), myoclonus were features. No enzyme defect was detected during life. A similar family was described by Poser *et al.* (1957).

Five sibs (unaffected parents) from a sibship of ten had an external

ophthalmoplegia, pyramidal signs, and extrapyramidal features (Staal *et al.*
1983). Only one had a cerebellar ataxia. The family is similar to that of
Ferguson and Critchley (1929), but neither amyotrophy, fasciculation or
areflexia occured. The age of onset was 20–50 years.

A post-mortem on one (with ataxia) showed demyelination of spinocerebellar tracts and olivocerebellar pathways. Purkinge cells and the cells of.
Clarke's column were in addition affected.

Counselling

This is a heterogeneous category. Most families will have the dominant form.

Familial ataxic diplegia with deficient cellular immunity

First described by Hagberg *et al.* (1970) in a brother and sister, the neurological picture is that of a mild spastic diplegia with marked cerebellar ataxia.
A further family is reported by Graham-Pole *et al.* (1975). Two out of four
sibs were affected with diplegia and extensor plantar reflexes but without
obvious ataxia. A T-lymphocyte deficiency was confirmed. There is clearly a
relationship between these families and those discussed under Ataxia
telangiectasia (see p. 188).

Charlevoix–Saguenay syndrome

A group of patients from a genetic isolate with a spastic ataxia are reported by
Barbeau (1980). Onset of the ataxia occurred early and no patient walked
normally. The disability was, however, not relentlessly progressive and
appeared to remain stable for many years. Muscle tone was markedly
increased in the lower limbs where the Babinski reflex was easily obtainable.
Posterior column sensory loss, dysarthria, nystagmus, *pes cavus*, and
atrophy of the small hand muscles were prominent. Unlike Friedreich's
ataxia urinary incontinence was an important sign. Intelligence was low
normal and though none of the echocardiographic features of Friedreich's
ataxia were recorded there was a suggestion that some patients had a mitral
prolapse. Inheritance was clearly recessive.

Electrophysiological studies confirm the presence of denervation with slow
conduction velocity (Bouchard *et al.* 1979).

(vii) Spastic paraplegia and skin lesions

Spastic paraplegia, oligophrenia, and ichthyosis (Sjögren–Larsson syndrome)

Sjögren (1956) and Sjögren and Larsson (1957) described 28 patients with
congenital ichthyosis, spastic paraplegia, and mental retardation. The onset

of the motor deficit occurred soon after birth and in only four patients was there evidence of progression. Of the other five patients, four died early and one had mental retardation without spasticity. Inheritance was clearly recessive.

The dermatological findings are distinct; scaling ichthyosis is present on the neck, trunk, axilla, and flexural areas of the extremities. Scalp hair is normal or thin, and moderate hyperkerotosis might be present on the face.

Richards *et al.* (1957) described a Greek Cypriot sibship with the disease, Richards (1960) an English girl who was the offspring of a first-cousin marriage, and Sayli *et al.* (1969) three Turkish children whose parents were first cousins. Fourteen individuals from eight sibships were from an isolated population in North Carolina (Witkop and Henry 1963). Twins (female of unknown zygosity) were reported by Guilleminault *et al.* (1973).

In the original publication by Sjögren and Larsson (1957) three affected persons were found to have retinitis pigmentosa and about one-third of patients have these changes. In the pedigree of Selmanowitz and Porter (1967) those affected had widely spaced teeth. These authors also comment on a speech deficit out of proportion to the mental retardation. This was also noticed by Witkop and Henry (1963). The Sjögren and Larsson patients came originally from Västerbotten in North Sweden where intermarriage is common. Members of 13 families were described and in eight consanguinity was found.

The paper by Heijer and Reed (1965) described four additional patients from Västerbotten and two from California. A fatty acid defect has been found in skin fibroblasts (Avigan *et al.* 1985).

Spastic paraplegia of adult onset and congenital ichthyosis

Two sibs were neurologically well until their 40s and then developed a progressive spastic paraparesis with urinary incontinence (McNamara *et al.* 1975). There were, in addition, features of a mixed peripheral neuropathy and/or myopathy.

Sjögren–Larsson syndrome with peripheral neuropathy

Two sibs, (a third died, possibly of the same condition) had ichthyosis (one at birth, the other only on biopsy) and developed a spastic quadriparesis (Maia 1974). No abnormal sensory signs were present, but ankle jerks were reduced. Histology showed the peripheral nerves to be affected (vacuolization and fragmentation of myelin).

Rud syndrome

The combination of oligophrenia, infantilism, ichthyosis, and epilepsy, known as the Rud syndrome. A spastic paraparesis is not part of the syndrome and is therefore distinguishable from Sjögren–Larsson disease.

Three sibs with Rud syndrome were reported by Marxmiller *et al.* (1985). The authors summarized the literature and found that ichthyosis, mental retardation, hypogonadism, and short stature were the most consistent features. Only a half have epilepsy and few have deafness.

There is a suggestion that some patients in this possibly heterogeneous group may have a deficiency of steroid sulphatase (Andria *et al.* 1984).

Sjögren–Larsson and Rud syndrome in the same family

It has been suggested that the Sjögren–Larsson and Rud syndrome constitute two varieties of the same disease. Kissel *et al.* (1973) report a mother with the Rud syndrome producing a child with Sjögren–Larsson syndrome. The father of the child was her own father.

HSP and hyperkeratosis

At least two families have been reported with this combination. In one (Fitz-simmons 1983), mental retardation was a feature and the pyramidal tract involvement had an onset within the first couple of years of life. It was uncertain whether the condition was X-linked (four males and their mother were affected), but subsequently there was a letter (Powell *et al.* 1983), about a three-generation family with male to male transmission. The hyperkeratosis involves the palms and soles.

(viii) Spastic paraplegia, oligophrenia, and microcephaly

The older literature is difficult to evaluate because syphilis was often a cause for spasticity and mental retardation. Böök (1953) described sibs, and Paskind and Stone (1933) reported three male sibs with microcephaly and spastic paraparesis. All three were severely retarded. At post-mortem pachygyria, agyria, and heterotopia were features.

Mild spasticity in the lower limbs is a frequent finding in microcephaly and is not specifically related to long tract involvement (see page 26).

X-linked spastic paraplegia and oligophrenia

Baar and Gabriel (1966) described 13 males in five generation affected with a progressive spastic paraparesis. The onset was at birth or in the first year of life and there was evidence of a bilateral facial palsy, a basal ganglia disorder (athetosis), and cerebellar dysfunction.

(ix) Spastic paraparesis and a sensory neuropathy (without retinal degeneration)

A pedigree reported by Khalifeh and Zellweger (1963) had 47 members affected over four generations. The spastic paraparesis began before the age of 10 years and later a peripheral sensory disturbance followed. Five cases, including one pair of sibs whose parents are cousins and one pair of half sibs (Cavanagh *et al.* 1979*a*), had a mutilating sensory neuropathy in addition to spasticity. Pain and temperature were most severely affected. In one sib pair the paraplegia and neuropathy developed within the first two years of life, whereas in the other it was much later. The pathology in one case was reminiscent of Denny–Brown's hereditary sensory neuropathy with extension to include pyramidal tract neurons. Recessive inheritance makes that particular designation unlikely.

Koenig and Spiro (1970) report a family in which a mother and all of her five children had a spastic paraplegia and a peripheral sensory neuropathy. The sensory changes involved pain touch or temperature/loss up to mid-calf level. Nerve conduction was normal.

(x) Familial spastic paraparesis and deafness

A large family with progressive spastic paraparesis affecting only males in three generations was reported by Wells and Jankovic (1986). The age of onset was towards the end of the first decade of life. Granularity of the retinal pigment and lens opacities occurred in half of the males.

Deafness was sensori-neural. Other features included short stature, hypogonadism, and a raised CSF protein. Four had a tremor (postural). X-linked inheritance is a possibility.

Hereditary spastic paraplegia, deafness, neurogenic bladder, and syndactyly

A three-generation family had nine affected members with the above combination. Syndactyly affected digits 4 and 5 and in three of the nine patients there was mild hearing loss (Opjordsmoen and Nyberg-Hansen 1980). The age of onset varied considerably, but occurred mostly in early adulthood.

(xi) Troyer syndrome

It is uncertain whether this is different from the other recessive syndromes (see above). It was first described in the Amish (Cross and McKusick 1967).

Onset is in early childhood, and, in addition to the distal wasting and spasticity, a dysarthria (monotonous and nasal) and mild cerebellar signs occurred in some. The severity of the spasticity and contractures made walking impossible by the third of fourth decade. Severe mental retardation was not a feature, but inappropriate laughter and crying occurred. Shortness of stature was a characteristic sign. A similar family (Neuhauser *et al.* 1976*a,b*) is reported in which two brothers from the Old Order Amish had a slowly progressive spastic paraplegia accompanied by mental and growth retardation. The mental retardation and non-progressive nature of the disease suggested to the authors that this might be a new syndrome.

(xii) MASA syndrome

This is an acronym for *M*ental retardation, *A*dducted thumbs, *S*huffling gait, and *A*phasia. An X-linked family (with one affected girl) was described by Bianchine and Lewis 1974). Reflexes in the lower limbs were pathologically brisk. Other families are reported by Gareis and Mason (1984) and Yeatman (1984). Linkage to X_q28 has been established (Kenwrick *et al.* 1986). (See p. 278.)

(xiii) Mast syndrome

This is characterized by an impassive face, a progressive spastic paraparesis and presenile dementia. Twenty cases in the Amish in a single pedigree were reported by (Cross and McKusick 1967). Onset was in the teens or twenties. Inheritance was recessive.

(xiv) Spastic paraparesis with cone-shaped epiphyses

This is a rare combination, reported only once (Fitzsimmons and Guilbert 1987) in male monovular twins. The onset was probably towards the end of the first year of life and there was considerable spasticity by 4 years. Their intelligence level was in the low/normal range. X-rays of the hands showed marked shortening of the metacarpals, especially the 3rd, 4th, and 5th. Cone-shaped epiphyses were present. The mode of inheritance of the condition has not yet been established.

16

Basal ganglia disease

Parkinson's disease

The role of inheritance in Parkinson's disease is uncertain and this is in line with the difficulty in interpreting occasional familial clustering in many common disorders.

Prevalence

Most figures suggest a prevalence of between 70 and 100 per 100 000, but there is a wide variation.

Family studies

There is conflicting evidence about the frequency of Parkinson's disease in near relatives of patients. The following studies show figures ranging from those in which there is no difference from controls to those in which the frequency is significantly greater.

Below 10 per cent	Scarpalezos (1948)
	Schwab and England (1958)
10–20 per cent	Pollock and Hornabrook (1966)—14 per cent
	Jenkins (1966)—13 per cent
	Kondo *et al.* (1973)—19.8 per cent
	Kurland (1958)—16 per cent
Above 20 per cent	Gudmundsson (1967)—22 per cent
	Mjönes (1949)—38 per cent
	Allan (1937)—62 per cent

Duvoisen *et al.* (1969), using rigorous criteria, found no difference in frequency of the disease between patient's sibs, and spouses' sibs.

In the study by Martin *et al.* (1973*b*), 130 index cases were analysed for the familial occurrence of Parkinson's disease, using the spouse as control. It was concluded that there was a higher frequency among parents and sibs when compared with the control group. When first-, second-, and third-degree relatives were considered together 26.8 per cent of probands and 14.8 per cent of controls reported at least one affected relative. They produced a table which could be utilized in genetic counselling.

Age of onset in proband	Affected sibs (%)	Risks for sibs
35–44	8.3	1 in 12
45–54	5.4	1 in 18
55–64	3.8	1 in 26
65–74	1.4	1 in 71

In those cases with an affected parent the risks are greater. From the age of 25 years until 50, sibs of probands with an affected parent have a steadily increasing risk of developing Parkinson's disease. The risk at 50 years is just over 5 per cent and at 60 it is 15 per cent.

If the age of onset in the proband is ignored the risks for sibs of probands whose parents are unaffected are also greater than the risks for controls, but is much less marked. At 35 years the risk is just under 2 per cent and at 60 it is still less than 5 per cent. However, in a large Finnish study Marttila and Rinne (1976) found no statistical difference between patients and controls in the frequency of Parkinson's disease in relatives, nor did they find that essential tremor was more frequent—a possible source of error in those studies in which a high proportion of affected relatives were found.

Recent studies have tended to point to the unimportance of genetic factors in Parkinson's disease. Marsden *et al.* (1986) having advertised in the Parkinson's disease Society journal, identified 22 pairs of twins of whom 11 were thought to be identical and 11 non-identical. Concordance was found in one pair from each group. 43 pairs of monzygotic twins were identified by Ward *et al.* (1983) and only one pair was concordant.

It is concluded (see also study by Bharucha *et al.* 1986*b*) that there is no difference in concordance rates between monozygotic and dizygotic twin pairs confirming the low genetic input.

Counselling

Risks to offspring are small

Parkinson's and dementia

There is considerable controversy about the frequency of dementia in Parkinson's disease. Figures between 20 and 40 per cent for severe or moderate dementia of the subcortical type are quoted (Brown and Marsden 1984). It should be noted that aphasia, apraxia and agnosia are not features

Lewy body disease (Lewy body dementia)

The clinical picture is that of dementia and Parkinsonian features especially rigidity, bradykinesia, and tremor. However, the dementia might precede the basal ganglia features by up to 18 months (Gibb *et al.* 1987). The Lewy bodies are found throughout the cortex and brainstem. Many cases have been reported from Japan (Yoshimura 1983) with a range in age of onset between 26 and 72 years.

Most cases are sporadic, but note the families of Morris *et al.* (1984) and Kim *et al.* (1981). Whether they belong to this group is not resolved.

In a study (Heston 1980) in which probands were selected because Parkinson's disease was accompanied by dementia, the frequency of Parkinson's disease in sibs (dementia only in one) was 23.3 ± 9 per cent. If a parent was additionally affected the risk was almost 30 per cent. The age of onset was 21–60 years (average 53.5) and the average survival was 7.1 years.

Whether these patients had Lewy body disease is uncertain.

Counselling

Only meaningful in the presence of a family history. Single cases are likely to be sporadic but note above.

Juvenile Parkinson's

Juvenile Parkinson's has an age of onset before 40 years (Yokochi *et al.* 1984) It is characterized by akinesia, often without tremor, diurnal variation, and a brisk response to L-dopa.

In Japan 10 per cent of idiopathic Parkinson's disease is juvenile in onset and 5 per cent is familial (Yokochi *et al.* 1984) It tends to be slowly progressive with dystonic posturing and an action tremor. A report from Portugal (Lima *et al.* 1987) is similar. There were 21 patients of whom two had an affected sibling. A family history of an essential tremor (relationship not stated) was found in three.

Yamamura *et al.* (1973) described five families in which sibs were affected. In two, the parents' marriage was consanguineous. The onset varied from 13 to 28 years. Diurnal fluctuation of symptoms was commented on and the clinical picture was that of bradykinesia tremor and rigidity with brisk reflexes. No pathology was available. Sibs have also been reported by Ota *et al.* (1968). Two sibs (onset at 9½ and 10 years) with juvenile Parkinson's disease responsed favourably to L-dopa (Sachdeva *et al.* 1977).

Incomplete penetrance in a dominant pedigree was described by Martin *et al.* (1971), but the pattern of inheritance is uncertain. There was no postmortem confirmation of the condition described as juvenile Parkinson's, disease in two sibs with an onset at 10 and 19 years. The unaffected father's father developed Parkinson's disease when treated with tranquilizers. A suggestion is made that he had subclinical disease and that the inheritance in this family is as a dominant with incomplete penetrance.

Note: some patients with striato-nigral degeneration might present clinically with Parkinson's disease (see page 297), but this is a separate disease.

Counselling

It must be concluded that juvenile Parkinson's is a separate genetic entity with an autosomal recessive mode of inheritance (in most). There might be a sporadic non-genetic form.

There is some evidence (Barbeau and Bourcher 1982) that juvenile Parkinson's can be divided into those in whom tremor is the first feature and those that begin with rigidity and akinesia. Most of the familial cases come from the latter group.

Toxicity to phenothiazine in children with a family history of Parkinson's disease

The induction of a Parkinsonian-like picture by appropriate drugs is occasionally genetically determined. One estimate is that 22 per cent of patients who manifest the syndrome in response to drugs have a family member with Parkinson's disease (Myrianthopoulos *et al.* 1969). The same phenomenon in a mother and infant has been reported by Holmes and Flaherty (1976), and Hill (1977).

Parkinsonism and cerebellar ataxia

See page 198.

Parkinsonism, cerebellar ataxia, and a peripheral neuropathy

See page 242.

Parkinsonism, depression, and taurine deficiency

Six affected members over three generations with an onset of symptoms in the fifth decade were described by Perry *et al.* (1975). Depression which was unresponsive to medication was the initial symptom. The concentrations of taurine were diminished in plasma, cerebrospinal fluid, and, at autopsy, in all regions of the brain.

Juvenile Parkinson's disease and anosmia (see Kallmann's syndrome —p. 41)

Monozygotic twin sisters developed Parkinson's disease and anosmia at the age of 39 (Kissel and André 1976). Both responded to L-dopa. Two previous reports (Yamamoto *et al.* 1966; Constantinidis and De Ajuriaguerra 1970) showed dominant inheritance.

Basal ganglia disease with a prominent movement disorder

Clinically, this is a group of conditions characterized by

(1) Complex extrapyramidal signs—chorea/athetosis/rigidity/dystonia;

(2) dementia;

(3) pyramidal involvement;

(4) progression.

'System' degenerations

There is considerable diversity amongst authors in the use of this term. Mostly, it refers to (see Oppenheimer 1976):

A. (1) autonomic nervous system failure (see p. 296);
    (2) olivo-ponto-cerebellar atrophy (OPCA; see p. 198);
    (3) striato-nigral degeneration;
    (4) dentato-rubro-pallido-luysian degeneration.

All might present with extrapyramidal features hence the inclusion here.

B. To the Oppenheimer classification might be added Jellingers categories:
   (1) pure-pallidal (including pallido-pyramidal);
   (2) pallido-luysian;
   (3) pallido-striato ± nigral degeneration.

Problem

There is a major problem about the conditions described in the literature as pure-pallidal and pallido-luysian degeneration. There have been very few recent case reports and of those that do exist (mostly 20 years ago) some might better be classified under striato-nigral degeneration (A_3).

Note: Those patients with striato/nigral degeneration might appear to have juvenile Parkinson's disease (see page 293).

It would be easy from a clinical viewpoint to divide patients into those with predominantly a Parkinsonian picture (but not juvenile Parkinson's with its diurnal fluctuation and biochemical response) and suggest that these will mostly turn out to have striato-nigral degeneration and those whose illness starts with an extrapyramidal movement disorder other than Parkinsonism and suggest that these patients are more likely to have pallidal involvement, but this seems not to be the case.

A1. Autonomic nervous system failure

This condition can be subdivided into:

 (i) familial orthostatic hypotension;
(ii) Shy–Drager.

The difference is that Shy–Drager is a multisystem disorder involving the striato-nigral system and is non-genetic.

Familial orthostatic hypotension

A family with orthostatic hypotension, upper and lower motor neuron signs in the lower limbs, Parkinsonism, cerebellar ataxia, and sphincter involvement, had members affected over three generations (Lewis 1964). The precise classification of this family is uncertain.

An orthostatic syndrome associated with a plasma concentration of bradykinin above the normal range has been described in families with dominant transmission (Streeten *et al*. 1972). Lightheadedness or syncope and facial erythema associated with an orthostatic fall in pulse pressure were features.

Shy–Drager syndrome

In 1960, Shy and Drager described a syndrome characterized by orthostatic hypotension, urinary and rectal incontinence, loss of sweating, iris atrophy, external ocular palsies, rigidity, tremor, akinesia, impotence, fasciculations, and distal wasting. Onset was in middle life. Most cases of Shy–Drager syndrome are single.

A2. *Olivo-ponto-cerebellar atrophy*

See p. 198.

A3. *Striato-nigral degeneration*

Since Adams *et al.* (1961) reported this condition, there have been about 28 patients with pathologically proven striato-nigral degeneration (Rosenberg *et al.* 1976). Clinically, they present with Parkinson's disease (at a relatively early age), and pathologically with bilateral degeneration of the corpus striatum and substantia nigra. The putamen is dark brown or green but the origin of the pigment is not known.

Oppenheimer (1976) had the following to say about striato-nigral degeneration: 'The disease is regarded by some authors as a distinct entity but it must be pointed out that many recorded patients with pontocerebellar atrophy have lesions in the striatum and pigmented nuclei similar to those described by Adams *et al.* (1961), and that in some of these cases the final diagnosis was of striato-nigral degeneration and olivopontocerebellar atrophy'.

Although difficulties in balance and rapid alternating movement have been described, they are extrapyramidal and not cerebellar in origin. However, in the case described by Fahn and Greenberg (1972) there were features at post-mortem of olivopontocerebellar degeneration in addition to the striatonigral degeneration.

The diagnosis is suggested when akinesia and tremor occur in a young person, but the disease has also been described at post-mortem in persons at an age where Parkinsons disease is prevalent and there is then no way of distinguishing the two conditions clinically (Trotter 1973; Koeppen *et al.* 1971). Striato-nigral degeneration has also been referred to as the rigid akinetic form of paralysis agitans. Neither chorea, athetosis, dystonia, dementia, nor myoclonus are part of the syndrome.

The seven cases (pathologically proven) of striato-nigral degeneration reported by Takei and Mirra (1973) all had an onset at a time when idiopathic Parkinsons disease would have been appropriate. All had rigidity and half had tremor. Six out of the seven were diagnosed as Parkinsons disease and one as orthostatic hypotension. It is rigidity rather than tremor that is the clinical hallmark of striato-nigral degeneration.

Most of the reported cases are sporadic (Gray and Rewcastle 1967; Sharpe *et al.* 1972).

Note: some early case reports of progressive pallidal atrophy belong to this group. The difficulty for the clinician is to distinguish this condition from the predominantly Japanese condition of juvenile Parkinson's (see p. 293). The lack of response to L-dopa and the absence of diurnal variation are important clues.

A4. Dentato-rubro-pallido-luysian atrophy (DRPLA)

This has been divided into three types, but some authors do not accept this subdivision because the groups can occur within the same family. Naito and Oyanagi (1982) call the condition 'familial myoclonus epilepsy'. The main signs are dementia, epilepsy, ataxia, involuntary movements, and myoclonus.

Note: it is difficult for the clinician to distinguish between DRPLA and OPCA (or cerebellar ataxia plus). The difficulty occurs when extrapyramidal features are part of the clinical picture of OPCA. The following should help.

OPCA (olivo-ponto-cerebellar atrophy) always involves cerebellum and brainstem, and less constantly the extrapyramidal system. More specifically it involves the pons, the cerebellar cortex and substantia nigra. Clinically, the basal ganglia features are usually Parkinsonian.

DRPLA (dentato-rubro-pallido-luysian atrophy) cerebellar deep nuclei and pallidum are involved and the extrapyramidal features are usually chorea/athetosis.

There are three clinical types of DRPLA (Iizuka *et al.* 1984), but they are likely to be the expression of the same gene.

(a) Ataxo-choreoathetosis

In the early stages there is a cerebellar ataxia which is subsequently replaced by choreoathetosis. Smith *et al.* (1958) reported a single case with a late onset. An ataxia was followed by choreoathetosis and a tendency to fall backwards. Grimacing facial movements and mild organic dementia occurred. Cases 1, 2, and 4 fall into this group. Verhaart (1958) reported a man in his early 40's who had become paranoid, demented, spastic, and dysarthric. Other cases are reported by Titica and Van Bogaert (1946), Maeshir *et al.* (1980) and Suzuki *et al.* (1985).

(b) Pseudo-Huntington's

Chorea and dementia predominate. Cerebellar involvement is mild or absent —this is rare.

(c) Myoclonus—epilepsy type

This is characterized by pronounced myoclonus, epilepsy and progressive mental deterioration (Naito *et al.* 1982).

Counselling in DRPLA

There is convincing evidence from the Iizuka *et al.* (1984) paper that this condition is autosomal dominant in inheritance. Whether all cases are genetic is uncertain.

Note: the similarity with the Ramsay-Hunt dominant syndrome (see p. 119).

B1. *Pure pallidal degeneration (Hunt–von Bogaert syndrome or pallidopyramidal degeneration.*

This is very rare. In 1917 Ramsay-Hunt published cases of juvenile Parkinsonism (one child was the offspring of a first-cousin marriage) in which a post mortem showed the presence of pallidal atrophy. In the Van Bogaert (1946) report two sibs had pure pallidal involvement. Jellinger (1968) described a single case showing torsion dystonia, athetosis ending in dystonic semiflexed contractures. Onset was in childhood.

Similarly, in 1954, Davison described five cases of what he called pallido-pyramidal degeneration. In two instances sibs were affected and in both of these two families the affected members were the offspring of consanguineous unions. Onset was in early adulthood and the clinical picture was that of progressive Parkinsonism (mask-like face, drooling, dysarthria, almost dystonic rigidity, pill rolling tremor). Extensor plantar reflexes were present. In one there was evidence of intellectual deterioration. A very similar sibship with an age of onset within the first decade is described by Horowitz and Greenberg (1975) in which akinesia, rigidity, tremor, and pyramidal tract signs occurred. This condition the authors called pallido-pyramidal syndrome, but there was no pathological confirmation.

In 1963, Lange and Poppe reported six sibs and a post-mortem (Lange *et al.* 1970) showed narrowing of the pallidum with normal putamen, caudate, and subgstantia nigra. Three other members were affected in two previous generations with transmission through an unaffected parent.

Counselling

In view of the sibs described by Davidson (1954) this is likely to be autosomal recessive.

B2. *Pallido-luysian degeneration*

Jellinger (1968) described a single case (sporadic) with athetosis, torsion dystonia, and a slight ballistic component. The age of onset was in the teens and progression was slow. The globus pallidus was bilaterally atrophic with moderate retrograde atrophy in the putamen and part of the corpus Luysii.

This, too, is a very rare entity with few recent reports and many believe that this too might be part of the following categories.

Counselling

Uncertain, very rare, and all are sporadic.

B3. *Pallido-striato-nigral degeneration*

Neumann (1959) described two isolated cases. Both had onset in adulthood. One showed constant involuntary movement of face, tongue, and eyes. The reflexes were hyperactive and the legs were in a scissor position. At autopsy there was extensive degeneration of pallido-putamonal connections. The other patient showed progressive rigidity and, whereas the patient with the dystonic features had degeneration of the putamen, the second patient with rigidity had prominent nigral changes.

There are some system degenerations which are difficult to classify. Gray *et al.* (1985) reported pallido-luysian-nigral degeneration with amyotrophic lateral sclerosis in a lady with onset at 32. The initial sign was dystonia in the limbs and brisk reflexes. Progression was rapid. Her face became expressionless, but with some grimacing. The tone was increased in addition to the severe and diffuse amyotrophy.

Mayer *et al.* (1986) reported two sibs with juvenile Parkinson's (onset 19 and 24 years) with areflexia and a retinal degeneration. Pathologically, there was pallido-luysian-nigral-dentate and dorsal column involvement.

Progressive supranuclear palsy (PSP)

Mata *et al.* (1983) reported three sibs who in their 20's developed a Parkinsonian illness including a postural tremor, pyramidal signs, and impairment of upward gaze. Within 10 years dementia occured. A CT scan showed diffuse atrophy. Pathologically, the cases resembled PSP (Progressive supranuclear palsy), but the onset was too early. The Parkinson–dementia complex of Guam was excluded on clinical grounds (non-Chamorra descent).

A recessively inherited extrapyramidal syndrome with dementia (Hallervorden–Spatz disease)

The onset is in early childhood. Most of the patients develop rigidity, hyperkinesia (dystonia or athetosis), spasticity, and progressive dementia (see p. 484). Inheritance is as a recessive.

Infantile symmetrical necrosis of neostriatum or corpora striata (or familial striatal degeneration)

This can be divided as follows.

1. Abrupt onset after systemic disease (non-genetic).
2. Sub-acute necrotizing encephalopathy (see p. 487).
3. Familial cases.

There have been only three convincing reports of the familial type: Paterson and Carmichael (1924), Miyoshi *et al.* (1969), and Roessmann and Schwartz (1973).

First described under the heading of holotopistic striatal necrosis (Miyoshi *et al.* 1969), this disease has an early onset (between 1 and 9 years) with sub-acute progression of mental deterioration, increased muscle tone (rigid extension), increased deep reflexes (sometimes with extensor plantars), hyperactivity or irritability, athetoid movement (occasionally a fine tremor or myoclonus), dysarthria and dysphasia. Post-mortem studies showed necrosis of caudate nucleus and putamen. Miyoshi *et al.* (1969) described five sibs in two unrelated families, and Roessman and Schwartz (1973) two further sibs (onset in infancy). Inheritance is probably recessive and the clinical picture, age of onset, is quite different from striatonigral degeneration. In one of the children reported by Miyoshi *et al.* (1969), fundoscopic examination revealed dark red spots in the central fovea. The duration of the illness lasted from days to a few years. Single cases have been reported by Verhaart (1938), and Hawke and Donohue (1950). Pathologically, the three cases described by Erdohazi and Marshall (1979) are similar, but in two the onset was later. A girl who deteriorated mentally from the age of 5 years developed akinetic-myoclonic seizures. She had an impassive face, an unsteady tip-toe gait, generalized rigidity, brisk jerks, and a fine rest tremor. Post-mortem showed striatal degeneration. One year after the death of his daughter, the father developed choreiform movements and was said to have Huntington's chorea (Erdohazi and Marshall 1979). The same authors describe another childhood-onset case, and one of infantile onset, both without family history.

Single cases are reported by Mito *et al.* (1986) with an onset at about 6–12 months, but these cases are probably different from those of Erdohazi and Marshall. The condition as a whole is probably not homogeneous, but note above the reports of sibs.

Rett syndrome

More than 1000 cases of this condition have been described—all have been female and with rare exceptions all have been single events. There are four clinical stages.

   I Stage of arrest or stagnation after a normal 6–12-month period. There is diminished eye contact and non-specific hand waving. Deceleration of skull growth occurs after initial normal measurement from birth to 6 months.

  II Rapid deterioration stage. Skills are lost and the ability to manipulate the hands becomes difficult. Patients are clumsy, even ataxic, and there are bouts of apnea followed by hyperventilation. Fits occur. Hand wringing begins.

III Pseudostationary phase. The girls are now retarded with a gait apraxia and truncal ataxia. Seizures can be troublesome.

IV Late stage (around the age of 20).

Further deterioration occurs, often resulting in death before 40 years.

Criteria for a diagnosis

(a) Female

(b) Normal for first 6 months often 12–18 months

(c) Normal head circumference at first

(d) Deceleration at 6 months–4 years

(e) Loss of abilities

(f) Loss of purposeful hand movements

(g) Hand wringing, clasping, washing 1–4 years

(h) Apraxic/ataxia 1–4 years

(i) Over-breathing

(Leiber 1985).

Counselling

Recurrence risks are small. They have to date been monozygotic affected twins, half sisters (the same mother), and one pair of female sibs. All the rest have been single females. Inheritance is as an X-linked dominant—most being fresh mutations (Opitz 1986).

Dystonia musculorum deformans

This progressive disorder is characterized by dystonic postures. The movements that occur are slow and sustained and predominate in the trunk and proximal parts of the limbs (Marsden and Parkes 1973).

In 1929, Mankowsky and Czerny postulated that there were two hereditary forms of the disease, but all the patients do not fit neatly into these two genetic groups, and non-genetic forms probably exist. The inheritance pattern of torsion dystonia in the Jewish population has still not been resolved. In a recent study of 47 patients in Israel, six were of Afro-Asian origin and 40 came from Eastern Europe (Zilber *et al.* 1984). There were four families (all of European origin). In three, more than one generation was affected and in one, cousins had the same disease. Transmission was not regular in any and either dominant inheritance with reduced penetrance or recessive inheritance (pseudodominance) is postulated.

The problem is highlighted by the family described by Eldridge *et al.* (1984). Twins aged 61 are reported (age of onset between 10 and 28 years) with variability in progression. Their first cousin once removed had the same condition. This was a Jewish family originally from different areas of Eastern Europe, but to postulate dominant inheritance, penetrance would have had to be totally reduced in at least four members.

Classification

(a) Autosomal recessive type—predominantly in Jews.

(b) Autosomal dominant type—predominantly in Jews.

(c) Autosomal dominant—non-Jewish type.

(d) X-linked.

(e) Sporadic-non-hereditary.

(a) Recessive

The recessive form has an age of onset under 20 years (mean age of onset 8.8 years in the Bundey *et al.* (1975) study). In another study (Menkes 1976) 84 per cent of patients had an onset at between 8 and 14 years. It is more severe than the dominant type and tends to occur more frequently among Ashkenazi Jews. In a United States study (Eldridge 1970; Eldridge *et al.* 1971) between 50 and 60 per cent of the group (where environmental causes and two-generation families were excluded) were Jewish. The average age of onset was 9.3 years; progression was rapid with a slowing down of the disease when adolescence was reached. In Bundey *et al.*'s (1975) study in the UK, only two out of the 17 were Ashkenazi Jews, but this does represent an unusually high incidence in this ethnic group as they only constitute 2–3 per cent of the population of South-East England. Onset is usually in a limb and only later are the neck and truck involved. Non-Jewish recessive pedigrees with typical early onset of limb involvement and slow progression are reported. A Spanish gipsy family with unaffected first-cousin parents and three further non-gipsy Spanish families with two affected were reported by Gimenez-Roldan *et al.* (1976). Eldridge (1970) has reported families of French Canadian origin and American Indians. In the early onset group only one-third to one-half are thought to be recessively inherited.

(b) Dominant form

Bundey *et al.* (1975) suggested that some of the early onset group without affected sibs or parents might be fresh dominant mutants, and found a significantly increased mean paternal age as evidence for this. The dominant form usually has an onset later than the recessive form but can range from 2 to 46 years. It is more variable in its clinical manifestation than the early onset recessive group. Torticollis, blepharospasm, and spinal dystonia, in addition to limb dystonia, occur and the trunk is often affected before the limbs. The disease is milder and there is no predeliction for Ashkenazi Jews, but of the four families with dominant transmission described by Eldridge (1970) two were Jewish. Pseudodominance cannot be excluded.

Johnson *et al.* (1962) described a large pedigree in whom 16 had an onset before the age of 20 years (usually 6–15 years). Some of the patients were minimally affected and would probably have gone unnoticed except for careful investigation. Spasmodic torticollis, writing difficulties, speech defects, an inversion or plantar flexion of the foot were minimal expressions of the disorder. Penetrance was reduced. These authors reviewed the literature and found 18 families in whom more than one generation was affected and in only one family (Santangelo 1934) did they consider recessive inheritance was possible.

Other dominant pedigrees were reported by Zeman *et al.* (1959) and Yanagisawa *et al.* (1972). In the first mentioned report, tremor was present and sometimes this was the only manifestation of a gene carrier. The tremor was aggravated by voluntary movement and postural stress. Non-penetrance was found in at least two individuals who had children with tremor.

In the Lewen family, first reported by Schwalbe in 1908 (see Zeman 1976), and later by Regensburg (1930) and Jankowska (1935) the inheritance was clearly dominant. Some of the gene carriers had a tremor. A pedigree with a skips (Hoefnagel *et al.* 1970) and of late onset (45–48 years) emphasizes how difficult counselling can be in the dominantly inherited group. The family of Zeman *et al.* (1960) showed not only skips but also variable expression of the disease. Some members had atypical dystonia, others had dyssynergia, and some members had contracture deformities, tics, and muscle spasm (also blepharospasm).

A large pedigree with 121 cases, reported by Larsson and Sjögren (1960), might represent another entity. Many of those affected had deformities and contractures of the spine or joints.

(c) A D—non-Jewish torsion dystonia

Analysis of 41 patients in 15 families (Burke *et al.* 1986) showed an age of onset (median) of 8 years (range 0.8–57 years). Legs and arms were more affected than other parts. Roughly half developed generalized dystonia within an average of 1 year. The earlier the age of onset, especially in the legs, the more likely it was for the condition to become generalized.

It is still uncertain whether Jewish and non-Jewish torsion dystonia are the same conditions. In both it can be mild and non-penetrant, and the age of onset is not significantly different in the two. Also axial structures are not more frequently involved at the onset in the non-Jewish type.

(d) X-linkage

There is an unusually high frequency of torsion dystonia in Panay in the Phillipines (Lee *et al.* 1976). Six sets of brothers were found and in two pedigrees two generations of males were affected with transmission through unaffected females. The mean age of onset was 31 years. Spasmodic eye blinking was an early symptom in some.

A pedigree with X-linkage had deafness and possibly dementia as additional features (Scribanu and Kennedy 1976). This was an Anglo-French family.

Counselling

It does not seem possible at the this stage to give accurate recurrence risks in single cases. If one-third to one-half of the early onset cases are recessively inherited then the risk to a sib of a single case with normal parents would be about 10 per cent. Some will be fresh dominants. Taken as a whole the risks are high enough for counselling those at risk to regard their position seriously.

Benign idiopathic dystonia

Sibs are reported by Willemse (1986). The onset was in the first year of life, but by the age of 18 months the movements had disappeared.

Torsion dystonia with translucencies in the basal ganglia

Under the title of idiopathic torsion dystonia PeBenito *et al.* (1984) reported two sibs who, at the age of 3 years in one and 8 months in the other, developed a slowly progressive torsion dystonia. The parents were unaffected and unrelated. Intelligence of the sibs was average. A CT scan showed symmetrical linear zones of decreased attenuation of the lenticular nuclei indistinguishable from Wilson's disease. The condition differs from familial striatal necrosis of Roesmann and Schwartz (1973) in that intellectual deterioration did not occur (see p. 301). This autosomal recessive condition is important in that it must be differentiated from idiopathic or genetic torsion dystonia of childhood in which the tranlucencis do not occur. For other causes see p. 338.

Writer's cramp

A father and son are reported with writer's cramp. Onset was in the 30's. There was a postural tremor and a mild finger-nose tremor, but no other dystonic features (Martinez-Martin and Pareja 1985). See also Cohen *et al.* (1987).

Myoclonic dystonia

A family with dominantly inherited dystonia affecting the arms, trunk, and face, but sparing the legs is reported by Quinn and Marsden (1984).

There was a dramatic response of the signs to alcohol. Onset was between 5 and 47 years. The myoclonus was present at rest, exacerbated by action and was superimposed on the dystonia. The condition had previously been described by Obesco *et al.* (1983).

There is some similarity of this condition to that described by Mahloudji and Piekielny (see p. 121).

Segawa disease or hereditary progressive dystonia with marked diurnal fluctuation

At least six families have been recorded with an onset between one and nine years of age of dystonia in a limb (Segawa *et al.* 1976). The trunk is only mildly affected. L-dopa dramatically relieved the symptoms.

In the original report by Segawa *et al.* (1976):

(a) symptoms worsened during the day;

(b) sleep was beneficial;

(c) there was a response to L-dopa.

A series of 20 European (as opposed to Japanese) children were reported by Deonna (1986). There was diurnal fluctation in 13, and an age of onset between 2 and 10 years. The signs were limited to the limbs in 13. None had significant abnormalities of trunk or neck. Scoliosis or opisthotonus, occured in five in addition to limb dystonia, but none had a predominant axial dystonia. Postural tremor occured in six and the condition always started in a limb.

In the study by Deonna (1986), like that of Segawa, there was autosomal dominant inheritance with reduced penetrance in 11 cases from five families.

Even when sibs have unaffected parents reduced penetrance seems more likely than recessive inheritance.

Nygaard and Duvoisin (1986) described members of a family with an onset between 16 months and 8 years of a gait disorder mostly due to dystonic posturing of the legs. Progression occured and the dystonia became more generalised and Parkinsonian features developed. There were diurnal fluctations. Tremor occured in many, but it was a postural or intention tremor and not a 'typical' Parkinsonian tremor. Neither IQ changes, pyramidal, sensory, or cerebellar signs were noted. They reviewed 20 cases from 13 families and suggest autosomal dominant inheritance with variable penetrance.

Autosomal recessive inheritance was reported in two sibs with an onset between 1 and 9 years (Ouvrier 1978), but it is possible that all are autosomal dominant AD with variable expression. There is usually a good response to L-dopa.

The condition differs from dystonia musculorum deformans in that there is a predominance of females, an absence of Jewish parentage, benefit from sleep, diurnal variation, and a response to L-dopa.

Dystonia and multiple contractures

Brothers with non-progressive contractures developed choreiform movements in infancy, and both had an EEG indicating a cerebral disorder (Fenichel *et al.* 1971).

GM₂, gangliosidosis with dystonia (see p. 440)

The onset of the disorder was at two and a half years with almost total immobility at 10 years. The patient had no seizures, but his development was retarded (Meek *et al.* 1984).

GM₁ gangliosidosis with dystonia (see p. 436)

Progressive extrapyramidal syndrome and retinitis pigmentosa

Two brothers were reported with an onset in the first decade of life of retinitis pigmentosa and progressive rigidity (Winkelman 1932). In one of the brothers no pyramial signs were present, whereas in the other reflexes were hyperactive.

No other cases have been reported. One family (see Mayer, p. 300) had two members with normal fundi, but absent ERG's. However, other features were present.

Huntington's chorea

Huntington's chorea has a prevalence of 4–7 per 100 000 with figures slightly higher in Tasmania and the Moray Firth area of Scotland, and low in Japan (Kishimoto *et al.* 1959) (0.4 per 100 000). Reed and Chandler (1958) in a nearly complete ascertainment for lower Michigan found a prevalence of 5.0 per 100 000. In an early study in East Anglia, Critchley (1934) calculated a prevalence of 1.2 per 100 000, but where cases were more energetically ascertained (Caro 1977) the figures for East Anglia rose to 9.24 per 100 000. The prevalence was unevenly distributed with a high rate in Lowestoft of 30 per 100 000 and a prevalence in Great Yarmouth, only 9 miles away, equal to that elsewhere in the United Kingdom. Of the early published series only Tasmania (Brothers 1964) has a higher prevalence than East Anglia (4.6 per 100 000). Other studies follow (prevalence rate per 100 000):

London	1938	1.8
Northampton	1970	6.3
Bedfordshire	1971	7.5
Essex	1967	2.5
Leeds	1972	4.3
Cornwall	1953	5.6

A pocket of high incidence is reported in Venezuela. According to Avila-Giron (1973) a local fruit, useful in the making of drugs, was the focus for trade with Germany in the 1860–70 period. All the local inhabitants suffering from Huntington's chorea can be traced back to a German sailor who had the disease.

Taken as a whole, most prevalence figures are in the region of 4–7 per 100 000. Areas of high prevalence (Lowestoft) are probably related to the founder effect, compounded by the fact that Huntington's chorea families are less mobile.

The heterozygote frequency within a population—that is the frequency in the population of those who carry the gene for Huntington's chorea irrespective of whether they are already clinically affected or not—is calculated as:

21.23×10^{-5} by Pearson *et al.* (1955)
10.1×10^{-5} by Reed and Chandler (1958)
10.69×10^{-5} by Stevens (1973)

Diagnosis

The diagnosis is based on the observation of choreiform movements and progressive dementia in the presence of a 'positive family history'. The average age of onset varies from series to series but lies between 35 and 45 years.
The average age of death is between 51 and 55 years.

Bell (1934)	35.5
Brothers (1964)	37.5
Bolt (1970)	42.5
Cameron and Venters (1967)	43.2
Heathfield (1967)	44.2
Mattsson (1974)	37.1
Panse (1942)	37.2
Reed and Chandler (1958)	35.3
Wendt and Drohm (1972)	43.4

Fifteen per cent of patients in Maryland, USA, reported to have Huntington's chorea did not have the condition (Folstein *et al.* 1987.) The commonest mistakes were with:

tardive dyskinesia with schizophrenia,
tardive dyskinesia alone,
psychosomatic illness,
Alzheimer's disease,
stroke, and
other dementias.

Looking at the problem the other way around, alcoholism, head injury and schizophrenia were diagnosed whereas Huntington's chorea was the correct diagnosis.

Chorea

The choreiform movements are fragmented, seemingly purposive, and disorderly. Frowning, pursing of the lips, smiling, and swallowing are common gestures. Movement of the whole head, limbs, and trunk develop later and are made worse by excitement and disappear during sleep. The patients and their spouses are often unaware of the movements and they are most easily confused with nervous tics and habit spasms.

Mental changes

It is well known that personality changes might preceed overt dementia as presenting manifestation of Huntington's chorea. The difficulty is to differentiate those behaviour problems which arise in a family disrupted by the disease, from those associated with the disease itself.

No specific pattern of dementia has been identified. A general decrease on the Wechsler Intelligence Test similar to normal aged people, without focal features and no selective impairment of immediate recall has been reported (Aminoff *et al*. 1975; Norton 1975). When patients are tested early in the disease, the IQ might be in the normal range. In a study by Caine *et al*. (1978) patients had insight into their dementia. They were overwhelmed by too much input, had difficulty in recalling material on demand, lacked spontaneity, and lost details in memory.

Psychiatric symptoms antedating the chorea are common (Dewhurst *et al*. 1970). Anxiety states, personality disorders, alcoholism, suicide, and self-mutilation might first draw attention to the disease. Violence, social deterioration, paranoid delusion, and depression, with or without suicidal attempts, were the common reasons for admission to mental hospitals. Of the married patients, 38 per cent were subsequently divorced or separated. Neglect of children, violence towards children, attempted incestuous sodomy, and exhibitionism are all problems. Assault, offences against property, breach of the peace, and financial chaos add to Huntington's chorea being a familial disease in its broadest sense.

Hans and Gilmore (1968) stated that whereas only a proportion of the children inherit the gene, all bear the stigmata of social deprivation and secondary personality change.

Childhood- and juvenile-onset Huntington's chorea

This group constituted about 5–12 per cent of all cases of Huntington's chorea (Bruyn 1973, 5 per cent; Dewhurst and Oliver 1970, 10–12 per cent).

In the early onset disease the choreiform movements (often mild) become apparent many years after a Parkinson-like picture has emerged. Unusual in adult-onset Huntington's chorea, but commonly present in this variant, are seizures, cerebellar ataxia, dystonic posturing, mutism, and rigidity. Beyers and Dodge (1967) found that two-third of their patients had fits, and Markam and Knox (1965) found that half of theirs did. All the patients show dementia, but choreiform movements might be difficult to detect in the presence of intense rigidity. Death occurs between the ages of 6 and 20 years. Oliver and Dewhurst (1969) suggested that the early onset cases were not rare, but may be erroneously diagnosed as Wilson's disease, Hallervorden–Spatz disease, or post-encephalitic Parkinsons. Goebel *et al.* (1978) extended the list to include subacute sclerosing panencephalitis, Batten's disease, and metachromatic leukodystrophy.

Taking the rigid akinetic form separately, Bittenbender and Quadfasel (1962) found a mean age of onset for 64 cases to be 22.2 years which was significantly lower than for the usual type of Huntington's chorea. Sixteen per cent had epilepsy, but the prognosis in terms of lifespan was not different. The relationship between the early onset rigid form and the sex of the parent remains controversial (see following section).

Paternal effect

It is to be expected that in single-gene inheritance, affected offspring have an equal chance of having an affected mother or father, (congenital myotonic dystrophy is an exception). This has not proved true for the juvenile-onset (under 20 years) Huntington's chorea. This was first suggested by Bruyn (1968), and confirmed by Merritt *et al.* (1969) and Barbeau (1970). Barbeau reported 33 cases of juvenile onset in whom the father was the affected parent in 26 cases (3:1 ratio). An early onset, rigidity, and a high frequency of affected fathers were closely associated in the report by Brackenridge and Teltscher (1975). Brackenridge (1980) has again explored this phenomenon. He concludes that the sex and the neurological signs of the affected parent influence the type of illness and the prognosis in the offspring. If a parent has the rigid type the chances of having a rigid child is five times greater than for parents with the choreiform type of Huntington's disease. Where the father was rigid the risk of a patient having the early onset rigid type was three times greater than when the mother was rigid. In addition, 35 per cent of the offspring of fathers who are not rigid will have the early onset type, whereas only 13 per cent if the mother is the affected non-rigid parent.

In a study by Bird *et al.* (1974), when the father was the affected parent, the age of death in his offspring was 10 years before his, and only 2 years before when the mother was the affected parent. These differences are highly significant, but not all authors agree with this interpretation of the results. A similar study on patients from the Low Countries (Vegter Van der Vlis *et al.* 1976) suggested that the apparent anticipation when father is the affected parent is largely an artefact due to sampling bias. When the bias is reduced the effect is much less marked.

Do early onset cases run in families?

It has been suggested (Oliver and Dewhurst 1969) that the rigid form tends to run true in certain families, but concordance for age of onset relates to sibs and large pedigrees will be found to have both types. A few representative families will be described. In the family of Bittenbender and Quadfasel (1962), a mother and her daughter had rigidity and dementia but at least three other members had the typical choreiform picture. Perrine and Goodman (1966) reported an unusual family in which rigidity (the proband's initial complaint was an inability to open his mouth) was a prominent sign. The age of onset was early in the three members with this type of picture. Two were sibs, the offspring of a father with probable Huntington's chorea, and the third a cousin once removed with a juvenile onset of personality change and choreiform movements whose mother was affected. Two sibs with the rigid, akinetic form of Huntington's chorea were found in a family in which other members had the more typical form (Enevoldsen and Albertsen 1970). The lesions at autopsy did not differ from those usually found in Huntington's chorea.

Bird and Omenn (1975) reported monozygotic male twins, their mother and a daughter of one of the twins, all with the akinetic, rigid form of Huntington's chorea. The mother had an onset at the usual age, the twins in their early twenties and the daughter at the age of five years. Other monozygotic twins with an onset in their early 20s were reported by Oepen (1973).

The rigid form of Huntington's chorea occurred in two members of a family (Bird and Paulson 1971) both of whom had an early onset of the disease (under 20 years). Chorea was absent and the condition was known as 'Knox County dystonia'. It was only when a member in another branch of the same family was examined and found to have the typical clinical picture that the diagnosis of Hungtington's chorea was made. Cambell *et al.* (1961) recorded this type of picture in two boys who were the offspring of a father with typical Huntington's chorea (with onset at 40 years). Similar cases were reported by Rotter (1932), Hempel (1938), and Lindenberg (1960). In one of the families reported by Jervis (1963) the father and his two children all had the rigid form. Other families with juvenile onset in more than one member were reported by Markham (1969), Oliver and Dewhurst (1969), and Liss *et al.* (1973).

If concordance for age of onset in the early rigid form of Huntington's chorea, especially when the father is the affected parent, proves to be correct then the data will be of use in reassuring unaffected sibs at an earlier age than is currently possible. At present, the data are not conclusive and the counsellor should proceed with caution.

The age of onset in sibs of those with an onset before 20 years was looked at by Hayden *et al.* (1985). The mean age of onset was found to be 26.8 years which is significantly less than the expected 39.8. The data might be useful in partially reassuring those who are well at 40.

The adult-onset rigid form of Huntington's chorea is infrequent, but families with more than one member with this form were reported by Samuel (1966) and Liss *et al.* (1973).

At least two kindreds are known where two relatives with Huntington's chorea have married and had affected offspring (Eldridge *et al.* 1973). In both kindreds an affected child had the usual onset, course, and severity of the disease, but in both families one infant died early and may have been homozygous for the Huntington's chorea gene. DNA studies have suggested that homozygotes are not more seriously affected.

Atypical Huntington's chorea

Those with dementia, but no chorea

Members in a family reported by Curran (1930) had only dementia, although an affected uncle had involuntary movements and two other close relatives were said to be ataxic. The validity of this type of clinical presentation remains uncertain. They could have had Gerstmann–Straussler disease—see p. 93.

Those with chorea and no dementia

Bruyn (1968) referred to a family in which the disease manifested itself in the form of chorea when the subjects were over the age of 60 years, and Gaule (1932) described a family where the proband had the full picture, but other members had only facial tremor at an advanced age. Senile chorea, although said to occur as a result of vascular lesions in the elderly, is rare and should be diagnosed with caution.

Fresh mutations

Few proven cases of fresh mutations in Huntington's chorea have been reported. The criteria suggested for their acceptance are stringent and can rarely be fully satisfied. These are:

(a) Both parents should have lived sufficiently long to have developed the disorder.

(b) Reliable evidence about their health is necessary.

(c) Paternity needs to be proven.

(d) Affected fresh mutants should be shown to have affected offspring.

Case reports partially satisfying these conditions have been reported by Stevens and Parsonage (1969) and Chiu and Brackenridge (1976). Reed and

Neel (1959) do not believe that specific instances of new mutations can be demonstrated.

In a large series (Caro 1977) there was a small proportion (2.8 per cent of all cases) in which even after an extensive search no family history could be substantiated. The figure was higher in Heathfield's (1967) study, but the search less extensive. In his study from a neurological hospital (the National Hospital, Queen Square, London) where unusual neurology tends to be referred, Heathfield found that 50 per cent of cases did not have a significant family history, but in most instances, the facilities for investigating relatives were not available. In general, the diagnosis of Huntington's chorea should still be seriously considered in those with progressive chorea and dementia even in the absence of a family history. Alzheimer's disease and chorea with acanthocytosis (see p. 323) must be excluded.

Mutation rate

In a Swedish series (Mattsson 1974) nine out of 362 patients were singletons —the parents were healthy and had passed the age of 60 years. The calculated a mutation rate of 0.7×10^{-6} when 0.97 is a measure of relative fertility, but he stated that illegitimacy could not be excluded.

In the Wendt and Drohn (1972) series, no new mutations could be detected in 4000 cases. All single cases were illegitimate or a parent died early. Other estimates of the mutation rate are:

	Direct	Indirect
Reed and Neel (1959)	5.4×10^{-6}	9.6×10^{-6}
Wendt and Drohm (1972)	—	1.5×10^{-6}
Mattsson (1974)	5.0×10^{-6}	0.7×10^{-6}

Fertility

It has been suggested that the fertility of patients with Huntington's chorea is greater than that of their normal sibs (Reed and Neel 1959; Wallace and Parker 1973), but when the fertility is compared with the general population, there is no difference. The most likely explanation is that the normal sibs have voluntarily limited their families (Shoulson and Chase 1975). Secondly, it is suggested that females with Huntington's chorea have more children than those fathered by the male with Huntington's chorea—and this holds good at all ages of onset (Jones 1973). It is possible that these observations relate to the dominant role of the female in determining in many instances the size of the family. When her inhibition fails due to the disease process she has more children.

Counselling

Members of the executive committee of 'Combat' in a letter to the *British Medical Journal* (1978), expressed the opinion that individuals at risk have a right to know the facts about their situation so that they can make informed choices about childbearing.

In a postal survey of 2600 members of the American Committee to Combat Huntington's chorea, just under a half chose to reply and of these more than 90 per cent of the respondents were able to state the risk correctly. There was no difference in their understanding whether they had received the information from a lay society or a physician. This high proportion might be expected · as all the respondents were members of a lay group committed to provide information to the membership. Where a whole population at risk is analysed, as in the study of Harper *et al.* (1979) in South Wales, 39 per cent had genetic information from a professional source outside the family itself.

A publicity drive on television evoked a response from 176 who had affected relatives. About half wanted information about inheritance, others about the prognosis, or someone to talk to about their predicament. Part of the function of the genetic counsellor is to assess what the consultand knows, and to explain the nature and clinical course of the disease. 'We knew Huntington's chorea was in our family but not what it meant for us and our children' is a fairly typical response. The family doctor is the ideal person to give advice and to support the family. Alternatively, the neurologist or genetic counsellor could provide counselling and long-term support. The purpose of continuing care for those at risk is to allay anxiety about symptoms that are unrelated to Huntington's chorea, to offer where needed practical support, and to provide information about recent research developments. A long-term programme would attract more family members to the clinic and non-directive genetic counselling should reduce the prevalence of the disease. That non-directive counselling is effective was shown by Carter and Evans (1979) in a letter to *The Lancet* about a follow-up study of patients seen at the National Hospital, Queen Square, London and the genetic clinic at the Hospital for Sick Children, Great Ormond Street, London. The mean number of children born to those at risk was slightly more than half the replacement rate. When those who had come to the clinic before having any children were analysed separately their mean number of children was less than half the replacement rate suggesting that counselling does substantially reduce the prevalence of the disease. Programmes aimed at providing information will be effective if full ascertainment of those at risk can be achieved—even if counselling is non-directive.

The genetic counsellor will need to provide information to the offspring of those at risk. In the pedigree shown in Fig. 16.1 the consultand is clinically

Fig. 16.1

well at the age of 55 years. Her mother is known to have developed Hunting-ton's chorea at the age of 40 years. To what extent can the 1 in 4 risk of the consultand having an affected child be altered?

The calculations for various problems of this type are shown below.

Age of consultand	Probability that a gene carrier will have developed the disorder by a certain age
30 years	0.1
40 years	0.35
50 years	0.6
60 years	0.85

Data from Harper *et al.* (1979).

At 30 years of age

	Carrier of the gene	Not a carrier
(a) Prior probability	$\frac{1}{2}$	$\frac{1}{2}$
(b) Conditional probability of carrying the gene, but not expressing it	$9/10$	1 (if he hasn't the gene he can't express it)
(c) Joint probability = a × b	$\frac{1}{2} \times 9/10 = 9/20$	$: \frac{1}{2} \times 1 = \frac{1}{2}$
(d) Final probability	$(9/20)/(9/20 + \frac{1}{2}) = 9/19$	
	$= 47$ per cent	

The risk at 30 would only fall from 50 to 47 per cent.
Risks to grandchildren of the affected person would be 23.5 per cent.

At 50 years of age

	Carrier of the gene	Not a carrier
(a) Prior probability	$\frac{1}{2}$	$\frac{1}{2}$
(b) Conditional probability	4/10	1
(c) Joint probability a × b	4/20	$\frac{1}{2}$
(d) Final probability	$(4/20)/(4/20 + \frac{1}{2})$	
	= 4/14	
	= 28 per cent	

The risk at 50 would drop from 50 to 28 per cent.
Faced at that stage with the problem, the risk to children would still be significant (about 1 in 7) and risks to grandchildren 1 in 14.

At 60 years of age

	Carrier of the gene	Not a carrier
(a) Prior probability	$\frac{1}{2}$	$\frac{1}{2}$
(b) Conditional probability	1.5/10	1
(c) Joint probability a × b	1.5/20	10/20
(d) Final probability	$(1.5/20)/(1.5/20 + \frac{1}{2})$	
	= 1/8	

Risk to children 1 in 16.
Risk to grandchildren 1 in 32.

The other statistic which could be used less easy to manipulate and is based on the finding of an intrafamilial correlation in the age of onset of the disease.

Bell (1942) recorded an age of onset correlation of 0.593 for parents and offspring from 153 pairs, and Panse (1942) a value of 0.66 from 103 pairs. Brackenridge (1972) found a correlation of 0.531. There is also a good correlation for age of death—0.472 (Bell 1942), and 0.456 (Brackenridge 1972). Sib-sib correlation was 0.465 in the Bell series (1942) and 0.587 from Brackenridge (1972). A much lower figure was calculated by Reed and Chandler (1958), namely 0.28 in their Michigan study.

It is, however, the insidiousness of onset that limits the usefulness of all the calculations that are based on the age of onset, denial of symptoms, and unrecognized early symptoms are common occurrences.

Most people at risk have probabilities of greater than 1 in 10 for developing Huntington's chorea and their options are:

(1) not to have children;
(2) to have one or two children;

(3) to disregard the risk completely;

(4) to use alternative methods of raising children (see following section);

(5) to have the predictive test on themselves;

(6) to have the predictive test pre-natally on a fetus at risk.

Sterilization

Those who decide not to have children need to take precautions. It is not unknown for contraception to be used half-heartedly and in such a way that if a mistake 'happens', the guilt could be channelled into 'failed contraception'. Those who are more certain of their decision might want to be sterilized. In general, it is easier to consider the procedure in patients who are older than 30 and more difficult when the consultand is in his or her late teens or early twenties. It is appropriate to advise the young to wait, on the grounds that therapy might become available. All well-meaning, but misguided attempts by parents to have their children sterilized should be resisted.

Realistic, but optimistic counselling

It is clear that the identification of those at risk for the purpose of helping them make an informed decision about family planning also identifies those who must then live with the knowledge that they have a serious risk of developing a progressive neurological disorder for which there is at present no treatment. A considerable proportion of the counselling session should therefore be devoted to helping individuals accept the situation, possibly by emphasizing that there is a decade or two before the period of maximum risk is entered and that in time there might well be useful therapy. Recent research ideas can be outlined and a feeling generated that doctors and scientists are interested in Huntington's chorea and that prospects for improved understanding of the mechanism of the disease are good.

To some extent the recent advance in the localization of the gene has helped to emphasize that research is progressing rapidly and recent television programmes have highlighted therapeutic advances, albeit in experimental animals.

If part of the usefulness of the follow-up clinic is to consider periodically all aspects of this truly familial disease then the relationship between those at risk and the counsellor will inevitably progress to one of mutual confidence and understanding. In order to be reassured that myoclonic jerks experienced at night are not choreiform and in no way different from normal experience then the person at risk will want to feel confident that the doctor is telling the truth. 'You will tell me if I do show signs of the disease?' is a common request and the doctor can only answer in the affirmative.

The man who is losing his job because his personality has changed sufficiently for him no longer to be able to cope will not be helped by untruths. On the other hand, the man who has decided not to have further children and has suspicious, but not definite signs will not be helped by a vague opinion. Intervention by the counsellor should usually await a crisis and the information only be dealt with if adequate support from the family, doctor, and social worker is available. Not all patients want to attend a follow-up clinic and prefer to cope by not thinking about the problem. It is the availability of a clinic, where patients can attend, albeit even irregularly, that is of prime importance.

Over the past 5 years the author has found it necessary on a number of occasions to intervene in a non-crisis situation. Those at risk have become more eloquent in their demands that they need to know the diagnosis as soon as it can be confidently entertained so that they can organize with due care the remaining years of their lives. In nearly all of these situations, the at risk people have thought through their options with great care and without exception their wishes have been adhered to. To date there has been one suicide in this group (out of 40).

Decisions about family planning

Most people faced with the high risk do not proceed to have children. When the disease has been identified after the birth of one child, some parents are unsure whether they should stop, or provide a hopefully normal playmate for their firstborn. One of the frequent justifications used by families for accepting the 1 in 4 risk and planning children is that the late onset of the disease would allow the children to have at least half a life. Most of these parents have lived with family members with an onset in the late 40s or early 50s. It is justified in this situation to gently point out that the age of onset does not run true within families and that an onset in the 30s is frequent enough to be seriously considered.

Alternative methods to having a child of one's own

Adoption, fostering, AID, and ovum implantation have all been considered by individuals at risk—AID where the father is at risk and ovum implantation where mother is at risk. They all have the disadvantage that there is a real possibility that the family will be disrupted by the disease so that couples need to consider their alternatives carefully. Fostering has been suggested as the best alternative, but the strain of short-term fostering would need to be assessed.

Predictive tests

There is no test that will, with certainty, indicate whether the offspring of those with Huntington's chorea are gene carriers or not.

However, the Huntington's chorea gene has now been localized to the

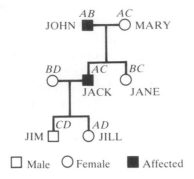

Fig. 16.2   An example of how the test works
Acknowledgments to Mrs Dalby of 'Combat'. Note—the current cross-over rate
is 2–3 per cent.

short arm of chromosome 4 (Gusella *et al.* 1983). This extraordinary achievement was to some extent serendipitous. The researchers chose a small number
of probes of known locality and found that one on the short arm of chromosome 4 segregated in a large Venezuelan pedigree with those who had the
disease. Sufficient polymorphism (variation) exists in the area alongside the
Huntington's gene for the two number 4 chromosomes to be differentiated
from each other in perhaps two-thirds of families.

The family tree in Fig. 16.2 is an example of how the test could work. John
has *HC*, and his two number 4 chromosomes are coded *A B*. His son, Jack, is
also affected, and his are coded *A C*. As he must have inherited one of the pair
from each parent, it can be seen that he inherited the *A* from John and the *C*
from his mother, Mary. The *HC* gene must therefore be travelling with *A*.
Jane, John's daughter, is coded *B C* and, therefore, has a low risk of carrying
the *HC* gene. Jack's son, Jim, has inherited the *C* from Jack, so that he also is
at low risk, but Jill has inherited the *A* and so she will be at high risk (98 per
cent).

There are, of course, many variations within families, It is essential that the
haplotypes, or coding letters, of the parents and grandparents are different so
that the chromosome 4 which comes from the affected side of the family can
be clearly identified.

Different haplotypes occur with varying frequency in populations. In any
one family one of these patterns can be seen as identifying the presence of the
HC gene, but it is not necessarily the same code in each family. Thus, in each
family group the letter code has to be identified, and blood tests from as
many key members as possible have to be obtained to do this.

If John had already died prediction in the case of Jack would be difficult
(as either the *A* or the *C* could have come from the surviving Mary).

Figure 16.3 shows a situation where *B* must have come from the affected
parent who is dead. *B* might not contain the mutation (50 per cent risk).

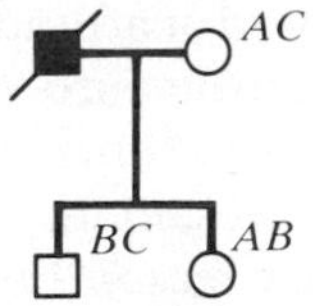

Fig. 16.3

Prenatal diagnosis might be considered. The result that can be expected is either that the fetus has a low risk (2 per cent) of having inherited the *HC* gene (the chromosome 4 from the affected grandparent has not been passed down)—which will occur half of the time—or that the fetus has a 50 per cent risk in that it has inherited the chromosome 4 from the parent which came originally from the affected grandparent. It will still not be known whether or not this carries the *HC* gene, but clearly the implication is that if the fetus has the same number 4 as the affected grandparent, then the risk will approach 50 per cent. If the parent does show symptoms of *HC* later on, then a child born with this '50' per cent result would automatically be at a very high risk.

Prenatal testing is likely to be by chorion biopsy rather than by amniocentesis, since the former can be performed at an earlier stage (10 weeks).

Example 3 shows how the test works. Jack has received the *A* from John, but does not know whether or not it is carrying the *HC* gene. Fetus 1 is clear, (2 per cent risk) not having inherited the *A*, but fetus 2 has the *A* and is thus at the same risk as Jack. A termination would be considered. Fetus 3 carries the *B* also originally from John and the same considerations apply. Fetus 4 is clear (± error rate).

Every effort needs to be made to collect and store blood samples from affected persons and elderly key relatives, even if no-one in the family is currently seeking prediction.

Also, diagnostic accuracy is now more essential than ever and if there is any doubt whatsoever, a post-mortem examination of the brain should be performed.

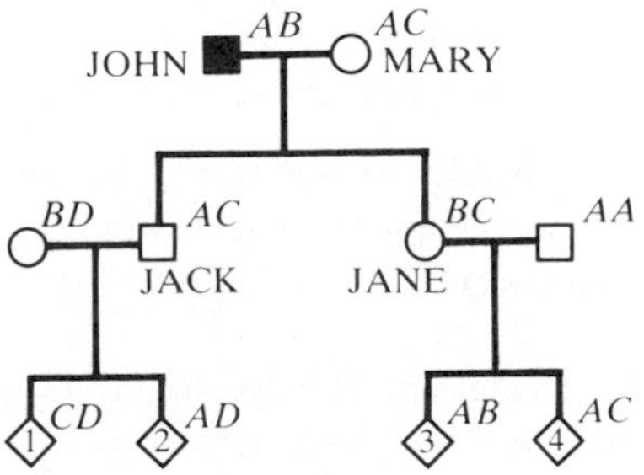

Fig. 16.4

A number of studies have looked at attitudes to presymptomatic testing in Huntington's chorea. One (Mastromauro *et al.* 1987) looked at over 100 persons who were at 50 per cent risk. Ninety-six per cent thought that the test should be available and 66 per cent said they would use it themselves. About 40 per cent said that the primary reason for taking the test was to end the uncertainty in their lives. About 15 per cent of those who wanted to be tested thought they might be at risk of suicide. In the Kessler *et al.* study about 2–6 per cent of those at risk might have severe psychiatric problems if the results were positive. Kessler *et al.* (1987) suggests that these issues could well be judiciously explained prior to testing and that the life of those at risk might be protected to some extent by enhancing the coping skills of caretakers. It is true to say that everyone has the 'right' to end their life, but as stated by Kessler, gestures, whether verbal or actual, might and should be seen as cries for help to find reasons to live.

Counselling for the predictive test

A typical situation might be as follows

1. A combined interview which would include the genetic counsellor in the presence of a 'personal counsellor' who could be a social worker/health visitor/psychologist, who would act as a link between the clinic and the patient and be responsible for further discussion. At the initial interview (after drawing up pedigree, confirming the diagnosis and the calculation of the risk) the test, the error rate, the time schedule, and how the results would be given are discussed. Blood is taken only from affected persons who are seriously ill.

2. A follow-up appointment by the personal counsellor where the above points are re-emphasized and matters arising from the initial interview can be discussed. It is important that at any time the patient should be able to opt out.

3. A second interview with the genetic counsellors follow after 3 months. If the patient still wants the test, blood is taken and the laboratory work is done to see whether the family is informative.

4. Results should be handled by the genetic counsellor and the personal counsellor as a team. If the result is negative, i.e. the risk for the fetus or for the patient is 2 per cent this is relatively straightforward 'good' news.

If the result is positive (usually 98 per cent), support to help the patient is essential. There is clearly a role for the social worker, other members of the family, or the lay organization 'Combat'. This must be arranged at the first visit.

Availability is the key. In general, the decision to take the test is the responsibility of the client and it will be very difficult to withold the test from any adult who wanted it. The person must have reached the age of majority.

Pre-natal and pre-symptomatic testing are now on offer in the UK and the United States. Those centres offering the test are proceeding cautiously and guidelines have been drawn up (similar to those above) to help recipients of the test through a process which might result in a period of anxiety.

A first trimester pre-natal diagnosis for Huntington's disease was published by Hayden *et al.* (1987). The fetus proved to have a 48 per cent risk (i.e. a positive result) and on this basis the parents elected to terminate. A study by Quarrell *et al.* (1987) looked at 55 couples who had a 50 per cent risk. In three of nine pregnancies the test was negative, in three the risk was ± 50 per cent and in two the test was uninformative. In the latter situation the family was evaluated after the pregnancy had commenced and hence an unsatisfactory situation emerged.

Registers in Huntington's chorea

Genetic registers have been recommended for the purpose of tracing and counselling those at high risk of transmitting a serious genetic disorder (Emery *et al.* 1978). Dominant and X-linked diseases are the most appropriate for consideration and Huntington's chorea is a good choice of a dominantly inherited condition.

Experience from the Department of Psychiatry, University of Melbourne, was reported by Chiu and Teltcher (1978). In 1978 there were 216 kindred listed. The suggested advantages of the register are assistance with diagnosis and ascertainment of those 'at risk' and recently an excellent microcomputer programme has been designed for this purpose (Sarfarazi *et al.* 1987).

Confidentiality is strict and information is released only to medical practitioners or fellow workers in the field. It was found that whereas acceptability of the register to the families at risk was achieved, this was less evident when professionals were consulted. Informed consent is mandatory and no individuals should be included without their knowledge and approval. The accountability of those involved with keeping the register must be considered and bodies such as the Genetic Society, Combat, or a Central Health Authority would be suitable to audit the operation.

Not all families would agree to allow their names to be included on a register, nor even in some instances for the geneticist to inform the general practitioner that they are at risk for developing Huntington's chorea. Some have had the experience that the 'at risk' status has counted against them when applying for life insurance and for a job. These wishes should be respected by the geneticist.

Familial chorea with acanthocytosis

Both dominant and recessive pedigrees are known, and there is no way of differentiating between the two types. There is in addition no constant relationship between the blood disorder and the neurological disease. Both can occur independently within a family or co-exist. Critchley *et al.* (1968) described a man with involuntary movements ('finger-snapping, grimacing, choreiform movements') with an onset in his early 20s. Eating became difficult and there was occasional inappropriate laughter. Reflexes were absent. Acanthocytes were present in the peripheral blood but serum lipids were normal. Three of the patient's sibs were known to have died in a mental institution and were said to have been psychotic. Two sibs had acanthocytosis but no neurological illness. A niece had Friedreich's ataxia and extensive myocardial involvement, but no involuntary movements. She had a few cells resembling acanthocytes in her bloodstream. Her mother (proband's sister) had acanthocytosis but, at the age of 49, was neurologically intact. Critchley *et al.* reported an English family in 1970.

A dominant pedigree was published by Levine *et al.* (1968). Nineteen members had neurological signs, and 15 had acanthocytosis. Only three had chorea and some were only very mildly affected. Aminoff (1972) described a dominant pedigree with weakness and wasting of late onset (58 years) and with involuntary orofacial movements. A sib was similarly affected. No intellectual deterioration was noted. There was a suggestion that members in the preceding generations might have had the same disease.

Three siblings, the offspring of parents who are first cousins once removed, had a clinical picture of an adolescent onset of chorea, mild to moderate dementia, decreased reflexes, and acanthocytosis (Bird *et al.* 1978). Serum beta-lipoproteins were normal. Post-mortem changes showed marked neuronal loss and gliosis of the caudate and putamen. The parents were alive at 87 and 80 years, respectively. A further point differentiating the condition from Huntington's chorea was the presence of decreased tendon reflexes. The family of Estes *et al.* (1967, same family as Levine *et al.* 1968) came from New England, and that of Critchley *et al.* (1968) from Kentucky.

A dominant pedigree was reported by Kito *et al.* (1980). They proposed that the syndrome be called 'amyotrophic chorea with acanthocytosis' and that the criteria originally suggested by Levine *et al.* (1968) to characterize the syndrome be adopted:

(a) extrapyramidal movement disorder;

(b) generalized muscle weakness;

(c) acanthocytosis (1–50 per cent);

(d) occasionally ataxia, epilepsy, schizophrenia;

(e) normal IQ

Two further sibs were reported by Vance *et al.* (1987). They emphasized that hyporeflexia, marked buccal-oral dyskinesia, a peripheral neuropathy, a normal mental state, and an increased CPK should alert the clinician to this diagnosis. An NMR scan might show caudate nuclear atrophy.

At least five autopsies have now been performed. All have shown changes in the caudate, the putamen, and pallidum.

There is also ample evidence of sensory change. At least one post-mortem showed a marked loss of large myelinated axons and their neuronal cell bodies in the ventral spinal outflow tracts (Sobue *et al.* 1986) Segmental de- and remyelination was also marked in both motor and sensory nerves. Two brothers were clinically similarly affected and parents were normal and non-consanguineous. Other brothers are reported by Gross (1985). Orofacial tics and seizures occured as well as neurogenic atrophy.

Spitz *et al.* (1985) reported tics, Parkinsonism and motor neuron disease in two brothers whose parents were distant cousins. Both had acanthocytosis. In one, the leg weakness and wasting started at 13 years and he went off his feet at 40. Facial tics began at 36 and thereafter Parkinsonian features and an ophthalmoplegia. Fasciculation was present in the legs.

Counselling

From the foregoing it must be concluded that both autosomal dominant and autosomal recessive forms exist. It does not seem possible in single cases to differentiate between the two.

Hereditary whispering dysphonia/dystonia/chorea

Members of an inbred population in North Queensland were able to shout and yell with emotion, but when they tried to speak normally only managed a whisper (Parker 1985). Despite the inbreeding it was concluded that the pattern was autosomal dominant with incomplete penetrance. The speech disturbance was often only the initial problem in a condition that developed into a torsion dystonia. A sister and brother within this family had bio-chemically proven Wilson's disease, but none of the other members were similarly affected.

The family was originally included in a survey of the atypical Westphal variant of Huntington's chorea. Some members in the family had violent jerking spasms of the trunk and limbs whereas others had transient torticollis, a spastic dysphonia, and champing movement of the jaw. The clinical signs differed from generation to generation. Indeed, one generation was remembered for the choreiform movements.

Wilson's disease

In 1911–12 Wilson described 'a familial nervous disease associated with cirrhosis of the liver'. The clinical picture is so variable that any child or young adolesent with basal ganglia disease or ataxia should be considered to have Wilson's disease (a treatable disease) until otherwise proven. Unexplained deterioration of school work, a speech disorder, or psychotic episodes are other early features.

A proportion of the homozygotes present with symptoms and signs of liver disease. Whether the presenting features are related to the liver or the nervous system is age dependent—the younger the age of onset, the more likely that the hepatic manifestations will precede the neurological signs. In a series (quoted by Cox *et al.* 1972) the liver was the organ involved in 14 of 15 patients who had an onset at 10 years or less. In six patients with primary neurological disease, the age of onset was greater than 23 years. The hepatic manifestations of the disease are not always cirrhotic. Seventeen patients presented with chronic active hepatitis (Scott *et al.* 1978) with a mean age of onset of 18 years and in a series of 200 patients with Wilson's disease (Sternlieb and Scheinberg 1972), about a dozen presented with chronic active hepatitis. Renal tubular acidosis, aminoaciduria, and renal stones have all been reported (Bearn 1960).

As stated by Sternlieb (1978) there are five main characteristic features of Wilson's disease.

(a) Kayser–Fleischer rings—an essential feature for the diagnosis of Wilson's disease in those with neurological disease, but not in those with only hepatic manifestations.

(b) A ceruloplasmin of less than 20 mg per 100 ml; this is true for 95 per cent of cases of those with neurological disease, but normal or even raised levels do not exclude the diagnosis.

(c) Marked diminution of incorporation of radioactive copper into ceruloplasmin (there is also an alteration in the rate of uptake of radioactive copper by the liver and a prolonged turnover in the body).

(d) A urinary excretion of more than 100 μg of radioactive copper in 24 hours (Sass-Kortsak 1975). Similar levels have been reported in primary biliary cirrhosis (La Russo *et al.* 1976).

(e) A liver copper in excess of 250 μg/g on liver biopsy and the histological presence of fine fat droplets, nuclear vacuoles, and a positive stain for copper (Scott *et al.* 1978).

The loading test using radioactive ^{64}Cu can be performed by most radio-isotope laboratories (Osborn and Walshe 1969), but some tests are more sophisticated requiring a whole body counter or ^{67}Cu (Vierling *et al.* 1978). Normally, the copper disappears from the serum within 4–6 hours with a slow secondary rise which follows the release of that copper incorporated into freshly synthesized ceruloplasmin secreted by the liver. In Wilson's disease, this secondary rise is absent.

Sibs were reported by Bearn (1960), Walshe (1967), and Bickel *et al.*

(1957). It is, however, still uncertain whether Wilson's disease is a single genetic entity or not. Cox *et al.* (1972) suggested a subdivision into those from Eastern Europe—the Slavic form—a Western European type which is also found in Orientals, and atypical cases. The Slavic form occurring in Jews and non-Jews is primarily a neurological disorder with an onset after 16 years. The Western European form has an earlier age of onset (less than 16 years), and the initial symptoms are usually hepatic, although between 10 and 16 years both hepatic and neurological manifestations can occur. There is also a suggestion from an Israeli study (Passwell *et al.* 1977), where the disease was found to occur in all ethnic groups, that heterogeneity exists. In the Arab patients the age of onset was earlier and the course more severe. Concordance for age of onset was found within sibships but not between Jews and non-Jews. The consanguinity rate was very high in the Arab patients. Similar families were reported from Japan (Arima and Sano 1968).

In a series of 29 patients (Dobyns *et al.* 1979) none were Jewish and parental consanguinity was not noted in any of these families. In an analysis of 18 Indian probands Dastur *et al.* (1969) found consanguinity in 27.7 per cent in a group where, for social reasons, cousin marriages were not encouraged. Neurological signs were the initial manifestations in 14, and hepatic in three, suggesting that the Indian type was of the Slavic form. In eight patients bony changes were prominent. These patients gave a history of early difficulty in walking, and an inability to run due to bone deformity or fractures. On X-ray there was generalized osteoporosis and active rickets. In three of five families, this type of deformity bred true.

Unusual findings included hypoparathyroidism (Carpenter *et al.* 1983) and hyperphosphaturia and rickets in association with a renal tubular acidosis.

A simple division between those patients with an early onset hepatic disease and those with the late onset neurological disorder is possibly no longer tenable. There are now on record many patients with neuro-psychiatric symptoms who present before the age of 10 years with intellectual deterioration, dystonia, choreoathetosis, lethargy, and poor school performance.

Areas of low density in the basal ganglia are useful diagnostic findings but cerebral and cerebellar atrophy have also been reported (Lingam *et al.* 1987).

Only 1.9 per cent of cases of fulminant hepatic failure have Wilson's disease (Nazer *et al.* 1983). Neither absence of Kayser–Fleischer rings nor a normal ceruloplasmin excludes the condition.

Counselling

The 1 in 4 recurrence risk after the diagnosis has been established in one child seldom has preventative value as most parents have completed their family by the time the diagnosis is made in the first child. More important in this potentially treatable disorder is the recognition of early biochemical and clinical evidence of Wilson's disease in sibs before overt manifestations so that treatment can be started early. This can be achieved by appropriate radioactive copper tests and urinary and blood copper estimations. It should be the responsibility of the genetic counsellor or neurologist to see that this service is offered in conjunction with a regular follow-up.

A possible linkage to esterase-D on chromosome 13 was suggested by Frydman *et al.* (1985) in a large inbred Israeli-Arab family.

Hypoceruloplasminaemia

Values of less than 21 mg/100 ml in males and 23 mg/100 ml in females who were clinically normal and had no evidence of Wilson's disease, were reported by Cox (1966), and a large pedigree over three generations by Edwards *et al.* (1979). The subjects are either heterozygous for Wilson's disease or have a gene (which lowers ceruloplasmin) distinct from that responsible for Wilson's disease.

Willvonseder syndrome

The combination of dementia, spastic dysarthria, paresis of vertical eye movement, a complex gait disturbance, and abnormal copper metabolism was reported in three brothers by Willvonseder *et al.* (1973).

One brother had reduced serum ceruloplasmin, but normal urinary copper excretion. Kayser–Fleischer rings were not a feature.

Familial inverted choreoathetosis

A family with infantile onset of progressive choreoathetosis occurring in 10 members over four generations was reported by Fisher *et al.* (1979). Dementia was not a feature.

Familial chorea and myoclonic epilepsy

See p. 119 for the family of Takahata *et al.* (1978). Many of the families described under this heading resemble the condition of hereditary essential myoclonus (see p. 121) and hereditary benign chorea (see p. 194; Kurlan *et al.* 1987).

Lesch–Nyhan syndrome

The diagnosis of Lesch–Nyhan syndrome rests initially on the observation of motor delay and involuntary movements (usually choreoathetosis) in a male infant towards the end of the first year of life. The diagnosis can be confirmed by a raised uric acid level and the absence of the enzyme hypoxanthine guanine phosphororibosyl transferase. Later, self-mutilation and seizures occur in most children. Males who have the disease have carrier mothers in more than two-thirds of cases and fresh mutations are less frequent than expected. Both carrier detection and prenatal diagnosis can be offered (Bakay *et al.* 1980; Halley and Heykels-Dully 1977). The condition has been localized to Xq26–7.

Lesch–Nyhan syndrome without self-mutilation

A patient is reported by Gottlieb *et al.* (1982), with uric acid nephrolithiasis, spasticity, and choreoathetosis. He was not mentally retarded.

Basal ganglia disease without dementia or progression

Familial essential benign chorea

An early childhood onset of chorea, not associated with dementia, seizures, rigidity, or ataxia and without marked progression has been described in families. Bird *et al.* (1976) described the condition in eight members of a family over four generations (a dominant). Haerer and Jackson (1967) described a similar condition in several members over three generations. Other families were reported by Pincus and Chutorian (1967), Sadjadpour and Amato (1973), Deonna and Voumard (1979), Burns *et al.* (1976), and Behan and Bone (1977). All were compatible with dominant inheritance.

Chun *et al.* (1973) reported a family of four sibs with unaffected parents suggesting that there might be a recessive variety. They also described two dominant pedigrees—with reduced penetrance in one. An unaffected mother had two children with early onset non-progressive chorea. Her brother, sister, and father had 'shaky hands'. Nutting *et al.* (1969) described another probably recessively inherited benign chorea in three sibs. The parents were unrelated and normal.

Several authors have noted that an intention tremor can occur in families with essential benign chorea (Pincus and Chutorian 1967; Haerer *et al.* 1967). It was emphasized that the tremor was not the disintegration of movement by sudden choreiform jerks. Harper (1978), in a review of the subject, suggested reduced penetrance as a possible explanation for the recessively inherited pedigree of Chun, and added two Welsh families to the list of dominant pedigrees.

However, the possibility of a recessive type remains. In another pedigree, two out of four sibs were affected (Damasio *et al.* 1977). Both parents were unaffected.

An association with intellectual impairment is described by Leli *et al.* (1984) in a four-generation family. Onset was in the first year of life. Movements increased during childhood but remained stable after adolescence. Intellectual achievement was in the low normal range.

There is some indication of a beneficial response to steroids (Robinson and Thornett 1985).

Hereditary non-progressive athetoid hemiplegia

A syndrome characterized by congenital left-sided hemiparesis with hemihypoplasia and hypertonicity was described in four members over three generations by Haar and Dyken (1977). Towards the end of the first decade athetoid movements developed in the affected hand, but there was little progression after childhood. Right-sided cerebral atrophy was present. Inheritance is as a dominant with incomplete penetrance.

Trembling of the chin

There have been a number of families described with isolated involuntary trembling of the chin (Frey 1930; Grossman 1957; Wadlington 1958). Two English families with many affected in three succesive generations were recorded by Laurance *et al.* (1968). Although embarrassing and sometimes distressing, it is usually benign. Onset is in early childhood.

Familial spasmodic torticollis

Both dominant and recessive pedigrees are known, but most cases are sporadic. Under the title 'A wry-neck family', Thompson (1896) reported spasmodic torticollis in two brothers and two sisters. The onset was in young adulthood and childhood. Tibbetts (1971), in a series of 71 cases, found one pair of sibs and one patient's father and brother were similarly affected. In a large series (92 patients) Armstrong *et al.* (1965) found one patient with a maternal grandmother who was similarly affected. Four cases of familial spasmodic torticollis (Gilbert 1977) were a man and his daughter and, in a separate family, two sibs. The age of onset in the sibs was 7 and 16 years, and in the father and daughter 57 and 19 years, respectively.

Torticollis occurs as a phenomenon in dystonia musculorum deformans, but seldom remains isolated. In all the cases cited above, no progression occurred to other parts of the body.

Gilles de la Tourette syndrome

The condition that now bears Tourette's name was first described by Itard (1825) when he reported a young French woman who, at the age of 7 years, developed tic-like movements of the hands, shoulders, and arms (Eldridge *et al* 1977). Within a few years she began to utter uncontrollable obscene words. Tourette syndrome has an age of onset between 2 and 15 years, it affects more males than females, and clinically, there are recurrent, involuntary rapid motor movements resulting in violent head jerking, shoulder twitching, and tongue thrusting, often accompanied by obscene shouting. In the early stages patients are able to voluntarily suppress the movements, but this becomes progressively more difficult.

The familial nature of the syndrome was noted by Tourette himself in 1885 when he observed that mild cases occurred in the same family as those with the classical clinical picture, but little more was said about the genetics until 1959 when Eisenberg confirmed the familial aggregation. The past 28 years has seen a steady unravelling of the genetic contribution and the progress made might be seen as a forerunner in our understanding of some of the commoner genetic disorders (Tourette syndrome has a prevalance of 0.1–0.5 per thousand).

It is the nature of the familial aggregation that has been difficult to explain. Within families more members are affected than could be accounted for by chance, but transmission is not regular and 'skips' within families have led to argument about the likelihood of polygenic inheritance versus the affect of a single gene of major importance, with incomplete penetrance. There has also been a suggestion that many cases of the Tourette syndrome are not genetically determined but are phenocopies (Comings *et al*. 1984). The relationship between isolated simple tics, multiple motor tics and the Tourette syndrome has been long debated (Pauls *et al*. 1984). Simple tics occur in 20 per cent of normal children, but multiple recurrent tics are much rarer and they occur more frequently in families with Tourette syndrome. It is now thought that they are likely to be a manifestation of the same gene.

Most of the earlier workers have used complicated mathematical models in an attempt to differentiate between multifactorial inheritance, and the effect of a single major gene locus. In a big study of 250 patients and their families,

Comings *et al.* (1984), provide strong evidence for a major semi-dominant gene with low heritability, but it was not until a more thorough assessment was undertaken, mostly by personal examination of the families that the pattern of inheritance became more understandable.

There had over the past 20 years been accumulating evidence that various psychiatric conditions were prevalent in the Tourette families. These have varied from hyperactivity and exhibitionism to obsessive compulsive behaviour (Montgomery *et al.* 1982) Comings and Comings (1984) noted that 62 per cent of Tourette syndrome male patients had an attention deficit disorder (ADD) with hyperactivity. An examination of the pedigrees suggested that the gene could be expressed as an attention deficit with hyperactivity alone, without tics. These authors also noticed that some patients referred for ADD with hyperactivity had on careful questioning Tourette syndrome-like symptoms. They listed echolalia, compulsive touching, inability to be able to relax, self-mutilation, and repetitive acts, as common features in Tourette patients.

In a recent paper, Pauls and Leckman (1986) have examined the specific genetic hypothesis that Tourette syndrome is a highly penetrant dominant trait. They first analysed Tourette syndrome within families, only scoring those with the full clinical complement and then re-analysed the families including chronic multiple tics as a gene manifestation. A third investigation looked at the same families using chronic tics or obsessive compulsive behaviour as part of the condition. What emerges is that the psychiatric manifestations and multiple tics are likely to be manifestations of the Tourette gene and the syndrome could be a simple dominant with high penetrance. The authors clearly state that they are not suggesting that all those with obsessive compulsive disorders have the Tourette gene. Interestingly, it had been previously noted that there might be two types of persons manifesting obsessive compulsive behaviour; those with a family history of tics and those without, so the new evidence is not altogether surprising.

The final part of the genetic jigsaw might have been filled in at the 7th International Congress on Human Genetics in Berlin. Comings *et al.* (1986) reported a family in which six members with varying manifestations of the Tourette syndrome all had a 7q22:18q22 balanced translocation. If it can be shown that the breakpoint went through the Tourette gene (they excluded chromosome 7) then DNA technology will finally solve the problem that complicated mathematics could not and that is that in most instances Tourette syndrome is inherited as a single gene disorder with simple transmission.

Whether Tourette syndrome is more prevalent in Ashkenazi Jews is still uncertain. Eldridge *et al.* (1977) looked at 21 selected patients from New York. In 18 there was a positive family history. 17 of the 21 were of Ashkenazi Jewish or other Eastern European ancestry. Despite the relatively high prevalence of the disorder in Ashkenazi Jews, incomplete dominance could not be excluded. In another series (Golden 1978) six out of 13 families were Ashkenazi Jews.

Wassman *et al.* (1978) studied the syndrome in a mid-western-American city, chosen specifically because it was less likely to have a large population of Ashkenazi Jews, and found that those affected were predominantly of that ethnic group. The pedigree analysis suggested incomplete dominant inheritance. It should, however, be noted that a recent study (Nee *et al.* 1980) does not support the evidence that the Tourette syndrome patients with a family history are predominantly Jewish. Those affected showed no ethnic predeliction.

For references the following pedigrees seemed to show dominant inheritance. One with incomplete penetrance showed seven affected members over two generations (Frost *et al.* 1976), and in the pedigree reported by Sanders (1973) and Friel (1973) two generations were affected. In the 15 patients examined by Moldofsky *et al.* (1974), five had a parent with the same disorder. In three families sibs only were affected. A remarkable pedigree with no fewer than 17 out of 43 members (with skips) was reported by Guggenheim (1979).

Five sets of twins were reported (Wassman *et al.* 1978) in which at least one twin had Gilles de la Tourette syndrome. Of the three monozygotic pairs, two were concordant and the one dizygotic twin pair was discordant.

Counselling

Autosomal dominant inheritance with variable expression and penetrance should be the model for the genetic counsellor.

Essential tremor (hereditary tremor, senile tremor)

This condition has been recognized for many years (Dana 1887; Gowers 1893) but only a few post-mortem reports have been published (Myle and Van Bogaert 1940; Herskovits and Blackwood 1969). No specific lesions were found.

The disorder may start at any age (10–70 years). It is only slowly progressive and involves mainly the upper limbs and head (Critchley 1949). The tremor is accentuated by emotion and often relieved by alcohol.

The upper extremities are affected more often than the lower, the head more often than the trunk (Critchley 1972). Speech is occasionally affected. Cerebellar signs are not usually present but the most severely affected members have gait ataxia. The condition is dominantly inherited.

A similar disorder was described by Hornabrook and Nagurney (1976) in a study in Papua New Guinea. Extrapyramidal signs were not a feature in 175 cases (although seven patients also had a 4–5 c.p.s. pill-rolling tremor). The onset was in middle life, penetrance was variable, and there was a strong female preponderance (only 27 per cent of the cases were males). Larsson and Sjögren (1960) described 121 members of a family over four generations. The variability of the clinical manifestations was pronounced in the same family. The age of onset was between 30 and 65 years. In their family, the degree of penetrance was high.

Another pocket of high incidence was reported by Rautakorpi and Rinne (1978) in two south-west Finnish communities. The prevalence in individuals over the age of 40 years was 55.7 per 1000. The mean age of onset was 42 years. Progress was slow and the disability small.

A large family, now members of the Mennonite religious sect of Lancaster County, Pennsylvania, had many members affected over 10 generations (Fogelson and Zwericki 1969). The first manifestation of the tremor was in the upper extremities and it then spread to the head and then the lower extremities.

A series of 16 patients were seen by Sutherland *et al.* (1975) of whom 10 had near relatives (parents or sibs) who were similarly affected. Age of onset in this series varied from 15 to 65 years.

Orthostatic tremor

In this condition tremulousness appears in the trunk and legs shortly after standing and diminishes upon walking, sitting and lying down. A family with essential tremor was described by Wee *et al.* (1986) in which some members had an orthostatic tremor.

Shuddering attacks in children

Six infants with attacks of shuddering described by the parents 'as if cold water had been poured down their spine' were reported by Vanasse *et al.* (1976). In five cases a parent had an essential tremor.

Familial basal ganglia calcification

Basal ganglia calcification was called Fahr's disease following a report in 1930 of patients with symmetrical calcification of the basal ganglia, dentate nucleus, and cerebral cortex. According to Melchior *et al.* (1960), it is now certain that this is not a single disease entity, and the eponym Fahr's disease is no longer tenable.

Genetic causes

A. Recessive

(a) Pseudohypoparathyroidism (including familial basal ganglia calcification of unknown cause).

(b) Hypoparathyroidism.

(c) Basal ganglia calcification with mental retardation and steatorrhoea.

(d) Cockayne syndrome, see p. 472

(e) Familial encephalopathy and basal ganglia calcification, see p. 336.

(f) Disorders of folate metabolism, see p. 502.

Many of the older reports do not differentiate between those sibships with normal and abnormal parathyroid function nor is it certain which familial cases are compatible with the diagnosis of pseudohypoparathyroidism. Pseudohypoparathyroidism is characterized by short stature, obesity, mental retardation, brachymetacarpia, abnormal facies, and subcutaneous calcification. The fourth and fifth metacarpals are short. Patients fail to respond to parathyroid hormone with phosphaturia. Pseudopseudohypoparathyroidism is clinically similar, but with normal calcium and phosphate metabolism. Families are known where the two co-exist. There remains a group in which there is no biochemical evidence of parathyroid disease nor clinical evidence of pseudohypoparathyroidism. This is called familial basal ganglia calcification, which can be both recessively or dominantly inherited.

(a) Pseudohypoparathyroidism (including familial basal ganglia calcification of unknown cause)

Two sisters, whose parents were second cousins, developed in middle life a Parkinsonian-like syndrome. Dense calcification of the basal ganglia was shown (a third sib was possibly affected; Matthews 1957). Because no response could be obtained to parathyroid hormones, the possibility of pseudohypoparathyroidism was raised, but dismissed by Mann *et al.* (1962).

Two families with a total of five affected sibs had members with mental deterioration, spasticity, contractures, and athetoid movements (Melchior *et al.* 1960). In the sibship described by Fritzsche (1935), and Sala and Savoldi (1959) parental consanguinity was recorded. Bruyn *et al.* (1964) suggested that the case of Fritzsche might be Albright's disease (pseudohypoparathyroidism).

Two female sibs (Bruyn *et al.* 1964) were retarded from early infancy, but then developed progressive dementia. They were unusually short.

Five out of nine sibs in two families from Western Norway developed progressive mental deterioration, pyramidal and extrapyramidal signs (choreoathetosis, rigidity, mask-like facies), and cerebellar ataxia (Nyland and Skre 1977). Dense symmetrical calcification was present in the basal ganglia, occipital cortex, and dentate nucleus. Calcium and phosphate were normal. The great-great-grandmothers were sisters. No physical stigmata of pseudohypoparathyroidism were found and the authors called this pseudo-pseudohypoparathyroidism.

Two brothers who were mentally slow (out of six sibs) developed a progressive dystonia in their teens (Caraceni *et al.* 1974). Tests for parathyroid function were normal.

(b) Basal ganglia calcification with hypoparathyroidism

A sibship described by Morse *et al.* (1961) showed five members with idiopathic hypoparathyroidism. Four were mentally defective and all five had partial odontia. The basal ganglia calcification was present in three out of the five. One member had steatorrhoea and four had a vitamin B_{12} deficiency. Other families were reported by Goldman *et al.* (1952) and Eaton *et al.* (1939).

In a review of 37 cases of basal ganglia calcification four cases had hypoparathyroidism and there were three cases of familial basal ganglia calcification (Ogata *et al.* 1987).

In one family the parents were cousins. Father and two daughters showed the lesion, both daughters were retarded with cerebellar dysfunction, short stature, obesity, seizures, and had the facial features of Albright's disease.

(c) Basal ganglia calcification, mental retardation, and steatorrhoea

This combination was found in four sibs (including non-identical twin girls) in a sibship of 16 (Cockel *et al.* 1973). Serum calcium and phosphate were normal. Inheritance is probably as a recessive.

(d) Familial encephalopathy with basal ganglia calcification

Eight infants, including two sibships, had an onset in infancy of failure of the head to grow, accompanied by spasticity, dystonia and abnormal eye movements (Aicardi and Goutieres 1984). Convulsions occurred occasionally. The c.s.f showed a mild lymphocytosis and the CT scan calcification of the basal ganglia, areas of reduced density in the white matter and enlarged ventricles. Death occurred in infancy. Inheritance is as a recessive trait.

A further case is reported by Giroud *et al.* (1986). The parents were first cousins. At 3 months he seemed to lose interest and was found to be hypotonic. The CSF showed a lymphocytosis and at 8 months he was microcephalic, had truncal hypotonia, bilateral pyramidal tract signs and chorea. A CT scan showed frontal atrophy, hypodensity of white matter and punctate calcification in the lenticular nuclei.

Note the similar family reported by Mehta *et al.* (1986).

B. Dominant

(a) Familial basal ganglia calcification with neurological features

Dense symmetrical calcification in the caudate, lenticular, and dentate nuclei associated with ataxia, spasticity, and intellectual impairment was described by Foley (1951). The patient's daughter had less marked calcification and no neurological signs. A pedigree in which nine members of a family in three generations had calcification of the basal ganglia, cerebral and cerebellar gyri, and the cerebellar dentate nucleus was reported by Boller *et al.* (1977).

From the same family described above (Boller *et al.* 1977) in which most members had asymptomatic intracranial calcification, Boller *et al.* (1973) reported a mother and son with compulsive repetition of a phrase or word, occurring with increasing rapidity and with decreasing volume. Dementia and choreiform movements were also features.

In a pedigree spanning five generations (English/Scottish; Larsen *et al.* 1985) 13 had dystonia dating from childhood or early adulthood involving the face, neck, and limbs. The calcification was not limited to the basal ganglia, but involved the cerebral white matter and cerebellum. Several patients had both dystonia and calcification but others had either dystonia or calcification (not all had CT scans).

A single case with mental retardation and hidrotic ectodermal dysplasia (generalized hypotrichosis, dystrophic nails, hyperkeratosis on palms and soles had calcification (striopallidodentatal) of the basal ganglia (Copeland *et al.* 1977).

Dementia with an onset in middle life associated with basal ganglia and dentate nucleus calcification occurred in a mother and two of her children (Adachi *et al.* 1968).

A dominant pedigree was reported in which five members had basal ganglia calcification (Roberts 1959). One had epilepsy, but neither extrapyramidal nor somatic features were prominent.

A family in which at least five members had basal ganglia calcification on X-ray examination is reported (Moskowitz *et al.* 1971) with autosomal dominant inheritance. The proband, a 47-year-old man, had a 10-year history of progressive choreoathetosis and possibly cerebellar ataxia. Two other members had a relatively early onset of Parkinson's disease. A metabolic study did not reveal parathyroid hormone unresponsiveness.

(b) Basal ganglia calcification with pseudohypoparathyroidism

Nichols *et al.* (1961) described nine members affected in three generations. All had the calcification and seven had calcium deficiency and tetany. The two remaining members were symptom-free. All were short, mentally dull and had short, stubby fingers. Mann *et al.* (1962) suggested that this might be Albright's hereditary osteodystrophy or pseudohypoparathyroidism. Inheritance was as a dominant.

Kuroiwa *et al.* (1982) reported a Japanese family with eight members affected over three generations. All had calcification of the basal ganglia. All the females had bilaterally short 4th metatarsals. Serum Ca and phosphorus were normal, and none had neurological abnormalities.

(c) Kearns-Sayre syndrome (see p. 380)

Robertson *et al.* (1979) reported cerebellar, brainstem, and basal ganglia calcification in a young girl aged 8 years. Strobos *et al.* (1957) described two brothers with retinitis pigmentosa and cerebellar ataxia. Both had basal ganglia calcification. Their sister had a similar clinical disorder without the calcification.

Similarly, Dewhurst *et al.* (1986) reported on a West Indian boy, (who had features of Kearns–Sayre without basal ganglia calcification) but had been treated for hypoparathyroidism since the age of 2 years. In the report by Seigal *et al.* (1979) intracranial calcification occurred in four out of eight patients, one of whom had definite hypoparathyroidism.

Translucencies in the basal ganglia

(a) Leigh's disease (classical) p. 487

(b) Mitochondrial disease p. 379

(c) Lang *et al.* (1984) reported four sibs with onset at 3–10 years of dystonia. In one, the first signs were writer's cramp and she was wheelchairbound at 29. Other signs included optic atrophy and nystagmus (not in all four). CT scans in two showed bilateral translucence of basal ganglia and one had raised lactate and pyruvate. A post-mortem in one showed 'honey-combing' of the lentiform nuclei.

A futher report of the same family (Marsden *et al.* 1986) indicated that the mother of the sibs had developed Parkinsonism at the age of 55. They described a second family with four affected, including cousins, in an inbred family. The cardinal features were dystonia and loss of visual function due to optic atrophy and some pyramidal involvement, but no dementia. There were symmetrical lucencies in the putamen. Unlike holotopistic necrosis of the striatum (see p. 301; scans could be similar) mental retardation was not a feature. Note similarities to Novotny *et al.* (1986; see p. 5) and to juvenile Leigh's disease.

(d) In Hallervorden–Spatz dementia is part of the clinical picture. Swaiman *et al.* (1983) described Hallervorden–Spatz disease with bilateral basal ganglia lucencies and seablue histiocytes. This is very atypical for Hallervorden–Spatz and the clinical diagnosis is uncertain.

(e) Wilson's disease (see p. 325).

(f) Dystonia as in dystonia musculorum deformans (see p. 306).

Three (single) patients were reported by Berkovic *et al.* 1987. These authors think that their patients might have had mitochondrial disease, but this was not proven.

Paroxysmal movement disorders

Familial paroxysmal dystonic choreo-athetosis (periodic dystonia)

An autosomal dominant condition in which the motor movements are in the form of paroxysmal choreo-athetosis is now a well established clinical entity. Starting in childhood and often provoked by movement, attacks last from a few minutes to a few hours, and although the violence of the movement can be great enough to throw the patient to the ground, consciousness in not lost. Stevens (1966) noted a refractory period after an attack, but attacks can recur many times a day.

There are three types. Goodenough *et al.* (1978 describe two of these).

(i) Kinesigenic group—provoked by movement—clinically, these are brief movements (choreic, athetoid, or dystonic), attacks occur daily, and respond to anticonvulsants.

(ii) Non-kinesigenic group—of Mount and Reback (1940)—these are longer duration movements which occur less frequently and rarely respond to anticonvulsants. Attacks can be induced by alcohol and stress. Grimacing is seldom seen and the attacks start with chorea and then develop into athetoid or dystonic postures.

(iii) The 2nd pedigree in the Lance paper (1977) and the family reported by Plant *et al.* (1984) are different. The attacks are 5–30 minutes in duration and are precipitated by prolonged exertion. In the cases reported by Plant (1983) and by Lance (1977) inheritance was as a dominant.

Type I pedigrees

Przuntek and Monninger (1983) reported a six generation family. In the kinesigenic form, half of the families reported had similar movement disorders in another member.

Autosomal dominant inheritance is reported by Hudgins and Corbin (1966), Jung *et al.* (1973), Kato and Araki (1969), Fukuyama and Okada (1968), Lance (1963), and sibs by Smith and Heersema (1941), Stevens (1966), Kertesz (1967), Whitty *et al.* (1964), and Lance (1977).

Note: this condition must be differentiated from hereditary hyper-explexia where a sudden startle-movement causes the person to become stiff and fall to the ground (see p. 112).

The condition differs from that described by Segawa *et al.* (1976) in that in the Segawa syndrome the dystonia comes on gradually and persists until the patient rests (p. 306).

Expression might be variable. Twelve members in a single family were affected by paroxysmal painful dystonic choreoathetosis (Kurlan *et al.* 1987). The attacks were often precipitated by cold and prolonged activity. Other members just experienced painful cramps without involuntary movement. The authors suggest that the cramping might be evidence of carrier status and if so the pedigree becomes a simple dominant.

Type II

In the non-kinesigenic group the family originally described by Mount and Reback (1940) had members with attacks lasting from minutes to hours, in which at least three generations were affected. Inheritance (although stated to be recessive) was clearly dominant.

It should be noted that in the Mount and Reback family the attacks of choreoathetosis were triggered by alcohol and anxiety. In other families, coffee and fatique have been the triggers. Attacks last from 5 minutes to 4 hours and are not influenced by antiepileptic medication. Other families are those of Boel and Casaer (1984), the family of Richards and Barnett (1968), and the first pedigree in the Lance paper (1977). The family reported by Mayeux and Fahn (1982) is similar, but two brothers had the attacks and there was, in addition, a family history of a cerebellar ataxia.

Ataxia with dystonic posturing

A recessively inherited ataxia (three sibs) with an age of onset between 4 and 8 years starting with dysarthria, nystagmus, ataxia, and brisk reflexes was described by Graff-Radford (1986). Attack of 'tightening' were reported lasting 30 seconds to several minutes during which the sibs couldn't speak.

The episodes of dystonia respond to antiepileptic drugs. This condition is similar to familial paroxismal dystonic choreoathetosis, but that condition is mostly dominant (see above). A CT scan showed atrophy of the vermis and basis pontis. The condition is also similar to that described by Mayeux and Fahn (1982) in two brothers (see previous section and review by Gancher and Nutt 1986).

Familial paroxysmal hypnogenic dystonia

A family with dominant inheritance of flexion spasms during sleep, occasionally of a painful nature, and beginning in childhood is described by (Lee *et al.* 1985). Some were dystonic in nature and probably not epileptic.

Dystonia/hypnagogic

There might be two groups.

1. Those that do not respond to carbamazapine (such as the family of Lee *et al.* (1985). It is uncertain what proportion are genetic.
2. Those that do respond—mostly sporadic.

Both types occur in non-REM sleep (Lugaresi *et al.* 1986).

17

Muscle disorders

A. Muscular dystrophies
   1. Duchenne muscular dystrophy
   2. Duchenne-type dystrophy in females
   3. Autosomal recessive Duchenne-like dystrophy
   4. Becker muscular dystrophy
   5. Emery–Dreifuss—X-linked
   6. Other X-linked muscular
      (i) Humero-pelvic dystrophy with contractures
      (ii) Scapulo-peroneal muscular dystrophy
      (iii) Mabry type, Wadia type
      (iv) Rigid spine syndrome
   7. Facioscapulo-humeral
   8. Hauptmann–Thannhauser muscular dystrophy
   9. Emery–Dreifuss autosomal dominant
  10. Quadriceps
  11. Limb girdle
  12. Scapuloperoneal–autosomal dominant
  13. Oculopharyngeal

B. Myopathies
  14. Distal—autosomal dominant
  15. Distal—autosomal recessive
  16. Others
      (a) Mitochondrial myopathies
      (b) Nemaline myopathy
      (c) Central core disease
      (d) Multicore
      (e) Centronuclear or myotubular myopathy
      (f) Reducing body
      (g) Fingerprint
      (h) Sarcotubular
      (i) With lysis of myofibrills
      (j) Loss of cross striations

    (k) Inclusion body
    (l) Type I fibre disproportion
    (m) Spheroid body
    (n) Excessive tubular aggregates

C. Muscle disease without specific pathology
    (a) Benign hereditary myopathy
    (b) Congenital muscular dystrophy
    (c) Myoglobinuria
    (d) Myositis ossificans
    (e) Myosclerosis
    (f) Congenital hypotonia
    (g) Inflammatory myopathy
    (h) Familial distal cramps
    (i) Arthrogryposis multiplex congenita
    (j) Malignant hyperpyrexia
    (k) Muscle hypertrophy
    (l) Others

D. Metabolic myopathies

E. Myasthenia

F. The myotonias

A. Muscular dystrophy

1. Duchenne muscular dystrophy

This is the commonest dystrophy with a birth incidence of one in 3000–3500 males. The approximate prevalence rate for the north of England (Walton and Gardner-Medwin 1974) is 2.8 per 100 000 for Duchenne compared with 0.12 for autosomal recessive muscular dystrophy of early onset and 0.21 for the milder Becker X-linked muscular dystrophy. The first symptoms occur before the age of five years. The early gait disturbance is due to either pelvic girdle weakness or shortening of the Achilles tendons and the disease progresses with subsequent inability of the males to climb stairs or to rise off the floor. The Gowers manoeuvre (climbing up the legs) occurs at this stage in the disease. Pseudohypertrophy is a frequent early finding occurring especially in the calf muscles. Weakness develops in both upper and lower limbs, with relative sparing of the deltoids and hamstrings, but eventually the weakness spreads to involve most muscles except those of the face, ocular, and bulbar muscles. Severe kyphoscoliosis and cardiac involvement are common, and mental retardation occurs in a significant number of those affected (possibly half). The most useful criterion for differentiating the Duchenne dystrophy from the milder forms is the age when patients become wheelchair bound. Only 3 per cent of Duchenne patients are still ambulant after 11.3 years (Emery and Skinner 1976).

Genetics

Where a woman is shown on genetic grounds to be a carrier, then half of her sons could be affected and half her daughters could be carriers. She has a 1 in 4 chance of having a child with Duchenne muscular dystrophy. In practice this is a high risk which few women are prepared to take. In the study of Emery *et al.* (1972) only two out of 41 high-risk women (risks greater than 1 in 20) elected to have further pregnancies after counselling. Of the five women with risks between 2.5 and 5 per cent, only one out of five elected to have further children. If the risk is unacceptable then until recently only two options were available: either not to have children or to have only girls. A small proportion of women opt for the latter alternative. Many decide not to, on the grounds of either a disinclination to accept therapeutic abortion in general or because of misgivings about termination of half normal male offspring who would not be affected by the dystrophy (only half the males are theoretically at risk). The possibility of putting half of their daughters into a position in which they at present find themselves—namely, of having to make a decision about childbearing—in the next generation, is more acceptable as it is reasonable to assume that within the next two decades the diagnosis of an affected male *in utero* (before 20 weeks) will be achieved. Most counsellors feel justified in pointing this out. All of this has now changed. The gene for Duchenne muscular dystrophy has now been localized to the short arm of the X-chromosome (Xp21) and both linked probes as well as probes for part of the gene itself have been developed.

The distance between the probes and the mutation in the gene is such that there is about 5 per cent recombination between probe and gene. The linked probe is a restriction length polymorphism (i.e. a short piece of radiolabelled DNA) which is geographically situated alongside the Duchenne gene. Enzymes (restriction enzymes) cut DNA at precise locations and there is sufficient variation in the non-coding sequences alongside the Duchenne gene for different cutting sites to be present on the two X-chromosomes. This generates fragments of different lengths which make it possible in most cases to be able to differentiate between the two X-chromosomes. To recapitulate, the radioactive labelled probe highlights an area alongside the gene. Because of the different lengths generated by the variability in cutting sites the two X-chromosomes can be distinguished on an electrophoretic 'blot'. This is possible because long fragments move less quickly down the gel than the short pieces.

The following terms used by molecular geneticists apply to Duchenne muscular dystrophy and Huntington's chorea.

Polymorphism	refers to variation of appreciable frequency in a population, i.e. not due to a recent new mutation.
RFLP	restriction fragment length polymorphism: refers to length variation which originates when DNA is cut at different sites using an enzyme. Although the enzyme will cut at precise points the variation in the base pair sequences between the two chromosomes might ensure that the lengths differ.
Probe	a single strand length of DNA of specific sequence. It is made radioactive and will only hybridize with the complementary sequence.
Linkage	if an RFLP and a gene are situated close enough on the same chromosome that they are inherited together, they are said to be linked.
Restriction enzymes	These are enzymes which cut DNA at precise recognizable base sequences, say, GAC/TAA. Both strands are cut.

There has been a further interesting development. Most clinicians expected Duchenne dystrophy to result from a simple point mutation (perhaps an aminoacid substitution), but it is now likely that in a significant number of cases deletions are responsible for the expression of the condition.

It was also surprising to find how big the Duchenne locus is. It probably contains 60 exons covering more than 2000 kilobases. Some of the more recently developed probes cover part of the gene itself (referred to as the dystrophin gene).

A probe identified by Kunkel (1986; PERT 87), was found to be deleted in 7 per cent of Duchenne patients. This suggested that the probe was at the site of the Duchenne gene. Surprisingly, the same sequence was found to recombine with the Duchenne locus in 5 per cent of meioses, suggesting that different mutations at the Duchenne locus were likely to be responsible for the same clinical disease. Other probes (XJ1.1) have also been found to be deleted (Thompson *et al.* 1986), but again there is a recombination frequency of 5 per cent in some families.

Not all patients deleted for PERT 87 are also deleted for XJ1.1 (Thompson *et al.* 1986), but approximately 10.7 per cent of Duchenne patients will be deleted for one or other probe. The most recent figures are well in excess of even this figure and it might be that 60–70 per cent are deletions as opposed to point mutations.

It should be noted that cytogenetically these deletions are not visible. Information about the localization of genes alongside the Duchenne gene has been furthered by observing patients with Duchenne muscular dystrophy plus other known genetic conditions. For instance one patient had in addition congenital adrenal hypoplasis, glycerol kinase deficiency and mental retardation (Dunger *et al*. 1986; Clarke *et al*. 1986). A boy has also been described with Duchenne muscular dystrophy, chronic granulomatous disease, retinitis pigementosa and mental retardation with a deletion at Xp21.1–p21.3 (Franke *et al*. 1985). This must mean that in these cases the deletion has extended beyond the Duchenne locus to involve other genes and this helps to identify other loci.

Carrier

An obligate or definite carrier of the gene for Duchenne muscular dystrophy is a woman with two affected sons by separate unrelated fathers, or an affected son and another close male relative affected (see Fig. 17.1). Possible carriers are those women who are the mothers, sisters, or other female relatives of a single affected male.

If after the birth of a male with Duchenne muscular dystrophy no evidence is found of another affected male in the family, carrier detection rests on biochemical, DNA, histological, and electromyographic investigations. Of these tests, the CPK estimation and the DNA analysis are the most useful.

The elevation of creatinine phosphokinase in some mothers of sons with Duchenne muscular dystrophy has long been known (Dreyfus *et al*. 1966;

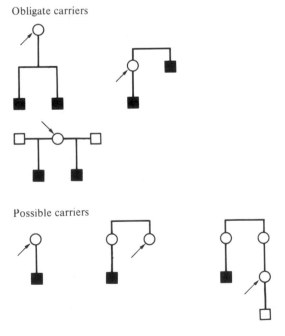

Fig. 17.1

Wilson *et al.* 1965; Hughes 1962). It is possible to alter significantly a genetic risk in about 60–70 per cent of possible carriers using this method. However, the relatively high (one-third) proportion of unhelpful tests limits the usefulness of creatinine phosphokinase estimations. Three estimates, preferably with weekly intervals, are more reliable than a single reading. If the creatinine phosphokinase values for both controls and carriers are logged, they have a near normal distribution (Wilson *et al.* 1965; Dennis and Carter 1978). Because the two curves overlap, a proportion of carriers cannot be detected as the values fall both into the tail of the carrier curve as well as into the normal control curve. The distributions are, however, far enough apart for each laboratory to be able to construct a graph whereby for a given creatine phosphokinase value (usually an average of three CPKs) the probability or odds of being a carrier can be calculated. The graph constructed by Dennis and Carter (1978) is shown in Fig. 17.2. The graph has been drawn presupposing a prior genetic probability of 1:1 or 1 in 2. Where the genetic odds are less than this, the biochemical odds should be multiplied by the prior genetic probability relating to the pedigree under consideration. If, for example, the genetic odds of being a carrier are 4:1 against, and the biochemical probability is 8:1 against, the final probability is 1:32 (1 in 33) of being a carrier and 1:64 (1 in 65) for having an affected son. In practice the CPK odds can only reduce the risk by a factor of about 10. This is usually satisfactory unless the consultand is the mother of the affected boy. When this is the situation the mother will have a two-thirds chance of being a

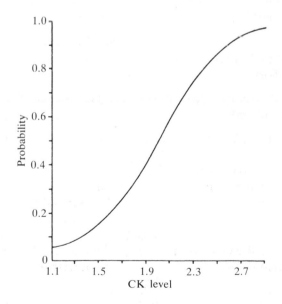

Fig. 17.2   Probability p^1 that a woman is a carrier, plotted against CK level.

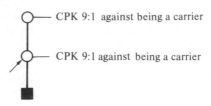

CPK 9:1 against being a carrier

CPK 9:1 against being a carrier

Fig. 17.3

carrier, and even if her biochemical odds are 10:1 against, the final odds of 1 in 6 might not be low enough for her to accept the risk (2:1 × 1:10 = 2:10 = 1:5 or 1 in 6).

If the consultand's mother's CPK can be used then the final risks might reach acceptable levels (first read how to calculate genetic risks on p. 352).

The consultand (arrowed) has a 1 in 9 chance of being a carrier which is a significant reduction from ⅔ the risk if no information about CPKs were available (Fig. 17.3).

The 1 in 9 chance is derived as follows starting with grandmother.

	Carrier		Not a carrier
(a) Prior probabily	4μ		$1-4\mu \approx 1$
(b) If a carrier the chances of having that CPK is	1/10	if not a carrier	9/10
(c) Joint probability = a × b	$4\mu/10$		9/10
(d) Final probability		$4\mu/10$	

$$\frac{4\mu/10}{4\mu/10 + 9/10}$$

New probability for $= 4\mu/9$
grandmother

Now halve this for next generation and add 2μ (chance of a fresh mutation from sperm or ovum)

(i) Prior probability for mother	$2\frac{4}{18}\mu$		$1-2\frac{4}{18}\mu \approx 1$
(ii) If a carrier there is half a ½ chance of having an affected son		if not a carrier μ boy must be fresh mutation (from ovum)	
(iii) Joint = i × ii	$40\mu/36$		μ
(iv) If a carrier the chance of having that CPK	1/10	the chance if not a carrier	$9/10$
(v) Joint	$\dfrac{40\mu}{360}$		$9/10\mu$

(vi) Final

$$\frac{40\mu/360}{40\mu/360 + 9/10\mu}$$
$$= 40/364$$
$$= 1/9$$

Precautions in using creatine phosphokinase

(a) Effect of exercise on creatine phosphokinase High creatine phospho-kinase levels can occur in normal female subjects after prolonged, strenuous exercise (well-trained athletes showed a creatine phosphokinase which was three times higher after a two hour run—Vassella *et al.* 1965). A 50 per cent increase in creatine phosphokinase activity occurred in six normal females after a three-mile walk (Hudgson *et al.* 1967), and the levels may remain elevated up to 72 hours after the exercise (Nuttall and Jones 1968). In those subjects who undergo a two to three week training programme, the elevation is less marked (Hunter and Critz 1971). There is no suggestion, however, that normal daily exercise has an effect on the creatine phosphokinase value so that no special precautions are needed in the hour prior to the testing of possible carriers. Violent muscular contractions as in generalized epilepsy, and muscle necrosis might raise levels, supporting the observation that it is the localized muscle response rather than a rise in body temperature that is responsible for the elevation (Vassella *et al.* 1965).

(b) Effect of pregnancy on creatine phosphokinase The creatine phospho-kinase levels of carrier females are lowered by pregnancy making carrier detection unreliable (Blyth and Hughes 1971; Emery and King 1971). However, significantly raised levels during early pregnancy could still be a useful indicator of carrier status. Fleisher *et al.* (1965), have found reduced creatine phosphokinase levels during pregnancy in non-carrier women and this has been confirmed by Bundey *et al.* (1979). The influence of oral contraceptive pills on the creatine phosphokinase levels is reported in two studies (Simpson *et al.* 1974; Perry and Fraser 1973) both with negative results. CPK rises in the post-menopausal woman, but not substantially.

(c) Effect of age on creatine phosphokinase It is well known that the levels are high in patients with Duchenne muscular dystrophy and that they reach a maximum at 10 years and then tail off. A decrease in creatine phosphokinase with advancing age in five definite and 14 possible carriers was reported by Moser and Vogt (1974). Other papers reporting a fall in individual carriers with age are those of Thompson *et al.* (1967) and Zatz *et al.* (1976*a*).

The creatine phosphokinase in normal females also varies with age (Bundey *et al.* 1979). The levels in healthy pre-menarchal girls became pro-gressively lower in the post-menarchal stage, and even lower in maturity and pregnancy. The levels rose again after menopause. Similar results were obtained by Nicholson *et al.* (1979). The only study to contradict this is that of Satapathy and Skinner (1979). In their females the creatine phosphokinase appeared to increase with age until 15 years and then remain fairly constant. However, it is pointed out by Bundey *et al.* (1979) that the normal female patients used were hospitalized and should not be compared with normal ambulant females.

Optimum time for testing possible carriers

Girls are usually tested when they are 17 or 18 years old. If seen earlier they are recommended to return for the three creatine phosphokinase estimations.

Muscle biopsy

About 10 per cent of carriers of Duchenne muscular dystrophy are manifesting carriers in that they have signs of the disease (Moser and Emery 1974). However a biopsy is not desirable.

Quantitative electromyography

In a report from Moosa *et al.* (1972) the carrier detection rate could be increased by combining this time-consuming technique with creatine phosphokinase estimations, but at present, few laboratories use this method. Previous reports (Gardner-Medwin *et al.* 1971) have concluded that quantitative electromyography can only make a marginal contribution to carrier detection.

The genetic risks using pedigree data

Five pedigrees were presented to a number of practising geneticists asking their participation in assessing the difficulties in using pedigree data to derive the genetic risk of being a carrier (Bundey 1978). Fewer than one-half were able to calculate the probabilities correctly but this method used in conjunction with the biochemical test is a potent way of reducing risks.

Clearly, if a possible carrier's sister has a son with Duchenne dystrophy and they have in addition eight normal brothers, then the chances that the mother is a carrier must be reduced and hence the risk of the consultand inheriting the disorder from her mother must likewise decrease.

Bayesian probability techniques are most useful in this situation as no single test is available that can unequivocally discriminate between carriers and non-carriers (Murphy 1968; Murphy and Mutalik 1969).

These methods show how opinions held prior to an event can be changed in the light of observations.

For instance if there are two bags of balls, one containing all black balls and the other an equal mix of black and red then the prior probability of entering the bag with only black balls is ½. Experience could help to change the probability. For instance, if 10 black balls are picked consecutively from a single bag, then the probability of being in the black bag must increase. Thus, as applied to Duchenne muscle dystrophy, information about the consultand's sibs, offspring, and antecedents and also the results of special investigations could be used to alter prior genetic probabilities.

In the following four pedigrees, a method of altering the probability is illustrated.

A prior probability of four times the mutation rate (i.e. 4μ is used as a starting point, as this is the probability of any woman in the general population being a carrier for an X-linked lethal condition.

Figure 17.4 illustrates how this can be derived.

It can be seen that starting at the top of the pedigree—say with Adam and Eve—the first generation of girls will have a 2μ chance of being a carrier—one μ is derived from a fresh mutation on the father's X chromosome and one μ on the X from the mother. As a daughter will inherit only one X from mother, the 2μ will halve in every generation. It can be seen that 4μ will be reached after several generations and that this figure will not be exceeded.

The following terms will be used and are defined as follows.

Prior probability—uses pedigree information anterior to the proband, e.g. the number of normal brothers the consultand has.

Conditional probability—this uses information about the offspring of the consultand, i.e. how many normal sons she has had as well as CPK information.

Joint probability—the prior probability × the conditional probability.

Final probability—the probability of being a carrier over the sum of all the probabilites.

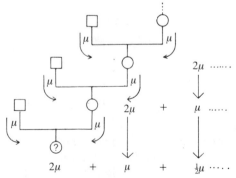

$$\text{Final Sum} = 2\mu\left(1+\tfrac{1}{2}+\tfrac{1}{4}+\tfrac{1}{8}\ldots\ldots\text{etc}\right)=4\mu$$

Fig. 17.4

Examples

1. The mother of an affected boy wants to know the risk of being a carrier. There is no information anterior to the consultand. As indicated above the prior probability of being a carrier is the population frequency which equals 4μ (Fig. 17.5).

Fig. 17.5

	Carrier	Not a carrier
(a) Prior probability	4μ	$1-4\mu \approx 1$
(b) Conditional probability	If she is a carrier the chance of having an affected son is $\frac{1}{2}$	If she is not a carrier the chance of having an affected son is 1μ (fresh mutation from mother)
(c) Joint probability (a × b)	2μ (i.e. $\frac{1}{2} \times 4\mu$)	1μ (i.e. $1 \times 1\mu$)
(d) Final probability	$\dfrac{2\mu}{2\mu + 1\mu}$	$= 2/3.$

2. Does the birth of a normal male lower the risk (Fig. 17.6)?

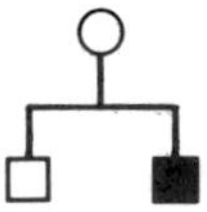

Fig. 17.6

	Carrier	Not a carrier
(a) Prior probability	4μ	$1-4\mu \approx 1$
(b) Conditional probability	The chance of having a normal son is 1/2	1
	The chance of having an affected son is 1/2	1μ (A fresh mutant from mother's ovum)

(c) Joint probability (a × b) $4\mu \times 1/4 = \mu$ $\mu \times 1 \times 1 = \mu$

(d) Final probability $\dfrac{1\mu}{1\mu + 1\mu}$ $= 1/2.$

A normal son decreases the risk from 2/3 to 1/2.

3. The chance of a consultand being a carrier depends on the probability of her mother being a carrier.

How much does the birth of three normal males lower her probability? The reasoning is as follows (Fig. 17.7).

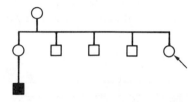

Fig. 17.7

The chance of the probands mother being a carrier is 4μ. Therefore:

	Carrier	Non-carrier
(a) Prior probability	4μ	$1-4\mu \approx 1$
(b) Conditional probability	The chance of having three normal boys is	
	$\frac{1}{2} \times \frac{1}{2} \times \frac{1}{2}$	1
The chance of having a carrier daughter		
	$\frac{1}{2}$	No real chance
The chance that this daughter will have an affected son		
	$\frac{1}{2}$	μ (fresh mutant from mother's ovum—father passes on his Y chromosome to his son)

There is, however, a further possibility which must be added into the calculation on the non-carrier side. Supposing the mother is not a carrier, her other daughter (proband sister) could be a carrier on the basis of having received a mutant gene on the X from either her mother's ovum or father's sperm and hence have a 2μ chance of being a carrier. This is an alternative (either/or situation) and must be added to the conditional probability on the non-carrier side.

Therefore:

$+ 2\mu$ and, as only half her sons would be affected, $\times \frac{1}{2}$

$= \mu$

(c) Joint probability (a × b) $4\mu \times \frac{1}{32} = \frac{1}{8}\mu$

$\mu + \mu = 2\mu$

(d) Final probability

$$\frac{\frac{1}{8}\mu}{\frac{1}{8}\mu + 2\mu}$$

$$= \frac{1}{17}.$$

This is the probability that mother is a carrier. Her daughter—our consultand—has $\frac{1}{34}$ of a chance of being a carrier and only a 1 in 68 chance of having an affected son.

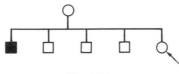

Fig. 17.8

4. The chance of the consultand—marked with an arrow—being a carrier is influenced by the probability that her mother is a carrier. The mother's prior probability of being a carrier is 4μ (Fig. 17.8).

	Carrier	Not a carrier
(a) Prior probability	4μ	$1-4\mu \approx 1$
(b) Conditional probability	If a carrier the probability of having three normal sons is	
	$\frac{1}{2} \times \frac{1}{2} \times \frac{1}{2}$	1
	If a carrier the probability of having an affected son is	
	$\frac{1}{2}$	μ (fresh mutation from mother's ovum)
(c) Joint probability (a × b)	$4\mu \times \frac{1}{16} = \frac{1}{4}\mu$	μ
(d) Final probability	$\dfrac{\frac{1}{4}\mu}{\frac{1}{4}\mu + \mu}$	
	$= \frac{1}{5}$	

Chance that consultand is a carrier $= \frac{1}{10}$.
Chance of having an affected son $= \frac{1}{20}$.

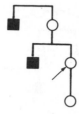

Fig. 17.9

It should be remembered that the prior probability is not 4μ when the pedigree is as shown in Fig. 17.9). The arrowed consultand has a prior probability of $\frac{1}{2}$ and her daughter of $\frac{1}{4}$. No other calculations are indicated.

Carrier detection using DNA (see pp. 344–6)

The following simple example illustrates how this is done.

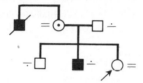

The arrowed female's mother is a carrier because she has an affected brother and son. Her daughter (arrowed) has $\frac{1}{2}$ a chance of being a carrier. Looking at the DNA analysis it can be seen that she received a lower band from her father and, therefore, an upper band (the one without the mutation) from her mother. The band (supposedly with the mutation) is identified as the lower band because the affected boy has that band and the unaffected boy has the upper band.

Prenatal diagnosis

Those women who choose to have fetal sexing at 8–10 weeks by chorion villus biopsy should be fully informed about the procedure. Chorion villus biopsy carries a 3 per cent risk of causing an abortion and the test will be followed by a short wait, which is the average time necessary for confirmation of the sex of the fetus by chromosomal analysis. If a male fetus is found, a termination is usually resorted to at about 12 weeks.

Fetal blood sampling

This has not proved reliable and has been abandoned.

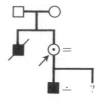

Using DNA analysis

The arrowed woman is an obligate carrier by having an affected brother and an affected son. If on using the linked probe she is heterozygous (two bands appear on the gel representing fragments of different length) it can be assumed that these bands represent her two X-chromosomes. Her son receives only one X and has received the lower band. It is therefore the lower band which is likely to contain the mutant gene and the upper band which does not. Given the current risk of recombination causing a 'false' result (5%), a male with an upper band would have a high probability (90 per cent) of being unaffected. If in the same family there had been an unaffected boy and he had the upper band then this strengthens the assumption that it is the lower band that is harbouring the mutation.

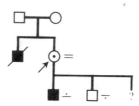

It might be that the lady in question doesn't want further children, but is more concerned about her daughter.

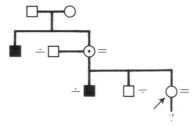

In this situation it is necessary to test her husband. If he is ÷ then his daughter must have received the lower band from him and the upper (the good news band) from her carrier mother. If the opposite occurred, then the daughter is likely to be a carrier and pre-natal tests can be offered during a subsequent pregnancy.

More often there is no family history. CPK's might help predict whether mother is a carrier or not but they might not substantially alter her 2/3 *a priori* risk of being a carrier. The new DNA tests might help in the following way.

1. There might be a detectable deletion in her affected boy. If this is the situation, the same deletion can be sought in a male fetus in a subsequent pregnancy.

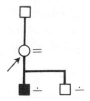

2. If not, 'exclusion' can be offered. Again it is uncertain whether the arrowed lady is a carrier or not. If she is, then it is reasonable to suggest that males with the lower band are at high risk and to counsel those with an upper band as likely to be unaffected. However, there is a chance that the affected boy is a fresh mutation and that unnecessary termination of males with the lower band will result. Nevertheless, this option will allow her the very reasonable choice of a normal son, and that is often acceptable. It might help to decide whether a mother is a carrier by looking at both her affected and unaffected sons. If both the affected and unaffected boys have received the same X-chromosome then, barring a crossover it is likely that the affected son has the condition as a fresh mutation.

In practice

There is now ample evidence that the tests as outlined above are very effective in substantially altering the reproductive options for those who have a close relative with Duchenne (or Becker) muscular dystrophy. The risk of being a carrier was lowered to 2.5 per cent or less in one-third of women in the report by Goodship *et al.* (1988) and these figures are improving.

Glycerol kinase deficiency

Infantile onset glycerol kinase deficiency can sometimes be difficult to distinguish from Duchenne (Kohlschutter *et al.* 1987). There is usually evidence of congenital adrenal hypoplasia with adrenal insufficiency, hyperglycerolaemia, and mental retardation.

2. Duchenne muscular dystrophy in females

Occasionally females develop a clinical picture of a progressive muscular dystrophy indistinguishable from Duchenne muscular dystrophy. The following possibilities exist:

(a) Females with Turner syndrome (a single X-chromosome; Ferrier *et al.* 1965), or XO/XX/XXX mosaic (Jalbert *et al.* 1966), or structurally abnormal X-chromosome (Berg and Conte 1974) are equivalent to hemizygous males and are susceptible to X-linked conditions.

(b) Random inactivation of one X-chromosome (Lyonization) does not always lead to the inactiviation of an equal number of maternally and paternally derived X-chromosomes and if more X-chromosomes with the normal allele are inactivated leaving a high proportion of Xs with the mutant gene, females may be affected. Evidence for this in Duchenne muscular dystrophy is reported by Gomez *et al.* (1977). One of a pair of female monozygous twins developed Duchenne muscle dystrophy. Their maternal uncle had typical Duchenne and their mother had large calves and a raised creatine phosphokinase.

A further case of a Duchenne-like muscular dystrophy in one of a pair of identical twin sisters was reported by Burn *et al.* (1986*a*). Laboratory studies suggested that in the affected girl only the maternal X-chromosome was active whereas in the other the active X was paternally derived. Both girls are therefore probably heterozygous and the difference is determined by a quirk in X-inactivation.

The distinction between manifesting carriers of an X-linked Duchenne dystrophy and those who have the genetically separate autosomal recessive variant can be difficult.

(c) When an affected male has an affected sister, the possibility of an autosomal recessive type of a Duchenne-like muscular dystrophy should be considered although there is no certain way of distinguishing between the two.

Six girls from five families are reported by Hazama *et al.* (1979). In one typical X-linked pedigree, two brothers and two uncles were affected. The girl had pseudohypertrophy and was in a wheelchair by the age of 10 years. The authors suggest unequal inactivition of the X chromosomes to explain the phenomenon. In another pedigree two sisters had a more prolonged course. Their brother was in a wheelchair by 11 and died at 26. The authors suggest that the girls could be manifesting carriers, but this is by no means certain. In a third pedigree a brother and sister were affected. Other possible pedigrees are described under limb-girdle muscular dystrophy (p. 372); family of Jackson and Strehler—where some members had an onset as early as 4 years of age).

3. Autosomal recessive variety of Duchenne-type muscular dystrophy

See also previous section. Penn *et al.* (1970) analysed all previous reported cases and selected 19 where the clinical features conformed, but in some of the cases laboratory investigations had not excluded spinal muscular atrophy, polymyositis, or chromosomal abnormalities. They included five cases of their own. Four of the five girls were still ambulant at an age when most (97 per cent) of Duchenne males would be in a wheelchair. The authors concluded that at least three of the girls could be fitted into the autosomal recessive limb girdle muscular dystrophy category, and not one had a typical Duchenne muscular dystrophy. One had an affected brother. The authors criticised the Lamy and de Grouchy (1954) report in which 11 girls were affected with what was called the Duchenne dystrophy, implying that 10 per cent of all Duchenne dystrophies are recessively inherited, on the grounds that no clinical description or progress reports were given. Four girls with a muscular dystrophy similar to Duchenne (Ionasescu and Zellweger 1974) had an onset at two years of moderate to mild weakness of pelvic girdle muscles, a normal electrocardiogram, and only slow progression. Some of the reports of severe muscular dystrophy in girls with recessive inheritance (Johnston 1964) were diagnosed clinically and cannot be accepted in the absence of more sophisticated tests.

However, the existence of an autosomal recessive childhood-onset muscular dystrophy has been clearly established. It is thought to have a slightly later age of onset and slower progression. Parental consanguinity and a normal algebraic sum of R and S waves in precordial leads on the electrocardiogram is suggestive of the recessive type (Jackson and Carey 1961). Two sisters with an early onset of weakness, including pseudohypertrophy, are reported by Dubowitz (1978).

Two sisters reported by Somer *et al.* (1985) were the product of a consanguineous union. Onset was at 6 and 7 years, and they were in wheelchairs by 11 and 12. Both had pseudohypertophy and mild facial weakness. Neither had a cardiomyopathy nor mental retardation.

Gardner-Medwin and Johnston (1984) reviewed 12 girls and two boys with a severe non-congential muscular dystrophy thought to be autosomal recessive muscular dystrophy. In no instance did the girls have a close male relative with Duchenne muscular dystrophy. They emphasized early toe-walking (before major walking difficulties), the milder course, relatively more deltoid muscle weakness than triceps, normal intelligence, and normal a ECG as features which help to differentiate the condition from Duchenne muscular dystrophy. It should be noted that slight facial weakness was present in four cases—a very unusual manifestation in Duchenne muscular dystrophy and the CK tended to be lower than in Duchenne males of the same age.

Muscle biopsy showed necrosis with fibrous tissue and fatty changes—less severe than Duchenne. Many showed both necrosis and regeneration. Hyaline fibres were prominent. Hypertrophy and splitting were seen in more than half and variation in fibre size was constant. The biopsy changes tended to be focal.

Duchenne and Becker muscular dystrophies

Linkage analysis has suggested that Duchenne and Becker muscular dystrophy are either very closely linked or allelic (Kingston *et al.* 1984). In general, families breed true, but there has been further evidence for occasional families with clinical evidence of both (Hausmanova-Petrusewicz and Borkowska 1978; Mostacciuolo *et al.* 1987; Jackson *et al.* 1974). Recently, Liechti–Gallati *et al.* (1987) and Hodgson *et al.* (1986) have found deletions in the Duchenne region in patients with Becker muscular dystrophy confirming that these two conditions are at the same locus.

Examples of the above variability are as follows. In three families described by Furakawa and Peter (1977), both the severe and benign types occurred. In a pedigree described by Shaw and Dreifuss (1969) two patients never walked whereas other members had the Becker type. Walton (1956) described a similar family. Four of five affected boys in a family reported by Hausmanowa-Petrusewicz and Borkowska (1978) had the mild Becker type of dystrophy. One boy was severely affected and had an onset at four years and was in a wheelchair by 13 years, whereas the other boys had an onset from 16 to 22 years. The severely affected male was mentally retarded. In general, the Becker and Duchenne muscular dystrophies breed true and parents should be counselled on this basis.

There is only one case on record of a man with Duchenne muscular dystrophy fathering children (Thompson 1978). He was one of 10 males affected in five generations. The proband was in a wheelchair by 12 years and data on the other nine showed that they all became wheelchair bound between 9 and 13 years. Of those dead, four died before the age of 20 years, and two at 24 and 27 years, respectively. The son of the affected male was unaffected and on creatine phosphokinase estimation his daughter (proband's daughter) had a high probability of being a carrier.

4. Becker muscular dystrophy

About 10 per cent of X-linked muscular dystrophy is of the mild variety and has a mean age of onset of 11.1 years (Emery and Skinner 1976). There is, however, considerable variation in the age of onset with some overlap into the Duchenne range. Emery and Skinner (1976) suggested that the best criterion for differentiating the two disorders is to take the age at which the male becomes chairbound. If the dividing line is taken at 11.2 years then 97 per cent of Duchenne sufferers are chairbound and only 3 per cent of those with Becker's dystrophy. Shaw and Dreifuss (1969) designate as mildly affected anyone who continues to walk beyond 16 years or dies after the age of 32 years. Clinically, the disease is similar to Duchenne dystrophy with a predominantly proximal myopathy and pseudohypertrophy of calf muscles. Early severe myocardial involvement is unusual although electrocardiographic changes can develop late in the disease. In the family reported by Markand *et al.* (1969), three out of four affected males had cardiac involvement. In the Emery and Skinner (1976) series, two out of 26 males had cardiac failure in middle life. Most patients are still ambulant until the third decade. Mental retardation does occur, but not to the same extent as in Duchenne muscular dystrophy (Emery and Skinner 1976; Ringel *et al.* 1977).

Almost all affected males have a significantly raised creatine phosophokinase, but more so than in the Duchenne dystrophy, the creatine phosphokinase varies in Becker's dystrophy with age. This is evident in the affected males and carrier women. As in Duchenne dystrophy only a proportion of carrier women have elevated creatine phosphokinase levels. Skinner *et al.* (1975) found that the risk could be altered in about 60 per cent of cases using this method, but an obligate carrier range for Becker's dystrophy must be used.

In a more recent study (Aston *et al.* 1984) obligate carriers had levels of CPK above the reference range upper limit in just under half of the cases and the ratios should be treated as for Duchenne.

The principles of genetic counselling in Becker's dystrophy do not differ from those in Duchenne muscular dystrophy. However, the male with Becker's dystrophy, if mildly affected, will be able to have children whereas this does not occur in Duchenne. The relative fertility of affected males is in the order of 67 per cent (Emery and Skinner 1976), and this figure can be used to calculate the frequency of female carriers·in the population.

Whereas the prior probability of being a carrier for an X-linked lethal condition (Duchenne) is 4μ—where there is no anterior information available—the same figure for Becker muscular dystrophy utilizing 70 per cent as the relative fertility is calculated to be 18μ (Emery 1976). This prior probability can be altered as shown in Fig. 17.10. It can be seen that negative anterior

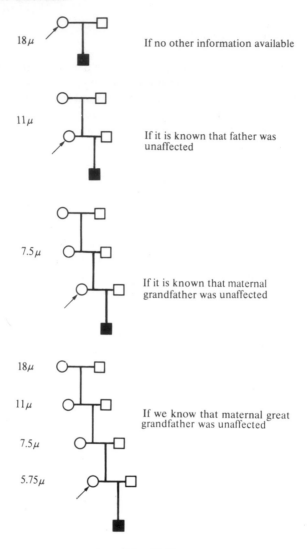

If no other information available

If it is known that father was unaffected

If it is known that maternal grandfather was unaffected

If we know that maternal great grandfather was unaffected

18μ

11μ

7.5μ

18μ

11μ

7.5μ

5.75μ

Fig. 17.10

information substantially reduces the probability that the affected male's mother is a carrier.

If she has unaffected brothers this can be utilized in the same way as illustrated in Fig. 17.8.

It should be noted that the probability of mother being a carrier will not fall below 2μ—which is the probability of receiving a fresh mutation on the X chromosome in sperm or egg cell.

The calculation of the probability that mother is a carrier is facilitated by

18μ becomes $9\mu + 2\mu = 11\mu$

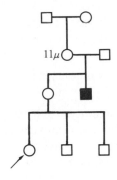

Fig. 17.11

starting as far back in the pedigree as possible and working forwards. For each generation in which the father is unaffected his daughter's probability is reduced by half to which must be added 2μ (a fresh mutant from mother or father; Fig. 17.10).

In the pedigree shown in Fig. 17.11 there is evidence from the family history that the affected man's parents and grandparents were unaffected. The prior probability that his mother is a carrier is 11μ. For the arrowed consultand to be a carrier her mother and grandmother must be a carrier.

The calculation proceeds as follows:

	Carrier	Not a carrier
(a) Prior probability	11μ	$1-11\mu \approx 1$
(b) Conditional probability	If a carrier, the chances of having an affected son is $\frac{1}{2}$	If not a carrier then the male is affected as a fresh mutation μ
(c) Joint probability	$11\mu \times \frac{1}{2}$ 5.5μ	$1 \times \mu$ μ
(d) Final probability	$\dfrac{5.5\mu}{5.5\mu + \mu}$ $= 11/13.$	

The new prior probability that the mother of the affected male is a carrier is 11/13.

For the next generation the probability must be halved

(e) Prior for mother of consultand	11/26	15/26

To this should be added 2μ but to add 2μ to a fraction will not for practical counselling purposes alter the risks.

(f) Conditional on having
    two normal sons $\qquad$ $\frac{1}{2} \times \frac{1}{2}$ $\qquad\qquad\qquad\qquad$ 1

(g) Joint (e × f) $\qquad\qquad\qquad$ 11/104 $\quad$: $\quad$ 60/104

$$= 11 \quad : \quad 60$$

$$= \quad 11/71.$$

Risk that mother of consultand is a carrier is 1/6.5.
Risk for consultand = 1/13.

    All the male offspring of a Becker dystrophy sufferer will be unaffected, but all of his female children will be obligate carriers. Many geneticists will not spontaneously offer affected males the option of having only male offspring on the grounds that by the time his carrier female offspring come to have children of their own, some of the problems might be resolved and, indeed, the recent advance in DNA analysis have ensured that these carriers can in most instances at least be offered the option of normal males. Parents who do not wish to take the high risk of having offspring with the severe type can be reassured that it is unusual for the severe and mild type to occur within the same family (Blyth and Pugh 1958).

    A family in which males had a clinical picture compatible with the Becker type of muscular dystrophy had two severely affected females in two previous generations (Aguilar *et al.* 1978*b*). Examination of the females showed the proximal muscles of upper and lower limbs to be affected with, in addition, pseudohypertrophy. This is most unusual and must be a very rare event. Other Becker pedigrees are reported by Ringel *et al.* (1977). The majority of these patients had an onset after 7 years and were walking beyond the age of 20 years.

DNA in carrier detection and for pre-natal diagnosis

As discussed, the same probes as in Duchenne muscular dystrophy (with the same recombination) are currently being used for carrier detection and prenatal diagnosis. Furthermore, in a study of 33 patients with Becker muscular dystrophy Hart *et al.* (1987) found two with clear deletions (in the Duchenne region). There did not seem to be a correlation between the extent of the deletion and the severity of the dystrophy.

5. Emery–Dreifuss type of X-linked dystrophy

Note: a number of X-linked pedigrees similar to Emery–Dreifuss have been published. In general, they are described as X-linked muscular dystrophies with contractures, a humero-pelvic distribution, and a cardiomyopathy. They will be referred to in the next section.

    In an article entitled 'An unusual type of benign X-linked muscular dystrophy', Emery and Dreifuss (1966) presented a large pedigree with eight males affected over three generations in which the onset of pelvic girdle weakness occured around the age of 4 or 5 years. Later the weakness spread to the proximal muscles of the upper limbs. This same family had previously been reported by Dreifuss and Hogan (1961) and subsequently by McKusick (1971). The dystrophy resembles that of Becker, but differs in that in Emery's cases flexion contracture of the elbows and ankles dated from early childhood. Pseudohypertrophy did not occur and the dystrophy was confined to proximal muscles. Cardiac conduction defects were present in all affected males and mental retardation was not a feature (a proportion of Becker dystrophy cases are mentally handicapped). Other cases are reported by Rowland *et al.* (1979) and Rotthauwe *et al.* (1972), and another family by Waters *et al.* (1975). As Rowland points out cases described in the literature as muscular dystrophy with permanent atrial paralysis might have this disease. He suggests that the condition Dubowitz (1973) describes as 'Rigid spine syndrome' might also be this disease (see below).

Recently, five further cases in a single family are discussed by Merlini *et al.* (1986). Variability within a family in terms of severity can be extreme.

Carrier detection has not been possible, but recently a linkage has been suggested by Boswinkel *et al.* (1985) using a probe close to Xq28. This has been confirmed by Yates *et al.* (1986; Xq27.3-qter).

6. Other conditions similar or identical to Emery–Dreifuss

(i) X-linked muscular dystrophy with contractures humero-pelvic distribution and a cardiomyopathy

It is uncertain whether this is any different from Emery–Dreifuss syndrome. Johnston and McKay (1986) described a family in which the onset was in infancy (delayed walking occured in one). In the full blown picture wasting and weakness occurred in the biceps, triceps, shoulder (forearm and hand were spared), face, (with ptosis), and the sternomastoids. The thighs were moderately weak and wasted. Toe walking and *pes cavus* occured, and the calf muscles, the peronei, and anterior tibial muscles were especially affected. Extensor digitorum brevis was preserved and there was no pseudohypertrophy. The proband needed a pacemaker and clearly had a cardiomyopathy.

There is a further article by Hopkins *et al.* (1981) on Emery–Dreifuss humero-peroneal muscular dystrophy with unusual contractures and bradycardia which is similar to that described by Johnston and McKay.

(ii) Scapuloperoneal syndrome (Fig. 17.12)—X-linked

It is uncertain whether X-linked Scapulo-peroneal dystrophy is the same as Emery–Dreifuss (see p. 374), although it probably is. In the original Emery–Dreifuss description there was sparing of the distal muscles, but most other

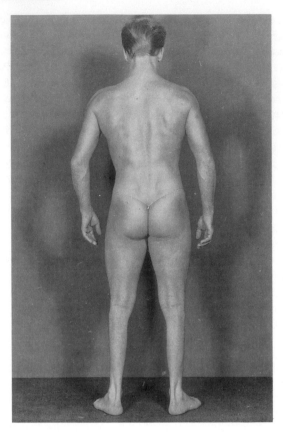

Fig. 17.12   Scapuloperoneal muscular atrophy: shoulder girdle and distal muscle wasting in lower limbs.
Courtesy of the Department of Medical Illustration, National Hospital for Nervous Diseases, Queen Square, London.

cases in the literature have predominantly distal wasting in the lower limbs. Emery (1987) has recently re-looked at the original Virginian family with this condition and suggests that there might be families with a scapulo–peroneal myopathy which are separate from the Emery–Dreifuss syndrome in which early (< 5 years) contractures do not occur, i.e. the family of Thomas *et al.* (1975).

(iii) Other types

Mabry *et al.* (1965) described another possible variant of the Becker type. The age of onset was between 11 and 13 years, pseudohypertrophy and myocardial involvement occured, but there were no flexion deformities. The authors suggested that the age of onset was earlier than usually found in Becker dystrophy and that the involvement of the myocardium places it in an intermediate position between Duchenne and Becker. Inheritance is X-linked recessive. This category has not been accepted by Becker (1972) nor by Bradley *et al.* (1978).

A benign X-linked muscular dystrophy with cardiac involvement is reported by Wadia *et al.* (1976). In the four cases on which data were available, the age of onset varied between 10 and 15 years, and those affected became chairbound around 20–30 years. One had severe contractures and two pseudohypertrophy. Another family with severe myocardial involvement is reported by Kuhn *et al.* (1979).

It is uncertain whether this category is different from the Emery–Dreifuss syndrome.

Note: for Emery–Dreifuss dominant see p. 370.

(iv) Rigid spine syndrome (Dubowitz)

This is not a homogeneous entity. Some reports are clearly of the X-linked Emery–Dreifuss syndrome and others are autosomal dominant or recessive. The condition was reviewed recently by Poewe *et al.* (1985) who reported on three single males and one male whose sister and mother were also affected. In this family there was early involvement of the spinal extensors causing scoliosis and a rigid back, followed by shoulder girdle muscle atrophy and a cardiomyopathy. One died at 13 years. The EMG was myopathic, and the muscle biopsy confirmed primary muscle disease.

A brother and sister were reported by Echenne *et al.* 1983. Both seemed normal in the neonatal period and hypotonia and weakness were first noticed at 10 months (proximal more than distal). At 7, in spite of extensive physiotherapy, the cervical spine became rigid. The IQ was low normal in both, but one showed hypodensity of white matter on the CT scan. In both, the muscle weakness progressed and joint contractures (in one) developed. (In view of the white matter changes this might be a separate condition.)

Most cases of the rigid spine syndrome are sporadic with an onset in infancy, with little or only mild progression and without cardiomyopathy. The condition is reviewed by Van Munster *et al.* (1986). Other single cases are those of Goto *et al.* (1979) and Seay *et al.* (1977).

One unusual case of the syndrome with a fatal cardiomyopathy is reported by Colver *et al.* (1981).

Counselling

The condition is heterogeneous. Some reports are clearly Emery–Dreifuss X-linked or Emery–Dreifuss autosomal dominant, but as seen above sibs have also been described suggesting autosomal recessive inheritance.

7. Facio-scapulo-humeral dystrophy (Fig. 17.13) (FSH)

The onset of facio-scapulo-humeral dystrophy occurs at any age from childhood to adulthood. Infantile cases are rare and might not belong to this group (Hanson and Rowland 1971). The diagnosis is based on the distribution of the affected muscles. The facial muscles are often involved first and might, in those with minimal manifestations, be the only signs of the disease. Inability to close the eyes tightly, whistle, or maintain the mouth shut, is evidence of facial weakness. The shoulder girdle involvement makes elevation of the arms difficult and it is especially the sternal head of the pectoralis, the spinati, latissimus dorsi, biceps, and triceps that are affected.

In a review of 200 cases, Kazakov *et al.* (1974) suggested that the facio-scapulo-humeral distribution was often only a stage in progressive disease. They distinguish two types:

(a) gradually descending type;

(b) 'Jump' type.

There were 162/200 cases in which the spread occurred gradually to muscles in the pelvic girdle, and 38/200 cases where a 'jump' took place to distal parts of the legs without involving the proximal muscles (see section on scapulo-peroneal dystrophy).

Large dominant pedigrees are reported by Tyler and Stevens (1950) and Pearson (1933). These and others were analysed by Walton (1955/56). The incidence of facio-scapulo-humeral dystrophy is calculated as 3.8/million in Wisconsin and 9.2/million for pooled data (Morton and Chung 1959). Geographical differences are indicated by a figure of 1 in 20 000 for South Baden (Becker 1964 *a,b*). A formal genetic analysis of facio-scapulo-humeral

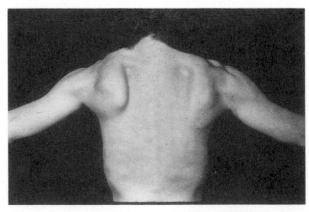

Fig. 17.13   Facio-scapulohumeral dystrophy showing shoulder girdle weakness.

dystrophy (Morton and Chung 1959) suggests that penetrance is usually complete.

Within families the variation in severity is great. Severely affected patients might have mildly affected offspring. There is some evidence (Kazakov *et al.* 1974) that the 'gradual' and 'jump' types breed true within families.

Recently, retinal vascular abnormalities have been described as a manifestation of FSH dystrophy, with or without deafness (Fitzsimons *et al.* 1987, Gieron *et al.* 1985). These authors suggest that retinal changes are an integral part of FSH and not a separate condition.

Genetic counselling in facio-scapulo-humeral dystrophy

Inheritance is clearly dominant and the main problem is the detection of those who are minimally affected and when to proclaim unaffected members unlikely to still develop signs.

Almost all cases can be detected by the late twenties, although family members need to be examined before pronouncing them unaffected. If those who want children are on clinical examination without signs at the age of 30 then risk to offspring should be small.

Early onset facio-scapulo-humeral dystrophy

Whether this entity with an onset in infancy is different from the later onset condition described above is uncertain. Many cases are sporadic and there is evidence that hearing loss might not be infrequent (Voit *et al.* 1986). Occasionally, death occurs in childhood (McGarry *et al.* 1983). Of the 10 patients reported by Voit *et al.* the pattern of inheritance was autosomal dominant in six.

Variant of facio-scapulo-humeral dystrophy

A family with dominantly inherited facio-scapulo-humeral dystrophy with slow progression and variation in severity is reported by Hurwitz *et al.* (1965). Segregating in the same family was an aminoaciduria involving the over-excretion of lysine, cystine, orinithine, and arginine.

Autosomal recessive inheritance of facio-scapulo-humeral dystrophy

Two publications are widely quoted as evidence for occasional autosomal recessive inheritance in facio-scapulo-humeral muscular dystrophy. Chung and Morton (1959) state simply that they had four families without giving any clinical details and Stevenson (1953/4) had nine families with a limb girdle dystrophy and facial involvement. The data on these nine families were compatible with recessive inheritance. However, in all nine, skeletal muscle weakness and wasting preceded recognition of facial involvement, and in seven of the nine the first symptoms were in the lower limbs—an unusual finding in facio-scapulo-humeral dystrophy. This mode of inheritance for FSH must be extremely rare and should be counselled with caution.

An autosomal recessive dysrtrophy in Manitoba Hutterites

The distribution of weakness and wasting in the families reported by Shokeir and Kobrinsky (1976) was of the facio-scapulo-humeral type. All those affected could be traced to one colony in South Dakota and the disease was probably introduced from Europe between 1874 and 1879.

The onset was between 1 and 9 years, and wasting was initially confined to the quadriceps and pelvic girdle muscles. The face became affected later in the course of the illness and eventually distal muscles were involved.

Subsequently, Shokeir and Rozdilsky (1985) have described three further individuals who belong to another branch of the Hutterites in Saskatchewan and comment that the dystrophy also occurs in the Hutterites of Alberta.

Facio-scapulo-humeral dystrophy/Coats disease/deafness/mental retardation

Small (1968) reported a condition very similar to the above except that mental retardation was a part of this condition. The onset of the muscle problem was within the first year of life and the family structure was different in that there were three affected sibs born to unaffected parents suggesting autosomal recessive inheritance.

8. Muscular dystrophy with contractures and cardiomyopathy (Hauptmann–Thannhauser)

See also Emery–Dreifuss—Autosomal dominant, which is probably the same condition. A French-Canadian family in whom many members were affected over three generations by an inability to extend the arms and legs had, in addition, atrophy and diminished power (Hauptmann and Thannhauser 1941). The disorder was not progressive. All those affected had short necks and they were unable to fully bend the head.

9. Emery–Dreifuss autosomal dominant

A condition similar to the Emery–Dreifuss dystrophy is described by Fenichel *et al.* (1982). The age of onset was between 2 and 10 years and the condition was relatively benign with only slow progression. Contractures occurred early. The distribution was humeropelvic and the cardiomyopathy was severe. It differed from Emery–Dreifuss dystrophy in the mode of inheritance—autosomal dominant (father and two children). It is similar to the family described by Chakrabarti and Pearce (1981), but in that family the progress was more rapid.

Similar cases are reported by Gilchrist and Leshner (1986). Clinically, the patients were like those with the Emery–Dreifuss syndrome, but the cardiac defect in Emery–Dreifuss is mostly an atrial conduction arrest, whereas in this group it was more variable and included atrial fibrillation, ventricular arrhythmia, cardiac failure, or a cardiomyopathy.

The autosomal-dominant dystrophy with humero-peroneal weakness and cardiomyopathy would, on present evidence, fall into this group.

In the patient reported by Galassi *et al.* (1986) there was a childhood onset of weakness. She could walk until 4 years of age. At 10 years she could not straighten her elbows or extend the wrists or thumbs nor flex her neck. At 30 she developed a tachychardia and throughout this illness the facial musculature was normal. On examination the weakness involved deltoids, biceps, triceps, but not the scapulae. The proximal leg muscles were normal, but the anterior tibial and peroneal muscles were affected resulting in bilateral footdrop. Her daughter had the same condition. An EMG confirmed a myopathy and a biopsy showed fibre I atrophy and predominance. An ECG revealed a hypertrophic cardiomyopathy in mother and a chronic bradycardia in her daughter.

Miller *et al.* (1985) described the Emery–Dreifuss syndrome in a father and daughter. The age of onset in the daughter was at 5 years with difficulty in running and in her father in his 30's—the first symptoms being weakness of the legs and limitation of neck flexion. At that stage he had evidence of a cardiomyopathy (his daughter developed cardiac involvement in her 30's and she had calf hypertrophy.

Counselling

This type of Emery–Dreifuss—autosomal dominant—can only be identified on pedigree evidence or when a female is affected. In the latter situation a manifesting heterozygote of the X-linked variety cannot be excluded. There is also a report by Takamoto *et al.* (1984) of a female whose parents were consanguineous.

10. Quadriceps myopathy

This chronic, slowly progressive myopathy has been only infrequently reported. Up to 1974 not more than a dozen cases had been published, mostly single, but one sibship (van Wijngaarden *et al.* 1968) had two affected brothers. Dominant pedigrees are also known. A family in which a man, his brother, and three of his daughters had a quadriceps myopathy with onset in adult life is reported by Espir and Matthews (1973). The hands became affected later in the course of the disease. In two other cases (Turner and Heathfield 1961; van Wijngaarden *et al.* 1968) there was evidence of wider involvement although the brunt of the disease was on quadriceps musculature. In the large dominant pedigree reported by Bacon and Smith (1971) the quadriceps distribution of wasting and weakness predominated, but progression to the arms occurred later. Onset was in middle life.

Counselling

Mostly dominant, but note the single report in brothers (see above).

11. Limb-girdle muscular dystrophy (Fig. 17.14)

Characterized by an onset in adolescence or early adulthood (sometimes later), both the hip and shoulder girdles are involved. Progress of the disease is variable. Enlargement of calf muscles is rare, but when it occurs and it can be difficult to distinguish this condition in the male from the Becker X-linked muscular dystrophy.

Recessive

This is the usual mode of inheritance. Stevenson (1953/4) found parental consanguinity in 19.4 per cent of his cases, although only in one instance were sibs affected. Sibs are reported by Dubowitz (1978).

   In a study of this type of muscular dystrophy in the Amish in the midwest of America, Jackson and Strehler (1968) found 37 affected in 14 families. Consanguinity was a feature in all cases. The onset was between four and 15 years and confinement to a wheelchair occurred between 12 and 44 years of age. Creatine phosphokinase levels were markedly raised in the affected.

   Recently complete ascertainment of adult onset limb girdle myopathy in the Lothian Region of Scotland was attempted (Yates and Emery 1985), and a prevalence of 0.9 per 100 000 proven cases was obtained. In 10 sibships there were 11 affected (one pair of male sibs). If those who might have Becker muscular dystrophy were excluded the prevalence figure was between 0.3 and 0.7 per 100 000. See p. 374 for counselling.

Limb girdle myopathy of late onset with diabetes mellitus

Swash *et al.* (1970) described four sisters with a late-onset (40–70 years) proximal myopathy, cataracts, and Dupytren's contracture. Two brothers

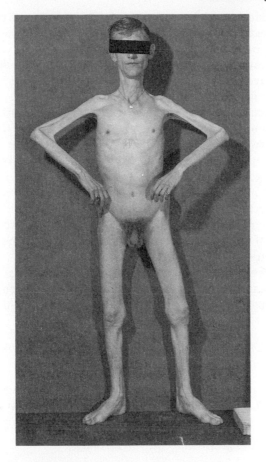

Fig. 17.14   Limb-girdle muscular dystrophy.
Courtesy of the Department of Medical Illustration, National Hospital for
Nervous Diseases, Queen Square, London.

were diabetic, one had cataracts, but neither had atrophy of the palmar fascia
nor muscle wasting.

Dominant limb-girdle muscular dystrophy

A large pedigree in which nine males were affected over four generations
(with male-to-male transmission) was presented by De Coster *et al.* (1974).
No member had either facial involvement or shoulder girdle atrophy and the
weakness and wasting was confined to pelvic girdle muscles. The onset was in
the fifth decade.

A large pedigree in which members had an adult onset of proximal weakness was reported by Schneiderman *et al*. (1969). The face was unaffected. A muscle biopsy revealed changes of a dystrophy on light microscopy and destruction of myofibrils on electron microscopy. In this family there was a suggested linkage with the Pelger–Huet anomaly (bipolar nuclei in polymorphonuclear leucocytes).

In the pedigree reported by Gilchrist *et al*. (1988), the arms were affected, but never without lower limb involvement. The CPK was elevated but not grossly so and two were in a wheelchair 20 years after onset.

A similar pedigree is reported by Chutkow *et al*. (1986). Six members were affected in two generation. The onset was in the 2nd decade and the distribution was initially pelvifemoral and then scapulohumeral. Proximal muscles were more affected than distal. The face was unaffected. At biopsy the muscle fibres were rimmed or had autophagic vacuoles. Two patients had a small number of ragged red fibres which were thought to be a non-specific reaction to injury.

The CPK's were normal or raised two-fold, the ECG was unremarkable and the EMG was myopathic. The family reported by Henson *et al*. (1967) in which only females were affected could be an example of this condition.

Limb-girdle muscular dystrophy with the Wolff-Parkinson–White syndrome

Two sibs had a limb girdle muscular dystrophy with onset in early adulthood (Weissleder *et al*. 1987). One of the sibs had the conduction defect.

Counselling in limb girdle muscular dystrophy

This remains difficult. The diagnosis should be considered in single females (or sisters) with the appropriate distribution of weakness and wasting. Mostly, the request is for offspring risks and these are likely to be small. It is at present not possible to make a diagnosis in a single male but it is accepted in this dilemma to counsel as if the patient has Becker dystrophy provided that the onset occurs before the age of 20 years.

Only an assessment of the pedigree will allow differentiation between the autosomal dominant and recessive types.

12. Scapuloperoneal syndrome

The following classification based on Mercellis *et al*. (1980) is useful.

1. Scapuloperoneal muscular dystrophy (autosomal dominant)

2. Scapuloperoneal muscular atrophy
 A. Without sensory loss
  (i) autosomal dominant
  (ii) autosomal recessive
 B. With sensory loss
  (i) autosomal dominant
  (ii) autosomal recessive

3. X-linked (could be the same as Emery—Dreifuss)

1. Scapuloperoneal muscular dystrophy (dominant)

Thomas *et al.* (1975) reported a family where a mother and daughter had an onset in adolescence, and Serratrice *et al.* described families in 1969 and 1979, and Scarlato *et al.* in 1978.

The difficulty in deciding whether the dystrophy of scapuloperoneal distribution is myogenic or neurogenic is clearly seen in the kindred K reported initially by Davidenkow (1939). For more than 40 years members of the kindred have been studied by various authors who have called it both scapulo-humeral, scapulo-humero-peroneal, and fascioscapulo-humeral dystrophy. The pedigree is clearly dominant (five generations) but more recently Kazakov *et al.* (1974, 1976) have shown that within the kindred, members could be individually classified as either facio-scapulo-humeral or (facio)-scapulo-peroneal syndromes. Often the (facio)-scapulo-humeral syndrome is a stage in the development of the facio-scapulo-peroneal muscular dystrophy.

A dominant pedigree with at least four affected over three generations is described by Feigenbaum and Munsat (1970) as a neuromuscular syndrome. They report that among the cases shown to be familial, there are more neurogenic than myopathic, but in some instances it is impossible to confirm the exact nature of the disorder.

In the cases collected by Ricker and Mertens (1968; 13 out of 212 patients with muscular dystrophy were thought to have the scapuloperoneal syndrome), dominant transmission occurred in at least five members who belonged to three families.

2. Scapuloperoneal muscular atrophy (neurogenic)

See p. 260.

3. X-linked scapuloperoneal myopathy

The six males who were affected over four generations (Thomas *et al.* 1972) had a myopathy, *pes cavus*, and contractures at the elbows. A cardiomyopathy developed in adult life. The authors pointed out some resemblance of this family to that described by Emery and Dreifuss (1966; see p. 364), but the distribution of muscle involvement was different. In the Thomas *et al.* (1972) pedigree there was linkage to the locus determining deutan colour blindness. Seventeen males with a myopathic scapuloperoneal syndrome occurred in three generations (Rotthauwe *et al.* 1972). In this family, as in the Thomas *et al.* (1972) pedigree, a cardiomyopathy was present. A similar family was reported by Hopkins and Karp (1976).

Most authors now accept that this condition is the same as the Emery–Dreifuss syndrome.

X-linked scapuloperoneal myopathy without contractures but with a severe cardiomyopathy

A severely affected family with a fatal cardiomyopathy and the appropriate distribution of muscle weakness and wasting is reported by Bergia *et al.* (1986) in three male cousins. All were normal until 5–6 years then the muscle and intellectual problems were noted. No contractures developed.

13. Oculopharyngeal myopathy (Fig. 17.15)

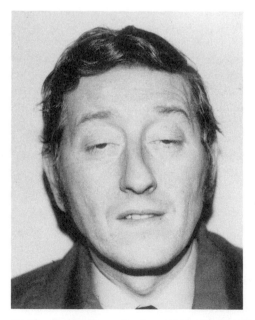

Fig. 17.15   Oculopharyngeal muscular dystrophy.

Dominant

In 1915, Taylor reported a French-Canadian family with ptosis and progressive dysphagia. Many of the subsequent families (Myrianthopoulos and Brown 1954; Peterman *et al*. 1964; Teasdall *et al*. 1964; Bray *et al*. 1965) were reported from Canada and the United States. Barbeau (1967) succeeded in tracing many of these families back to a French-Canadian ancestor who is known to have had 249 affected descendants. Aarlie (1969) reported a dominant pedigree as did Victor *et al*. (1962). The age of onset in this group is more advanced than in most of the other myopathies with the exception of the distal myopathies.

When the limbs are affected proximal musculature is more vulnerable than distal muscles. Bilateral facial weakness occurred in members of the three families reported by Bray *et al*. (1965).

In the family reported by Julien *et al*. (1974), electron microscopy of muscle showed 'finger print' inclusions. Inheritance was as a dominant but this may be a separate condition (see p. 392).

Recessive

Genetic heterogeneity of oculopharyngeal myopathy is suggested by an analysis of the pedigree of Fried *et al*. (1975), in which two sibs of Ashkenazi Jewish origin were affected. The unaffected parents were first cousins once removed and third cousins. The clinical picture in these sibs is indistinguishable from the dominant type although the age of onset (mid-30s) is less than the usual 45–60 years in the dominant oculopharyngeal dystrophy. Recessive inheritance in the group as a whole is rare.

Oculopharyngodistal myopathy

Four Japanese families (Satoyoshi and Kinoshita 1977) with this distribution are described. Seventeen members were affected. Because of the distal distribution of weakness (foot drop) the authors suggested that this condition it is not related to the oculopharyngeal myopathies where wasting is predominantly proximal.

In Welander's (1951) series of 249 patients with a distal myopathy only six had ocular signs and none had swallowing difficulties.

Oculopharyngeal myopathy with distal distribution and cardiomyopathy

A Japanese man had this combination (Goto *et al*. 1977). His son had cardiac involvement and the patient's sister had ptosis. No mitochondrial abnormalities were found.

B. Myopathies

14. Distal myopathy

In 1951, Welander described over 200 cases of a distal myopathy in Sweden. Her cases were characterized by dominant inheritance, an onset between 20 and 77 years, and slow progression. Initially, the intrinsic muscles of the hands and feet were involved.

Out of 14 cases described by Edström (1975), in 10 there was evidence of dominant inheritance. Biopsy findings in the distal myopathies show striking similarities in the early stages to myotonic dystrophy (but markedly different from Charcot–Marie–Tooth disease and the spinal muscular atrophies).

Fewer than 10 cases have been reported elsewhere. Markesbery *et al.* (1974) described a dominant pedigree (three generations affected) with onset in middle life and gradual progression. Distal myopathy has long been associated with the name of Gowers, but it seems likely that the young male Gowers described had myotonic dystrophy (Markesbery *et al.* 1974).

An early onset group also occurs. Biemond's family (1955) had 19 affected over five generations. Only one patient had an electromyogram and this was stated to be typical of muscular dystrophy. The onset was earlier than in Welander's cases (5–15 years), progression was slow and the disease started simultaneously in hands and feet.

A father and two sons have been reported with an early onset distal myopathy (Van der Does de Willebois *et al.* 1968). Bilateral foot drop occurred at 2 years. Magee and de Jong (1965) had one case with a similar onset, and Bautista *et al.* (1978) described a three-generation pedigree. The distal weakness was relatively mild and only the most severely affected member in the Bautista pedigree sought orthopaedic advice. It is suggested (Bautista *et al.* 1978) that the selective hypotrophy of type I fibres found in his probands might indicate that these families belong to the group called fibre type I disproportion (see p. 394). An indication that the late and the early onset distal type might be manifestations of the same gene defect is suggested by the pedigree described by Sumner *et al.* (1971). There was a striking difference in the age of onset within this family (15 to 50 years).

A possible separate dominantly inherited distal myopathy was reported by Edström *et al.* (1980). The onset was in middle life, involved the thenar muscles and hand flexors early on in the course of the disease, and was more malignant than the types mentioned above. A muscle biopsy showed sarcoplasmic bodies.

15. Distal muscular dystrophy—autosomal recessive (ARDMD)

Seventeen cases had an onset in young adulthood. The distribution was distal in the legs, but the weakness then spread to the thighs and glutei. Early in childhood there was difficulty in standing on tip-toe (not on heels), then of climbing stairs. The forearms were mildly affected (Miyoshi *et al.* 1986). The CPK was increased and an EMG was myopathic. A biopsy showed changes similar to Duchenne muscular dystrophy. Seven were the offspring of consanguineous marriages.

Note: same as Kuhn and Schroder (1981) who reported pseudodominance in three families. Inheritance is probably recessive.

16. Others

(a) Mitochondrial disease

It now seems certain that many of the so-called ocular myopathies described in the older literature would now be found to be mitochondrial myopathies. Whether this is always the case is uncertain. The diagnosis of mitochondrial disease is usually made by finding 'ragged red fibres' on muscle biopsy, but it is increasingly unlikely that this feature is always present.

Some of the entities under this heading will be described separately, but in summary:

(1) mitochondrial disease has multisystem involvement, but nearly always includes muscles;

(2) abnormal mitochondrial function is most often diagnosed by finding ragged red fibres on muscle biopsy;

(3) proven or postulated defects of aerobic oxidation metabolism are an increasingly frequent finding.

Biochemically the following deficiencies may be found (data from Nishizawa *et al.* 1987 and Shapira *et al.* 1977)

Complex I	NADH-coenzyme Q-reductase,
Complex II	Succinate-coenzyme Q-reductase
Complex III	Coenzyme Q cytochrome C-reductase
Complex IV	Cytochrome C-oxidase

Clinically the main entities are:

(1) Kearns–Sayre

(2) MELAS (Pavlakis *et al.* 1984; see Di Mauro *et al.* 1985)

(3) MERRF (Fukuhara *et al.* 1980)

(4) Progressive external ophthalmoplegia

(5) Chronic myopathy

(6) Myopathy with excercise intollerance

(7) Fatal infantile encephalopathy

(8) Leigh's disease

(9) Some patients have cataracts (might be separate)

(1) Kearns–Sayre syndrome or progressive external ophthalmoplegia atypical retinal pigmentation, and heart block.

These are the chief criteria required for a diagnosis of this group. Mental retardation (40 per cent of cases), cerebellar ataxia, deafness, short stature, and elevated cerebrospinal fluid protein have all been described (Berenberg *et al.* 1977). Familial cases are few and Berenberg could not find any convincing cases in his review of the literature. However, Bastiaensen *et al.* (1978) reported two propositi who had members in three generations affected with ptosis, whereas only the proband had the full syndrome. Other possibly inherited cases are those of Lambert and Fairfax (1976) and Schnitzler and Robertson (1979). In the former report the proband had heart block and a dystrophy of the scapuloperoneal distribution, and his sister and daughter had the same distribution of weakness without heart block. In a more convincing report a father developed weakness in his legs in middle age and then developed ataxia, dysarthria, and ophthalmoplegia (Jankowicz *et al.* 1977). His intelligence was below normal. His son had the cardiac abnormality and external ophthalmoplegia. Both had retinitis pigmentosa. Muscle biopsy showed a mitochondrial myopathy.

The differentiation between the group called ophthalmoplegia 'plus' and the Kearns–Sayre syndrome is not always clear. The brother and sister reported by Tamura *et al.* (1974) had ocular palsies, growth impairment, deafness, facial weakness, dysphagia, dysarthria, a proximal myopathy, and high cerebrospinal fluid protein. Another sib had only the progressive ophthalmoplegia. Onset was in early adulthood. Muscle biopsy showed a marked abnormality of the mitochondria. The patient of Okamura *et al.* (1976) had an affected sister and consanguineous parents. Single cases include the seven reported by Olsen *et al.* (1972) and those of Zintz and Villiger (1967), Morgan-Hughes and Mair (1973), Schotland *et al.* (1976), and Hülsmann *et al.* (1967).

Focal neurological signs, vomiting, myoclonus, and seizures are rare unless accompanied by hypoparathyroidism (Pavlakis *et al.* 1984).

Counselling   See p. 385.

(2) MELAS

M   Mitochondrial myopathy
E   Encephalopathy
L   Lactic

A   Acidosis

S   Stroke-like episodes

MELAS can be differentiated from MERRF by

(a) the stroke-like episodes

(b) episodic vomiting;

(c) absence of cerebellar ataxia.

Other features include normal early development, short stature, seizures, hemianopia, hemiparesis, and cortical blindness. The onset is usually between 3 and 11 years (Pavlakis *et al.* 1984).

A sib pair with MELAS was reported by Hart *et al.* (1977). A brother and sister with this combination began to dement in the middle of the first decade of life. An external ophthalmoplegia was present in the male sib.

A similar family is reported by Monnens *et al.* (1975). Two first cousins were affected.

The patient described by Kobayashi *et al.* (1982) had all the features of MELAS and the Kuriyama *et al.* (1984) patient had MELAS (not both MELAS and MERRF).

(3) MERRF (myoclonic epilepsy ragged red fibres)

This is characterized by action myoclonus, myoclonic epilepsy, cerebellar ataxia, an intention tremor, weakness, and often a foot deformity—*pes cavus*. Muscle weakness need not be present.

Early development is normal and symptoms begin between 5 and 42 years. Other features include seizures, dementia, and hearing loss (Holliday *et al.* 1983; Rosing *et al.* (1983). In the latter pedigrees transmission was always through females (to both sex offspring) suggesting mitochondrial transmission.

A family was reported by Tsairis *et al.* (1973), Those affected had myoclonic epilepsy.

A combination of growth failure, neural deafness, and a myopathy was found in two sisters (Hackett *et al.* 1973). Light microscopy showed large abnormal mitochondria. Hyperalaninaemia, lactic acidosis, and hyper-pyruvicacidaemia were features.

A girl of 6 years with marked weakness of the muscles of her trunk, shoulder, and pelvic girdle was reported by Tarlow *et al.* (1973). Muscle biopsy showed bizarre mitochondrial changes. A lactic acidosis and raised levels of pyruvic acid and alanine in the blood were part of the disease. The mother was clinically normal, but had a raised plasma lactate and pyruvate; other unaffected members had a raised creatine phosphokinase level.

The patients of Hackett and Tarlow are not easy to classify. They have been referred to as juvenile mitochondrial myopathy (Rowland *et al.* 1983) which is characterized by lactic acidemia, growth failure, and a myopathy.

Other cases are those of Fitzsimons *et al.* 1981. The mother and three maternal aunts were also affected (Sasaki *et al.* 1983) and Riggs *et al.* (1984) reported the condition in sibs. See also Byrne *et al.* 1985*b*.

Some patients are difficult to classify and within this group the clinical picture can fluctuate before leading to profound dementia. In the two isolated cases reported by Morgan-Hughes *et al.* (1982), one was found to have optic atrophy at 9 years, but did not have further problems until 34. He then developed a stroke-like episode from which he recovered and 10 years later he developed dyspraxia and generalized dystonia without muscle weakness. In their other case myoclonus preceeded the ataxia, weakness, and confusion by many years. This second patient could have had MERRF.

Combination of MELAS and MERRF

Several cases have features of both MELAS and MERRF see Holliday *et al.* (1983), and Pavlakis *et al.* (1984). It should also be noted that in MERRF stroke-like episodes have been described (Fukuhara *et al.* 1980).

Counselling   See p. 385.

(4) Chronic progressive external ophthalmoplegia

This is not a single entity and as the signs might progress to involve other muscles, group-designation is difficult and the precise diagnosis needs to be considered over a period of time. There are at least three groups (the oculopharyngeal dystrophy is dealt with elsewhere).

(a) Ptosis and ophthalmoplegia.

(b) Ptosis, ophthalmoplegia, and weakness of other muscles.

(c) Ophthalmoplegia plus (ophthalmoplegia with peripheral or central nervous system involvement). See Kearns–Sayre p. 380.

In general those with only eye muscle involvement are less likely to be mitochondrial than those with additional features. Ophthalmoplegia can also occur in:

(d) MELAS (p. 380)

(e) MERRF (p. 381)

(f) Progressive muscle weakness alone (p. 383)

Croft *et al.* (1977) reviewed 122 papers (1868–1971), and found a positive family history in about 37 per cent of 335 cases. In their own series, seven out of 13 had another close affected relative. The inheritance in these cases was as a dominant.

(a) Ptosis and ophthalmoplegia   A father and son with this combination are reported by Croft *et al.* (1977), and other families by Beckett and Netsky (1953) and Danta *et al.* (1975)—who reported three mother-daughter pairs.

(b) Ptosis, ophthalmoplegia, and involvement of other muscles   Dominant pedigrees are recorded by Croft *et al.* (1977), Barrie and Heathfield (1971), and Danta *et al.* (1975). Variation in expression might necessitate personal examination, especially for ptosis, in parents and offspring before excluding the familial form. If groups (a) and (b) are considered together and taking into account the variability in expression, then about one-third of cases are dominantly inherited. This suggests that the risk to offspring of single cases is probably greater than 10 per cent. Genetic counselling is made difficult by the problem of assessment of ptosis and the late onset of the disease.

Under the title 'familial progressive external ophthalmoplegia and ragged red fibres' Iannaccone *et al.* (1974) reported a four-generation family with progressive ptosis, ophthalmoplegia, and facial and shoulder-girdle muscles. Two patients with ragged red fibres are reported by Byrne *et al.* (1985*a*). Both were single cases.

(5) Myopathy

Mitochondrial myopathies might present, especially in adult practice, as a pure muscle disease. In an analysis of 67 patients (Petty *et al.* 1986) about a half had only ptosis and limb weakness, whereas only 3/67 had multisystem problems.

A family reported by Hudgson *et al.* (1972) and later by Worsfold *et al.* (1973) had six members affected over four generations. The pattern of muscle involvement was compatible with a clinical diagnosis of facio-scapulo-humeral dystrophy. There was some variation within the family in that in one of the six the face was spared, and the clinical picture resembled a limb-girdle dystrophy.

Quite another clinical picture was found in three members affected over three generations (Lapresle *et al.* 1972). Inheritance was as a dominant, and progression was slow. Mild distal weakness in the legs was probably present from birth, and the calf muscles were hypertrophied. Spiro *et al.* (1970) reported this condition in a father and son.

Sibs with slowly progressive proximal weakness were reported by Shy *et al.* (1966).

(6) Myopathy with excercise intolerance and weakness

There have been a number of children and adults with this combination (Morgan-Hughes *et al.* 1985). The patients reported by Land *et al.* (1981) had in addition an ophthalmoplegia and the one reported by Arts *et al.*

(1983*b*) presented as a limb girdle myopathy. There was no central nervous system involvement and most have had Complex I (NADH-Coenzyme Q) reductase deficiency.

(7) Acute encephalopathy of infancy or fatal infantile mitochondrial disease

This is a neonatal illness presenting with vomiting, hypotonia, weakness, ptosis, areflexia, and sometimes a lactic acidosis accompanied by the de Toni–Fanconi–Debre Syndrome (renal tubular acidosis). Sibs are reported by Van Biervliet *et al.* (1977) and by Michom *et al.* (1983). Single patients of Di Mauro *et al.* (1980) and of Zeviani *et al.* 1985 had the same condition. Some have a cardiomyopathy (Rimoldi *et al.* 1982; and Sengers *et al.* 1984) and most will have Complex IV or Cytochrome C-Oxidase deficiency.

Two patients have been described with an Alpers-like clinical picture (Gabreels *et al.* 1984). Prick *et al.* (1981) reported an infantile onset Alpers-like disease with a defect of NADH oxidation in liver and muscle.

(8) Leigh's disease

Subacute necrotizing encephalomyelopathy   Under the title mitochondrial cytopathy or Leigh's syndrome, Egger *et al.* (1984) reported a boy who developed an anaemia at 8 months, but who only had neurological signs at 6 years and then developed sensorineural hearing loss, a pigmentary retinopathy, ptosis, ophthalmoplegia, ataxia, and distal muscular weakness. A CT scan showed low density areas in the basal ganglia and a muscle biopsy showed ragged red fibres. At post-mortem the pathological features were indistinguishable from Leigh's disease. Clinically, the boy did not have Leigh's disease and it might be concluded that low attenuation in the basal ganglia can be found in a Kearns–Sayre-type clinical picture.

The single case reported by Crosby and Chou (1974) had a cardiac defect and could fall into this group.

(9) Mitochondrial myopathy (skeletal muscle and heart), congenital cataract, and lactic acidosis

Seven out of 22 children in three unrelated families had this combination (Sengers *et al.* 1975*b*). Storage of lipids and glycogen was found in muscle. The onset was early in life with delayed motor development, but normal intelligence.

Affected members in a family had progressive external ophthalmoplegia, congenital cataract, somatic weakness (and gonadal dysgenesis in one member; Barron *et al.* 1979). Muscle biopsy showed ragged red fibres and mitochondrial inclusions. A similar family is reported by Pepin *et al.* (1980).

The clinical features were cataract, bilateral ophthalmoplegia, and facial, pharyngeal, and limb weakness.

Biochemical findings in mitochondrial disease

These will not be discussed because the clinical variability, given a defect in any of the known enzyme deficiencies, is so great that at present it does not help the geneticist to establish a diagnosis nor recurrence risks. It should, however, be stated that defects in the respiratory chain have been found in four of the five respiratory complexes.

Genetic counselling in mitochondrial cytopathy

Despite the enormous strides that have been made in understanding the biochemistry and the clinical boundaries of mitochondrial disease, counselling remains extremely difficult.

A few general points are pertinent.

1. Cytoplasmic/maternal transmission is still only suggestive—on pedigree evidence—in a small number of families. In six families studied by Egger *et al.* (1983), maternal transmission occurred. They reviewed families from the literature and found 21 with maternal transmission and three families where male transmission occurred. However, in two of the latter three families, mothers passed on the condition many more times than fathers did. It is proposed that cytoplasmic inheritance is the likely mode of inheritance in mitochondrial cytopathy.

The following families were included in the review by Egger *et al.* (1983): Hudgson *et al.* (1972); Tsairis *et al.* (1973); Iannaccone *et al.* (1974); Shapira *et al.* (1975); Monnens *et al.* (1975); Hart *et al.* (1977); Berenberg *et al.* (1977); Barron *et al.* (1979); Pepin *et al.* (1980); Spiro *et al.* (1970).

The following observations would be expected if cytoplasmic inheritance were operative.

(a) Transmission would be, in the vast majority of cases, through an affected mother rather than father. Occasional paternal transmission would not exclude cytoplasmic inheritance as there is a small amount of cytoplasm around the nucleus in sperm.

(b) Transmission would affect sons and daughters in more or less equal proportions.

(c) There would be no transmission down the paternal line.

(d) The ratio in a sibship of those affected to those unaffected would not be those expected by simple Mendelian inheritance. It is possible that all sibs could be affected.

It remains uncertain whether some, or all, of the clinical subdivisions are transmitted in this way.

2. Most cases of Kearns–Sayre are sporadic, but a 5 per cent offspring and sib-sib risk is appropriate.

3. MERRF—see pedigree of Rosing *et al.* (1983). This subdivision seems to fit best the criteria for maternal transmission. Sib recurrence risks are high (greater than 1 in 10).

4. MELAS. There have been sufficient reports of affected sibs to counsel single figure risks. It is probably not worthwhile to do a muscle biopsy on a clinically normal mother.

5. Myopathies. In the Petty *et al.* (1986) paper a family history was obtained in 13 instances. There were three sib pairs with normal parents and six instances of female to offspring transmission. In one case an unaffected mother had an affected brother and daughter (?maternal transmission), but there were two affected cousins with intervening normal males. Risks to sibs and offspring are about 5 per cent.

6. Acute encephalopathy. Single figure sib risks (> 1 in 10).

7. Recently the mitochondrial hypothesis for at least some of the mitochondrial myopathies has been confirmed by the finding of deletions in muscle mitochondria of patients with this group of conditions (Holt *et al.* 1988).

X-linked mitochondrial disease

There have been two reports of an X-linked disease involving heart muscle, skeletal muscles and accompanied by neutropenia. Most of those affected were dead before their 2nd year. In the pedigree presented by Barth *et al.* (1983) the cause of death was septicaemia and heart failure. The other report was by Neustein *et al.* (1979).

Non-progressive external ophthalmoplegia (Congenital)

A family in which seven members in three generations had congenital ptosis, myopia, and varying degrees of paralysis of cranial nerves III, IV, and VI was reported by Mace *et al.* (1971). A similar family was described by Bradburne in 1912 and Holmes (1956). In a dominant pedigree reported by Slatt (1965) the clinical findings were similar, but only the muscles innervated by the third cranial nerve were affected.

(b) Nemaline myopathy

Rod-like bodies have been observed in different categories of muscle disease, including toxic and inflammatory myopathies. In the genetically determined nemaline myopathies the age of onset varies from early childhood to middle life. Deterioration can be either gradual over many years or respiratory problems might be severe enough in infancy to cause early death. The facial muscles can be markedly affected, but ophthalmoplegia is not usually a feature. Dysmorphic features reminiscent of Marfan's syndrome are suggested by the high-arched palate, thin face, prominent jaw with malocclusion of the teeth, and pectus carinatum.

Dominant (Adult)

The first cases were described by Shy *et al.* (1963) and Connen *et al.* (1963). The mother of Shy's case had proximal weakness and a 'myopathic' electromyogram, but most of the more than 50 cases that have been published are sporadic (Arts *et al.* 1978*a*). All the (approximately six) adult-onset nemaline myopathies have been single cases. In a few families more than one generation has been affected. The mother of Engel and Gomez's (1967) case had slight weakness of the sternomastoid, facial, triceps, and intrinsic hand muscles. Two of the patient's sibs were thought to have died from a similar disorder.

Spiro and Kennedy (1965) reported the presence of a non-progressive nemaline myopathy in mother and daughter. In the 10-year-old daughter the clinical picture was that of moderate weakness of shoulder and pelvic girdles with slight weakness of sternocleidomastoid muscles. A maternal male cousin, aunt, and grandmother allegedly had varying degrees of weakness but they were not studied. The mother and daughter described by Hopkins *et al.* (1966) again suggest dominant inheritance, as did the mother and daughter pairs of Afifi *et al.* (1965) and Shafiq *et al.* (1967).

It can be difficult to interpret individual pedigrees; Gonatas *et al.* (1966) reported sibs with biopsy proven nemaline myopathy. However, it became clear that the patients were relatives of the mother and daughter described in 1963 by Shy *et al.*, and the mode of transmission was dominant.

Some patients with nemaline myopathy develop multiple joint webbing and are clinically indistinguishable from the multiple pterygium syndrome (Papadia *et al.* 1987).

Recessive (severe)

Sibs in a single generation have been described by Nienhuis *et al.* (1967) and a sister of the proband in the Karpati *et al.* (1971) report is said to have had the same disease. Both children were severely affected and died before the age of five years. In the Neustein (1973) pedigree, three out of four sibs were affected. All three affected children died under the age of one year. The onset was from birth (there were even reduced movement *in utero*) and only the three affected showed histological changes.

An infant with a fatal outcome was the product of a consanguineous union (Tsujihata *et al.* 1983).

Counselling

In a review (Arts *et al.* 1978*a*) six families were studied. One showed dominant inheritance. In the other five families, recessive inheritance was suggested. The authors were unable to distinguish clinically between the dominant and recessive pedigrees. The minor changes found in the parents were interpreted as manifestations of a heterozygote. In one of their families two sibs (out of four) were affected and both parents had nemaline rods on biopsy. In one single case, the same was found. In one family, only one parent showed the rods. The parents of another single case showed an increased number of fibres with central nuclei and this was also taken as evidence for heterozygosity. Until the significance of these changes in the parents is elucidated, genetic counselling will remain difficult. In a review of nemaline myopathy in infancy, Norton *et al.* (1983) included two sibships from the literature out of a total of six reports. Those were the families of Neustein (1973) and Gillies *et al.* (1979).

At this stage it must be concluded that the severe neonatal form is recessive. Two single cases, both with consanguineous parents (Arts and de Groot 1983) support this supposition.

(c) Central core disease

So-called because the muscle histology shows the central part of the muscle to be occupied by one or several cores running along the fibre (Gardner-Medwin 1977).

The disease is usually non-progressive. Weakness is moderate and the onset of walking is often delayed (Gonatas *et al.* 1965; Shy and Magee 1956). The lower limbs are affected more often than the upper, and proximal muscles more often than the distal. In the original report (Shy and Magee 1956) five members were involved over three generations—clearly a dominant pedigree. Since then Dubowitz and Roy (1970) and Bethlem *et al.* (1966) have described dominant pedigrees. In the report of Bethlem *et al.* (1966) the motor milestones were normal, and the clinical picture was that of stiffness and fatigue rather than weakness. In a single case (Telerman-Toppet *et al.* 1973) bilateral *pes cavus* was the only manifestation. Other dominant pedigrees are those of Palamucci *et al.* (1978) and Armstrong *et al.* (1971). Congenital dislocation of the hip is frequently found.

The only possible alternative mode of inheritance was suggested by Dubowitz and Platts (1965) who described an affected brother and sister whose parents were first cousins. In the woman, focal wasting of the right shoulder girdle was present from birth and only after an episode of uremia at the age of 40 years did generalized weakness occur. Her brother had a congenital focal atrophy of the right shoulder but biopsy was refused. It is thought that the central cores develop during the course of the myopathy.

Central core disease is probably not a homogeneous muscle disorder. Morgan-Hughes *et al*. (1973) reported a family in which a mother and two children had a non-progressive myopathy. Typical central cores were found in mother, but not in her children, although they had morphologically abnormal muscle. Support for the idea that central core disease might not be a pure entity also comes from Afifi *et al*. (1965) who found central cores and nemaline bodies in the same family. Variation in expression of the condition is well demonstrated. In a family in which four members had a congenital non-progressive myopathy, two had the adult-onset progressive type (Patterson *et al*. 1979).

In a family in which the disease has been traced back for five generations (Isaacs *et al*. 1975), members had muscular weakness, hypotonia, kyphoscoliosis, and *pes cavus*. The pelvic muscles were particularly affected and proximal shoulder girdle less so. No ocular muscles were involved.

Single cases have also been described (Bethlem *et al*. 1971; Wynne-Davies and Lloyd-Roberts 1976). There is one report in which one of monozygotic twins was affected (Cohen *et al*. 1978). This patient had arthrogryposis multiplex congenita.

Counselling Autosomal dominant.

(d) Multicore disease (minicore disease)

This is characterized by the presence of numerous lesions similar to those found in central core disease, but which are eccentrically placed and smaller than central cores. The first case was described by Engel and Gomez (1966) with an onset in infancy. Bonnett *et al*. (1974) reported a case with an onset at 33 years. Out of a total of five cases to date (Heffner *et al*. 1976) three were families with more than one affected member. In Heffner's families, there were identical twins with an onset in infancy of a non-progressive generalized weakness. Proximal more than distal muscles were involved—especially the sternomastoids. The twin's sister suffered from a neuromuscular disease (no biopsy), but as with her brothers, improvement took place and she showed no signs of weakness. Engel *et al*. (1971) reported two cases. One was sporadic but in the other a maternal grandmother and maternal uncle were said to have a neuromuscular disease, but no details were given. In Bonnette *et al*.'s (1974) late-onset case, the weakness has been progressive. The genetics are, as yet, uncertain. More recently, Brownell (1979) has reported multicore disease in a mother and child, but Lake *et al*. (1977) reported sibs (brothers) with this condition. Both were floppy in the newborn period, but whereas one seemed to have little weakness at one year, the other at five years had a high-pitched voice, mild ptosis, bilateral facial weakness, and generalized muscle wasting.

Another sibship with normal parents is reported by Pages *et al*. (1985).

Clinically, some of those affected have a Marfanoid habitus with scoliosis and arachnodactyly.

A sibship (normal parents) was reported by Swash and Schwartz (1981). There was an onset in infancy of hypotonia, delayed motor development and ptosis. At the end of the first decade proximal weakness of the arms and legs developed with facial weakness and an ophthalmoplegia. CK was normal and the EMG was myopathic. A biopsy showed multicores with loss of cross-striation (see p. 393).

Counselling

Still uncertain. Both dominant and recessive families have been reported.

(e) *Centronuclear or myotubular myopathy*

Classification of myotubular myopathy (Centronuclear) (Heckmatt *et al.* 1985)

1. Severe neonatal—X-linked.
2. Less severe infantile—floppy infant or severe weakness in childhood. Most are isolated, but sibs are reported by Raju *et al.* (1977).
3. Milder juvenile and adult forms—both recessive and dominant.

(1) X-linked

Van Wijngaarden *et al.* (1969) reported a pedigree with six affected males.

In a second family with X-linked myotubular myopathy (Barth *et al.* 1975) two males were affected. In addition, mother had two brothers who died during the neonatal period. In five obligatory female carriers, muscle biopsy revealed changes compatible with the carrier status. The clinical picture in this family was characterized by respiratory problems and generalised muscle weakness during the newborn period with no improvement and early death.

Meyers *et al.* (1974) reported two boys (sibs) with a myotubular myopathy. The mother had two stillborn male sibs and her mother had a stillborn male by another marriage. Brothers were reported by Bradley *et al.* (1970) and more recently two male sibs are reported by Askanas *et al.* (1979).

There have now been a number of additional cases of X-linked myotubular myopathy (Heckmatt *et al.* 1985). The onset is usually at birth with respiratory and swallowing difficulties, and marked floppiness. Most have a facial diplegia. In the Heckmatt *et al.* (1985) study, post-haemorrhageic hydrocephalus was an added problem, but not all infants die.

(2 + 3) Less severe infantile and late onset forms

These will be taken together as they can occur within the same family. The clinical picture is that of congenital slowly progressive weakness. Motor milestones are delayed, but most patients walk, although with difficulty, in their second year (Gardner-Medwin 1977). Late-onset cases with the first symptoms between two and 15 years have been described (Bradley *et al.* 1970). Ophthalmoplegia, including ptosis, is the most characteristic clinical feature, but this is occasionally absent. Schockett *et al.* (1972) believe that the clinical course is downhill and the progression is more serious than in central core disease or the mitochondrial myopathies. The weakness is generalized, both proximal and distal muscles are affected. When the facial and eye muscles are affected a picture not remarkably different from the Möbius syndrome results.

Recessive

The brother and sister reported by Ortiz de Zarate and Maruffo (1970) might have a recessive form of the disease as both parents were clinically normal. The mother was said to be 'pathologically weak' during childhood but recovered fully. Biopsies on both parents revealed changes which the authors thought might be significant in father. but not in mother. The same paper reported an affected child, the offspring of first-cousin parents. However, a maternal aunt, aged 45 years, had ptosis and difficulty with her gait.

Dominant

Sher *et al.* (1967) reported an 18-year-old Negro female with an onset of a myotubular myopathy in infancy. At 18 years, she had bilateral ptosis, external ophthalmoplegia, and marked atrophy of her facial muscles, shoulder girdle, pelvic girdle, and distal musculature. Her 16-year-old sister was similarly affected. Their mother had only mild bilateral ptosis. A biopsy taken from mother's gastrocnemius showed 39 per cent of nuclei in a central position as compared with 92 and 80 per cent in her two affected daughters (normally 1–3 per cent of fibres can contain central nuclei). The authors suggested that inheritance is as an autosomal recessive, but as the father could not be examined, dominant inheritance could not be excluded.

Two adult cases had the typical histological picture of a centronuclear myopathy (Pepin *et al.* 1976). Other members were suspected of being involved suggesting dominant inheritance with variable expression and penetrance.

The unaffected mother of the single case described by Munsat *et al.* (1969*a*) had 50 per cent central nuclei. The father was not tested.

A mother and her 11-month-old daughter were affected (Kinoshita *et al.* 1975*b*), but whereas the daughter had typical myotubules, the mother had selective atrophy of type I fibres without central nuclei (see p. 394).

McLeod *et al.* (1972) reported a dominant pedigree. Sixteen were affected over five generations. The mother in the Karpati *et al.* (1970) family clinically resembled a limb-girdle dystrophy except that the onset was in very early life.

Had it not been for mother, the daughter's problem might not have been detected. A family with dominant inheritance showed reduced penetrance (Pavone *et al.* 1980).

A three-generation family with a congenital myotubular myopathy is reported by Reske-Nielsen *et al.* (1987). Serious problems began in childhood but the progress of the illness was interrupted by long periods in which there was very little progression. Bilateral ptosis and facial involvement occurred.

Another mother and daughter pair was reported in Schockett *et al.*'s (1972) survey of the literature.

Very rarely, centronuclear myopathy is accompanied by calf muscle hypertrophy as was shown in a man and his two sons (Bill *et al.* 1979).

Counselling

Many cases are single and cannot be put into the two groups mentioned above. This makes genetic counselling difficult. In general, the dominantly inherited category seems commoner than the recessive.

Single cases are reported by Bill *et al.* (1979), Heckmatt *et al.* (1985), and Moosa and Dawood (1987) in four black Africans. Bergen *et al.* (1980) reported a single case. Onset was in late childhood (no precise age given) and at 38 he had a mild scoliosis, ophthalmoplegia with severe ptosis, and diffuse muscle weakness.

Myotonia in myotubular myopathy

Two sisters developed symptoms in infancy (Gil-Peralta *et al.* 1978). In adulthood the distribution was that of a limb-girdle myopathy. Both had percussion and grip myotonia.

Cardiac involvement in myotubular myopathy

Two sibs with a cardiomyopathy and a myotubular myopathy are reported by Verhiest *et al.* (1976). Other cases are reported by Bethlem *et al.* (1969) and Shafiq *et al.* (1972).

(f) Reducing body myopathy

Two unrelated infants with progressive weakness and an unusual sulphydryl-containing compound within the muscle fibres are reported by Brooke and Neville (1972). Both were single cases.

(g) Fingerprint body myopathy

The independent existence of this type of congenital myopathy in which there are subsarcolemmal fingerprint inclusions on electron microscopy is controversial. The weakness first noted in early infancy is non-progressive (Curless *et al*. 1978). First described in a single male infant by Engel *et al*. (1972) and then by Gordon *et al*. (1974), the only familial cases have been half-brothers (Fardeau *et al*. 1976). Four of the six reported cases have been mentally retarded. Some of the cases also show fibre type I preponderance.

Fingerprint bodies have also been described in dermatomyositis (Carpenter *et al*. 1972), in myotonic dystrophy (Tomé and Fardeau 1973), and in an adult-onset oculopharyngeal muscular dystrophy (Julien *et al*. 1974).

(h) Sarcotubular myopathy

A benign familial congenital myopathy involving two brothers from the Hutterite community was reported by Jerusalem *et al*. (1973*b*). The patient's parents were third cousins. The weakness was proximal in both upper and lower limbs. Neck flexion and facial muscles were affected. The axial muscles were involved to such an extent that chest excursion was reduced and, in one sib, the cough was feeble (see also p. 395).

(i) Familial myopathy with lysis of myofibrils

The clinical picture of early floppiness (sat at 10 months and walked at 18–20 months), waddling gait, generalized weakness more marked proximally, but without hypertrophy was described by Cancilla *et al*. (1971). A muscle biopsy revealed alteration of type I fibres with an accumulation of a fine granular material. There is some resemblance of this case to familial focal loss of cross striation (see below). However, in Cancilla *et al*.'s (1971) report, the areas of abnormality were always located in the subsarcolemmal regions, and vacuoles were present.

(j) Familial focal loss of cross-striation

In 1967, Engel described a 14-year-old girl with a flaccid, proximal muscle weakness from birth. She had slight ptosis and an ophthalmoplegia. Muscle biopsy revealed focal lesions with loss of cross striation. Type I fibres were small but predominant. Sibs with this disorder were reported by Van Wijngaarden *et al*. (1977). This condition might be the same as mini or multi-core disease (see p. 389).

(k) Familial myofibrillar inclusion body myopathy

A family with dominant transmission of a myopathy showed cytoplasmic myofibrillar inclusions in three symptomatic and seven asymptomatic family members (Clark *et al*. 1978). A single case is recorded by Jerusalem *et al*. (1979) in a 22-year-old-woman.

(l) Type I fibre disproportion

In 1973, Brooke drew attention to an entity in which there was a disparity in fibre size between type 1 and 2 fibres, with consistently larger type 2 fibres, (selective atrophy of type 1). The clinical picture was of a non-progressive congenital hypotonia and weakness complicated by hip dislocation and contractures. Of his 12 cases (reported in 1973) nine had another affected member (on history alone) and the pattern suggested an incomplete dominant. Eleven of the 12 patients were of small stature.

A selective atrophy of the type I muscle fibres occurs in many different conditions. It is common in nemaline myopathy (Gonatas *et al.* 1966; Radu and Ionescu 1972) and in centronuclear myopathy (Engel *et al.* 1968).

A family in which mother and son (and possibly grandmother) had atrophy or hypotrophy of type I and hypertrophy of type II fibres was reported by Kinoshita *et al.* (1975*a*). The authors considered the atrophy to be the essential pathological feature and suggested their cases are the same as those described under 'type I fibre disproportion' by Dubowitz and Brooke (1973) and Brooke (1973). Those affected have an unusually thin face, high arched palate, and scoliosis with moderate weakness of neck and limb muscles, more marked proximally than distally. In a review of nine patients (Cavanagh *et al.* 1979*b*) all were single, although in one case the parents were second cousins.

Taken as a whole, the genetics of the condition or group of conditions are still uncertain but dominant inheritance with or without reduced penetrance accounts for many reports. A family illustrating the reduced penetrance is described by Eisler and Wilson (1978). Fibre type disproportion was found in three sibs and their mother. Only two of the sibs had a progressive proximal muscular disease starting at the age of 2 and 6 years. Neither child was floppy at birth.

Two sisters with neonatal hypotonia (Fardeau *et al.* 1975) were the offspring of a father who had biopsy changes of fibre type disproportion, but was clinically unaffected. Sibs are reported by Jaffe *et al.* (1988).

Identical twins with slow motor development and hypotonia both had type I fibre atrophy on muscle biopsy (Curless and Nelson 1977). Their father did not walk until $2\frac{1}{2}$ years, and the paternal grandfather had weakness and an abnormal gait from infancy. A male sib of the twins was similarly affected. The affected members had macrocephaly.

Some reports of fibre type disproportion have shown, in addition, central nuclei. These are the cases of Bethlem *et al.* (1969), Engel *et al.* (1968)—sibs; Karpati *et al.* (1970)—mother and daughter; and Inokuchi *et al.* (1975)—a single case.

Further evidence for heterogeneity is presented by Sulaiman *et al.* (1983).

A little girl had marked proximal weakness, first noted at 2½, but then improved and at 4½ she had only mild weakness. A CPK was elevated, but an EMG suggested neurogenic change. Her father had neuromuscular problems since infancy. A diagnosis of muscular dytrophy was made at 18 months and he was wheelchair bound at 30. He had multiple flexion contractures and a rigid spine. The little girl had fibre type disproportion but her father had anterior horn cell atrophy.

It should be noted that one of the original cases of Brooke (data from Sulaiman *et al.* 1983) developed a rigid spine.

It remains uncertain if this group as a whole is a single entity.

Fibre type disproportion/ophthalmoplegia

These are rare cases of fibre type I hypotrophy, with facial weakness, ophthalmoplegia, central nuclei, and lysis of type I fibres. The first case was reported by Bender and Bender (1977).

There is no fibre necrosis, regeneration or endomysial connective tissue proliferation. CPK is normal. Whether this is neurogenic or myopathic is unclear. Onset is in infancy. All cases (two) have thus far been single.

(This condition differs from 'familial myopathy with lysis of myofibrils' in that the lysis is peripheral in that condition but central in the report above—see p. 393.)

(m) Spheroid body myopathy

This is characterized by adolescent onset of proximal weakness, fatigue, and pain in the lower limbs accompanied by slowly progressive atrophy (Goebel *et al.* 1978). The distinguishing feature is the finding of spheroid bodies in type I fibres.

(n) Myopathy with excessive tubular aggregates

Some of the cases (families) reported to have this condition might have a myotubular myopathy. For instance, a dominant family is reported by Rohkamm *et al.* (1983). Onset was in the 20s and the initial complaint was difficulty in climbing stairs. Later the arms (proximally and distally) became affected. Neither atrophy nor pain were experienced and the face was not involved. Sixty to ninety per cent of all muscle fibres contained tubular aggregates. Marked atrophy of type II fibres occurred. It should be noted that aggregates might be found in intoxication with alcohol or diazepam and the condition is not homogenous.

The subject is reviewed by Niakan *et al.* (1985). Most are single cases. Myalgia is a common feature.

C. Muscle disease without specific pathology

(a) Benign hereditary myopathy

This is an unsatisfactory designation. Some authors call this minimal change myopathy. This category includes those myopathies where neither clinical nor histological examination suggest a specific known disease entity. Although predominantly a congenital myopathy, the age of onset is sometimes delayed to four or five years after birth (see also next section).

(i) Recessive

Turner and Lees (1962) reported a family in which six out of 13 sibs were affected. Hypotonia was present from birth and walking was delayed. As the children grew older the hypotonia was replaced by localized atrophy and weakness, especially in the muscles of the neck, shoulders, and hands. Later, proximal muscles in the pelvic girdles were affected and one of the sibs died from the condition.

(ii) Dominant

Under the title 'benign hereditary myopathy', Bethlem and Van Wijngaarden (1976) described three large pedigrees with dominant inheritance of muscular weakness beginning around the fifth year of life. Progression was slow, but symptoms worsened in middle life. Weakness and atrophy were generalized, but proximal more than distal, and moderate in severity. Many patients developed flexion contractures of the elbows, and plantar flexion contractures of the ankles. None showed contracture of the wrists, metacarpophalangeal joints, or of the thumbs. Four out of 28 cases had congenital torticollis. A marked variation in fibre size and an increase in fatty tissue were the main histological findings.

A pedigree of Polish descent with six affected over four generations (Arts *et al.* 1978*b*) was clinically similar to the other known pedigrees. However, creatine phosphokinase was elevated in all of the patients (only one out of nine in Bethlem and Van Wijngaarden's (1976) paper had a raised creatine phosphokinase).

Counselling

Because the entity is unclear and might be very heterogeneous it is not surprising to find that both autosomal recessive and dominant families have been reported. The distribution of weakness is mainly proximal and it is not impossible that some patients might be described as having a mild limb girdle myopathy.

(b) Congenital muscular dystrophy (CMD)

There seems to be at least three distinct types

1. Fukuyama type with central nervous system involvement, but normal vision.
2. Congenital muscular dystrophy with central nervous system malformations and a retinal dysplasia (COD-MD)
3. Congenital muscular dystrophy—type Donner—which has neither central nervous system nor retinal malformation.
4. Congenital muscular dystrophy more severe in its progression than Donner (no central nervous system malformation). One case of Kihira and Nonaka (1985) falls into this group.

Pathologically, the diagnosis can be difficult. There is an overlap with congenital non-progressive myopathy. Some authors regard endomysial fibrosis as a cardial pathological features of CMD, but mostly it is not easy to differentiate from minimal change myopathy (Nonaka *et al.* 1983). (See p. 396.)

For instance, in a series of 10 Japanese patients (Kihira and Nonaka 1985) there were nine females and one male. The cases were all single except for one pair of sibs. Muscle biopsy findings varied from mild myopathic to advanced dystrophic changes. Fibre hypertrophy (with occasional splitting) was assumed to reflect a chronic dystrophic process.

1. Fukuyama congenital muscular dystrophy

In Japan this type of muscular dystrophy is a frequent cause of the floppy infant syndrome. Fukuyama *et al.* (1960) reported 15 cases and found the frequency of retardation and epileptic seizures so high that he suggested the name 'congenital cerebromuscular dystrophy'. Micropolygyria was found at post-mortem. Kamoshita *et al.* (1976) reviewed reports from Japan. First-cousin marriages occurred in two, and sibs were affected in one.

In a review of 25 patients, Yoshioka *et al.* (1980) found low density areas in the white matter in 56 per cent. Cerebral atrophy was found in 80 per cent.

The ophthalmological findings were myopia (36 per cent) optic atrophy either confirmed or suspected in 36 per cent, but none had obvious retinal changes.

Fukuyama and Ohsawa (1984) looked at 153 families with this condition. Consanguinity occured in 26.8 per cent. No parent was affected. The segregation ratio obtained was very close to the expected 25 per cent. Two Dutch sibs were reported by Peters *et al.* (1984). They differed from classical Fukuyama only in that they had choroidal hypoplasia and a histopathological proven delay in myelination but these differences are probably not sufficient to call this a separate disorder.

Congenital muscular dystrophy with CNS malformation (normal eyes)

A sib pair reported by Egger *et al.* (1983) were thought to be different from Fukuyama disease in that the cerebral pathology was more severe. There were areas of low density in the white matter, but it is likely that this sib pair is not different from Fukuyama muscular dystrophy.

2. *Cerebro-ocular-dysplasia muscular dystrophy or COD-MD*

Muscle, brain, eye disease was first described by Santavuori *et al.* in 1978. They reported 13 children and five adults, but the eye abnormalities were discussed in only 10. Severe myopia occured in six, but eight out of the ten had congenital glaucoma, seven had hypoplasia of the choroid, and seven had optic nerve atrophy. Unfortunately, the retina could not be studied. Later Korinthenberg *et al.* (1984), reported two males who were alive at 9 and 8 years with a similar condition.

Towfighi *et al.* (1984) discussed in more detail the neuropathological features in four patients. Important findings are agyria, cerebellar hypoplasia, gliomesodermal proliferation, hypoplasia of the pyramidal tracts, hydrocephalus, an encephalocele, fusion of the frontal lobes, and abnormal olfactory bulbs. In the Korinthenberg patients there were areas on the CT scan of white matter hypodensity and the eye signs included cataracts, corneal opacities, immaturity of the anterior chamber, hypoplasia of the optic nerve, microphthalmia, and a retinal dysplasia. It is not certain whether this condition is different from HARD ± E or the Walker Warburg syndrome (see p. 63).

In both, the inheritance pattern is recessive.

3. *Congenital muscular dystrophy (Donner)*

The clinical picture is that of marked hypotonia and generalized muscular weakness present at birth, often with arthrogryposis multiplex congenita. Muscle biopsy shows non-specific changes with necrosis of a severe degree out of all proportion to the relative mildness of the clinical picture. Undue pessimism about the prognosis is unwarranted as some of the children survive many years. In many cases, the condition is static and in some muscle power improves (Dubowitz 1978). Inheritance is probably as a recessive.

Sibs are reported by Zellweger *et al.* (1967), Pearson and Fowler (1963), Banker *et al.* (1957), and Donner *et al.* (1975).

In 1930, Ullrich described two patients with muscle weakness at birth, hypotonia, contractures of proximal joints, hyperextensibility of distal joints, a high arched palate, hyperhidrosis, and posterior protrusion of the calcaneus. There was no progression.

They described this condition as 'congenital hypotonic-sclerotic muscular dystrophy', but this condition is probably not different from congenital muscular dystrophy type-Donner.

Congenital muscular dystrophy with cerebral lesions, but normal intelligence

Two females sibs with congenital muscular dystrophy (hypotonia, areflexia, raised CPK and, on muscle biopsy, involvement of type I and II fibres) were reported by Nogen (1980). One sib had a communicating hydrocephalus and the CT scan showed lucencies in the white matter. The IQ was in the normal range. The other sib had mild to moderate white matter changes and a normal IQ.

This condition is unlike Fukuyama disease where there is usually severe mental retardation and epilepsy. However, a Japanese boy is reported by Yoshioka *et al.* (1987) in which a CT scan revealed low density areas in the white matter and the muscle biopsy changes were compatible with a dystrophy. The CK and intelligence level were normal.

(c) Familial myoglobinuria and idiopathic rhabdomyolysis

This rare and often lethal condition has two forms (Savage *et al.* 1971). Type I is precipitated by exertion and Type II by infection. Death might result from hyperkalaemia, renal tubular necrosis, or respiratory muscle paralysis. Sibs have been reported by Bowden *et al.* (1956), Wheby and Miller (1960), and by Savage *et al.* (1971). All have been type II, and in each of these sibships the disease was severe. In a type I pedigree (Hed 1955), three brothers were affected, but only in one did the disease start in childhood, and in none was it fatal. Consanguinity is reported in the case of Tavill *et al.* (1974).

(d) Progressive myositic ossification (Munchmeyer disease)

Characterized by ectopic calcification in the connective tissue of muscles, this condition is dominantly inherited (Smith *et al.* 1976). The onset is usually within the first decade of life and the first signs are the detection of hot, painful lumps. Then follows stiffness of shoulder-girdle muscles, upper arm, pelvis, and neck (Illingworth 1971). Microdactyly of the great toe, abnormal teeth, hypogenitalism, and deafness are other features. The inheritance is thought to be as a dominant. McKusick (1960) quoted a family of five males affected in three generations. Remissions and exacerbations occur.

(e) Familial myosclerosis

The characteristic clinical feature is the hard, woody texture of muscles to palpation. Bradley *et al.* (1973) questioned the continued usefulness of this term as the condition is the end result of diverse diseases including anterior horn cell disease. They discussed two sibs with spinal muscular atrophy and myosclerosis. Other familial cases (some with primary muscle disease) were reported by Cordier *et al.* (1952) and Löwenthal (1954). The latter author described two families, one with four affected sibs and the other with four members affected over three generations.

(f) Congenital hypotonia-sclerotic muscular dystrophy

Two sibs, the offspring of a consanguineous mating, and a single girl with cousin parents are described by Furukawa and Toyokura (1977). Intelligence was above average and all those affected were susceptible to recurrent respiratory infections. This entity could be the same as congenital muscular dystrophy-type.

(g) Inflammatory myopathy

Inflammatory reactions have been noted in what appears to be familial muscular disease. Munsat *et al.* (1972) described polymyositis-like changes in a family with autosomal dominant facio-scapulo-humeral dystrophy.

   Jennekens *et al.* (1975) described an inflammatory myopathy present in two Dutch families with a scapulo-ilio-peroneal distribution of weakness and wasting. Transmission was as a dominant. A total of 26 members were affected in these two apparently unrelated families. In a later phase of the disease a cardiomyopathy appeared. The onset of the neuromuscular conditions varied between 17 and 42 years.

A father and his daughter with myositis was reported by Lewkonia and Buxton (1973). The daughter's illness resembled childhood dermatomyositis and father had adult-onset polymositis without systemic involvement.

(h) Hereditary persistent distal cramps

A disease characterized by painful muscle spasms affecting both hands and feet had an age of onset that varied from the second decade in the third generation, to childhood in the fourth and fifth generations (Jusic *et al.* 1972). Inheritance is as a dominant and the cramps occur both on exertion and during full relaxation. They are persistent during spinal anaesthesia and peripheral nerve block. A similar condition called adolescent familial cramps was described in two males and their male first cousins. Creatine phosphokinase levels were raised but this was also a feature in their clinically normal sisters (Hanson and Mincy 1975). The relationship of this condition to that in which malignant hyperpyrexia is a feature is unclear. At least seven members were affected over three generations in the family reported by Van der Bergh *et al.* (1980). The onset was in childhood, and most of the affected were asymptomatic after the age of 20 years. Myokymia was seen in proximal musculature—see p. 410 (myo-adenylate deficiency).

(i) Myopathic arthrogryposis multiplex congenita

Two main groups of arthrogryposis can be recognized—namely the neurogenic and myopathic variants. Both types are occasionally inherited. However, in a study of 66 cases only sporadic occurrence was found (Wynne-Davies and Lloyd-Roberts 1976). A birth incidence of about 1 in 50 000 for Scotland is calculated, and most are non-genetic.

Strehl and co-workers (1985) examined 22 children with AMC. The final diagnosis was neurogenic in 10 (spinal in five, cerebral in two, mixed in three) and myopathic in nine. Three had a congenital muscular dystrophy, three congenital myotonic dystrophy, two a non-specific myopathy, and one fibre-type disproportion.

Two infants with the myopathic type, from an inbred Palestinian Moslem family were reported by Der Kaloustian *et al.* (1972).

Another recessive pedigree (Lebenthal *et al.* 1970) included 17 males and six females with clinical manifestation of the disease. Analysis of this Arab pedigree revealed a very high rate of consanguineous marriage.

The same family was also reported by Weissman *et al.* (1963). Six of those affected had congenital heart lesions.

For other myopathic types see under Congenital muscular dystrophy, page 397.

Concordance for arthrogryposis multiplex congenita in identical twins was reported by Lipton and Morgenstern (1955) and discordance by Hillman and Johnson (1952) in two sets of identical twins.

Counselling

This is extremely difficult. Clinically, if the recessive Pena-Shokeir syndrome (neonatal death from lung hypoplasia) can be excluded risks are likely to be small. It is also clear that the pathological differentiation between end stage muscle or anterior horn cell disease can be difficult. If the pathologist is certain that it is a 'congenital muscular dystrophy', recurrence risks might be 5 per cent (experience is that most cases are single). Prenatal monitoring of fetal movement might be a useful procedure.

Amyoplasia congenita

This condition has a characteristic distribution of muscle weakness and joint contractures and it should be possible to diagnose it on clinical grounds. The weakness is at the shoulder girdle which in both infancy and adulthood is internally rotated with fixed flexion at the elbows. The arms cannot be elevated above the head and the distal joints in the hands and the feet are frequently involved. The condition is mostly sporadic (Hall *et al.* 1983).

(j) Malignant hyperpyrexia (King syndrome)

The usual effect of anaesthesia is to lower body temperature. In 1960, Denborough and Lovell reported 10 hyperpyrexial deaths in the same family following an anaesthetic. Inheritance seemed to be as a dominant; see also Denborough *et al.* 1962). Since the first report in 1960 over 600 cases have been reported (Kaplan *et al.* 1977). Three groups of individuals are known to be susceptible.

(a) Those with a dominantly inherited myopathy.

(b) Those with myotonia congenita.

(c) A phenotype characterized by short stature, pectus carinatun, a spinal deformity, cryptorchidism, mandibular hypoplasia, antimongoloid eye slant, ptosis, and low-set ears (King *et al.* 1972). The similarities to Noonan syndrome have been noted by Hunter and Pinsky (1975).

The main clinical signs during anaesthesia are early rigidity of the jaw and progressive rigidity of the limbs. The association of malignant hyperpyrexia with muscle diseases was fully reviewed by Barlow and Isaacs in 1970. They added a family with three hyperpyrexial deaths during anaesthesia and Britt *et al.* (1969) reported a family with dominant inheritance.

In five patients with a malignant hyperpyrexial myopathy (Harriman *et al.* 1973) clinical signs prior to the anaesthetic were minimal. Muscle biopsy showed abnormalities, but the changes were non-specific. Creatine phosphokinase estimations were not a reliable guide to those at risk. At present those at risk can only be satisfactorily investigated using *in vitro* tests on muscle tissue from a biopsy specimen (Ellis and Halsall 1980).

Isaacs and Barlow (1973) showed pedigrees of three such families. In two, death from malignant hyperpyrexia occurred in a single generation, but if a raised creatine phosphokinase level is taken as an indication of muscle disease then other members were at risk and the pedigree could be interpreted as a dominant. It is again pointed out by Kaplan *et al.* (1977) that creatine phosphokinase estimation is not a reliable marker of the condition.

Of the congenital myopathies, central core disease is the one most frequently associated with malignant hyperpyrexia.

Evidence for heterogeneity is provided by McPherson and Tayler (1982) in a report on 12 Wisconsin families and a review of the literature. They looked at 133 reports of familial malignant hyperpyrexia. Patients came from 93 separate kindred. Just over half were clearly dominant. In 12 per cent siblings only were affected, but in 20 per cent the condition appeared to be sporadic.

As the authors state, whether malignant hyperthermia is a symptom of many conditions or whether malignant hyperthermia is a heterogeneous condition is mainly semantic. Unfortunately, no single laboratory test can exclude the condition. The CPK might be normal, but the *in vitro* pharmacological responses of biopsied muscles are useful in families to determine who is at risk. The frequency of the condition is about 1 in 15 000 anaesthetics administered in childhood (Nelson and Flewellan 1983).

(k) Kocher-Debre-Semelaigne syndrome (muscular hypertrophy and hypothyroidism)

Sibs are reported by Cross *et al.* (1968), Najjar (1974), and Abdurrahman and Gassmann (1976). Four sibs were affected, including twins. When the thyroid deficiency is treated the hypertrophy disappears.

(l) Others

Hypertrophia musculorum vera

Muscle hypertrophy in the absence of thyroid disease or myotonia is a rare disorder. A large dominant pedigree in which all affected members showed enlargement of the calves and masseters was reported by Poch *et al.* (1971). No muscle weakness or wasting occurred. Many cases described in the past, may have had myotonia congenita (Walton and Gardner-Medwin 1981).

Ocular myopathy with curare sensitivity

Some families with a progressive ocular myopathy are curare sensitive. A South Indian family in which parents were second cousins had three affected children. In another branch of the same family first-cousin parents who were partially affected themselves had three out of seven affected children (Matthew *et al.* 1970).

Absence of muscle groups

(a) Poland's syndrome

The association of congenital absence of the sternocostal head of the pectoralis major and minor, hypoplasia of breast, nipple, and rib, pectus excavatum, or carinatum and scoliosis with ipsilateral syndactyly is known as Poland's syndrome (Ravitch 1977).

Most cases are single. In a series of 43 consecutive cases all were sporadic (Ireland *et al.* 1976). An association between Poland's anomaly and the Moebius syndrome is now well established.

(b) Absent gluteal muscles

Two sibs had absent gluteal muscles and spina bifida occulta (Carnevale *et al.* 1976). Both parents and two other sibs had sacral spina bifida occulta and two further sibs died, one with anencephaly and the other with probable spina bifida.

(c) Universal muscular hypoplasia

Also known as muscular infantilism, the weakness and hypoplasia is present at birth and no progression takes place. Histologically the condition closely resembles type I fibre disproportion. No patient in the inbred family reported by Pelias and Thurmon (1979) had contractures, hip dislocation, or kyphoscoliosis. Inheritance was recessive.

(d) Familial hypoplasia of thenar eminence

A family (mother and two children) with an absence of the abductor pollicis brevis (in all three) and in two, in addition, an absence of the extensor pollicis longus is reported by Selmar *et al.* (1986).

(e) Familial absence of the trapezius muscles

Two brothers with a congenital absence of trapezius, pectoralis, supraspinatus and serratus anterior are reported by Gross-Kieselstein and Shalev (1987). The weakness was not progressive.

(f) Familial absence of pectoralis major, serratus anterior and latissimus dorsi

Single members in three generations had this combination (David and Winter 1985). This condition should not be confused with Poland's anomaly which is nearly always sporadic.

D. Metabolic myopathies (mostly lipid storage myopathies)

Two genetic abnormalities of fatty acid metabolism are known.

(a) Carnitine palmityl transferase deficiency

Engel *et al*. (1970) reported twins with muscle cramps and myoglobinuria. In an editorial in the same journal Bressler (1970) suggested that the twins had a deficiency in carnitine metabolism. This was subsequently proven.

The typical clinical picture is that of recurrent myoglobinuria from childhood occurring three or four times per year, often after exercise preceded by fasting, or the ingestion of fatty food (Reza *et al*. 1978). Later, episodes of cramp, muscle weakness, and acute renal failure occur. Sibs have been recorded. Di Mauro and Di Mauro (1973) reported the condition in one of two brothers, but both were subsequently found to be affected (Bank *et al*. 1975). The brothers had recurrent myoglobinuria and renal failure, but no permanent muscle weakness. Metabolic studies showed a disorder of lipid metabolism and an absence of carnitine palmityl transferase. A high fasting plasma triglyceride is a useful clue to the diagnosis. Another sib pair is recorded by Isaacs *et al*. (1976).

A very unusual presentation is reported by Carey *et al*. (1987). The patient was well until her teens when she began to experience cramps in her leg muscles related to exercise. Mild ptosis developed in her twenties, and at 70 there was increasing weakness of arms and legs. On examination she had an external ophthalmoplegia and weakness more proximal than distal. A biopsy

showed ragged red fibres and abnormal mitochondria on electron microscopy. The enzyme was deficient in muscle.

Note: It initially seemed likely that there were two genetically distinct
enzyme deficiencies of carnitine palmityl transferase called CPT I and CPT
II; these enzymes are active on either side of the mitochondrial inner
membrane and both are needed for transfer of fatty acids. A patient with type
II deficiency was reported by Scholte *et al.* (1979) and type I by Hostetler
et al. (1978). Doubt has been cast on the difference (Zietz and Enger 1987).

(b) Carnitine deficiency

The division into systemic and muscle carnitive deficiency might no longer be
tenable.

Indeed, some cases of systemic deficiency present with weakness (Scarlato
et al. 1978), and differentiation between the systemic and muscle carnitine
deficiency is not possible (Scholte *et al.* 1979). These authors reported two
sibs with a marked difference in the severity of the disease. Blood carnitine in
the severely affected female was normal. Low levels were found at post-
mortem in muscle and heart, but not in liver; whereas her mildly affected
sister had decreased levels in blood and muscle. It is possible that the normal
blood and liver levels in the severely affected girl were due to the profound
muscle wasting and that both sibs had the systemic type of carnitine
deficiency.

Type I—muscle

The carnitine content is low in skeletal muscles but normal in plasma and
liver. The onset of weakness mostly occurs during the first decade of life,
although floppiness can be detected at birth. Single cases are those of Engel
and Angelini (1973), but recessive inheritance was strongly suggested by the
presence of the disease in two sibs (consanguineous mating) reported by
Willner *et al.* (1979). Furthermore, the mother and father of an affected boy
(Van Dyke *et al.* 1975*a*) had reduced muscle carnitine content suggesting
autosomal recessive inheritance of the condition.

Type II—systemic

This condition presents in childhood with vomiting and an encephalopathy as
well as a myopathy. Lipid accumulation takes place in muscle, heart, liver,
and plasma. The distribution of weakness in Karpati *et al.*'s (1975) case was
that of a proximal myopathy of upper and lower limbs, with atrophy of facial
muscles. Intermittent liver insufficiency and ketoacidosis occurred in both
Karpati *et al.*'s (1975) and Boudin *et al.*'s (1976) cases. Sibs have been

recorded by Cornelio *et al.* (1977) from an isolated area in North Italy where consanguineous matings were frequent in the past.

Some authors now believe that systemic carnitine deficiency might not exist as a primary condition. It is usually secondary to excessive loss or decreased liver synthesis, secondary to organic acidurias, defects of beta-oxidation and defects in the respiratory chain enzymes.

Two sisters with systemic carnitine deficiency are reported by Cruse *et al.* 1984, one was asymptomatic and the other had had an encephalopathic episode before developing weakness. She was thought initially to have Reye syndrome (she was ketotic with hypoglycaemia).

Lipid storage myopathy with normal carnitine levels

Not all lipid storage myopathies are associated with carnitine deficiency. An adult with a striking lipid excess in muscle (Jerusalem *et al.* 1975) had normal carnitine levels.

Mitochondrial changes were not prominent but they have been reported previously (Jerusalem *et al.* 1973*a*). These authors reported a 7-week-old girl with severe weakness sparing only the ocular muscles with macroglossia and an enlarged liver. By 22 months improvement had taken place.

Glycogen storage disease

Only type II and type V have significant muscle involvement.

Glycogenosis type I

A Japanese girl with proven Von Gierke disease had a lipid storage myopathy. Her affected sibs only had the liver disease (Yamaguchi *et al.* 1978).

Glycogenosis type II

This autosomal recessive disorder characterized biochemically by the lysosomal accumulation of glycogen is due to a deficiency of acid alpha-glucosidase (acid maltase). It was first described in 1932 by Pompe, but it was Hers (1963) who identified the enzyme defect that has led to clinical recognition even in the less severe cases.

Clinically, the disorder has been subdivided on the basis of age of onset, organ involvement, and progression of the disease.

(a) An infantile form in which the heart, liver, and tongue are enlarged. Respiratory and skeletal muscle weakness is severe and death occurs before two years. Two infantile cases reported by Engel *et al.* (1973) had affected sibs. Prenatal diagnosis is available.

(b) A juvenile form (3–19 years) with slower progression is recognized. The enlarged organs and respiratory problems are less constant and death occurs before the age of 20 years. Cardiac enlargement occurs in fewer than half the reported cases and the clinical picture is that of a progressive dystrophy. Sibs were reported by Zellweger *et al.* (1965), Engel (1970), and Mehler and Di Mauro (1977; two sets of sibs).

(c) An adult form occurs where the initial diagnosis is often of a muscular dystrophy or polymyositis (Engel 1970). Weakness is more pronounced proximally than distally in a limb-girdle distribution. Engel (1970) reported four cases. All were isolated. However, non-identical adult twins with acid maltase deficiency (male and female) confirm the autosomal recessive nature of this condition; sibs were reported by Mehler and Di Mauro (1977).

In general, acid maltase deficiency in adults has an age of onset between 20 and 35 years. Patients are thin, tall, and weak with wasted paraspinal and gluteal muscles, and with predominantly lower limb weakness. All show vacuolated lymphocytes filled with glycogen, and an increased CPK. Two of the patients were sibs concordant for adult onset disease (Trend *et al.* 1985).

The groups may not be genetically distinct. Busch *et al.* (1979) reported a 16-week-girl with the infantile form. Her paternal grandfather was well until 53 years when he developed a proximal myopathy. Acid maltase was grossly deficient. In a report of an adult myopathy due to acid maltase deficiency (Hudgson *et al.* 1968) a sib died at 4 years of age.

Other adult onset cases have been described by Martin *et al.* (1976), and McComas *et al.* (1983). In this latter report are three sibs of whom two had respiratory muscle disease due to weakness of the diaphragm and intercostal muscles.

Glycogenosis type III (Cori-Forbes disease)

Muscle involvement is unusual, but five patients were reported by Di Mauro *et al.* (1979). In addition to the hepatic and cardiac involvement all five had an adult onset of weakness and distal wasting. In two instances a sib had cirrhosis, but did not have muscle weakness.

Glycogenosis type V (Mcardle's disease)

Classical muscle fatigability with an onset in childhood or adolescence, intermittent dark urine, muscle cramps on exertion, with persistent weakness and wasting in adulthood (fourth-fifth decade) are the main clinical features.

Of the 47 cases reported up to 1974 (Chui and Munsat 1976), about 60 per cent had affected relatives and all the familial cases were compatible with recessive inheritance. Consanguinity was recorded in the cases of Schmid and Hammaker (1961), Rowland *et al.* (1963), and Cochrane *et al.* (1973). An additional sibship was reported by Rowland *et al.* (1966).

An interesting family is reported by Schmidt *et al.* (1987). Mother and son (age of onset 7 with exercise induced cramps and pigmenturia) had the condition, but it was milder in mother who was thought to be a manifesting heterozygote.

There are, however, four unusual clinical presentations. (DiMauro *et al.* 1986).

a. Very mild tiredness without cramps or myglobinuria.

b. Late onset weakness (6–7th decade) no cramps or myoglobinuria.

c. Mild congenital weakness.

d. Severe, progressive, and fatal.

(a) Mild tiredness without cramp or myoglobinuria

In an exceptional family described by Chui and Munsat (1976) there were six affected members over four generations. Those affected showed the typical clinical triad of muscle weakness, exercise intolerance, and the 'second wind' phenomenon (repetitive muscle exercise helps to avoid further pain), but the course was more benign than in the recessive form. It is also suggested that there are two autosomal recessive forms: one in which the enzyme is lacking on both biochemical and immunological tests whereas in the other the enzyme deficiency can only be demonstrated immunologically.

Of the cases reported up to 1974 there have been twice as many affected males as females.

(b) Late form with permanent weakness

Two sibs with an onset at 50 years had deficient skeletal muscle phosphorylase (Engel *et al.* 1963). One had severe symptoms and no skeletal muscle phosphorylase and the other had mild symptoms and 35 per cent of normal phosphorylase.

(c) Presenting as mild congenital weakness

In the case reported by Cornelio *et al.* (1983), problems with sucking occured soon after birth. Motor milestones were delayed and at 3 years he walked with a waddling gait and tired easily. CPK was raised and there was moderate proximal weakness. A muscle biopsy was morphologically normal but the histochemical stain for phosphorylase showed no reaction. The enzyme defect was confirmed biochemically.

(d) Fatal infantile form of muscle phosphorylase deficiency

A single case was reported with rapid progress of weakness beginning at four weeks of age resulting in death at 13 weeks (Di Mauro and Hartlage 1978).

Glycogenosis type VII (Muscle phosphofructokinase deficiency (Tarui)

Clinically, this is very similar to McArdle disease in that there is pain (cramps) provoked by intense exercises and relieved by rest. Myoglobinuria is less frequent.

Few cases of Type VII have been recorded. Partial activity of the enzyme in both parents of a male with the disorder suggests recessive inheritance (Layzer *et al.* 1967)

(a) A fatal form of the type VII

An infant died at 7 months from respiratory failure, limb weakness, seizures, cortical blindness, and corneal opacification (Servidei *et al.* 1986). The muscle showed marked decrease in phospho-fructo-kinase.

(b) Adult muscle phosphorylase 'b' kinase deficiency

There are three types (Abarbanel *et al.* 1986).

 (i) X-linked—enzyme is lacking in liver.
 (ii) Autosomal recessive type—enzyme is lacking in liver, muscle, and erythrocytes.
(iii) Confined to muscle—presenting as a childhood or adulthood myopathy.

Only type (iii) concerns us here. The adult onset muscle type is described by Abarbanel *et al.* (1986). There was severe excercise intolerance, cramps, and one episode of pigmenturia.

Glycogenic myopathy

Adult male twins with severe muscle fatigue after effort were reported by Lehoczky *et al.* (1965). No atrophy or hypertrophy was present. Muscle glycogen was found to be increased.

Myo-adenylate deaminase deficiency

Three sibs with muscle weakness and post-exercise cramping were reported by Fishbein *et al.* (1978). Histochemical staining showed only mild type I fibre atrophy in one sib. The adenylate deaminase stain was negative and assay of myo-adenylate deaminase showed a very low level. The authors suggested this is a common cause for some of the non-progressive myopathies. Inheritance now seems dominant with incomplete penetrance (Kelemen *et al.* 1982).

E. Myasthenic syndromes

Myasthenia gravis

Infantile or congenital (Table 17.1)

Myasthenia gravis has a prevalence of 2–4 per 100 000 (Kurland and Alter 1961). In Finland the estimate is 5.6 per 100 000 (Pirskanen 1977). Myasthenia gravis occurring in infancy accounts for between 1 and 2 per cent of all cases.

If figures from Bundey (1972) and Namba *et al.* (1971) are combined, then 4 per cent of all myasthenics have an affected near relative and 42 per cent of the familial cases have an onset before two years (quoted from Fenichel 1978).

Some authors suggest that there is a clinical difference between familial infantile and familial congenital myasthenia in that the congenital form has prominent involvement of extra-ocular muscles at or soon after birth, whereas the infantile form has severe life-threatening respiratory or feeding

Table 17.1 Classification of childhood-onset myasthenia gravis (from Fenichel 1978)

	Onset	Initial symptoms and signs	Progression	Antibodies to AChRP
Juvenile	mostly after 10 years, but between 6–12 and adolescence	ptosis	remission in 20%	+
Neonatal	1–3 days	feeding difficulties	last 2–3 weeks	+ transient
Congenital	birth	extra-ocular muscle palsies	persists	–
Familial infantile	birth	apnoea	improves	–

problems, but no limitation in eye movement. For a confident diagnosis of the familial form, myasthenia should be absent in the mother (to exclude transient myasthenia) and another sib should be affected.

The clinical findings in congenital myasthenia differ from the adult group in that they are milder, only slowly progressive or static, and without fluctuation. (Extra-ocular and facial muscles, particularly orbicularis oculi are affected.)

Counselling

Autosomal recessive inheritance is most likely in both infantile and congenital types.

The first cases of the infantile type in sibs were described by Greer and Schotland (1960) by Conomy *et al.* (1975), and Gieron *et al.* (1985). These latter authors followed three sibs to adulthood and two have severe respiratory problems in adulthood.

The congenital type with extra-ocular palsies

Sibs are described by Whitely *et al.* (1976), Gath *et al.* (1970), Walsh and Hoyt (1959), and Dick *et al.* (1974), and twins by McLean and McKone (1973).

The unaffected parents in one sibship with three affected with the congenital form (Warrie and Pillai 1967) were first cousins. However, inheritance is uncertain as the brother and sister of the unaffected parent had the disease with an onset in childhood.

Juvenile and adult type

Both dominant and recessive inheritance have been reported.

The prevalence of myasthenia gravis in the first-degree relatives of probands in this group is about 1–2 per cent (Bundey 1972).

Out of five index patients with an onset between 6 and 18 months (two with first-cousin parents) one had an affected sib (Bundey 1972) and the data suggested autosomal recessive inheritance. All of these patients presented with ptosis.

In the study by Namba *et al.* (1971) mother and child were affected in four families and in one, symptoms began in adult life in both mother and daughter. The mother in the family reported by Foldes and McNall (1960) had myasthenia gravis for 56 years with an onset at 17 years. Her two daughters developed myasthenia gravis in adult life.

Myasthenia gravis in father and child occurred in six families (Namba *et al.* 1971). The onset was in adult life in four fathers (not mentioned in two). The age of onset in their offspring was between 8½ and 60 years.

Most studies confirm an earlier age of onset in those families with recessive inheritance. The age of onset was before the age of two years in a family of four sibs, and in two other families the onset was in childhood or early adulthood (Namba *et al*. 1971). Where the relationship was more remote (second cousins or uncle and niece) the onset was in adulthood. There were, however, four sib pairs with onset in adulthood.

A population study in Finland (Pirskanen 1977) ascertained 264 patients. Nineteen cases or 7.2 per cent were familial and came from eight families. Seventeen of the 19 were female. In four families there were two affected sibs. A parent and child were affected in two families, a niece and aunt in one, and cousins (four first cousins and their second cousin) in one. The age of onset in the familial cases varied from 4.5 to 35 years and was not different from the non-familial cases. The series comprised of five sets of twins all of which were discordant. One pair was probably monozygotic. That the familial cases might be based on the inheritance of susceptibility genes is given some impetus from the family of Honeybourne *et al*. (1982) in which grandmother and grandchild had myasthenia and thyroid disease. The intervening mother had thyroid disease alone, as did grandmother's sister. Her daughter had myasthenia and a thymoma. Grandmother and granddaughter had HLA-A_1B_8, known to be associated with myasthenia gravis, but the granddaughter did not derive the haplotype from grandmother.

There are other reports of familial thymoma. (Pascuzzi 1986). They report brothers with adult onset myasthenia gravis both with thymomas.

In a recent study of myasthenia (Kerzin-Storrar *et al*. 1988) there was a family history of auto-immune disease in one-third of patients, but only in one instance was it myasthenia. The study also confirms the increase frequency of the haplotype HLA B_8Dr_3 in myasthenia gravis, especially in females under the age of 40 who do not have a thymona. Identical twins with myasthenia gravis are reported by Murphy and Murphy (1986). The age of onset was in late childhood.

Counselling

Not all studies have found secondary cases. In a study of 70 consecutive patients with myasthenia gravis, Jacob *et al*. (1968) found no additional cases and no parental consanguinity. The data from the Pirskanen (1977) study suggest that the risk of a child developing myasthenia gravis when one parent is affected is 0.6 per cent (a low risk) and that the risk of having another affected sib is also 0.6 per cent. As the authors point out the risks are minimal even though they are 100 times those of the general population.

HLA-A and myasthenia gravis

A number of reports indicate that there is a higher frequency of HLA-B_8 in patients with myasthenia gravis than in the general population (Behan *et al*.

1973; Fritze *et al.* 1974; Pirskanen 1976). In the latter study the association was strongest in females with an onset before 35 years and in patients with thymus hyperplasia. In the familial cases—mother and son, cousins, two sib pairs—there was no consistent association with HLA-B$_8$. Note report of Kerzin-Storrer *et al.* 1988—see above.

Familial limb-girdle myasthenia

The combination of a dystrophy and myasthenia as a single entity has long been disputed. Walton *et al.* (1956) used the term myasthenic myopathy to refer to patients with myopathic wasting and weakness accentuated by exercise and improved by neostigmine. It is well known that wasting can occur in longstanding myasthenia, but usually only in severely affected muscles. Myopathies with myasthenic features are therefore difficult to categorize.

In the family of McQuillen (1966) onset was in childhood and the distribution of weakness of the limb-girdle type. A brother and sister were affected and their father had weakness (deltoid, biceps, and radialis), but no myasthenia. A second family is known in which four out of eight sibs were affected (Johns *et al.* 1973). Onset was in the second decade of life and progression was slow.

Type I fibre hypoplasia, tubular aggregates, cardiomyopathy, and myasthenia

This category is probably the same as that described above. Three sisters with this combination are reported (Dobkin and Verity 1976). Parents were unrelated. Muscle weakness was proximal and non-progressive. The onset was in childhood and the weakness was made worse by fatigue and responded to anticholinesterase medication. A single case was reported by Rowland and Eskenazi (1956).

(F) The myotonias

The myotonias can be classified as follows.

(a) Myotonia congenita—Thomsen's disease (dominant).

(b) Myotonia congenita (recessive).

(c) Myotonic dystrophy.

(d) Paramyotonia congenita with cold paralysis (Eulenburg).

(e) Paramyotonia congenita without cold paralysis (de Jong).

(f) Paralysis periodica paramyotonica.

(g) Adynamia episodica hereditaria.

(h) Others.

Myotonia congenita

(a) Dominant type

A Danish physician, Julius Thomsen (1876), described a disease which affected himself and many other members of his family over five generations. It is said (Winters and McLaughlin 1970) that Thomsen's description was prompted by his affected son being accused of malingering to avoid military

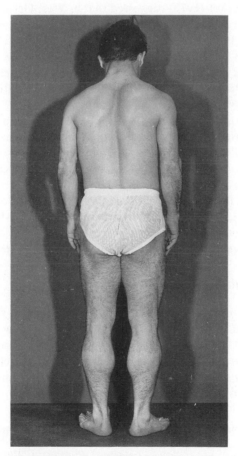

Fig. 17.16 Herculean build in myotonia conqenita.
Courtesy of the Department of Medical Illustration, National Hospital for Nervous Diseases, Queen Square, London.

service. The cardinal features are myotonia and muscle hypertrophy. Patients with severe involvement can look 'Herculean' (Fig. 17.16) Prognosis is usually good and lifespan not altered. Degenerative changes are thought not to occur. Affected infants have delayed facial relaxation after crying and a sudden movement in the older child might induce myotonia and a fall. Percussion myotonia of skeletal muscles and the tongue is invariably present. Dominant pedigrees are reported by Isaacs (1959), Penders and Delwaide (1972), and Harel *et al.* 1979. Reduced penetrance is rare, but this was reported in a family by Boltshauser *et al.* (1980).

(b) Recessive type

Recessively inherited myotonia congenita is much less common in most countries. However, Becker (1977) suggested that in West Germany the recessive form might be only slightly less common than the dominant (1 in 46 000 as opposed to 1 in 23 000 for the dominant form). He has described a number of families in which the clinical features differ from the dominant type. Myotonia is severe and widespread and symptoms are absent in infancy and early childhood. Progression of the disorder takes place until at least puberty and mild weakness is a feature. Harper and Johnston (1972) described the recessive type in three out of five sibs. The unaffected parents were first cousins. The description of the condition in dizygotic twins with unaffected parents (Ron and Pearce 1972) is also presumptive of recessive inheritance. A family with four affected children with unrelated unaffected parents, and another with three affected sibs, was reported by Crews *et al.* (1976). In this survey of patients (from the USA) onset was in early childhood or infancy and the moderate proximal weakness tended to improve with repetitive activity. Females were less severely affected.

It is uncertain whether the following two entities are separate categories.

Myotonia congenita and muscle irritability

A family with dominant hereditary myotonia (distinct from Thomsen's disease) showed muscular hypertrophy and increased muscular irritability (Torbergsen 1975). The disorder differs from Thomsen's disease only in the distribution of the myotonia. It does not occur in muscle innervated by cranial nerves, but distally, and instead of percussion of muscles causing an indentation, it causes a localized swelling.

Myotonia congenita with painful muscle contractions (dominant)

A large pedigree with myotonia congenita was described by Sanders (1976). Fourteen family members were affected over four generations. No affected member had weakness or wasting. In the adult members, the involuntary contractions were painful. In one of Becker's (1971) dominant pedigrees muscle cramp was a feature after exercise. A similar family was reported by Stöhr *et al.* (1975). Two of the patients in the Sanders (1976) family developed cramps when they developed hypothyroidism but improved on thyroid replacement theraphy.

(c) Myotonic dystrophy (Fig. 17.17)

The clinical features of this disease are myotonia, weakness, and atrophy affecting the face and sternomastoids, and distal rather than proximal muscle weakness, frontal baldness, lens opacities, gonadal atrophy, cardiomyopathy, mild endocrine and bony changes, and mild dementia. Changes in the skull X-ray (thick calvarium) and peripheral nerves can occur. Of the biochemical changes, low serum immunoglobulins (G and M) due to hypercatabolism, an excessive insulin response to a glucose load, abnormality of erythrocyte shape, membrane deformability, and abnormality of membrane lipids have all been reported (see below).

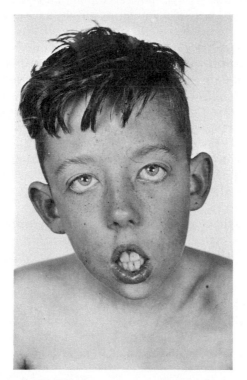

Fig. 17.17 Dystrophia myotonica.
Courtesy of the Department of Medical Illustration, Great Ormond Street Hospital for Sick Children, London.

The disorder is inherited as a dominant, but genetic counselling is difficult because of the late onset of the disease, the problem of identification of gene carriers because of minimal manifestations, and the observation that affected mothers might have a child with an early onset, severe form of the disease. The prevalence calculated by Klein (1958) was 4.9 per 100 000 and by Todorov *et al*. (1970) as 13.5 per 100 000 births and more recently, by Grimm (1975), 5.5 per 100 000.

Congenital myotonic dystrophy

This is characterized by respiratory distress and poor feeding after birth. A typical facial appearance due to bilateral facial paralysis and ptosis is present in the newborn, and hypotonia and talipes equinovarus are additional features. The mean I Q for those cases with neonatal respiratory distress was not different from those without distress, suggesting that the mental retardation is unlikely to be related to anoxia (Harper 1975*a,b*). The diagnosis is at times difficult as the more characteristic manifestations of the disease (myotonia and cataracts) might not develop until much later.

Many reports of congenital myotonic dystrophy have now been published. Fifty-four sibships containing 70 patients were analysed by Harper (1975*a,b*). The inheritance of the congenital form differs markedly from orthodox Mendelian ratios in that mother was affected in 51 out of the 54 families and father only once (the affected parents were not identified in the other two instances). The tendency for the mother to be affected was first noted by Vanier in 1960. Harper (1975*a,b*) suggested that a maternal intrauterine factor is responsible for the congenital form, but the precise nature of this factor is not yet evident. Bundey and Carter (1972) presented evidence that there might be two genetic types of myotonic dystrophy—one with early onset and one with late onset. The alternative hypothesis that the infantile cases are born to mothers because of the low fertility of affected males (Bundey and Carter 1972) is not accepted by Harper.

The incidence of the congenital type of myotonic dystrophy remains uncertain as the severe cases might succumb neonatally from respiratory complications and not be recognized. Harper's (1975*a,b*) survey suggested that at least five children with the severe congenital type are born in Great Britain each year.

Myotonic dystrophy presenting with severe dysphagia was described in two brothers (Pruzanski 1965). The mother had classical myotonic dystrophy.

Detection of heterozygotes of myotonic dystrophy

A reliable predictive detection test for those at risk who want to have children, but have not yet passed their period of maximum risk, is not available although DNA markers might be useful (see next section). A thorough clinical examination is mandatory. Asymptomatic first-degree relatives were found to have the disease in 13.6 per cent of those examined by Bundey *et al.* (1970) and 17.6 per cent in Harper's (1973) study.

The study of Polgar *et al.* (1972) suggested that electromyography was more useful than other tests. Whereas 85 per cent of clinically affected patients had changes on slit lamp examination, all the patients had electromyographic changes. These authors did have one family in whom characteristic lens opacities were found in the absence of electromyographical myotonia.

Genetics

Thomasen (1948) found that in all the families he studied one parent was affected. Other studies which have confirmed dominant inheritance are those of Klein (1958) and Bundey (1974). The average age of onset in index cases is in the third decade. When the offspring were investigated the onset was earlier, presumably because the knowledge that the disease is in the family facilitates early diagnosis. This, and the observation that if a family (and not the population as a whole) is investigated at a particular time late-onset cases will be missed, i.e. those who have not yet developed clinical manifestations. This accounts for the phenomenon of 'anticipation' (with each succeeding generation the onset appears to be earlier). Most workers believe that anticipation is an error of sampling (Penrose 1948).

Counselling

The risk to offspring of those affected is 1 in 2. If (Bundey *et al.* 1970) those at risk reach the age of 30 years and are clinically normal then the risk is reduced by half. Bundey (1977) suggested that if the consultand who has an affected parent and an affected sib (with an age of onset between 1 and 19 years) waits until the age of 30 before starting a family, then the risk of being a gene carrier if the consultand is clinically normal is 1 in 6 and, if slit lamp and electromyographic examinations are normal, 1 in 12. If the onset in the sib is after 20 years then the risk at 30 is 1 in 3 if clinically normal, and 1 in 4 if the special tests are negative.

The risk that an affected mother might have a severely affected infant with mental retardation is not known. A reasonable figure is between 1 in 10 and 1 in 15 but this is based on small numbers (Bundey 1977). After the birth of one infant with the neonatal type the risk for a subsequent similarly affected child might be as high as 1 in 3 (Harper 1975*a,b*). If a child born after an infant with the severe congenital form is clinically normal and remains so through adolescence, the likelihood of still developing the disease after the age of 20 is small, as the congenital type seems to breed true within sibships (Harper's data).

Recent figures (Glanz and Fraser 1984) suggest the following. The risk to sibs of having the severe illness after a neonatally severely affected infant is 22 per cent i.e. 1 in 5 (collected from published data). Their own figure was 29 per cent, 1 in 3. Therefore there is a half a chance of inheriting the gene and two-thirds of gene carriers will have the severe neonatal form. The risk to the offspring of mothers who have myotonic dystrophy who have as yet no children ± 9 per cent, i.e. 1 in 10. The risk of the severe neo-natal type after the birth of an affected child with the delayed onset form is plus or minus 5 per cent (1 in 20).

The gene for myotonic dystrophy has also been found to be linked to the locus for peptidase D on chromosome 19. Unfortunately, polymorphism at that locus is not common. Most units have now stopped using the linkage between the myotonic dystrophy gene and the secretor locus in favour of using linked RFLP (restriction fragment length polymorphisms). There is close linkage with apolipoprotein C11 and using Bam 1 and Tac 1. Five families were shown to be closely linked (Bird *et al.* 1987).

In informative families, haplotype determination on the fetus might give 96–98 per cent chances of the foetus not having inherited the gene.

Another probe LDR152 (D19519) has recently (Bartlett *et al.* 1987) been shown to be tightly linked to the myotonic dystrophy gene. No recombinations were found.

Myotonic dystrophy and HMSN

The association was noted by Ziegler and Rogoff (1956), and Wald *et al.* (1962).

A remarkable family is reported by Spaans *et al.* (1986). There were 13 members with HMSN, eight of whom also showed prominent signs of myotonic dystrophy. The disorder segregated with markers known to be linked to myotonic dystrophy on chromosome 19. The authors conclude that the syndrome could be caused by an allelic form of the myotonic dystrophy gene or by a closely linked gene on 19.

A syndrome resembling myotonic dystrophy with external ophthalmoplegia (and minimal myotonia)

Hawkes and Absolon (1975) described a 29-year-old male with onset of an illness before the age of 2 years. When seen in adulthood, he had bilateral ptosis and a myopathic facies. No myotonia was present on clinical examination but prominent myotonic-like discharges were found on EMG. Cataracts were present on slit lamp examination. Unlike myotonic dystrophy, there was an external ophthalmoplegia, and no frontal balding. The proband's father was found to have congenital ptosis and stippled lens opacities on slit lamps examination. Muscle biopsy revealed evidence of a myotubular myopathy. A

similar disorder was described by Schotland and Rowland (1964). They called
it myotonic dystrophy without apparent myotonia. Ophthalmoplegia was
again a feature.

(d) Paramyotonia congenita

It is uncertain if this condition is genetically distinct from adynamia episodica
hereditaria. In Eulenburg's original description (1886) myotonia was evident
on cooling and relieved by warmth. On exposure to cold the facial expression
becomes fixed and finger flexion, especially of the ring and little fingers, is
prominent. Myotonia of the tongue, orbicularis oris, lid lag, and delayed
relaxation after 'tap' are characteristic. Onset is in early childhood and
mechanical myotonia of the tongue is the most constant sign in all age groups
(De Jong *et al.* 1973). Muscle weakness is mainly proximal and may be
precipitated (with or without cold) by activity. Paradoxical myotonia—that
is, myotonia made worse and not better by activity—was a feature, even in
Eulenburg's original cases (Magee 1963, 1966). Five to ten repetative fist
clenches elicit the paradoxical myotonia (De Jong *et al.* 1973). Muscle
wasting is not usually present nor does muscle hypertrophy occur. The condition is non-progressive and even improves in adulthood (Drager *et al.* 1958).
Affected members seldom consult a doctor for their muscle disease. The
disease has been studied many times, in particular by Becker (1970), and
according to Lundberg *et al.* (1974) most cases described in the literature
originate from 'one and the same mutant occurring before the second half of
the seventeenth century'. De Jong *et al.* (1973) followed a huge pedigree over
19 years inheritance was clearly dominant. Other dominant pedigrees were
represented by Thrush *et al.* (1972)—eight affected in three generations, and
Baxter and Dyck (1961)—four members in two generations.

Two large families (clearly autosomal dominant) are reported by Wegmuller *et al.* (1979). Their patients all had signs in infancy—the parents
noticed that if their children were exposed to cold, crying distorted their
facial features (especially after a wash in cold water). Speaking or laughing
caused a grimace. They had narrow palpebral fissures and the mouth was
twisted or tightly shut. Repeated innervation made it worse. Warmth allowed
the muscles (fingers became stiff, clumsy, and flexed) to relax. It should be
noted that muscle stiffness caused by chilling was not relieved by repeated
contractions. Myotonia could be evoked by percussion and action, but it was
paradoxical—after repeated hand clasps muscle tension increased although
never to the extent of provoking paralysis.

The following two entities might not be distinct categories.

(e) Paramyotonia congenita without cold paralysis

In 1958, De Jong described a family with 37 affected members with paramyotonia without cold paralysis (also reported by Brüngger and Kaeser

1977). These authors added another family with 26 affected members. The myotonia occurred on percussion of the tongue, thenar, and other muscles; paradoxical myotonia occurred on repeated opening of the fists, on eye closure and on chewing, but myotonia was not precipitated by exposure to cold or potassium load. The onset of the hand stiffness occurred between 4 and 5 years of age.

(f) Paralysis periodica paramyotonica

A father and daughter (and possibly the father's brother) were affected by weakness (often lasting days) which could be provoked by exercise, especially in cold weather but also in a warm environment (Appenzeller and Amick 1972). Paradoxical myotonia of the eyelids and a myotonic lid lag were present. Percussion myotonia in the tongue and shoulder-girdle muscles could be elicited. The authors believe that this dominantly inherited condition can be distinguished from paramyotonia congenita by the presence of paralysis independent of myotonia and cold exposure. Becker (1970) suggested the name paralysis periodica paramyotonica for this disorder and considered it to be distinct from paramyotonia congenita. A father and son, probably with the same disorder (Lundberg *et al.* 1974), had three cardinal symptoms: myotonia, abnormal fatiguability, and periods of weakness. Myotonia increased after exposure to cold and was made worse by repeated movement.

(g) Adynamia episodica hereditaria (hyperkalaemic periodic paralysis

A number of families have been described (Tyler *et al.* 1951; Stevens 1954; Armstrong 1962) in which the paralysis has responded unfavourably to potassium.

Gamstorp (1962) described pedigrees with dominantly inherited periodic paralysis and suggested that this is a separate entity and called it adynamia episodica hereditaria. Weakness is provoked by rest after exercise and is associated with a normal or rising serum potassium during attacks. A high intake of carbohydrates protects the patient. Onset is during the first decade and, although attacks might occur every day, they do not last for longer than an hour or two (Bradley 1969). There is some confusion about the occurrence of myotonia in this condition, and this relates to the fact that Gamstorp, in her original study (1956) of two families, did not note its presence. However, she subsequently reported that some patients did develop it (Gamstorp 1962). To add to the confusion, Drager *et al.* (1958) considered that Gamstorp's cases were but a variant of Eulenberg's paramyotonia congenita (1886). These authors presented a large pedigree with 30 affected members. Myotonia is now accepted as a frequent feature of the condition—lid lag being the most characteristic finding, but myotonia is not always present.

Some members experienced facial spasms only when exposed to the cold (Egan and Klein 1959). Persistent weakness is known to occur. Inheritance is clearly dominant. A family with 31 individuals affected over six generations was reported by Thomas *et al.* (1978). Attacks lasted longer than usual in this family (from 1 to 4 days) and some patients had permanent proximal weakness. In a father and two sons (Saunders *et al.* 1968) with myotonic lid lag and facial myotonia, permanent weakness occurred in the extensor group of muscles of the forearm. In one son paralysis could be provoked by either potassium or cold.

Becker (1971) distinguishes between adynamia episodica hereditaria, in which weakness can be precipitated by potassium, but not by cold, and paralysis periodica paramyotonia, where both potassium as well as cold can precipitate the weakness.

(h) Others

Chondrodystrophia myotonia (Schwartz–Jampel syndrome)

First fully described by Aberfeld *et al.* (1965), the syndrome consists of myotonic blepharophimosis, skeletal deformities, muscle weakness, and short stature. The skeletal abnormalities are coxa vara or valga, pectus carinatum, flattening of vertebrae, and kyphoscoliosis. The clinical picture is not too dissimilar from Morquio's disease (Huttenlocher *et al.* 1969*a*). In the children described by this author, the facial muscle involvement was striking, leading to a 'pinched face'. The family had originally been seen and described by Schwartz and Jampel (1962) and the syndrome now bears their names. Inheritance is as a recessive. Up to 1977 there were at least 15 affected families described with worldwide distribution. Sibs have been reported by Schwartz and Jampel (1962), Mereu *et al.* (1969), Beighton (1973), and by Pavone *et al.* (1978). Consanguinity was present in the families of Pavone *et al.* (1978), Saadat *et al.* (1972), Beighton (1973), and Greze *et al.* (1975).

A recent review of the literature (Farrell *et al.* 1987) shows that about one-third of the children have muscle hypertrophy.

Hypokalemic periodic paralysis

The onset of this disorder is in the second decade and consists of periodic weakness of proximal muscles, with only rare involvement of eye muscles. Respiratory muscles are not affected during an attack. The condition is of later onset than hyperkalaemic periodic paralysis (second decade instead of first), attacks usually last several hours and are provoked by rest following exercise and a heavy meal. During an attack speech is often affected unfortunately, many of those affected (Gardner-Medwin 1977) develop a progressive myopathy. Adequate therapy with a potassium supplement or acetazolamide, and the avoidance of carbohydrate foods might delay the myopathy. In the family reported by Buruma and Bots (1978) the evidence suggests that permanent weakness might be independent of the severity and frequency of the periodic attacks.

The family was again examined by Buruma *et al.* (1985). Indeed, it was the same family as that originally described by Biemond and Daniels (1934), thereby enabling Buruma and co-workers to observe a 50-year follow-up. There were 28 affected members in four generations. Permanent muscle weakness was common in the family, and although rest following exertion and carbohydrate rich meals provoked an attack most frequently, cold was also a precipitating factor.

Patients are usually normal between attacks, but Dyken *et al.* (1969) collected 33 cases with permanent weakness and some were described by Pearson (1964), and Engel *et al.* (1965).

Myotonia of the eyelids—more commonly associated with hyperkalemic periodic paralysis—can be present in hypokalemic paralysis (Resnick and Engel 1967).

McArdle (1962) reported a large pedigree in which a myotonic phenomenon could be elicited, but other than lid lag myotonia is not a significant part of the clinical picture.

Many dominant pedigrees with or without skips are reported (Corbett and Nuttall 1975—a large Negro pedigree; Cusins and Van Rooyen 1963; Kobayashi 1975). It is important to note (Brooke 1977) that the condition is often not expressed in females and may appear to be recessive.

Normokalemic periodic paralysis

One large family described by Poskanzer and Kerr (1961) might represent a separate entity. Inheritance is dominant. Paralysis was severe, lasting days or weeks, and attacks were triggered by rest after exercise. Acetazolamide and hydrocortisone could prevent attacks.

18

Multiple sclerosis

The occurrence of familial multiple sclerosis is found in most consecutively collected series. The frequency of this varies considerably. In the two English studies of Pratt *et al.* (1951) and Schapira *et al.* (1963), the mean frequency of familial multiple sclerosis was in the region of 6 per cent. The prevalence varies from country to country and is usually regarded in terms of high, medium, and low. A prevalence of 30–80 per cent 100 000 occurs in high-frequency areas, and 5 per 100 000 in low-frequency areas.

High risk areas exist in Northern Europe (British Isles, central France, the low countries, Germany, Poland, Czechoslovakia, Northern Switzerland, Sweden, and Finland), Northern USA between the latitudes of 38 and 50°, Southern Canada, New Zealand, and probably the Southern Antartic.

Exceptionally high prevalence figures (1 in 1000) have been recorded in the Orkneys and Shetlands (Fog and Hyllested 1966).

The change in frequency in the Faroe Islands is instructive (situated in the North Atlantic). Prior to the 2nd World War there were no recorded cases (data of Kurtzke and Hyllasted 1986 summarized by McDonald 1987), but between 1943 and 1973 there were 32 'cases'. The only significant change recorded in the environment was the use of the island by British troops in 1940. Whether this is a clue to the environmental 'factor' remains to be seen.

Medium frequency areas occur in Asia, the Pacific Islands, Africa, Latin America, Alaska, Greenland, Scandinavia above 65°N, Southern Europe, and the USA between 30 and 33° latitudes.

The frequency of multiple sclerosis is low in the European descendants of settlers in South Africa, Western Australia, Queensland, and Hawaii (Acheson 1977). It is also rare in Japan and Equatorial Africa.

Kurtzke (1975) pointed out that all the high and medium prevalence zones are in Europe or those areas colonized by Europeans, and suggested that the disease was transmitted elsewhere from Europe.

Migration and its effect on prevalence

A decline in risk to that of intermediate prevalence status possibly occurs when migration from a high-risk country of origin to a low-risk country of settlement takes place (Kurtzke 1975). Dean and Kurtzke (1971) showed that when children under the age of 15 years migrate they are protected from the high risk of their country of origin. In support of these conclusions, Dean (1967) in South Africa calculated a prevalence of 3 per 100 000 for Afrikaners, 11 per 100 000 for English speaking natives, and 50 per 100 000 for immigrants from Europe (not born in South Africa). No cases among the black Africa population had as yet been reported and there was a deficiency of immigrants who had migrated to South Africa before they were 15.

Looking at the reverse situation, that is migration from a country of low risk to that of high risk, Dean *et al.* (1976) found few immigrants from Africa, West Indies, India and Pakistan—all low-risk countries—admitted with multiple sclerosis to London hospitals. The frequency in peoples from high-risk countries showed figures only slightly lower than native Londoners.

Recently, Elian and Dean (1987) studied a group of children of West Indian immigrant parents. The children were born in England and at the time of the study were between 20 and 29 years of age. The incidence of multiple sclerosis in this group was equal to that for people in Northern Ireland and possibly the same as in south-east England.

Those who moved from one high prevalence area to another retained the high risk, but even here it is difficult to distinguish between genetic and environmental factors.

For instance, the population in the areas of high prevalence in North America came from high prevalence European areas such as Scandinavia. They did however, settle in areas which were environmentally the closest to their place of origin thereby bringing together both genetic and environmental factors (McDonald 1987)

Family studies

Studies have established that familial aggregation exists (Table 18.1). Six per cent of probands will have an affected first-degree relative. The risk of a consultand having an affected sib, parent, or child is about 10–20 times the general population risk. There is no evidence of an increased conjugal incidence of multiple sclerosis although this might only mean that the significant exposure takes place long before marriage (at the age of 10–15 years, with a latency of 20 years).

In most studies, females are affected more frequently than males. In the general population, the ratio of female to male cases is 3 to 2 and not 1 to 1. In the Pratt *et al.* (1951) study transmission from father to son occurred in three cases, whereas from father to daughter it was observed in 11. Transmission from mother to son occurred in 10 cases, but from mother to daughter in 27. It can be seen that in various studies the proportion of cases in sibs and parents of index cases far exceeds the frequencies expected in the general

Table 18.1 Taken from Kuwert (1977)

Familial incidence	Country	Cases of multiple sclerosis	Families with more than one	Familial incidence (%)
Curtius and Speer (1937)	Germany	106	10	9.4
Pratt et al. (1951)	England	310	20	6.5
Muller (1953)	Sweden	750	27	3.6
Hadley (quoted by Kuwert 1977)	Scotland	150	5	3.3
Millar and Allison (1954)	N. Ireland	668	44	6.58
Sutherland (1956)	Nova Scotia	127	14	11
Abb and Schaltenbrand (1956)	Germany	472	29	6.5
Hyllested (1956)	Denmark	2731	73	2.6
MacKay and Myrianthopoulos (1958)	U.S.A. and Canada	54	18	22
Alter et al. (1962)	Israel	282	0	0
Fog and Allison (quoted in Kuwert 1977)	Orkney, Shetland Is	60	7	11.6
Allison (1963)	Faroe Islands	16	2	12.5
Schapira et al. (1963) England	Durham North	607	35	5.8

population. If the study of MacKay and Myrianthopoulas (1958) is excluded, the prevalence rate in sibs and parents is 20–40 times that expected in the general population. This is confirmed in a review by Drachman *et al.* (1976) who found 12 patients with another affected first-degree relative. The prevalence in first-degree relatives was calculated at 800 per 100 000 that is 15–20 times the general population figure for Chicago (40 per 100 000).

The prevalence ratio is sibs and parents of index cases in other studies can be seen in Table 18.2. The figures far exceed those expected.

Twin studies

When a disease is due to a single dominant mutant gene concordance in monozygous twins should be 100 per cent and in non-identical twins 50 per cent if a parent is affected and 0 if not. When a disease is due solely to environmental factors, concordance in identical twins should be no different from that in fraternal twins. In multifactorial inheritance, concordance in identical twins will be incomplete but greater than in non-identical twins.

In 1930, Goldfiam reported concordance in twins (zygosity unknown), and four years later Legras (1934) reported concordance in uni-ovular twins. Most of the older literature on the subject is not reliable concerning the zygosity of the twins. Monozygotic twins discordant for multiple sclerosis are also known (Gardner-Thorpe and Foster 1975).

McKay and Myrianthopoulos (1966) collected 90 monozygous and 85 dizygous twins. They found concordance in 24/90 and 11/85 respectively. In a later study, Cendrowski (1968) found three pairs of twins out of 300 patients with multiple sclerosis during an epidemiological survey in Western Poland. All three pairs were dizygous and all were discordant. In the Bobowick *et al.* (1978) analysis of 12 pairs of twins only one (possibly two) of five pairs of monozygotic twin pairs were concordant.

More recently Williams *et al.* (1980) found concordance in half of the monozygotic twin pairs and in three out of 14 dizygotic pairs, but all twin studies are biased towards family concordance in twin pairs. (Of the three concordant dizygotic paris only one pair was HLA identical.)

In a Canadian twin study (Ebers *et al.* 1986), concordance rates in dizygotic twins (2.3 per cent was similar to that found in siblings 1.9 per cent whereas in monozygotic twins concordance was 25.9 per cent. In a Finnish twin study there were 11 MZ pairs and 10 DZ pairs (Kinnunin *et al.* 1987). There was one concordant MZ pair.

An identical twin of a patient with multiple sclerosis has a 20–25 per cent chance of developing the disease, that is, 300 times the general population risk (Eldridge *et al.* 1978). The risk for a non-identical twin is somewhere between 2.5 and 15 per cent.

Table 18.2 From Kuwert (1977)

	Prevalence (per 10 000)	Sibs			Parents	
		Observed	Ratio of O : E		Observed	O : E
Curtius and Speer (1937)	2.3	4/414	42 to 1		1/212	21 to 1
Pratt et al. (1951)	5.0	6/538	22 to 1		3/620	10 to 1
Millar and Allison (1954)	5.8	34/2939	20 to 1		11/1336	14 to 1
Sutherland (1956)	6.7	7/547	19 to 1		2/254	12 to 1
MacKay and Myrianthopoulos (1958)	5.0	12/204	118 to 1		1/62	32 to 1
Schapira et al. (1963)	5.0	25/2151	23 to 1		7/1204	12 to 1

Histocompatibility antigens

HLA-A$_3$

Most studies examining histocompatibility antigens in multiple sclerosis have found an increased frequency of HLA-B$_7$ and, to a lesser degree, of HLA-A$_3$. In the study of Naito *et al.* (1972) the antigen HLA-A$_3$ was found in 40.4 per cent of those affected compared with 23.5 per cent of normal Caucasian controls, and in the study of Paty *et al.* (1977) HLA-A$_3$ was found in 37.5 per cent of patients compared with 25.30 per cent of controls. Other studies are those of Arnason *et al.* (1974) where HLA-A$_3$ was 52 per cent versus 23 per cent of controls, and in a second series 35.5 per cent versus 23 per cent controls. Stewart *et al.* (1977) observed 46 per cent versus 28 per cent, and Thorsby *et al.* (1977) found 68 per cent versus 30.2 per cent. However, in the studies of Möller *et al.* (1975) and Bertrams *et al.* (1975), the difference was less pronounced.

HLA-B$_7$

Jersild *et al.* (1975) found 39.7 per cent of patients with multiple sclerosis to have HLA-B$_7$ as compared to 26.8 per cent of controls.

Other studies are those of Platz *et al.* (1975), 41.07 per cent versus 19.92 per cent for controls, and Stewart *et al.* (1977), 46 per cent versus 29 per cent.

HLA-Dw$_2$ antigens

If the frequency of HLA Dw$_2$ antigens from patients with multiple sclerosis are compared with controls, they are found in 60 per cent of patients as compared to 11 per cent of controls (Jersild *et al.* 1975). No association with HLA-C antigens has been noted.

Paty *et al.* (1977) found HLA-Dw$_2$ to be present in 48 per cent of multiple sclerosis patients versus 19 per cent of controls. Data from the seventh Workshop (Bodmer 1978) found HLA-Dw$_2$ to be present 53 per cent of multiple sclerosis patients and 21 per cent of controls.

Ia antigens—now HLA-DR antigens

The antigen Ag-7a was found with a high frequency in a small group of patients with multiple sclerosis (Winchester *et al.* 1975). Subsequently, Compston *et al.* (1976) looked at six B-lymphocyte antigens in 59 patients with multiple sclerosis. They found that BT 101 occurred in 83 per cent of patients with multiple sclerosis compared to 33 per cent of normals. The relationship between the B-lymphocyte alloantigens and the products of the HLA-D locus is still uncertain. It is known that BT 102 shows a significant association with Dw_4 so that the possibility of linkage disequilibrium with a D-locus gene is possible.

Recently, Francis *et al.* (1980) looked at a class II HLA antigen, DQwl and found it overrepresented in the multiple sclerosis population. This study took place in the Grampian region in Scotland in a known high frequency region. The HLA-DR_2 frequency was 49.4 per cent which is similar to figures in other European studies of MS, but there was no difference when this figure was compared with the local non-MS population.

These findings accord well with the study of Compston (1981) who suggested that MS is more common in those areas with a high DR_2 frequency in the population.

The association with DR_2 decreases as one moves south in Europe and north within the UK, but so does the frequency of DR_2 in the general population. In the study of Swingler and Compston (1986) multiple sclerosis seemed to be more prevalent in those populations where the background frequency of DR_2 was higher.

As summarized by McDonald (1987) there is not always an association of MS with DR_2 as in the Japanese and Gulf Arabs, and in the Hungarian gypsies where the frequencies of DR_2 is particularly high the frequency of MS is low. It is most likely a marker for another gene or genes which is more intimately involved with the clinical occurrence of multiple sclerosis.

Family studies of HLA-antigens

If a patient has multiple sclerosis because of susceptibility associated with his HLA status, then family studies should show that affected members share his haplotype and those unaffected do not. The problem is, as stated parental haplotype will occur in the affected offspring with a probability not significantly different from 50 per cent. Large pedigrees are usually needed and these are rare in multiple sclerosis.

Of the familial cases reported there have been only two with affected members in three generations making simple dominant inheritance unlikely (Drachman *et al.* 1976; Bird 1975). However, family studies have identified situations where affected members share a common haplotype, either identical with or different from the expected HLA-A_3, HLA-B_7, and HLA-Dw_2 haplotypes.

Alter *et al.* (1976) analysed the segregation of HLA haplotypes in 10 families in which there were at least two cases of multiple sclerosis. In nine of the 10, a significant association between the disease and one parental haplotype was found, but the specific type was different in nearly every family.

A similar result was found in a study of eight families from Southern Sweden (Olsson *et al.* 1976) in which two or more members had multiple sclerosis. All eight families showed affected members to share one major histocompatible haplotype within a family. In only two families was the haplotype $HLA-A_3$, $HLA-B_7$, or $HLA-Dw_2$.

In the Olsson *et al.* study (1976) the authors found that 18 unaffected members had the 'multiple-sclerosis-susceptible' haplotype and 10 unaffected did not. The difference was not significant.

In the single family described by Jersild *et al.* (1973), both affected sibs carried $HLA-Dw_2$ although an unaffected sib had the same haplotype. The family described by Bird (1975) showed affected members to share a common haplotype $HLA-A_{11}$ and $HLA-W_{16}$, but insufficient information was given on unaffected sibs to establish a significant relationship. The sharing of a haplotype in sibs with multiple sclerosis was also shown by Zander *et al.* (quoted by Batchelor *et al.* 1978). They studied 28 sib pairs, both of whom had multiple sclerosis. Twenty-four out of the 28 shared a common haplotype. In seven pairs it was $HLA-A_3$, $HLA-A_7$, or $HLA-Dw_2$.

Not all studies confirm HLA-haplotype segregation within families.

Thirteen families with at least two affected were selected by Hens and Carton (1978). They found that the segregation of HLA haplotypes in affected sib pairs was not significantly different from that expected.

A summary of the family date was given by Eldridge *et al.* (1978). They selected seven pedigrees of their own in which there were 20 affected members. They also analysed the data from 28 other families reported in the literature and concluded that these data do not confirm the presence of a single gene responsible for susceptibility. Families used were those of Alter *et al.* (1976), Bird (1975), Jersild *et al.* (1973), Drachman *et al.* (1976), and Olsson *et al.* (1976). No consistent segregation of multiple sclerosis with a specific haplotype was noted. In their own families there was an increase in $HLA-Dw_2$, $HLA-A_3$, and $HLA-B_7$, but unaffected members of these families also had an increased frequency of these antigens. As emphasized by Drachman *et al.* (1976), no causal relationship has been found between an HLA trait and multiple sclerosis. No haplotype is necessary or sufficient for multiple sclerosis to occur. Their data suggest that the prevalence rate of multiple sclerosis is the same for all family members whether or not they have inherited a particular HL-A haplotype. Whether multiple sclerosis is linked to a locus in the HLA region or is only associated with a HLA antigen is still not resolved.

Visscher *et al.* (1979) examined 13 families in which more than one affected member occurred. In four, more than one generation was affected and in the others, sibs only. Within each family all cases of multiple sclerosis shared at least one common HLA haplotype.

The probability of this occurrence in the absence of linkage is less than

1/1000, but it presumes that only 5 per cent of those with the hypothetical multiple sclerosis susceptible gene or genes would have the disease. In a study by Haile *et al.* (1980) to test the hypothesis that the multiple sclerosis gene or genes are linked to the HLA loci the data was again compatible with linkage at low penetrance (5 per cent) levels, assuming an autosomal dominant mode of inheritance. The linkage analysis is, however, still uncertain as it depends on the difficult estimate of penetrance. In addition, the patients used in this study were drawn from two different areas and only in one were the Lod scores suggestive of linkage.

There is as yet no acceptable explanation for the diverse HLA findings in the various family studies and most investigations have not provided information on Dw_2 antigens. To account for the strong association with Dw_2 and the weak linkage (20 centimorgans in the Haile *et al.* (1980) study) it is possible that multiple sclerosis is heterogeneous, with HLA-linkage in some families but not in others.

Familial multiple sclerosis involving many members of a pedigree remains rare. However, three such pedigrees are reported by Hartung *et al.* (1988) and by Kinnunen *et al.* (1987). All the affected individuals, who were typed, in the Hartung pedigree, were $HLA-DR_2$, and when looked for $HLA-DQw_1$.

Genetic counselling

In Britain the prevalence of multiple sclerosis is 1 in 1600. An increase by 20 times (the risk to the offspring of a proband) would give a risk of 1 in 80. All studies agree that this risk is low and parents and those affected can at present be reassured. Where two first-degree relatives are affected the risk could be increased on an empiric basis to between 3 and 5 per cent, and to 2 per cent where the relationship between those affected is more remote.

Recently, Sadovick and Macleod (1981) calculated empiric risks as follows:

Male index risk to son	0.6%	Female index risk to son	0.8%
daughter	2.6%	daughter	0.4%
brother	2.6%	brother	1.9%
sister	2.0%	sister	2.2%

Some figures, especially those relating to the sib risk of female index cases are higher than those quoted in the above table. For instance, Sweeney *et al.* 1986 calculated a risk of 3.9 per cent in a British Columbian population where the background risk is about 0.1–0.2 per cent.

Kuwert (1977) suggested that individuals carrying the antigen $HLA-A_3$ and $HLA-B_7$ have a 50 per cent higher risk of developing multiple sclerosis than persons without these antigens. Even if this was found to be correct, the risk in Britain would still only be 1 in 1000—a very low risk. Although in some families the possession of a specific haplotype might confer on first-degree relatives a slightly higher risk; the data remain inconclusive.

The strongest association is with Dw_2 antigens where the relative risk is ninefold (Compston, personal communication). For genetic counselling purposes this would mean an increase in the risk from about 1 in 1600 to 1 in 170—a very small risk that could not be used profitably in genetic counselling.

19

Degenerative illnesses of childhood

A. The gangliosidosis and related disorders

 1. GM_1 gangliosidosis
 2. GM_2 gangliosidosis
 3. GM_3 gangliosidosis
 4. Sandhoff's disease
 5. A B variant
 6. Niemann–Pick disease
 7. Gaucher's disease
 8. Ceroid lipofuscinosis
 9. Mucopolysaccharidosis
10. Sialidosis
11. Mucolipidosis
12. Mannosidosis
13. Fucosidosis
14. Farber's disease

1. GM_1 gangliosidosis

In 1968 Okada and O'Brien described a case with beta-galactosidase deficiency in liver, brain, and fibroblasts. Since then at least three distinct phenotypes have been distinguished by age of onset and the degree of visceral and skeletal involvement. The diagnosis is suggested by the presence of foamy mononuclear cells in the urinary sediment or vacuolated peripheral lymphocytes. Foam cells in the marrow appear in all types, although vacuolization of peripheral lymphocytes is more prominent in the juvenile form (O'Brien *et al.* 1972). There is no difference in the degree of beta-galactosidase deficiency in the three types.

Type I (infantile form) 'pseudo-Hurler syndrome' (Landing's disease)

This is characterized by an onset soon after birth of a progressive motor and psychomotor deterioration with bony deformities, an enlarged spleen and liver, and peripheral oedema. The facial features are coarse and consist of frontal bossing, a flat nasal bridge, large low-set ears, and gingival ridge hyperplasia. A cherry red spot at the macula is found, but the corneas are clear. The infant is dull and hypoactive and problems arise with poor feeding and swallowing.

The CT scan appearance of the brain shows areas of reduced density of white matter which is unusual for a storage disorder (Curless 1984).

A recent report by Giugliani *et al.* (1985) suggest that birthweight may be significantly increased with a tendency to macrosomy. Hypotonia in the early months was a prominent feature. Death usually occurs before 2 years. X-rays of the bones show changes suggestive of Hurler syndrome and the pituitary fossa is elongated and appears shoe shaped. A type-I pedigree is described by Singer and Schafer (1972). Inheritance is recessive. Presumed monozygotic twins are reported by Ginsburg and Long (1977).

Some cases fall clinically between type I and type II in that they have an intermediate age of onset. Pinsky *et al.* (1974) described a girl who, at 5 months, developed generalized seizures. On examination she was noted to have minimal hepatosplenomegaly, and spasticity in her lower limbs. Bone-marrow examination revealed vacuolated lymphocytes and at 8 months her peripheral blood leucocytes showed the beta-galactosidase activity to be one-tenth of normal.

Type II (juvenile form)

Symptoms occur between 7 and 16 months and the onset is thus later than type I. Deterioration is less rapid and visceral enlargement is absent. Weak-ness and ataxia appear first. Hearing and vision are normal and skeletal changes are minimal. Facial changes are not pronounced and the fundi are normal. Seizures develop later in the illness, and death occurs between three and 10 years. O'Brien *et al.* (1972) described the features in five members of two families (recessive inheritance in both), and other pedigrees have been described by Derry *et al.* (1968), Wolfe *et al.* (1970), Chou *et al.* (1974), and Lowden *et al.* (1974). Inheritance is recessive.

Type III (adult and chronic GM₁ gangliosidosis some with dystonia)

Facial coarseness, spondyloepiphyseal dysplasia, but no macular red spots nor visceromegaly were the features in two brothers with beta-galactosidase deficiency (Stevenson *et al.* 1978). These were adult males with an onset at three years. Both deteriorated intellectually.

A Japanese patient (parents 2nd cousins) had an onset at 13 years of a mild dysarthria (Ushiyama *et al.* 1985). At 16 years she began to drag one foot and at 22 had facial grimacing and dystonia (segmental). Her intellect was within normal limits. Her ganglion cells in Meissners plexus had osmiophillic lamellar inclusions similar to membranous cytoplasmic bodies. Goldman *et al.* (1981) emphasized the basal ganglia pathology and extra pyramidal clinical presentation. In their case the age of onset was 4 years and at the age of 27 intellectual deterioration was minimal. Sibs with a similar clinical picture had an onset in one at 19 years and later in the other (Nakano *et al.* 1985).

Two sibs are reported by Wenger *et al.* (1980). They had ataxia, mild intellectual deterioration, slurred speech, and mild vertebral changes.

Prenatal diagnosis has been achieved by Nadler and Gerbie (1970) and Lowden *et al.* (1973).

Adult onset GM_1 gangliosidosis with ataxia/myoclonus

Three sibs with adult onset ataxia, myoclonus and pyramidal signs had coarse facial features (one had cataracts) and optic atrophy (two) (Mutoh *et al.* 1986). Corneal clouding occured in all three. Early development was normal, but school performance was below average and a mild dementia occurred. None had bony lesions and one had a cherry red spot. Neuraminidase was normal.

The authors note the clinical difference between this type and the usual adult onset form. The sibship is unique and might be a separate condition at the same beta-galactosidase locus.

2. GM_2 gangliosidosis

Nomenclature of the GM_2 gangliosidosis

Hexosaminidase A is composed of one alpha- and two beta-subunits whereas Hexosaminidase B comprises four beta-subunits. The alpha-chain is coded for on chromosome 15 and the beta-subunit on chromosome 5.

Mutations at the beta-locus will affect Hex A and B, and cause Sandhoff disease. Those patients with Tay–Sachs disease have alpha-locus mutations.

Classical infantile Tay-Sachs disease

Initially the disease is characterized by hypotonia and the baby shows an abnormal startle response. Unlike the normal Moro reflex, the startle response to sudden sounds (rattling of a key) is a sudden extension of arms, head, and trunk. The children do not learn to sit, but despite the floppiness, reflexes are often brisk. The cherry red macular spots can develop as early as two months (Stephens, personal communication). Towards the end of the first year the child becomes disinterested, seizures develop, and the child assumes a frog-like position. At 18 months, the head appears to increase in size, bulbar palsy develops and death ensues usually before the third year.

Inheritance is as a recessive and in the large series of Aronson *et al.* (1959, 1960), 90 per cent of those affected in New York City were of Jewish descent. The condition has been described in Caucasians, Chinese, Egyptians, Hindus, Japanese, Negroes, and Singhalese. The frequency of carriers in the Jewish population of New York City is about 1 in 30 with a carrier rate in non-Jewish Americans of 1 in 300 (Aronson *et al.* 1960). Hexosaminidase A is absent.

The distribution of the ancestral origin of Tay–Sachs disease in North America is heavily weighted towards Ashkenazi Jews of Polish and Russian origin (Petersen *et al.* 1983), but what is suprising are the numbers from Austria, Hungary, and Czechoslovakia. The overall numbers are small and need to be interpreted with caution (all data from Peterson *et al.* 1983), but the significant numbers from Austria, Hungary and Czechoslovakia cannot be overlooked in theories about a founder effect. Indeed, the higher carrier frequency from Middle European Jews rather than American Jews from Poland/Russia has led to suggestions that the Tay–Sachs gene proliferated among the antecedents of the Jews before migration to Poland and Russia.

Control and prevention of Tay–Sachs disease

Since the 1971 pilot scheme programme for the screening of possible carriers of Tay–Sachs disease in the Jewish communities of Baltimore, Maryland and suburban Washington, more than 100 000 Ashkenazi Jews have been tested (Kaback *et al.* 1977). In the original study the tests were organized on a voluntary basis with the help of the leaders of the religious community. The organization of resources took 14 months. The test results were communicated in the original pilot study by an evening telephone call, when only preliminary questions were answered and an appointment for further genetic counselling was arranged. At this session the implications for close relatives were discussed and parents tested first in order to determine on what side of the family carriers should be looked for. Distant relatives were contacted by the carrier, if this is what he or she chose, but the programme provided a clearly written letter and brochure to the carriers for their use in contacting relatives. Effective public education for both the medical and ethnic communities by trained volunteers must be intensive before the screening test is instigated as the communication of the results can cause misunderstanding and confusion (Kaback *et al.* 1977). The consequence of the testing, such as amniocentesis, termination, and what it means to be a carrier, is included as part of the programme. There have been doubts expressed about the benefit of the Tay–Sachs screening programme (Kuhr 1975). To prevent one child with Tay–Sachs disease, it would be necessary to identify 72 people as carriers and this price, in terms of the psychiatric burden, might be too high to pay. It is, however, argued that with proper planning and education, it is possible to dispel unreasonable fears. The consequences of refusing to make information available with regard to carrier status might cause greater anguish.

The screening programme has not been favourably viewed by all those directly concerned. In a recent report, Steele (1980), only 15 per cent of adult North American Jews of reproductive age wanted to be screened. The stress associated with the screening programme was suggested in a follow-up of carriers identified in Toronto. Seventeen out of 19 have not attempted subsequent pregnancies. A fear was expressed that the programme might be adding to the already steady decline in numbers of North American Jews. Steele (1980) suggests that people want treatment not prevention.

Juvenile GM₂ gangliosidosis

Bernheimer and Seitelberger (1968) first described the juvenile variant in which the hexosaminidase A deficiency is partial. The onset is late (between 2 and 6 years) and progression is slower than in Tay–Sachs disease. Death occurs in the second decade of life. The patients are frequently of non-Jewish origin. The initial symptoms (Suzuki and Suzuki 1970) are personality changes and intellectual deterioration, followed by seizures, a cerebellar ataxia, and pyramidal and extrapyramidal signs. A decerebrate state is reached between 5 and 14 years. The preservation of sight is in contrast to the early loss of sight in the infantile variety and in ceroid lipofuscinosis (Battens disease). In the case described by Buxton *et al.* (1972) a younger sister of the proband had the biochemical defect, but was asymptomatic at 3 years of age.

Cherry red spots at the macula were seen in three out of the eight cases described by Brett *et al.* (1973), but these spots were unlike those found in Tay–Sachs disease. Among the 14 cases in the literature up to 1973 (Brett *et al.*) there have been four pairs of affected sibs. Consanguinity is reported in the parents of the two Lebanese sibs (Andermann *et al.* 1974) and in a further, possibly unrelated, Lebanese child, by the same authors. The unrelated parents of the affected child reported by Brandt *et al.* (1977) had enzyme levels in the carrier range.

There is some indication that it might be possible to subdivide the group into late infantile and juvenile cases. The younger groups more frequently have the startle reaction and the cherry red spots but this is not a constant finding. Unfortunately, the division does not tally with the degree of deficiency of hexosaminidase A (Brett *et al.* 1973).

Tay–Sachs with intermediate levels

A patient with an unusual Hex A mutation is described by O'Brien *et al.* (1978). Development was normal until 4 months when deterioration was noted and at 17 months the patient was decerebrate and had bilateral optic atrophy and cherry red spots. Hex A activity measured with synthetic substrates was normal, but the *in vivo* activity was markedly decreased. It is suggested that the patient had an allelic compound Hex A.

Normal subjects with absent Hex A

A healthy 28-year-old male was tested when he volunteered for a Tay–Sachs screening programme and Hex A was found to be absent (O'Brien 1978*b*). It was proposed that the proband had an allellic compound which leads to a reduction of Hex A activity to the synthetic substrate.

Navon *et al.* (1973, 1976) described a deficiency of hexosaminidase A in healthy members of families in which there were affected individuals with Tay–Sachs disease. It was proposed that these persons are compound heterozygotes consisting of an allele for Tay–Sachs and another for a mutant allele at the same locus. Heterozygotes for the second mutant cannot be distinguished from Tay–Sachs heterozygotes using artificial substrates. Compound heterozygotes of the allele for Tay–Sachs cannot be differentiated from the Tay–Sachs homozygote. A healthy non-Jewish female with a low level of hexosaminidase was presumed by Kelly *et al.* (1976) to be a compound of the two mutant alleles. They also showed that the second mutant gene segregated in the family according to Mendelian laws.

The importance of this finding is that before prenatal diagnosis is undertaken parental hexosaminidase levels should be looked at in order to circumvent aborting a normal fetus with deficient enzyme levels because of compound heterozygosity.

Adult GM$_2$ gangliosides

Rapin *et al.* (1976) described an atypical spinocerebellar degeneration in three Jewish sibs. They presented between the ages of 2–3 years with a gait disturbance. Inco-ordination progressed and distal weakness and wasting resulted in severe pes cavus and foot drop. Sensation was intact except for the unpleasant nature of light touch. Speech was slow and laboured and was affected early in the disease. Convergence was poor and there was a progressive loss of evoked nystagmus—both vestibular and optokinetic. The fundi and vision were normal. Extrapyramidal features such as dystonia, facial grimacing, and episodic involuntary vertical eye movements occurred. Intellect was not seriously affected. No seizures or myoclonic jerks occurred.

Another similar family is reported by Johnson *et al.* (1977). Onset was at two and a half. At four the patient was found to have a cherry red spot and a cerebellar ataxia. Seizures and dementia were not at that stage present. Note also the clinical picture of juvenile progressive dystonia and a cerebellar type of ataxia (Meek *et al.* 1984) have been described in GM$_2$ gangliosidosis. Spinal muscular atrophy (Navon *et al.* 1980) and motor neurone disease (Yaffe *et al.* 1979) have also been reported.

An Ashkenazi Jewish brother and sister developed a cerebellar ataxia, proximal weakness, and wasting (neurogenic) and a progressive supranuclear ophthalmoplegia. Onset was in the 30's. Other features were occasional facial twitching, a brisk jaw jerk with a pout reflex, and extensor plantar responses. Hexosaminidase A was significantly reduced (Harding *et al.* 1987).

Two sibs are reported by Mitsumoto *et al.* (1985). One presented with juvenile onset of amyotrophic lateral sclerosis. Subsequently, this evolved into a mild dementia, ataxia and an axonal neuropathy, whereas the 2nd sib had a 'pure' spinal muscular atrophy.

Intrafamilial variability was commented on by Argov and Navon (1984). Intermittent psychosis leading to a diagnosis of schizophrenia was common, but within families some had amyotrophic lateral sclerosis or spinal muscular atrophy and some (often a sib) spinocerebellar degeneration.

Two brothers with an onset in middle life of ataxia, fasciculation and a neurogenic EMG, had clinical picture which was said to resemble motor neuron disease (Frederico *et al.* 1986).

Note: the adult onset cases have been summarized by Navon *et al.* (1986). The main systems involved are cerebellar, pyramidal and lower motor neurones. Within families there can be considerable variation. One of the patients referred to by Navon *et al.* had a spinocerebellar syndrome whereas another member of the same family presented with lower motor neuron disease. Psychosis is common.

3. GM$_3$ gangliosidosis

Maclaren *et al.* (1976) and Max *et al.* (1974) reported a male Jewish infant with features of pseudo-Hurler's syndrome (type I GM$_1$) with an accumulation of the GM$_3$ molecule. He died at 3½ months. The infant was retarded from birth, the corneas were clear and bony abnormalities were absent. There was circumstantial evidence that a maternal uncle had died at 2½ years from a similar disorder and the possibility of X-linked recessive inheritance is suggested by the authors. The dysmorphic features were those of a coarse facies, dry loose hirsute skin, macroglossia, pronounced gingival hypertrophy, low-set malformed hirsute ears, a depressed nasal bridge with prominent maxilla, and underdeveloped mandible. The liver and spleen were enlarged.

4. Sandhoff's disease

The disease presents by the age of nine months with blindness, progressive mental and motor deterioration, cherry red spots, and macrocephaly. The clinical picture is indistinguishable from that of Tay–Sachs disease except in the ancestry of the patient. Most of the families have been non-Jewish. Inheritance is recessive. Okada *et al.* (1972) described sibs in two families. The parents were British and unrelated in one family and were Mexican-American in the other. A sibship (two affected) in which the parents were first cousins is reported by Lowden *et al.* (1978). In this pedigree carrier detection was achieved. Prenatal diagnosis has been successful (Desnick *et al.* 1973; Harzer *et al.* 1975).

In contrast to Tay–Sachs disease, both hexosaminidase A and B are absent.

Experience from the Lebanon (Der Kaloustian *et al.* 1981) suggests that Sandhoff in that community is relatively frequent compared with Tay–Sachs. Of the five families that they reported four were Moslem and one Christian.

Variants of Sandhoff's disease (including juvenile onset)

A Negro infant with a clinical picture of Sandhoff's disease had reduced hexosaminidase A levels in the serum and to a lesser extent in fibroblasts, but higher than in other reported cases (Spence *et al.* 1974). Hexosaminidase B activity was absent. The parents were 'indirectly related'. A female paternal cousin was diagnosed as having amaurotic familial idiocy. She had psycho-motor deterioration, optic atrophy, and retinal pigmentation. Two sibs of this cousin had a similar condition. Death occurred between 3 and 4 years.

A single case (male) with an onset at five years developed a progressive cerebellar ataxia and dementia (MacLeod *et al.* 1977). Hexosaminidase A and B were absent in the serum. This is probably juvenile Sandhoff's disease.

Two adult sisters developed a spinocerebellar degeneration and had bilateral pes cavus, dysarthria, and facial grimaces (Oonk *et al.* 1979). Peripheral dimunition of vibration and position sense were found and reflexes were brisk, except for the ankle and triceps jerks. Both sisters had a deficiency of hexosaminidase A and B.

A single juvenile case of both A and B deficiency has been described by Goldie *et al.* (1977).

A normal adult with absent activity of Hex A and Hex B was thought by the authors to be a compound of the allele for Sandhoff's disease and a mutant allele with activity for natural but not synthetic substrate (Dreyfus *et al.* 1975).

Another variant of Sandhoff's disease, but possibly a new disorder, was reported by Johnson and Chutorian (1978) in a single case with juvenile onset of ataxia and a cherry red spot. On starch-gel electrophoresis both Hex A and Hex B were absent, but an A-like substance was present.

5. Activitor mutant

AB variant (infantile)

This subdivision is characterized by a clinical picture similar to Tay–Sachs disease, but milder. When tested with synthetic substrates Hex A and B are normal. All the patients with the AB variant have so far been non-Jewish. The cases reported by Sandhoff *et al*. (1971) and Kolodny *et al*. (1973) were sibs in a Scottish-Irish family, and a black infant was reported by De Baecque *et al*. (1975). Seizures, myoclonus to sounds, and hypotonia developed at 9 months; spasticity, dementia, and death followed. A cherry red spot was noted in De Baecque *et al*.'s (1975) case and in the case of Kotagel *et al*. (1986), GM_2 gangliosides were detected in the CSF.

AB variant (adult)

An adult-onset single (non-Jewish) AB variant of GM_2 gangliosidosis is reported by O'Neill *et al*. (1978). Activity of both hexosaminidase A and B were markedly increased. The clinical picture was that of a normal pressure hydrocephaly with seizures and dementia.

6. Niemann–Pick disease

Niemann–Pick disease is divided into the following types.

A. Acute neuropathic form.

B. Chronic form without central nervous system involvement.

C. Chronic neuropathic form—late onset (after 3 years).

Those affected develop ataxia and seizures, and lose their speech. Hepato-splenomegaly might not be striking.

D. Nova Scotia variety.

E. Adult—non-neuropathic form.

Type A. Acute neuropathic form

Classical infantile form

This has an onset during the first year of life with mental and motor deterioration. Death ensures before three years. Recurrent vomiting and bouts of fever, with failure to thrive and emaciation, are prominent early symptoms. On examination, there is a protuberant abdomen, due to the enlarged liver and spleen. Cherry red spots were found in four cases of type A Niemann–Pick disease (Walton *et al*. 1978) and this lesion possibly occurs in 50 per cent of cases. The four patients had, in addition, coarse corneal clouding. It is the classic infantile type that predominantly affects Jews— about 40 per cent of cases in the United States (Fredrickson and Sloan 1972), and the deficient enzyme is sphingomyelinase. The characteristic pathologic change is a ballooning of neurons with membranous cytoplasmic bodies. *In utero* diagnosis of the infantile type was achieved in 1971 by Epstein *et al*. and using the chromogenic method by Patrick *et al*. in 1977. Inheritance is autosomal recessive.

Type B. Chronic form

This group does not involve the central nervous system and will not be discussed.

Type C. Niemann–Pick/juvenile dystonic lipidosis

Most of the cases of progressive supranuclear ophthalmoplegia and juvenile dystonic lipidosis or ophthalmoplegic neurolipidosis have now been reclassified as Niemann–Pick Type C.

It is characterized by a delayed onset, protracted course and only moderate-hepatosplenomegaly. There is a modest increase of sphingomyelin content in the spleen and kidneys. A typical report (Anzil *et al*. 1973) was of a 14-year-old boy who was well up to 8 years. He developed epilepsy at 9 years and then followed a long illness with dementia and spasticity as the main signs. The parents were Yugoslavians and non-Jewish. No consanguinity was recorded.

The condition is highlighted by an accumulation of sphingomyelin and other lipids in the liver and spleen but normal sphingomyelinase activity. In the brain there is neuronal ballooning in the deep cortical layers, the basal ganglia, substantia nigra, reticular formation, and motor nerve nuclei. Electron microscopy shows polymorphous cytoplasmic bodies associated with axonal spheroids.

Other cases are those of Lowden *et al*. (1967) and the sibs noted by Forsythe *et al*. (1959). Two brothers with this type were mentally retarded (Sogawa *et al*. 1978). Sibs (some of mother's ancestors were Jewish) had an onset after the first two years of life (Philippart *et al*. 1969). Both had seizures, cherry red spots, cerebellar ataxia, spasticity, mental deterioration, and hepatosplenomegaly. Antenatal diagnosis has been achieved (Harzer *et al*. 1978).

All previously recorded cases of progressive supranuclear ophthal-moplegia and Niemann–Pick disease are now regarded as Type C.

Two sibs with clumsiness and intellectual decline between the age of 6 and 10 years had foamy, lipid-laden histiocytes ('sea-blue histiocytes') in the bone marrow and were thought to have Niemann–Pick type C or D (Dunn and Sweeney 1971). A similar condition in two brothers (Wherrett and Rewcastle 1969) and in nine Spanish-American children (Wenger *et al.* 1977) have been reported. All the patients had a progressive supranuclear palsy. Out of three sisters with dementia and dystonia within the first decade, one had a vertical supranuclear ophthalmoplegia (de Leon *et al.* 1969). Two brothers (Neville *et al.* 1973) with a supranuclear ophthalmoplegia, ataxia, and dementia fall into the same category. A brother and sister with an onset at 8 and 6 years of dystonia, involuntary movements, facial grimacing, gait difficulty, and a supranuclear ophthalmoplegia did not have intellectual deterioration or epilepsy (Karpati *et al.* 1977). Both had enlarged spleens.

Two cases are reported by Frank and Lasson (1985). The cardinal features were splenomegaly, psychomotor retardation, clumsiness, dystonia (in one), seizures, ataxia, and slow vertical eye movements which were limited. Sudden rapid vertical eye movements were not possible although the eyes could be made to move upward by passive flexion of the head. Both patients and their carrier relatives showed sea-blue histiocytes

The storage cells are different from those found in classic Niemann–Pick disease in their strongly positive acid-phosphatase reaction. In the two sibs reported by Hagberg *et al.* (1978) there were peculiar foam cells from those found in Gaucher and Niemann–Pick diseases. They were thought to be distinct, but they probably fall into this group. The two sisters had a splenomegaly from infancy and at two to four years developed a progressive encephalopathy. One sib probably had a supranuclear ophthalmoplegia.

It should also be noted that some patients present with juvenile dystonia alone and do not develop a supranuclear ophthalmoplegia (Frederico *et al.* 1986).

Type D. Nova Scotia variant

This was first described by Crocker and Farber (1958) in a French-Canadian Catholic subgroup in Nova Scotia. Early development is normal and neuro-logical manifestations begin in middle childhood, with death between 12 and 20 years. Diagnosis was based on the clinical picture and the presence of Niemann–Pick cells. All known cases were traced in Nova Scotia to a couple born in the late 1600s (Winsor and Welch 1978). Type D patients have shown sphingomyelinase levels in skin fibroblasts of 100 per cent, in bone marrow of 58 per cent, and in the spleen of 18 per cent. Types C and D are indistingui-shable except for the Nova Scotia ancestry in type D.

Type E. Adult form

This is non-neuropathic and will not be described.

7. Gaucher's disease

(a) Infantile type or type 2

Progressive deterioration begins between 4–6 months and the signs are those of hypertonicity, opisthotonus, choking spells, and cyanosis. The liver and spleen are enlarged. Death usually occurs before the age of 1 year (Drukker *et al.* 1970). The brain is reduced in size. Lipid-laden Gaucher cells, which are large macrophages with an eccentric nucleus, are found dispersed throughout the reticulo-endothelial system.

Whereas the chronic form of Gaucher's disease is frequently found in Ashkenazi Jews, this has not been found to apply to the acute infantile form, although Jewish ancestry has occasionally been reported (Bernstein and Shelden 1959). One infantile case in a Sephardic Jewish child was reported by Drukker *et al.* (1970). Single cases were reported by Sengers *et al.* (1975*a*), Verity and Montasir (1977), and in a black infant by Forster *et al.* (1978); identical twins were reported by Espinas and Faris (1969).

(b) Juvenile type or type 3

Six cases had an onset between six months and a year (Herrlin and Hillborg 1962). Mental retardation and a behaviour disorder, progressive dementia, generalized stiffness, and defective co-ordination occurred. Particularly noticeable was the inability to adduct the eyes. Epilepsy occurred in three. All the patients came from a small area in Northern Sweden and were related. A disabling myoclonus was present in the sibs reported by Nishimura and Barranger (1980).

Three West Indian children, the offspring of a consanguineous mating, had juvenile Gaucher's disease (Tripp *et al.* 1977). Myoclonic and generalized epilepsy, intellectual deterioration, and a defect of lateral horizontal gaze were the main neurological features.

(c) Adult type

A distinguishing characteristic of adult Gaucher's disease is the absence of neurological involvement. However, an extraordinary family is reported by Miller *et al.* (1973) in which a brother and sister had biochemically and histologically proven Gaucher's disease and three other sibs are said to have died from the same disorder. All had neurological involvement. The clinical picture was that of seizures, mental deterioration, an extra-ocular movement disorder, diffuse muscle wasting, finger, nose, and heel-shin ataxia, and a wide-based gait. Position sense and the perception of vibration were impaired. The onset of neurological symptoms occurred in the late 30s in one sib and in the 40s in the other. Hepatosplenomegaly was a feature in both.

An Ashkenazi Jewish male, who at 39 years developed a progressive myoclonic epilepsy and intellectual deterioration, was found to have Gaucher's disease (King 1975). His parents were first cousins.

Two sibs with adult-onset Gaucher's disease had predominantly psychiatric symptoms (depression, irritability, and confusion, Neil *et al.* 1979). Both had an enlarged liver and spleen. Generalized seizures, myoclonus, facial immobility, rigidity, and a cerebellar gait developed later.

An adult case with a retinopathy was reported by McKeran *et al.* (1985). The onset of splenic enlargement was in infancy but the neurological signs (mainly Parkinsonian) began in middle life.

8. Ceroid lipofuscinosis

The diagnosis is based on the demonstration of excessive amounts of auto-fluorescent lipopigment in cells of the nervous system and certain visceral cells.

The neuronal ceroid lipofuscinoses can be divided into various subtypes based on clinical criteria. The group is also known as Batten's disease.

(a) Infantile—Santavuori.

(b) Late infantile—Jansky–Bielchowsky.

(c) Juvenile—Spielmeyer–Vogt or Spielmeyer–Sjögren.

(d) Adult—Kuf's

In an attempt to add histological criteria to the clinical classification, Carpenter *et al.* (1977) suggested the following electron microscopic differences: the late infantile group have curvilinear membranous inclusions of tubular shape whereas the infantile group has granular osmophilic deposits (rarely, adults have them too). In the juvenile form, granular or granulo-vesicular osmophilic inclusions are seen, some of which are crystalline and resemble human fingerprints.

Infantile ceroid lipofuscinosis (Santavuori type)

The infantile type of Batten's disease has an onset at about 1 year of age with slowness of development, ataxia, loss of vision, and myoclonic jerks (Santavuori 1973). Most infants develop normally until 8 months of age. Santavuori *et al.* (1974) reviewed the clinical features. Mental retardation was noted at 12–18 months, and vision started to deteriorate from 12 months.

All were virtually blind at 2 years. All but one of their patients had myo-clonus. Acne, hirsutism, and precocious puberty appear in some of the patients. Microcephaly occurred in all and the majority (all except four) had dystophic maculae and optic atrophy. This type of Batten's disease is common in Finland, but Haltia *et al.* (1973) have reported Swedish patients with the disease. The clinical picture did not differ in the Finnish and Swedish patients, and the latter possibly stemmed from Finnish ancestors in the province of Ostrobothnia where the disease is particularly prevalent. Jervis and Donahue (1975) described a 6-year-old boy with profound mental retardation and quadriplegia. His parents were German and non-Jewish. The onset was at 18 months.

The EEG is of diagnostic importance. It shows a generalized slowing with abundant theta and delta-activity which initially is of enlarged amplitude. Thereafter, there is a marked diminution in amplitude the record becoming almost isoelectric. Seen at 3–4 years this is very characteristic and it is often combined with a marked loss of amplitude in the ERG.

At post-mortem there is a selective loss of cortical neurons with large amounts of neuronal lipopigment. Lymphocytes and fibroblast may show granular deposits before the first sign appears (Haynes *et al.* 1988). It is possible that some cases included under the diagnosis of Alper's disease are examples of the infantile ceroid lipofuscinosis. Under the title 'An infantile form of neuronal "storage" disease', Pampiglione and Harden (1974) described two sib pairs with features of the Santavuori infantile type of Batten's disease. The early milestones were unremarkable, but at 12–21 months, a loss of skills occurred. Optic atrophy was an early feature, and spasticity or jerkiness of the limbs developed between 2 and 3 years. A vegetative state was reached by 3 years, and death by 9 years.

The progressive disappearance of phasic activity on electroencephalogram, and absent visual-evoked responses is quite distinct from young children with other types of amaurotic family idiocy. This type of ceroid lipofuscinosis is rare in Britain.

Late infantile ceroid lipofuscinosis (Jansky–Bielchowsky type)

This is characterized by an onset between 1 and 4 years of convulsions, myoclonic jerks, mental deterioration, cerebellar signs, optic atrophy, and blindness. The visual loss may occur relatively late in the disease. Death occurs between within 2–6 years of onset.

Single cases are reported by Gonatas *et al.* (1968), Duffy *et al.* (1968), Herman *et al.* (1971), and sibs by Richardson and Bornhofen (1968); Andrews *et al.* (1971) described two sibships.

Prenatal diagnosis has been successful despite the absence of knowledge about the specific enzyme deficiency. This has been achieved by studying uncultured amniocytes and finding on standard electron microscopy, curvilinear bodies (MacLeod *et al.* 1985).

An early juvenile form of Batten's disease

Five cases of Batten's disease with onset between five and eight years of ataxia and drops attacks had a late onset of visual loss (Lake and Cavanagh 1978). The cases differ from the juvenile form in that vacuolated lymphocytes are absent and the VER shows large responses. Inheritance is autosomal recessive.

Juvenile ceroid lipofuscinosis (Spielmeyer–Sjögren or Batten's disease)

The onset is typically between the ages of 5 and 10 years. Visual loss, intellectual deterioration, a retinal pigmentary degeneration, extrapyramidal signs, and spasticity develop. Seizures occur late and the children might survive 10 years and more after the first symptoms (Gordon *et al.* 1972). In contrast to the late infantile form, dementia and visual problems are early features. In a series of 13 patients reported by Elze *et al.* (1978), the initial symptoms were visual in all of the patients and the neurological state was characterized by a progressive hypertonic–hypokinetic syndrome. Pyramidal signs were not a feature but all had a cerebellar ataxia. When the children were first seen by an ophthalmologist, seven were diagnosed as having macular degeneration and four as having retinitis pigmentosa. Motor conduction velocity was reduced in one-third of patients.

The sibs reported by Michielsen *et al.* (1984) both with visual loss due to a pigmentary retinopathy followed by mental deterioration and seizures. Both seemed to have the 'sick sinus' syndrome and had cardiac problems. The series contained one pair of sibs. Twenty-three patients from 16 families were recorded by Lyon (1975). Consanguinity was reported in a single case of Nakano *et al.* (1979), and sibs by Gadoth *et al.* (1975).

Adult ceroid lipofuscinosis (Kuf's type)

This is also called 'Adult amaurotic idiocy'. This category has an onset in adulthood and presents as a cerebellar or extrapyramidal syndrome with only moderate dementia and no visual symptoms (Kufs 1925).

Dominant

A pedigree reported by Boehme *et al.* (1971) had members with a cerebellar ataxia, myoclonic jerks, and a moderate dementia. No sensory disturbance or retinopathy were found. Azurophilic hypergranulation of polymorphonuclear neurophils was found in all living patients and was absent in non-carriers. The age of onset was in the 30–40-year range and *grand mal* epilepsy was the first sign. Four generations were affected.

Another dominant pedigree was reported by Dumon-Radermecker (1965).

The clinical picture was that of a late-onset dementia and a cerebellar syndrome. The pedigree of Brodner *et al.* (1976) was said to be dominant but no details were given. Single cases have been reported by Fine *et al.* (1960), Pallis *et al.* (1967), and Kornfeld (1972).

A dominant pedigree with incomplete penetrance is reported by Maere and Muyle (1938) and different branches of the same family by Van Bogaert and Klein (1955) and by Van Bogaert (1961). In the family of Ferrer *et al.* (1980; a dominant with reduced penetrance) those affected had dementia and involuntary movements of the face and neck.

Recessive

Chou and Thompson (1970) described a man who was well up to the age of 17 years when he developed seizures, mental deterioaration, slurred speech, and athetoid-like movements. His sister died from a similar condition, and another sister had akinetic episodes, myoclonic jerks, and cerebellar disease. The parents were first cousins. The neuropathological findings on the proband were characteristic of the adult type of cerebral lipidosis. Sibs reported by Kraft (1970) were said to have Kuf's disease. Diagnosis was confirmed by cortical biopsy in one. The clinical picture was of a diffuse cerebral deterioration. Three brothers with adult onset (30s) ceroid lipofuscinosis were reported by Vercryssen *et al.* (1982). This is the same family as Dom *et al.* (1979). Diagnosis remains a problem. Somatosensory evoked potentials were found by Vercruyssen to be abnormal.

A family is reported by Iseki *et al.* (1987) in which the pre-morbid diagnosis was familial amyotrophic lateral sclerosis, but after autopsy the condition was clearly adult neuronal ceroid lipofuscinosis. Two sibs born to consanguineous parents had the disease. The onset was in their twenties with dementia and progressive muscle atrophy. Pyramidal signs were additionally present. A further Japanese sibship was reported by Tobo *et al.* (1984).

Counselling

It can be seen that adult onset ceroid lipocuscinosis is heterogeneous—both dominant and recessive pedigrees are reported. Only pedigree analysis allows a differentiation between the (at least) two types and this is unhelpful in the absence of a family history.

Pigment variant of lipofuscinosis

This term was first introduced by Seitelberger and Simma (1962) and up to 1978 there have been six reports (Jervis and Pullarkat 1978). Three of the six have been Ashkenazi Jews and this includes affected sibs. The age of onset was from one to 31 years. All showed mental and neurological deterioration (some have ataxia, weakness, and tremor). The pigment resembled that found in Hallervorden–Spatz disease but no neuroaxonal dystrophy or unusual glial cells were present. The cases are those of Jervis (1952), Zeman and Scarpelli (1958), Moschel (1954), Simma (1957), and Jakob and Kolkmann (1973).

Ceroid lipofuscinosis storage disease and sea-blue histiocytes with posterior column dysfunction

A large pedigree in which at least five members had sea-blue histiocytes in the bone marrow was ascertained because two sisters, from early adulthood, had a progressive gait disturbance due to defective position sense (Swaiman *et al.* 1975). Reflexes were brisk. Some members had a postural tremor of the hands and one had no abnormal neurological signs.

Unusual clinical manifestations in ceroid lipofuscinosis

Single cases in the late infantile age range with chorea as a prominent symptom have been described (Dal Canto *et al.* 1974). The case reported by Elfenbein (1968) had a juvenile onset, normal fundi, and no myoclonus, whereas dystonia, dementia, ataxia, seizures, and a paralysis of upward gaze were prominent features.

9. Mucopolysaccharidosis

Hurler's disease (mucopolysaccharidosis type I-H)

Mental and physical retardation have their onset during the first year of life, but the characteristic clinical picture is often only recognizable during the second to third years. It is then that the coarse facial appearance, characterized by macrocephaly, thickened skin, flat nasal bridge, prominent tongue and lips (often resulting in an open mouth), occur. Corneal opacities develop, growth is stunted, and the bent appearance due to kyphoscholiosis and protuberant abdomen is seen. Spastic quadriparesis due to C1–C2 subluxation has been reported (Brill *et al.* 1978). Liver, spleen, and heart are enlarged, and the bone dysplasia and restricted joint mobility complete the picture. Inheritance is as a recessive and sibs are reported by Lowry and Renwick (1971).

Occasionally, a patient with classical Hurler's might be found to have intellectual functions which are unusually well preserved (Watts *et al.* 1986). These authors postulate that the occasional patient has low residual alpha-L-iduronidase activity in neuronal cells alone.

Antenatal diagnosis is available. The deficient enzyme is alpha-L-iduronidase.

Scheie disease (mucopolysaccharidosis type I–S)

The absence of Hurler-like facial features and mental retardation differentiates type I–S from Hurler's syndrome. Originally designated type V it was subsequently found to share an alpha-L-iduronidase deficiency with Hurler's syndrome and the two entities are thought to be caused by different alleles at the same locus. Limitation of joint movement, corneal opacities, aortic incompetence, and moderate bone changes are present. There is often an increase in body hair. Enlargement of the liver, and less often the spleen, occurs. Affected sibs are reported by Scheie *et al.* (1962), Koskenoja and Suvanto (1959), and Soodan and Goel (1976).

Antenatal diagnosis is available.

For the Hurler–Scheie compound see page 455.

Hunter's syndrome (mucopolysaccharidosis type II)

Clinical heterogeneity has been recognized in Hunter's syndrome, and on this basis it has been divided into a mild and a severe form. The main criteria used to differentiate between the two are the severity of the mental retardation and the course of the disease. In the severe form, retardation is marked by the age of 3 years and death from congestive heart failure or respiratory infection occurs before the age of 15 years. Typical X-linked pedigrees are those of Beebe and Formal (1954), Millman and Whittick (1952), and Nja (1946). In the mild type, survival beyond 20 years and relatively satisfactory mental development are found. Six cases of the mild form in five generations are reported by Hobolth and Pedersen (1978). Other cases have been reported by Lichtenstein *et al.* (1972). Biochemically, iduronate sulphatase activity in serum and in cultured cells is low.

In the majority of instances, the two types breed true. An exception to this is the report of Yatziv *et al.* (1977) in which both severe and mild types occurred together in two separate sibships. The authors stressed that despite the previous acceptance of presumed clinical heterogeneity of the two types of mucopolysaccharidosis II, biochemical heterogeneity had not been proven. These two families emphasize the need for caution in counselling and indicate the possible role of modifying genes or environmental factors. The clinical picture is similar to that of Hurler's disease, but less severe. Corneal opacities are unusual and this is the most useful clinical feature in distinguishing it from Hurler's syndrome.

To date there is no totally reliable carrier-detection test. Heterozygosity has been established by clonal analysis of cultured fibroblasts (Migeon *et al.* 1977) or by hair-root analysis (Fluharty *et al.* 1977), but neither method is totally reliable.

Carrier detection using enzyme assays can, using serum and lympho-cytes, detect 86 per cent of obligate carriers. (Zlotogora and Bach 1984). Pre-natal diagnosis has been achieved on many occasions (Archer *et al.* 1984).

Mossman *et al.* (1983) reported a girl with clinical features of Hunter syndrome. On chromosome analysis she had an apparent balanced transloca-tion with the breakpoint on the X at q27 or q26. The localization was supported by linkage analysis in the study of Upadhyaga *et al.* (1985).

Hunter's syndrome in females

Two unrelated girls, one with Ashkenazi Jewish parents (one of whom has adult Gaucher's disease) and the other whose parents were second cousins, had profound iduronate sulphatase deficiency (Neufeld *et al.* 1977). An autosomal recessive type of Hunter's syndrome remains a possibility. Chromosome analysis was normal.

Sanfillipo A and B (mucopolysaccharidosis type III)

In the Sanfillipo syndrome, early development is normal but by the age of two to four years, rapid intellectual and motor deterioration begin. Speech is lost, the gait becomes ataxic, muscular atrophy and mild joint stiffness develop. Those affected are often tall initially. The coarse facial features of the mucopolysaccharide disorders are present by the third year and certain of the osseous abnormalities, such as thickened calvareum, and vertebral body changes, occur, but are mild. The vertebral bodies are convex but do not have the anterior 'hook' as seen in Hunter's syndrome. The head circumference is often increased and seizures occur by the sixth to ninth year. The liver and spleen are enlarged and the cardiac ventricular system is dilated. Corneal clouding is not a feature. Sibs (three) are reported by Greenwood *et al.* (1978).

The clinical severity varies considerably. Gordon and Thursby-Pelham (1969) described a child who reached primary school, whereas some patients die before their tenth year. Most of the variation is interfamilial but in the Van der Kamp *et al.* (1976) paper, cases in two inter-related families had marked differences. In one the onset was late and the patient was still able to communicate at 24 years, whereas another lost speech at the age of six years. A similar sibship is reported by Andria *et al.* (1979) in which two severe and one mild case occurred in the same family (type B).

Biochemically, there are two types of Sanfillipo's disease, but clinically they cannot be distinguished from one another. Sanfillipo A has a deficiency of heparan sulphate sulphamidase and type B of alpha-N-acetylglucos-aminidase. In type A it is heparan sulphate that accumulates in cultured skin fibroblasts and peripheral blood leucocytes. Sibs are reported by Farriaux *et al.* (1974) and Wallace *et al.* (1966).

Three male sibs with Sanfillipo A were reported by Wisniewski *et al.* 1985). They demonstrated ceroid lipofuscine storage in the brain similar to that found in Batten's disease.

Antenatal diagnosis has been achieved in Sanfillipo A (Harper *et al.* 1974).

Sanfillipo C

Two related Greek families had members with typical Sanfillipo syndrome. When fibroblasts from those affected were cultured with fibroblasts from patients who have Sanfillipo A or Sanfillipo B, the metabolic defect was corrected. Inheritance is recessive (Kresse *et al.* 1976).

Morquio's syndrome (mucopolysaccharidosis IV)

This spondyloepiphyseal dysplasia syndrome is unlikely to be seen in neurological practice. The main clinical features are short stature, normal intelligence, hypermobile prominent joints, short neck, hypoplasia of the odontoid, and a pectus carinatum deformity. Outside of the skeletal system, corneal clouding, deafness, impaired aortic valve function, and a thin enamel layer on the teeth are features. For the neurologist it is important to note that atlantoaxial subluxation is a potential hazard (Beighton and Craig 1973). Biochemically, deficiency of chondroitin 6-sulphate sulphatase is responsible (Matalon *et al.* 1974). Keratan sulphate excretion in the urine is excessive.

A milder form was first described by Dale in 1931 and more recently by Arbisser *et al.* (1977) and Groebe *et al.* (1980).

Sibs are reported by Morquio (1929)—the parents were consanguineous —and in a large study Gadbois *et al.* (1973) found 48 cases in 27 kindred in Quebec.

Maroteaux–Lamy syndrome (mucopolysaccharidosis type VI)

These patients with short stature have Hurler-like skeletal abnormalities, joint movement limitation, and corneal opacities but a relatively mild degree of facial coarseness—indeed, this might be hardly noticeable. The liver and spleen are enlarged, the heart is affected, intelligence is near normal, and survival into late childhood and early adulthood is common. The urine contains excess mucopolysaccharides with a preponderance of dermatan sulphate and aryl sulphatase B activity is deficient in skin, liver, kidney, spleen, and brain (Stumpf *et al.* 1973). Inheritance is autosomal recessive. Neurological complications are rare. Two brothers with the mild variant (type B) are described by Pilz *et al.* (1979).

Sly syndrome (mucopolysaccharidosis type VII)

This was first recognized by Sly *et al.* (1973) in a patient with short stature, a large liver and spleen, skeletal deformities including a thoraco-lumbar gibbus, possible mental retardation, and beta-glucuronidase deficiency. The parents and some sibs were shown to have intermediate enzyme levels suggesting recessive inheritance. Onset is within the first decade and most cases reported after the initial one have had cloudy corneas. A further patient is reported by Beaudet *et al.* (1975). In a review of the subject by Danes *et al.* (1977) it was noted that of the seven families reported to date, one had beta-glucuronidase deficiency, three others showed deficiency of alpha-L-iduronidase, and three others no deficiency at all.

Di Ferrante syndrome (mucopolysacchariodosis type VIII)

This syndrome had the clinical and biochemical features of both Morquio and Sanfillipo syndrome. A 5-year-old male with short stature, mental retardation, coarse hair, large liver, hypoplasia of the odontoid process, but only mild bony changes was reported by Ginsberg *et al.* (1977). The enzyme involved is glucosamine 6-sulphate sulphatase and a partial deficiency in both parents in Ginsburg *et al.*'s (1977) case suggests autosomal recessive inheritance.

Genetic heterogeneity

That identical or similar phenotypes might be the result of diverse mutations at separate loci has been suggested for many genetic conditions. The mucopolysaccharidoses are subdivided on the basis that each has a specific enzyme deficiency, but the classification was revised when it was found that the two clinically distinct entities Hurler and Scheie syndrome shared the same enzyme deficit. At that stage it was not evident whether more than one locus coded for the enzymes or whether two mutant allelic genes at the same locus were involved. If the two genes that were responsible were allelic, then compounds of the two would be expected to occur. A compound—that is a double heterozygote of these two mutant alleles at the same locus—is theoretically possible, and there have been various claims for this.

The initial report was that of McKusick *et al.* (1972) who described seven possible cases. The patient described by Winters *et al.* (1976) was short and mentally retarded with a coarse face, hypertelorism, flat nasal bridge, corneal clouding, restriction of movement of neck and arms, a bilateral carpal tunnel syndrome, and multiple bony abnormalities. The phenotype did not resemble either Hurler or Scheie syndromes. The patient did not present to her doctor until she was 25 years old and then with paranoia. Two

Japanese brothers, aged 18 and 21 years, had Hurler-like clinical features and a normal intelligence (Kajii *et al.* 1974). Both were short, had corneal opacities, hearing defect, cardiac enlargement, clawed hands, and multiple bony defects. The two sibs described by Horton and Schimke (1970) were also of normal intelligence.

All of these cases are thought to be compound heterozygotes.

10. Sialidosis

Sialidase deficiency has been reported in mucolipidosis I and this has necessitated, re-classifying it as a sialidosis. The sialidoses have the following clinical subdivisions:

Type I.   Normosomatic (cherry red spot—myoclonus syndrome, normal beta-galactosidase)

Type II.   Dysmorphic group—(a) infantile onset usually normal galactosidase; (b) juvenile onset, absent beta-galactosidase.

Note: the subdivision into those with both galactosidase deficiency and neuraminidase deficiency (localized to chromosome 20) and those with neuraminidase deficiency alone (localized to chromosome 10; Mueller *et al.* 1984).

Type III.   Salla disease with elevate free sialic acid

Type IV.   Sialuria and sialic acid storage disease (acute and severe). Free sialic acid accumulation in tissue and urine (possibly separate from III).

Sialidosis type I—normosomatic

An adult developed at the age of 29 years proximal weakness of upper and lower extremities with occasional myoclonic jerking, causing difficulty with fine motor movement (Thomas *et al.* 1978*a*). No obvious abnormality in sensation nor in mental status was observed. A cherry red macular spot was present. A brother died of a similar illness. In particular, no Hurler-like features were observed in either sibling. Fibroblast culture showed cells with markedly reduced sialidase (alpha-N-acetyl neuraminidase) activity.

A black girl with an onset at six years of progressive ataxia, and focal and *grand mal* seizures, had a cherry red spot (Justice *et al.* 1977), and the 17-year-old girl with progressive myoclonus and a cherry red spot reported by Goldstein *et al.* (1974) falls into this category. Vacuolated foam cells were seen in bone marrow and liver biopsies. Other single cases are those of Hambert and Petersen (1970), and Tittarelli *et al.* (1966).

Rapin *et al.* (1978) reported Italian sisters with an onset at eight years of an incapacitating, painful, slowly progressive myoclonus often precipitated by movement, touch, or sound but with relatively normal or low normal intelligence. Absent were corneal clouding, joint stiffness, enlarged liver and spleen, or skin lesions. Visual loss was progressive. Consanguinity was denied.

Anderson *et al.* (1958) published the clinical association of cherry red spots and myoclonus in two sisters. The onset was in the teens. A white halo surrounded the red spots.

In the pedigree reported by Durand *et al.* (1977) two sibs had bilateral cherry red spots, punctate opacity of the lens, and only minimal neurological signs and symptoms. Sibs were reported by O'Brien (1978).

Thomas *et al.* (1979) reported brothers with this condition. Their parents were first cousins.

Sialidosis type II—infantile

This neurodegenerative condition (Spranger *et al.* 1977) is characterized by Hurler-like features with short stature, coarse facies, gingival hyperplasia, wide-spaced teeth, large tongue, thoracic kyphosis, cherry red macular spots, fine corneal opacities, and mental retardation. The neurological changes are those of muscle hypotonia, myoclonus in 75 per cent of cases. (Lowden and O'Brien 1979), cerebellar ataxia, and choreoathetosis. Excessive amounts of sialic-acid-containing compounds are found in fibroblasts, leucocytes, and urine. Cultured fibroblasts show diminished alpha-N-acetylneuraminidase (sialidase). Beta-galactosidase is usually normal. Ultrastructural changes resemble GM_1 gangliosidosis (Yamano *et al.* 1985).

An increase in sialic acid and a specific deficiency of acid neuraminidase are the biochemical hallmarks of the disorder. Cases are reported by Spranger *et al.* (1968), and Kelly and Graetz (1977). All described cases are single with unrelated parents, but intermediate values of the enzyme in the parents of these cases is suggestive of recessive inheritance.

The differentiation between the types is imperfect. A patient with consanguineous parents was clinically of type II (Winter *et al.* 1980). He manifested a coarse facies, mental retardation, dysostosis multiplex, and had myoclonus and a cherry red spot. The onset was before the first year of life and the beta-galactosidase was normal (suggestive of the infantile form). The patient, however, survived into adult life, and did not have the enlarged liver and spleen that is characteristic of the infantile form.

The infantile form of sialidosis type II is a rare disease. A patient with this condition and 21-hydroxylase deficiency was reported by Oohira *et al.* (1985) in a Japanese child—the offspring of first cousin parents. A severely affected infant was reported by Aylsworth *et al.* (1980). The infant had ascites, hepatosplenomegaly, the nephrotic syndrome, growth failure, inguinal hernia, dysostosis multiplex, and a pericardial effusion. He died at

24 months. Neuraminidase was deficient, but beta-galactosidase was not done. A 3-month-old Japanese infant (Okada *et al.* 1983) had severe psychomotor delay, a coarse face, a big liver and spleen, dysostosis multiplex, bilateral inguinal hernia and vacuolated lymphocytes. There were numerous vacuoles in ballooned glomerular and tubular epithelial cells. This is also called nephrosialidosis.

Type II—Juvenile Goldberg's disease

Lowden and O'Brien (1979) summarize the clinical features of this type as an onset between 8 and 15 years of joint stiffness with coarse facial features and decreased visual acuity. Hepatomegaly is found only in the infantile group, whereas mental retardation is found in both the juvenile and infantile types.

Note: Most type II patients have neuraminidase and beta-galactosidase deficiency. The condition is characterized by a juvenile onset, coarse face, corneal clouding, angiokeratoma, ataxia, and myoclonus. There are vacuolated lymphocytes and foam cells in the bone marrow (Sakuraba *et al.* 1983). The condition affects mostly Japanese and has a juvenile or adult onset, slow progression and long survival (Loonen *et al.* 1984).

Three Japanese cases (Suzuki *et al.* 1977) had a similar clinical course and all three had macular cherry red spots, myoclonus, a cerebellar ataxia, and corneal opacities. All three had gargoyle-like facies, two were deaf, and two had generalized seizures. The parents in two of the three cases showed normal activities of beta-galactosidase. The deficiency of this enzyme is not thought to be the primary defect but is secondary to a more fundamental biochemical defect. Other Japanese cases are reported by Kuriyama *et al.* (1975), sibs by Fukinaga *et al.* (1976), and single cases by Orii *et al.* (1972), Okada *et al.* (1978), and Koster *et al.* (1976).

A late juvenile form (onset at 18 years) is reported by Miyatake *et al.* (1979).

Not all cases are Japanese. The condition was first reported by Goldberg and co-authors (1971). The proband was of Mexican ancestry and the product of a consanguineous marriage. He sat at 6 months, walked at 16 months, and talked at 2 years. At the age of 2–3 years, he was thought to be a slow learner. His facial appearance was moderately coarse. At 8 years, he was noted to have a cherry red spot at the macula and at 10 he developed generalized seizures. A dental X-ray showed multiple impacted teeth with thin enamel. The vertebral bodies showed beaking. Vacuolated lymphocytes were not seen. He was at 12 years moderately mentally retarded, but his half-sister was well until 13 years. At 22 she could not work, had frequent seizures, and looked like a gargoyle with a kyphoscoliosis and keel chest. She died at 22 from left ventricular heart failure. Beta-galactosidase was deficient in fibroblasts (15 per cent of normal).

Type III—Salla disease

Sialic acid (N-acetylneuraminic acid) is a constituent of numerous glyco-proteins and glycolipids. Salla disease (most, but not all cases have Finnish origin) is due to a deficiency of oligosaccharide-specific neuraminidase which leads to lysosomal storage of sialic acid and oligosaccharides (Renlund *et al.* 1983; Baunkotter *et al.* 1985). Clinically, it is characterized by mental retardation, myoclonus, ataxia (in some) increased muscle tone, and delayed walking and speech. There are coarse facial features and small stature. Onset is usually in the first year of life and deterioration occurs between the ages of 16 and 30 years, mainly in the form of athetosis, spasticity, and loss of speech. Three sibs reported by Wolburg-Buchholz *et al.* (1985) were without Finnish ancestry. They lost motor control of the head before the age of 1 year and both were still alive, but demented at 9 and 17 years. Another sib had died at 8 years. Lysosomal vacuolation was found in skin, muscle, nerve, and liver.

In a review of a number of patients with Salla disease (Renlund 1984) the diagnosis rested on an increased amount of free sialic acid in the urine. All patients had delayed motor development, ataxia, poor speech, and mental retardation. None had hepatosplenomegaly.

A 5-year-old boy reported by Echenne *et al.* (1986) was a native of France. He presented with hypotonia and delayed milestones and did not develop coarse facial features (although his voice was hoarse). The diagnosis was made by conjunctival biopsy and by finding a 10-fold increase in free sialic acid in the urine.

Type IV—Severe infantile neuraminic acid storage disease

There are biochemical similarities of Salla disease with infantile neuraminic acid storage disease, but in this condition, deterioration is much earlier and more rapid (see Baunkotter *et al.* 1985). The coarse facial features are promi-nent as is the organomegaly. The cases of Tondeur *et al.* (1982), Stevenson *et al.* (1983) and Gillan *et al.* (1984) fall into this group. In this group is the Caucasian boy described by Paschke *et al.* (1986). There was an early onset of retardation with coarse facial features, pronounced visceromegaly and vacuolated lymphocytes. Occasionally, there are punctate areas of calcifica-tion in tarsal bones, proximal femoral epiphyses, and in the spine. In the cases reported by Tondeur infants had trunk hypotonia, but spasticity in the lower limbs. Infections are frequent and fetal hydrops might be a prominent feature.

A newborn was reported with non-immune hydrops and hepatospleno-megaly (Beck *et al.* 1984). She died at 6 months. The diagnosis was confirmed by enzyme assay, beta-galactosidase was normal.

Similar cases are those of Laver *et al.* (1983) and Johnson *et al.* (1980, four cases).

11. Mucolipidosis

There are three known mucolipidoses:

Type II.   I-cell disease.
Type III.   Pseudopolydystrophy (pseudo-Hurler's disease).
Type IV.

Type II mucolipidosis—I-cell disease

In the neonatal period (Spritz *et al.* 1978) growth failure, multiple areas of bone destruction in the metaphyseal segments of the long bones, coarse wizened facies, bulbous nose, thick doughy skin, and downy hair over most of the body are the main features. Limitations of joint mobility, gingival hypertrophy, hypotonia, long thin fingers, hepatomegaly, and microcephaly complete the clinical picture. An elevated activity of multiple lysosomal enzymes in the serum is the biochemical hallmark of this disease.

Some of the clinical features are similar to Hurler's syndrome, but with normal excretion of mucopolysaccharides. In addition, the marked widening of the long bones occurs early and whereas Hurler's syndrome is usually diagnosed towards the end of the first year of life I-cell disease presents in the first six months (Milla 1978). The term I-cell disease, or inclusion cell disease, refers to the cytoplasmic inclusions seen with the phase-contrast microscope in skin fibroblasts (Leroy *et al.* 1971).

Single cases were reported by Tondeur *et al.* (1971) and Wiesmann *et al.* (1971).

Two sibs are described by Whelan *et al.* (1983). Infants seldom survive more than 2 years. It is also interesting to note that both sibs had punctate calcification around the calcaneus and knees.

Prenatal diagnosis has been achieved by Matsuda *et al.* (1975) and Aula *et al.* (1975). High lysosomal hydrolase activity in amniotic fluid and low activity in cultured fetal cells were significant (confirmed by examination of the fetus).

Type III

This is characterized clinically by mild mental retardation, corneal opacity, joint stiffness, valvular heart disease, short stature, and minimal to moderate coarseness of facial features (Taylor *et al.* 1973). The severity of the joint involvement differentiates type III from type II. The clinical picture resembles that of Maroteaux—Lamy disease (MPS type VI).

Twelve cases reviewed by Kelly *et al.* (1975) included four sib pairs. There was no consanguinity and nine of the parents were of Irish extraction. The disorder became manifest at 2–4 years of age with joint stiffness, and the hands were clawed by the age of 6 years. Mild coarseness of the facial features then developed. X-rays at that stage showed dysostosis multiplex. In the 12 patients a slit-lamp examination of the cornea showed fine peripheral infiltrates, but these were not as severe as in mucopolysaccharidosis type I and VI. Hand muscle wasting and sensory changes were common. School performance was fair to poor.

There is a deficiency of nearly all the lysosomal enzymes, although some might be totally absent (beta-galactosidase and alpha-L-fucosidase), others are partially reduced. The I-cell phenomenon has also been demonstrated in this group (Thomas *et al.* 1973). The mucolipidoses type II and III share a reduced activity of lysosomal enzymes yet they are clinically different. The only biochemical difference is that the residual acidity of beta-galactosidase is three to five times greater in the patients with type II than in type III, suggesting that beta-galactosidase may play a role in the pathophysiology of these conditions (Leroy and O'Brien 1976). Several sibships have been reported (Robinow 1974; Starrenveld and Ashenhurst 1975).

Mucolipidosis type IV

This is characterized by corneal opacities in infancy, full rather than coarse facial features and psychomotor retardation (Kohn *et al.* 1977). Most cases thus far have been Ashkenazi Jews. The enzyme deficit is unknown but multiple cytoplasmic storage bodies can by demonstrated by electron microscopy.

Antenatal diagnosis, based on inclusion bodies in amniotic cells, has been attempted and was found to be positive (Kohn *et al.* 1977).

A typical case report is that of Berman *et al.* (1974) who describes corneal clouding in a male Jewish infant. His liver was minimally enlarged and at five months mild motor retardation was noted. Grossly abnormal storage cells were found in the liver, conjuctiva, and cultured fibroblasts. No abnormal bone, visceral, or facial features were present. Mucopolysaccharides were not present in the urine but an unusual storage material was noted in bone marrow histocytes.

The 23-year-old male described by Newell *et al.* (1975) falls into this group. The onset of mental retardation was before one year and corneal clouding was noted at 2 years. The neurological picture consisted of spasticity, ataxia, optic atrophy, and retinal degeneration. There were no skeletal or facial changes.

A pathological examination of the eye in a 23-year-old male with this condition (Riedel *et al.* 1985) showed early cataracts, marked outer retinal degeneration, and optic atrophy.

These cases differed from the sialidoses in the absence of obvious skeletal manifestations, and from I-cell disease where retardation and skeletal changes are noticed early. In type III (pseudo-Hurler polydystrophy), corneal clouding is mild and is usually only recognized from the age of three years onwards.

12. Mannosidosis

This condition is characterized by a low activity of alpha-mannosidase and the accumulation of mannose rich oligosaccharides in nervous tissue, liver, spleen, leucocytes, and urine (Kistler *et al.* 1977). Clinically, the retarded patients have a mild Hurler-like appearance with gingival hypertrophy. Some have recurrent infections, hearing loss, corneal clouding, and mild liver and spleen enlargement. Most patients are diagnosed between two and 10 years. In contrast to many of the storage diseases, the course is relatively mild. Sibs have been reported by Loeb *et al.* (1969), Norden *et al.* (1973), Autio *et al.* (1973), and Vidgott *et al.* (1977). Up to 1976, about 23 reports had appeared (Desnick *et al.* 1976).

A mild variant is reported by Warner *et al.* (1984). At 15 years the patient had an I Q just within normal limits, a mild hearing deficiency and little else. She had had an enlarged liver, but that was no longer present. Foam cells were found in the marrow. Mannose containing oligosaccharides were excreted in the urine and detected with high performance liquid chromatography.

13. Fucosidosis

This neurovisceral storage disease was first described, by Durand *et al.* (1966), in two sibs with consanguineous parents. The clinical picture was that of a Hurler phenotype. There was an absent alpha-fucosidase resulting in an accummulation of glycolipid material.

Different phenotypes have been noted by Patel *et al.* (1972*b*) and Schafer *et al.* (1971), but the usual clinical picture is an onset within the first two years of life of a progressive neurological deterioration with spasticity and dementia. Hepatosplenomegaly, vacuolized lymphocytes, angiokeratoma corporis diffusum, cardiomegaly, thick skin, and abundant sweating have all been described.

Prenatal diagnosis was attempted by Poenaru *et al.* (1976). Tests were negative and the fetus was unaffected.

A single case was reported by Durand *et al.* (1967). The parents in Durand's family originated from the region of Calabria in Southern Italy. A second set of two affected sibs, again with consanguineous parents, came from the same village (Borrone *et al.* 1974), and Romeo *et al.* (1977) have shown that the sibships are related. Another Calabrian family was reported by the latter authors, but it has been difficult to prove common ancestry with the other pedigree.

Fucidosis type II

Two sibs developed normally until the second year of life when psychomotor delay was noted for the first time (Kouseff *et al.* 1976). The skin lesions developed in one sib at six months and in the other between 4 and 5 years of age. In four other patients with type II, the skin lesions developed between 4 and 5 years. It is suggested that type II patients have angiokeratoma, but milder retardation and neurological deficit, longer survival, and normal salinity in sweat. However, both the severe and mild forms have been found in different branches of the same family (Romeo *et al.* 1977). A single case with onset at 7½ years had no skin lesions. In this patient there was a reduction of both alpha-L-fucosidase and arylsulphatase A.

14. Farber's disease (ceramidosis or lipogranulomatosis)

Characterized by joint swelling, subcutaneous nodules, and growth and psychomotor retardation. About 17 patients have been reported (Dulaney *et al.* 1976). Sibs are reported by Farber *et al.* (1957) and Amirhakimi *et al.* (1976). Prenatal diagnosis has been achieved.

B. Leucodystrophies and related disorders

1. Krabbe's disease
2. X-linked leucodystrophy
   (a) Adrenoleucodystrophy
       (i) Adrenomyeloneuropathy
       (ii) Infantile sudanophilic leucodystrophy
   (b) Pelizaeus–Merzbacher
       Type I: classic
       Type II: connatal
       Type III: transitional
       Type IV: Lowenberg and Hill
       Type V: atypical
       Type VI: Cockayne
3. Metrachromatic leucodystrophy
   1. Late infantile
   2. Juvenile

    3. Adult

    4. Austin variant

4. Alexander's disease

5. Glial and glioneuronal degeneration
    (a) Spongy degeneration of Van Bogaert or Bertrand/Canavan
    (b) Alper's disease
    (c) Glioneuronal degeneration—including the cerebrohepatic degenerations

6. Menkes disease

7. Hallervorden–Spatz

8. Infantile neuro-axonal dystrophy

9. Leigh's disease

10. Reye syndrome

1. Krabbe's disease (globoid cell leucodystrophy) (Fig 19.1)

First described in 1916 by Krabbe, it was Collier and Greenfield (1924) who used the word 'globoid' for the typical phage cells found in white matter. Similar changes are now known to occur in peripheral nerve myelin (Lake 1968; Joosten *et al.* 1974). It is believed that the globoid cell material is derived from myelin breakdown and that the underlying metabolic defect is related to an absence or severe reduction of a beta-galactosidase (cerebrosidase). This substance can be measured in leucocytes, cultured skin fibroblasts, and cells derived from amniotic fluid.

    Hagberg (1984) and Loonen *et al.* (1985) suggest the following subdivision (data combined).

(a) Early infantile—classic irritative—hypertonic, onset <6 months

(b) Late infantile—onset 6–18 months

(c) Late onset 1½–15 years

(a) Early infantile

There is an onset at about 4–6 months, characterized by progressive mental and motor deterioration, tonic seizures, optic atrophy, deafness, spasticity, nystagmus, and bouts of fever. A peripheral neuropathy with the slowing of nerve conduction velocity is a frequent feature. Deterioration is rapid and death occurs between the ages of 1½ and 2 years.

(b) Late infantile

This subdivision has an onset between 6 and 18 months. A Japanese patient is

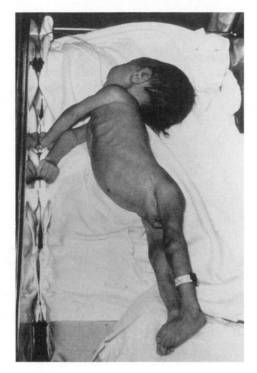

Fig. 19.1 Krabbe's disease.

reported by Kurokawa *et al.* (1987). She was irritable, had mild dysphagia, and a mild spastic diplegia. At 23 months she had grossly deteriorated.

Prolonged floppy infant variant (Hagberg)

A patient with delayed onset (at 12 months) was reported by Vos *et al.* (1983). His initial signs were a loss of skills (he walked at a year) and when seen at 17 months he was floppy rather than spastic. He died at 28 months. The authors point out that unlike the usual early onset cases there was no visual failure nor gait disturbance.

The child reported by Lieberman *et al.* (1980) was seen at 5 weeks because a sib had died from Krabbes disease. Prenatal diagnosis on this infant had been positive, but the pregnancy proceeded. Development was normal for 2 months and then progressive loss of tone was noted. Even before this, at 5 weeks, there was evidence of peripheral nerve involvement. By 9 months muscle tone and movement were absent and he died at 17 months.

(c) Late onset

A girl who was well up to the age of 3½ years and then suffered a progressive neurological deterioration is reported by Loonen *et al.* (1985). The authors review the clinical differences between the late infantile and the late childhood forms. In the latter group failing vision was the commonest early symptom with subsequent hemiparesis or ataxia. The course is protracted and death occurs from 10 months to 7 years after onset. In the late infantile group irritability, motor deterioration, stiffness, and ataxia seem to appear early. The pathological changes are identical.

Young *et al.* (1972) described beta-galactocerebrosidase deficiency in a brother and sister with an onset in the brother 2½ years and in his sister at 5½ years. The onset in the sister (still alive at 11 years) was with defective vision and in her brother as a progressive cerebellar syndrome. He died at 5 years. It is likely that the late-onset form is genetically distinct from the infantile form, but the difference cannot as yet be explained on the basis of varying degrees of enzyme deficiency.

Other late-onset cases were collected by Crome *et al.* (1973). The cases of Neubürger (1922), Collier and Greenfield (1924), Christensen *et al.* (1961), and Liu (1970) had an onset between 2 and 5 years, and in this group there is good evidence for recessive inheritance (sibs affected in all of these families). Clinically, the late-onset (including adult-onset) disease is characterized by failing vision due to both cortical blindness and optic atrophy (Crome *et al.* 1973). Peripheral neuropathy was not invariably present (even electrically) and pyramidal signs might predominate. Kelly (1973) suggested that the late-onset variety could be due to a mutation in a different portion of the same gene, i.e. an allelic mutation.

Counselling

Recessive inheritance is firmly established and sibs with the disease have been reported by Farrell *et al.* (1973) and Nelson *et al.* (1963). Parental consanguinity occurred in the cases of Van Gehuchten (1956). Suzuki and Suzuki (1971), Harzer (1977), and Farrell *et al.* (1978) reported successful prenatal diagnosis of Krabbe's disease with enzymatic and morphologic confirmation in the fetal tissue. It had also been achieved in one of dizygotic twins by Suzuki *et al.* (1978). It is important that the enzyme study be done on cultured amniotic fluid fibroblasts as the cell-free amniotic fluid estimation of cerebroside beta-galactosidase can be normal, whereas the enzyme is profoundly deficient in the cultured cells (Harzer 1977).

Recently, Harzer *et al.* (1987) reviewed their experience of 58 pregnancies tested, 23 were positive (and confirmed). The test on a chorion villus sample is still not reliable enough to use this method only.

2. X-linked leucodystrophy

There are two types of X-linked leucodystrophy:

(a) adrenoleucodystrophy;

(b) Pelizaeus–Merzbacher.

Adrenoleucodystrophy, Siemerling–Creutzfeldt disease, sudanophilic leucodystrohpy, melanodermic leucodystrophy

Adrenoleucodystrophy is a combination of adrenocortical insufficiency (Addison's disease) with cerebral sclerosis (Schilder's disease). It is now apparent that many males previously described as having Schilder's disease had adrenoleucodystrophy.

Up to 1975 there were 50 reported cases (Schaumburg *et al.* 1975) and all have been males except for the one case of Pilz and Schiener (1973). The disease is characterized by the presence of skin hyperpigmentation which often precedes the first neurological signs by 1–3 years. The age of onset of the cerebral deterioration is between 5 and 8 years, and consists of progressive spasticity, resulting in disturbance of joint movement and speech. Seizures, both myoclonic and generalized, occur with repeated bouts of vomiting, abdominal pain, and shock. Deteriorating vision is often an early complication in children and adults, and blindness may result.

The CT scan appearance is very suggestive of the diagnosis. There are symmetrical low density areas in the periventricular white matter, the trigones, occipital, and proximal temporal horns of the lateral ventricles. They are often confluent and continuous across the midline via the corpus callosum.

The pathology is that of severe demyelination, the main brunt of the disease falling on parietal and occipital lobes. The fresh lesions contain so-called 'glitter cells' with sudanophilic droplets which are perivascularily situated. The axons are preserved. Inclusion bodies have also been observed in the adrenal cortex, Leydig cells of the testis, and Schwann cells.

A long lifespan in those affected has been reported by Ropers *et al.* (1975) who described two members of a large pedigree who died from the disease during their 30s and 40s. In one of their two pedigrees, two males, as well as one illegitimate male, were affected making an X-linked disorder virtually certain. Because of the longevity of some of the affected males, Ropers *et al.* (1975) were able to observe an affected male who fathered three children. His daughter was an obligate carrier and she had an affected son. The evidence for adrenoleucodystrophy being X-linked is conclusive (this was first suggested by Fanconi *et al.* in 1963). X-linked pedigrees are those of Hoefnagel *et al.* (1962), Sharr (1975), and Hormia (1978).

The diagnosis is sometimes difficult, as shown by the pedigree of Gagnon and Le Blanc (1959), where some members had Addison's disease alone and others (all males) had the neurological deficit; and in the report by Schaumburg *et al.* (1975), of 17 cases, one patient with the full picture had two brothers and a cousin with Addison's disease. Likewise, the mother of an affected boy had a brother with Addison's disease alone (Forsyth *et al.* 1971). Affected males are reported by Powers and Schaumburg (1973), Strunk and Struck (1965)—seven males over four generations; Powell *et al.* (1975)—two brothers; and Domagk *et al.* (1975). ACTH stimulation in 10 of Schaumburg *et al.*'s (1975) series of patients showed an abnormal result in nine, suggesting that this test could be useful in investigating adrenal function in patients with an obscure neurological disorder.

It has now been conclusively shown (Moser *et al.* 1981) that very long chain fatty acids, mainly C_{25} and C_{26} are responsible for the interference with myelin formation.

The increase in very long chain fatty acid is best expressed by the C_{26}/C_{22} ratio. It is also now evident that males may be biochemically affected but remain clinically well for many years (Naidu *et al.* 1988).

Gene localization

The gene has been localized to the Xq28 region and a large kindred of 17 (Migeon *et al.* 1981) members subsequently studied by Aubourg *et al.* (1987) showed no recombination.

Prenatal diagnosis

In 11 out of 54 pregnancies at risk for adrenoleucodystrophy, amniocytes were found to have abnormal amounts of very long chain fatty acids (Moser *et al.* 1984; Watkins *et al.* 1987). A post-mortem revealed lamellar inclusion in adrenal cortical cells in the fetus, confirming the diagnosis.

Carrier detection

Obligate carriers have plasma C_{26} concentration in the intermediate range (Moser *et al.* 1984, 1983). Of 76 obligate carriers nine could not be detected in this way. Of these nine, fibroblast very long chain fatty acids were abnormal in four. Thus, more than 90 per cent of carriers can be detected in this way (Watkins *et al.* 1987).

A carrier with a chronic non-progressive foot weakness (pyramidal) was a carrier for adrenoleucodystophy. Her eldest son was affected (Noetzel *et al.* 1987). A significant number of carrier women will show spasticity.

Neonatal form of adrenoleucodystrophy

The clinical features might be indistinguishable from Zellweger syndrome and this includes the facial dysmorphism. Infants are hypotonic from birth, show slow development, and then regress within the first 4 years of life. The condition has been comprehensively reviewed by Kelley *et al.* (1986).

Peroxysomes are absent or markedly reduced in the liver. Differentiation from Zellweger syndrome can be difficult but neither stippled epiphyses nor renal cysts have been reported in the neonatal form of adrenoleuco-dystrophy.

Hiperpipecolic acidaemia occurs in both conditions (Kelley and Moser 1984).

Inheritance

Autosomal recessive.

Adrenomyeloneuropathy

This is a variant of adrenoleucodystrophy. Two sisters with Addison's disease were reported by Hewitt (1957). One had spasticity in the legs. In the case reported by Gumbinas *et al.* (1976) the parents were first cousins.

Five men with a slowly progressive spastic paraparesis, distal symmetrical polyneuropathy, Addison's disease, and hypogonadism were reported by Schaumburg *et al.* (1977). A single adult male with Addison's disease and a spastic paraparesis (Budka *et al.* 1976) had no evidence of cerebral deterioration. He died in an Addisonian crisis.

A report of a pedigree in which adrenoleucodystrophy and adrenomyelo-neuropathy co-existed suggests that the two conditions might be due to the same mutant gene (Davis *et al.* 1979). The adult-onset and childhood-onset syndromes have also been reported in a single family by Griffin *et al.* (1977) and Weber *et al.* (1980). In this latter family a male had an onset at 20 years of a spastic paraplegia and was found to be adrenocortical deficient. Four other males had a severe adrenoleukodystrophy. Three brothers had a typical childhood onset adrenoleukomyeloneuropathy (Marsden *et al.* 1982, confirmed in one at post-mortem). A maternal uncle had an onset at 12 of a gait disturbance with epilepsy. He subsequently demented and developed a cerebellar dysarthria, spastic ataxic gait (pathologically brisk jerks), mild optic atrophy, and a denervating peripheral neuropathy in his feet. He had a raised plasma ACTH.

Infantile sudanophilic leucodystrophy

In those cases of sudanophilic leucodystrophy with onset under the age of one year seizures can be severe. Male sibs with the clinical picture were reported by Bignami *et al.* (1966) and Coleman *et al.* (1977). The latter authors found raised levels of serotonin in the blood and central nervous system.

Infantile Refsum

This often presents in the first months of life with vomiting and hepato-megaly. Later, development will be noted to be retarded and a sensorineural deafness (Scotto *et al.* 1982) and visual loss due to a pigmentary retinopathy develop. The facial features can be similar to Zellweger syndrome.

(b) Pelizaeus–Merzbacher disease

Pelizaeus–Merzbacher disease is one of the leucodystrophies. Pathologically, there is widespread hypomyelination with some preservation of islands of myelin giving a tigroid pattern (Konishi and Kamoshita 1975). In some cases with an onset in infancy the myelin sheaths are rare or absent (Watanabe *et al.* 1973). As reviewed by Stark (1972), there are six types. The diagnosis remains difficult. There is some evidence (Garg *et al.* 1983) that brainstem evoked responses are abnormal from an early age—Wave I I has a prolonged latency and all of the subsequent waves are absent—but this might not be universally true. It is also difficult to distinguish between types I and II.

Type I—classic (mostly X-linked)

This was described by Pelizaeus (1885) and Merzbacher (1910). Many males were affected over four generations and the disorder was characterized by gross nystagmus and head tremor with an onset in the infantile or late infantile period. There is slowness in motor development and later cerebellar ataxia, choreoathetosis (especially face and upper limbs), spasticity of lower limbs, and dysarthria develop. Seizures are infrequent, but dementia occurs resulting in death after 5–6 years. The condition is probably inherited as an X-linked recessive. A huge pedigree with many affected in six generations was reported by (Watanbe *et al.* 1973) and, another family with type I Pelizaeus–Merzbacher disease had 11 affected males with X-linked inheri-tance; abnormal eye movements were not a feature (Johnson *et al.* 1978). A male and his two half brothers by the same mother had the classic type I disease (Niakan *et al.* 1979). Four isolated affected girls have, however, been reported including one by Konishi and Kamashita (1975) and one by Liano *et al.* (1974). A CT scan shows atrophy but NMR scan shows the gross myelin deficiency.

Type II—Seitelberger, connatal, or congenital type (both X-linked and recessive)

This condition is more severe than type I. From the start there is little developmental progress. Nystagmus, gross involuntary movement, and spasticity of the lower limbs are the main features. In an X-linked pedigree (Ulrich and Herschkowitz 1977) the male proband had continuous symmetrical nystagmus from birth with failure of motor development. In previous reports (Seitelberger 1970: Watanabe *et al.* 1969; Adachi *et al.* 1970; Zeman *et al.* 1964; Nisenbaum *et al.* 1965; Vuia 1978) all were X-linked except Nisenbaum's pedigree where three girls were affected. In Vuia's (1978) report microcephaly, pachygyria, and cystic degeneration were present.

Two cousins (the offspring of sisters) developed stridor at 3 weeks. At 4½ weeks the stridor had increased, with vomiting, poor sucking, and intercostal indrawing (Renier *et al.* 1981). At 8 weeks one was noted to be delayed and had developed nystagmoid eye movements. Motor retardation was severe, and grey optic discs and pyramidal tract signs were features. A CT scan was normal. The diagnosis was confirmed by brain biopsy.

These authors also suggest that the Plott syndrome (see p. 24) especially the family of Watters and Fitch could have this condition.

In a recent review of this condition Cassidy *et al.* (1987) summarize the clinical picture. The age of onset is at birth or soon thereafter and death occurs between the ages of 4 months and 7 years. Most died before 3 years. Nystagmus (roving, horizontal, or rotary) and a movement disorder (head nodding, involuntary jerking of head, eyes, or limbs), and seizures of all types were common. Optic atrophy has occured in 50 per cent of patients and spasticity develops. At post-mortem most show an absence of CNS myelin. Cassidy *et al.* (1987) describe a girl with this condition which is likely to be heterogeneous. Only pedigree analysis allows the geneticist to distinguish between the autosomal recessive and X linked types.

Type III—transitional type between I and II

All cases have been sporadic. Both girls as well as boys have been described. The main clinical features are nystagmus (not in all the cases), dementia, ataxia, spasticity, and a movement disorder (chorea or athetosis). The onset in the girl described by Jellinger and Seitelberger (1969) was after the age of 3 years when she developed what was thought to be a meningoencephalitis, but this was followed by gradual mental deterioration and a relapsing neurological disorder. At 13 years, she had gross spasticity, ataxia, and optic atrophy. During the course of the illness she had choreiform movements. The patchy demyelination of the cortex was characteristic. Other cases in this group are listed by Seitelberger (1970), Jacobi (1947), Blackwood and Cumings (1954), Thulin *et al.* (1968), and Hallervorden (1957).

Type IV—adult type of Lowenberg and Hill (1933)

Onset is in the fifth decade and the chief signs are a stiff gait, tremor, ataxia, generalized hyperreflexia, and progressive intellectual deterioration.

Inheritance is autosomal dominant and a family with four generations affected was reported by Peiffer and Zerbin-Rüdin (1963). The onset was at 40 years, initially with a slowly progressive paraparesis followed by dementia. A patchy demyelination of cerebrum and cerebellum was found. Unlike types I and I I there might be considerable damage to axons. A sibship of three with a much earlier age of onset of dwarfism, microcephaly, and tapetoretinal degeneration (Lüthy and Bischoff 1961) looked clinically more like type V I.

Type V

Atypical cases are grouped by Seitelberger (1970) in this category. Only six cases were listed by him in which demyelination was patchy. Sibs have been reported by Diezel and Huth (1963).

Type V I—Cockayne's syndrome

This is characterized by microcephaly, dwarfism, mental retardation, ataxia, retinal pigmentation, bilateral cataracts, large ears and nose, and progressive contracture of joints. Early development is usually normal. Hypersensitivity to sunlight followed by scarring, carious teeth, and intracranial calcification have all been reported.

It is the sunken eyes, large ears, prominent beaked nose, and wizened facial appearance that is suggestive of the clinical diagnosis. Optic atrophy might develop during the course of the illness. The age of onset is usually after the first year of life and the eye signs and deafness might not be early manifestations. A CT scan might show basal ganglia calcification.

Pathologically, there is widespread involvement of white matter with a tigroid pattern of demyelination. A peripheral neuropathy suggestive of segmental demyelination (Moosa and Dubowitz 1970) lends support to the syndrome being a leucodystrophy. This view was supported by Rowlatt (1969), and Norman and Tingey (1966). Sibs are reported by Gerstl *et al.* (1965), Sugarman *et al.* (1977), and Neill and Dingwall (1950) and in twins by McIntyre and Brown (1965). A report of sibs and a review of the literature was provided by Soffer *et al.* (1979).

Of diagnostic importance is that cultured skin fibroblasts have been shown to be hypersensitive to ultraviolet light and this has been used is prenatal diagnosis (Sugita *et al.* 1982).

Early-onset Cockayne syndrome Occasional patients present in infancy with failure to grow and this involves both linear growth and head circumference. Cataracts develop early as does skin sensitivity to sunlight. Most patients show the same chromosomal breakage as do the late onset Cockayne patients. Inheritance is recessive (Lowry 1982).

Dermatoleucodystrophy with neuroaxonal spheroids

A condition in two Japanese sibs with an onset within the first year of life of mental deterioration, spasticity, and tremor was reported by Matsuyama *et al*. (1978). The skin lesions consisted of thickening and wrinkling were noted at birth. At post-mortem there was a leucodystrophy with multiple axonal spheroids and granules resembling ceroid-lipofuscin. The condition, resembles Pelizaeus–Merzbacher disease.

3. Metachromatic leucodystrophy

This is characterized by diffuse demyelination of peripheral and central white matter and by the accumulation of cerebroside sulphate (sulphatide) due to a deficiency of arylsulphatase. Five clinical types have been described.

A. Allelic types
    (1) Late infantile
    (2) Juvenile
    (3) Adult onset
    (4) Variant form
    (5) Pseudo-A.S.A.-deficiency (P D)

There are three biochemical subtypes of arylsulphatase, namely types A, B, and C.

B. Non-allelic forms
    (1) Cerebroside sulphatase activator deficiency
    (2) Multiple sulphatase deficiency

  The enzyme arysulphatase A is deficient in the first three groups. It is possible that the residual enzyme activity can be correlated with age of onset and, therefore, with the above classification.

A1. Late infantile metachromatic leucodystrophy

The majority of cases have their onset between the ages of 1–2 years. Classically, there are three stages of the disease (Hagberg 1962). Irritability and poor eating are early non-specific symptoms. The child is floppy and begins to lose the ability to maintain an erect posture, walk, sit, or crawl. Mild optic atrophy may be present. Protein in the cerebrospinal fluid is moderately raised. The second stage shows further decline but the muscle tone changes to that of hypertonicity in the legs with hypotonicity of trunk and head still being evident. Electroencephalographic and electromyographic changes consist of marked slowing of cerebral rhythm and evidence of peripheral nerve involvement. Finally, extensive spasticity with opisthotonus become prominent and myoclonic jerks occur. Reflexes are depressed but the plantar responses might be extensor. When the onset is slightly later (between 2 and 3 years), a gait disturbance is often the first symptom (Percy *et al.* 1977).

In some cases the presentation is that of a polyneuropathy. An Indian child, the product of first-cousin parents, showed this (De Silva and Pearce 1973), as did a case of Tasker and Chutorian (1969).

In Sweden, the incidence of the late infantile form of metachromatic leuco-dystrophy has been estimated at 1 in 40 000 (Gustavson and Hagberg 1971).

Monozygotic twins were reported by Hashimoto *et al.* (1978).

A2. Juvenile metachromatic leucodystrophy

The onset is between 3 and 21 years, and a gait disturbance (incoordination) or behaviour problems are often the first signs (Gordon 1978). Most develop the peripheral neuropathy but not invariably so. Austin *et al.* (1968) described two brothers with an onset at about 2–6 years (still alive at 61) in the one, and 30 years in the other. The initial picture was that of subtle mental deterioration with minor lapses of memory and lack of concentration. Unusual movements developed late in the disease described as 'generalized spasms accompanied by athetoid posturing of the limbs'. In the case of Sourander and Svennerholm (1962), movements were choreic in type. His patient had an onset at 16 years. A study of the late infantile and juvenile forms (Schutta *et al.* 1966) included 13 with an onset between one and two years with death in the first decade, and another four with ages of onset between four and 10 years. The proportion of sibs affected suggested auto-somal recessive inheritance and they found that within families the clinical course tended to breed true. Müller *et al.* (1969) described a brother and sister, and other sibs (pathology in one only) were reported by Harzer and Recke (1975) and Haltia *et al.* (1980). Single cases were reported by Suzuki and Mizuno (1974), Haberland *et al.* (1973), Clark *et al.* (1979), and Warner (1978). Three siblings, the offspring of related Iranian parents, presented with dystonia which progressed to involve neck, spine, and limbs (Yatziv and Russell 1981). Onset was between 3 and 6 years. Mental deterioration was not a feature. Metachromatic material was seen on sural nerve biopsy.

A3. Adult metachromatic leucodystrophy

Pilz (1972) suggested that adult onset should include all cases that manifest after the end of the second decade. No clear distinction can be made between those of juvenile and those of adult onset. In the nine other 'adult'-onset metachromatic leucodystrophy cases (autopsy proven) in the literature, the age of onset was mostly in the second decade. Preclinical diagnosis of the adult metachromatic leucodystrophy is sometimes possible (Pilz and Hopf 1972). A normal sib of an adult patient had a deficiency of the enzyme arylsulphatase A in leucocytes, and an increased excretion of sulphatides and metachromatic substance in the urine sediment. The urinary excretion of sulphatides was higher in the unaffected sib than in her brother. A nerve-conduction study showed her to have a reduced conduction velocity but clinically she was well. A subsequent report (Pilz *et al.* 1977) showed that the sister did indeed develop a progressive neurological disorder consisting of dementia, optic atrophy, and spasticity. Two adult patients were diagnosed at 37 and 34 years. Profound enzyme deficiency (homozygous levels) was found in three clinically normal relatives aged 13, 39 and 40 years.

It is, however, well documented that low enzyme levels of arylsulphatase A can be found in clinically healthy people who do not develop a neurological disorder. Dubois *et al.* (1977) reported a family in which enzyme assays revealed low arylsulphatase A in a clinically healthy father and three of his children (one other child had the late infantile form of metachromatic leucodystrophy). The parents were first cousins. The father must be heterozygous for the metachromatic leucodystrophy gene, but to account for the biochemical findings, it is postulated that he has another mutant gene which affects the expression of enzyme activity.

Other reports of the adult-onset metachromatic leucodystrophy in sibs are those of Percy *et al.* (1977)—three sibs; Hoes *et al.* (1978)—two brothers; and Hirose and Bass (1972). The clinical picture in these sibships was that of a progressive cerebral deterioration with optic atrophy and pyramidal, cerebellar, and peripheral neuropathic signs.

Occasionally, adult-onset metachromatic leucodystrophy presents very late in life. A man with progressive mental deterioration beginning at the age of 62 years in association with a polyneuropathy was found to have significantly reduced arylsulphatase A (Bosch and Hart 1978). This was a single case.

In adults the diagnosis should be considered in cases of presenile dementia in which the VER is delayed or peripheral nerve conduction velocities are slowed (Wulff and Trojaborg 1985).

Psychotic problems might be the first manifestation. The patient reported by Cerizza *et al.* (1987) seemed, as the initial sign, to neglect her family and personal affairs. Four sibs with an onset of a progressive dementia between 15 and 21 years are reported by Alves *et al.* 1986. A nerve biopsy confirmed the diagnosis. The patient reported by Finelli, (1985) presented with schizophrenic thought disorder and the adult patient of Skomer *et al.* (1983) presented with a dementia at the age of 20 years. The low density areas on CT were multifocal, well defined and symmetrical. They were however much less extensive than those seen in the late infantile form of MLD.

A4. Variant form

Pfeiffer *et al.* (1984) reported a boy who developed at the age of 3 years a gait ataxia, dysmetria, dysarthria, and a spastic paresis. Nerve conduction was normal. A CT scan showed only atrophy although at post-mortem there was demyelination.

Biochemical studies on the patient and his family suggested a pseudo-ASA deficiency, but he had sulphatide accumulation in the brain. The authors suggest that the patient is either heterozygous or homozygous for a variant form of ASA.

A5. Pseudo-deficiency

Healthy individuals may show deficient aryl-sulphatase A (Kihara *et al.* 1980) and this might be demonstrated using either the synthetic or natural substrates. This phenomenon is best explained by supposing that there are three alleles.

ASAp
ASA –
ASA +

Pseudo-deficient individuals are ASA-/ASAp and ASAp might be frequent in the population. Pseudo deficiencies might raise difficulty with prenatal diagnosis, but it can be resolved if both parents are investigated. If one is a compound heterozygote then the absence of enzyme in the fetus might be unreliable (Zlotogara and Bach 1983).

A small number of patients with non-progressive mental retardation have ASA deficiency (Butterworth *et al.* 1978). It must be concluded that these patients have pseudo-deficiency.

B1. Activator mutant in metachromatic leucodystrophy

Two sibs, the product of consanguineous Mexican-American parents, were reported by Shapiro *et al.* 1979 and again by Stevens *et al.* 1981. Clinically they had juvenile MLD but the ARSA activity in leucocytes and fibroblasts were half normal. Hydrolysis of cerebroside sulphate by growing fibroblasts was markedly attenuated and supplementation of the growing fibroblast with sulphatase activator normalized the biochemical response in a loading test (data from Stevens *et al.* 1981).

B2. The Austin variant or variant O (multiple sulphatase deficiency)

In 1965, Austin described two siblings, a brother and sister, who presented with late infantile metachromatic leucodystrophy associated with features of the mucopolysaccharidoses. Onset occurred at about one year, and death between 3 and 12 years. Biochemically, the Austin variant is characterized by a multiple deficiencies of arylsulphatases—namely A, B, and C (Moser *et al.* 1972). Other sibs were reported by Mossakowski *et al.* (1961), and single cases by Thieffrey *et al.* (1966), Rampini *et al.* (1970), and Raynaud *et al.* (1975). The clinical features are those of the late infantile metachromatic leucodystrophy and no juvenile- or adult-onset cases have as yet been described. Retinal atrophy may be a feature and flared ribs or a prominent or depressed sternum suggest a mucopolysaccharidosis—however, no other signs of Hurler's disease appear.

4. Alexander's disease

The pathological hallmark of this disease is the presence of refractile bodies in relation to astrocytes (Russo *et al.* 1976). These bodies are thought to be the same as Rosenthal fibres.

Clinically, infantile, juvenile, and adult cases have been described.

Infantile

The infantile group is characterized by megalencephaly and spastic quadriparesis.

Juvenile

This subgroup has on onset between 7 and 14 years, mostly with progressive bulbar symptoms and spasticity. Seizures and cerebral deterioration are less prominent and the average duration of the illness is 8 years (Borrett and Becker 1985).

Adult

The onset is from 20 to 70 years. The adult-onset patients have a combination of intermittent ataxia and spastic quadriparesis (Russo *et al.* 1976). This last group has been described as multiple sclerosis with Rosenthal fibres (Ogasawara 1965). A sister and three brothers (Wohlwill *et al.* 1959) had the syndrome. Histological evidence of the disease was proven in one.

Counselling

Diagnosis during life is difficult and the majority of cases are isolated. Genetic counselling remains difficult. There is one recent report in the Italian literature (Barbieri *et al.* 1980) of affected sibs, but further reports are needed before being certain that this is a recessive disorder.

Alexander's disease—focal in the medulla/upper cord

In the patient described by Goebel (1981) the first symptom was an inability to feed on solid food which developed at about 18 months. At 9 years she was severely malnourished and on examination her voice was aphonic, the vocal cord paralysed (unilateral), and there was tongue atrophy with fasciculations. She died at 14 years. At post-mortem there was a circumscribed lesion in the enlarged medulla, proliferation of astrocytes, incomplete demyelination, perivascular infiltrates, and abundant Rosenthal fibres.

5. Glial and glioneuronal degenerations

This is a group of conditions with an onset early in life of mental deterioration, myoclonus, severe seizures (often status epilepticus), abnormal movements, nystagmus, blindness, and deafness (Jellinger and Seitelberger 1970).

The cases can be classified as follows (Jellinger and Seitelberger 1970).

(a) When the primary pathology affects the astroglia; this type is represented by spongy degeneration in infancy, e.g. Van Bogaert–Bertrand disease and Canavan disease.

(b) Where the primary pathology affects both glial tissue and neurons; glioneuronal degeneration, Alpers' disease.

(c) Gliovasal. When the primary disorder involves the astroglia, with secondary increased permeability which gives rise to a mesodermal reaction and hypervascularity, e.g. hepatocerebral disease.

(a) Spongy degeneration (Van Bogaert and Bertrand or Canavan disease)

First described by Canavan in 1931 but suggested as a distinct entity by Van Bogaert and Bertrand (1949), spongy degeneration has its onset between 2 and 5 months, with death before the fourth year. The clinical features are those of poor head control, cortical blindness, optic atrophy, megalencephaly, increased tone, and hyperactive reflexes. Deterioration is gradual and seizures are not prominent. At post-mortem the cortical and subcortical vacuoles are characteristic giving a spongy appearance on light microscopy. Intramyelin oedema similar to Van Bogaert–Bertrand disease has been described in hexachloraphine intoxication (Towfighi and Gonatas 1976) and occasionally in the aminoacidopathies.

Up to 1973, 78 instances out of a total of 135 had been reported in 49 families (Adachi *et al.* 1973). The disease is inherited as an autosomal recessive and Jews from Lithuania and the Western Ukraine are more frequently affected (Crome 1965). This author described spongy degeneration in three unrelated non-Jewish infants and there have been reports of an Irish-American family (Hogan and Richardson 1965), a report from India (quoted in Andermann *et al.* 1969) and four Iranian sibs with first-cousin parents (Mahloudji *et al.* 1970). In the latter family, three had been diagnosed as having familial hydrocephalus until the megalencephaly was shown to be the cause of the large head, and the pathology to be a spongy degeneration. Consanguinity in the 26 sibships recorded by Andermann *et al.* (1969) was 12 per cent and in their series of cases from the literature 66 per cent of the total were Jews. Banker *et al.* (1964) traced the geographic origins of their Jewish patients to present-day Lithuania and an adjacent area of Eastern Poland. In a study of dizygotic twins concordant for the disease, histological examination has suggested that the primary pathology might be in the astrocytic mitochondria (Adornato *et al.* 1972).

Six cases in Chinese children (Luo and Huang 1984) had an age of onset between 4 and 25 months. Initial symptoms were lethargy and vomiting followed by blindness (mostly optic atrophy) convulsions and psychomotor decline. Megalencephaly was a feature. Three of the cases were sisters and all were of Han origin—the major nationality of Chinese.

Note: a deficiency in aspartoacylase resulting in *N*-acetylaspartic aciduria has been noted (Matalon *et al.* 1988)

Juvenile type of spongy degeneration

There have been infrequent reports of this type. Onset is after the age of five years (Jellinger and Seitelberger 1969). A Japanese woman with an onset at 11 years had growth retardation, ptosis, and progressive external ophthalmoplegia. Then followed blindness, partial deafness, tremor, ataxia, loss of reflexes, and generalized weakness (Goodhue *et al.* 1979).

(b) Alper's–progressive neuonal degeneration of childhood

In 1931 Bernard Alpers described an infant with cortical neuronal loss. Later Blackwood *et al.* (1963) described five patients in three families and noted in two siblings, in one of the families, the presence of liver cirrhosis. He called the condition 'diffuse cerebral degeneration' whereas the same condition was called spongy glioneuronal dystrophy by Jellinger and Seitelberger (1970). The onset was between infancy and 15 months with failure to thrive and poor development.

This is succeeded by generalized seizures, myoclonus, deafness, blindness (sometimes optic atrophy), spasticity, ataxia, and choreoathetosis. Sibs were reported by Jellinger and Seitelberger (1970), Ulrich and Cunz (1966), and Klein and Dichgans (1969)—in the latter family onset was later than usual, namely 5½ and 13 years. Three sibs with progressive mental deterioration, seizures, and rigidity had an onset in early infancy (Sandbank and Lerman 1972). Post-mortem on one showed severe neuronal loss with microgliosis and astrogliosis. The perinuclear mitochondria in the neurons were disorganized. Two mentally retarded sibs (unrelated parents) had an early onset of convulsions, followed by a progressive cerebral deterioration, blindness, and kyphoscoliosis (Liu and Sylvester 1960). The authors called this 'familial diffuse progressive encephalopathy'. Similar pathology with minor differences was described in two pairs of sibs (Laurence and Cavanagh 1968). The sibs were microcephalic and had no seizures before the retardation was diagnosed. Two had an enlarged liver and spleen.

Twin brothers had in addition to the spongy degeneration, microcephaly, internal hydrocephalus and pontocerebellar atrophy (Vuia 1977). Both the degeneration and malformations were thought to be based on glial atrophy and the mitochondrial abnormality that was found could be the primary defect.

For other possible cases, see the paper entitled 'familial type of leucodystrophy with microcephaly' (Vuia and Gutermuth 1973), the sibs described by Gambetti *et al.* (1969), and the family described under the title 'A protracted form of spongy degeneration' by Adachi and Volk (1968).

Alpers, in 1960, reviewed the condition pointing out that most cases are sporadic. However, sibs have been reported by Ford *et al.* (1951) and Kramer (1953). In all the reported cases the onset was in infancy and death followed between 3 months and 8 years later. Very early development is normal. For an excellent review see Egger *et al.* (1987).

The VER is abnormal in most cases and the CT scan shows progressive atrophy particularly in occipital areas. The EEG is unusual and characteristic (Boyd *et al.* 1986).

Post-mortem changes vary from mild and superficial atrocytosis and spongy change to loss of neurons and extensive gliosis through the depth of the cortex, especially involving the visual cortex. The liver shows fatty degeneration, extensive cell loss, bile duct proliferation and fibrosis (Harding *et al.* 1986).

The cases described by Wefring and Lamvik (1967) and Huttenlocher *et al.* (1976) probably had this condition. Wefring and Lamrik (1967) reported two sibs who developed convulsions at 11 and 14 months of age, became hypotonic, demented, and died. Shortly before their death, jaundice appeared and at post-mortem there were spongy cirrhotic changes. The parents were unrelated.

Two more sibships—two affected in each family—were reported by Huttenlocher *et al.* (1976). The onset was between 11 and 31 months. The clinical progression was that of vomiting, failure to thrive, fever, seizures (multifocal), dementia, and hypotonia, leading to stupor and death. Cortical blindness, ascites, and jaundice completed the picture. Liver involvement can be late or only found at post-mortem. Increased blood ammonia should be looked for.

Counselling

Autosomal recessive. Prenatal diagnosis is not possible.

(c) Spongy glioneuronal dystrophy (and other conditions affecting brain and liver)

This term was suggested by Jellinger and Seitelberger (1970) for the border-line cases with prominent histological changes of vacuolation. A single case is reported by Hopkins and Turner (1973), but a sib could have been affected. In this report there were both hepatic and cerebral lesions, and the same association occurred in one case of Jellinger and Seitelberger (1970). The similarity of this condition with that described below (Zellweger syndrome) is evident.

Other conditions affecting brain and liver

Infantile spongy degeneration with glycogen storage disease

A four-month-old Japanese child had this combination (Takei and Solitare 1972). She was the offspring of first-cousin parents.

Zellweger syndrome (cerebro-hepato-renal syndrome)

The outstanding features are severe hypotonia from birth, a high forehead and occiput, open sutures, a large fontanelle, hypertelorism, a flat face, and low, sometimes dysplastic, ears. Developmental milestones are delayed and seizures occur. Nystagmus is sometimes a feature. Hepatic dysfunction is either found during the child's brief life or liver changes are found at post-mortem.

Other organs, including the kidneys and bone, may be involved. At post-mortem the main finding is that of micropolygyria and subcortical hetero-topia suggesting a disturbed neuronal migration (Volpe and Adams 1972).

Sibs are reported by Bowen *et al.* (1964) and sibs with first-cousin parents by Brun *et al.* (1978). Inheritance is recessive.

X-rays of the skeleton show punctate epiphyseal calcification especially involving the patellae—an unusual localization in Conradi–Hunermann disease (the acute form of rhizomelic chondrodysplasia) in which punctate calcification in similar areas also occurs.

There are many new findings in Zellweger syndrome which can be listed as follows.

1. An increase in very long-chain fatty acids.

2. Absence of peroxisomes from the liver.

3. A decrease in plasmalogens.

4. A decrease in DHAP-AT, i.e. dihydroxyacetone phosphate acyltransferase.

5. It is also now clear that Zellweger syndrome is but one of a number of closely related peroxisomal disorders. These can be classified as follows.

Classifications from Monnens and Heymans (1987)—'abridged'

Group I   *Impairment of one peroxisomal function*
             Adult Refsum disease
             X-linked adrenoleucodystrophy

Group II   *General impairment with reduced number of peroxisomes*
             Zellweger syndrome
             Infantile Refsum's disease
             Neonatal adrenoleucodystrophy
             Acute rhizomelic chondrodysplasia

Group III   *General impairment with normal numbers of peroxisomes*
             Pseudo-Zellweger
             Hyperpipelocic acidemia

Biochemically some of the conditions are characterized as follows

X-linked ALD   Increased $C_{26} : C_{22}$ ratio
                    Normal phytanic acid oxidase

Classical Refsum   Increased phytanic acid, and phytanic acid oxidase

Zellweger   Increased $C_{26} : C_{22}$ ratio
                    Abnormal bile acids
and   Normal or increased phytanic and pipecolic acids

Neonatal ALD   Increased phytanic acid oxidase

Infantile Refsum

Only complementation studies differentiate these conditions

Recessive cerebral dysgenesis

Two sibs with severe mental retardation, spastic cerebral palsy, and seizures died at 3-4 years. At post-mortem there was an almost complete absence of myelin in cerebrum, brain stem, cerebellum, and spinal cord (Neühauser *et al.* 1977). The authors suggested that the condition represents an inability of oligodendrocytes to form myelin. This condition is similar to that reported by Waggoner *et al.* (1942).

6. Menkes syndrome (kinky-hair disease)

Steely-hair disease (Menkes *et al.* 1962) is characterized by pili torti, mental retardation, abnormalities of body temperature, arterial tortuosity, and bony changes. The metaphyses of some bones may be widened and have a wavy irregularity. Danks *et al.* (1972) showed that the syndrome was caused by a deficiency of copper. Inheritance is X-linked with a birth frequency of about 1 in 35 000 live births (Danks *et al.* 1972). These authors presented five families in which 14 males were affected.

The symptoms start at about 2 months of age and death occurs usually before three years. Seizures are often the first symptom (Gordon 1974) and the electroencephalogram at five months is hypsarrhythmic (Friedman *et al.* 1978).

The primary hair (fetal) is usually normal, but the secondary hair is stubbly and white (Gordon 1974). Microscopically the hair is twisted around its long axis.

Serum copper and ceruloplasmin levels are low. A high copper content of the duodenal mucosa has been found (Danks *et al.* 1973) and this may in part explain the deficient body copper stores. It is, however, certain that a simple deficiency is not the whole explanation for the manifestation of this disease. Membrane transport defects might make copper unavailable to important intracellular sites (especially neural tissue) during fetal development (Horn *et al.* 1978).

The gene for Menkes syndrome has been localised to a position between X cen and Xp113, i.e. close to the centromere (Wicacker *et al.* 1983). The carrier status can be determined using multiple hairs from different parts of the scalp and antenatal diagnosis has been successful (Horn 1983).

A review of the Danish experience was published by Tonnescn *et al.* (1987). They looked at 27 high risk pregnancies using direct copper measurement on chorionic villi. Two males were definitely affected. Six normal test results were confirmed to be normal. The authors warn against some canulas which might cause contamination. Alternatively, affected males can be diagnosed by increased uptake of labelled copper by amniocytes. Carrier detection is less exact, but can be achieved in most instances.

Haas *et al.* (1981) reported an X-linked condition which might (see McKusick 1988) be allelic to Menkes. Four males were affected over two generations and had early hypotonia, but later developed hyperreflexia. Seizures did not occur at an early stage and the hair was normal. Retardation was severe, but there was slow achievement of some milestones. Athetoid and choreiform movements of the head and feet developed. Serum copper and ceruloplasmin were low, and there was some evidence of a block in gut copper absorption.

7. Infantile neuro-axonal dystrophy (INAD)—Seitelberger disease and Hallervorden–Spatz disease

There is controversy about the relationship of infantile neuro-axonal dystrophy to Hallervorden-Spatz disease. Seitelberger *et al.* (1963) called infantile neuro-axonal dystrophy a late infantile form of Hallervorden–Spatz disease, only for Cowen and Olmstead (1963) to separate the two. Authors who suggest that infantile neuro-axonal dystrophy and Hallervorden–Spatz disease are the same disease are Seitelberger and Gross (1957), Sacks *et al.* (1966), Rodzilsky *et al.* (1971), and Gilman and Barrett (1973).

Martin and Martin (1974) suggested that there is a clinical difference between the two diseases. Hallervorden–Spatz disease has a later onset (childhood) and three distinct phases occur. The initial spasticity is replaced by dystonia and later rigidity. In contrast those patients with infantile neuro-axonal dystrophy develop normally during the first year of life and then regress. Most of the children learn to sit, but few learn to walk. Death occurs between 4 and 10 years.

Axonal spheroids are typical of both diseases, but they differ in their distribution. In infantile neuro-axonal dystrophy they are ubiquitous whereas in Hallervorden-Spatz disease they are found predominantly in the pallido-reticular area.

Seitelberger *et al.*'s (1963) case did not have the pigmentary changes so typical of the Hallervorden–Spatz syndrome, but he stated that those cases with early onset, i.e. infantile Hallervorden–Spatz syndrome, might not have had time to develop the pigmentary changes. Since then, Indravasu and Dexter (1968) have described three patients with a late infantile Hallervorden–Spatz syndrome with pigmentation in the appropriate places.

Hallervorden–Spatz—early childhood (classical)

The clinical picture is that of a progressive extrapyramidal disorder and dementia beginning after early childhood (Dooling *et al.* 1974). The extrapyramidal features are dystonic postures, ridigity, and choreoathetosis. Occasionally, lower motor neuron lesions with wasting and reduced reflexes are seen (Amctani 1974). Hyperpigmentation of the pallidoreticular zone, and the presence of axonal spheroids in the pallidum and zone reticularis of the substantia nigra are typical.

In the original sibship, described by Hallervorden and Spatz in 1922, five girls were affected. Increasing rigidity, dysarthria, and dementia were the cardinal features. Optic atrophy was not recorded. However, in a family described by Kramer (1914) optic atrophy occurred in the three affected male children.

Inheritance is recessive. In Dooling *et al.*'s (1974) review of 64 cases of

Hallervorden–Spatz syndrome, about half of the probands had another affected sib. The data do not permit the calculation of a segregation ratio, but recessive inheritance is most likely. The majority of reported cases have come from central Europe (Elejalde *et al.* 1979). According to Dooling *et al.* (1974), the following sibships have both the clinical and pathological features of Hallervorden–Spatz disease—Kalinowsky (1927), Winkelman (1932; see p. 182), Urechia *et al.* (1950), Simma (1957), Grcevic (1966), Myle and Fadilogu (1967), Rodzilsky *et al.* (1971), Radermecker and Martin (1972), and Swisher *et al.* (1972).

Five out of 11 sibs in a non-consanguineous family (Elejalde *et al.* 1979) had an age of onset during the second year of life. Gait difficulties because of increasing rigidity; facial grimacing; dysphagia; dystonic movements of face, neck, and limbs resembling a torsion dystonia; and mental deterioration were the main findings. The course of the illness was further complicated by generalized muscular atrophy. The earliest signs were dystonia and rigidity of the lower limbs. Psychomotor development was normal for the first two years.

A CT scan shows bilateral increased density in the globus pallidus region —the caudate nuclei were normal (Boltshauser *et al.* 1987).

Hallervorden–Spatz—infantile

A sibship (three sibs) with the infantile Hallervorden–Spatz disease was reported by Martin and Martin (1974). Two had been previously reported (Martin and Martin 1972) and Fadiloglu (1971) reported sibs with an early onset.

A single case described by Perry *et al.* (1985) had an onset between 10 and 20 months. He was initially clumsy then ataxic, dyskinetic, and developed choreoathetoid movements. He could still walk at 8½ years, but then deteriorated. Fundal examination showed optic atrophy and a pigmentary retinopathy. At post-mortem there was a small, firm brown globus pallidus with many neuroaxonal swellings.

Hallervorden–Spatz (restricted to pallidal nuclei)

A single case (no family history) of a man with an onset at 10 years of a hypokinetic-rigid motor disturbance—looking clinically like a severe dystonia is reported (Kessler *et al.* 1984). At post-mortem, only the pallidal nuclei were affected without involvement of zona reticularis and the cerebral cortex.

Hallervorden–Spatz—late onset

Three sibs are described with an age of onset at 55, 47, and in early adolescence—in whom the features were dementia, rigidity, bradykinesia, mild tremor a shuffling gait, dystonia, blepharospasm, and anarthria (Jankovic *et al.* 1985). At post-mortem iron deposits were found in the globus pallidus and caudate nuclei, and spheroids were present in those areas and in the spinal cord. Three single cases (Eidelberg *et al.* 1987) are reported with an age of onset between 31 and 42 years. One of the patients had previously been thought to be mildly mentally handicapped. The authors note that the neuropathological features in their patient have been produced experimentally by giving heavy metals, especially iron and alluminium and noted further the similarities between the histopathology of this condition and the Chamorro ALS/PD disease (see p. 271).

Two sibs, the product of first cousin parents, developed a Parkinsonian syndrome in their twenties (Alberca *et al.* 1987). The post-mortem findings in one confirmed the diagnosis.

8. Infantile neuroaxonal dystrophy

(A) Infantile

The onset of this autosomal recessive condition is in late infancy. Towards the end of the first year of life, motor and mental deterioration occur with a clinical picture of either spasticity or hypotonia (or both), optic atrophy (occasionally), pendular nystagmus, deafness, and cranial nerve palsies. Those affected do not learn to walk or talk. Convulsions and extrapyramidal features rarely occur. Death results between the ages of 4 and 10 years. Pathologically, spheroid bodies occur—thought to be swellings of the axon —everywhere in grey matter with bulbospinal predominance. They have also been seen in peripheral nerves and autonomic nerve ganglia (Shimono *et al.* 1976; Yagishita *et al.* 1978). The pallidum is not abnormally pigmented. Sibs have been reported by Ametani (1974), Berard-Badier *et al.* (1971)— parents were first cousins, and by Duncan *et al.* (1970). A case presenting at 20 months with uncontrollable seizures (Butzer *et al.* 1975) is unusual. Cowen and Olmstead (1963), Huttenlocher and Gilles (1967), Martin and Martin (1972), and Jellinger and Jirasek (1971) all emphasized the rarity of seizures. When seizures are present the clinical presentation is difficult to differentiate from the infantile form of ceroid lipofuscinosis, but the pathology is different.

Two infants (single cases) had an onset of regression in the 2nd year of life. Both showed evidence of denervation on EMG (Ferrer *et al.* 1983). The diagnosis was confirmed by conjunctival biopsy which showed tubulo-vesicular aggregates in some amyelinic and rarely myelinic axons.

Eight cases are reviewed by Ramaeckers *et al.* (1987). The onset was between 6 months to 2 years, with loss of skills, early visual loss (nystagmus, squint, optic atrophy) and pyramidal signs in conjunction with marked hypotonia. There was loss of ankle jerks and an unsteady, ataxic gait. The diagnosis can be made by brain or conjunctival biopsy and cases may be missed on peripheral nerve biopsy alone. Clinical diagnosis is difficult because of the very variable presentation.

After 2 years prominent fast activity is present on the EEG, but this might not be seen in the very early or very late stages. Other investigations reveal an abnormal VEP (consistent and progressive), and a normal ERG and EMG.

Juvenile neuroaxonal dystrophy

Dorfman *et al.* (1978) proposed the name juvenile neuro-axonal dystrophy (JNAD) for the syndrome in their two male sibs, in whom there were no obvious basal ganglia signs, but a dementing illness. They called the disease juvenile neuro-axonal dystrophy because of the numerous spheroids found at post-mortem.

Dystonia musculorum deformans needs to be considered as an alternative diagnosis (see Elejalde *et al.* 1979), but, the postural torsion is less severe and the prognosis is more favourable than in Hallervorden–Spatz disease.

Neonatal form of neuro-axonal dystrophy (rare)

An infant had a poor suck in the neonatal period. In addition there was severe hypotonia, a pendular nystagmus, and feeble tendon reflexes. Seizures developed at 18 months and death occured at 34 months (Nagashima *et al.* 1985).

INAD and renal tubular acidosis (rare)

Three sibs had a neurological problem (Maccario *et al.* 1983). Two died within the first year of life (one was certainly normal up till 4 months). Both were hypotonic, irritable, and had a renal tubular acidosis. A CT scan on one showed a hyperdense area in the thalamus. The other sib was retarded (at 7 years) and had a similar CT finding. He had a stiff, broad-based gait and a rest tremor of the hands. He also had a renal tubular acidosis. The authors suggest that his clinical picture was Hallervorden–Spatz and that the other two had INAD!

9. Leigh's encephalopathy (subacute necrotizing encephalopathy)

This condition might not be a single entity, but there is evidence that some cases represent a defect in the oxidation of pyruvate. There can be a marked variability in the symptoms of feeding difficulty, weakness, visual deterioration, ataxia, spasticity of lower limbs, convulsions, deafness, and a pigmentary retinopathy with an onset usually before the age of 2 years and after the age of 6 months. The diagnosis needs to be considered where children have recurrent episodes of ataxia, choreo-athetosis, hypotonus, and vomiting (Gordon *et al.* 1974*a*).

Characteristically, there are symmetrical lesions in the basal ganglia and brain stem. Pathologically, there are soft brown areas with blood vessel proliferation and spongy necrosis of cells, but nerve cells and axons are preserved. In muscle, ragged red fibres have been reported twice (Egger *et al.* 1982*b*)—suggesting heterogeneity. The CT scan appearance in Leigh's showed low attenuation areas in the basal ganglia in three patients (Campistol *et al.* 1984).

In three of the 23 families reviewed by Pincus (1972) the age of onset within the family did not vary by more than 6 months, but the course of the disease can vary considerably, and acute and chronic types have occurred within the same family.

A typical pedigree with three affected sibs out of five was reported by Howard and Albert (1972). Concentrating on the ocular findings, they described nystagmus, strabismus secondary to ocular muscle palsy, and optic atrophy as common manifestations of the disease. Three sibs were affected in a pedigree reported by Gordon *et al.* (1974), and other sibships were reported by Montpetit *et al.* (1971), Feigin and Wolf (1954), Ebels *et al.* (1965), Dayan *et al.* (1970), Murphy (1973), and Saudubray *et al.* (1976).

(a) Leigh's disease with NADH deficiency

Some cases of Leigh's have now been shown to have an oxidative phosphorylase defect. One of two sibs reported by Van Erven *et al.* (1986) had evidence in muscle of a disturbed oxidation of NADH. Lactate and pyruvate were intermittently elevated in serum. No ragged red fibres were seen.

(b) Pyruvate dehydrogenase

Evidence that Leigh's disease is caused by a defect in the pyruvate dehydrogenase complex is at present unconvincing. Wallace (1985) reviewed the literature to date. Of the three cases in which there was post-mortem confirmation, the enzyme was never absent and the reduction was only in some of the tissues that were tested.

However, twenty-eight patients with lactic acidosis were reviewed by Miyabayashi *et al.* (1985). Two had autopsy proven Leigh's, and both had pyruvate dehydrogenase activities of less than 10 per cent of the norm.

The significance of reduced pyruvate dehydrogenase deficiency in Leigh's disease is still unknown.

(c) Pyruvate carboxylase deficiency

A severely retarded infant with congenital lactic acidosis had repeated vomiting and convulsions soon after birth. She failed to thrive, was microcephalic, hypotonic, and apathetic. Pyruvate and lactate were raised in blood, and CSF and the enzyme pyruvate carboxylase was 5 per cent of controls in liver and fibroblasts (Tsuchiyama *et al.* 1983).

The association of this condition with renal tubular acidosis and cystinuria has also been noted (Oizumi *et al.* 1983).

Many authors are uncertain whether this deficiency can be a cause of Leigh's disease as the measurement of pyruvate carboxylase can be technically very difficult.

Note: pyruvate carboxylase and dehydrogenase enzyme systems and the respiratory chain are located in mitochondira, and a deficiency might give rise to abnormalities of mitochondrial metabolism.

(d) Cytochrome—C-oxidase deficiency

One patient reported by Miyabayashi *et al.* (1985) had Leigh's disease at post-mortem and showed a reduction of cytochrome—C-oxidase in liver and brain. In a living sib the enzyme was reduced in muscle and fibroblasts. Neither had ragged red fibres.

Genetics of group as a whole

There is good evidence for autosomal recessive inheritance in most instances (see review by Van Erven *et al.* 1987). It might be that there is occasional maternal transmission as in the family reported by Berkovic *et al.* (1987), seven members in two generations had a mitochondrial encephalopathy. Their age of onset was very diverse. One, at post-mortem, had Leigh's disease, others had a system degeneration. It should be noted that subunits of cytochrome—C-oxidase are coded for in mitochondrial DNA.

Acute neonatal Leigh's

Leigh's disease with a severe neonatal onset and rapid clinical course was described by Seitz *et al.* (1984). The infant was floppy from birth and then developed a metabolic acidosis within the first few days of life. On examination there was also a hypertrophic cardiomyopathy and at post-mortem the brain showed changes typical of Leigh's disease involving the cortex, basal ganglia, brain stem, and cord. In addition, a leucoencephalopathy of the hemispheres with sudanophilic tissue degradation advanced neuronal loss, and reactive gliosis were found. The heart showed on electronmicrosopy a primary mitochondrial cardiomyopathy, but similar changes were not detected in the CNS nor in the peripheral nervous system. The inheritance is probably recessive.

Juvenile Leigh's

This has an onset between 4 and 16 years and was reviewed by Van Erwen *et al.* 1987. These authors added a case of their own (not autopsy proven). Development was normal until 14 years. The patient then had exercise intolerance, weakness, an intellectual decline, Parkinsonian features, shortness of breath, and pyramidal signs.

In summary the characteristic features are pyramidal tract involvement, jerky eye movements, extra pyramidal and cerebellar signs, hyperventilation, and exercise intolerance. No ragged red fibres are found and the mitochondria look, on electronmicroscopic examination, normal. NADH dehydrogenase deficiency was demonstrated in the case described above.

Adult-onset Leigh's disease

A mother and her two sons had a degenerative disease of adult onset characterized by optic atrophy and a central scotoma. This was followed by a quiescent period (Kalimo *et al.* 1979). At the age of 50 an ataxia and seizures (both myoclonic and *grand mal*) developed with spasticity and slight dementia. A post-mortem showed lesions characteristic of Leigh's disease.

Leigh's disease with ragged red fibres

See Egger *et al.* 1982*b*.

10. Reye syndrome

Familial aggregation in Reye syndrome is known to occur, but simple inheritance is not thought to be operative. In a survey of 85 families, affected sibs were seen in three (Hilty *et al.* 1979). The development of signs in the second sib occurred within five days of the first in the sibship reported by Hilty, and similar short interval was found in at least 12 other families in the literature (Norman 1968; Huttenlocher *et al.* 1969*b*; Glick *et al.* 1970; Olson *et al.* 1971). The interval was longer or not stated in eight other families (e.g. Hochberg *et al.* 1975; Van Caillie *et al.* 1977). It is, therefore, suggested that a virus might be responsible.

Fatty changes in the liver are crucial for the diagnosis, but in a review of liver histology in children dying a traumatic death, fatty metamorphosis was common so that this feature might not be as specific as first thought.

Various suggestions about the etiology include defects in fatty acid metabolism especially abnormalities in medium chain fatty acid oxidation. In the latter condition the diagnosis can be made during an acute attack as a pathognomonic pattern of dicarboxylic acid and suberylgycinic can be shown in the urine.

Multiple system atrophy with intranuclear hyaline inclusions

Up to 1985, 10 cases had been reported with this condition (Patel *et al.* 1985). In two instances sibs were affected. The age of onset has varied from 1½ years to 45 and it is interesting to note that the sibs reported by Janota (1979) were concordant for their age of onset which was 35 and 45. The sibs reported by Schuffler *et al.* (1978) were both affected in childhood.

The clinical spectrum includes pain and muscle spasm, ataxia, choreoathetosis, mental deterioration, cranial nerve palsies, and pyramidal tract involvement. In those with an early age of onset the prognosis is poor. Friedreich's ataxia had been diagnosed in four instances. Diagnosis in life is difficult, but at post-mortem inclusions are present throughout the central and peripheral nervous system.

(C) Porphyria

The following porphyrias have been found

(a) Intermittent acute porphyria
(b) Variegate porphyria
(c) Hereditary coproporphyria
(d) Porphyria cutanea tarda
(e) Protoporphyria
(f) Congenital erythropoietic protoporphyria

Only types (a), (b), and (c) are likely to be seen by the neurologist.

(a) Intermittent acute porphyria

In the Swedish type there is a high urinary excretion of porphobilinogen G and delta-amino levulinic acid during the acute attacks, and for many months thereafter. On standing the urine becomes dark red.

In Northern Sweden and particularly Lapland the prevalence of heterozygotes is 1 in 900 and for the rest of Sweden 1 in 20 000 (Wetterberg 1967). Inheritance is as a simple dominant, but the identification of gene carriers can be difficult.

Waldenström (1956) emphasized the rarity of clinical or biochemical findings in acute intermittent porphyria prior to puberty.

Screening for acute intermittent porphyria using urinary porphobilinogen as a diagnostic test is difficult in those with latent or quiescent disease, but since the demonstration of deficient activity of uroporphyrinogen-I-synthetase, rapid screening has become possible (Tishler *et al*. 1976). There is, however, an overlap between the gene carriers and normal controls, so that most laboratories will give a likelihood of falling into one or other group.

McColl *et al*. (1982) suggested that by utilizing the estimation of both white cell delta amino-levulinic acid synthase and erythrocyte porphobilinogen deaminase the accuracy for the detection of gene carriers has improved when compared to measuring porphobilinogen deaminase alone. However, there have been families (Mustajoki 1981) with increased excretion of porphobilinogen and delta-aminolevulinic acid in the urine, but with normal red blood cell deaminase (mostly there is a 50 per cent depression of activity; data from Goldberg 1985). Recently, polyclonal antibodies against human erythrocyte porphobilinogen deaminase have been developed and used in 22 families by Anderson *et al*. (1981).

Those authors have identified two groups based on the amount of immunologically cross-reacting enzyme protein (CRIM), but there are likely to be several subtypes even with these groups.

Within families each affected member had the same CRIM subtype.

The neurological involvement in the acute disease is a motor neuropathy, probably due to a functional, but reversable block of peripheral nerve activity. In those with a prolonged block, degeneration of the axon occurs. Sensory symptoms might appear but, if so, they are mild. The motor weakness ascends and can result in a quadriplegia and respiratory failure. The autonomic nervous system and lower cranial and peripheral nerves are primarily involved.

(b) Variegate porphyria

The enzymatic defect interferes with the conversion of protoporphyrin to heme. Variegate porphyria differs from acute intermittent porphyria in the presence of skin lesions (abnormal fragility) and the continuous faecal excretion of protoporphyrin IX and coproporphyrin (Romeo *et al*. 1977). The condition is common in South Africa, where more than 5000 cases have been detected. Elsewhere fewer than 300 patients have been reported (Mustajoki 1980). During an acute attack the excretion of porphyrin precursors in the urine is elevated.

The frequency in South African Afrikaners is 3/1000 (Dean 1967), and this considerable prevalence can be accounted for by a founder effect. There are clinical differences between the South African type of variegate porphyria and that found elsewhere. About 80 per cent of patients in South Africa have the skin lesion, whereas in Finland only 50 per cent do (Mustajoki and Koskelo 1976). It is uncertain whether this difference is due to the effect of sunlight or genetic heterogeneity. Inheritance is a dominant and penetrance is almost complete (Dean 1971).

Faecal porphyrin excretion in prepubertal children is usually normal, and in a study of children of a porphyria parent, no case of an elevated faecal porphyrin was found (Mustajoki 1978).

In the family study of Fromke *et al.* (1978) porphyria was difficult to diagnose when there was a slight or borderline increase in faecal porphyrin, especially in the teenage or younger group of patients. Part of the reason for the lack of penetrance is that a trigger is needed to precipitate symptoms.

The majority of patients have a clear-cut elevation in adulthood, but in some, the values might be only slightly abnormal and hence difficult to interpret. It is concluded that porphyria cannot invariably be diagnosed by current methods, but the frequency of gene carriers who cannot be detected after puberty is low.

It should be stressed that most patients with variagate porphyria are symptom free throughout their lives and the mortality during an acute attack mostly relates to experiences before 1960.

The neurological features are the same as those described under acute intermittent porphyria, and the skin lesions are similar to those found in porphyria cutanea tarda, namely bullous ulceration and hyperpigmentation, especially in exposed areas.

(c) Hereditary coproporphyria

This is characterized by the excretion of large quantities of coproporphyrin III in the stools and urine. The neurological signs and symptoms are not different from those seen in the two more common varieties described above. Seizures have been described, but it is difficult to ascertain the exact relationship between the seizures, anticonvulsant therapy, and the disease itself. Houston *et al.* (1977) described a 9-year-old child with mental deterioration and epilepsy who had coproporphyria. The seizures started at 2½ years and seizure control worsened during an acute attack.

(D) Aminoacidopathy

This group of conditions is characterized by the following cascade of events. Infants are initially normal but this is followed rapidly by:

lethargy;

drowsiness;
irritability;
allerations in tone;
convulsions;
loss of reflexes.

The diagnostic possibilities must include:

(a) organic acidamia;
(b) hyperammonamias—especially urea cycle defects;
(c) non-ketotic hyperglycaenemia.

If disorders of calcium, magnesium, and sodium can be excluded, then the first step in arriving at the diagnosis is to differentiate between those conditions causing a reducing substance in the urine, hypoglycaemia, ketosis, metabolic acidosis, respiratory alkalosis, and hyperammonenia.

The following groups will be considered:

I   Organic acidaemias

II   Branched-chain aminoacidurias

III Disorders of the urea cycle

IV   Disorders of folate metabolism

V   Others

I. Organic acidaemia

This group of conditions is characterized by the presence of small molecular weight organic acids in the blood and urine. They include:

(1) propionic acidaemia

(2) methylmalonic aciduria

(3) isovaleric acidaemia

(4) beta-methylcrotonylglycinuria

(5) glutaric aciduria
     Type I
     Type II

(6) glutathione synthetase deficiency

(7) multiple carboxylase deficiency

(8) beta-ketothiolase deficiency

Clinically

This group presents with repeated infections—sudden death, intermittent

coma, failure to thrive, vomiting, hypotonia, and difficult to treat hypoglycaemia (Brandt 1984). Cerebral palsy occurs most commonly in the
branched chain aminoacidurias. Convulsions are common, and if the infant
survives, the picture might be confused by the presence of ataxia and
athetosis. Mental retardation might occur, but not always. Intelligence can
be normal in isovaleric acidaemia, 2 methylacetoacetyl Co-A thiolase deficiency, methylmalonic acidaemia and glutaric aciduria type I. Brandt (1984)
suggests that it is unlikely in the organic acidaemias for pure mental retardation to occur uncomplicated by other symptoms or signs.

(1) Propionic acidaemia (glycinaemia or ketotic hyperglycinaemia)

The symptoms are those described under the general heading, but they can be
either severe or mild. There is some suggestion that these two types are part of
a single disorder and both can occur within the same family (Wolf *et al.*
1979). Prenatal diagnosis has been achieved (Gompertz *et al.* 1975). Inheritance is recessive (sibs are reported by Brandt *et al.* (1974*a*) and Sweetman
et al. (1979)).

The enzyme involved is propionyl-CoA carboxylase and the disorder
according to a newborn screening programme has a frequency of 1 in 350 000
births. (Wolf and Feldman 1982). Initially, the symptoms might be vomiting,
lethargy, and hypotonia, and failure to treat will result in seizures, coma, and
death. Treatment with protein restriction and biotin may be successful.

(2) Methylmalonic aciduria

There are probably at least six forms of this condition, including a benign
type. Two brothers with the adult benign type are reported by Giorgio *et al.*
(1976).

Matsui *et al.* (1983) analysed information on a number of patients with this
disorder. They found that the most frequent manifestations were lethargy,
failure to thrive, recurrent vomiting, dehydration, respiratory distress,
hypotonia, and developmental delay. About one-third had hepatomegaly.
There was not a large difference in presentation between the different
mutants (they analysed four). There was, however, a difference in the age of
presentation. Important for the geneticist is that if the fetus has a subtype
which responds to cobalamin therapy then it is important to know this.
Therapy might need to be started in utero (Matsui *et al.* 1983).

(3) Isovaleric acidaemia

An unusual and strikingly offensive odour resembling sweaty feet is characteristic if present. Those children who survive are mentally retarded. Sibs are

reported by Levy *et al.* (1973). A cerebellar ataxia might be part of the clinical picture.

An 11-day-old infant with this condition had diffuse cerebral oedema and a massive cerebellar haemorrhage. At post-mortem there was spongiosus of white matter with reactive gliosis, and the cells in white and grey matter showed focal clustering and degeneration (Fischer *et al.* 1981).

(4) Beta-methylcrotonylglycinuria

A typical case had, at 11 weeks, difficulty with feeding, tachypnoea, irritability, lethargy, and infantile spasms (Finnie *et al.* 1976). Some patients previously thought to have this condition were subsequently found to have multiple carboxylase deficiency.

(5) Glutaric aciduria, glutaryl Co-A dehydrogenase deficiency type I

Inborn errors of lysine and tryptophane metabolism were reported in six patients with progressive choreoathetosis, cerebellar ataxia, and dysarthria (Brandt *et al.* 1978). Some of the patients were mentally retarded. Inheritance is recessive and sibs are reported by these authors and by Gregersen *et al.* (1977).

The diagnosis is made by the finding of glutaric acid during organic acid screening. The age of onset is early—below one year—but the outcome is variable; some die early, while some survive and have mild mental retardation.

Glutaric aciduria type I can be mistaken for Leigh's disease (Stutchfield *et al.* 1985). These authors report a male with hypotonia and poor head control at 7 months, dystonia and choreiform movements at 11 months, and marked athetosis and dystonia at 2–3 years. Initially, he had an acidosis (raised lactate and pyruvate) and an encephalitic-like illness with convulsions. A CT scan showed cerebral atrophy. Athetosis is also reported by Leibel *et al.* 1980.

Patients with glutaric aciduria type I differ from those with type II in that they do not have hypoglycaemia nor do they accumulate other organic acids.

Pre-natal diagnosis is possible (Goodman *et al.* 1980).

Type II glutaric aciduria

Neonatal acidosis, hypoglycaemia, and hyperammonaemia are the main metabolic consequences. Male sibs are reported by Sweetman *et al.* (1980) and a single male by Przyrembel *et al.* (1976). Autosomal recessive inheritance is likely.

An acrid sweaty feet odour has been noted in at least two patients.

Vomiting, seizures, lethargy and coma are the main features. There are two types.

(a) With congenital abnormality—mostly renal cysts. Patients may have a Potter-like face and there is fatty infiltration of the liver and heart at post-mortem (Goodman and Frerman 1984).

(b) Without congenital abnormality, presenting as above.

(c) There might be a third type with an onset at any time from the neo-natal period to adulthood with recurrent vomiting, acidosis and hypoglycaemia.

In this late onset type of glutaric aciduria type II, the onset can be in early adulthood with a lipid myopathy. (de Visser *et al.* 1986). The main features were a Gowers sign, a waddling gait and minimal facial weakness. The creatinine level in muscle is low (secondary phenomenon) and the enzyme acyl-CoA dehydrogenase is deficient in muscle. It is of note that the clinically normal sibs had the same biochemical defect.

Some Type II cases have episodes of fatigue, hypoglycaemia and hepatic dysfunction associated with a metabolic acidosis (Dusheiko *et al.* 1979).

(6) Glutathione synthetase deficiency (5-oxoprolinuria)

An over-production of pyroglutamate can cause an acidaemia and neuro-logical disease. Mental retardation, an ataxic gait, and dementia have all been described (Marstein *et al.* 1976). Inheritance is recessive. Two sisters with this condition had a chronic metabolic acidosis with an increased rate of haemolysis (Larsen *et al.* 1985). Both initially had an IQ in the middle of the normal range, but lost 30 points over about 7–10 years. Both had an abnormal ERG.

(7) Multiple carboxylase deficiency

There is a deficiency of propionyl-coenzyme A (CoA) carboxylase, 3-methylcrotonyl CoA carboxylase, and pyruvate carboxylase. In addition, there might be acetyl-CoA carboxylase deficiency. Patients accumulate lactic acid and a number of other organic acids, and develop hypotonia, seizures, ketosis, and coma. Alopecia and a skin rash are part of the clinical picture.

There are two forms:

(a) neonatal in which there is a deficiency in holocarboxylase synthetase;

(b) late onset—biotinidase deficiency.

Other features include developmental delay, persistent keratoconjunctivitis, periods of acidosis, convulsions, sometimes within the first year of life, hyperventilation/stridor, optic atrophy, and sensorineural deafness (Thuy *et al*. 1986).

As discussed by Wolf *et al*. (1985), the condition can be diagnosed rapidly by using a simple rapid semiquantative colorimetric procedure.

(8) Acetoacetyl-Co A thiolase deficiency (or beta-ketothiolase deficiency)

A patient with this biochemical defect and delayed development was recorded by De Groot *et al*. (1977) and sibs by Robinson *et al*. (1979). The defect is in isoleucine metabolism and the ketoacidosis may be provoked by infection.

Clinical findings vary considerably. There might be persistent vomiting within days of birth or there might be a single episode of acidosis as late as 7 years (Schutgens *et al*. 1982). Psychomotor development is often normal.

II. Branched-chain keto-aciduria (maple syrup urine disease)

Infants often appear normal at birth and symptoms begin on day 3–5. Difficulty with feeding, a high-pitched cry, and abnormal eye movements occur, in addition to the more usual manifestations of alternating periods of hypo- and hypertonicity, convulsions, coma, apnoea, and death (Snyderman 1975). A smell of maple syrup or curry can be a helpful diagnostic sign. Inheritance is autosomal recessive.

There are different forms. The classic type, an intermediate type (Schulman *et al*. 1970) an intermittent type (Goedde *et al*. 1970) and a thiamine responsive type (Scriver *et al*. 1971).

In the classical type the onset is neonatal and the course is acute, whereas in the intermediate type the development is often slow, there is failure to thrive, irritability, and an intolerance to different formulas (Gonzalez-Rios *et al*. 1985). The patient described by these authors presented at 10 months in a ketoacidotic coma. The authors postulated that although their patients had the intermediate variety there was biochemical evidence for a unique mutation not described previously.

III. Disorders of urea cycle—hyperammonaemia

Infants are well for a few hours or a day then deteriorate, feed poorly, vomit, hyperventilate with grunting, and have seizures, leading eventually to posturing and coma (levels above 400 mM/l). Levels above 500 mM/l result in increased intra-cranial pressure (based on Breningstall 1986).

1. Levels of 500–2000 result in hyperammonaemic coma.

2. Levels in maple syrup urine disease are mostly less.
3. In urea cycle disorders ketosis and acidosis are absent and an early alkalosis might occur.
4. In organic acid and pyruvate metabolic defects ketosis would be expected.
5. Hyperglycinaemia and hyperglycinuria occur in organic acidaemias.
6. Hypoglycaemia occurs mostly in organic acidaemia.

Therefore, do the acid base status, look for ketones, measure serum glucose, and quantify the amino acids.

Urea cycle defect causing hyperammonaemia (based on Breningstall 1986)

		Orotic acid
Carbamyl-phosphate-synthetase deficiency	↑ Glutamine alanine ↓ citrulline, arginine	decreased
N-acetyl-glutamate-synthetase deficiency	↑ Glutamine, alanine ↓ citrulline, arginine	decreased
Ornithine transcarbamylase deficiency	↑ Glutamine alanine ↓ citrulline, arginine	+ + +
Arginino-succinic-acid-synthetase deficiency	Citrulline + + +	⎫
Arginino-succinic-acid lyase deficiency	+ + + Arginino-succinic acid	⎬ small rise
Arginase deficiency	+ + + Arginine	⎭

1. Carbamoyl-phosphate synthetase deficiency

The clinical presentation is that of recurrent vomiting seizures and coma. Both a neonatal form with complete deficiency and a later onset form with partial deficiencies exist. Differentiation from ornithine-transcarbamoylase deficiency is made by the absence of orotic acid in the urine (Aleck and Shapiro 1978). This has now been mapped to the short arm of chromosome 2 (Adcock and O'Brien 1984).

2. Ornithine-transcarbamylase deficiency

Three males in one family were reported by Cambell *et al.* (1973) and in a study of four kindreds eight males died in the neonatal period (Short *et al.* 1973). Females can be mildly affected and have elevation of blood ammonia. There is evidence of sex-linked inheritance as the disease is invariably more severe in the males.

Most males have severe hyperammonaemia and die rapidly if not treated, but some might only present between 6 and 18 months of age. The diagnosis can be suspected when hyperammonaemia occurs without an increase in the blood citrulline, arginino-succinic acid, or arginine. Orotic acid is increased in the urine. Carrier detection has been achieved by measuring urinary excretion of orotic acid following a protein load, but some obligate carriers cannot be detected in this way. Recombinant DNA techniques have allowed the development of an OTC probe (Rozen *et al.* 1985) and there are now a small number of families in which this has proved useful in detecting carriers.

Prenatal diagnosis using the technique of fetal liver biopsy has been successful (Rodeck *et al.* 1982).

However, there is a small risk of abortion in performing fetal liver biopsy; hence, the development of prenatal diagnosis using gene specific probes (Pembrey *et al.* 1985).

3. Argininosuccinic acid synthetase deficiency (citrullinaemia)

After an initial normal period there is anorexia, vomiting, lethargy, and convulsions leading to death. Citrullinaemia might present in late childhood after many years of uncontrollable epilepsy (Origuchi *et al.* 1984). Incidence is in the vicinity of 1 in 70 000 live births.

The clinical spectrum of this disorder includes patients with an acute and fatal course, and those with a near normal development (Van der Zee *et al.* 1971). In a family reported by Burgess *et al.* (1978) two sibs had the acute course, and the third, who was biochemically affected, was clinically normal. The milder group might mental have retardation and poor growth. The diagnosis is based on the presence of arginino-succinic acid in plasma and urine.

4. Argininosuccinic acid lyase deficiency (argininosuccinicacidaemia)

This is characterized by convulsions, anorexia, vomiting, and lethargy. About 30 well documented families are known, and both severe and mild types occur. Short, brittle hair occurs in about half of the patients (Ccderbaum *et al.* 1973). Patients seem hairless, but there is a stubble showing trichorexis nodosa.

5. Arginase deficiency

Argininaemia had been reported in nine patients up to 1979 (Snyderman *et al.* 1977). All were retarded and the neurological syndrome included a marked spastic diplegia. Sibs were recorded by the above authors and by Terheggen *et al.* (1975); inheritance is recessive. Other features include opisthotonus, convulsions, microcephaly, vomiting, and a clumsy ataxia. The onset is in the neonatal period or later in the first year of life with toe walking, scissoring, increasing spasticity, ataxia, chorea, and athetosis. Prenatal diagnosis has been difficult as the enzyme cannot be detected in fibroblasts. Many cases have been in Spanish or those of Spanish-American origin (Jorda *et al.* 1986) and the infants these authors describe had severe protein intolerance of early onset.

Defects of fatty acid oxidation (Dicarboxylic aciduria)

The symptoms relate to episodic non-ketotic hypoglycaemia mimicking Reye's syndrome. The episodes are precipitated by fasting, vomiting, and increased caloric requirement.

There are seven different types (Vianey-Liaud *et al.* 1987).

1. Systemic carnitine deficiency.
2. Hepatic carnitine deficiency.
3. Long chain acyl-CoA dehydrogenase deficiency.
4. Medium chain acyl-CoA dehydrogenase deficiency.
5. Short chain acyl-CoA dehydrogenase deficiency.
6. Riboflavin responsive multiple deficiency.
7. Multiple or glutaric aciduria type I I.

These conditions can only be diagnosed during an acute attack or by using fasting or loading tests.

Organic acids especially dicarboxylic acids and acylglycines appear in the urine, and can be identified by gas chromatography and mass spectrometry.

Long chain disease is similar to medium chain disease but more severe and the first attack often occurs early (before 18 months) and death during an attack is more common.

Medium chain—see below.

Short chain—rare, only two patients are known. One was a middle aged woman who presented with muscle weakness and lactic acidosis after exercise (Turnbull *et al.* 1984), the other was a 21-month-old infant who became progressively weak, neonatally (Coates *et al.* 1986).

Medium chain acyl-CoA dehydrogenase deficiency

More than 50 cases have been reported. There is usually a history of one or more metabolic crises associated with infection, fever or diarrhoea with reduced level of consciousness, lethargy, and coma. A large liver is found in one-quarter of cases and a biopsy shows fatty infiltration. Hypoglycaemia is always found (above data from Vianey-Liaud *et al.* 1987).

Riboflavin responsive

A remarkable family of seven affected infants in one sibship was reported by

Harpey *et al.* (1983). They were floppy at birth and had sweaty feet. A riboflavin rich diet late in pregnancy and in infancy prevented the syndrome.

IV. Folate metabolism

There are five known disorders either due to defective uptake or utilization of folate (Erbe 1979). Only those associated with mental retardation will be discussed.

(a) Congenital malabsorption

The two unrelated patients reported by Lanzkowsky *et al.* (1970) had severe mental retardation. One had seizures, athetosis, and punctate calcification of the basal ganglia. Another patient had first-cousin parents (Santiago-Borrero *et al.* 1973).

(b) Glutamate formiminotransferease deficiency

Some patients with the defect have been found to be retarded with seizures whereas others have been normal. This biochemical abnormality might be benign.

(c) Methylenetetrahydrofolate reductase deficiency or folic acid non-responsive homocystinuria

Proximal weakness, a waddling gait, and seizures were recorded in a 16-year-old boy (Shih *et al.* 1972). Four sibs of Irish ancestry were mentally retarded and spastic. Two mentally retarded sisters with catatonia, hallucinations, and peripheral neuropathy were described by Mudd *et al.* (1972) and Freeman *et al.* (1975), and a further family by Wong *et al.* (1977a,b).

The clinical features include severe mental retardation, spasticity and seizures (Haan *et al.* 1985). Biochemically, there is homocystinaemia and, homocystinuria. Infants might present as a fatal neonatal encephalopathy or even a non-progressive severe retardation with spasticity (Wong *et al.* 1977a,b). A progressive peripheral neuropathy has also been described (Singer *et al.* 1980).

V. Other inborn errors causing hyperammonaemia

Ornithinaemia

Nightblindness due to gyrate atrophy of the choroid and retina can be part of the syndrome. Nine patients with this abnormality included two sister pairs.

The urinary, cerebrospinal fluid, and aqueous humour ornithine was 10–20 times higher than controls (Simell and Takki 1973). Two sisters reported by Bickel *et al.* (1968) were mentally retarded. Onset is usually between 5 and 10 years. Patients develop progressive constriction of the visual fields and blindness. Myopia can be severe (Fukuda *et al.* 1983).

Hyperlysinuria and hyperammonaemia

This is a disorder of intestinal absorption reported by Brown *et al.* (1972). The mentally retarded male was the illegitimate son of a mentally retarded girl. Dibasic aminoaciduria with defective absorption of lysine and severe post-prandial hyperammonaemia were the main biochemical features.

Hyperlysinaemia

Twelve patients in eight families have been described in the English literature. In three lysine ketoglutarate reductase deficiency was found in skin fibroblasts (Cederbaum *et al.* 1979). Sibs are recorded by Woody *et al.* (1966).

Mental retardation, convulsions, lax ligaments, and in some cases subluxation of the lens are features (Woody and Pupene 1971).

More recently, ten patients were identified on the newborn screening or family surveys (Dancis *et al.* 1983). Their ages ranged from 2 to 24 years. Two had been treated. Mental development was normal in nine (including the two treated individuals) and only mildly subnormal in three. The authors conclude that the condition might be benign even without treatment, but still advocate at least a low protein diet.

Hyperlysinaemia with saccharopinuria

This is probably the same as above. Cederbaum *et al.* (1979) suggested that lysine ketoglutarate reductase and saccharopinuria dehydrogenase deficiency frequently occur together.

Others

Hyper-beta-alaninaemia

A male infant with drowsiness and seizures was reported by Scriver *et al.* (1966). Onset was at 2 weeks of age.

Carnosinaemia

The original case reported by Perry *et al.* (1967), had a progressive neurological disorder with myoclonic seizures. Since then two further sibs have been described with seizures beginning at 2–4 months followed by progressive cerebral deterioration, flexion deformities, muscle wasting, and hyperactive reflexes (Terplan and Cares 1972). At post-mortem the findings were reminiscent of an infantile neuroaxal dystrophy (see p. 486).

Hyperprolinaemia—type II (pyrroline-5-carboxylic acid dehydrogenase deficiency)

Three sibs were reported by Similä and Visakorp (1967). Two sibs had convulsions and mental retardation. A further sib also had a high proline level, but was clinically normal.

Hyperprolinaemia type I (proline oxidase deficiency)

The association of renal disease and deafness with hyperprolinaemia and mild mental retardation has been established. Inheritance is recessive (Efron 1965).

Methionine malabsorption (oasthouse disease)

White hair, convulsions, mental retardation, an odour of hops, and hyperpnoea are the main features (Smith and Strang 1958). A second case was reported by Hooft *et al.* (1968).

Sulphite oxidase deficiency

Three children with severe neurological damage are known. One was reported by Shih *et al.* (1971*a*). The patient was normal until 18 months and then developed an acute hemiplegia, progressive cerebral degeneration, seizures, and dislocated lenses. A similar patient was reported by the author as a personal communication from Carton.

Phenylketonuria

Phenylketonuria is not a single genetic disease, but a number of defects in the hydroxylation of phenylalanine to tyrosine. The three main types are:

1. Classical infantile form (phenylalanine hydroxylase deficiency).
2. Dihydropteridine reductase deficiency or defective regeneration of BH_4
3. Biopterin deficiency, or defective synthesis of BH_4

(1) PKU type I   The classical infantile form is associated (in the untreated) with vomiting, microcephaly, mental retardation, a musty odour, and dermal hypopigmentation. The distinction between this condition and the

benign persistent hyperphenylalaninaemia cannot be made until the child is older than 2½ years.

The disease has a prevalence of about 1 in 10 000 births in Western Europe and 1 in 50 in that population is a carrier. The gene has now been cloned and localized to chromosome 12q (Woo *et al.* 1984 *b*). Prenatal diagnosis using a gene specific probe has also been achieved (Woo *et al.* 1984*a*).

(2) PKU type II Hyperphenylalaninaemia due to dihydropteridine reductase deficiency (Smith *et al.* 1975; Kaufman *et al.* 1975) is a separate condition associated with gross retardation and death despite dietary control. The activity of phenylalanine hydroxylase in liver is 20 per cent of normal, but there is no activity of dihydropteridine reductase present in the liver, brain, or in fibroblasts. The parents in the Kaufman case were first cousins.

Another possible variant was reported by Watts *et al.* (1979). The patient was retarded and had phenylketonuria, but phenylalanine intolerance was less severe than in the classical disease. Phenylalanine tolerance was further impaired by cotrimoxazole, a compound which increases intolerance in normals, but not in classical phenylketonuria.

About 1–3 per cent of all hyperphenylalaninaemias are due to tetrahydro-biopterin deficiency (BH$_4$). Known causes are dihydropteridine reductase deficiency, dihydrobiopterin synthetase deficiency, and guanosine triphosphate cyclohydrolase deficiency. (Longhi *et al.* 1985). These same authors report two unrelated children with the reductase deficiency.

Prognosis is poor if therapy is begun after the first few months of life.

(3) Biopterin deficiency (PKU type III) This is characterized by progressive neurological deterioration. The untreated patient of Fukuda *et al.* 1985 had at 4 months episodic stiffening of the extremities and opisthotonus. At 6 months he was grossly hypotonic. Other features are stridor and convulsions. Sibs were reported by Niederwieser *et al.* (1986).

Some patients have an onset at 1–2 years of delayed development accompanied by deterioration, dementia, ataxia (spinal) immobility (Parkinsonian), a Parkinsonian tremor, and rigidity. More peripherally there is muscle wasting and fasiculations. The pyramidal tracts can also be involved, and myoclonus and generalized seizures occur. Urine shows elevated homocysteine and folate is reduced in CSF, serum, and the red blood cells. A CT brain scan shows atrophy and low density areas in the white matter.

In another case a baby girl with a negative Guthrie test on day 6 had, at 6 months, severe muscle hypotonia and mental retardation. The serum phenylalanine was raised and diet was commenced with little improvement (Schaub *et al.* 1978). The activity of phenylalanine 4-hydroxylase and dihydropteridine reductase in liver were normal, but phenylalanine 4-hydroxylase was only 2 per cent of normal in the *in vivo* test using denatured phenylalanine. A defective dihydropterin biosynthesis is postulated and the biochemical abnormality responded to BH_4 (biopterin) the coenzyme of phenylalanine 4-hydroxylase.

An attempt to treat the condition by using L-dopa, carbidopa, and 5-hydroxy tryptophan gave promising results (Kaufman *et al.* 1978).

Carrier detection

Oral or intravenous loading tests, with measurement of the phenylalanine/ tyrosine ratio under fasting conditions, have been used to detect heterozygotes, but discrimination between the normal and the heterozygote is not perfect. Carrier detection tests are, however, still useful when treated homozygotes want to know the carrier status of their partners, and to a lesser extent to detect the carrier status of normal sibs of homozygotes and their spouses. Gene tracking using DNA probes might be useful.

Mass screening

In 1953, Bickel *et al.* showed that a low phenylalanine diet could prevent severe mental retardation in children with phenylketonuria.

Newborn infants in the United Kingdom are tested at 6–14 days and those with the disease are treated. Not all homozygotes will be mentally retarded (1 in 6 or 7 untreated homozygotes have normal intelligence), but those who are treated will, with few exceptions, have an IQ above 70.

Offspring of homozygotes

The offspring of homozygotes with phenylketonuria will all be heterozygous, but should be unaffected. If the homozygote is female and not on treatment during pregnancy the foetus will face the hazard of coping with the maternal metabolic disturbance. Reported offspring have been microcephalic and severely mentally retarded.

Homocystinuria

The classical clinical presentation is that of seizures, mental retardation (about one-third have normal intelligence), fair hair, malar flush, downward-displaced dislocated lenses (often not initially present), glaucoma, Marfan-like chest deformity, osteoporosis, and venous or arterial thrombosis. The biochemical defect is a deficiency of cystathione synthetase which normally converts homocystine to cystathionine. Heterogeneity is evident and Fowler *et al.* (1978) suggest three distinct types (one severe and two mild).

A comprehensive review of 629 patients by Mudd *et al.* (1985), gives an idea of the frequency of individual features. The chance of dislocated lenses in untreated B6 responsive disease at 10 years was 55 per cent as opposed to untreated B6—non-responsive patients—82 per cent. The average IQ in B6 responsive patients was 79 whereas the IQ in those with B6 unresponsiveness was 57.

The chances of having thromboembolic event at 15 years was 12 per cent for the responsive type vrs 27 per cent for the unresponsive type and for osteoporosis it was 36 per cent vrs 64 per cent.

Mental retardation as the reason for ascertainment was 4 per cent, but was a contributing factor in 51.7 per cent. Taking the figures together 55.7 per cent were retarded.

Prenatal diagnosis

Aminoaciduria

Achieved in:
    Argininosuccinic aciduria
    Citrullinaemia
    Homocystinuria
    Maple syrup urine (severe infantile form)
    Methylmalonic acidaemia
    Propionic acidaemia
    Ornithine-transcarbamylase deficiencies
    PKU

Possible in:
    Hypervalinaemia
    Ornithinaemia

Lipid metabolism

Tay–Sachs and related gangliosides
Gaucher's disease
Niemann–Pick
GM_1 gangliosidosis
Metachromatic leucodystrophy
Krabbe's disease
Farber's lipogranulomatosis
Fabry's disease
Batten's disease
Mucopolysaccharidoses/mucolipidoses
    Hurler/Scheie
    Hunter
    Sanfilippo A and B

Maroteaux–Lamy
Adrenoleucodystrophy
Mucolipidosis II, III, IV
Mannosidosis
Fucosidosis
Sialidosis

Carbohydrate metabolism

Glycogen storage type II, IV
Galactosaemia

Others

Lesch–Nyhan syndrome
Menkes disease
Xeroderma pigmentosa

Ultrasonography

Total agenesis of the corpus callosum
Neutral tube defects
Hydrocephaly, hydranencephaly
Microcophaly
Severe limb reduction deformities
Porencephaly
Joubert syndrome

Condition with vacuolated lymphocytes

Small   Niemann–Pick disease
        Pompe's disease

Large   GMI gangliosidosis
        Mannosidosis
        Fucosidosis
        I-cell disease
        Mucolipidosis III
        Salla disease
        Aspartylglycosaminuria
        Juvenile Batten's

Bibliography

Aagenaes. O. (1959). Hereditary spastic paraplegia. *Acta psychiat. scand.* **34**. 489–94.

Aarlie J.A. (1969). Oculopharyngeal muscular dystrophy. *Acta neurol. scand.* **45**, 484–92.

Aase, J.M. and Smith, D.W. (1970). Facial asymmetry and abnormalities of palms and ears. A dominantly inherited developmental syndrome. *J. Pediat.* **76**, 928–30.

Abarbanel, J.M., Bashan, N., Potashnik, R., *et al.* (1986). Adult muscle phosphorylase "b" kinase deficiency. *Neurol.* **36**, 560–2.

Abb, L. and Schaltenbrand. G. (1956). Statistische Untersuchengen zum Problem der Multiplen sklerose II. Mitteilung. *Dt. Z. NervHeilk.* **174**, 199–218.

Abdallat, A., Davis, S.M., Farrage, J., and McDonald, W.I. (1980). Disordered pigmentation, spastic paraparesis and peripheral neuropathy in three siblings: a new neurocutaneous syndrome. *J. Neurol. Neurosurg. Psychiat.* **43**, 962–6.

Abdul-Karim, R., Iliya, F., and Iskandar, G. (1964). Consecutive hydrocephalus: report of two cases. *Obst. Gynec, N. Y.* **24**, 376–8.

Abdurrahman, M.B. and Gassmann, L. (1976). Kocher–Debre–Semelaigne syndrome in presumed identical twins. *Niger. J. Pediat.* **3**, 57–61.

Aberfield, D.C. and Namba, T. (1969). Progressive ophthalmoplegia in Kugelberg–Welander disease—report of a case. *Arch Neurol., Chicago* **20**, 253–6.

Aberfeld, D.C. and Rao, K.R. (1981). Familial arteriovenous malformation of the brain, *Neurol.,* **31**, 184–6.

Aberfeld, D.C., Hinterbuchner, L.P. and Schneider, M. (1965). Myotonia dwarfism, diffuse bone disease and unusual ocular and facial abnormalities (a new syndrome). *Brain* **88**, 313–22.

Acheson, E.D. (1977). Epidemiology of multiple sclerosis. *Br. med. Bull.* **33**, 9–14.

Acosta Rua, G.J. (1978). Familial incidence of ruptured intracranial aneurysms. *Arch Neurol., Chicago* **35**, 675–7.

Adachi, M. and Volk, B.W.(1968). Protracted form of spongy degeneration of the central nervous system (van Bogaert and Bertrand type). *Neurol., Minneap.* **18**, 1084–92.

Adachi, M., Wellman, K.F., and Volk, B.W. (1968). Histochemical studies on the pathogenesis of idiopathic non-arteriosclerotic cerebral calcification. *J. Neuropath. exp. Neurol.* **27**, 483–99.

Adachi, M., Volk, B.W., and Torii, J. (1970). Histochemical, ultrastructural and biochemical studies of a case with leukodystrophy due to congenital deficiency of myelin. *J. Neuropath. exp. Neurol.* **29**, 149–50.

Adachi, M., Schneck, L., Cara, J., and Volk, B.W. (1973) Spongy degenaration of the central nervous system (Van Bogaert and Bertrand type; Canavan's disease). *Hum. Path.* **4**, 331–47.

Adams, H.P., Subbiah, B., and Bosch, E.P. (1977). Neurologic aspects of hereditary hemorrhagic telanglectasia. *Archs Neurol., Chicago* **34**, 101–4.

Adams, J., Scaravilli, F., and Spokes, E. (1979). Clinical and neuropathological studies of a family with cerebral amyloidoses and dementia. *Neuropath. appl. Neurobiol.* **5**, 320.

Adams, J.H., Blackwood, W., and Wilson, J. (1966). Further clinical and pathological observations on Leber's optic atrophy. *Brain* **89**, 15–26.

Adams, R.D., Van Bogaert, L., and Van der Eecken, H. (1961). Dégénéréscences nigro-striées et cérébello nigro-striées. *Psychiatria Neurol.* **142**, 219–59.

Adcock, M.W. and O'Brien, W.E. (1984). Molecular cloning of cDNA for rat and human carbamyl phosphate synthetase. *J. Biol. Chem.* **259**, 13471–6.

Adie, W.J. (1932). Tonic pupils and absent tendon reflexes: a benign disorder genesis: its complete and incomplete forms. *Brain* **55**, 98–113.

Adler, E. (1961) Familial cerebral palsy. *J. chron. Dis.* **13**, 207–14.

Adolfsson, R., Forsell, A., and Johansson, G. (1978). Hereditary polycystic osteodysplasia with progressive dementia in Sweden. *Lancet* **i**, 120–10.

Adornato, B.T., O'Brien, J.S., Lampert, P.W. Roe T.F., and Neustein, H.B. (1972). Cerebral spongy denegeration of infancy—a biochemical and ultrastructural study of affected twins. *Neurol., Minneap.* **22**, 202–10.

Adornato, B.T., Afifi, A.K., der Kaloustian, V.M., and Mire, J.J. (1972). Muscular abnormality in xeroderma pigmentosum. High-resolution light microscopy and electron microscopic observations. *J. neurol Sci.* **17**, 435–42.

Afifi, A.K., Smith, J.W. and Zellweger, H. (1965). Congenital nonprogressive myopathy. Central core disease and nemalinc myopathy in one family. *Neurology* **15**, 371–81.

Aggerbeck, L.P., McMahon L.P., and Scanu, A.M. (1974). Hypobetalipoproteinemia: clinical and biochemical description of a new kindred with 'Friedreich's ataxia'. *Neurol, Minneap.* **24**, 1051–63.

Agamanolis, D.P., *et al.* (1986). Lipoprotein disorder, cirrhosis, and olivopontocerebellar degeneration in two siblings. *Neurol.*, **36**, 674–81.

Aguilar, L., Lisker, R., Hernandez-Peniche, J., and Martinez-Villar, C.

(1978*a*). A new syndrome characterised by mental retardation, epilepsy palpebral conjunctival telangiectasias and IGA deficiency. *Clin. Genet.* **13**, 154–8.

Aguilar, L. and Ramos, G.G. (1978*b*). Unusual inheritance of Becker type muscular dystrophy. *J. med. Genet.* **15**, 116–18.

Aicardi, J., Lefebvre, J., and Lerique-Koechlin, A. (1965). A new syndrome: spasms in flexion callosal agenesis, ocular abnormalities. *Electroenceph. clin. Neurophysiol.* **19**, 609–10.

Aicardi, J. and Goutieres, F. (1984). A progressive familiar encephalopathy in infancy with calcification of the basal ganglia and chronic cerebrospinal fluid lymphocytosis. *Ann. Neurol.* **15**, 49–54.

Airaksinen, E.M. (1984). Famillial porencephaly, *Clin. Genet.,* **26**, 236–8.

Akelaitis, A.J. (1938). Hereditary form of primary parenchymatous atrophy of cerebellar cortex associated with mental deterioration. *Am. J. Psychiat.* **94**, 1115–40.

Alajouanine, Th. and Nick, J. (1959). Sur'trois cas familiaux de sclerose laterale amyotrophique (forme commune: forme bulbaire a evolution aiguë: forme a type de poliomyelite anterieure chronique) survenus dans la meme fratrie. *Rev neurol.* **100**, 490–2.

Alajouanine, Th., Aubry, M., and Nehlil, J. (1943). Sur une affection familiale carcterisée par un syndrome de desequilibration avec importantes pertubations vestibulaires centrales. *Rev neurol.* **75**, 252.

Alberca, R., *et al.*. (1980). Progressive bulbar paralysis associated with neural deafness. *Arch. Neurol.,* **37**, 214–16.

Alberca, R., *et al.* (1987). Late onset Parkinsonian syndrome in Hallervorden–Spatz disease. *J. Neurol. Neurosurg. Psychiat.,* **50**, 1665–8.

Albers, J.W., *et al.* (1983). Juvenile progressive bulbar palsy, diagnostic findings. *Arch. Neurol.,* **40**, 351–3.

Albert, E. (1976). Ischaemic infarct of the brain stem combined with bisymptomatic Klippel-Trenaunay–Weber syndrome and cutis laxa. *J. Neurol. Neurosury. Psychiat.,* **39**, 581–5.

Aleck, K.A. and Shapiro, L.J. (1978). Genetic-metabolic considerations in the sick neonate. *Pediat. Clins N. Am.* **25**, 431–51.

Alexander, M.P., Emery, E.S., Koerner, F.C. (1976). Progressive bulbar paresis in childhood. *Arch. Neurol., Chicago* **33**, 66–8.

Alexander, W.S. (1966). Phytanic acid in Refsum's syndrome. *J. Neurol Neurosurg. Psychiat.* **29**, 412–16.

Alfonso, I. *et al.* (1986). Spinal cord involvement in encephalocutaneous lipomatosis. *Ped. Neurol.,* **2**, 380–4.

Allan, W. (1928). Inheritance of migraine. *Arch. intern. Med.* **42**, 590–9.

Allan, W. (1937). Inheritance of the shaking palsy. *Arch. intern. Med.* **60**, 424–36.

Allan, W. (1938). Familial occurence of tic Douloureux. *Arch. Neurol. Psychiat., Chicago* **40**, 1019-20.

Allan, W. (1939). Relation of hereditary pattern in clinical severity as illustrated by peroneal atrophy. *Arch. Intern. Med.*, **63**, 1123-31.

Allanson, J.E. (1987). Syndrome of the month: Noonan syndrome. *J. Med. Genet.*, **24**, 9-13.

Allen. H.B. and Parlette, H.L. (1973). Coat's disease. *Arch. Derm.* **108**, 413-15.

Allen, N. (1964). In *Pediatric neurology* (ed. T.W. Farmer), p. 162. Hoeber, New York.

Allison, R.S. (1963). Some neurological aspects of medical geography. *Proc. R. Soc. Med.* **56**, 71-6.

Allport, R.B. (1971). Mental retardation and spastic paraparesis in four of eight siblings. *Lancet* ii, 1089.

Alpers, B.J. (1931). Diffuse progressive degeneration of the gray matter of the cerebrum. *Arch. Neurol. Psychiat.* **25**, 469-505.

Alpers, B.J. (1960). Progressive cerebral degeneration of infancy. *J. nerv. ment. Dis.* **130**, 442-8.

Alström, C.H. (1950). A study of epilepsy in its clinical, social and genetic aspects. *Acta psychiat. neurol. scand.* Suppl. 63, 1-284.

Alström, C.H. Hallgren, B., Nilsson, L.B., and Asander, H. (1959). Retinal degeneration combined with obesity, diabetes mellitus and neurogenous deafness. A specific syndrome (not hitherto described) distinct from the Laurence-Moon-Bardet-Biedl syndrome. *Acta psychiat. neurol. scand.* **34** Suppl. 129, 1-35.

Alter M. (1963). Familial aggregation of Bell's palsy. *Arch. Neurol., Chicago* **8**, 557-64.

Alter, M. and Kennedy, W. (1968). The Marinesco-Sjögren syndrome. *Minn. Med.* **51**, 901-6.

Alter, M. and Schaumann, B. (1976). Hereditary amyotrophic lateral sclerosis. A report of two families. *Eur. Neurol.* **14**, 250-65.

Alter, M., Morariu, M., and Cohen-Libman, J. (1974). Hereditary amyotrophic lateral sclerosis. (Abstract.) *Neurol., Minneap.* **24**, 356.

Alter, M., Talbert, O.R., and Croffead, G. (1962). Cerebellar ataxia, congenital cataracts and retarded somatic and mental maturation. Report of cases of Marinesco-Sjögren syndrome. *Neurol., Minneap.* **12**, 836-47.

Alter, M., Harshe, M., Anderson, V.E., and Yunis, E.J. (1976). Genetic association of multiple sclerosis and H.L.-A determinants. *Neurol., Minneap.* **26**, 31-6.

Alves, D., *et al.* (1986). Four cases of late onset metachromatic leucodystrophy in a family: clinical, biochemical and neuropathological studies. *J. Neurol. Neurosurg. Psychiat.* **49**, 1417-22.

Alzheimer, A. (1910-11). Uber eigenartige Krankheitafälle des spateren Alters. *Z. ges. Neurol. Psychiat.* **4**, 356-85.

Ambler, M., Pogacar, S., and Sidman, R. (1969). Lhermitte–Duclos disease (granule cell hypertrophy of the cerebellum): pathological analysis of the first familial cases. *J. Neuropath. exp. Neurol.* **28**, 622–47.

Ametani, T. (1974). Infantile neuro-axonal dystrophy (INAD). Light and electron microscopic observation of an autopsy case. *Neuropaediatrie* **5**, 63–70.

Aminoff, M.J. (1972). Acanthocytosis and neurological disease. *Brain* **95**, 749–60.

Aminoff, M.J., Marshall, J., Smith, E.M., and Wyke, M.A. (1975), Pattern of intellectual impairment in Huntington's chorea. *Psychol. Med* **5**, 169–72.

Amirhakimi, G.H., Haghighi, P., Ghalambor, M.A., and Honari, S. (1976). Familial lipogranutomatosis (Faber's disease). *Clin. genet.* **9**, 625–30.

Amit, R. (1987). Familial juvenile onset Bell's palsy. *Eur. J. Pediat.* **146**, 608–9.

Ammermann, O. (1940). Isolierte Schädigung der unteren Oliven bei Myoklonus Epilepsie. *Arch. Psychiat. Nevenkh. Kr* **III**, 213–32.

Amolg, C. and Tal, E. (1968). A family with Kugelberg-Welander syndrome. *Confin. Neurol., Basel* **30**, 313–24.

Amuso, S.J. and Mankin, H.J. (1967). Hereditary spondylolisthesis and spina bifida. Report of a family in which the lesion is transmitted as an autosomal dominant through three generations. *J. Bone Jt Surg.* **49A**, 507–13.

Andermann, E., Remillard, G.M., Goyer, C., Blitzer, L., Andermann, F., and Barbeau. A. (1976). Genetic and family studies in Friedrcich's ataxia. *Can. J. neurol. Sci.* **3**, 287–301.

Andermann, E., *et al.* (1977). Agenesis of the corpus callosum with sensorimotor neuropathy. *Vth. Int. Conf. Birth Defects.* Montreal.

Andermann, F., Metrakos, J.D., Andermann, F., and Carpenter, S. (1969). Spongy degeneration of the central nervous system in infancy. *Prog. Neurogenet.* 794–808.

Andermann, F., Joubert, M., Karpati, G., Carpenter, S., and Melancon, D. (1972). Familial agenesis of the corpus callosum with anterior horn cell disease. A syndrome of mental retardation, areflexia and paraplegia. *Trans. Am. neurol. Ass.* **97**, 242–4.

Andermann, F., Andermann, E., Carpenter, S., Karpati, G., and Wolfe, L. (1974). Late onset GM_2 gangliosidosis in two Lebanese families. *Am. J. hum. Genet.* **26**, 10A.

Andersen, B. (1965). Marinesco–Sjögren syndrome: spinocerebellar ataxia, congenital cataract, somatic and mental retardation. *Devl med. Child Neurol.* **7**, 249–57.

Anderson, B., Margolis, G., and Lynn, W.S. (1958). Ocular lesions related to disturbances of fat metabolism, *Am. J. Ophthal.* **45**, Suppl., 23–41.

Anderson, F.M. and Geiger, L. (1965). Craniosynostosis. *J. Neurosurg.* **22**, 229–40.

Anderson, L.G., Cook, A.J., Coccaro, P.J., Coro, C.J., and Bosma, F. (1972). Familial osteodysplasia. *J. Am. med. Ass.* **220**, 1687–93.

Anderson, P.M., Reddy, R.M., Anderson, K.E., and Desnick, R.J. (1981). Characterization of the porphobilinogen deaminase deficiency in acute intermittent porphyria: immunologic evidence for heterogeneity of the genetic defect. *J. Clin. Invest.* **68**, 1–12.

Andersson, R. (1970). Hereditary amyloidosis with polyneuropathy. *Acta med. scand.* **188**, 85–94.

Anderson, R. and Hofer, P.A. (1974). Primary amyloidosis with polyneuropathy: some aspects on the histopathological diagnosis antemortum based on studies of specimens from 30 familial and non-familial cases. *Acta med. scand.* **196**, 115–20.

Andrade, C. (1952). A peculiar form of peripheral neuropathy. Familial atypical generalized amyloidosis with special involvement of the peripheral nerves. *Brain* **75**, 408–27.

Andrade, C. (1963). Clinique de la paramyloidose du type portugaise C. Andrade. *Acta neuropath.* Suppl. 2, 3–11.

Andrade, C. Canijo, M., Klein, D., and Kaelin, A. (1969). The genetic aspect of familial amyloidotic polyneuropathy. Portugese type of paramyloidosis. *Humangenet.* **7**, 163–5.

Andreas–Zietz, A. *et al.* (1986). DR2-negative narcolepsy. *Lancet* **iii**, 684–5.

Andrews, J.M., Sorenson, V., Cancilla, P.A., Price, H.M., and Menkes, J.H. (1971). Late infantile neurovisceral storage disease with curvilinear bodies. *Neurol., Minneap.* **21**, 20–17.

Andria, G., Di Natale, P., Del Giudice, E., Strisciuglio, P., and Murino, P. (1979). Sanfillipo B syndrome (MPS IIIB): mild and severe forms within the same sibship. *Clin. Genet.* **15**, 500–4.

Andria, G. *et al.* (1984). Steroid sulphate deficiency is present in patients with the syndrome 'Ichthyosis and male hypogonadism and with 'Rud syndrome'. *J. Inherit. Met. Dis.* **7**, Suppl. 2, 159–60.

Andriola, M. and Stolfi, J. (1972). Sturge–Weber syndrome. Report of an atypical case. *Am. J. Dis. Child.* **123**, 507–10.

Androp, S. (1941). Leber's primary optic atrophy with other central nervous system involvement. *Psychiat. Q.* **15**, 215–23.

Angelman, H. (1965). 'Puppet' children, a report on three cases. *Dev. Med. Child Neurol.* **7**, 681–8.

Annegers, J.F., Hauser, W.A., Elveback, L.R., Anderson, V.E. and Kurland, L.T. (1976). Seizure disorders in offspring of parents with a history of seizures—a maternal-paternal difference? *Epilepsia* **17**, 1–9.

Anyane-Yeboa, K., Gunning, L., and Bloom, A.D. (1980). Baller–Gerold syndrome: craniosynostosis-radial aplasia syndrome. *Clin. Genet.* **17**, 161–6.

Anzil, A.P., Blinzinger, K.M., Mehraein, P., and Dozic, S. (1973). Niemann–Pick disease type C: case report with ultrastructural findings. *Neuropaediatrics* **4**, 207–25.

Appenzeller, O. and Amick, L. (1972). Paralysis with paradoxic myotonia. *Trans. Am. neurol. Ass.* **97**, 245–7.

Appenzeller, O., Kornfeld, M., and Snyder, R. (1976). Acromutilating, paralyzing neuropathy with corneal ulceration in Navajo children. *Archs Neurol., Chicago* **33**, 733–8.

Araki, S., Mawatari, S., Ohta, M., Nakajima, A., and Kuroiwa, Y. (1968). Polyneuritic amyloidosis in a Japanese family. *Archs. Neurol., Chicago* **18**, 593–602.

Arbisser, A.I., *et al.* (1977). Morquio-like syndrome with beta galactosidase deficiency and normal hexosamine sulfatase activity. *Am. J. med. Genet.* **1**, 195–205.

Arbizu, T., Santamaria, J., Gomez, Quibez, A., and Serra, J.P. (1983). A family with adult spinal and bulbar muscular atrophy, X-linked inheritance and associated testicular failure. *J. Neurol. Sci.* **59**, 371–82.

Archer, I.M. Kingston, H.M. and Harper, P.S. (1984). Prenatal diagnosis of Hunter syndrome. *Prenatal Diagnosis* **4**, 195–200.

Argov, Z. Navon, R. (1984). Clinical and genetic variations in the syndrome of adult GM$_2$ gangliosidosis resulting from hexosamidase A deficiency. *Ann. Neurol.* **16**, 14–20.

Arima, M. and Sano, I. (1968). Genetic studies of Wilson's disease in Japan. *Birth Defects.* **4**, 54–9.

Armendares, S. (1970). On the inheritance of craniostenosis. Study of 13 families. *J. Génét. hum.* **18**, 121–34.

Armstrong, D., Pickrell, K., Fetter, B., and Pitts, W. (1965). Torticollis: an analysis of 271 cases. *Plastic reconstr. Surg.* **35**, 14–25.

Armstrong, F.S. (1962). Hyperkalaemic familial periodic paralysis. (adynamia episodica hereditaria). *Ann. intern. Med.* **57**, 455–61.

Armstrong, R.M. and Hanson, C.W. (1969). Familial gliomas. *Neurol., Minneap.* **19**, 1061–3.

Armstrong, R.M., Fogelson, M.H., and Silberberg, D.H. (1966). Familial proximal spinal muscular atrophy. *Archs Neurol., Chicago* **14**, 208–12.

Armstrong, R.N., Koenigsberger, R., Mellinger, J., and Lovelace, R.E. (1971). Central core disease with congenital hip dislocation: study of two families. *Neurol., Minneap.* **21**, 369–76.

Arnason, B.G.W., Fuller, T.C., Lehrich, J.R., and Wray, S.H. (1974) Histocompatibility types and measles antibodies in multiple sclerosis and optic neuritis. *J. neurol. Sci.* **22**, 419–28.

Aronson, S.M., Aronson, B.E., and Volk, B.W. (1959). A genetic profile of infantile amaurotic idiocy. *Am. J. Dis. Child.* **98**, 50–65.

Aronson, S.M., Valsamis, M.P., and Volk, B.W. (1960). Infantile amaurotic family idiocy. Occurence, genetic considerations and pathophysiology in the non-Jewish infant. *Pediatrics, Springfield* **26**, 229–42.

Arseni, C., Nereantiu, F., and Nicolescu, P. (1973). The infantile form of diffuse sclerosis with meningeal angiomatosis. *Neurol., Minneap.* **23**, 1297–301.

Arts, W.F. and de Groot, C.J. (1983). Congenital nemaline myopathy: two patients with consanguineous parents, one with a progressive course. *J. Neurol.* **230**, 123–30.

Arts, W.F., Bethlem, J., Dingemans, K., and Eriksson, A.W. (1978*a*). Investigations on the inheritance of nemaline myopathy. *Archs. Neurol., Chicago* **35**, 72–7.

Arts, W.F., Bethlam, J., and Volkers, W.S. (1978*b*). Further investigations on benign myopathy with autosomal dominant inheritance. *J. Neurol.* **217**, 201–6.

Arts, W.F.M., *et al.*. (1983*a*). Hereditary neuralgic amyotrophy. Clinical, genetic, electrophysiological and histopathological studies. *J. Neurol. Sci.* **62**, 261–79.

Arts, W.F.M, *et al.* (1983*b*). NADH-CoQ reductase deficient myopathy: successful treatment with riboflavin. *Lancet* **ii**, 581–2.

Asch, A.J. and Myers, G.J. (1976). Benign familial megalencephaly. *Pediatrics. Springfield* **57**, 535–9.

Aschoff, J.C., Becker, W., and Rettelbach, R. (1976). Voluntary nystagmus in five generations. *J. Neurol. Neurosurg. Psychiat.* **39**, 300–4.

Asher, P. and Schonell, F.E. (1950). A survey of 400 cases of cerebral palsy in childhood. *Arch. Dis. Childh.* **25**, 360–79.

Askanas, V., *et al.* (1979). X-linked recessive congenital muscle fibre hypotrophy with central nuclei. *Arch. Neurol., Chicago* **36**, 604–9.

Assmann, G., Simantke, O., Schaefer, H.E., and Smootz, E. (1977). Characterizations of high density lipoproteins in patients heterozygous for Tangier disease. *J. Clin. Invest.* **60**, 1025–35.

Aston, J.F. *et al.* (1984). Plasma pyruvate kinase and creatinine kinase activity in Becker muscular dystrophy. *J. Neurol. Sci.* **65**, 307–14.

Attal, C., Robain, O., and Chapnis, G. (1975). Familial nerve trunk paralyses. *Devl. med. Child Neurol.* **17**, 787–92.

Aubourg, P.R., *et al.* (1987). Linkage of adrenoleukodystrophy to a polymorphic DNA probe. *Ann. Neurol.* **21**, 349–52.

Auerbach, S.H., Dipiero. T.J., and Mejlszenkier, J. (1981). Familial recurrent peripheral facial palsy. *Arch. Neurol.* **38**, 463.

Aughton, D.J. and Cassidy, S.B. (1987). Hydrolethalus syndrome: Report of an apparent mild case, literature review and differential diagnosis. *Am. J. Med. Genet* **27**, 935–42.

Aula, P., Rapola, J., Autio, S., Raivio, K., and Karjalainen, O. (1975).

Prenatal diagnosis and fetal pathology of I-cell disease. (mucolipidosis type II). *J. Pediat.* **87**, 221–6.

Auld, A. W. and Bauermann, A. (1965). Trigeminal neuralgia in six members of one generation. *Arch. Neurol., Chicago* **13**, 194.

Austin, J., *et al.* (1968). Metachromatic leukodystrophy (M.L.D.) VIII M.L.D. in adults: diagnosis and pathogenesis. *Arch. Neurol., Chicago* **18**, 225–40.

Austin, J.H. (1965). Metachromatic leukodystrophy. In *Medical aspects of mental retardation* (ed. C.C. Carter) p. 768. Thomas, Springfield III.

Austin, J.H. and Stears, J.C. (1971). Familial hypoplasia of both internal carotid arteries. *Arch. Neurol., Chicago* **24**, 1–16.

Autio, S., Norden, N.E., Ockerman, P.A., Riekkinen, P., Rapola, J., and Louhimo, T. (1973). Mannosidosis, clinical fine-structural and biochemical findings in three cases. *Acta pediat. scand.* **62**, 555–65.

Avigan, J., Campbell, B.D., Yost, D.A., *et al.* (1985). Sjogren-Larsson syndrome: delta 5- and delta 6-fatty acid desaturases in skin fibroblasts. *Neurology* **35**, 401–3.

Avila-Giron, R. (1973). Medical and social aspects of Huntington's chorea in the State of Zulia, Venezuela. *Adv. Neurol.* **1**, 261–6.

Axelrod, F.B., Nachtigal, R., and Dancis, J. (1974). Familial dysautonomia: diagnosis, pathogenesis and management. *Adv. Pediat.* **21**, 75–96.

Axelrod, F.B., Pearson, J., Tepperberg, J., and Ackerman, B.D. (1983). Congenital sensory neuropathy with skeletal dysplasia. *J. Pediat.* **102**, 727–30.

Aylsworth A.S. Thomas, G.H. and Hood, J. (1980). A severe infantile sialidosis: clinical, biochemical, and microscopic features. *J. Pediat.* **96**, 662–8.

Baar, H.S. and Gabriel, A.M. (1966). Sex linked spastic paraplegia. *Am. J. ment. Defic.* **71**, 13–18.

Bacon, P.A. and Smith, B. (1971). Familial muscular dystrophy of late onset. *J. Neurol. Neurosurg. Psychiat.* **34**, 93–7.

Bader, J.L. and Miller, R.W. (1978). Neurofibromatosis and childhood leukemia. *J. Pediat.* **92**, 925–9.

Baier, W.K. and Doose, H. (1985). *Petit-mal* absences of childhood onset: familial prevelance of migraine and seizures. *Neuropediat.* **16**, 84–91.

Baird, P.A. and De Jong, B.P. (1972). Noonans syndrome (XX and XY Turner phenotype) in three generations of a family. *J. Pediat.* **80**, 110–14.

Bakay, B., Tucker-Pian, and Seegmiller, J.E. (1980). Detection of Lesch-Nyhan syndrome carriers; analysis of hair roots for HPRT by agarose gel electrophoresis and autoradiography. *Clin. Genet.* **17**, 369–74.

Baker, R.S. Ross, F.A., and Baumann, R.J. (1987). Neurologic complications of the epidermal nevus syndrome. *Arch. Neurol.* **44**, 227–32.

Bakwin, H. (1973). Reading disability in twins. *Devl med. Child Neurol.* **15**, 184–7.

Bale, A.E., *et al.* (1987). Linkage analysis in spinopontine atrophy: correlation of HLA linkage with phenotypic findings in hereditary ataxia. *Am. J. Med. Genet.* **27**, 595–602.

Ball, M.J. (1980). Features of Creutzfeldt-Jakob disease in brains of patients with familial dementia of Alzheimer type. *Can J. neurol. Sci.* **7**, 51–7.

Baller, F. (1950). Radiusaplasie and Inzucht. *Z. mensch. Vererb.-u. Konstitlehre* **29**, 782–90.

Ballet, G. and Rose, F. (1904). Un cas d'amyotrophie du type Charcot–Marie avec atrophie des deux nerfs optiques. *Revue neurol.* **12**, 522–4.

Bamezai, R., *et al.* (1987). Cerebellar ataxia and total albinism. *Clin. Genet.* **31**, 178–81.

Bank, W.J. and Morrow, G. (1972). A familial spinal cord disorder with hyperglycinemia. *Arch. Neurol., Chicago* **27**, 136–44.

Bank, W.J., Di Mauro, S., Bonilla, E., Capuzzi, D.M., and Rowland, L.P. (1975). A disorder of muscle lipid metabolism and myoglobinuria – absence of carnitine palmityltransferase. *N. Engl. J. Med.* **292**, 443–9.

Banker, B.Q., Victor, M., and Adams R.D. (1957). Arthrogryposis multiplex due to congenital muscular dystrophy. *Brain* **80**, 319–34.

Banker, B.Q., Robertson, J.T., and Victor, M. (1964). Spongy degeneration of the central nervous system in infancy. *Neurol., Minneap.* **14**, 981–1001.

Bannerman, R.M., Graf, C.J., and Upson, J.F. (1967). Ehlers–Danlos syndrome. *Br. med. J.* **iii**, 558.

Bannerman, R.M., Ingall, G.B., and Graf, ·C.J. (1970). The familial occurence of intracranial aneurysms. *Neurol. Minneap.* **20**, 283–92.

Bannister, R. and Oppenheimer, D.R. (1972). Degenerative diseases of the nervous system associated with autonomic failure. *Brain* **95**, 457–74.

Baraitser, M. (1977). Genetics of Möbius syndrome. *J. med. Genet.* **14**, 412–17.

Baraitser, M. (1982). The hypertelorism microtia clefting syndrome. *J. med. Genet.* **19**, 387–9.

Baraitser, M. and Parkes, J.O. (1978). Genetic study of narcoleptic syndrome. *J. med. Genet.* **15**, 254–9.

Baraitser, M., Bowen-Bravery, M., and Saldana-Garcia, P. (1980). Pitfalls of genetic counselling in Pfeiffer's syndrome. *J. med. Genet.* **17**, 250–6.

Baraitser, M., Brett, E.M., and Piesowicz, A.T. (1983*a*). Microcephaly and intracranial calcification in two brothers. *J. med. Genet.* **20**, 210–12.

Baraitser, M. Winter, R.M. and Brett, E.M. (1983*b*). Greig cephalopolysyndactyly: Report of 13 affected individuals in three families. *Clin. Genet.* **24**, 257–65.

Baraitser, M., Patton, M.A., Lam, S.T.S., Brett, E.M. and Wilson, J.

(1987). The Angelman (happy puppet) syndrome: is it autosomal recessive? *Clin. Genet.* **31**, 323–30.

Barbeau, A. (1967). Oculopharyngeal muscular dystrophy in French Canada. *2nd Int. Cong. Neuro-genetics and Neuro-opthalmology,* Montreal, p. 76.

Barbeau, A. (1970). Parental ascent in the juvenile form of Huntington's chorea. *Lancet* **ii**, 937.

Barbeau, A. (1978). Friedreich's ataxia 1978. An overview. *Can. J. neurol. Sci.* **5**, 161–5.

Barbeau, A. (1980). Distribution of ataxia in Quebec. In *Spino-cerebellar degeneration* (ed. I. Sobue), pp. 120–2. Tokyo University Press.

Barbeau, A., *et al.* (1984). Recessive ataxia in Acadians and Cajuns. *Can. J. Neurol. Sci.* **11**, 526–33.

Barbeau, A. and Bourcher E., (1982). New data on the genetics of Parkinsons disease. *Can. J. Neurol. Sci.* **9**, 53–60.

Barbeau, A. and Giroux, J.-M. (1972). Erythrokeratodermia with ataxia. *Trans. Am. neurol. Ass.* **97**, 55–6.

Barbieri, F., Filla, A., de Falco, F.A., and Buscaino, G.A. (1980). Alexanders disease. A clinical study with computerised tomographic scans of the first two Italian cases. *Ital. Acta neurol.* **35**, 1–9.

Bardach, M. (1925). Systematisierte Naevusbildungen bei einem einerligen Zwillingspaar. *Z. Kinderheilk.* **39**, 542–50.

Bardet, G. (1920). Sur un syndrome d'obcsite infantile avec polydactylie et retinite pigmentaire. These de Paris, No 470.

Bargeton, E., Nezelot, Cl., Guran, P., and Job, J.-C. (1961). Etude anatomique d'un cas d'arthrogrypose multiple congenitale et familiale. *Rev. neurol.* **104**, 479.

Barker, D., *et al.* (1987). Gene for Von Recklinghausen neurofibromatosis is in the pericentromeric region of chromosome 17. *Science* **236**, 1100–2.

Barlow, M.B. and Isaacs, H. (1970). Malignant hyperpyrexial deaths in a family. Report of three cases. *Br. J. Anaesth.* **42**, 1072–6.

Barnard, R.O. and Lang, E.R. (1964). Cerebral and cerebellar gliomas in a case of von Recklinghausen's disease with adrenal phaeochomocytomas. *J. Neurosurg.* **21**, 506–11.

Barolin, G.S., Hodkewitsch, E., Höfinger, E., Scholz, H., Bernheimer, H., and Molzer, B. (1979). Klinisch-biochemiche Verlaufsuntersuchungen bei heredopathia atactica polyneuritiformis (Morbus Refsum). *Fortschr. Neurol. Psychiat.* **47**, 53–66.

Barraquer-Forré, L. and Barraquer-Bordas, L. (1953). De la semiologic ganglior adiculaire postérieure dans l'amyotrophie de Charcot–Marie–Tooth: troubles trophiques, douleurs fulgurantes, troubles sensitifs. *Acta neurol. psychiat. belg.* **53**, 55–70.

Barre, J.A. and Reys, L. (1924). Syringomyelie chez le frére et la soeur. *Rev. Neurol.* **11**, 521–30.

Barre, R.G., Suter, C.G., and Rosenblum, W.I. (1978). Familial vascular malformation or chance occurence? Case report of two affected family members. *Neurol., Minneap.* **28**, 98–100.

Barrie, M. and Heathfield, K. (1971). Diagnosis of ocular myopathy. *Br. J. Ophthal.* **55**, 633–8.

Barron, S.A., Heffner, R.R., and Zwirecki, R. (1979). A familial mitochondrial myopathy with central defect in neutral transmission. *Arch. neurol., Chicago.* **36**, 553–6.

Barth, P.G. (1987). Disorders of neuronal migration. *Can. J. Neurol. Sci.* **14**, 1–16.

Barth, P.G., Van Wijngaarden, G.K., and Bethlem, J. (1975). X-linked myotubular myopathy with fatal neonatal asphyxia. *Neurol., Minneap.* **25** 531–6.

Barth, P.G. Mullaart, R. Stam, F.C. and Sloop, J.L. (1982). Familial lissencephaly with extreme neopallial hypoplasia. *Brain Dev.* **4**, 145–51.

Barth, P.G. *et al.* (1983). A X-linked mitochondrial disease affecting cardiac muscle, skeletal muscle and neutrophil leucocytes. *J. Neutrol. Sci.* **62**, 327–55.

Bartlett, R.J. *et al.* (1987). A new probe for the diagnosis of myotonic muscular dystrophy. *Science* **235**, 1648–50.

Bartley, J.A. and Hall, B.D. (1978). Mental retardation and multiple congenital abnormalities of unknown etiology: frequency of occurence in similarly affected sibs of the proband. *Birth Defects* **14**, 127–37.

Bastiaensen, L.A.K. *et al.* (1978). Ophthalmoplegia- plus: a real nosological entity *Acta neurol. scand.* **58**, 9–34.

Batchelor, J.R., Compston, A., and McDonald, W.I. (1978). The significance of the association between HLA and multiple sclerosis. *Br. med. Bull.* **34**, 279–84.

Battersby, R.D.E., Ironside, J.W., and Maltby, E.L. (1986). Inherited multiple meningiomas: A clinical, pathological and cytogenetic study of an affected family. *J.N.N.P.* **49**, 362–7.

Bauman, M.L. and Hogan, G.R. (1973). Laurence–Moon–Biedl syndrome. *Am. J. Dis. Child.* **126**, 119–26.

Baunkotter, *et al.* (1985). N-Acetylneuraminic acid storage disease. *Hum. Genet.* **71**, 155–9.

Bautista, J., Rafel, E., Castilla, J.M., and Alberca, R. (1978). Hereditary distal myopathy with onset in early infancy. Observation of family. *J. Neurol. Sci.* **37**, 149–58.

Baxter, D.W. and Dyck, P.J. (1961). Paramyotonia congenita. *Can. med. Ass. J.* **85**, 113–18.

Bay-Nielsen, E. and Cohn. J. (1969). Hereditary defect of the sacrum and coccyx with anterior sacral meningocele. *Acta paediat., Stockh.* **58**, 268–74.

Bearn, A.G. (1960). A genetical analysis of thirty families with Wilson's disease (hepatolenticular degeneration). *Ann. hum. Genet.* **24**, 33–43.

Beaudet, A. L., Di Ferrante, N. M., Ferry, G. D., Nichols, B. L., Jr, and Mullins, C. E. (1975). Variation in the phenotype expression of beta-glucuronidase deficiency. *J. Pediat.* **86**, 388–94.

Beaumont, P. J. V. (1968). The familial occurence of berry aneurysm. *J. Neurol. Neurosurg. Psychiat.* **31**, 399–402.

Beaussart, M. and Loiseau. P. (1969). Hereditary factors in a random population of 5,200 epileptics. *Epilepsia* **10**, 55–63.

Becak, W., Becak, M. L., and Andrade, J. D. (1964). A genetical investigation of congenital analgesia. *Acta genet. Statist. med.* **14**, 133–42.

Beck, M., *et al.* (1984). Neuraminidase deficiency presenting as non-immune hydrops fetalis. *Eur. J. Pediat.* **143**, 135–9.

Becker, P. E. (1964*a*). Myopathien. In *Becker Humangenetic,* Vol. 3, Part 1, pp. 411–550. Springer, Stuttgart.

Becker, P. E. (1964*b*). Atrophia Musculorum Spinalis Pseudomyopathica Hereditäre neurogene proximale amyetrophie von Kugelberg und Welander. *Z. mensch. Vererb.-u KonstitLLehre* **37**, 193–220.

Becker, P. E. (1970). Paramyotonia congenita (Eulenburg). In *Advances in human genetics* (ed. P. E. Boiler, W. Leiz, F. Vogel, and G. G. Wendt) Vol. 3, pp. 1–134. Thieme, Stuttgart.

Becker, P. E. (1971). Genetic approaches to the nosology of muscle disease: myotonias and similar diseases. *Birth Defects* **7**, 52–62.

Becker, P. E. (1972). Neues zur Genetik und Klassifikation der Muskeldystrophien. *Hum. Genet.* **17**, 1–22.

Becker, P. E. (1977). Syndromes associated with myotonia. In *Pathogenesis of human muscular dystrophies* (ed. L. P. Rowland), pp. 699–703, Excerpta Medica, Amsterdam.

Becker, P. E. and Wieser, S. (1964). Zur Genetik der essentiellen Myoklonie. *Hum. Genet.* **1**, 14–23.

Beckett, R. S. and Netsky, M. G. (1953). Familial ocular myopathy and external ophthalmoplegia. *Archs Neurol. Psychiat., Chicago* **69**, 64–72.

Beebe, R. T. and Formal, P. F. (1954). Gargoylism. Sex- linked transmission in nine males. *Trans. Am. clin. clim. Ass.* **66**, 199–204.

Begleiter, M. L. and Harris, D. J. (1980). Holoprosencephaly and endocrine dysgenesis in brothers. *Am. J. Med. Genet.* **7**, 315–8.

Behan, P. O. and Bone, I. (1977). Hereditary chorea without dementia. *J. Neurol. Neurosurg. Psychiat.* **40**, 687–91.

Behan, P. O., Simpson, J. A. and Dick, H. (1973). Immune response genes in myasthenia gravis. Letter *Lancet* **ii**, 1033.

Behan, W. M. H. and Maia, M. (1974). Strümpell's familial spastic paraplegia: genetics and neuropathology. *J. Neurol. Neurosurg. Psychiat.* **37**, 8–20.

Behr, C. (1909). Die komplizierte hereditär-familiäre Optikusatrophie des Kindesalters. *Klin. Mbl. Augenheilk.* **47**, 138–60.

Behse, F. and Buchthal, F. (1977). Peroneal muscular atrophy (PMA) and related disorders. II. Histological findings in sural nerves. *Brain* **100**, 67–85.

Behse, F., Buchthal, F., Carlsen, F., and Knappeis, G.G. (1972). Hereditary neuropathy with liability to pressure palsies. Electrophysiological and histopathological aspects. *Brain* **95**, 777–94.

Beighton, P.H. (1971). Recessively inherited Charcot–Marie–Tooth syndrome in identical twins. *Clin. Determ. Birth Defects* **7**, 105.

Beighton,P.H. (1973). The Schwartz syndrome in Southern Africa. *Clin. Genet.* **4**, 548–55.

Beighton, P.H. (1978). *Inherited disorders of the skeleton.* Churchill Livingstone, Edingurgh.

Beighton, P.H. (1988) Sclerosteosis. *J. med. Genet.* **25**, 200–3.

Beighton, P.H. and Craig, J. (1973). Atlantoaxial subluxation in the Morquio syndrome. *J. Bone Jt Surg.* **55B**, 478–81.

Beighton, P.H. and Lindenberg, R. (1971). Alzheimer's disease in multiple members of a family. *Birth Defects.* **7**, 232–3.

Beighton, P.H., Durr, L., and Hamersma, H. (1976). The clinical features of sclerosteosis. A review of the manifestation in 25 affected individuals. *Ann. intern. Med.* **84**, 393–7.

Beighton, P.H., Horan, F., and Hamersma, H. (1977). A review of the osteopetroses. *Post-grad. med. J.* **53**, 506–15.

Beighton, P.H., Hamersma, H., and Horan, F. (1979*a*). Craniometaphyseal dysplasia – variability of expression within a large family. *Clin. Genet.* **15**, 252–8.

Beighton, P.H., Hamersma, H., and Raad, M. (1979*b*). Oculodento-osseous dysplesia: heterogeneity or variable expression. *Clin. genet.* **16**, 169–77.

Bell, J. (1934). Huntington's chorea. In *Treasury of human inheritance,* (ed. L.S. Penrose) Vol. 4, Part I, pp. 1–67. Cambridge University Press, London.

Bell, J. (1935). On the peroneal type of progressive muscular atrophy. In *Treasury of human inheritance,* (ed. L.S. Penrose). Vol. 4, Part VI. Cambridge University Press, London.

Bell, J. (1942). On the age of onset and age of death in hereditary muscular dystrophy with some observations bearing on the question of antedating. *Ann. Eugen.* **11**, 272–89.

Bell, J. (1958). The Laurence–Moon syndrome. In *Treasury of human inheritance,* (ed. L.S. Penrose). Vol. V, Part III, pp. 51–96. Cambridge University Press, London.

Bell, J and Carmichael. E.A. (1939). On hereditary ataxia and spastic paraplegia. In *Treasury of human inheritance,* (ed. L.S. Penrose). Vol. IV, Part 3, pp. 141–281. Cambridge University Press, London.

Belmaric, J. and Chau, A.S. (1969). Medulloblastoma in newborn sisters. *J. Neurosurg.* **30**, 76–9.

Benda, C. E. (1954). The Dandy–Walker syndrome or the so-called atresia of the foramen of Magendie. *J. Neuropath, exp. Neurol.* **13**, 14–29.

Bender, A. N. and Bender, M. B. (1977). Muscle fiber hypotrophy with intact neuromuscular junctions. A study of a patient with congenital neuromuscular disease and ophthalmoplegia. *Neurol* **27**, 206–12.

Benhaiem–Sigaux, N. *et al.*. (1985). A retromedullary arteriovenous fistula associated with the Klippel–Trenaunay–Weber Syndrome. A clinicopathological study. *Acta Neuropathol.* **66**, 318–24.

Benjamins, D. (1980). Progressive bulbar palsy of childhood in siblings. *Ann. Neurol.* **8**, 203.

Benke, P. J. and Cohen, M. M. Jr. (1983). Recurrence of holoprosencephaly in families with a positive history. *Clin. Genet.* **24**, 324–8.

Bennett, J. H., Rhodes, F. A., and Robson, H. N. (1959). A possible genetic basis for Kuru. *Ann. J. hum. Genet.* **11**, 169–87.

Benson, M. D. and Cohen, A. S. (1977). Generalized amyloid in a family of Swedish origin: a study of 426 family members in seven generations of a new kinship with neuropathy, nephropathy, and central nervous system involvement. *Ann. intern. Med.* **86**, 419–24.

Bentley, S. J., Cambell, M. J., Kaufmann, P. (1975). Familial syringomyelia. *J. Neurol. Neurosurg. Psychiat.* **38**, 346–9.

Bentsen, K. G., Dalsgaard–Nielsen, E., and Möller, H. U. (1938). A propos de la maladie de Sturge–Wever, considérée surtout au point de vue des symptômes opthalmoloques. *Acta ophthal.* **16**, 279–94.

Bentzen, N. (1972). Familial incidence of intra-cranial aneurysms: case reports. *N.Z. med. J.* **75**, 153–5.

Berant, W, and Berant, N. (1973). Radioulnar synostosis and craniosynostosis syndromes. *Birth Defects.* **2**, 137–89.

Berard-Badier, M., Gambarelli, D., Pinsard, N., Hassoun, H., and Toga, M. (1971). Infantile neuro-axonal dystrophy or Seitelberger's disease II. Peripheral nerve involvement: electron microscopic study in one case. *Acta neuropath. Suppl.* V, 30–9

Berenberg, R. A., *et al.* (1977). Lumping or splitting ? "Opthalmoplegia-plus" or Kearns-Sayre syndrome? *Ann. Neurol.* **1**, 37–54.

Berg, B. O. and Conte, F. (1974). Duchenne muscular dystrophy in a female with structurally abnormal X-chromosome. *Neurology, Minneap.* **24**, 1356.

Berg, H. (1913). Vererbung der tuberösen Sklerose durch zwei bzw drei Generationen. *Z. ges Neurol. Psychiat.* **19**, 528–39.

ter Berg, H. W. M., *et al.* (1986). Familial Association of intracranial aneurysms and multiple congenital anomalies. *Arch Neurol.* **43**, 30–3.

ter Berg, H. W. M., Bijlsma, J. B., and Willemse, J. (1987). Familial occurence of intracranial aneurysms in childhood: A case report and review of literature. *Neuropaediat.* **18**, 227–30.

Berg, R. A., Aleck, K. A., and Kaplan, A. M. (1983). Familial Porencephaly. *Arch. Neurol.* **40**, 567–9

Bergen, B.J., Carry, M.P., Wilson, W.B., Barden, M.T., and Ringel, S.P. (1988). Centronuclear myopathy: extraocular and limb muscle findings in an adult. *Muscle and Nerve* 3, 165–71.

Bergia, B., Sybers H.D., and Butler, I.J. (1986). Familial lethal cardiomyopathy with mental retardation and scapuloperoneal muscular dystrophy. *J.neurol. Neurosurg. Psychiat.* 49, 1423–26.

Berginer, V.M., and Abeliovich, D. (1981). Genetics of cerebrotendinous xanthomotosis (CTX): an auosomal recessive trait with high gene frequency in Sephardim of Moroccan orign. *Am J Med Genet.* 10, 151–7.

Bergland, R.M. (1968). Congenital intraspinal extradural cyst. *J. Neurosurg.* 28 495–9

Bergstedt, M.I., Johansson, S., and Müller, R. (1962). Hereditary spastic ataxia with central retinal degeneration and vestibular impairment. *Neurology, Minneap.* 12 124–32.

Bergstrom, L.V., Neblett, L.M., and Hemenway, W.G. (1972). Otologic manifestations of acrocephalosyndactyly. *Archs Otolar.* 96, 11–23.

Berkovic, S.F., *et al.* (1987). Progressive dystonia with bilateral putaminal hypodensities. *Arch.Neurol.* 44, 1184–7

Berlyne, G.M. and Berlyne, N. (1960). Anaemia due to 'blue-rubber-bleb' naevus disease. *Lancet* II 1275–7.

Berman, E.R., Livni, N., Shapira, E., Merin, S., and Levij, I.S. (1974). Congenital corneal clouding with abnormal systemic storage bodies: a new variant of mucolipidosis. *J. Paediat.* 84 519–26

Berman W., Haslam, R.H.A., Konigsmark, B.W., Capute, A.J., and Migeon, C.J. (1973). A new familial syndrome with ataxia, hearing loss and mental retardation. *Arch Neurol., Chicago* 29, 258–61.

Bernheimer, H. and Seitelberger, F. (1968). Über das Verhalten der Ganglioside im Gehirn bei z fällern von spätinfantiler amaurotischer Idiotic. *Wein klin. Wochenschr.* 80 163–4.

Bernstein, J. and Shelden, W.E. (1959). A note on the development of Gaucher cells in a newborn infant. *J. Pediat.* 55, 577–81.

Berry, S.A. Pierpont, M.E., and Gorlin, R.J. (1984). Single central incisor in familial holoprosencephaly. *J. Pediat.* 104, 877–80.

Bertrams, J., Grosse-Wilde, H., Netzel, B., Mempel., W., and Kuwert, E. (1975). A Caucasian histocompatibility super haplotype with increased susceptibility to multiple sclerosis. *Histocompatibility testing* (ed. F. Kissmenger-Nielson), Report of the Sixth International Workshop, Arhus, 1975, pp. 782–7. Munksgaard, Copenhagen.

Bejsovec, M., Kullenda, Z., and Ponca, E. (1967). Familial intrauterine convulsions in pyridoxine dependency. *Arch. Dis. Child.* 42, 201–7.

Bethlem, J. and Van Wijngaarden, G.K. (1976). Benign myopathy with autosomal dominant inheritance. *Brain* 99, 91–100.

Bethlem, J., Van Gool, J., Hülsmann, W.C., and Meijer, A.E.F.H. (1966).

Familial non-progressive myopathy with muscle cramps after exercise. *Brain* **89**, 569–88.

Bethlem, J., Van Wijngaarden, G.K. Meijer, A.E.F.H., Hülsmann, W.C. (1969). Neuromuscular disease with type I fiber atrophy, central nuclei, and myotube- like structure. *Neurol., Minneap.* **19**, 705–10.

Bethlem, J, Van Wijngaarden, G.K., Meijer, A.E.F.H., and Fleury, P. (1971). Observations on central core disease. *J. neurol. Sci.* **14**, 293–9.

Beyers, R. and Dodge, J. (1967). Huntington's chorea in children: report of four cases. *Neurol., Minneap.* **17**, 587–96.

Bharucha, N.E., Bharucha, E.F., and Bhabha, S.K. (1986*a*). Machado–Joseph–Azorean disease in India. *Arch. Neurol.* **43**, 142–4.

Bharucha, N.E. *et al.* (1986*b*). A case-control study of twin pairs of discordant for Parkinsons disease: a search for environmental risk factors. *Neurology* **36**, 284–8.

Bianchi, E. *et al.* (1985). A family with Saethre–Chotzen syndrome. *Am. J. Med. Genet.* **22**, 649–58.

Bianchine, J.W. and Lewis, R.C. (1974). The Masa syndrome: a new heritable mental retardation syndrome. *Clin. Genet.* **21** 123–9.

Bickel, H., Gerrard, J., and Hickman, E.M. (1953). Influence of phenylalanine uptake on phenylketonuria. *Lancet* **ii**, 812.

Bickel, H., Neale, F.C., and Hall, G. (1957). A clinical and biochemical study of hepatolenticular degeneration (Wilson's disease). *Q. J. med.* **26**, 527–58.

Bickel, H., Feist, D., Muller, H., and Quadbeck, G. (1968). Ornithinämie, eine weiters Aminosäurenstoffwechelstörung mit Hvinschädigung. *Dt. med. Wschr.* **93**, 2247–51.

Bickers, D.S. and Adams, R.D. (1949). Hereditary stenosis of the aqueduct of Sylvius as a cause of congenital hydrocephalus. *Brain* **72**, 246–62.

Bickerstaff, E.R. (1950). Hereditary spastic paraplegia. *J. Neurol. Neurosurg. Psychiat.* **13**, 134–45.

Bickerstaff, E.R. (1961). Basilar artery migraine. *Lancet* **i**, 15–17.

Bicknell, J.M., Carlow, T.J., Kornfeld, M., Stovring, J., and Turner, P. (1978). Familial cavernous angiomas. *Archs Neuroll., Chicago* **35**, 746–9.

Biedl, A. (1922). Ein Geschwisterpaar mit adiposogenitaler dystrophi. *Dt. med. Wschr.* **48**, 1630.

Biemer, J.J. and McCammon, R.E. (1975). The genetic relationship of abetalipoproteinemia and hypobetalipoproteinemia: report of the occurence of both diseases within the same family. *J. Lab. clin. Med.* **85**, 556–65.

Biemond, A. (1934*a*). Brachydactylie, Nystagmus en cerebellaire Ataxie als familiair syndroom. *Ned. Tijdschr. Geneesk.* **78**, 1432–31.

Biemond, A. (1934*b*). Het syndroom van Laurence–Biedl en een aanverwant, niuew syndroom. *Ned. Tijdsch. Geneesk.* **78**, 1801–9.

Biemond, A. (1951). Les degenerescences spino-cerebelleuses. *Folia psychiat. neurol. neurochir. neerl.* **54**, 216–23.

Biemond, A. (1955). Myopathia distalis juvenilis hereditaria. *Acta psychiat. neurol. scand.* **30**, 25–38.

Biemond, A. (1963). Paramyoclonus multiplex (Friedreich). Clinical and genetic aspects. *Folia psychiat. neurol. neurochir. neerl.* **66**, 270–6.

Biemond, A., and Beck, W. (1955). Neural muscle atrophy with degeneration of the substantia nigra. *Confinia neurol.* **15**, 142–53.

Biemond, A. and Daniels A.P. (1934). Familial periodic paralosis and its transition into spinal muscular atrophy. *Brain* **57**, 91–108.

Biemond, A. and Sinnege, J.L.M. (1955). Tabes of Freidreich with degeneration of the substantia nigra, a special type of hereditary parkinsonism. *Confinia neurol* **15**, 129–42.

Bignami, A., MacCagnami, F., Zappella, M., and Tingey, A.H. (1966). Familial infantile spasms and hypsarrhythmia associated with leukodystrophy. *J. Neurol. Neurosurg. Psychiat.* **29**, 129–34.

Bill, P.L.A., Cole, G., and Proctor, N.S.F. (1979). Centronuclear myopathy. *J. Neurol. Neurosurg. Psychiat.* **42**, 548–56.

Bille, B. (1962). Migraine in school children. *Acta paediat., Stockh.* **51**, Suppl. 136, 1–151.

Bing, R. (1905). Eine kombinierte Form der heredofamiliaren Nervenkrankheiten. *Dt. Arch. klin. Med.* **85**, 199.

Bird, A.V. and Krynauw, R.A. (1953). Lindau's disease in a South African family. *Br. J. Surg.* **40**, 433–7.

Bird, E.D., Caro, A.J., and Pilling, J.B. (1974). A sex-related factor in the inheritance of Huntington's chorea. *Ann. hum. Genet.* **37**, 255–9.

Bird, M.T. and Paulson, G. (1971). The rigid form of Huntington's chorea. *Neurology, Minneap.* **21**, 271–6.

Bird, T.D. (1975). Apparent familial multiple sclerosis in three generations. Report of a family with histocompatibility antigen typing. *Archs Neurol., Chicago* **32**, 414–16.

Bird, T.D. and Kraft, G.H. (1978). Charcot–Marie–Tooth disease: data for genetic counselling relating age to risk. *Clin. Genet.* **14**, 43–9.

Bird, T.D. and Omenn, G.S. (1975). Monozygotic twins with Huntington's disease in a family expressing the rigid variant. *Neurology, Minneap.* **25**, 1126–9.

Bird, T.D. and Shaw, C.M. (1978). Progressive myoclonus and epilepsy with dentatorubral degeneration in a clinicopathological study of the Ramsay Hunt syndrome. *J. Neurol. Neurosurg. Psychiat.* **41**, 140–9.

Bird, T.D. Carlson, C.B., and Hall, J.G. (1976). Familial essential (benign) chorea. *J. med. Genet.* **13**, 357–62.

Bird, T.D., Cederbaum, S., Valpey, R.W., and Stahl, W.L. (1978). Familial degeneration of the basal ganglia with acanthocytosis: a clinical neuropathological and neurochemical study. *Ann. Neurol.* **3**, 253–8.

Bird, T.D., *et al.* (1987). The use of apolipoprotein CII as a genetic marker for myotonic dystrophy. *Arch. Neurol.* **44**, 273–5.

Bird, T.D. *et al.* (1983). Genetic linkage evidence for heterogeneity in Charcot–Marie–Tooth neuropathy (HMSN Type I). *Ann. Neurol.* **14**, 679–84.

Bittenbender, J. and Quadfasel, F. (1962). Rigid and akinetic forms of Huntington's chorea. *Archs. Neurol., Chicago.* **7**, 275–88.

Bixler, D., Christian, J.C., and Gorlin, R.J. (1969). Hypertelorism, microtia and facial clefting: a new inherited syndrome. The Clinical Delineation of Birth Defects, Part II Malformation Syndromes. *Birth Defects.* Original Article Series V, 77–81.

Bjerre, I. and Cornelius, E. (1968). Benign familial neonatal convulsions. *Acta pediat., Stockh.* **57**, 557–61.

Björk, A., Lindblom, U., and Wadensten, L. (1956). Retinal degeneration in hereditary ataxia. *J. Neurol. Neurosurg. Psychiat.* **19**, 186–93.

Blackwood, W. and Cumings, J.N. (1954). A histological and chemical study of three cases of diffuse cerebral sclerosis. *J. Neurol. Neurosurg. Psychiat.* **17**, 33–49.

Blackwood, W., Buxton, P.H., Cumings, J.N., Robertson, D.J., and Tucker, S.M. (1963). Diffuse cerebral degeneration in infancy (Alper's disease). *Arch. Dis. Childh.* **38**, 193–204.

Blandfort, M., Tsuboi, T., and Vogel, F. (1987). Genetic counselling in the epilepsies. *Hum. Genet.* **76**, 303–31.

Blank, C.E. (1960). Aperts syndrome (a type of acrocephalosyndactyly) – observation of a British series of 39 cases. *Ann. hum. Genet.* **24**, 151–64.

Blass, J.P., Kark, R.A.P., and Engel, W.K. (1971). Clinical studies of a patient with pyruvate decarboxylase deficiency. *Archs Neurol., Chicago* **25**, 449–60.

Blau, J.N. and Whitty, C.W.M. (1955). Familial hemiplegic migraine. *Lancet* **ii**, 1115–16.

Blethen S.L. and Weldon V.V. (1985). Hypopituitarism and septooptic dysplasia in first cousins. *Am. J. Med. Genet.* **21**, 123–9.

Blumel, J. and Kniker, W.T. (1959). Laurence–Moon–Bardet–Biedl syndrome. Review of the literature and a report of five cases including a family group with three affected males. *Tex. Rep. Biol. Med.* **17**, 391–410.

Blumel, J., Evans, E.B., and Eggers, G.W.N. (1957). Hereditary cerebral palsy. A preliminary report. *J. Pediat.* **50**, 454–8.

Blyth, H. and Hughes, B.P. (1971). Pregnancy and serum CPK levels in potential carriers of severe X-linked muscular dystrophy. Letter. *Lancet* **i**, 855–6.

Blyth, H. and Pugh, R.J. (1958). Muscular dystrophy in childhood. The genetic aspect. *Ann. hum. Genet.* **23**, 127–63.

Bobowick, A. R., Kurtzke, J. F., Brody, J. A., Hrubec, Z., and Gillespie, M. (1978). Twin study of multiple sclerosis: an epidemiologic enquiry. *Neurol., Minneap.* **28**, 978–87.

Bodmer, W. F. (1978). The H. L. A. system: introduction. *Br. med. Bull.* **34**, 213–16.

Boehme, D. H., Cottrell, J. C., Leonberg, S. C., and Zeman, W. (1971). A dominant form of neuronal ceroid-lipofuscinosis. *Brain* **94**, 745–60.

Boel, M. and Casaer, P. (1984). Paroxysmal kinesigenic choreoathetosis. *Neuropediat.* **15**, 215–17.

Boghen, D. and Peyronnard, J.-M. (1976). Myoclonus in familial restless legs syndrome. *Arch. Neurol., Chicago* **33**, 368–70.

Bolger, G. B. *et al.* (1985). Chromosome translocation 14:22 and oncogene (c-sis) variant in a pedigree with familial meningioma. *N. Engl. J. Med.* **312**, 564–7.

Boller, F. and Segarra, J. M. (1969). Spino-pontine degeneration. *Eur. Neurol.* **2**, 356–73.

Boller, F., Boller, M., Denes, G., Timberlake, W., Zieper, I., and Albert, M. (1973). Familial palilalia. *Neurol. Minneap.* **23**, 1117–25.

Boller, F., Boller, M. and Gilbert, J. (1977). Familial idiopathic cerebral calcifications. *J. Neurol. Neurosurg. Psychiat.* **40**, 280–5.

Bollinger, A. (1961). Dic meralgia paraesthetica. *Suisses Arch. Neurolchir. Psychiat.* **87**, 58–102.

Bolt, J. M. W. (1970). Huntington's chorea in the West of Scotland. *Br. J. Psychiat.* **116**, 259–70.

Boltshauser, E. and Isler, W. (1977). Joubert syndrome: episodic hyperpnoea, abnormal eye movements, retardation and ataxia associated with dysplasia of the cerebellar vermis. *Neuropaediatrie* **8**, 57–66.

Boltshauser, E. *et al.* (1987). Computed tomography in Hallervorden-Spatz disease. *Neuropediat.* **18**, 81–3.

Boltshauser, E., Meyer, M., Metaxas, M., Mahler, M., and Schiller, H. (1980). Dominant myotonia congenita. Pedigree with skipping of one generation. *J. Neurol.* **222**, 235–8.

Bonduelle, M., Escourolle, R., Bouygues, P., Lormeau, G., Dumas, J.-C., and Merland, J.-J. (1971). Maladie de Creutzfeldt–Jakob familiale. *Revue meurol.* **125**, 197–209.

Bonduelle, M., Escourolle, R., Bouygues, P., Lormeau, G., and Gray, F. (1976). Atrophie olivo-ponto-cerebelleuse familiale avec myoclonies les limites de la dyssynergie cerebelleuse myoclonique (syndrome de Ramsay Hunt). *Rev. neurol.* **132**, 113–24.

Bone, I., Johnson, R. H. and Ferguson-Smith, M. A. (1976). Occurence of familial spastic paraplegia in only one of monozygotic twins. *J. Neurol. Neurosurg. Psychiat.* **39**, 1129–33.

Bonnet, P., Dechaume, J., and Blanc, E. (1938). L'aneurisme cirsoide de la retine (aneurisme racemeux) ses relations avec l'aneurisme cirsoide du cerveau. *Bull. Soc. fr. Ophthal.* **51**, 521–4.

Bonnette, H., Roelofs, R., and Olson, W.H. (1974). Multicore disease: report of a case with onset in middle age. *Neurology Minneap.* 24, 1039–44.

Böök, J.A. (1953). A genetic and neuropsychiatric investigation of a North-Swedish population. *Acta genet. Statist. med.* 4, 1–100, 345–414.

Book, J.A., Schut, J.W., and Reed, S.C. (1953). A clinical and genetical study of microcephaly. *Am. J. ment. Defic.* 57, 637–60.

Bornstein, B. (1961). Restless legs. *Psychiatria Neurol.* 141, 165–201.

Borrett, D. and Becker, L.E. (1985). Alexander's disease. A disease of astrocytes. *Brain* 108, 367–85.

Borrone, C., Gatti, R., Trias, X., and Durand, P. (1974). Fucosidosis: clinical biochemical, immunologic and genetic studies in two new cases. *J. Pediat.* 84, 727–30.

Bosch, E.P. and Hart, M.N. (1978). Late adult-onset metachromatic leukodystrophy. *Archs Neurol., Chicago* 35, 475–7.

Bosch-Baneyeras, J.M. *et al.* (1984). Poland-Mobius syndrome associated with dextrocardia. *J. Med. Genet.* 21, 70–1.

Boswinkel, E. *et al.* (1985). Linkage analysis using eight DNA polymorphisms along the length of the X chromosome locates the gene for Emery–Dreifuss muscular dystrophy to distal Xq. *Cytogenet. Cell Genet.* 40, 586 (abst.).

Bouchard, J.P., Barbeau, A., Bouchard, R., and Bouchard, R.W. (1979). Electromyography and nerve conduction studies in Friedrcich's ataxia and autosomeal/recessive spastic ataxia of Charlevoix–Saguenay (ARSACS). *Can. J. Neurol. Sci.* 6, 185–9.

Boucher, B.J. and Gibberd, F.B. (1969). Familial ataxia hypogonadism and retinal degeneration. *Acta neurol. scand.* 45, 507–10.

Boudin, G., *et al.* (1971). Cas familial de paralysie bulba-pontine chronique progressive avec surdité. *Rev. Neurol.* 124, 90–2.

Boudin, G., *et al.* (1976). Fatal systemic carnitine deficiency with lipid storage in skeletal, muscle, heart, liver and kidney. *J. Neurol. Sci.* 30, 313–25.

Bouldin, T.W., Riley, E., Hall, C.D., and Swift, M. (1980). Clinical and pathological features of an autosomal recessive neuropathy. *J. neurol. Sci.* 46, 315–23.

Bouwsma, G. and Van Wijngaarden, G.I.C. (1980). Spinal muscular atrophy with hypertrophy of the calves. *J. neurol. Sci.* 44, 275–9.

Boveri, P. (1910). De le nevrite hypertrophique familiale (type I Pierre Marie). *Sem. med.* 13, 145–58.

Bowden, D.H., Fraser, D., Jackson, S.H., and Walker, N.F. (1956). Acute recurrent rhabdomyolysis (paroxysmal myohaemoglobinuria). *Medicine, Baltimore* 35, 335–53.

Bowen, P., Lee, C.S.N., Zellweger, H., and Lindenberg, R. (1964). A familial syndrome of multiple congenital defects. *Bull. Johns Hopkins Hosp.* 114, 402–14.

Boyd, M.C., Steinbok, P. and Paty, D.W. (1985). Familial arteriovenous malformations. Report of four cases in one family. *J. Neurosurg.* **62**, 597–9.

Boyd, S.G., *et al.* (1986). Progressive neuronal degeneration of childhood with liver disease (Alpers disease): characteristic neurophysiological features. *Neuropediat.* **17**, 75–80.

Boyd, S.G., Harden, A. and Patton, M.A. (1988). The EEG in early diagnosis of Angelman's (happy puppet) syndrome. *Eur. J. Ped.* **147**, 503–13.

Boynton, R.C. and Morgan, B.C. (1973). Cerebral arteriovenous fistula with possible hereditary telangiectasia. *Am. J. Dis. Child.* **125**, 99–101.

Boysen, G., Galassi, G., Kamieniecka, Z., Schlaeger, J., and Trojaborg, W. (1979). Familial amyloidosis with cranial neuropathy and corneal lattice dystrophy. *J. Neurol. Neurosurg. Psychiat.* **42**, 1020–30.

Brackenridge, C.J. (1972). Familial correlations for age at onset and age at death in Huntington's disease. *J. med. Genet.* **9**, 23–32.

Brackenridge, C.J. (1980). Parental factors associated with rigidity in Huntington's chorea. *J. med. Genet.* **17**, 112–14.

Brackenridge, C.J. and Teltscher, B. (1975). Estimation of the age of onset of Huntington's disease from factors associated with the affected parent. *J. med. Genet.* **12**, 64–9.

Bradburne, A.A. (1912). Hereditary ophthalmoplegia in five generations. *Trans. Opthhal. Soc. U.K.* **32**, 142–53.

Bradley, W.G. (1969). Adynamia episodica hereditaria. *Brain* **92**, 345–78.

Bradley, W.G., Price, D.L., and Watanabe, C.K. (1970). Familial centronuclear myopathy. *J. Neurol. Neurosurg. Psychiat.* **33**, 687–93.

Bradley, W.G., Hudgson, P., Gardner-Medwin, D., and Walton, J.N. (1973). The syndrome of myosclerosis. *J. Neurol. Neurosurg. Psychiat.* **36**, 651–60.

Bradley, W.G., Thrush, D.C. and Campbell, M.J. (1975). Recurrent brachial plexus neuropathy. *Brain* **98**, 381–98.

Bradley, W.G. Madrid, R., and Davis, C.J.F. (1977). The peroneal muscular atrophy syndrome. Clinical, genetic electrophysiological and nerve biopsy studies. III. Clinical, electrophysiological and pathological correlations. *J. neurol. Sci. Armst.* **32**, 123–36.

Bradley, W.G., Jones, M.Z., and Fawcett, P.R.W. (1978). Becker-type muscular dystrophy. *Muscle Nerve.* **1**, 111–32.

Bradshaw, J.P.P. (1954). A study of myoclonus. *Brain* **77**, 1381–57.

Bradshaw, P. and Parsons, M. (1965). Hemiplegic migraine, a clinical study. *Q. J. Med.* **34**, 65–85.

Brait, K., Fahn, S., and Schwartz, G.A. (1973). Sporadic and familial parkinsonism and motor neuron disease. *Neurol., Minneap.* **23**, 990–1002.

Brandon, M.W.G., Kirman, B.H., and Williams, C.E. (1959). Microcephaly in one of monozygous twins. *Arch. Dis. Childh.* **34**, 56–9.

Brandt, N. J. (1984). Symptoms and signs in organic acidurias. *J. Inherit. Met. Dis.* Suppl. 1, 23–7.

Brandt, N. J., Brandt, S. Christensen, E., Gregersen, N., Rasmussen, K. (1978). Glutamic aciduria in progressive choreo-athetosis. *Clin. Genet.* **13**, 77–80.

Brandt, S. (1949). Hereditary factors in infantile progressive muscular atrophy. *Am. J. Dis. Child.* **78**, 226–36.

Brandt, S. Carlsen, N., Glenting, P., and Helweg-Larsen, J. (1974). Encephalopathia myoclonica infantilis (Kinsbourne) and neuroblastoma in children. A report of three cases. *Devl Med. child. Neurol.* **16**, 286–94.

Brandt, S., *et al.*. (1977). Juvenile neurolipidosis of Bernheime–Seitelberger type: histopathological and biochemical findings. *Acta neurol. scand.* **56**, 587–602.

Brasfield, R. D. and Das Gupta. T. K. (1972). Von Recklinghausen's disease: a clinicopathological study. *Ann. Surg.* **175**, 86–104.

Braughman, F. A. Jr, List, C. F., Williams, J. R., Muldoon, J. P., Segarra, J. M., and Volker, J. S. (1969). The glioma-polyposis syndrome. *N. Engl. J. Med.* **281**, 1345–6.

Bray, G. M., Kaarsoo, M., and Ross, R. T. (1965). Ocular myopathy with dysphagia. *Neurol. Minneap.* **15**, 678–84.

Bray, P. F. and Wiser, W. C. (1964). Evidence for a genetic etiology of temporal-central abnormalities in focal epilepsy. *N. Engl. J. Med.* **271**, 926–33.

Bray, P. E., Ziter, F. A., Lahey, M. E., and Meyers, G. G. (1969). The coincidence of neuroblastoma and acute ccrcbellar encephalopathy. *J. Pediat.* **75**, 983–90.

Breitner, J. C. S. and Folstein, M. (1984a). Familial Alzheimer dementia: a prevalent disorder with specific clinical features. *Psychol. Med.* **14**, 507–15.

Breitner, J. D. S. Folstein, M. F. (1984b). Familial nature of Alzheimer's disease. (Letter). *N. Engl. J. Med.* **311**, 192.

Breningstall, G. N. (1986). Neurologic syndromes in hyperammonemic disorders. *Pediat. Neurol.* **2**, 253–62.

Bressler, R. (1970). Carnitine and the twins. *N. Engl. J. Med.* **282**, 745–6.

Brett, E. M., Ellis, R. B., Haas, L., Ikonne, J. U., Lake, B. D., Patrick, A. D., and Stephens, R. (1973). Late onset of GM_2 gangliosidosis. Clinical, pathological and biochemical studies on eight patients. *Archs Dis. Childh.* **48**, 775–85.

Bretz, G. W., Baghdassarian, A., Graber, J. D., Zackerle, B. J., Norum, R. M., and Blizzard, R. M. (1970). Co-existence of diabetes mellitus and insipidus and optic atrophy in two siblings. *Am. J. Med.* **48**, 398–403.

Brewis, M., Poskanzer, D. C., Rolland, C. and Mille, H. (1966). Neurological disease in an English city. *Acta neurol. scand.* **42**, Suppl. 24, 1–89.

Brill, C.B., Rose, J.S., Godmilow, L., Sklower, S., Willner, J., and Hirschhorn, K. (1978). Spastic quadriparesis due to C_1–C_2 subluxatioin and Hurler syndrome. *J. Pediat.* **92**, 441–3.

Brisman, R. and Abbassioun, K. (1971). Familial intracranial aneurysms. *J. Neurosurg.* **34**, 678–82.

British Medical Journal (1978). Predictive tests in Huntington's chorea. Editorial. *Br. med. J.* **i**, 528–9.

Britt, B.A., Cocher, W.G., Kalow, W. (1969). Hereditary aspects of malignant hyperthermia. *Can. Anaesth. Soc. J.* **16**, 89.

Brodner, R.A. Noh, J.M., and Fine, E.J. (1976). A dominant form of adult neuronal ceroid-lipofuscinosis (Kurf's disease) with associated occipital astrocytoma: early diagnosis by cortical biopsy. *J. Neurol. Neurosurg. Psychiat.* **39**, 231–8.

Brodribb, A.J.M. (1970). Vertebral aneurysm in a case of Ehlers–Danlos syndrome. *Br. J. Surg.* **57**, 148–51.

Brodrick, J.D. (1974). Hereditary optic atrophy with onset in early childhood. *Br. J Ophthal.* **58**, 817–22.

Brooke, M.H. (1973). Congenital fiber type dysproportion. In *Proc. 2nd Int. Congr. Muscle Disease,* Perth, Australia, Part 2 *Clinical studies in myology* (ed. BA. Kakulas) I CS No. 295, Vol. 1, pp. 147–59. Excerpta Medica, Amsterdam.

Brooke, M.H. (1977). *A clinician's view of neuromuscular disease.* Williams and Wilkins, Baltimore.

Brooke, M.H. and Neville, H.E. (1972). Reducing body myopathy. *Neurology, Minneap.* **22**, 829–40.

Brothers, C.R.D. (1964). Huntington's chorea in Victoria and Tasmania. *J. neurol. Sci.* **1**, 405–20.

Brown, B.J., Lewis, L.A., and Mercer, R.D. (1974). Familial hypobetali-poproteinemia report of a case with psychomotor retardation. *Pediatrics, Springfield* **54**, 111–13.

Brown, J.H., Fabre, L.F., Farrell, G.L., and Adams, E.D. (1972). Hyperlysinuria with hyperammonemia. *Am. J. Dis. Childh.* **124**, 127.

Brown, J.W. and Coleman, R.F. (1966). Hereditary spastic paraplegia with ocular and extrapyramidal signs. *Bull. Los Ang. neurol. Soc.* **31**, 21–34.

Brown, M.R. (1951). Wetherbee Ail: the inheritance of progressive muscular atrophy as a dominant trait in two New England families. *New Engl. J. Med.* **245**, 645–7.

Brown, P., Cathala, F., Sandowsky, D., and Gajdusek, D.C. (1979). Creutzfeldt–Jakob disease in France: II. Clinical characteristics of 124 consecutive verified cases during the decade 1968–1977. *Ann. Neurol.* **6**, 430–7.

Brown, R.G. and Marsden, C.D. (1984). How common is dementia in Parkinson's disease? *Lancet* **i**, 1262–5.

Brown, S. (1892). On hereditary ataxy with a series of twenty-one cases. *Brain* **xv**, 250–68.

Browne, T.R., Adams, R.D., and Roberson, G.H. (1976). Hemangioblastoma of the spinal cord. Review and report of five cases. *Archs Neurol., Chicago* 33, 326–32.

Brownell, A.K.W. (1979). Familial multicore disease. (Abstract). *Ann. Neurol.* 6, 155.

Brownell, B. and Oppenheimer, D.P. (1965). An ataxic form of subacute presenile polioencephalopathy (Creutzfeldt–Jacob disease). *J. neurol. neurosurg. Psychiat.* 28, 350–61.

Brownstein, S., Bernado, A.I., Uprato, S., and Salim, I. (1974). Neurofibromatosis with the eye fly *Siphunculina funicola* in an eyelid tumor. *Can. J. Ophthal.* 11, 261–6.

Brucher, J.M., Dom R., Lambaert A., *et al.* (1981). Progressive pontobulbar palsy with deafness; clinical and pathological study of two cases. *Arch. Neurol.* 38, 186–90.

Bruens, J.H. Guazzi, G.C., and Martin, J.J. (1968). Infantile form of meningeal angiomatosis with sudanophilic leucodystrophy associated with complex abiotrophies. *J. Neurol. Sci.* 7, 417–25.

Brun, A., Gilboa, M., Meeuwisse, G.W. and Nordgren, H. (1978). The Zellweger syndrome: subsellular pathology, neuropathology and the demonstration of pneumocystis carinii pneuomonitis in two siblings. *Eur. J. Pediat.* 127, 229–45.

Brüngger, U. and Kaeser, H.E. (1977). Paramyotonia congenita without cold paralysis and myotonia levior. *Eur. Neurol.* 15, 2–4.

Brunt, P.W. and McKusick, V.A. (1970). Familial dysautonomia. A report of genetic and clinical studies with a review of the literature. *Medicine, Baltimore* 49, 343–74.

Brust, J.C.M., Lovelace, R.E. and Devi, S. (1978). Clinical and electrodiagnostic features of Charcot–Marie–Tooth syndrome. *Acta neurol. scand.* 58, Suppl. 68–150.

Bruyn, G.W. (1968). Huntington's chorea. *Handbook of clinical neurology.* Vol. 6 (ed. P.J. Vinken and G.W. Bruyn) pp. 298–377. North Holland, Amsterdam.

Bruyn, G.W. (1973). Clinical variants and differential diagnosis. In *Advances in neurology,* Vol. 1. Raven Press, New York.

Bruyn, G.W. and Went, L.N. (1964). A sex-linked heredodegenerative neurological disorder associated with Leber's optic atrophy, Part I. Clinical Studies. *J. neurol. Sci.* 1, 59–80.

Bruyn, G.W., Bots, G.T.H., and Staal, A. (1964). Familial bilateral vascular calcification in the central nervous system. *Folia psychiat. neurol. neurochir. neerl.* 67, 342–76.

Bucheit, W.H., Burton, C., Haag, B., and Shaw, D. (1969). Familial papilledema and idiopathic intracranial hypertension. *N. Engl. J. Med.* 280, 938–42.

Buchthal, F. and Behse, F. (1977). Peroneal muscular atrophy (PMA) and related disorders, I. Clinical manifestations as related to biopsy findings, nerve conduction and electromyography. *Brain* **100**, 41–66.

Buchthal, F. and Olsen, P. Z. (1970). Electromyography and muscle biopsy in infantile spinal muscle atrophy. *Brain* **93**, 15–30.

Budka, H., Sluga, E., and Heiss, W. D. (1976). Spastic paraplegia associated with Addisons disease: adult variant of adreno-leukodystrophy. *J. Neurol.* **213**, 237–50.

Buge, A., *et al.* (1978). Maladie de Creutzfeldt–Jakob familiale. *Revue neurol.* **134**, 165–81.

Brucher, J. M. *et al.* (1981). Progressive pontobulbar palsy with deafness. Clinical and Pathological study of two cases. *Arch. Neurol. Vol.* **38**, 186–90.

Bull, J., Gammal, T. E., and Popham, M. (1969). A possible genetic factor in cervical spondylosis. *Br. J. Radiol.* **42**, 9–16.

Bull, J., Nixon, W., and Pratt, R. T. C. (1955). The radiological criteria and familial occurence of primary basilar impression. *Brain* **78**, 229–47.

Bundey, S. (1972). A genetic study of infantile and juvenile myasthenia gravis, *J. Neurol. Neurosurg. Psychiat.* **35** 41–51.

Bundey, S. (1974). Detection of heterozygotes for myotonic dystrophy. *Clin. Genet.* **5**, 107–9.

Bundey, S. (1977). Disorders of muscle: genetic counselling. *Hosp. Med.* **17**, 342–9.

Bundey, S. (1978). Calculation of genetic risks in Duchenne muscular dystrophy by geneticists in the United Kingdom. *J. med. Genet.* **15**, 249–53.

Bundey, S. and Carter, C. O. (1972). Genetic heterogeneity for dystrophia myotonia. *J. med. Genet.* **9**, 311–15.

Bundey, S., and Carter, C. O. (1974). Recurrence risks in severe undiagnosed mental deficiency. *J. ment. defic. Res.* **18**, 115–34.

Bundey, S. and Evans, K. (1969). Tuberous sclerosis: a genetic study. *J. Neurol. Neurosurg. Psychiat.* **32**, 591–603.

Bundey, S. and Griffiths, M. I. (1977). Recurrence risks in families of children with symmetrical spasticity. *Devl Med. Child. Neurol.* **19**, 179–91.

Bundey, S. and Lovelace, R. E. (1975). A clinical and genetic study of chronic proximal spinal muscular atrophy. *Brain* **98**, 455–72.

Bundey, S., Carter, C. O., Soothill, J. F. (1970). Early recognition of heterozygotes for the gene for dystrophia myotonica. *J. Neurol. Neurosurg. Psychiat.* **33**, 279–93.

Bundey, S., Doniach, D., and Soothill, J. F. (1972). Immunological studies in patients with juvenile-onset myasthenia gravis and in their relatives. *Clin. exp. Immun.* **11**, 321–32.

Bundey, S., Harrison, M. J. G., and Marsden, C. D. (1975). A genetic study of torsion dystonia. *J. med. Genet.* **12**, 12–19.

Bundey, S., Griffiths, M.I., and Young, J.A. (1978). Recurrence risk of symmetrical spastic cerebral palsy. Letter. *Devl Med. child Neurol.* 20, 390–1

Bundey, S., Crawley, J.M., Edwards, J.H., and Westhead, R.A. (1979). Serum creatine kinase levels in pubertal, mature, pregnant, and post-menopausal women. *J. med. Genet.* 16, 117–21.

Burck, U., *et al.* (1981). Neuropathy and vitamin E in man. *Neuropediatrics.* 12, 267–78.

Burdick, A.B. Owens, L.A., and Peterson, C. (1981). Slowly progressive autosomal dominant spastic paraplegia with late onset, variable expression and reduced penetrance: A basis for diagnosis and counselling. *Clin. Genet.* 19, 1–7.

Burgess, E.A., Oberholzer, V.G., Semmens, J.M., and Stern, J. (1978). Acute neonatal and benign citrullinaemia in one sibship. *Arch. Dis. Childh.* 53, 179–82.

Burke, R.E., Brin, M.F. Fahn, S., Bressman, S.B. and Moskowitz, C. (1986). Analysis of the clinical course of non-Jewish, autosomal dominant torsion tyslonia. *Movement disorders* 1, 163–78.

Buruma, O.J.S. and Bots, G. Th. A.M. (1978). Myopathy in familial hypokalaemia periodic paralysis independent of paralytic attacks. *Acta neurol. scand.* 57, 171–9.

Burn, J., *et al.* (1986*a*). Duchenne muscular dystrophy in one of monozygotic twin girls. *J. Med. Genet.* 23, 494–500.

Burn, J., *et al.* (1986*b*). A syndrome with intracranial calcification and microcephaly in two sibs, resembling intrauterine infection. *Clin. Genet.* 30, 112 1.

Burns, J., Neuhauser, G., and Tomasi, L. (1976). Benign hereditary non-progressive chorea of early onset. Clinical genetics of the syndrome and report of a new family. *Neuropaediatrie* 7, 431–8.

Burton, B.K. (1979). Recurrence risks for congenital hydrocephalus. *Clin. Genet.* 16, 47–53.

Buruma, O.J.S. Bots, G.T.A.M. Went, L.N. (1985). Familial hypokalemic periodic paralysis. A large family. *Arch. Neurol.* 42, 28.

Busch, H.F.M., Koster, J.F., and Van Weerden, T.W. (1979). Infant and adult onset acid maltase deficiency occuring in the same family. *Neurol., Minneap.* 29, 415–16.

Busis, N.A. (1988). Peripheral neuropathy, high serum IgM, and paraproteinemia in mother and son. *Neurol.* 38, 679.

Busis, N.A. and Hochberg, F.H. (1985). Familial syringomyelia. *J. neurol. neurosurg. Psychiat.* 48, 936–8.

Bussone G. *et al.* (1984). Divry–van Bogaert syndrome, clinical and ultrastructural findings. *Arch. Neurol.* 41, 560–2.

Butterworth, J., *et al.* (1978). Low arylsulphatase. An activity in a family without metachromatic leukodystrophy. *Clin. Genet.* 14, 213–8.

Butzer, J.F., Schochet, S.S., and Bell, W.E. (1975). Infantile neuro-axonal dystrophy. *Acta neuropath., Ber.* **31**, 35–43.

Buxton, P., *et al.* (1972). A case of GM₂ gangliosidosis of late onset. *J. Neurol. Neurosurg. Psychiat.* **35**, 685.

Byers, R.K. and Banker, B.Q. (1961). Infantile muscular atrophy. *Arch. Neurol., Chicago.* **5**, 140–64.

Byrne, E., Thomas, P.K. and Zilkha, K.J. (1982). Familial extrapyramidal disease with peripheral neuropathy. *J. Neurol. Neurosurg. Psychiat.* **45**, 372–4.

Byrne, E., *et al.* (1984). Ataxia-without-telangiectasia. Progressive multi-system degeneration with IgE deficiency and chromosomal instability. *J. Neurol. Sci.* **66**, 307–17.

Byrne, E., *et al.* (1985*a*). Partial cytochrome oxidase (AA3). Deficiency in chronic progressive external ophthalmoplegia. Histochemical and biochemical studies. *J. Neurol. Sci.* **71**, 257–71.

Byrne, E., *et al.* (1985*b*). Mitochondrial myoneuropathy with respiratory failure and myoclonic epilepsy. A case report with biochemical studies. *J. Neurol. Sci.* **71**, 273–81.

Cable, W.J.L., *et al.* (1982). Fabry disease: significance of ultrastructual localization of lipid inclusions in dermal nerves. *Neurol.* **32**, 342–53.

Cagianut, B., Rhyner, K., Furrer, W., and Schnebly, H.P. (1981). Thiosulphate-sulphur transferase (Rhodanese) deficiency in Lebers hereditary optic atrophy. *Lancet,* **ii**, 981–2.

Caine, E.D., Hunt, R.D., Weingartner, H., and Ebert, M.H. (1978). Huntington's dementia. *Arch. gen. psychol.* **35**, 377–84.

Caldwell, J.B., Howard, R.O., and Riggs, L.A. (1971). Dominant juvenile optic atrophy. A study in two families and review of hereditary disease in childhood. *Archs Ophthalm., N.Y.* **85**, 133–47.

Cambell, A.M.G. and Hoffman. H.L. (1964). Sensory radicular neuro-pathy associated with muscle wasting in two cases. *Brain* **87**, 67–74.

Cambell, A.M.G., Corncr, B., Norman, R.M., and Urich, H. (1961). The rigid form of Huntington's disease. *J. Neurol. Neurosurg. Psychiat.* **24**, 71–7.

Cambell, A.M.G., Rosenberg, L.E., Snodgrass, P.J., and Nuzum, C.T. (1973). Ornithine transcarbamylase deficiency: a cause of lethal neonatal hyperammonemia in males. *New Engl. J. Med.* **288**, 1–6.

Cameron, D. and Venters, E.A. (1967). Some problems in Huntington's chorea. *Scott. med. J.* **12**, 152–6.

Campbell, S., Tsannatos, C., and Pearce, J.M. (1984). The prenatal diagnosis of Joubert's Syndrome of familial agenesis of the cerebellar vermis. *Prenatal Diagnosis* **4**, 391–5.

Campistol, J. Alvarez, E.F. and Cusi, V. (1984). CT scan appearances in subacute necrotizing encephalomyelopathy. *Dev. Med. Child Neurol.* **26**, 509–27.

Canale, D., Bebin, J., and Knighton, R.S. (1964). Neurologic manifestations of von Recklinghausen's disease of the nervous system. *Confinia neurol.* **24**, 359–403.

Canavan, M.M. (1931). Schilder's encephalitis periaxialis diffusa. *Archs Neurol. Psychiat., Chicago* **25**, 299–308.

Cancilla, P.A., Kalyanaraman, K., Verity, M.A., Munsat, T., and Pearson, C.M. (1971). Familial myopathy with probably lysis of myofibrils in type I fibre. *Neurol. Minneap.* **21**, 579–85.

Cantoni, A., *et al.* (1985). Seven hereditary syndromes with pigmentary retinopathy. *Clin. Pediat.* **24**, 578–83.

Cantu, J.M, Urrusti, J., Rosales, G., and Rojas, A. (1971). Evidence for autosomal recessive inheritance of costovertebral dysplasia. *Clin. Genet.* **2**, 149–54.

Cantu, J.M., *et al* (1977). Autosomal recessive microcephaly associated with chorioretinopathy. *Hum. Genet.* **36**, 243–7.

Cantu, J.M., Fragoso, R., Garcia-Cruz, D., and Sanchez-Corona, J. (1978). Dominant inheritance of holoprosencephaly. *Birth Defects* **14**, 215–20.

Cao, A., Cianchetti, L., Calisti, L., and Tangheroni, W. (1976). A family of juvenile, proximal spinal muscular atrophy with dominant inheritance. *J. med. Genet.* **13**, 131–5.

Cao, A., Cianchetti, L., Signorini, E., Loi, M., Sanna, G., and De Virgilis, S. (1977). Agenesis of the corpus callosum, infantile spasms, spastic quadriplegia, microcephaly and severe mental retardation in three siblings. *Clin. Genet.* **12**, 290–6.

Carenini, L., Finoccharo, G., Di Dancito, S., Visciani, A., and Negri, S. (1984). Electromyography and nerve conduction study in autosomal dominant olivopontocerebellar atrophy. *J. Neurol.* **231**, 34–7.

Caraceni, T., and Giovannini, P. (1977). Familial syringomyelia: a report of four cases. *Arch. Psychiat. Neurol.* **224**, 331–40.

Caraceni, T., Broggi, G., and Avanzini, G. (1974). Familial idiopathic basal ganglia calcification exhibiting 'dystonia musculorum deformans' features. *Eur. Neurol.* **12**, 315–9.

Carey, J.C., Laub, J.M., and Hall, B.D. (1979). Penetrance and variability in neurofibromatosis: a genetic study of 60 families. *Birth Defects* **15**, 271–81.

Carey, M.F. *et al.* (1987). Carnitine palmitoyl transferase deficiency with a typical presentation and ultrastructural mitochondrial abnormalitirs. *J. Neurol. Neurosurg. Psychiat.* **50**, 1060–2.

Carnevale, A., del Castillo, V., Sotillo, A.G., and Larrondo, J. (1976). Congenital absence of gluteal muscles. Report of two sibs. *Clin. Genet.* **10**, 135–8.

Carney, R.G. (1976). Incontinentia pigmenti—a world statistical analysis. *Arch. Derm.* **112**, 535–42.

Caro, A.J. (1977). A genetic problem in East Anglia. Huntington's chorea. Ph.D. Thesis, University of East Anglia.

Carpenter, G. (1901). Two sisters showing malformations of the skull and other congenital abnormalities. *Rep. Soc. Study. Dis. Child.* **1**, 110.

Carpenter, S. and Schumacher, G.A. (1966). Familial infantile cerebellar atrophy associated with retinal degeneration. *Arch. Neurol., Chicago* **14**, 82–94.

Carpenter, S., Karpati, G., Eisen, A., Andermann, F., and Watters, G. (1972). Childhood dermatomyositis and familial collagen disease. (Abstract). *Neurol., Minneap.* **22**, 425.

Carpenter, S., Karpati, G., Andermann, F., Jacob, J.C., and Andermann, E. (1977). The ultrastructural characteristics of the abnormal cytosome in Batten–Kufs disease. *Brain* **100**, 137–56.

Carpenter, T.O., Carnes, D.L. and Anast, C.S. (1983). Hypoparathyroidsm in Wilsons disease. *N. Engl. J. Med.* **309**, 873–7.

Carr, R.D. (1966). Is the Melkersson–Rosenthal's syndrome hereditary? *Arch. Derm.* **93**, 426–7.

Carroll, W.M., and Mastaglia, F.L.(1979). Leber's optic neuropathy: a clinical and visual evoked potential study of affected and asymptomatic members of a six-generation family. *Brain* **102**, 559–80

Carroll, W.M., Kriss, A., Baraitser, M., Barratt, G., and Halliday, A.M. (1980). The incidence and nature of visual pathway involvement in Friedreich's ataxia. *Brain* **103**, 413–34.

Carson, M.J., Slager, H.T., and Steinberg, M. (1977). Simultaneous occurence of diabetes mellitus, diabetes insipidus and optic atrophy in a brother and sister. *Am. J. Dis. Child.* **131**, 1382–5.

Carter, C.O. and Evans, K. (1973a). Spina bifida and anencephalus in Greater London. *J. med. Genet.* **10**, 209–34.

Carter, C.O. and Evans, K. (1973b). Children of adult survivors with spina bifida cystica. *Lancet* **ii**, 924–6.

Carter, C.O. and Evans, K. (1979). Counselling and Huntington's chorea. (Letter). *Lancet* **ii**, 470–1.

Carter, C.O., David, P.A., and Laurence, K.M. (1968). A family study of major central nervous system malformations in South Wales. *J. med. Genet.* **5**, 81–106.

Carter, C.O., Evans, K., and Till, K. (1976). Spinal dysraphism: genetic relation to neural tube malformations. *J. med. Genet.* **13**, 343–50.

Carter, C.O., *et al.* (1982). A family study of craniosynostosis with probable recognition of a distinct syndrome. *J. Med. Genet.* **19**, 280–5.

Carter, H.R. and Sukavajana, C., (1956). Familial cerebello-olivary degeneration with late development of rigidity and dementia. *Neurol., Minneap.* **6**, 876–84.

Cartier, L. Galvez, S. Gajdusek, D.C. (1985). Familial clustering of the ataxic form of Creutzfeldt–Jakob disease with Hirano bodies. *J. Neurol. Neurosurg. Psychiat.* **48**, 234–8.

Cartlidge, N.E.F. and Bone, G. (1973). Sphincter involvement in hereditary spastic paraplegia. *Neurol., Minneap.* **23**, 1160–3.

Carton, D. (1978). Benign familial neonatal convulsions. *Neuropaedriatrie* **9**, 167–71.

Carvalho, J., Coimbra, A., and Andrade, C. (1976). Peripheral nerve fibre changes in asymptomatic children of patients with amyloid polyneuropathy. *Brain* **99**, 1–10.

Casaer, P., *et al.* (1985). Variability of outcome in Joubert syndrome. *Neuropediatric* **16**, 43–5.

Cassidy, S.B., *et al.* (1987). Connatal Pelizaeus–Merzbacher disease: An autosomal recessive form. *Ped. Neurol.* **3**, 300–5.

Cassie, R. and Boon, A.R. (1977). Sex-linked hydrocephalus. *J. Med. Genet.* **14**, 72–3.

Castaner-Vendrell, E. and Barraquer-Bordas, L. (1949). Six membres de la meme famille avec tic douloureux du trijumeau. *Mschr. Psychiat, Neurol.* **118**, 77–80.

Castroviejo, I.P., Roderigues-Costa, T., and Castillo, F. (1973). Spondylothoracic dysplasia in three sisters. *Dev. med. Child. Neurol.* **15**, 348–54.

Cavanagh, J.B. and Meyer, A. (1956). Aetiological aspects of ammon's horn sclerosis associated with temporal lobe epilepsy. *Br. med. J.* **ii**, 1403.

Cavanagh, N.P.C. (1978). Cerebellar ataxia in infancy and childhood related to a disturbance of pyruvate and lactate metabolism. *Dev. med. Child. Neurol.* **20**, 672–4.

Cavanagh, N.P.C., Eames, R.A., Galvin, R.J., Brett, E.M., and Kelly, R.E. (1979*a*). Hereditary sensory neuropathy with spastic paraplegia. *Brain* **102**, 79–94.

Cavanagh, N.P.C., Lake, B.D., and McMeniman, P. (1979*b*). Congenital fibre type disproportion myopathy. A histological diagnosis with an uncertain clinical outlook. *Arch. Dis. Childh.* **54**, 735–43.

Cawthorne, T. and Haynes, D.R. (1956). Facial palsy. *Br. med. J.* **ii**, 1197–200.

Caylor, G.G. (1969). Cardiofacial syndrome. *Arch. Dis. Childh.* **44**, 69–75.

Cederbaum, S.D., *et al.* (1973). Argininosuccinic aciduria. *Am J. ment. Defic.* **77**, 395–404.

Cederbaum, C.D., Blass, J.P., Minkoff, N., Brown, W.J., Cotton, M.E., and Harris S.H. (1976). Sensitivity to carbohydrate in a patient with familial intermittent lactic acidosis and pyruvate dehydrogenase deficiency. *Pediat. Res.* **10**, 713–20.

Cederbaum, S.D., Shaw, K.N.F., Dancis, J., Hutzler, J., and Blaskovics, J.C. (1979). Hyperlysinemia with saccharopinuria due to combined lysine – ketoglutarate reductase and saccharopine dehydrogenase deficiencies presenting as cystinuria. *J. Pediat.* **95**, 234–8.

Cendrowski, W.S. (1968). Multiple sclerosis: discordance in three pairs of dizygotic twins. *J. med. Genet.* **5**, 266–8.

Cerizza, M., Nemni, R., and Tamma, F. (1987). Adult metachromatic leukodystrophy: an underdiagnosed disease? *J. Neurol. Neurosurg. Psychiat.* **50**, 1710–2.

Chaco, J. (1969). Marinesco–Sjögren syndrome with myopathy. *Confinia neurol.* **31**, 349–51.

Chakrabarti, A. and Pearce, J.M.S. (1981). Scapoloperoneal syndrome with cardiomyopathy – report of a family with autosomal dominant inheritance and unusual features. *J. Neurol. Neurolsurg. Psychiat.* **44**, 1146–52.

Chakravorty, B.G. and Gleadhill, C.A. (1966). Familial incidence of cerebral, aneurysms, *Br. med. J.* **i**, 147–8.

Chalhub, E.G. (1976). Neurocutaneous syndromes in children. *Pediat. Clins N. Am.* **23**, 499–516.

Challa, V.R., Goodman, H.O., and Davis, C.H. (1983). Familial brain tumors: Studies of two families and review of recent literature. *Neurosurg.* **12**, 18–23.

Chalmers, N. and Mitchell, J.D. (1987). Optico-acoustic atrophy in distal spinal muscular atrophy. *J. Neurol. Neurosurg. Psychiat.* **50**, 238–9.

Chamberlain, S. *et al.* (1988). Mapping of mutation causing Friedreich's ataxia to human chromosome 9. *Nature.* **334**, 248–9.

Chambers, W.R., Harper, B.F., and Simpson, J.R. (1954). Familial incidence of congenital aneurysms of cerebral arteries. Report of cases of ruptured aneurysms in father and son. *J. Am. med. Ass.* **155**, 358–9.

Chance, P.F., Murray, J.C., Bird, T.D., and Kochin, R.S. (1987). Genetic linkage relationship of Charcot–Marie–Tooth disease (HMSN 1b) to chromosome 1 markers. *Neurol.* **37**, 325.

Chandler, J.H. and Bebin, J. (1956). Hereditary cerebellar ataxia – olivo-pontocerebellar type. *Neurology, Minneap.* **6**, 187–95.

Chandra, R.K. (1963). Congenital analgia in two siblings. *J. Ind. med. Ass.* **41**, 25–6.

Chao, D.H.-C. (1959*a*). Congenital neurocutaneous syndromes in childhood. II. Tuberous sclerosis. *J. Pediat.* **55**, 447–59

Chao, D.H.-C. (1959*b*). Congenital neurocutaneous syndromes in childhood. Sturge–Weber disease. *J. Pediat.* **55**, 635–49.

Chapman, R.C. and Diaz-Perez, R. (1962). Pheochromocytoma and cerebellar hemangioblastoma. *J. Am. med. Ass.* **182**, 1014–17.

Chapman, R.C., Kemp, V.E., and Talliaferro, I. (1959). Pheochromocytoma associated with multiple neurofibromatosis and intracranial hemangioma. *Am. J. Med.* **26**, 883–90.

Charcot, J.-M. and Marie, P. (1886). Sur une forme particuliere d'atrophie musculaire progressive, souvent familiale, debutant par les pieds et les jambes, et atteignant plus tard les mains. *Revue méd.* **6**, 97–138.

Chazot, G., Guard, O., Seytier, A., Robert, J.M., and Schott, B. (1975). Syndrome de Ramsay Hunt et aniridie chez deux jumelles monozygotes. *Rev. neurol.* **131**, 43–8.

Chemke, J., Czernobilsky, B., Mundel, G., and Barishak, R. (1975). A familial syndrome of central nervous system and ocular malformation. *Clin. Genet.* **7**, 1–7.

Chemke, J., Katznelson, K., and Zucker, G. (1985). Familial glioblastoma multiforme without neurofibromatosis. *Am. J. Med. Genet.* **21**, 731–5.

Cheney, W. D. (1965). Acro-osteolysis. *Am. J. Roentg.* **94**, 595–607.

Childs, B., Nyhan, W. L., Borden, M., Bard, L., and Cooke, R. E. (1961). Idiopathic hyperglycinemia and hyperglycinuria. A new disorder of amino acid metabolism. *Pediatrics, Springfield* **27**, 522–38.

Chio, A. *et al.* Phenotypic and genotypic heterogeneity of dominantly inherited amyotrophic lateral sclerosis. *Acta Neurologica Scandinavia* **75**(4), 277–282.

Chiu, E. and Brackenridge, C. J. (1976). A probable case of mutation in Huntington's disease. *J. med. Genet.* **13**, 75–7.

Chiu, E. and Teltcher, B. (1978). Huntington's disease: the establishment of a national register. *Med. J. Aust.* **2**, 394–6.

Chotzen, F. (1932). Eine eigenartige familiäuare Ertwicklungsstörung. (Atrocephalosyndaktylie, dysostosis craniofacialis und hypertelorismus.) *Mschr. Kinderheilk.* **55**, 97.

Chou, L., Kaye, C. I. and Nadler, H. L. (1974). Brain β-galactosidase and GM_1 gangliosidosis. *Pediat. Res.* **8**, 120–5.

Chou, S. M. and Thompson, H. G. (1970). Electron microscopy of storage cytosomes in Kuf's disease. *Arch. Neurol., Chicago.* **23**, 489–501.

Christensen, E., Melchior, J. C., and Negri, S. (1961). A comparative study of 16 cases of diffuse sclerosis with special reference to the histopathological findings. *Acta neurol. scand.* **37**, 163–207.

Christian, J. C., Bixler, D., Blythe, S. C., and Merritt, A. D. (1969). Familial telecanthus with associated congenital anomalies. In *The clinical delineation of birth defects.* Vol. II. *Malformation Syndromes,* pp. 82–5. National Foundation, New York.

Christian, J. C. Andrews, P. A., Conneally, P. M., and Muller, J. (1971*a*). The adducted thumbs syndrome. *Clin. Genet.* **2**, 95–103.

Christian, J. C., Dexter, R. N., and Donohue, J. P. (1971*b*). Hypogonadtrophic hypogonadism with anosmia: the Kallmann syndrome. *Birth Defects.* **vii**, 166–71.

Christoferson, A., Gustafson, M. B., and Petersen, A. G. (1961) Von Hippel–Lindau's disease. *J. Am. med. Ass.* **178**, 280–2.

Chiu, E. and Brackenridge, C. J. (1976). A probable case of mutation in Huntington's disease. *J. med. Genet.* **13**, 75–7.

Chui, L. A. and Munsat. T. L. (1976). Dominant inheritance of McArdle syndrome. *Arch. Neurol., Chicago* **33**, 636–41.

Chiu, E. and Teltscher, B. (1978). Huntington's disease: the establishment of a national register. *Med. J. Aust.* **2**, 394–6.

Chun, R., Daly, R., Mansheim, B., and Wolcott, G. (1973). Benign familial chorea with onset in childhood. *J. Am. med. Ass.* **225**, 1603–7.

Chung, C.S. and Morton, N.E. (1959). Discrimination of genetic entities in muscular dystrophy. *Am. J. hum. Genet.* **11**, 339–59.

Chutkow, J.G., *et al.* (1986). Adult onset autosomal dominant limb-girdle muscular dystrophy. *Ann. Neurol.* **20**, 240–8.

Ciccarelli, E.C. and Vesell, E.S. (1961). Laurence–Moon–Biedl syndrome. *Am. J. Dis. Childh.* **101**, 519–24.

Claireaux, A.E. (1972). Multicystic encephalomalacia. *Dev. Med. Child. Neurol.* **14**, 662.

Clark, J.R. d'Agostino, A.N., Wilson, J., Brooks, R.R., and Cole, G.C. (1978). Autosomal dominant myofibrillar inclusion body myopathy. (Abs.) *Neurol., Minneap.* **28**, 399.

Clark, J.R., Miller, R.G., and Vidgoff, J.M. (1979). Juvenile-onset metachromatic leukodystrophy: biochemical and electrophysiologic studies. *Neurol., Minneap.* **29**, 346–53.

Clark, J.V. (1970). Familial occurence of cavernous angiomata of the brain. *J. Neurol. Neurosurg. Psychiat.* **33**, 871–6.

Clarke, A., *et al.* (1986). Duchenne muscular dystrophy with adrenal insufficiency and glycerol kinase deficiency: high resolution cytogenetic analysis with molecular, biochemical and clinical studies. *J. Med. Genet.* **23**, 510–8.

Clarke, J.M. (1910). On recurrent motor paralysis in migraine. *Br. med. J.* **i**, 1534–8.

Coates, P.M., Hale, D.E., and Winter, S.C. (1986). Short-chain acyl CoA dehydrogenase deficiency associated with severe skeletal muscle weakness, and muscle carnitine deficiency. *Pediat. Res.* **20**, 328A.

Cochrane, P., Hughes, R.R., Buxton, P.H., and Yorke, R.A. (1973). Myophosphorylase deficiency (McArdle's disease) in two interrelated families. *J. Neurol. Neurosurg. Psychiat.* **36**, 217–24.

Cockel, R., Hill, E.E., Rushton, D.I., Smith, B., and Hawkins, C.F. (1973). Familial steatorrhea with calcification of the basal ganglia and mental retardation. *Q. J. Med.* **42**, 771–83.

Codish, S.D., Kraszeski, J., and Pratt, K. (1973). C.N.S. developmental anomaly in the basal cell nevus syndrome: another congenital neuro-cutaneous syndrome? *Neuropaediatrics* **4**, 338–43.

Cogan, D.G. (1952). A type of congenital ocular motor-apraxia presenting jerky head movements. *Trans. Am. Acad. Ophthal. Oto- lar.* **56**, 853–6.

Cohen, L.G., Hallett, M., and Sudarsky, L. (1987). A single family with writers cramp, essential tremor and primary writing tremor. *Movement Disorder* **2**, 109–16.

Cohen, M.E., Duffner, P.K., and Heffner, R. (1978). Central core disease in one of identical twins. *J. Neurol. Neurosurg. Psychiat.* **41**, 659–63.

Cohen, M.M. Jr (1979). Craniosynostosis and syndromes with craniosynostosis: incidence, genetics, penetrance, variability, and new syndrome updating *Birth Defects:* Original Article Series **XV**, 13–63.

Cohen, M.M. Jr. (1986). Syndromes with craniosynostosis. *Diagnosis, Evaluation, and Management* (ed. M.M. Cohen, Jr.), Raven Press, New York, pp. 417–418.

Cohen, M.M. Jr. and Gorlin, R.J. (1969). Genetic considerations in a sibship of cyclopia and clefts. The clinical delineation of birth defects. II. Malformation syndromes. *Birth Defects*. Original Series. **V**, 113–18.

Cohen, M.M. Jr., Sedano, H.O., Gorlin, R., and Jirasek, J.E. (1971). Frontonasal dysplasia (median cleft face syndrome): comments on etiology and pathogenesis. *Birth Defects:* Original Article Series **VII**, 117–19.

Cohen, N.M.J., Shaham, M., Dagan, J., Shmueli, J., and Kohn, G. (1975). Cytogenetic investigations in families with ataxia-telagiectasia. *Cytogenet.* **15**, 338–56.

Cohn, R. and Kurland, L.T. (1958). Mirror movements and the general phenomenon of synkinesia. *Trans. Am. neurol. Ass.* **84**, 40–4.

Coimbra, A. and Andrade, C. (1971). Familial amyloid polyneuropathy: an electron microscope study of the peripheral nerve in five cases. I. Interstitial changes. *Brain* **94**, 199–206.

Colan, R.V. Snead, C.C. and Ceballos, P. (1981). Olivopontocerebellar atrophy in children: a report of seven cases in two families. *Ann. Neurol.* **10**, 355–63.

Cole, E.C., and Meek, D.C. (1985). Juvenile non-ketotic hyperglycinaemia (NKH) in 3 siblings. *J. Inher. Metab. Dis.* **8**, 123.

Coleman, M., Hart, P.N., Randall, J., Lee, J., Hijada, D., and Bratenahl, CH. G. (1977). Seratonin levels in the blood and central nervous system of a patient with sudanophilic leukodystrophy. *Neuropaediatrics* **8**, 459–66.

Collier, J. and Greenfield, J.G. (1924). The encephalitis periaxiales of Schilder; a clinical and pathological study with an account of two cases, one of which was diagnosed during life. *Brain* **47**, 489–519.

Collins, D.L. and Schimke, R.N. (1982). Moebius syndrome in a child and extremity defect in her father. *Clin. Genet.* **22**, 312–14.

Colver, A.F. *et al.* (1981). Rigid spine syndrome and fatal cardiomyopathy. *Arch. Dis. Childh.* **56**, 148–51.

Comings, D.E. and Amromin, G.D. (1974). Autosomal dominant insensitivity to pain with hyperplastic myelinopathy and autosomal dominant indifference to pain. *Neurology, Minneap* **24**, 838–48.

Comings, D.E. and Comings, B.G. (1984). Tourette's syndrome and attention deficit disorder with hyperactivity. Are they genetically related? *J. Am. Acad. Child Psychiat.* **23**, 138–46.

Comings, D.E., Comings, B.G., Devor, E.J. and Cloninger, C.R. (1984). Detection of major gene for Gilles de la Tourette syndrome. *Am. J. Hum. Genet.* **36**, 586–600.

Comings, D.E., *et al.* (1986). Evidence the Tourette syndrome gene is at

18q22.1. 7th International Congress of Human Genetics, Berlin (Abst.) Part II, 620.

Compston, D.A.S. (1981). Multiple sclerosis in the Orkneys (letter). *Lancet* **2**, 98.

Compston, D.A.S., Batchelor, J.R., and McDonald, W.I. (1976). B-lymphocyte alloantigens associated with multiple sclerosis. *Lancet* **ii**, 1261–5.

Connen, P.E., Murphy, E.G., and Donohue, W.L. (1963). Light and electron microscopic studies of 'myogranules' in a child with hypotonia and muscle weakness. *Can. med. Ass. J.* **89**, 983–6.

Connor, J.M., Stephenson, J.B.P., and Hadley M.D.M. (1986). Non-penetrance in tuberous sclerosis. *Lancet* **ii**, 1275.

Conomy, J.P., Levinsohn, M., and Fanaroff, A. (1975). Familial infantile myasthenia gravis: a cause of sudden death in young children. *J. Pediat.* **87**, 428–9.

Conrad, K. (1937). Erbanlas und Epilepsie IV. Ergebnisse einer Nachkommenschaftsuntersuchung an Epileptikern. *Z. ges. Neurol. Psychiat.* **159**, 521–81.

Constantinidis, J. (1969). Forme hereditaire de la maladie de Pick sans lésions neutronales specifiques. *Acta neurol. psychiat. Lellen.* **8**, 226–32.

Constantinidis, J. (1978). Is Alzheimer's disease a major form of senile dementia? Clinical, anatomical and genetic data. In *Alzheimer's disease: senile dementia and related disorders,* Vol. 7, Ageing, (ed. R. Katzman, R.D. Terry, and K.L. Bick), pp. 15–25. Raven Press, New York.

Constantinidis, J. and De Ajuriaguerra, J. (1970). Syndrome familial avec tremblement parkinsonien et anosmie et se therapeutique par la l-DOPA associeée á un inhibiteur de la carboxylase. *Sem. Hôp. Ther., Paris* **46**, 263–9.

Constantinidis, J., Garrone, G., and de Ajuriaguerra, J. (1962). L'heredite des demences de l'age avancé. *Encephale.* **51**, 301–44.

Copeland, D., Lamb, W. Klintworth, G. (1977). Calcification of basal ganglia and cerebellar roof nuclei in mentally defective patient with hidrotic ectodermal dysplasta. *Neurol.* **27**, 1029–33.

Corbett, V.A. and Nuttall, F.Q. (1975). Familial hypokalemic periodic paralysis in Blacks. *Ann. intern. Med.* **83**, 63–5.

Cordero, J.F. and Holmes, L.B. (1978). Phenotypic overlap of the BBB and G syndromes. *Am. J. med. Genet.* **2**, 145–52.

Cordier, J., Lowenthal, A., Rademecker, M.A., and Van Bogaert, L. (1952). Sur une forme congenitale et hereditaire de sclerose musculaire generalisee. La famille Duc. *Acta neurol. psychiat. belg.* **52**, 422–32.

Coria, F, *et al.* (1983). Occipital dysplasia and Chiari type I deformity in a family clinical and radiological study of three generations. *J. Neurol Sci.* **62**, 147–58.

Cornelio, F., *et al.* (1977). Fatal cases of lipid storage myopathy with carnitine deficiency. *J. Neurol. Neurosurg. Psychiat.* **40**, 170–8.

Cornelio, F., Bresolin, N., Dimauro, S. Mora, M. and Balestrini, M. R. (1983). Congenital myopathy due to phosphorylase deficiences. *Neurology.* **33**, 1383–5.

Cornell, J. Sellars, and Beighton, S. (1984). Autosomal recessive inheritance of Charcot–Marie–Tooth disease associated with sensorineural deafness. *Clin. Genet.* **25**, 163–5.

Corston, R. N. and Godwin-Austen, R. B. (1982). Transient global amnesia in four brotters. *J. Neurol. Neurosurg. Psychiat.* **45**, 375–7.

Cosgrove, G. R. Leblanc and Meagher-Villemure, K. (1985). Cerebral amyloid angiopathy. *Neurol.* **35**, 625–31.

Cottrill, C., Glueck, C. J., Leuba, V., and Millett, F. (1974). Familial homozygous hypobetalipoproteinemia. *Metabolism* **23**, 779–91.

Courten-Myers, G. and Mandybur, T. I. (1987). Atypical Gerstmann–Straussler syndrome or familial spinocerebellar ataxia and Alzheimer's disease. *Neurol.* **37**, 269–75.

Coutinho, P. and Andrade, C. (1978). Autosomal dominant system degeneration in Portuguese families of the Azores Islands. A new genetic disorder involving cerebellar, pyramidal extapyramidal and spinal cord motor functions. *Neurol. Minneap.* **28**, 703–9.

Cowchock, F. S., Duckett, S. W. and Stretz, L. J. (1985). X-linked motor-sensory neuropathy type II with deafness and mental retardation: a new disorder. *Am. J. Med. Genet.* **20**, 307–15.

Cowen, D. and Olmstead, E. V. (1963). Infantile neuro-axonal dystrophy. *J. Neuropath. exp. Neurol.* **22**, 175–236.

Cox, D. W. (1966). Factors influencing ceruloplasmin levels in normal individuals. *J. Lab. clin. Med.* **68**, 893–904.

Cox, D. W., Fraser, F. C., and Sass-Kortsak, A. (1972). A genetic study of Wilson's disease: evidence for heterogeneity. *Am. J. hum. Genet.* **24**, 646–66.

Craig, W., Wagener, H. P., and Keinshan, J. W. (1941). Lindau–Von Hippel disease: a report of four cases. *Acta neurol. psychiat.* **466**, 36.

Crawford, A. H. (1978). Neurofibromatosis in the pediatric patient. *Orthop. Clin. N. Am.* **9**, 11–23.

Crawford, C. (1960). Report of a family showing 'mirror' movements. *Australas. Ann. Med.* **9**, 176–9.

Crawford, M. D. and Sarner, M. (1965). Ruptured intracranial aneurysm: community study. *Lancet* ii, 1254–7.

Crawfurd, M. d'A., Harcourt, R. B., and Shaw, P. A. (1979). Non-progressive cerebellar ataxia, aplasia of pupillary zone of iris, and mental subnormality (Gillespie's syndrome) affecting three members of a non-consanguineous family in two generations. *J. med. Genet.* **16**, 373–8.

Crawfurd, M. d'A., Jackson, P., and Kohler, H. G. (1978). Meckel's syndrome (dysencephalia splanchno-cystica) in two Pakistani sibs. *J. med. Genet.* **15**, 242–5.

Cremers, C.W.R.J., Wijdeveld, P.G.A.B., and Pinckers, A.J.L.G. (1977). Juvenile diabetes mellitus, optic atrophy, hearing loss, diabetes insipidus, atonia of the urinary tract and bladder and other abnormalities (Wolfram syndrome). *Acta pediat. scand.* Suppl. 264. 1–16.

Crews, J., Kaiser, K.K., and Brooke, M.H. (1976). Muscle pathology of myotoniacongenita. *J. neurol. Sci.* **28**, 449–57.

Critchley, E. (1972). Clinical manifestation of essential tremor. *J. Neurol. Neurosurg. Psychiat.* **35**, 365–72.

Critchley, E.M.R., Clark, D.B., and Wikler, A. (1968). Acanthocytosis and neurological disorder without betalipoproteinemia. *Archs Neurol., Chicago.* **18**, 134–40.

Critchley, E.M.R., Betts, J.J., Nicholson, J.T., and Weatherall, D.J. (1970). Acanthocytosis, normolipoproteinacemia and multiple tics. *Post-grad. med. J.* **46**, 698–701.

Critchley, M. (1934). Huntington's chorea and East Anglia. *J. St. Med.* **42**, 575–87.

Critchley, M. (1949). Observations on essential (heredofamilial) tremor. *Brain* **72**, 113–39.

Critchley, M. (1970). *The dyslexic child.* Thomas, Springfield, III.

Crocker, A.C. and Farber, S. (1958). Niemann Pick disease: a review of 18 patients. *Medicine, Baltimore* **37**, 1–95.

Croft, P.B., Cutling, J.C., Jewesbury, E.C.O., Blackwood, W., and Mair, W.G.P. (1977). Ocular myopathy (progressive external ophthalmoplegia) with neuropathic complications. *Acta neurol. scand.* **55**, 169–97.

Crome, L. (1965). Spongy degeneration of the brain. *Dev. med. Child. Neurol.* **7**, 322–3.

Cromc, L., Hanefeld, F., Patrick, D. and Wilson, J. (1973). Late onset globoid cell leucodystrophy. *Brain* **96**, 841–8.

Crosby, T.W. and Chou, S.M. (1974). 'Ragged-red' fibers in Leigh's disease. *Neurol., Minneap.* **24**, 49–54.

Cross, H.E. and McKusick, V.A. (1967). The Troyer syndrome. A recessive form of spastic paraplegia with distal muscle wasting. *Arch. Neurol., Chicago.* **16**, 473–85.

Cross, H.E., Mekusick, V.A., and Breen, W. (1967). A new oculocerebral syndrome with hypopigmentation. *J. Pediat.* **70**, 398–406.

Cross, H.E. Hollander, C.S., Rimoin, D.L., and McKusick, V.A. (1968). Familial agoitrous cretinism accompanied by muscular hypertrophy. *Pediatrics, Springfield* **41**, 413–20.

Crowe, F.W. (1964). Axillary freckling as a diagnostic aid in neurofibromatosis. *Ann. intern. Med.* **61**, 1142–3.

Crowe, F.W., Schull, W.J., and Neel, J.V. (1956). *A clinical, pathological and genetic study of multiple neurofibromatosis.* Thomas, Springfield, III.

Cruse, R.P., Conomy, J.P., Wilbourn, A.S., and Hanson, M.R. (1977). Hereditary hypertrophic neuropathy combining features of tic douloureux, Charcot–Marie–Tooth disease and deafness. *Cleveland Clin. Q.* **44**, 107–11.

Cruse, R.P. *et al.* (1984). Familial systemic carnitine deficiency. *Arch. Neurol.* **41**, 301–5.

Crutchfield, C.A. and Gutmann, L. (1973). Hereditary aspects of accessory deep peroneal nerve. *J. Neurol. Neurosurg. Psychiat.* **36**, 989–90.

Cruz-Martinez, A., Perez Conde, C., Ramon y Cajal, S., and Martinez, A. (1977). Recurrent familiar polyneuropathy with liability to pressure palsies. Special regards to electrophysiological aspects of twenty-five members from seven families. *Electromyography.* **17**, 102–24.

Curatolo, P., Libutti, G., and Dallapiccola, B. (1980*a*). Aicardi syndrome in a male infant. *J. Pediat.* **96**, 286–7.

Curatolo, P., Mercuri, S., and Cotroneo, E. (1980*b*). Joubert syndrome: a case confirmed by computerised tomography. *Dev. Med. child Neurol.* **22**, 362–78.

Curless, R.G. (1984). Computed tomography of GM$_1$ Gangliosidosis. *J. Pediat.* **105**, 964–8.

Curless, R.G. and Nelson, M.B. (1977). Congenital fibre type disproportion in identical twins. *Ann. Neurol.* **2**, 455–9.

Curless, R.G., Payne, C.M., and Brinner, F.M. (1978). Fingerprint body myopathy: a report of twins. *Devl Med. child Neurol.* **20**, 793–8.

Curran, D. (1930). Huntington's chorea without choreiform movement. *J. Neurol Psychopath.* **10**, 305–10.

Currie, S. (1970). Familial oculomotor palsy with Bell's palsy. *Brain* **93**, 193–8.

Currier, R.D., Glover, G., Jackson, J.F., and Tipton, A.C. (1972). Spinocerebellar ataxia: study of a large kindred. *Neurology, Minneap.* **20**, 1040–3.

Curth, H.O. and Warburton, D. (1965). The genetics of incontinentia pigmenti. *Arch. Derm.* **92**, 229–35.

Curtius, F. and Speer, H. (1937). Multiple sklerose und Erbanlage. z Mitteilung. *Z. ges. Neurol. Psychiat.* **160**, 226–45.

Cushing, H. and Bailey, P. (1928). Hemangiomas of the cerebellum and retina (Lindau's disease) with a report of a case. *Arch. Opthal., N. Y.* **57**, 447–63.

Cusins, P.J. and Van Rooyen, R.J. (1963). Familial periodic paralysis (7 cases in a Durban family). *S. Afr. med. J.* **37**, 1180–3.

D'Agostino, A.N., Kernohan, J.W., and Brown, J.R. (1963*a*). The Dandy–Walker syndrome. *J. Neuropath. exp. Neurol.* **22**, 450–70.

D'Agostino, A. N., Soule, E. H., and Miller, R. H. (1963*b*). Sarcomas of the peripheral nerves and somatic soft tissues associated with multiple neurofibromatosis. *Cancer, N. Y.* **16**, 1015–27.

Dalby, M. A. (1969). Epilepsy and 3 cps second spike and wave rhythms. A clinical, electroencephalographic and prognostic analysis of 346 patients. *Acta neurol. scand.* Suppl. 40.

Dal Canto, M. C., Rapin, I., and Suzuki, K. (1974). Neuronal storage disorder with chorea and curvilinear bodies. *Neurology, Minneap.* **24**, 1026–32.

Dale, T. (1931). Unusual forms of familial osteochondrodystrophy. *Acta radiol.* **12**, 337–58.

Dallaire, L., Clarke-Fraser, F., and Wigglesworth, F. W. (1971). Familial holoprosencephaly. *Birth Defects:* Original Series **VII**, 136–42.

Dalsgaard-Nielson, T. (1965). Migraine and heredity. *Acta neurol. scand.* **41**, 287–300.

Daly, D. and Bickford, R. G. (1951). Electroencephalographic studies of identical twins with photoepilepsy. *Electroenceph. clin. Neurophysiol.* **3**, 245–9.

Daly, D., and Yoss, R. E. (1959). A family with narcolepsy. *Proc. Staff Meet. Mayo Clin.* **34**, 313–19.

Daly, D., Siekert, R., and Burke, E. C. (1959). A variety of familial light sensitive epilepsy. *Electroenceph. clin Neurophysiol.* **11**, 141–5.

Daly, R. F., and Forster, F. M. (1975). Inheritance of reading epilepsy. *Neurol. Minneap.* **25**, 1051–4.

Daly, R. E. and Sajor, E. F. (1973). Inherited tic douloureux. *Neurol. Minneap.* **23**, 937–9.

Damasio, H., Antunes, L., and Damasio, A. R. (1977). Familial non-progressive involuntary movements of childhood. *Ann. Neurol.* **1**, 602–3.

Damaske, N. M., Cohen, D. N., Gutman, F. A., and Schumacher, O. P. (1975). Optic atrophy, diabetes mellitus and diabetes insipidus. *J. pediat. Ophthal.* **12**, 16–21.

Dambska, M., Wisniewski, K., Sher, J. H. (1984). An autopsy case of hemimegalencephaly. *Brain Dev.* **6**, 60–4.

Dana, C. L. (1887). Hereditary tremor, a hitherto undescribed form of motor neurosis. *Am. J. med. Sci.* **94**, 386–93.

Dancis, J., *et al.* (1983). The prognosis of hyperlysinemia: an interim report. *Am. J. Hum. Genet.* **35**, 438–42.

Danes, B. S., Rottell, B. K., Eviatar, L., and Stolzenberg, J. (1977). Genetic heterogeneity within the chondroitinsulphaturias. *J. med. Genet.* **14**, 103–7.

Danforth, H. B. (1964). Familial Bell's palsy. *Ann. Otol. Rhinol. Lar.* **73**, 179–83.

Danks, D. M., Campbell, P. E., Stevens, B. J., Mayne, V. and Cartwright, E.

(1972). Menkes kinky hair syndrome. An inherited defect in copper absorption with widespread effects. *Pediatrics, Springfield* **50**, 188–201.

Danks, D.M., Dartwright, E., Stevens, B.-J., and Townley, R.R.W. (1973). Menkes kinky hair disease: further definition of the defect in copper transport. *Science, N. Y.* **179**, 1140–2.

Danks, D.M., Mayne, C., and Kozlowski, K. (1974). *A precocious autosomal recessive type of osteodysplasty.* The clinical delineation of birth defects, No. 19. Williams & Wilkins, Baltimore.

Danner, R., Shewmon, A., and Sherman, M.P. (1986). Seizures in an atelencephalic infant. *Arch. Neurol.* **42**, 1014–16.

Danon, M.J. and Carpenter, S. (1985). Hereditary sensory neuropathy: Biopsy study of an autosomal dominant variety. *Neurol.* **35**, 1226–9.

Danta, G. (1975). Familial carpal tunnel syndrome with onset in childhood. *J. Neurol. Neurosurg. Psychiat.* **38**, 350–5.

Danta, G., Hilton, R.C., and Lynch, P.G. (1975). Chronic progressive external ophthalmplegia. *Brain* **98**, 473–92.

Daras, M. Tuchman, A.J. and David, S. (1983). Familial spinocerebellar ataxia with skin hyperpigmentation. *J. Neurol. Neurosurg. Psychiat.* **46**, 743–4.

Darras, B.T. *et al.* (1986). Familial amyloidosis with cranial neuropathy and corneal lattice dytrophy. *Neurology.* **36**, 432–50.

Dastur, D.K., Singhal, B.S., Gootz, M., and Seitelberger, F. (1966). A typical inclusion bodies with myoclonic epilepsy. *Acta neuropath.* **7**, 16–25.

Dastur, D.K., Manghani, D.K., and Wadia, N.H. (1969). Wilson's disease in India. In *Progress in neurogenetics* (cd. A. Barbeau and J.-R. Brunette), pp. 615–21. Excepta Medical, Amsterdam.

Daube, J.R. and Chou, S.M. (1968). Lissencephaly: two cases. *Neurol., Minneap.* **16**, 179–91.

Daube, J.R. and Peters, H.A. (1966). Hereditary essential myoclonus. *Arch. Neurol., Chicago* **15**, 587–94.

David, T.J. and Winter, R.M. (1985). Familial absence of the pectoralis major, serratus anterior and latissimus dorsi muscle. *J. Med. Genet.* **22**, 390–2.

Davidenkow, S. (1939). Scapuloperoneal amyotrophy. *Arch. Neurol. Neurosurg. Psychiat.* **41**, 694–701.

Davidson, E.A. and Robertson, F.E. (1955). Alzheimer''s disease with acne rosacea in one of identical twins. *J. Neurol. Neurosurg. Psychiat.* **18**, 72–7.

Davidson, S. and Falconer, M.A. (1975). Outcome of surgery in 40 children with temporal lobe epilepsy. *Lancet* **i**, 1260–3.

Davidson, S. and Watson, C.W. (1956). Hereditary light sensitive epilepsy. *Neurol. Minneap.* **6**, 235–61.

550   *Bibliography*

Davies, D.M. (1954). Recurrent peripheral-nerve palsies in a family. *Lancet* **ii**, 266–8.

Davis, C.J.F., Bradley, W.G., and Madrid, R. (1978). The peroneal muscular atrophy syndrome (clinical, genetic, electrophysiological and nerve biopsy studies). *J. Génét. hum.* **26**, 311–49.

Davis, L.E., Snyder, R.D., Orth, D.N., Nicholson, W.E., Kornfeld, M., and Seelinger, D.F. (1979). Adrenoleukodystrophy and adrenomyeloneuropathy associated with partial adrenal insufficiency in three generations of a kindred. *Am. J. Med.* **66**, 342–7.

Davidson, C. (1954). Pallido-pyramidal disease. *J. Neuropath. exp. Neurol.* **13**, 50–9.

Day, R.E. and Schutt, W.H. (1979). Normal children with large heads – benign familial megalencephaly. *Archs. Dis. Childh.* **54**, 512–17.

Dayan, A.D., Ockenden, B.G., and Crome, L. (1970). Necrotizing encephalomyelopathy of Leigh. Neuropathological findings in 8 cases. *Archs Dis. Childh.* **45**, 39–48.

Dean, G. (1967). Annual incidence, prevalance and mortality of multiple sclerosis in white South-African-born and in white immigrants to South Africa. *Br. med. J.* **ii**, 724.

Dean, G. (1971). Screening tests for porphria. *Lancet* **i**, 86–7.

Dean, G. and Kurtzke, J.F. (1971). On the risk of multiple sclerosis according to age and immigration to South Africa. *Br. med. J.* **iii**, 725–7.

Dean, G., McLoughlin, H., Brady, R., Adelstein, A.M., and Tallett-Williams, J. (1976). Multiple sclerosis among immigrants in Greater London. *Br. med. J.* **i**, 861–4.

De Baecque, C.M., Susuki, K., and Rapin, I. (1975). Gm$_2$-gangliosidosis A.B. variant: clinicopathological study of a case. *Acta neuropath.* **33**, 207–26.

Debicka, A. and Adamczak, P. (1979). Przypadek dziedziczenia zespolu Sturge a–Webera. *Klin. oczna.* **81**, 541–2.

De Coster, W., De Reuck, J., and Thiery, E. (1974). A late autosomal dominant form of limb-girdle muscular dystrophy: a clinical, genetic and morphological study. *Eur. Neurol.* **12**, 159–72.

De Groot, C.J., Haan, G.I., Hulstaert, C.E., and Hommes, F.A. (1977). A patient with severe neurological symptoms and acetoacetyl CoA thiolase deficiency. *Pediat. Res.* **11**, 1112.

De Haene, A. (1955). Agénésie partielle du vermis du cervelet á caractére familial. *Acta neurol. psychiat. belg.* **55**, 622–8.

Dejerine, J. and Sottas, J. (1893). Sur la nevrite interistitielle, hypertrophique et progressive de l'enfance. *C.R. Séanc. Soc. Biol.* **5**, 63–96.

De Jong, J. G. Y. (1950). Peripheral facial palsy. *J. Am. med. Ass.* **144**, 1586.

De Jong, J. G. Y. (1958). Een familie met ean ganz bijzondere vorme van myotonie. *Ned. Tijdschr. Geneesk.* **102**, 890.

De Jong, J. G. Y., Slooff, J. L., Van der Eerden, A. A. J. J. (1973). A family with paramytonia congenita with the report of an autopsy. *Acta neurol. scand.* **49**, 480–94.

De Jong, J. G. Y., *et al.* (1976). Agenesis of the corpus callosum, infantile spasms, ocular anomalies (Aicardi's syndrome). *Neurology, Minneap.* **26**, 1152–8.

De Jong, J. G. Y., van Gent, C. M., and Delleman, J. W. (1977). Cerebrotendinous cholestanolosis in relation to other cerebral xanthomatosis. *Clin. Neurol. Neurosurg.* **79**, 253–72.

Delay, J. and Pichot, P. (1946). La phacomatose angiomateuse familiale. *Revue neurol.* **78**, 151–4.

de Leon, G. A., Kaback, M. M., and Elfenbein, I. B. (1969). Juvenile dystonic lipidosis. *Johns Hopkins med. J.* **125**, 62–77.

Delwaide, P. J. and Schoenen, J. (1976). Non-hypertrophic familial neuropathy associated with intention tremor. A variety of Charcot–Marie–Tooth's disease? *J. neurol. Sci.* **27**, 59–69.

De Myer, W. (1972). Megalencephaly in children. *Neurol., Minneap.* **22**, 634–43.

De Myer, W. (1986). Megalencephaly: Types: clinical syndromes and management. *Ped. Neurol.* **2**, 321–7.

De Myer, W. and Zeman, W. (1963). Alobar holoprosencephaly (arhinencephaly) with median cleft lip and palate. Clinical, nosologic and electroencephalographic considerations. *Confinia neurol.* **23**, 1–36.

De Myer, W., Zeman, W., and Palmer, C. G., (1963). Familial alobar holoprosencephaly (archinencephaly) with median cleft lip and palate. *Neurology Minneap* **13** 913–18.

De Myer, W., Zeman, W., and Palmer, G. G., (1964). The face predicts the brain. Diagnostic significance of median face anomalies for holoprosencephaly (arhinencephaly). *Pediatrics, Springfield,* **34**, 256–63.

Denborough, M. A. and Lovell, R. R. H. (1960). Anaesthetic deaths in a family. *Lancet* **ii**, 45.

Denborough, M. A., Forster, J. F., Lovell, R. R., Maplestone, P. A., and Villiers, J. D. (1962). Anaesthetic death in a family. *Br. J. Anaesth.* **34**, 395–66.

Dennis, J. and Bower, B. D. (1972). The Aicardi syndrome. *Dev. med. Child Neurol.* **14**, 382–90.

Dennis, N. R. and Carter, C. O. (1978). Use of overlapping normal distributions in genetic counselling. *J. med. Genet.* **15**, 106–8.

Denny-Brown, D. (1951). Hereditary sensory radicular neuropathy. *J. Neurol. Neurosurg. Psychiat.* **14**, 237–52.

Denslow, G. T. and Robb, R. M. (1979). Aicardi's syndrome: a report of four cases and review of the literature. *J. pediat. ophthal. Strabismus.* **16**, 10–15.

Deonna, T. (1986). Dopa-sensitive progressive dystonia of childhood with fluctuations of symptoms – Segawa's syndrome and possible variants. *Neuropediat.* **17**, 81–5.

Deonna, Th. and Voumard, C. (1979). Benign hereditary (dominant) chorea of early onset. *Helv. paediat. Acta* **34**, 77–83.

Der Kaloustian, V. M., Afifi, A. K., and Mire, J. (1972). The myopathic variety of arthrogryposis multiplex congenita: a disorder with autosomal recessive inheritance. *J. Pediat.* **81**, 76–82.

Der Kaloustian, V. M., *et al.* (1985). Familial spinocerebellar degeneration with corneal dystrophy. *Am. J. Med. Genet.* **20**, 325–39.

Der Kaloustian, V. M., Khoury, M. J., Hallal, R., *et al.* (1981). Sandhoff disease: a prevalent form of infantile gm2 gangliosidosis in Lebanon. *Am. J. Hum. Genet.* **33**, 85–9.

Derry, D. M., Fawcett, J. S., Andermann, F., and Wolfe, L. S. (1968). Late infantile systemic lipidosis. Major monosialogangliosidoses: delineation of two types. *Neurol., Minneap.* **18**, 340–8.

De Rudolf, G. (1936). Tonic pupils with absent tendon reflexes in mother and daughter. *J. Neurol. Psychopath.* **16**, 367–8.

Deshaies, Y., Rott, H. D., Wissmuller, H. F., Schwanitz, G., Le Marec, B. and Koch, G. (1979). Microcephalie recessive liee a l'X. *J. Génét. hum.* **27**, 221–36.

De Silva, K. L. and Pearce, J. (1973). Neuropathy of metachromatic leokodystrophy. *J. Neurol. Neurosurg. Psychiat.* **36**, 30–3.

Desnick, R. J., Klionsky, B., and Sweeley, C. C. (1978). Fabry's disease. In *The metabolic inheritance of disease* (ed. J. B. Stanbury, J. B. Wyngaarden, and D. S. Fredrickson) pp. 810–40. McGraw Hill, New York.

Desnick, R. J., Krivil, W., and Sharp, H. R. (1973). *In utero* diagnosis of Sandhoff's disease. *Biochem. Biophys. Res.* **51**, 20–4.

Desnick, R. J., Walling, L. L. and Anderson, P. M. (1976). Mannosidosis: studies of the alpha-mannosidase isozymes in health and disease. *Adv. exp. Med. Biol.* **68**, 277–99.

Destunis, G. (1944). Die olivo-ponto-cerebellare heredoataxie. *Z. ges. Neurol. Psychiat.* **177**, 683.

Devos, J. (1957) Late cerebellar cortical atrophy of dominant heredity with ocular paralysis; anatomo-clinical study. *Psychol. Neurol.* **133**, 46–62.

de Weerdt, C. L. (1972) and Went, L. N. (1971). Neurological studies in families with Leber's optic atrophy. *Acta neurol. scand.* **47**, 541–54.

Dewhurst, K. and Oliver, J. E. (1970). Huntington's disease of young people. *Eur. Neurol.* **3**, 278–89.

Dewhurst, K., Oliver, J. E., and McKnight, A. L. (1970). Socio-psychiatric consequences of Huntington's disease. *Br. J. Psychiat.* **116**, 225–8.

Dewhurst, A.G., *et al.* (1986). Kearns–Sayre syndrome, hypoparathyroidism and basal ganglia calcification. *J. Neurol. Neurosurg. Psych.* **49**, 1323–4.

Dick, A.P. and Stevenson, C.J. (1953). Hereditary spastic paraplegia. Report of a family with associated extrapyramidal signs. *Lancet* **i**, 921–3.

Dick, H.M., Behan, P.D., Simpson, J.A., and Durward, W.F. (1974). The inheritance of HL-A antigens in myathesia gravis. *J. Immunogenet.* **1**, 401–12.

Diebler, C. and Dulac, O. (1987). *Pediatric neurology and neuroradiology*. Springer Verlag, Berlin, Heidelberg and New York.

Diebold, K., Kastner, M., and Penin, H. (1974). Über 2 Geschwisterfälle mit progressiver Myoklonusepilepsie und 5 Falle mit Dyssynergia cerebellaris myoclonica in mehreren Generationen einer Sippe—eine klinische und genetische Studie. *Nervenarzt* **45**, 595–601.

Dieker, H., Edwards, R.H., Zurheim, G., Chou, S., Hartman, H.A. and Opitz, J.M. (1969). The lissencephaly syndrome. In *Malformation syndromes*. Part II (ed. D. Bergsma), pp. 53–64. The National Foundation–March of Dimes, New York.

Diezel, P.B. and Huth, K. (1963). Pelizaeus-Merzbachersche Erkrankung mit familiärem Befall. *Dt. Z. Nervenheilk.* **184**, 264–87.

Diliberti, J.H. *et al.* (1984). The fetal valpruate syndrome. *Am. J. Med. Genet.* **19**, 473–81.

Di Mauro, S. and Di Mauro, P.M.M. (1973). Muscle carnitine palmityltransferase deficiency and myoglobinuria. *Science, N.Y.* **182**, 929–31.

Di Mauro, S. and Hartlage, P.L. (1978). Fatal infantile form of muscle phosphorylase deficiency. *Neurology, Minneap.* **28**, 1124–9.

Di Mauro, S., *et al.* (1979). Debrancher deficiency. Neuromuscular disorder in 5 adults. *Ann. Neurol.* **5**, 422–36.

Di Mauro, S., *et al.* (1980). Fatal infantile mitochondrial myopathy and renal dysfunction due to cytochrome-C-oxidase deficiency. *Neurology* **30**, 795–804.

Di Mauro, S., *et al.* (1985). Mitochondrial myopathies. *Ann. Neurol.* **17**, 521–38.

Di Mauro, S., *et al.* (1986). Metabolic myopathies. *Am. J. Med. Genet.* **25**, 635–51.

Ditlefsen, E.M.L. and Tonjum, A.M. (1960). Intracranial aneurysms and polycystic kidneys. *Acta med. scand.* **168**, 51–4.

Divry, P. and Van Bogaert, L. (1946). Une maladie familiale characterisée par une angiomatose diffuse cortico-meningee non calcifiante et une demyelinisation progressive de la substance blanche. *J. Neurol. Neurosurg. Psychiat.* **9**, 41–54.

Djindjian, M., Djindjian, R., Hurth, M., Rey, A. and Houdart, R. (1977).

Spinal cord arteriovenous malformations and the Klippel–Trenaunay–Weber syndrome. *Surg. Neurol.* **8**, 229–37.

Dobkin, B.H. and Verity, M.A. (1976). Familial progressive bulbar and spinal muscular atrophy. Juvenile onset and late morbidity with ragged-red fiber. *Neurol., Minneap.* **26**, 754–63.

Dobkin, B. and Verity, M.A. (1978). Familial neuromuscular disease with type I fiber hypoplasia, tubular aggregates, cardiomyopathy, and myasthenic features. *Neurol.* **28**, 1135–40.

Dobyns, W.B., Goldstein, N.P. and Gordon, H. (1979). Clinical spectrum of Wilson's disease. *Proc. Staff Meet. Mayo Clin.* **54**, 35–42.

Dobyns, W.B., Stratton, R.F. and Greenberg, F. (1984). Syndromes with lissencephaly. I: Miller–Dieker and Norman–Roberts syndromes and isolated lissencephaly. *Am. J. Med. Genet.* **18**, 509–26.

Dobyns, W.B., *et al.* (1987). Familial cavernous malformations of the central nervous system and retina. *Ann. Neurol.* **21**, 578–83.

Dom, R., Brocher, J.M., Ceuterick, C., Carton, C. and Martin, J.J. (1979). Adult ceroid-lipofuscihosis (Kuf's disease) in two brothers. Retinal and visceral storage in one, diagnostic muscle biopsy in the other. *Acta Neuropathol. (Berlin).* **45**, 67–72.

Domagk, J., Linke, I., Argyrakis, A., Spaar, F.W., Rahlf, G., and Schulte, F.J. (1975). Adrenoleukodystrophy. *Neuropaediatrics* **6**, 41–64.

Dominok, G.W. and Kirchmair. H. (1961). Familiare Haufung von Fehbildungen der Arhinencephalie Gruppe. *Z. Kinderheilk.* **85**, 19–30.

Donaghy, M., *et al.* (1987). Hereditary sensory neuropathy with neurotrophic keratitis: Description of an autosomal recessive disorder with a selective reduction of small myelinated nerve fibres and a discussion of the classification of *Brain* **110**, 563–83.

Donat, J. and Auger, R. (1979). Familial periodic ataxia. *Archs. Neurol., Chicago.* **36**, 568–9.

Donnai, D. and Farndon, P.A. (1986). Syndrome of the month: Walker-Warburg syndrome (Warburg syndrome, hard + / – e syndrome). *J. Med. Genet.* **23**, 200–3.

Donner, M., Rapola, J., and Somer, H. (1975). Congenital muscular dystrophy a clinicopathological and follow-up study of 15 patients. *Neuropaediatrics* **6**, 239–58.

Donohue, W.L. and Uchida, I. (1954). Leprechaunism: a euphimism for a rare familial disorder. *J. Pediat.* **45**, 505–19.

Dooling, E.L., Schoens, W.C., and Richardson, E.P. Jr. (1974). Hallervorden–Spatz syndrome. *Arch. Neurol., Chicago.* **30**, 70–83.

Doose, H. and Baier, W.K. (1987). Epilepsy with primarily generalized myoclonic-astatic seizures: a genetically determined disease. *Eur. J. Pediat.* **146**, 550–4.

Doose, H. and Gerken, H. (1973). On the genetics of EEG anomalies in childhood. IV. Photoconvulsive reaction. *Neuropaediatrics* **4**, 162–71.

Doose, H., Petersen, C.E., Völzke, E., and Herzberger E. (1966). Fieberkrampfe und Epilepsie. *Arch. Psychiat. NervKrankh.* **208**, 400–32.

Doose, H., Gerken, H., Hien-Völpel, K.-F., and Völzke, E. (1969). Genetics of photosensitive epilepsy. *Neuropaediatrics* **1**, 56–73.

Doose, H., Gerken, H., Leonhardt, R., Völzke, E., and Völz, C. (1970). Centrencephalic myoclonic-astatic petit mal. *Neuropaediatrics* **2**, 59–78.

Doose, H., Gerken, H., Horstmann, T., and Völzke, E. (1973). Genetic factors in spike-wave absences. *Epilepsia* **14**, 57–75.

Doose, H., Gerken, H., Kigfer, R. and Völzke, E. (1977). Genetic factors in childhood epilepsy with focal sharp wave. II. E.E.G. findings in patients and siblings. *Neuropaediatrics* **8**, 10–20.

Dorfman, L.J., Pedley, T.A., Tharp, B.R., and Scheithauer, B.W. (1978). Juvenile neuroaxonal dystrophy: clinical, electrophysiological and neuropathological features. *Ann. Neurol.* **3**, 419–28.

Drachman, D.A., Davison, W.C., and Mittal, K.K. (1976). Histocompatibility (H.L.A.) factors in familial multiple sclerosis. *Arch Neurol., Chicago.* **33**, 406–13.

Drager, G.A., Hammill, J.F., and Shy, G.M. (1958). Paramyotonia congenita. *Arch. Neurol. Psychiat., Chicago.* **80**, 1–9.

Dreifuss, F.E. and Hogan, G.R. (1961). Survival in X-chromosomal muscular dystrophy. *Neurol., Minneap.* **11**, 734–7.

Dreschfeld, J. (1886). On some of the rarer forms of muscular atrophies. *Brain* **9**, 178–95.

Dretakis, E.K. and Kondoyannis, P.N. (1974). Congenital scoliosis associated with encephalopathy in five children of two families. *J. Bone Jt Surg.* **56A**, 1747–50.

Dreyfus, J.C., Schapira, F., Demos, J., Rosa, R., and Schapira, G. (1966). The value of serum enzyme determinations in the identification of dystrophic carriers. *Ann. N.Y. Acad. Sci.* **138**, 304–14.

Dreyfus, J.C., Poenaru, L., and Svennerholm, L. (1975). Absence of hexosaminidase A and B in a normal adult. *N. Engl. J. Med.* **292**, 61–3.

Drukker, A., Sacks, M.I. and Gatt, S. (1970). The infantile form of Gaucher's disease in an infant of Jewish Sephardi origin. *Pediatrics, Springfield* **45**, 1017–23.

Dubi, J., Regli, F., Bischoff, A., Schneider, C., and de Crousaz, G. (1979). Recurrent familial neuropathy with liability to pressure palsies. *J. Neurol.* **220**, 43–55.

Dubois, G., Harzer, K., and Baumann, N. (1977). Very low arylsulfatase A and cerebroside sulfatase activities in leukocytes of healthy members of metachromatic leukodystrophy family. *Am. J. hum. Genet.* **29**, 191–4.

Dubowitz, V. (1964). Infantile muscular atrophy—a prospective study with particular reference to a slowly progressive variety. *Brain* **87**, 707–18.

Dubowitz, V. (1965). Familial low birth weight dwarfism with an unusual facies and a skin eruption. *J. med. Genet.* **2**, 12.

Dubowitz, V. (1973). Rigid spine syndrome: a muscle syndrome in search of a name. *Proc. R. Soc. Med.* **66**, 219–20.

Dubowitz, V. (1976). Screening for Duchenne muscular dystrophy. *Arch. Dis. Childh.* **51**, 249–51.

Dubowitz, V, (1978). *Muscular disorders in childhood.* Saunders, Philadelphia.

Dubowitz, V. and Brooke, M.H. (1973). *Muscle biopsy—a modern approach.* Saunders, London.

Dubowitz, V. and Platts, M. (1965). Central core disease of muscle with focal wasting. *J. Neurol. Neurosurg. Psychiat.* **28**, 432–7.

Dubowitz, V. and Roy, S. (1970). Central core disease of muscle: clinical, histochemical and electron microscopic studies of an affected mother and child. *Brain* **93**, 133–46.

Duffy, P., Wolf, J., Collins, G., Devoe, A.G., Streeten, B., and Cowen, D. (1974). Possible person-to-person transmission of Creutzfeldt–Jakob disease. *N. Engl. J. Med.* **290**, 692–3.

Duffy, P.E., Kornfeld, M., and Suzuki, K. (1968). Neurovisceral storage disease with curvilinear bodies. *J. Neuropath. exp. Neurol.* **27**, 351–70.

Dulaney, J.T., Milunsky, A., Sidbury, J.B., Hobolth, N. and Moser, H.W. (1976). Diagnosis of lipogranulomatosis (Farber disease) by use of cultured fibroblasts. *J. Pediat.* **89**, 59–61.

Dumon, J., Macken, J., and De Barsy, Th. (1971). Concordance for amyotrophic lateral sclerosis in a pair of dizygous twins of consanguineous parents. *J. med. Genet.* **8**, 113–16.

Dumon-Radermecker, M. (1965). Formes tres tardies d''idiotie amaurotique dans une souche d'oligophrenie congenitale avec epilepsie. *Acta neurol. psychiat. belg.* **65**, 778–807.

Duncan, C., Strub, R., McGarry, P., and Duncan, D. (1970). Peripheral nerve biopsy as an aid to diagnosis in infantile neuro-axonal dystrophy. *Neurology, Minneap.* **20**, 1024–32.

Duncan, P.A., Klein, R.M., Wilmott, P.I., and Sharpiro, L.R. (1979). Greig's cephalopolysyndactyly syndrome. *Am. J. Dis. Childh.* **133**, 818–21.

Dunger, D.B., *et al.* (1986). Deletion on the X chromosome detected by direct DNA analysis in one of two unrelated boys with glycerol kinase deficiency, adrenal hypoplasia, and Duchenne muscular dystrophy. *Lancet* **i**, 585–7.

Dunn, H.C., and Sweeney, V.P., (1971). Progressive supranuclear palsy in an unusual juvenile variant of Niemann-Pick disease. (Abstract). *Neurol. Minneap.* **21**, 442.

Dunn, H.G., Daube, J.R., and Gomez, M.R. (1978). Heredofamilial brachial plexus neuropathy (hereditary neuralgic amyotrophy with brachial predeliction) in childhood. *Dev. med. Child Neurol* **20**, 28–46.

Dupuis, M. J. M., Pierre, Ph. and Gonsette, R. E. (1987). Transient global amnesia and migraine in twin sisters. *J. Neurol. Neurosurg. Psychiat.* **50**, 816-24.

Durand, P. and Belotti, B. M. (1957). Un caso di indifferenza congenita al dolore—'algoatarassia Primo contributo della letteratura Italiana. *Helv. pediat. Acta* **12**, 116-26.

Durand, P., Borrone, C., and Della Cella, G. (1966). A new mucopolysaccharide lipid storage disease? *Lancet* **ii**, 1313-14.

Durand, P., Philippart, M., Borrane, C., and Della Cella, G. (1967). A new glycolipid storage disease. (Abstract). *Pediat. Res.* **1**, 416.

Durand, P., *et al.* (1977). Sialidosis (mucolipidosis I) *Helv. Paediat. Acta* **32**, 391-100.

Durham, D. G. (1958). Congenital hereditary Horner's syndrome. *Archs Ophithal. N. Y.* **60**, 939-40.

Dusheiko, G., *et al.* (1979). Recurrent hypoglycemia associated with glutaric aciduria type II in an adult. *N. Engl. Med. J.* **301**, 1405-9.

Duvoisen, R. C., Gearing, F. R., Schweitzer, M. D., and Yahr, M. D. (1969). A family study of Parkinsonism. In *Progress in neuro-genetics* (ed. A. Barbeau and J. R. Brunette), pp. 492-96. Excerpta Medica, Amsterdam.

Dwulet, F. E. and Benson, M. D. (1984). Primary structure of an amyloid prealbumin and its plasma precusor in a heredofamilial polyneuropathy of Swedish origin. *Proc. Nat. Sci.* **81**, 694-8.

Dwulet, F. E. and Benson, M. D. (1986). Characterization of a transthyretin (prealbumin) variant associated with familial amyloidotic polyneuropathy type II (Indiana/Swiss). *J. Clin. Invest.* **78**, 880.

Dyck, P. J. (1975). Inherited neuronal degeneration and atrophy affecting peripheral motor, sensory and autonomic neurons. In *Peripheral neuropathy,* Vol. II (ed. P. J. Dyck, P. K. Thomas, and E. H. Lambert), pp. 825-667. Saunders, Philadelphia.

Dyck, P. K. and Gomez, M. R. (1968). Segmental demyelination in Dejerine–Sottas disease: light, phase contrast, and electron microscopic studies. *Proc. Staff Meet. Mayo Clin.* **43**, 280-96.

Dyck, P. K. and Lambert, E. H. (1968*a*). Lower motor and primary sensory neuron diseases with peroneal muscular atrophy. I. Neurologic, genetic and electrophysiological findings in hereditary polyneuropathies. *Arch. Neurol., Chicago* **80**, 603-18.

Dyck, P. K. and Lambert, E. H. (1968*b*). Lower motor and primary sensory neurone disease with peroneal muscular atrophy. II. Neurologic, genetic, and electrophysiological findings in various neuronal degenerations. *Arch. Neurol., Chicago.* **18**, 619-25.

Dyck, P. K. and Lambert, E. H. (1969). Dissociated sensation in amyloidosis. *Arch. Neurol., Chicago* **20**, 490-507.

Dyck, P. K. and Ohta, M. (1975). Neuronal atrophy and degeneration predominantly affecting peripheral sensory neurons. In *Peripheral neuropathy* (ed. P.J. Dyck, P.K. Thomas and E.H. Lambert), pp. 791–824. Saunders, Philadelphia.

Dyck, P.K., Kennel, A.J., Magal, I.V., and Kraybill, E.N. (1965). A Virginia kinship with hereditary sensory neuropathy: peroneal muscular atrophy and pes cavus. *Proc. Staff Meet. Mayo Clin.* **40**, 685–94.

Dyck, P.J. *et al.* (1983). Neurogenic arthropathy and recurring fractures with subclinical inherited neuropathy. *Neurology.* **33**, 357–67.

Dyken, M., Zeman, W., Rusche, T. (1969). Hypokalaemia periodic paralysis. Children with permanent myopathic weakness. *Neurol., Minneap.* **19**, 691–9.

Dyken, P. and Kolar, O. (1968). Dancing eyes, dancing feet: infantile polymyoclonia. *Brain* **91**, 305–19.

Earl, C.J., Fullerton, P.M., Wakefield, G.S., and Schutta, H.S. (1964). Hereditary neuropathy with liability to pressure palsies. *Q. J. Med.* **33**, 481–98.

Eaton, L.M., Camp, J.D., and Love, J.G. (1939). Symmetric cerebral calcification, particularly of the basal ganglia, demonstrable roentgenographically. *Arch. Neurol. Psychiat. Chicago* **41**, 921–42.

Ebels, E.J., Blokzijl, E.J., and Troelstra, J.A. (1965). A Wernicke-like encephalomyelopathy in children (Leigh) an inborn error of metabolism. *Helv. paediat. Acta* **20**, 310.

Ebers, G.C., *et al.* (1986). A population-based study of multiple sclerosis in twins. *N. Engl. J. Med.* **315**, 1638–42.

Echenne, B., *et al.* (1983). Congenital muscular distrophy and rigid spine syndrome. *Neuropediatrics* **14**, 97–101.

Echenne, B., Vidal, M., Maire, I., and Michalski, J. *et al.* (1986). Salla disease in one non-Finnish patient. *Eur. J. Pediat.* **145**, 320–2.

Edelsohn, L., Caplan, L., and Rosenbaum, A.E. (1972). Familial aneurysms and infundibular widening. *Neurol., Minneap* **22**, 1056–60.

Edström, L. (1975). Histochemical and histopathological changes in skeletal muscle in late-onset hereditary distal myopathy (Welander). *J. neurol. Sci.* **26**, 147–57.

Edström, L., Thornell, L.-E., and Eriksson, A. (1980). A new type of hereditary distal myopathy with characteristic sarcoplasmic bodies and intermediate (skeleton) filaments. *J. neurol. Sci.* **47**, 171–90.

Edwards, C.Q., Williams, D.M., and Cartwright, G.E. (1979). Hereditary hypoceruloplasminemia. *Clin. Genet.* **15**, 311–16.

Edwards, J.H. (1961). The syndrome of sex-linked hydrocephalus. *Arch. Dis. Childh.* **36**, 486–93.

Edwards, J.H. Norman, R.M., and Roberts, J.M. (1961). Sex linked hydrocephalus. Report of a family with 15 affected members. *Arch. Dis. Childh.* **36**, 481–5.

Edwards, J.H., Sethi, P.K., Scoma, A.J., Bannerman, R.M., and Frohman, L.A. (1976). A new familial syndrome characterized by pigmentary retinopathy, hypogonadism, mental retardation, nerve deafness and glucose intolerance. *Am. J. Med.* **60**, 23–32.

Efron, M.L. (1965). Familial hyperprolinemia. *N. Engl. J. Med.* **272**, 1243–54.

Egan, T.J. and Klein, R. (1959). Hyperkalaemic familial periodic paralysis. *Pediatrics, Springfield* **24**, 761–73.

Egger, J., Bellman, M., Ross, E.M. and Baraitser, M. (1982*a*). Joubert–Boltshauser syndrome with polydactyly in siblings. *J. Neurol. Neurosurg. Psychiat.* **45**, 737–9.

Egger, J., Wynne-Williams, C.J.E. and Erdohazi, M. (1982*b*). Mytochondrial cytopathy or Leigh's syndrome? Mitochondrial abnormalities in spongiform encephalopathies. *Neuropediatrics* **13**, 219–24.

Egger, J., *et al.* (1983). Involvement of the central system in congnital muscular dystrophy. *Dev. Med. Child. Neurol.* **25**, 32–42.

Egger, J., *et al.* (1984). Cortical subacute necrotizing encephalopathy. A study of two patients with mitochodrial dysfunctions. *Neuropediatrics.* **15**, 150–8.

Egger, J., *et al.* (1987). Progressive neuronal degeneration of childhood (PNDC) with liver disease. *Clin. Pediatr.* **26**, 167–73.

Eidelberg, D., *et al.* (1987). Adult onset Hallcrvorden-Spatz disease with neurofibrillary pathology: a discrete clinicopathological entity. *Brain* **110**, 993–1013.

Eisenberg, L., Ascher, E., and Kanner, L. (1959). A clinical study of Gilles de la Tourette's disease (Maladie des tics) in children. *Am. J. Psychiat.* **115**, 715–23.

Eisler, T. and Wilson, J.H. (1978). Muscle fiber-type disproportion. Report of a family with symptomatic and asymptomatic members. *Arch. Neurol., Chicago* **35**, 823–6.

Ek, J., *et al.* (1986). Peroxisomal dysfunction in a boy with neurologic symptoms and amaurosis (Leber disease): clinical and biochemical findings similar to those observed in Zellweger syndrome. *J. Pediat.* **108**, 19–24.

Ekbom, K.A. (1960). Restless legs syndrome. *Neurol., Minneap.* **10**, 868–73.

Ekbom, K.A. (1975). Hereditary ataxia, photomyoclonus, skeletal defor-maties and lipoma. *Acta neurol. scand.* **51**, 393–404.

Eldridge, R. (1970). The torsion dystonias. *Neurol., Minneap.* **20**, Suppl. 2, 1–78.

Eldridge, R., Edgar, A., and Cooper, I.S. (1971). Genetics geography and intelligence in the torsion dystonias. *Birth Defects:* Original Article Series **7**, 167–77.

Eldridge, R., O'Mearn, K., Chase, T.N., and Donelly, E.F. (1973). Offspring of consanguineous parents with Huntington's chorea *Adv. Neurol.* **1**, 211–21.

Eldridge, R., Sweet, R., Lake, R., Ziegler, M., and Shapiro, A.K. (1977). Gilles de la Tourette syndrome: clinical, genetic, psychologic, and biochemical aspects in 21 selected families. *Neurol., Minneap.* **27**, 115–24.

Eldridge, R., McFarland, H., and Sever, H. (1978). Familial multiple sclerosis: clinical, histocompatibility and viral serological studies. *Ann. Neurol.* **3**, 72–80.

Eldridge, R., Iivanainen, M., Stern, R., Koeber, T., and Wilder, B.J. (1983). 'Baltic' myoclonic epilepsy: hereditary disorder of childhood made worse by phenytoin. *Lancet* **ii**, 838–42.

Eldridge, R., *et al.* (1984). Dystonia in 61-year old identical twins: Observation over 45 years. *Ann. Neurol.* **16**, 356–8.

Elejalde, B.R., De Elejalde, M.J., and Lopez, F. (1979). Hallervorden–Spatz disease. *Clin. Genet.* **16**, 1–18.

Elejalde, B.R., Giraldo, C., Jimenez, R., and Gilbert, E.F. (1977). Acrocephalopolydactylous dysplasia. *Birth Defects:* Original Series **XIII**, 532–67.

Elfenbein, I.B. (1968). Dystonia juvenile idiocy without amaurosis. A new syndrome *Johns Hopkins med. J.* **123**, 205–21.

Elian, M. and Dean, G. (1987). Multiple sclerosis among the United Kingdom children of immigrants from the West Indies. *J. Neurol. Neurosurg. Psychiat.* **50**, 500–1.

Ellis, F.R. and Halsall, P.J. (1980). Malignant hyperpyrexia. *Br. J. hosp. Med.* **24**, 318–27.

Elwood, J.H. and Nevin, N.C. (1973). Factors associated with anencephalus and spina bifida in Belfast. *Br. J. Prev. Soc. Med.* **27**, 73–80.

Elze, K.-L., Koepp, P., Langenstein, I., Steinhausen, H.-C., Colmant, H.J., and Schwendemann, G. (1978). Juvenile type of generalized ceroid-lipofuscinosis (Spielmeyer–Sjögren syndrome). *Neuropaediatrics* **9**, 3–27.

Emery, A.E.H. (1971). Review. The nosology of the spinal muscle atrophics. *J. med. Genet.* **8**, 481–95.

Emery. A.E.H. (1976). *Methodology in medical genetics.* Churchill Livingstone, Edinburgh.

Emery, A.E.H. (1987). X-linked muscular dystrophy with early contractures and cardiomyopathy (Emery–Dreifuss type). *Clin. Genet.* **32**, 360–7.

Emery, A.E.H. and Dreifuss, F.E. (1966). Unusual type of benign X-linked muscular dystrophy. *J. Neurol. Neurosurg. Psychiat.* **29**, 338–42.

Emery, A.E.H., Hausmanowa-Petrusewicz, I., Davie, A.M., Holloway, S., Skinner, R., and Borkowska, J. (1976). International collaborative study of the spinal muscular atrophies, Part I. *J. Neurol. Sci.* **29**, 83–94.

Emery, A.E.H. and King, B. (1971). Pregnancy and serum creatine kinase levels in potential carriers of Duchenne X-linked muscular dystrophy. Letter. *Lancet* **i**, 1013.

Emery, A.E.H. and Skinner, R. (1976). Clinical studies in benign (Becker type) X-linked muscular dystrophy. *Clin. Genet.* **10**, 189–201.

Emery, A.E.H., Watt, M.S., and Clack, E.R., (1972). The effects of genetic counseling in Duchenne muscular dystrophy. *Clin. Genet.* **3**, 147–50.

Emery, A.E.H., *et al.* (1973). The nosology of the spinal muscular atrophies. In *Proceedings of the 2nd International Congress on muscle Diseases* (ed. B.A. Kakucas). *Exerpt. Nel. Amsterdam* pp. 439–49.

Emery, A.E.H., Brough, C., Crawfurd, M., Harris, R., and Oakshott, G. (1978). A report on genetic registers. Based on the Report of the Clinical Genetics Society Working Party. *J. med. Genet.* **15**, 435–42.

Emery, E.S., Fenichel, G.M., and Eng. G. (1968). A spinal muscular atrophy with scapuloperoneal distribution. *Arch. Neurol., Chicago* **18**, 129–33.

Enell, H. and Pehrson, M. (1958). Studies on osteopetrosis I. Clinical report of three cases with genetic considerations. *Acta paediat., Stockh.* **47**, 279–87.

Enevoldsen, E. and Albertsen, K. (1970). The rigid and akinetic type of Huntington's chorea. Two case reports. *Acta neurol. scand.* **46** Suppl. 43, 227–8.

Engel, A.G. (1970). Acid maltase deficiency in adults: studies in four cases of a syndrome which may mimic muscular dystrophy or other myopathies. *Brain* **93**, 599–616.

Engel, A.G. and Gomez, M.R. (1966). Congenital myopathy associated with multifocal degeneration of muscle fibers. *Trans. Am. neurol. Ass.* **91**, 222–3.

Engel, A.G., Lambert, E.H., Rosevear, J.W., and Tauxe, W.N. (1965). Clinical and electromyographic studies in a patient with primary hypokalaemic periodic paralysis. *Am. J. Med.* **38**, 626.

Engel, A.G., and Angelini, C. (1973). Carnitine deficiency of human skeletal muscle with associated lipid storage myopathy—A new syndrome. *Science, N.Y.* **179**, 899–901.

Engel, A.G., Gomez, M.R., Groover, R.V. (1971). Multicore disease. A recently recognized congenital myopathy associated with multifocal degeneration of muscle fibers. *Proc. Staff Meet. Mayo Clin.* **46**, 666–81.

Engel, A.G., Angelini, C., and Gomez, M.R. (1972). Fingerprint body myopathy. A newly recognized congenital muscle disease. *Proc. Staff Meet. Mayo Clin.* **47**, 377–88.

Engel, A.G., Gomez, N.R., Seybold, M.E., and Lambert, E.H. (1973). The spectrum and diagnosis of acid maltase deficiency. *Neurol., Minneap.* **23**, 95–106.

Engel, K., Kurland, L.T., and Klatzo, I. (1959). An inherited disease similar to amyotrophic lateral sclerosis with a patern of posterior column involvement: an intermediate form? *Brain* **82**, 203–20.

Engel, W. K. (1967). Muscle biopsies in neuromuscular disease: *Pediat. Clin. N. America,* **14**, 963–95.

Engel, W. K., Eyerman, E. L., and Williams, H. E. (1963). Late-onset type of skeletal muscle phosphorylase deficiency. *N. Engl. J. Med.* **268**, 135–7.

Engel, W. K., Dorman, J. D., Levy, R. I., and Fredrickson, D. S. (1967). Neuropathy in Tangier disease: α-lipoprotein deficiency manifesting as familial recurrent neuropathy and intestinal lipid storage. *Archs. Neurol., Chicago* **17**, 1–9.

Engel, W. K., Gold, G. N., and Karpati, G. (1968). Type I fiber hypotrophy and central nuclei: a rare congenital muscle abnormality with a possible experimental model. *Arch. Neurol., Chicago* **18**, 435–44.

Engel, W. K., Vick, N. A., Glueck, C. J., and Levy, R. I. (1970). A skeletal muscle disorder associated with intermittent symptoms and a possible defect of lipid metabolism. *N. Engl. J. Med.* **282**, 697–704.

England, A. C., and Denny-Brown, D. (1952). Severe sensory changes and trophic disorder in peroneal muscular atrophy (Charcot–Marie–Tooth type). *Arch. Neurol. Psychiat., Chicago* **67**, 1–22.

English, W. H. (1942). Alzheimers disease. *Psychiat. Q.* **16**, 91–106.

Epstein, C. J., Brady, R. O., Schneider, E. L., Bradley, R. M., and Shapiro, D. (1971). *In utero* diagnosis of Niemann–Pick disease. *Am. J. hum. Genet.* **23**, 533–5.

Erbe, R. W. (1979). Genetic aspects of folate metabolism. In *Advances in human genetics.* Vol. 9 (ed. H. Harris and K. Hirschorn), pp. 293–354. Plenum. New York.

Erdohazi, M. and Marshall, P. (1979). Striatal degeneration in childhood. *Arch. Dis. Childh.* **54**, 85–91.

Erickson, R. P. (1972). Leber's optic atrophy, a possible example of maternal inheritance. *Am. J. hum. Genet.* **24**, 348–9.

Erickson, T. C., Paul, L. W., and Suckle, H. M. (1950). Clinical observation on basilar impression of the skull. *Trans. Am. neurol. Ass.* **75**, 180–3.

Ervin, F. R. and Sternback, R. A. (1960). Hereditary insensitivity to pain. *Trans. Am. neurol. Ass.* **85**, 70–4.

Escobar, V. and Bixler, D. (1977). Are the acrocephalosyndactly syndromes variable expressions of a single gene defect? In *Natural history of specific birth defects* (ed. D. Bersma and R. B. Lowry), pp. 139–54. Alan R. Liss for the National Foundation–March of Dimes, New York.

Espinas, O. E. and Faris, A. A. (1969). Acute infantile Gaucher's disease in identical twins. *Neurol., Minneap* **19**, 133–40.

Espinosa, R. E., Okihiro, M. M., Mulder, D. W., and Sayre, G. P. (1962). Hereditary amyotrophic lateral sclerosis. A clinical and pathological report with comments on classification. *Neurol., Minneap.* **12**, 1–7.

Espir, M. L. E. and Matthews, W. B. (1973). Hereditary quadriceps myopathy. *J. Neurol. Neurosurg. Psychiat.* **36**, 1041–5.

Essen-Möller, E. (1946). A family with Alzheimer's disease. *Acta psychiat. neurol. scand.* **21**, 233–44.

Estes, J.W., Morley, T. J., Levine, I. M., and Emerson, C. P. (1967). A new hereditary acanthocytosis syndrome. *Am. J. Med.* **42**, 868–81.

Estrin, W.J. (1977). Amyotrophic lateral sclerosis in Dizygotic twins. *Neurol. Minneap.* **27**, 692–4.

Eulenburg, A. (1866). Ueber eine familiäre, durch 6 Generationen Verfolgbare form congenitaler Paramyotonie. *Neurol. Zentbl.* **5**, 265–72.

Euziere, J., Lafon, R., Mercadier, Cazaban, R., and Barjon, P. (1952). Hérédo-ataxie cérébelleuse et paralysie familiale de la verticalité du regard. *Rev. neurol.* **87**, 323–9.

Evans, A.T.G. (1954). Essential atrophy of the choroid with paraplegia and a strong family history of similar conditions. *Trans. ophthal. Soc. U.K.* **74**, 215–17.

Evans, P. J. (1950). Five cases of familial retinal abiotrophy. *Trans. ophthal. Soc. U.K.* **70**, 96.

Fadiloglu, S. (1971). Sur les formes infantiles pecoces de l'atrophie pigmantarie pallido-reticulee. *Acta neurol. belg.* **71**, 392–406.

Fahn, S. and Greenberg, J. (1972). Striatonigral degeneration. *Trans. Am. neurol. Ass.* **97**, 275–7.

Fahr, T. (1930). Idiopathische Verkalkung der Hirngefasse. *Zentbl. allg. Path. path. Anat.* **50**, 129–33.

Fairburn, B. (1973). 'Twin' intracranial aneurysms causing subarachnoid haemorrhage in identical twins. *Br. med. J.* **i**, 210–11.

Fairburn, B. and Urich, H. (1971). Malignant gliomas occuring in identical twins. *J. Neurol. Neurosurg. Psychiat.* **34**, 718–22.

Faivre, J., Lemarec, B., Bretagne, J., and Pecker, J. (1976). X-linked hydrocephalus with aqueductal stenosis, mental retardation and adduction-flexion deformity of the thumbs. *Child. Brain* **2**, 226.

Falconer, M.A. (1971). Genetic and related aetiological factors in temporal lobe epilepsy. *Epilepsia* **12**, 13–31.

Falden, A.I. and Townsend, J.J. (1976). Myoclonus in Alzheimer's disease—a confusing sign. *Arch. Neurol. Chicago* **33**,278–80.

Falls, H.F., Kruse, W.T., and Cotterman, C.W. (1949). Three cases of Marcus–Gunn phenomenon in two generations. *Am. J. Ophthal.* **32**, Suppl., 53–9.

Fanconi, A., Prader, A., Isler, W., Lüthy, F., and Siebenmann, R. (1963). Morbus Addison mit Hirnsklrose im Kindesalter. *Helv. paediat. Acta* **18**, 480–501.

Fanconi, G., and Ferrazzini, F. (1957). Kongenitale analgie (kongenitale generalisierte schmerzindifferenz). *Helv. paediat. Acta.* **12**, 79–115.

Farber, S., Cohen, J., and Uzman, L. L. (1957). Lipogranulomatosis: a new lipoglycoprotein storage disease. *J. Mt Sinai Hosp.* **24**, 816–37.

Fardeau, M., Harpey, J.P., and Caille, B. (1975). Disproportion congenitale des differents types de fibre musculaire avec petitesse relatives des fibres de type I. *Rev. neurol.* **131**, 745–66.

Fardeau, M., Tomé, F.M.S., and Derambure, S. (1976). Familial fingerprint body myopathy. *Arch. Neurol., Chicago* **33**, 724–5.

Farlow, M.R., De Myer, W., Dlouay, S.R., and Hodes, M.E. (1987). X-linked recessive inheritance of ataxia and adult onset dementia. *Neurol.* **37**, 602–7.

Farmer, T.W. and Allen, J.N. (1969). Hereditary proximal amytotrophic lateral sclerosis. *Trans. Am. neurol. Ass.* **94**, 140–4.

Farmer, T.W. and Mustian, V.M. (1963). Vestibulocerebellar ataxia. *Arch. Neurol., Chicago,* **8**, 471–80.

Farpour, H. and Mahloudji, M. (1975). Familial cerebrotendinous xanthomatosis. *Arch. Neurol., Chicago* **32**, 223–5.

Farrell, D.F., Perry, A.K., Kaback, M.M., and McKhann, G.M. (1973). Globoid cell (Krabbe) leukodystrophy heterozygote detection in cultured skin fibroblasts. *Am. J. hum. Genet.* **25**, 604–9.

Farrell, D.F., Clack, A.F., Scott, C.R., and Wennberg, R.P. (1975). Absence of pyruvate decarboxylase activity in man: a cause of congenital lactic acidosis. *Science. N.Y.* **187**, 1082–4.

Farrell, D.F., Sumi, S.M., Scott, C.R., and Rice, G. (1978). Ante-natal diagnosis of Krabbe's leukodystrophy: enzymatic and morphological confirmation in an affected fetus. *J. Neurol. Neurosurg. Psychiat.* **41**, 766–82.

Farrell, S.A., Davidson, R.G., and Thorp, P. (1987). Neonatal mani-festations of Schwartz–Jampel syndrome. *Am. J. Med. Genet.* **27**, 799–805.

Farriaux, J.P., Dhondt, J.L., Blankhaert, D., and Fontaine, G. (1974). Etude comparative des aspects clinques radiologiques, biochimiques et genetiques de la maladie de Sanfilippo de type A et de type B. *Helv. paediat. Acta* **29**, 349–70.

Faulkner, S.H. (1939). Familial ptosis with ophthalamoplegia externa starting late in life. *Br. med. J.* **ii**, 854.

Fazio, M. (1892). Ereditarieta della paralsi bulbare progressiva. *Rif. Med.* **8**, 327.

Fedrizzi, E., D'Angelo, A., Negri, S., and Ermacora, E. (1972). Peripheral sensory neuropathy in childhood. *Dev. med. Child. Neurol.* **14**, 501–7.

Feigenbaum, J.A. and Munsat, T.L. (1970). A neuromuscular syndrome of scapuloperoneal distribution. *Bull. Los Ang. neurol. Soc.* **35**, 47–57.

Feigin, I. and Wolf, A. (1954). A disease in infants resembling chronic Wernicke's encephalopathy. *J. Pediat.* **45**, 243–63.

Feiling, A. and Ward, E. (1920). A familial form of acoustic tumour. *Br. med. J.* **i**, 496–7.

Feinberg, A.P. and Leahy, W.R. (1977). Infantile spasm: case report of sex-linked inheritance. *Dev. med. Child Neurol.* **19**, 524–6.

Feldman, G.L., Weaver, D.D., and Lovrein, E.W. (1977). The fetal trimethadione syndrome. *Am. J. Dis. Childh.* **131**, 1389–92.

Feldman, R.G., Chandler, K.A., Levy, L.L., Glaser, G.H. (1963). Familial Alzheimer's disease. *Neurol., Minneap.* **13**, 811–24.

Fenichel, G.M. (1978). Clinical syndromes of myasthenia in infancy and childhood. *Arch. Neurol., Chicago* **35**, 97–103.

Fenichel, G.M., Emery, E.S., and Hunt, P. (1967). Neurogenic atrophy simulating facioscapulo humeral dystrophy. *Arch. Neurol., Chicago* **17**, 257–60.

Fenichel, G.M., Olson, W.H., and Kilroy, A.W. (1971). Hereditary dystonia associated with unique features in skeletal muscle. *Adv. Neurol.* **25**, 552–9.

Fenichel, G.M., Chul Sui, Yi., Kilroy, A.W., and Blouin, R. (1982). An autosomal dominant dystrophy with humeropelvic distribution and cardiomyopathy. *Neurol., Minneap.* **32**, 1399.

Ferguson, F. and Critchley, M. (1929). A clinical study of an heredo-familial disease resembling disseminated sclerosis. *Brain* **52**, 203–25.

Ferrer, I., Arbizu, T., Pena, J., and Serra, J.P. (1980). A golgi ultrastructural study of a dominant form of Kufs disease. *J. Neurol.* **222**, 183–90.

Ferrer, I., Fabregues, I., Pineda, M., and Alvarez, E. (1983). Diagnosis of infantile neuroaxonal dystrophy by conjunctival biopsy. *Neuropediatrics.* **14**, 53–5.

Ferrier, P., Bamatter, F., and Klein, D. (1965). Muscular dystrophy (Duchenne) in a girl with Turner's syndrome. *J. med. Genet.* **2**, 38–46.

Feuerstein, R.C. and Mims, L.C. (1962). Linear nevus sebaceus with convulsions and mental retardation. *Am. J. Dis. Childh.* **104**, 675–9.

Fickler, A. (1911). Klinische und pathologische anatomische Beitrag zu den Erkrankungen des Kleinhirns. *Dt. Z. Nervenheilk.* **41**, 306–75.

Field, C.E. (1939). Albers–Schönberg disease. A typical case. *Proc. R. Soc. Med.* **32**, 3220–4.

Fienman, N.L. and Yakovac, W.C. (1970). Neurofibromatosis in childhood. *J. Pediat.* **76**, 339–46.

Fine, D.I., Barrow, K.D., and Hirano, A. (1960). Central nervous system lipidosis in an adult with atrophy of cerebellar granular layer. *J. Neuropath. exp. Neurol.* **19**, 355–69.

Fine, R.M., Derbes, V.J., and Clark, W.H. Jr (1961). Blue rubber bleb nevus. *Arch. Derm., N.Y.* **84**, 802–5.

Finelli, P.F. (1985). Metachromatic leukodystrophy manifesting as a schizophrenic disorder: computed tomographic correlation. *Ann. Neurol.* **18**, 94–5.

Finer, N.N., Bowen, P., and Dunbar, L.G. (1978). Caudal regression anomalad (sacral agenesis) in siblings. *Clin. Genet.* **13**, 353–8.

Finlayson, M.H., Guberman, A., and Martin, J.B. (1973). Cerebral lesions in familial amyotrophic lateral sclerosis and dementia. *Acta neuropath.* **26**, 237–46.

Finnie, M. D. A., *et al.* (1976). Massive excretion of 2-oxoglutaric acid and 3-hydroxy isvaleric acid in a patient with a deficiency of 3-methyl-crotonyl-CoA-carboxylase. *Clin. Chim. Acta.* **73**, 513–19.

Finucci, J. M., Guthrie, J. T., Childs, A. L., Abbey, H., and Childs, B. (1976). The genetics of specific reading disability. *Ann. hum. Genet.* **40**, 1–23.

Finucci, J. M. and Childs, B. (1983). Dyslexia: Family studies. *Genetic aspects of speech and language disorders.* pp. 157–66. Academic Press, New York.

Fiorilli, M., *et al.* (1985). Variant of ataxia-telangiectasia with low level radiosensitivity. *Hum. Genet.* **70**, 274–7.

Fischbeck, K. H. *et al.* (1986). X-linked neuropathy: Gene localization with DNA probes. *Ann. Neurol.* **20**, 527–32.

Fischer, A. Q., *et al.* (1981). Cerebellar hemorrhage complicating isovaleric acidemia. *Neurology.* **31**, 746–8.

Fishbein, W. N., Armbrustmacher, V. M., and Griffin, J. L. (1978). Adenylate deaminase deficiency: a new disease of muscle. *Clin. Res.* **26**, 545–8.

Fisher, J. A., Burn, J., Alexander, F. W. and Gardner-Medwin, D. (1987). Angelman, (happy puppet) syndrome in a girl and her brother. *J. Med. Genet.* **24**, 294–8.

Fisher, M., Sargent, J., and Drachman, D. (1979). Familial inverted choreoathetosis. *Neurol., Minneap.* **29**, 1627–31.

Fisher, R. L. and Russman, B. S. (1974). Genetic syndromes associated with cerebral palsy. *Clin. Orthop.* **99**, 2–11.

Fishman, M. A. (1979). Febrile seizures: the treatment controversy. *J. Pediat.* **94**, 177–84.

Fitzpatrick, D. B., Hooper, R. E., and Seife, B. (1976). Hereditary deafness and sensory radicular neuropathy. *Arch. Otol.* **102**, 552–7.

Fitzpatrick, T. B., Szabó, G., Hori, Y., Simone, A. A., Reed, W. B., and Greenberg, M. H. (1968). White leaf-shapcd macules: earliest visible signs of tuberous sclerosis. *Arch. Derm., N. Y.* **98**, 1–6.

Fitzsimmons, J. S. and Guilbert, P. R. (1987). Spastic paraplegia associated with brachydactyly and cone shaped epiphyses. *J. Med. Genet.* **24**, 702–5.

Fitzsimmons, J. S., Fitzsimmons, E. M. Barrow, M. (1982). Fronto-metaphyseal dysplasia. Further delineation of the clinical syndrome. *Clin. Genet.* **22**, 195–205.

Fitzsimmons, J. S. *et al.* (1983). Four brothers with mental retardation. Spastic paraplegia, and palmoplantar hyperkeratosis. A new syndrome. *Clin. Genet.* **23**, 329–35.

Fitzsimons P. B. Clifton-Bligh, P., and Wolfender, W. H. (1981)., Mitochodrial myopathy and lactic acidemia with myoclonic epilepsy, ataxia, and hypothalamic infertility: a variant of Ramsay-Hunt syndrome? *J. Neurol. Neurosurg. Psychiat.* **44**, 79–82.

Fitzsimons, R.B., Gurwin, E.B. and Bird, A.C. (1987). Retinal vascular abnormalities in facioscapulohumeral dystrophy: a general association with genetic and theraputic implications. *Brain* 110, 631–48.

Flannery, D.B., *et al.* (1983). Non-ketotic hyperglycinemia in two retardated adults: a mild form of the infantile non-ketotic hyperglycinemia. *Neurology.* 33, 1064–6.

Fleck, H. and Zurrow, H.B. (1967). Familial amyotrophic lateral sclerosis. *N.Y. St. J. Med.* 67, 2368–73.

Fleisher, G.A., McConahey, W.M., and Pankow, M. (1965). Serum creatine kinase lactic dehydrogenase and glutamic oxalcacetic transaminase in thyroid disease and prognosis. *Proc. Mayo Staff Meet. Clin.* 40, 300–11.

Fleiszar, K.A., Daniel, W.L., and Imrey, P.B. (1977). Genetic study of infantile spasms with hypsarrhythmia. *Epilepsia.* 18, 55–62.

Fleury, P. and Hageman, G. (1985). A dominantly inherited lower motor neuron disorder presenting at birth associated with arthrogryposis. *J. Neurol. Neurosurg. Psychiat.* 48, 1037–48.

Fluharty, A.L., Yutaka, T., Stevens, R.L., and Kihara, H. (1977). Hair root analysis of sulfoiduronate sulfatase in a Hunter syndrome heterozygote. (Abstract.) *Am. J. hum. Genet.* 29, 43A.

Flynn, P. and Aird, R.B. (1965). A neuroectodermal syndrome of dominant inheritance. *J. neurol. Sci.* 2, 161–82.

Fog, M. and Hyllested, K. (1966). Prevalence of disseminated sclerosis in the Faroes, the Orkneys and Shetland. *Acta neurol. scand.* Suppl. 19, 6–11.

Fogelson, M.H. and Zwericki, R.J. (1969). Phenotypic variations of essential tremor in a large kindred. In *Progress in neurogenetics* (ed. A. Bardeau and J.-R Brunette), pp. 406–10. Excerpta Medica, Amsterdam.

Fois, A., *et al.* (1985). Giant axonal neuropathy. Endocrinological and histological studies. *Eur. J. Pediat.* 144, 274–80.

Foldes, F.F. and McNall, P.G. (1960). Unusual familial occurence of myasthenia gravis. *J. Am. med. Ass.* 174, 418–20.

Foley, J. (1951). Calcification of the corpus striatum and dentate nuclei occuring in a family. *J. Neurol. Neurosurg. Psychiat.* 14, 253–61.

Folstein, S.E., *et al.* (1987). Huntington's disease in Maryland: clinical aspects of racial variation. *Am. J. Hum. Genet.* 41, 168–79.

Font, R.L. and Ferry, A.P. (1972). Ocular and adnexal tumours. *Int. Ophthal. Clins* 12, 1–50.

Fontaine, G. Francois, P., Razemon, P., Farriaux, J.P., and Walbaum, R. (1974). Une observation familiale de ptosis avec blepharophimosis et epicanthus inversus. *J. Génét. hum.* 22, 233–41.

Foo, D., Chang, Y.O. and Rossier, A.B. (1980). Spontaneous cervical epidural hemorrhage. Anterior cord syndrome and muscular malformation. *Neurology.* 30, 1253–4.

Ford, F.R., Livingstone, S., and Pryles, C.V. (1951). Familial degeneration of the cerebral gray matter in childhood, with convulsions, myoclonus, spasticity, cerebellar ataxia, choreoathetosis, dementia, and death in status epilepticus; differentiation of infantile and juvenile types. *J. Pediat.* **39**, 33–43.

Forssman, B. (1971). Hereditary studies of congenital nystagmus in a Swedish population. *Ann. hum. Genet.* **35**, 119–47.

Forster, J., Chambers, J. P., Peters, S. P., Lee, R. E., and Glew, R. H. (1978). Acute neuropathic Gaucher disease in a black infant. *J. Pediat.* **93**, 823–4.

Forsyth, G. G., Forbes, M., and Cumings, J. N. (1971). Adrenocortical atrophy and diffuse cerebral sclerosis. *Arch. Dis. Childh.* **46**, 273–84.

Forsythe, W. I. (1955). Congenital hereditary vertical nystagmus. *J. Neurol. Neurosurg. Psychiat.* **18**, 196–8.

Forsythe, W. I., McKeown, E. F., and Neill, D. W. (1959). Three cases of Niemann–Pick's disease in children. *Arch. Dis. Childh.* **34**, 406–9.

Fowler, B., Kraus, J., Packman, S., and Rosenberg, L. E. (1978). Homocystinuria. Evidence for three distinct classes of cystathionine beta-synthase mutants in cultured fibroblasts. *J. clin. Invest.* **61**, 645–53.

Fowler, C. P., Harrison, M. J. G., and Snaith, M. L. (1986). Familial carpal and tarsal tunnel syndrome. (Letter). *J. Neurol. Neurosurg. Psychiat.* **49**, 717–18.

Fowler, G. W. and Williams, L. P. (1973). Technetium brain scans in tuberous sclerosis. *J. nucl. med.* **14**, 215–18.

Fowler, M., Dow, R., White, T. A., and Greer, C. H. (1972). Congenital hydrocephalus—hydranecephaly in five siblings with autopsy studies: a new disease. *Dev. med. Child. Neurol.* **14**, 173–88.

Fox, H. (1972). Neurocutaneous melanosis. In *Handbook of Clinical neurology* (ed. P. J. Vinken and G. W. Bruyn) Vol. 14, pp. 425–8.

Fox, J. W., Golden, G. T., and Edgerton, M. T. (1976). Frontonasal dysplasia with alar clefts in two sisters. *Plastic reconstr. surg.* **57**, 553–61.

Fracassi, T. and Parachu, L. (1935). The vascular abnormalities and tumours of the spinal cord and its membranes. *Rev. argent. Neurol.* **1**, 58.

Franceschetti, A. (1966). L'atrophie optique infantile compliquée (Maladie de Behr). *Génét. hum.* **15**, 322–31.

Franceschetti, A. and Klein, D. (1948). Hérédo-ataxies par dégénérescence, spino-pontocerebelleuse. Les manifestations tapeto-retiniennes. *Rev. Oto-Neuro-Ophtal.* **20**, 109–66.

Franceschetti, A., Konig, H., and Klein, D. (1953). L'importance du facteur heredodegenaratif dans l'hemiatrophie faciale progressive. *Suisses Arch. Neurol. Psychiat.* **71**, 311–16.

Francis, D. A., *et al.* (1987). Multiple sclerosis in North East Scotland. An association with HLA-DQWI. *Brain* **110**, 181.

Francois, J. (1974). L'heredite des degenerescences maculaires. *Ophthalmological, Basel* **168**, 417–45.

Francois, J. (1975). Les difficultés du conseil genetique dans les phacomatoses. *J. Génét. hum.* **23**, 17–24.

Francois, J. (1976). Les atrophies optiques hereditaries. *J. Génét. hum.* **24**, 183–200.

Frank, U. and Lassun, U. (1985). Ophthalmoplegic neurolipidosis—storage cells in heterozygotes. *Neuropediatrics.* **16**, 3–6.

Franke, U., *et al.* (1985). A minor Xp 21 chromosome deletion in a male associated with expression of Duchenne muscular dystrophy, chronic granulomatous disease, retinitis pigmentosa and McLeod syndrome. *Am. J. Hum. Genet.* **37**, 250–67.

Frantzen, E., Lennox-Buchthal, M., and Nygaard, A. (1968). Longitudinal EEG and clinical study of children with febrile convulsions. *Electroenceph. clin. Neurophysiol.* **24**, 197.

Frantzen, E., Lennox-Buchtal, N., Nygaard, A., and Stene, J. (1970). A genetic study of febrile convulsions. *Neurol., Minneap.* **20**, 909–17.

Fraser, D. (1880). Defect of the cerebellum occuring in a brother and sister. *Glasg. med. J.* **xiii**, 199–210.

Fraser, F.C. and Gunn, T. (1977). Diabetes mellitus, diabetes insipidus and optic atrophy. An autosomal recessive syndrome? *J. med. Genet.* **14**, 190–3.

Frederico, A., *et al.* (1986). Juvenile dystonia without vertical gaze paralysis: Niemann–Pick type C disease. *J. Inher. Met. Dis.* **9**, 314–16, Suppl.

Freundlich, E., Scatter, M., and Yatziv, S. (1981). Familial pellagra-like skin rash with neurological manifestations. *Arch. Dis. Childh.* **56**, 146–8.

Fredrickson, D.S. and Sloan, H.R. (1972). Sphingomyelin lipidosis Niemann–Pick disease. In *The metabolic basis of inherited disease,* 3rd edn (ed. J.B. Stanbury, J.B. Wyngaarden, and D.S. Fredrickson), p. 783. McGraw-Hill, New York.

Fredrickson, D.S., Altrocchi, P.H., Avioli, L.V., Goodman, D.S., and Goodman H.C. (1961). Tangier disease. *Ann. intern. Med.* **55**, 1016.

Freeman, J.M., Finkelstein, J.D., and Mudd, S.H. (1975). Folate responsive homocystinuria and 'schizophrenia': a defect in methylation due to deficient 5, 10-methylenetetrahydrofolate reductase activity. *N. Engl. J. Med.* **292**, 491–6.

Frey, E. (1930). Ein streng dominant erbliches Kinnmuskelzittern. *Dt. Z. Nervenheilk.* **115**, 9–26.

Frias, J.L., Felman, A.H., Rosenbloom, A.L., Finkelstein, S.N., Hoyt, W.F., and Hall, B.D. (1978). Normal intelligence in two children with Carpenter's syndrome. *Am. J. med. Genet.* **2**, 191–9.

Fried, K. and Emery, A.E.H. (1971). Spinal muscular atrophy type II. *Clin. Genet.* **2**, 203–9.

Fried, K. and Mundell, G. (1977). High incidence of spinal muscular atrophy type I (Werdnig–Hoffmann disease) in the Karaite community in Israel. *Clin. Genet.* **12**, 250–1.

Fried, K. and Sanger, R. (1973). Possible linkage between Xg and the locus for a gene causing mental retardation with or without hydrocephalus. *J. med. Genet.* **10**, 17–18.

Fried, K, Arlozorov, A., and Sira, R. (1975). Autosomal recessive oculopharyngeal muscular dystrophy. *J. med. Genet.* **12**, 416–18.

Friede, R. L. and Boltshauser, E. (1978). Uncommon syndrome of cerebellar vermis aplasia I: Joubert syndrome. *Dev. med. Child Neurol.* **20**, 758–63.

Friede, R. L. and de Jong, R. N. (1964). Neuronal enzyme failure in Creutzfeldt–Jakob disease: a familial study. *Arch. Neurol.* **10**, 181–95.

Friede, R. L. and Magee, K. R. (1962). Alzheimer's disease. *Neurol., Minneap.* **12**, 213–222.

Friedman, A. P. (1972). Current concepts in the diagnosis and treatment of chronic recurring headache. *Med. Clins. N. Am.* **56**, 1257–71.

Friedman, E., Harden, A., Koivikko, M., and Pampiglione, G. (1978). Menke''s disease: neurophysiological aspects. *J. Neurol. Neurosurg. Psychiat.* **41**, 505–10.

Friedman, M., Hatcher, G., and Watson, L. (1967). Primary hypo-magnesaemia with secondary hypocalcaemia in an infant. *Lancet* **i**, 703–5.

Friedreich, N. (1881). Paramyoklonus multiplex. *Virch. Arch. path. Anat. Physiol.* **86**, 421–30.

Friel, P. B. (1973). Familial incidence of Gilles de la Tourette disease with observations on aetiology and treatment. *Br. J. Psychiat.* **122**, 655–8.

Fritze, D., Herrman, C., Naeim, F., Smith, G. S., and Walford, R. L. (1974). HLA antigens in myasthenia gravis. *Lancet* **i**, 240–2.

Fritzsche, R. (1935). Eine familiar auftretende Form von Oligophrenie mit roentgenologisch nachweisbaren symmetrischen Kalkablagerungen im Gehirn, besonders in den Stammganglien. *Schweiz. Arch. Neurol. Psychiat.* **35**, 1–29.

Froment, J., Bonnet, P., and Colrat, A. (1937). Heredodegenerations retiniennes et spino-cerebelleuses. Variantes opthalmoscoriques et neurologiques par trois generations successive. *J. med. Lyon.* **18**, 153–62.

Fromke, V. L., Bossenmaier, I., Cardinal, R., and Watson, C. J., (1978). Porphyria variegata. *Am. J. Med.* **65**, 80–8.

Frost, N., Feighner, J., and Schuckit, M. A. (1976). A family study of Gilles de la Tourette Syndrome. *Dis. nerv. Syst.* **37**, 537–8.

Frydman, M., *et al.* (1985). Assignment of the gene for Wilson disease to chromosome 13: linkage to esterase D locus. *Pro. Nat. Acad. Sci.* **82**, 1819–21.

Frydman, M., Kauschansky, A., and Elian, E. (1984). Trigonocephaly: A new familial syndrome. *Am. J. Med. Genet.* **18**, 55–9.

Fryer, A.E. and Osborne, J.P. (1987). Tuberous Sclerosis—A clinical appraisal. *Pediat. Rev.* **1**, 239–55.

Fryer, A.S. *et al.* (1987). Evidence that the gene for tuberous sclerosis is on chromosome 9. *Lancet* **i**, 659–61.

Fukinaga, H., Hirose, K., Beppu, H., Uowo, M., and Suzuki, Y. (1976). Two siblings with mucolipidosis. *Clin. Neurol.* **16**, 566–73.

Fukuda, K., *et al.* (1983). Free amino acid concentration in blood cells of two brothers with gyrate atrophy of the choroid and retina with hyperornithinaemia. *J. Inherit. Metab. Dis.* **6**, 137–42.

Fukuda, K., Tanaka, T., Hyodo, S., Kobayashi, Y., and Usui, T. (1985). Hyperphenylalaninaemia due to impaired dihydrobiopterin biosynthesis: leukocyte function and effect of tetrahydrobiopterin therapy. *J. Inherit. Metab. Dis.* **8**, 49–52.

Fukuhara, N., *et al.* (1980). Myoclonus epilepsy associated with ragged-red fibres (mitochondrial abnormalities): disease entity or a syndrome. *J. Neurol. Sci.* **47**, 117–33.

Fukuyama, Y. and Ohsawa, M. (1984). A genetic study of the Fukuyama type congenital muscular dystrophy. *Brain Dev.* **6**, 373–90.

Fukuyama, Y. and Okada, R. (1968). Hereditary kinasthetic reflex epilepsy: report of five pedigrees with seizure induced by movement and review of literature. *Proc. Aust. Ass. Neurol.* **5**, 583–7.

Fukuyama, Y., Kawazura, M., and Haruna, H. (1960). A peculiar form of congenital progressive muscular dystrophy. *Pediatrica* **44**, 5–8.

Furman, J.M, *et al.* (1985). Infantile cerebellar atrophy. *Ann. Neurol.* **17**, 399–402.

Furukawa, T. and Peter, J.B. (1977). X-linked muscular dystrophy. *Ann. Neurol.* **2**, 414–16.

Furukawa, T. and Toyokura, Y. (1976). Chronic spinal muscular atrophy of facio-scapulo humeral type. *J. med. Genet.* **13**, 285–9.

Furukawa, T. and Toyokura, V. (1977). Congenital, hypotonic-sclerotic muscular dystrophy. *J. med. Genet.* **14**, 426–9.

Furukawa, T., Nakao, K., Sugita, H., and Tsukagoshi, H. (1968). Kugelberg–Welander disease with particular reference to sex- influenced manisfestations. *Arch. Neurol.* **19**, 156–62.

Furukawa, T., Igata, A., Toyokura, Y., and Ikeda, S. (1970). Sturge–Weber and Klippel–Trenaunay syndrome with nevus of Ota and Ito. *Arch. Derm. N.Y.* **102**, 640–5.

Furukawa, T., Akagami, N., and Maruyama, S. (1977). Chronic neurogemc quadriceps amyotrophy. *Ann. Neurol.* **2**, 528–30.

Fye, K.H., Jacobs, R.P., and Roe, R.L.(1975). Vascular manifestations of von Recklinghausen's disease. *West J. Med.* **122**, 110–16.

Gabreels, F.J.M. *et al.* (1984). Defects in citric acid cycle and the electron transport chain in progressive poliodystrophy. *Acta Neurol. Scan.* **70**, 145–54.

Gabreels–Festen A.A.W.M. and Hageman, A.T.M. (1986). Chronic inflammatory demelinating polyneuropathy in two siblings. *J. N. N. P.,* **49**, 152–6.

Gadbois, P., Moreau, J., and Laberge, C. (1973). La maladie de Morquio dans la province de Quebeck. *Un. méd. Can.* **102**, 602–7.

Gadoth, N., O'Croinin, P., and Butler, I.J. (1975). Bone marrow in Batten–Vogt syndrome. *J. neurol. Sci.* **25**, 197–203.

Gafni, J., Fischel, B., Reif, R., Yaron, M. and Pras, M. (1985). Amyloidotic polyneuropathy in a Jewish family. *Q. J. Med.* **55**, 33.

Gagnon, J. and Le Blanc, R. (1959). Sclerose cerebrale diffuse avec melanodermie et atrophie surrenale. Maladie d'Addison–Scholz. *Un. med. Can.* **88**, 392–412.

Gainer, J.V. Jr, Chou, S.M., and Chadduck, W.M. (1975). Familial cerebral sarcomas. *Arch. Neurol., Chicago* **32**, 665–8.

Gaist, G. and Piazza, G. (1959). Meningiomas in two members of the same family. *J. Neurosurg.* **16**, 110–13.

Gajdusek, D.C. (1962). Kuru: an appraisal of five years of investigation *Eugen. Q.* **9**, 69–74.

Gajdusek, D.C. and Gibbs, C.J. (1975). Familial and sporadic chronic neurological degenerative disorders transmitted from man to primates. In *Primate models of neurological disorders* (ed. B.S. Meldrum and C.D. Marsden), pp. 219–317. Raven Press, New York.

Gajdusek, D.C., Gibbs, G.J., and Alpers, M. (1966). Experimental transmission of a Kuru-like syndrome to chimpanzees. *Nature. Lond.* **209**, 794–6.

Gal, A., *et al.* (1985). Charcot–Marie–Tooth disease: suggestion of linkage with a cloned DNA sequence from the proximal Xq. *Hum. Genet.* **70**, 38–42.

Galassi, G., Modena, M.G., Benassi, A., *et al.* (1986). Autosomal-dominant dystrophy with humeroperoneal weakness and cardiopathy: a genetic variant of Emery–Dreifuss disease? *Ital. J. Neurol. Sci.* **7**, 125–32.

Gallai, V., Hockaday, J.M., Hughes, J.T., *et al.* (1981). Ponto-Bulbar palsy with deafness (Brown–Vialetto–Vanlaere syndrome). A report on three cases. *J. Neurol. Sci.* **50**, 259–75.

Gallemaerts, V., Kleyntjens, F., and Cloetens, W. (1939). Hérédo-ataxie cérébelleuse de P. Marie. *J. belge Neurol. Psychiat.* **39**, 667–75.

Gambetti, P., Mellman, W.J., and Gonatas, N.K. (1969). Familial spongy degeneration of the central nervous system (van Bogaert–Bertrand type). *Acta Neuropath., Berl.* **12**, 103–15.

Gamstorp, I. (1956). Adynamia episodica hereditaria. *Acta paediat., Stockh.* Suppl. 108, 1.

Gamstorp, I. (1962). A study of transient muscular weakness. *Acta neurol. scand.* **38**, 3–19.

Gamstorp, I. (1967). Progressive spinal muscular atrophy with onset in infancy or early childhood. *Acta paediat. scand.* **56**, 408–23.

Gancher, S.T., and Nutt, J.G. (1986) Autosomal dominant episodic ataxia: a heterogeneous syndrome. *Movement Disorder,* **1**, 239–53.

Garcia, C.A., Dunn, D., and Trevor, R. (1978). The lissencephaly (Agyria) syndrome in siblings. *Arch. Neurol., Chicago.* **35**, 608–11.

Garcin, R., Raverdy, Ph., Delthil, S., Man, H.X., and Chimenes, H. (1961). Sur une affection heredo-familiale associant cataracte, atrophie optique, signes extra pyramidaux et certains stigmates de la maladie de Friedreich. *Rev. neurol.* **101**, 373–9.

Gardner, J.H. and Feldmahn, A. (1966). Hereditary adult motor neuron disease *Trans. Am. neurol. Ass.* **91**, 239–41.

Gardner, J.H and Maloney, W. (1968). Hereditary brachial and cranial neuritis, genetically linked with ocular hypotelorism and syndactyly. *Neurol., Minneap.* **18**, 278.

Gardner, W.J. and Frazier, C.H. (1930). Bilateral acoustic neurofibromas. A clinical study and field survey of a family of five generations with bilateral deafness in 31 members. *Arch. Neurol. Psychiat., Chicago* **23**, 266–302.

Gardner, W.J. and Turner, O. (1940). Bilateral acoustic neurofibromas. Further clinical and pathologic data on hereditary deafness and Recklinghausen's disease. *Arch. Neurol. Psychiat., Chicago* **44**, 76–99.

Gardner-Medwin, D. (1977). Children with genetic muscular disorders. *Brit. J. Hosp. Med.* **17**, 314–340.

Gardner–Medwin, D. and Johnston, H.M. (1984). Severe muscular dystrophy in girls. *J. Neurol. Sci.* **64**, 79–86.

Gardner–Medwin, D., Hudgson, P., and Walton, J.N. (1967). Benign spinal muscular atrophy arising in childhood and adolescence. *J. neurol. Sci.* **5**, 121–58.

Gardner–Medwin, D., Pennington, R.J., and Walton, J.N. (1971). The detection of carriers of X-linked muscular dystrophy genes. *J. neurol. Sci.* **13**, 459–74.

Gardner–Thorpe, C. and Foster, J.B. (1975). Monozygous twins discordant for multiple sclerosis. *J. neurol. Sci.* **26**, 361–75.

Gareis, F.J. and Mason, J.D. (1984). X-linked mental retardation associated with bilateral clasp thumb anomaly. *Am. J. Med. Genet.* **17**, 333–8.

Garg, B.P, *et al.* (1983). Evoked response studies in patients with adrenoleukodystrophy and heterozygous relatives. *Arch. Neurol.* **40**, 350–9.

Garland, H.G. and Astley, C.E. (1950). Hereditary spastic paraplegia with amyotrophy and pes cavus. *J. Neurol. Neurosurg. Psychiat.* **13**, 130–3.

Garland, H.G. and Moorhouse, D. (1953). An extremely rare recessive hereditary syndrome including cerebellar ataxia oligophrenia, cataract and other features. *J. Neurol. Neurosurg. Psychiat.* **16**, 110–16.

Garvie, J.M. and Woolf, A.L. (1966). Kugelberg–Welander syndrome (hereditary proximal spinal muscular atrophy). *Br. med. J.* **i**, 1458–61.

574   *Bibliography*

Gass, J.D.M. (1971). Cavernous hamangioma of the retina : a new–oculo–cutaneous syndrome. *Am. J. Ophth.* **71**, 799–814.

Gath, I., Kayan, A., Leegaard, J., and Sjaastad, O. (1970). Myasthenia congenita: electromyographic findings. *Acta neurol. scand.* **46**, 323–30.

Gaule, A. (1932). Das Auftreten der Chorea Huntington in einer Familie der Nordostschweiz. *Schweiz. Arch. Neurol. Psychiat.* **29**, 90–112.

Gayral, L. and Gayral, J. (1966). Une souche familiale de onze cas de syndrome de Marinesco–Sjogren. *J.Génet. hum.* **15**, 63–9.

Gee, S. (1889). Hereditary infantile spastic paraplegia. *St. Bart's Hosp. Rep.* **25**, 81–3.

Geiger, L.R., Mancall, E.L., Penn, A.S., and Tucker, S.H. (1974). Familial neuralgic amyotrophy. Report of three families with review of the literature. *Brain* **97**, 87–102.

Gelardi, J.A.M. and Brown, J.W. (1967). Hereditary cataplexy. *J. Neurol. Neurosurg. Psychiat.* **30**, 455–7.

Gellis, S.S. and Feingold, M. (1976). Spondylothoracic dysplasia (costo-vetebral dysplasia, Jarcho–Levin syndrome). *Am. J. Dis. Childh.* **130**, 513–14.

Gelman, M.I. (1977). Autosomal dominant osteosclerosis. *Radiology* **125**, 289–96.

Gemignani, F. (1986). Spinocerebellar ataxia associated with localised amyotrophy of the hands, sensorineural deafness and spastic paraparesis in two brothers. *J.Neurogenet.* **3**, 125–33.

Gerken, H. and Doose, H. (1973). On the genetics of EEG anomalies. Part 3. (spikes and waves). *Neuropaediatrics* **4**, 88–91.

Gerken, H., Kiefer, R., Doose. H., and Volzke, E. (1977). Genetic factors in childhood epilepsy with focal sharp waves..I. Clinical data and familial morbidity for seizures. *Neuropaediatrics* **8**, 3–9.

Gerstenbrand, F. and Weingarten, K. (1962). Beitrag zum Problem der systematrophien des Kleinhirns. *Wien. klin. Wschr.* **74**, 702.

Gerstl, B., Malamud N., Hayman, R.B., and Bond, P.R. (1965). Morphological and neurochemical study of Pelizaeus–Merzbacher's disease. *J. Neurol. Neurosurg. Psychiat.* **28**, 540–7.

Giannelli, F., Avery, J.A., Pembrey, M.E., and Blunt, S. (1981). Prenatal exclusion of ataxia-telangiectasia. In *Ataxia-telangiectasia* (ed. D.G. Harnden and B.A. Bridges). Wiley, New York.

Gibb, W.R.G. Esiri, M.M, and Lees, A.J. (1987). Clinical and pathological features of diffuse Lewy body disease. *Brain* **110**, 1131– 53.

Gibberd, F,B. (1966). The clinical features of petit mal. *Acta neurol. scand.* **42**, 176–90.

Gibbs, C.J., *et al.* (1968). Creutzfeldt–Jakob disease (spongiform encephalopathy) transmission to the chimpanzee. *Science, N. Y.* **161**, 388–9.

Gibbs, J.L, (1985). The heart and tuberous sclerosis. *Br. Heart. J.* **54**, 596–9.

Gieron, M.A., Korthals, J.K., Kousseff, B.G. (1985). Facioscapuluhumeral dystrophy with cochlear hearing loss and tortuosity of retinal vessels. *Am. J. Med. Genet.* **22**, 143–7.

Gilbert, G. J, (1977). Familial spasmodic torticollis. *Neurol., Minneap.* **27**, 11–13.

Gilbert, G. J., McEntee, W. J., and Glaser, G. H. (1963). Familial myoclonus and ataxia. Pathophysiologic implications. *Neurol., Minneap.* **13**, 365–72.

Gilchrist, J. M. and Leshner, R. T. (1986). Autosomal dominant humer-operoneal myopathy. *Arch. Neurol.* **43**, 734–5.

Gilchrist, J. M. *et al.* (1988). Clinical and genetic investigation in autosomal dominant limb–girdle muscular dystrophy. *Neurology* **38**, 5–9.

Gillan, J. E. *et al.* (1984). Congenital ascites as a presenting sign of lysosomal storage disease. *J. Pediat.* **104**, 225–31.

Gillerot, Y., *et al.* (1987). Prenatal diagnosis of a dup (3p) with holoprosencephaly. *Am. J. Med. Genet.* **26**, 225–7.

Gillespie, F. D. (1965). Aniridia, cerebellar ataxia and oligophrenia in siblings. *Arch.Ophthal., N. Y.* **73**, 338–41.

Gillies, C., Raya, J., Vasan, U., Hart, W. E., and Goldblatt, P. J. (1979). Nemalin (rod) myopathy: a possible cause of rapidly fatal infantile hypotonia. *Arch. Pathal. Lab. Med.* **103**, 1–6.

Gilly, R., Cotton, J., Farouz, S., Noiret, A., and Maclet, M. (1971). Hydrocephalie congenitale et anomalie bilateral des pouces: syndrome malformatif liae au chromosome X. *Pédiatrie* **26**, 365–78.

Gilman, S. and Barrett, R. E. (1973). Hallervorden–Spatz disease and neuroaxonal dystrophy. Clinical characteristics and nosological consid-erations. *J.neurol. Sci.* **19**, 189–205.

Gilman, S. and Horenstein, S. (1964). Familial amyotrophic dystonic paraplegia. *Brain* **87**, 51–66.

Gil-Peralta, A., Rafel, E., Bautista, J., and Alberca, R. (1978). Myotonia in centronuclear myopathy. *J. Neurol. Neurosurg. Psychiat* **41**, 1102–8.

Gimenez-Roldan, S. and Estaban, A. (1977). Prognosis in hereditary amyotrophic lateral sclerosis. *Arch. Neurol., Chicago* **34**, 706–8.

Gimenez-Roldan, S., Peraita, P., Lopez–Agreda, J. M., Abad, J. M., and Estaban, A. (1971). Myoclonus and photic–induced seizures in Alzheimer's disease. *Eur. Neurol.* **5**, 215–24.

Gimenez-Roldan, S., Lopez-Fraile, I. P., and Estaban, A. (1976). Dystonia in Spain: study of a gypsy family and general survey. *Arch. Neurol., Chicago.* **14**, 125–36.

Gimenez-Roldan, S., Benito, C., and Mateo, D. (1978). Familial communicating syringomyelia *J.neurol. Sci.* **36**, 135–46.

Ginsburg, C. M. and Long, C. G. (1977). GM$_1$ ganglisidosis type I in twins. *J.med.Genet.* **14**, 132–4.

Ginsberg, L., Di Ferrante, D. T., Caskey, C. T., and Di, Ferrante, N. (1977).

Glucosamine-6-SO$_4$ sulfatase deficiency: a new mucopolysaccharidosis. *Clin. Res.* **25**, 471A.

Ginter, D.N., Konigsmark, B., and Abbot, M.H. (1974). X-linked spinocerebellar degeneration. In *Clinical delineation of birth defects*. XVI. Urinary system and others (ed. D. Bergsma) *Birth Defects*: Original Article Series, 334–6.

Giorgio, A.J., Trowbridge, M., Boone, A.W., and Pattern, R.S. (1976). Methylmalonic aciduria without Vit B$_{12}$ deficiency in an adult sibship. *N. Engl J. Med.* **295**, 310–13.

Girard, P.F., Trillet, M., Confaureux, C., and Chazot, G. (1977). Meningomatose multiple et familiale. Un syndrome voisin de la neurofibromatose de Recklinghausen. *Rev. neurol.* **133**, 359–62.

Girdany, B.R. (1959). Engelmann's disease (progressive diaphyseal dysplasia)—a non-progressive familial form of muscular dystrophy with characteristic bone changes. *Clin. Orthop.* **14**, 102–9.

Giroud, M. *et al.* (1986). A case of progressive familial encephalopathy in infancy with calcification of the basal ganglia and chronic cerebrospinal fluid lymphocytosis. *Child Nerv. System* **2**, 47–8.

Giroux, J.M. and Barbeau, A. (1972). Erythrokeratodemia with ataxia. *Arch. Derm* **106**, 183–8.

Giugliani, R., Dutra, J.C., Pereira, M.L.S. (1985). Gangliosidosis: clinical and laboratory findings in eight families. *Hum. Genet.* **70**, 347–54.

Glanz, A. and Frazer, P.C. (1984). Risk estimates for neonatal myotonic dystrophy. *J. Med. Genet.* **21**, 186–8.

Glick, T.H., Likosky, W.H., Levitt, L.P., Mellin, H., and Reynolds. D.W. (1970). Reye's syndrome: an epidemiologic approach. *Pediatrics, Springfield* **46**, 371–7.

Glista, G.G., Mellinger, J.F., and Rooke, E.D. (1975). Familial hemiplegic migraine. *Proc. Staff Meet. Mayo Clin.* **50**, 307–11.

Go, R.C.P., *et al.* (1984), Segregation and linkage analyses of Von Hippel Lindau disease among 220 descendants from one kindred. *Am. J. Hum. Genet.* **36**, 131–42.

Godeano, D., Winter, S.T., and Dar, H. (1973). Familial holoprosencephaly with median cleft lip. *J. Génét. hum.* **21**, 223–8.

Goebel, H.H., Heipertz, R., Scholz, W., Iqbal, K., and Tellez–Nagel, I. (1978). Juvenile Huntington's chorea: clinical ultrastructural and biochemical studies. *Neurol. Minneap.* **28**, 23–31.

Goebel, H.H., Bode, G., Caesar, R. and Kohlschutter (1981). Bubar palsy with Rosenthal fiber formation in the medulla of a 15 year-old girl. *Neuropediatrics* **12**, 382–91.

Goedde, H.W., *et al.* (1970). Clinical and biochemical–genetic aspects of intermittent branched–chain ketoaciduria. Report of two Scandinavian families. *Acta Pedsiat. Scand.* **59**, 83–8.

Gold. G.N. and Hogenhuis, L.A.H. (1968). Hypertrophic interstitial neuropathy and cataracts *Neurology, Minneap.* **18**, 526–33.

Goldberg. A, (1985). Molecular genetics of acute intermittent porphyria. *Br. Med. J.* **291**, 499–500.

Goldberg, M. F., Cotlier, E., Fichenscher, L. G., Kenyon, K., Enat, R., and Borowsky, S. A. (1971). Macular cherry-red spot, corneal clouding and β-galactosidase deficiency. *Arch. intern. Med.* **128**, 387–98.

Goldberg, R. and Jampal, R. (1962). Voluntary nystagmus in a family. *Arch. Ophthal., N. Y.* **68**, 32–5.

Goldberg, R. E., Pheasant, T. R., and Shields, J. A. (1979). Cavernous haemangioma of the retina : a four generation pedigree with neuro-cutaneous manifestations and an example of bilateral retinal involvement. *Arch. Ophthal.* **97**, 2321–4.

Golden, G. S. (1978). Familial occurrences of Tourette syndrome. (Abstract.) *Clin. Res.* **26**, 74A.

Golden, G. S. and French, J. H. (1975). Basilar artery migraine in young children. *Pediatrics. Springfield.* **56**, 722–6.

Golden, S. E. and Kaplan, A. M. (1986). Hypomelanosis of Ito: Neurological complications. *Pediat. Neurol.* **2**, 170–4.

Goldflam, S. (1930). *Die diagnostische Bedeutung des rosslimoschen Reflexes bei Erkkrankungen des Zenbralnervensystems.* Karger, Berlin.

Goldgaber, D, *et al.* (1987). Characterization and chromosomal localization of cDNA encoding brain amyloid of Alzheimer's disease. *Science* **235**, 877–80.

Goldie, W. D., Holtzman, D., and Suzuki, K. (1977). Chronic hexosaminidase A and B deficiency. *Ann. Neurol.* **2**, 156–8.

Goldman, J. E., Katz. D., Rapin, I, Adachi, M., Suzuki, K. and Suzuki, K. (1981). Chronic G. M. I gangliosidosis presenting as dystonia. I. clinical and pathological features. *Ann. Neurol.* **9**, 465–75.

Goldman, R., Reynolds, J. L., Cummings, H., and Bassett, S. H. (1952). Familial hypoparathyroidism. *J. Am. med. Ass.* **150**, 1104–6.

Goldstein, H. (1921). Meralgia parasthetica (Roth's or Bernhart's disease) with the report of five cases: three cases occurring in the same family. *Am. J. med. Sci.* **162**, 720–35.

Goldstein, J. L. and Fialkow, P. J. (1973). The Alström syndrome. Report of three cases with further delineation of the clinical, pathophysiological and genetic aspects of the disorder. *Medicine, Baltimore* **52**, 53–71.

Goldstein, M. L., Kolodny, E. H., Gascon, G. C., and Gilles, F. H. (1974). Macular cherry-red spot, myoclonic epilepsy and neurovisceral storage in a 17-year-old girl. *Trans. Am. neurol. Ass.* **99**, 110–12.

Gollop, T. R. and Fontes, L. R. (1985). The Greig cephalopolysyndactyly syndrome: report of a family and review of the literature. *Am. J. Med. Genet.* **22**, 59–68.

Gomez, M. (1979). *Tuberous sclerosis,* pp. 16–17. Raven Press, New York.

Gomez, M., Clermont, V., and Bernstein, J. (1962). Progressive bulbar paralysis in childhood (Fazio–Londe's disease). *Arch. Neurol.* **6**, 317–23.

Gomez, M.R., Mellinger, J.F., and Reese, D.F. (1975). The use of computerized transaxial tomography in the diagnosis of tuberous sclerosis. *Proc. Staff Meet. Mavo Clin.* **50**, 553–6.

Gomez, M.R., Engel, A.G., Dewald, G., and Peterson, H.A. (1977). Failure of inactivation of Duchenne dystrophy X- chromosome in one of female identical twins. *Neurol., Minneap* **27**, 537–41.

Gompertz, D., *et al.* (1975). Prenatal diagnosis and family studies in a case of propionicacidemia. *Clin. Genet.* **29**, 378–88.

Gonatas, N.K., Perez, M.C., Shy, G.M., and Evangelista, I. (1965). Central 'core' disease of skeletal muscle. Ultrastructural and cytochemical observations in two cases. *Am. J. Path.* **57**, 503–24.

Gonatas, N.K., Shy, G.M., and Godfrey, E.H. (1966). Nemaline myopathy. The origin of nemaline structure. *N. Engl. J. Med.* **274**, 535–9.

Gonatas, N.K., Gambetti, P., and Baird, H. (1968). A second type of late infantile amaurotic idiocy with multilamellar cytosomes. *J.Neuropath. exp. Neurol.* **27**, 371–89.

Gonzales-Rios, M.D.C, *et al.* (1985). A distinct variant of intermediate maple syrup urine disease. *Clin. Genet.* **27**, 153–9.

Goodbody, R.A. and Gamlen, T.R. (1974). Cerebellar hemanglioblastoma and genitourinary tumours. *J. Neurol. Neurosurg. Psychiat.* **37**, 606–9.

Goodenough, D.J., Fariello, R.G., Annis, B.L., and Chun, R.W. (1978). Familial and acquired paroxysmal dyskinesias. A proposed classification with delineation of clinical features. *Arch. Neurol., Chicago* **35**, 827–31.

Goodhue, W.W., Couch, R.D., and Namiki, H. (1979). Spongy degeneration of the central nervous system. An instance of the rare juvenile form. *Arch. Neurol. Chicago.* **36**, 481–4.

Goodman, R.M., Caren, J., Ziprkowski, M., Padeh, B., Ziprowski, L., and Cohen. B.E. (1971). Genetic considerations in giant pigmented hairy naevus. *Br J. Derm.* **85**, 150–7.

Goodman, R.M, Tadmor, R., Zaritsky, A., and Becker, S.A. (1975). Evidence for an autosomal recessive form of cleidocranial dysostosis. *Clin. Genet.* **8**, 20–9.

Goodman, R.M., Sternberg, M., Shem–Tov, Y., Katznelson, M.B.M., Hertz, M., and Rotem, Y. (1979). Acrocephalopolysyndactyly type IV. A new genetic syndrome in three sibs. *Clin. Genet.* **15**, 209–14.

Goodman, S.I. and Frerman, F.E. (1984). Glutaric acidaemia type II (multiple acyl–Co A dehydrogenase deficiency). *J. Inherit. Met. Dis.* Suppl 1, 33–7.

Goodman, S.I., *et al.* (1980). Antenatal diagnosis of glutaric acidemia. *Am. J. Hum. Genet.* **32**, 695–69.

Goodship, J. *et al.* (1988). Service experience using DNA analysis for genetic prediction in Duchenne muscular dystrophy. *J. Med, Genet.* **25**, 14–19.

Goodwin, J.F. and Lawson, C.W. (1947). Status epilepticus complicating pregnancy. *Br. med. J.* **ii**, 332–3.

Gordon, A. (1933). Familial degenerative disease. *Med. J. Rec.* **138**, 13–16.

Gordon, A.M., Capute, A.J., and Konigsmark, B.W. (1976). Progressive quadriparesis, mental retardation, retinitis pigmentosa and hearing loss—report of two sibs. *Johns Hopkins Med. J.* **138**, 142–5.

Gordon, A.S., Rewcastle, N.B., Humphrey, J.G., and Stewart, B.M. (1974). A chronic benign congenital myopathy: fingerprint body type. *Can. J. Neurol. Sci.* **1**, 106–13.

Gordon, H. (1959). Craniostenosis. *Br. med. J.* **ii**, 792–5.

Gordon, H. and Gordon, W. (1970). Incontenentia pigmenti: clinical and genetical studies of two familial cases. *Dermatological.* **140**, 150–68.

Gordon, N. (1968). Juvenile spinal muscular atrophy. *Dev. med. Child. Neurol.* **10**, 617–20.

Gordon, N. (1974). Menkes kinky-hair (steely-hair) syndrome. *Dev. med. Child. Neurol.* **16**, 827–9.

Gordon, N. (1978). The insidious presentation of the juvenile form of metachromatic leucodystrophy. *Post-grad. med. J.* **54**, 335–7.

Gordon, N. and Thursby–Pelham, D. (1969). The Sanfillipo syndrome: an unusual disorder of mucopolysaccharide metabolism. *Dev. med. Child. Neurol.* **11**, 485–92.

Gordon, N., Marsden, H.B., and Lewis, D.M. (1974). Subacute necrotising encephalomylopathy in three siblings. *Dev. med. Child. Neurol.* **16**, 64–78.

Gordon, N.S., Marsden, H.B., and Noronha, M.J. (1972). Neuronal ceroid lipofuscinosis (Batten's disease). *Arch. Dis. Childh.* **47**, 285–91.

Goren, H. Steinberg, M.C. and Farboody, G.H. (1980). Familial oculoleptomeningeal amyloidosis. *Brain* **103**, 473–95.

Gorlin, R.J. (1960). Craniofacial dysostosis, patent ductus arteriosus, hypertrichosis hypoplasia of labia majora, dental and eye anomolies—a new syndrome? *J. Pediat.* **56**, 778–85.

Gorlin, R.J. and Cohen, M.M. Jr (1969). Frontometaphyseal dysplasia. A new syndrome. *Am. J. Dis. Child.* **118**, 487–94.

Gorlin, R.J. and Glass, L. (1977). Autosomal dominant osteosclerosis. *Radiology* **125**, 547–8.

Gorlin, R.J. and Goltz, R.W. (1960). Multiple nevoid basal-cell epithelioma, jaw cysts and bifid rib: a syndrome. *N. Engl. Med. J.* **262**, 908–12.

Gorlin, R.J. and Winter, R.B. (1980). Frontometaphyseal dysplasia—evidence for X-linked inheritance. *Am. J. med. Genet.* **5**, 81–4.

Gorlin, R.J., Yunis, J., and Anderson, V.E. (1968). Short arm deletion of chromosome 18 in cebocephaly. *Am. J. Dis. Child.* **115**, 473–6.

Gorlin, R.J., Anderson, R.C., and Blaw, M. (1969). Multiple lentigenes syndrome, complex comprising multiple lentigines, electrocardiographic conduction abnormalities, ocular hypertelorism, pulmonary stenosis, abnormalities of the genitalia, retardation of growth, sensorineural deafness and autosomal dominant hereditary pattern. *Am. J. Dis. Child.* **117**, 652–62.

Goto, I., Kanazawa, Y., Kobayashi, T., Murai, Y., and Kuroiwa, Y. (1977). Oculopharyngeal myopathy with distal and cardiomyopathy. *J.Neurol. Neurosurg. Psychiat.* **40**, 600–7.

Goto, I, *et al.* (1979). Rigid spine syndrome. *J. Neurol. Neurosurg. Psychiat.* **42**, 276–279.

Goto, I., Tobimatsu, S., Ohta, M., Hosakawa, S., Shipaski, H., and Kuriowa, Y. (1982). Dentatorubropallidoluysian degeneration: clinical, neuro-ophthalmological, biochemical, and pathalogical studies on autosomal dominant form. *Neurology* **32**, 1395–9.

Gottlieb, R. P. *et al.* (1982). Hyperuricaemia and choreoathetosis in a child without mental retardation or self–mutilation—a new HPRT variant. *J. Inherit. Metab. Dis.* **5**, 183–6.

Gourie–Devi, M. Suresh, T.G., and Shankar, S.K. (1984). Monomelic amyotrophy. *Arch. Neurol.* **41**, 388–94.

Gowers, W.R. (1893). *A manual of diseases of the nervous system.* 2nd edn. Churchill, London.

Graf, C.J. (1965). Spontaneous carotid-cavernous fistula: Ehlers–Danlos syndrome and related conditions. *Arch. Neurol., Chicago* **13**, 662–72.

Graf, C.J. (1966). Familial intracranial aneurysms. Report of four cases. *J. Neurosurg.* **25**, 304–8.

Graff–Radford, N.R. (1986). A recessively inherited ataxia with episodes of dystonia. *J. Neurol. Neurosurg. Psychiat.* **49**, 591–4.

Gragg, G.W., Fogelson, M.H., and Zwirecki, R.J. (1971). Juvenile amyotrophic lateral sclerosis in two brothers from an inbred community. *Birth Defects*: Original Artical Series **VII**, 222–5.

Graham–Pole, J., Ferguson, A., Gibson, A.A.M., and Stephenson, J.B.P. (1975). Familial dysequilibrium-diplegia with T-lymphocyte deficiency. *Arch. Dis. Childh.* **50**, 927–33.

Gray, F., Eizenbaum, J.F., Gherardi, R., Degos, J.D., and Poirier, J. (1985). Luyso–pallido–nigril alropty and amyotrophic lateral sclerosis. *Acta New, path.* **66**, 78–82.

Gray, R.C. and Oliver, C.P.(1941). Marie's hereditary cerebellar ataxia (olivoponto cerebellar atrophy). *Minn. Med.* **24**, 327–35.

Gray, T. and Rewcastle, N.B. (1967). Parkinsonism and striatonigral degeneration. *Can. med. Ass. J.* **97**, 240.

Grcevic, N. (1966). Hallervorden–Spatz disease. Histological and histo-chemical study of a new case. *Proc. 5th Int. Congr. Neuropath., Zurich,* 1965, pp. 170–2. Excerpta Medica, Amsterdam.

Grebe, H. (1954). Familienbefunde bei letalen Anomalien der Koperform. *Acta Genet. med. Gemell.* **3**, 93–111.

Green, J.B., (1960). Familial amyotrophic lateral sclerosis occuring in 4 generations. Report of a case. *Neurol. Minneap.* **10**, 960–2.

Greenberg, R. and Schraufnager, D. (1979). The G syndrome. A case report. *Am. J. med. Genet.* **3**, 59–64.

Greenfield, J.G. (1912). Case of peroneal atrophy with signs of Friedreich's disease. *Proc. R. Soc. Med.* **5**, 75.

Greenwood, R.S., Hillman, R.E., Alcala, H., and Sly, W.S. (1978). Sanfillipo. A syndrome in the fetus. *Clin . Genet.* **13**, 241–50.

Greer, M. and Schotland, M. (1960). Myasthenia gravis in the newborn. *Pediatrics, Springfield.* **26**, 101–8.

Gregersen, N., Brandt, N.J., Christensen, C., Grøn, I., Rasmussen, K., and Brandt, S. (1977). Glutaric aciduria: clinical and laboratory findings in two brothers. *J. Pediat.* **90**, 740–5.

Gregoriou, M., Matsaniotis, N., and Papadakis, G. (1969). Progressive bulbar palsy. *Devl med. Child. Neurol.* **11**, 630–2.

Greig, D.M. (1922). A case of meningeal nevus associated with adenoma sebaceum. *Edinb. med. J.* **28**, 105–11.

Greig, D.M. (1924). Hypertelorism. A hitherto undifferentiated congenital cranio-facial deformity. *Edinb. med. J.* **31**, 560.

Greig, D.M. (1926). Oxycephaly. *Edinb. med. J.* **33**, 189.

Greitzer, L.J., Jones, K.L., Schnall, B.S., and Smith, D.W. (1974). Craniosynostosis—radial aplasia syndrome. *J.Pediat.* **84**, 723–7.

Greze, J., Baldet, P., Dumas, R., Cadilhac, J., Pages, A., and Jean, R. (1975). Dystrophic osteo–chondo–musculaire de Schwartz–Jampel. Deux cas familiaux. *Arch. fr. Pediat.* **32**, 59–75.

Gricpentrog, F. and Pauly, H. (1957). Intra und extrakranielle fruhmanifeste Medulloblastome bei erbgleichen Zwillingen. *Zentbl. Neurochir.* **17**, 129–40.

Griffin, J.W., Goren, E., Schaumburg, H., Engel, K., and Loriaux, L. (1977). Adrenomyeloneuropathy: a probable variant of adrenoleukodystrophy I. *Neurology, Minneap.* **27**, 1107–13.

Griffiths, M.I. and Barrett, N.M. (1967). Cerebral palsy in Birmingham. *Dev. med. Child. Neurol.* **9**, 33–46.

Griffiths, R.G, Mortimer, T.F., Oppenheimer, D.R. (1982). Congophilic angiopathy of the brain: a clinical and pathological report on two siblings. *J. Neurol. Neurosurg. Psychiat.* **45**, 396–408.

Griggs, R.C., Moxley, R.T., LaFrance, R.A., and McQuillen, J. (1978). Hereditary paroxysmal ataxia: response to acetazolamide. *Neurol. Minneap.* **28**, 1259–64.

Grimm, T. (1975). The ages of onset and at death in dystrophia myotonica. *J. Génét. hum.* **23**, Suppl., 172.

Grinker, R.R. Serota, H., and Stein, S.I. (1938). Myoclonic epilepsy. *Arch. Neurol. Psychiat. Chicago* **40**, 968–80.

Groebe, H., *et al.* (1980). Morquio syndrome (mucopolysac charidosis IV) associated with β-galactosidase deficiency. *Am. J. hum. Genet.* 32, 258–72.

Groen, J.J. and Endtz, L.J (1982). Pick's disease: hereditary second re-examination of a large family with discussion of other hereditary cases with particular reference to electroencephalography and computerized tomography. *Brain* 105, 443.

Gross. (1985). Familial amyotrophic chorea with acanthocytosis new clinical and laboratory investigations. *Arch. Neurol.* 42, 753–56.

Grosshans, E.M., Stoebner, P., Bergoend, H., and Stoll, C. (1971). Incontinentia pigmenti achromians (Ito). Etude clinique et histo-pathologique. *Dermatologica. Basel* 142, 65–78.

Gross-Kieselstein, E. and Shalev, R.S. (1987). Familial absence of the trapezius muscle with associated shoulder girdle abnormalities. *Clin. Genet.* 32, 145–7.

Grossman, B.J. (1957). Trembling of the chin: an inheritable dominant character. *Pediatrics, Springfield* 19, 453–5.

Grunthal, E. (1930). Uber ein Bruderpaar mit Pickscher Krankheit. *Z. Neurol. Psychiat.* 129, 350–75.

Guazzi, G.C. and Martin, J.J. (1967). La forme infantile de la leucodystrophie soudanophile avec angiomatose meningee non calcifante. *Acta neurol. belg.* 67, 463–74.

Gudmundsson, G., Hallgrimsson, J., Jonasson, T.A., and Bjarnason, O. (1972). Hereditary cerebral haemorrhage with amyloidosis. *Brain* 95, 387–404.

Gudmundsson, K.R. (1967). A clinical survey of parkinsonism in Iceland. *Acta neurol. scand.* 43, Suppl. 33.

Guggenheim, M.A. (1979). Familial Tourette syndrome. *Ann. Neurol.* 5, 104.

Guibaud, P., Larbre, F., Freycon, M.-T., and Genoud, J. (1972). Osteopetross of acidose renale tubulaire duex cas de cette association dans une fratrie. *Arch. fr. Pédiat.* 29, 269–86.

Guilleminault, C.G., Harpey, J.P., and Lafourcade, J. (1973). Sjögren–Larsson syndrome. Report of two cases in twins. *Neurol., Minneap.* 23, 367–73.

Guillozet, N. and Mercer, R.D. (1973). Hereditary recurrent brachial neuropathy. *Am. J. Dis. Child.* 125, 884–7.

Guiloff, R.J., Thomas, P.K., Contrenas, M., Armitage, S., Schwartz, G., and Sedgewick, E.M. (1982). Linkage or autosomal dominant type I hereditary motor and sensory neuropathy to the Duffy locus on Chromosome I. *J. Neurol. Neurosurg. Psychiat.* 45, 669.

Guinter, R.H., Hernried, L.S., and Kaplan, A.M. (1977). Infantile neurogenic muscular atrophy with prolonged survival. *J. Pediat.* 90, 95–7.

Gumbinas, M., Liu, H.M., Dawson, G., Larsen, M., and Green, O. (1976).

Progressive spastic paraparesis and adrenal insufficiency. *Arch. Neurol., Chicago* **33**, 678–80.

Gunderson, G.H., Greenspan, R.H., Glazer, G.H., and Lubs, H.A. (1967). The Klippel–Feil syndrome: genetic and clinical re-evaluation of cervical fusion. *Medicine, Baltimore* **46**, 491.

Gunn, T., Bortolussi, R., Little, J.M., Andermann, F., Frazer, F.C., and Belmont, M. (1976). Juvenile diabetes mellitus, optic atrophy, sensory nerve deafness and diabetes insipidus – a syndrome. *J.Pediat.* **89**, 565–70.

Gunter, M. and Penrose, L.S. (1935). Genetics of epiloia. *J. Genet.* **31**, 413.

Gurland, J.E., Tenner, M., Hornblass, A., and Wolintz, A.H. (1976). Orbital neurofibromatosis. *Arch. Opthal., N.Y.* **94**, 1723–5.

Gusella, J.F., Tenner, M., Hornblass, A., *et al.* (1983). A polymorphism DNA marker genetically linked to Huntington's disease. *Nature* **306**, 234–8.

Gustavson, K.-H. and Hagberg, B. (1971). The incidence and genetics of metachromatic leucodystrophy in northern Sweden. *Acta pediat. scand.* **60**, 585–90.

Gustavson, K.-H., Hagberg, B. and Sanner, G. (1969). Identical syndromes of cerebral palsy in the same family. *Acta paediat. scand.* **58**, 330–40.

Guzetta, F.Ferriere, G., and Lyon, G. (1982). Congenital hypomyelineation polyneuropathy. pathological findings compared with polyneuropathies starting later in life. *Brain* **105**, 395–415.

Gwinn, J.L., Barnes, Jr, G.R., Tucker, A.S., and Johnson, C. (1966). Radiological case of the month. *Am. J. Dis. Child.* **122**, 583–4.

Haan, E.A., *et al.* (1985). 5,10-methylenetetrahydrofolate reductase deficiency: clinical and biochemical features of a further case. *J. Inherit. Met. Dis.* **8**, 53–7.

Haar, F. and Dyken, P. (1977). Hereditary nonprogressive athetotic hemiplegia: a new syndrome. *Neurol., Minneap.* **27**, 849–54.

Haas, D.C. and Sovner, R.D. (1969). Migraine attacks triggered by mild head trauma and their relation to certain post-traumatic disorders of childhood. *J. Neurol. Neurosurg. Psychiat.* **32**, 548–54.

Haas, L.F., Austad, W.I., and Bergin, J.D. (1974). Tangier disease. *Brain* **97**, 351–4.

Haas, R.H., *et al.* (1981). An X-linked disease of the nervous system with disordered copper metabolism and features differing from Menkes disease. *Neurology.* **31**, 852–9.

Haberland, C., Brunngraber, E., Whitting, L., and Daniels, A. (1973). Juvenile metachromatic leukodystrophy. *Acta neuropath.* **26**, 93–106.

Haberland, W.F. (1961). Aspects genetiques de la sclerose laterale amyotrophique. *Wld. Neurol.* **2**, 356–65.

Hackenbruch Y., Meerhof, E., Besio, R., and Cordoso, H. (1975). Familial bilateral optic nerve hypoplasia. (*Am. J. Ophthal*). **79**, 314–20.

Hackett, T. N., Patrick, P. F., Bray, F. A., Ziter, F. A., Nyhan, W. L., and Creer, K. M. (1973). A metabolic myopathy associated with chronic lactic acidemia, growth failure, and nerve deafness. *J. Pediat.* **83**, 426–31.

Haddow, J. E., Shapiro, S. R., and Gall, D. G. (1970). Congenital sensory neuropathy in siblings. *Pediatrics, Springfield* **45**, 651–5.

Haerer, A. F. and Currier, R. D. (1966). Mirror movement. *Neurol., Minneap.* **16**, 757–60.

Haerer, A. F. and Jackson, J. F. (1967). Hereditary non-progressive chorea of early onset. *N. Engl. J. Med.* **276**, 1220–4.

Hagberg, B. (1962). Clinical symptoms, signs and tests in metachromatic leukodystrophy. In *Brain, lipids and lipoproteins and the leukodystrophies* (ed. J. Folch-Pi and H. Bauer), pp. 134–46. Elsevier, Amsterdam.

Hagberg, B. (1984). Krabbe's disease: clinical presentation of neurological variants. *Neuropediatr.* **15**, 11–15.

Hagberg, B. and Westerberg, B. (1983). The nosology of genetic peripheral neuropathies in Swedish children. *Dev. Med. Child. Neurol.* **25**, 3–18.

Hagberg, B., Hansson, O., Liden, S., and Nilsson, K. (1970). Familial ataxic diplegia with deficient cellular immunity. A new clinical entity. *Acta paediat. scand.* **59**, 545–50.

Hagberg, B., Sanner, G., and Steen, M. (1972). The dysequilibrium syndrome in cerebral palsy. *Acta paediat. scand.* Suppl. 226. 1–63.

Hagberg, B., Haltia, M., Sourander, P., Svennerholm, M., Vanier, M.-T., and Ljunggren, C.-G, (1978). Neurovisceral storage disorder simulating Niemann–Pick disease. *Neuropaediatrics.* **9**, 59–73.

Hagler, W. S., Hyman, B. N., and Waters, W. C. (1971). Von Hippel's angiomatosis retinae and pheochromocytoma. *Trans. Am. Acad. Ophthal.* **75**, 1022–34.

Haile, R. W., *et al.* (1980). Genetic susceptibility to multiple sclerosis: a linkage analysis with age of onset corrections. *Clin. Genet.* **18**, 160–7.

Hakamada, S., *et al.* (1983). Congenital hypomyelination neuropathy in a newborn. *Neuropediat.* **14**, 182–3.

Hakola, H. P. A. (1972). Neuropsychiatric and genetic aspects of a new hereditary disease characterized by progressive dementia and lipomembranous polycystic osteodysplasia. *Acta psychiat. scand.* Suppl. 232, 1.

Hakola, H. P. A., Jarvi, D. H., and Sourander, P. (1970). Osteodysplasia polycystica hereditaria combined with sclerosing leucoencephalopathy: a new entity of the dementia praesenilis group. *Acta neurol. scand.* **46**, Suppl. 43, 78.

Halal, F., *et al.* (1983). Intracranial aneurysm: a report of a large pedigree. *Am. J. Med. Genet.* **15**, 89–95.

Hall, B., Noad, K. B., and Latham, O. (1941). Familial cortical cerebellar atrophy. *Brain* **64**, 178–94.

Hall, J.G., Reed, S.D., and Driscoll, E.P. (1983). Amyoplasia: a common sporadic condition with congenital contractures. *Am. J. Med. Genet.* **15**, 571–90.

Hallervorden, J. (1957). Die degenerative diffuse sklerose. In *Handbuch spez. path. Anat. Histol.* Bd XIII/1A, (ed. F. Henke and O. Lubarsch), pp. 716–82. Springer, Berlin.

Hallervorden, J. and Spatz, H. (1922). Eigenartige Erkrankung in extrapyramidalen system mit besonderer Beteiligung des Globus pallidus und der Substantia nigra. *Z. ges Neurol. Psychiat.* **79**, 254–302.

Halley, D. and Heukels-Dully, M.J. (1977). Rapid prenatal diagnosis of the Lesch–Nyhan syndrome. *J. med. Genet.* **14**, 100–2.

Hallgren, B. (1959). Retinitis pigmentosa combined with congenital deafness: with vestibular-cerebellar ataxia and mental abnormality in a proportion of cases. A clinical and geneticostatistical study. *Archs Psychiat. Scand.* Suppl. 138, 1–101.

Halliday, A.M. (1967). The clinical incidence of myoclonus. In *Modern trends in neurology,* Vol. 4 (ed. D. Williams), pp. 69–105. Butterworths, London.

Halliday, J., Chow, C.W., Wallace, D., and Danks, D.M. (1986). X-linked hydrocephalus: a survey of a 20 year period in Victoria, Australia. *J. Med Genet* **23**, 23–31.

Halmagyi, G.M. Rudge, P., and Gresty, M.A. (1983). Downbeat nystagmus. *Arch, Neurol.* **40**, 777–84.

Halsey, J.H., Scott, T.R., and Farmer, T.W. (1967). Adult onset hereditary cerebello retinal degeneration. *Neurol., Minneap.* **17**, 87–90.

Haltia, T., Kristensson, K., and Sourander, P. (1969). Neuropathological studies in three Scandinavian cases of progressive myoclonus epilepsy. *Acta neurol, scand.* **45**, 63–77.

Haltia, T., Rapola, J., Santovuori, P., and Keränen, A. (1973). Infantile type of so-called neuronal ceroid-lipofuscinosis. Part 2. Morphological and biochemical studies. *J. neurol. Sci.* **18**, 269–85.

Haltia, M., Kovanen, J., Van Crevel, H., Bots, G. Th. A.M., and Stefanko, S. (1979). Familial Creutzfeldt–Jakob disease. *J. neurol. Sci.* **42**, 381–9.

Haltia, T., Palo, J., Haltia, M., and Icen, A. (1980). Juvenile metachromatic leukodystrophy. *Arch. Neurol., Chicago* **37**, 42–6.

Hambert, O. and Petersen, I. (1970). Clinical, electroencephalographical and neuropharmacological studies in syndromes of progressive myoclonus epilepsy. *Acta neurol. scand.* **46**, 149–86.

Hanhart, E. (1962). Die genealogie der 6 sicheren und 4 wahrschein lichen Falle von neurogene proximale amyotrophie (Kugelberg–Welander) in einer Sippe aus dem Isolat 1 (Kanton Schwyz). *Arch. Klaus-Sift. Veserb-Forsch.* **37**, 175.

Hans, M.B. and Gilmore, T.H. (1968). Social aspects of Huntington's chorea. *Br. J. Psychiat.* **114**, 93–8.

Hansen, R.L., Marx, J., Ptacek, L.J., and Roberts, R.C. (1977). Immunological studies on an aberrant form of ataxia telangiectasia. *Am. J. Dis. Child.* **131**, 518–21.

Hanson, J.W. and Smith, D.W. (1975). The fetal hydantoin syndrome. *J. Pediat.* **87**, 285–90.

Hanson, P.A. and Mincy, J.E. (1975). Adolescent familial cramps. *Neurol., Minneap.* **25**, 454–8.

Hanson, P.A. and Rowland, L.P. (1971). Mobius syndrome and facioscapulohumeral muscular dystrophy. *Arch. Neurol., Chicago.* **24**, 31–9.

Haque, I.U. and Glasauer, F.E. (1969). Hydranencephaly in twins. *N. Y. J. Med.* **69**, 1210–14.

Harding, A.E. (1981*a*). Genetic aspects of autosomal dominant late-onset cerebellar ataxia. *J. med. Genet.* **18**, 436–41.

Harding, A.E. (1981*b*). Early-onset cerebellar ataxia with retained tendon reflexes: a clinical and genetic study of a disorder distinct from Friedreich's ataxia. *J. Neurol. Neurosurg. Psychiat.* **44**, 503–8.

Harding, A.E. (1981*c*). Friedreich's ataxia: a clinical and genetic study of 90 families with a analysis of early diagnostic criteria and intrafamilial clustering of clinical features. *Brain* **104**, 589–620.

Harding, A.E. (1981*d*). 'Idiopathic' late-onset cerebellar ataxia: a clinical and genetic study of 36 families. *J. neurol. Sci.*

Harding, A.E. (1981*e*). Hereditary 'pure' spastic paraplegia: a clinical and genetic study of 22 families. *J. Neurol. Neurosurg, Psychiat.*

Harding, A.E. (1981*f*). The clinical features and classification of the late-onset autosomal dominant cerebellar ataxias: a study of 11 families, including descendants of 'the Drew family of Walworth'.

Harding, A.E. (1981*g*). Hereditary 'pure' spastic paraplegia: a clinical and genetic study of 22 families. *J. Neurol. Neurosurg. Psychiat.* **44**, 871–83.

Harding, A.E. and Thomas, P.K. (1980*a*). Distal spinal muscular atrophy: a report of 34 cases and a review of the literature. *J. neurol. Sci.* **45**, 337–48.

Harding, A.E. and Thomas, P.K. (1980*b*). The clinical features of hereditary motor and sensory neuropathy types I and II. *Brain* **103**, 259–80.

Harding, A.E. and Thomas, P.K. (1980*c*). Genetic aspects of hereditary motor and sensory neuropathy types I and II. *J. med. Genet.* **17**, 329–36.

Harding, A.E. and Thomas, P.K. (1980*d*). Autosomal recessive forms of hereditary motor and sensory neuropathy. *J. Neurol. Neurosurg. Psychiat.* **43**, 669–78.

Harding, A.E. and Thomas, P.K. (1984). Peroneal muscle atrophy with pyramidial features. *J. Neurol. Neurosurg. Psychiat.* **47**, 168–72.

Harding, A.E. and Zilka, K.J. (1981). Pseudodominant inheritance in Friedreich's ataxia. *J. med. Genet.* **18**, 285.

Harding, A. E., *et al.* (1985). Spinocerebellar degeneration associated with a selective defect of vitamin E absorption. *N. Engl. J. Med.* **313**, 32–5.

Harding, A. E., Young, E. P. and Schon, F. (1987). Adult onset supranuclear ophthalmoplegia, cerebellar ataxia and neurogenic proximal muscular weakness in a brother and sister: another hexosaminidase A deficiency syndrome. *J. Neurol. Neurosurg. Psychiat.* **50**, 687–90.

Harding, B. N., *et al.* (1986). Progressive neuronal degeneration of childhood with liver disease. *Brain* **109**, 181–206.

Harding, B. N., *et al.* (1988). Familial olivopontocerebellar atrophy with neonatal onset: a recessively inherited syndrome with systemic and biochemical abnormalities. *J. Neurol. Neurosurg. Psychiat.* **51**, 385–90.

Harel, S., Chui, L. A., and Shapira, Y. (1979). Myotonia congenita (Thomsen's disease). *Acta pediat, scand.* **68**, 225–7.

Harenko, A. and Toivakka, E. I. (1961). Myoclonus epilepsy (Unverricht–Lundborg) in Finland. *Acta neurol. scand.* **37**, 282–96.

Hariga, J. (1959). Hérédo-ataxie du type spastique avec ophthalmoplegie chronique progressive. *J. Génét. hum.* **8**, 61–71.

Harper, C., and Hockey, A. (1983). Proliferative vasculopathy and a hydranencephalic-hydrocephalic syndrome: a neuropathological study of two siblings. *Dev. Med. Child. Neurol.* **25**, 232–44.

Harper, P. S. (1971). Sturge–Weber syndrome with Klippel–Trenaunay–Weber syndrome. In *The clinical delineation of birth defect,* Vol. XII *Skin, hair and nails,* (ed. D. Bergsma) p. 314. Williams and Wilkins, Baltimore.

Harper, P. S. (1973). Pre-symptomatic detection and genetic counselling in myotonic dystrophy. *Clin. Genet.* **4**, 134–40.

Harper, P. S. (1975a). Congenital myotonic dystrophy in Britain 1. Clinical aspects. *Arch. Dis. Childh.* **50**, 505–13.

Harper, P. S. (1975b). Congenital myotonic dystrophy in Britain II. Genetic basis. *Arch. Dis. Childh.* **50**, 514–21.

Harper, P. S. (1977). Mendelian inheritance or transmissable agent?—the lesson of Kuru and the Australia antigen. *J. med. Genet.* **14**, 389–98.

Harper, P. S. (1978). Benign hereditary chorea. *Clin. Genet.* **13**, 85–95.

Harper, P. S. and Johnston, D. M. (1972). Recessively inherited myotonia congenita. *J. med. Genet.* **9**, 213–15.

Harper, P. S., Laurence, K. M. Parkes, A., Wusteman, F. S., Kresse, H., Von Figura, K., Ferguson-Smith, M. A., Duncan, D. M., Logan, R. W., Hall, F., and Whiteman, P. (1974). Sanfilippo A disease in the fetus. *J. med. Genet.* **11**, 123–32.

Harper, P. S., Walker, D. A., Tyler, A., Newcombe, R. G., and Davies, K. (1979). Huntington's chorea. The basis for long-term prevention. *Lancet* **ii**, 346–9.

Harper, P. S., Marks, R., Dykes, P. J., and Young, I. D. (1980). Ichthyosis hepatosplenomegaly and cerebellar degeneration in a sibship. *J. med. Genet.* **17**, 212-15.

Harpey, J.P. Charpewtier, C., Goodman, S.I., *et al.* (1983). Multiple acyl–CoA dehydrogenase deficiency occurring in pregnancy and caused by a defect in riboflavin metabolism in the mother. Study of a kindred with seven deaths is infancy. Value of riboflavin therapy is preventing this syndrome. *J. Ped.* **103**, 394-8.

Harriman, D.G.F. and Millar, J.H.D. (1955). Progressive familial myoclonic epilepsy in three families: its clinical features and pathological basis. *Brain* **78**, 325-49.

Harriman, D.G.F., Sumner, D.W., and Ellis, F.R. (1973). Malignant hyperpyrexia myopathy. *Q. J. Med.* **42**, 639-64.

Harris, W. (1936). Bilateral trigeminal tic: its association with hereditary and disseminated sclerosis. *Ann. Surg,* **103**, 161-72.

Hart, K.A., *et al.* (1987). DNA deletions in mild and severe Becker muscular dystrophy. *Hum. Genet.* **75**, 281-5.

Hart, M.N., Malamud, N., and Ellis, W.G. (1972). The Dandy–Walker syndrome: a clinico-pathological study based on 28 cases. *Neurol., Minneap.* **22**, 771-80.

Hart, Z.H., Chang, C.-H., Perrin, E.V.D., Neerunjun, J.S., and Ayyar, R. (1977). Familial poliodystrophy mitochondrial myopathy and lactate acidemia. *Arch. Neurol., Chicago* **34**, 180-5.

Hartung, H.-P., *et al.* (1988). Familial multiple sclerosis. *J. Neurol. Sci.* **83**, 259-68.

Harvald, B. (1954). *Hereditary in epilepsy. An electroencephalographic study of relatives of epileptics.* Egnar, Munksjaard, Copenhagen.

Harvald, B. and Hauge, M. (1956). On the heredity of glioblastoma. *J. Nat. Cancer Inst.* **17**, 289-96.

Harvald, B. and Hauge, N. (1965). Hereditary factors elucidated by twin studies. In *Genetics and the epidemiology of chronic diseases* (ed. J.V. Neel, M.W. Shaw, and W.J. Shull), pp. 61-76. Washington.

Harzer, K. (1977). Prenatal diagnosis of globoid cell leukodystrophy (Krabbe's disease) *Hum. Genet.* **35**, 193-6.

Harzer, K. and Recke, A.S. (1975). Sulfatide excreting heterozygous carrier of juvenile metachromatic leukodystrophy of asymptomatic patient of adult metachromatic leukodystrophy. *Hum. Genet.* **29**, 299-307.

Harzer, K., Schlote, W., Peiffer, J., Benz, H.U., and Anzil, A.P. (1978). Neurovisceral lipidosis compatible with Niemann–Pick disease type C: morphological and bichemical studies of a late infantile case and enzyme and lipid assays in a prenatal case of the same family. *Acta neuropath.* **43**, 97-104.

Harzer, K., *et al.* (1975). Prenatale diagnose der Gm_2-gangliosidose Type 2. *Dt. med. Wschr.* **100**, 106-8.

Harzer, K., *et al,* (1987). Prenatal enzymatic diagnosis and exclusion of

Krabbe's disease (globoid-cell leukodystrophy) using chorionic villi in five at risk pregnancies. *Hum. Genet.* **77**, 342–4.

Haschinski, C., Lasson, N.A., and Marshall, J. (1974). Multi-infarct dementia. A case of mental deterioration in the elderly. *Lancet* **ii**, 207.

Hashimoto, I. (1977). Familial intracranial aneurysms and cerebral vascular anomalies. *J. Neurosurg.* **46**, 419–27.

Hashimoto, T., Minato, H., Kuroda, Y., Toshima, K., Ohara, K., and Miyao, M. (1978). Monozygotic twins with presumed metachromatic leukodystrophy. *Arch. Neurol.* **35**, 689–91.

Hascovec, V. (1935). Picksov choroba. *Rev. teheque. Neurol. Psychiat.* **31**. 42–72. [*Abs. Ann. Med. Psychol.* **93**, 534 (1935).]

Haslam, R.H.A. and Smith, D.W. (1979). Autosomal dominant microcephaly. *J. Pediat.* **95**, 701–5.

Haukipuro, K., Keränen, N., Koivisto, E., Lindholm, R., Norio, R., and Punto, L. (1978). Familial occurence of lumbar spondylolysis and spondylolisthesis. *Clin. Genet.* **13**, 471–5.

Hauptmann, A. and Thannhauser, S.J. (1941). Muscular shortening and dystrophy. A heredofamilial disease. *Arch. Neurol. Psychiat.* **46**, 654–64.

Hausmanowa-Petrusewicz, I. (1978). *Spinal muscular atrophy.* US Department of Commerce, Springfield.

Hausmanowa-Petrusewicz, I. and Borkowska, J. (1978). Intrafamilial variability of X-linked progressive muscular dystrophy. Mild and acute form of X-linked muscular dystrophy in the same family. *J. Neurol.* **218**, 43–50.

Hausmanowa-Petrusewicz, I. and Zielinska, S. (1962). Zur nosologischen Stellung des scapulo-peronealen syndrome. *Dt. Z. Nervenhfeilk.* **183**, 377–82.

Hausmanowa-Petrusewicz, I., *et al.* (1984). Chronic proximal spinal muscular atrophy of childhood and adolescent: sex influence. *J. Med. Genet.* **21**, 447–50.

Havener, W.H. (1951). Cerebellar-macular abiotrophy. *Arch. Ophthal., N.Y.* **45**, 40.

Hawke, W.A. and Donohue, W.L. (1950). Bilateral symmetrical necrosis of the corpora striata. Report of a fatal case and reference to a possible syndrome of the corpora striata. *J. nerv. ment. Dis.* **113**, 20–39.

Hawkes, G.H. and Absolon, M.J. (1975). Myotubular myopathy associated with cataract and electrical myotonia. *J. Neurol. Neurosurg. Psychiat.* **38**, 761–4.

Hayden, M.R., *et al.* (1987). First-trimester prenatal diagnosis for Huntington's disease with DNA probes. *Lancet* **i**, 1284–5.

Hayden, M.R., Soles, J.A., and Ward, R.H. (1985). Age of onset of siblings of persons with juvenile Huntington's disease. *Clin. Genet.* **28**, 100–5.

Hayman, L.A., Evans, R.A., Ferrell, R.E., Fahr, L.M., Ostow, P. and

Riccardi, V.M. (1982). Familial cavernous angiomas: natural history and genetic study over a five-year period. *Am. J. med. Genet.* **ii**, 147–60.

Hazama, R., Tsujihata, M., Mori, M., and Mori, K. (1979). Muscular dystrophy in 6 young girls. *Neurol., Minneap.* **29**, 1486–91.

Heathfield, K.W.G. (1967). Huntington's chorea – investigations into the prevalence of this disease in the area covered by the N.E. Metropolitan Regional Hospital Board. *Brain* **90**, 203–32.

Heck, A.F. (1964). A study of neural and extraneural findings in a large family with Friedreich's ataxia. *J. neurol. Sci.* **1**, 226–55.

Heckmatt, J.Z., Sewry, C.A., Hodes, D., and Dubowitz, V. (1985). Congenital centronuclear (myotubular) myopathy. A clinical, pathological and genetic study in eight children. *Brain* **108**, 941–64.

Hed, R. (1955). Myoglobinuria in man with special reference to a familial form. *Acta med. scand.* Suppl. 303, 1–107.

Heffner, R.R., Cohen, M., Duffner, P., and Daigler, G. (1976). Multicore disease in twins. *J. Neurol. Neurosurg. Psychiat.* **39**, 602–6.

Heijbel, J., Blom, S., and Rasmuson, M. (1975). Benign epilepsy of childhood with centrotemporal EEG foci: a genetic study. *Epilepsia* **16**, 285–93.

Heijer, A. and Reed, W.B. (1965). Sjögren–Larsson syndrome. *Arch. Derm.* **92**, 545–52.

Heller, I.H. and Robb. P. (1955). Hereditary sensory neuropathy. *Neurol., Minneap.* **5**, 15–29.

Hellsing, G. (1930). Hereditärer facialiskrampf. *Acta med. scand.* **73**, 526–37.

Hempel, H.-C. (1938). Ein Beitrag zur Huntingtonschen Erkrankung. *Z. ges. Neurol. Psychiat.* **160**, 563–97.

Hens, L. and Carton, H. (1978). H.L.-A determinants and familial multiple sclerosis. H.L.-A typing of 13 families with at least two affected members. *Tiss. Antigen* **11**, 75–80.

Henson, T.E., Muller, J., and De Myer, W.E. (1967). Hereditary myopathy limited to females. *Arch. Neurol.* **17**, 238–47.

Herlitz, G. (1941). Studien uber die sogenannten initialen Fieberkrampfe bei Kindren. *Acta paediat. scand.* **229**, Suppl. 1.

Herman, M., Rubinstein, L.J., and McKhann, G.M. (1971). Additional electron microscopic observations on two cases of Batten–Spielmeyer–Vogt disease, (neuronal ceroid-lipofuscinosis). *Acta neuropath.* **17**, 85–102.

Hermann, K. (1956). Congenital word blindness. *Acta psychiat. neurol. scand.* Suppl. 103, 1–138.

Hermanussen, M. and Sippell, W.G. (1985). Heterogeneity of Kallmann's syndrome. *Clin. Gen.* **28**, 106–11.

Herrlin, K.-M. and Hillborg, P.O. (1962). Neurological signs in a juvenile form of Gaucher's disease. *Acta paediat. scand.* **51**, 137–54.

Herrmann, C., Aguilar, M.J., and Sacks, O.W. (1964). Hereditary photomyoclonus associated with diabetes mellitus, deafness, nephropathy and cerebral dysfunction. *Neurol., Minneap.* **14**, 212–21.

Herrmann, J. and Opitz, J.M. (1969). An unusual form of acrocephalosyndactyly. *Birth Defects* **5**, 39–42.

Hermann, J., Pallister, P.D., and Opitz, J.M. (1969). Craniosynostosis and craniosynostosis syndromes. *Rocky Mount, med. J.* **66**, 45–56.

Hers, H.G., (1963). α-Glucosidase deficiency in generalised glycogen storage disease (Pompe's disease). *Biochem. J.* **86**, 11–16.

Hersh, J.H., Podruch, P.E., Weisskopf, B. (1982). Pigmentary retinopathy, hearing loss, mental retardation, and dysmorphism in sibs: a new syndrome. *BDOAS* **18**, 175–82.

Herskovits, E. and Blackwood, W. (1969). Essential (familial hereditary) tremor, a case report. *J. Neurol. Neurosurg. Psychiat.* **32**, 509–11.

Herva, R., Von Wendt, L., Von Wendt, G., *et al.* (1987). A syndrome with juvenile cataract, cerebellar atrophy, mental retardation and myopathy. *Neuroped.* **18**, 164–9.

Herzberg, J.J. and Wiskemann, A. (1963). Die fuenfte Phakomatose. Basalzellnaevus mit familaeref Belastun und Medulloblastom. *Dermatologica.* **126**, 106–23.

Herzberg, L. (1980). Familial trigeminal neuralgia. *Arch. Neurol.* **37**, 285–6.

Heston, L.L. (1978). The clinical genetics of Picks disease. *Acta psychiat. scand.* **57**, 202–6.

Heston, L.L. (1980). Dementia associated with Parkinson's disease: a genetic study. *J. Neurol. Neurosurg. Psychiat.* **43**, 846–8.

Heston, L.L. and Mastri, A.R. (1977). The genetics of Alzheimers disease. *Arch. gen. Psychiat.* **34**, 976–81.

Hewitt, P.H. (1957). Addison's disease occuring in sisters. *Br. med. J.* **ii**, 1530–1.

Hicks, E.P. (1922). Hereditary perforating ulcer of the foot. *Lancet* **i**, 319–21.

Hicks, A.M. (1943). Congenital paralysis of lateral rotators of eyes with paralysis of muscles of face. *Arch. Ophthal.* **30**, 38–42.

Hidano, A., Kajima, H., Ikeda, S., Mizutani, Miyasato, H., and Niimura, M. (1967) Natural history of nevus of Ota. *Arch. Derm.* **95**, 187–95.

Hierons, R. (1956). Familial peroneal muscular atrophy and its association with the familial ataxias and tremor and longevity. *J. Neurol. Neurosurg. Psychiat.* **19**, 155–60.

Hilat, S.K., Solomon, G.E., Gold, A.P., and Carter, S. (1971). Primary cerebral arterial occlusive disease in children. I. Acute acquired hemiplegia. *Radiology* **99**, 87–93.

Hill, R.M. (1977). Toxicity of phenothizine derivatives in children with family history of Parkinson's disease. (*J. Pediat*). **90**, 667.

Hill, W. and Shérman, H. (1968). Acute intermittent familial cerebellar ataxia. *Arch. Neurol.* **18**, 350–7.

Hillman, J.W. and Johnson, J.T.H. (1952). Arthrogryposis multiplex congenita in twins. *J. Bone Jt Surg.* **34A**, 211–14.

Hilty, M.D., McClung, H.J., Haynes, R.E., Romshe, C.A., and Sherard, E.S. (1979). Reye syndrome in siblings. *J. Pediat.* **94**, 576–9.

Hintz, R.L., Menking, M., and Sotos, J.F. (1968). Familial holoprosencephaly with endrocrine dysgenesis. *J. Pediat.* **72**, 81–6.

Hirajama, K. (1972). Juvenile non-progressive muscular atrophy localised in the hand and forearm: observations in 38 cases. *Clin. Neurol.* **12**, 313–24.

Hirano, A., Kurland, L.T., and Sayre, G.P. (1967). Familial amyotrophic lateral sclerosis. *Arch. Neurol.* **16**, 232–43.

Hirose, G. and Bass, N.H. (1972). Metachromatic leukodystrophy in the adult: a biochemical study. *Neurol., Minneap.* **22**, 312–20.

Hirschowitz, B.I., Groll, A., and Ceballos, R. (1972). Hereditary nerve deafness in three sisters with absent gastric motility, small bowel diverticulitis and ulceration, and progressive sensory neuropathy. The Clinical Delineation of Birth Defects XIII. G.I. tract including Liver and Pancreas. *Birth Defects: Original Article Series* **VIII**, 27–41.

Hobolth, N. and Pedersen, C. (1978). Six cases of a mild form of the Hunter syndrome in five generations. Three affected males with progeny. (Abst.) *Clin. Genet.* **13**, 121.

Hochberg, F.H., Nelson, K., and Janzen, W. (1975). Influenza and type B related encephalopathy. The 1971 outbreak of Reye''s syndrome in Chicago. *J. Am. med. Ass.* **231**, 817–21.

Hockaday, T.D.R. (1966). Hypogonadism and life-long anosmia. *Postgrad. med. J.* **42**, 572–4.

Hodgson, S., *et al.* (1986). DNA deletion in boy with Becker muscular dystrophy. *Lancet* i, 918.

Hodgson, S.V., *et al.* (1985). Two cases of X/autosome translocation in females with incontinentia pigment. *Hum. Genet.* **71**, 231.

Hoefnagel, D. and Biery, B. (1968). Spasmus nutans. *Dev. med. Child Neurol.* **10**, 32–5.

Hoefnagel, D, Van den Noort, S., and Ingbar, S.H. (1962). Diffuse cerebral sclerosis with endocrine abnormalities in young males. *Brain* **85**, 553–68.

Hoefnagel, D., Allen, F.H., and Falk, C. (1970). Hereditary dystonia musculorum deformans. *Clin. Genet.* **1**, 258–62.

Hoes, M.J., Lamers, K.J., Hommes, O.R., and ter Haar, B. (1978). Adult metachromatic leokodystrophy. Arylsulphatase A values in four generations of one family with some reflections about the genetics. *Clin. Neurol. Neurosurg.* **80**, 174–88.

Hoffman, H.N. and Fredrickson, D.S. (1965). Tangier disease (familial high density lipoprotein deficiency): clinical and genetic features in two adults. *Am. J. Med.* **39**, 582–93.

Hoffman, P.M., Stuart, W.H., Earle, K.M., and Brody, J.A. (1970). Hereditary cerebello-olivary degeneration of late onset. (Abst.) *Neurol., Minneap.* **20**, 400.

Hogan, G.R. and Bauman, M.L. (1977). Familial spastic ataxia: occurence in childhood. *Neurol., Minneap.* **27**, 520–6.

Hogan, G.R. and Richardson, E.P. Jr (1965). Spongy degeneration of the nervous system (Canavan's disease) Report of a case in an Irish-American family. *Pediatrics. Springfield.* **35**, 284–94.

Holliday, F.L., *et al.* (1983). Mitochondrial myopathy and encephalopathy: three cases, a deficiency of NADH-COQ dehydrogenase. *Neurology* **33**, 1619–22.

Hollis, R.J., Kennaugh, A.A., Butterworth, S.V., and Taylor, A.M.R. (1987). Growth of large chromosomally abnormal T cell clones in ataxia telangiectasia patients is associated with translocation at 14q11. *Hum. Genet.* **76**, 389–95.

Holmes, C. and Flaherty, R.J. (1976). Trimethobenzamide HCI (Tigan)-induced extrapyramidal dysfunction in a neonate. *J. Pediat.* **89**, 669–70.

Holmes, G. (1905). Familial spastic paralysis associated with amyotrophy. *Rev. Neurol. Psychiat.* **3**, 256–63.

Holmes, G. (1907*a*). A form of familial degeneration of the cerebellum. *Brain* **30**, 466–89.

Holmes, G. (1907*b*). An attempt to classify cerebellar disease with a note on Marie's hereditary cerebellar ataxia. *Brain* **30**, 545–67.

Holmes, G.L. and Shaywitz, B.A. (1977). Strümpell's pure familial spastic paraplegia: cases study and review of the literature. *J. Neurol. Neurosurg. Psychiat.* **40**, 1003–8.

Holmes, L.B. and Schepens, C.L. (1972). Syndrome of ocular and facial anomalies, telecanthus and deafness. *J. Pediat.* **81**, 552–5.

Holmes, L.B., Nash, A., ZuRheim, G.M., Levin, M., and Opitz, J.M. (1973). X-linked aqueductal stenosis: clinical and neuropathological findings in two families. *Pediatrics, Springfield* **51**, 697–704.

Holmes, L.B., Driscoll, S., and Atkins, L. (1974). Genetic heterogeneity of cebocephaly. *J. med. Genet.* **11**, 35–40.

Holmes, W.J. (1956). Hereditary congenital ophthalmoplegia. *Am. J. Ophthal.* **41**, 615–18.

Holt, I.J., Harding A.E., and Morgan-Hughes, J.A. (1988). Deletions of muscle mitochondrial DNA in patients with mitochondrial myopathies. *Nature,* **331**, 717–19.

Honeybourne, D., Dyer, P., and Mohr, P.D. (1982). Familial myasthenia gravis. *J. Neurol. Neurosurg. Psychiat.* **45**, 854–6.

Hooft, C., *et al.* (1968). Further investigations in the methionine malabsorption syndrome. *Helv. paediat. Acta.* **23**, 334–9.

Hootnick, D. and Holmes, L.B. (1972). Familial polysyndactyly and craniofacial anomalies. *Clin. Genet.* **3**, 128–34.

Hopf, H.C. and Port, F. (1968). Friedreichsche Ataxie mit Beteiligung der peripheren Nerven nach Art der Neuralen Muskelatrophie (Myatrophische ataxie). *Dt. Z. Nervenheilk.* **194**, 1–16.

Hopf, H.C. and Volles, E. (1972). Spinocerebellar ataxia with neural myatrophy. *Neuropaediatrie* **12**, 97–105.

Hopkins, I.J. and Turner, B. (1973). Spongy glio-neuronal dystrophy: a degenerative disease of the nervous system. *J. Neurol. Neurosurg. Psychiat.* **36**, 50–6.

Hopkins, I.J., Linsey, J.R., and Ford, F.R. (1966). Nemaline myopathy. A long term clinicopathologic study of affected mother and daughter. *Brain* **89**, 299–310.

Hopkins, L.C. and Karp, H.R. (1976). X-linked recessive humero-peroneal neuromuscular disease associated with atrial paralysis and sudden death. (Abstract.) *Arch. Neurol., Chicago* **33**, 386.

Hopkins, L.C., Jackson, J.A., Elsas, L.J. (1981). Emery-Dreifuss humeroperoneal muscular dystrophy, an X-linked myopathy with unusual contractures and bradycardia. *Ann. Neurol.* **10**, 230–7.

Horan, F.T. and Beighton, P.H. (1978). Osteopathia striata with cranial sclerosis. An autosomal dominant entity. *Clin. Genet.* **13**, 201–6.

Hormia, M. (1978). Diffuse cerebral sclerosis, melanoderma and adrenal insufficiency (adreno-leukodystrophy) *Acta neurol. scand.* **58**, 128–33.

Horn, N. (1983). Menkes X-linked disease: prenatal diagnosis and carrier detection. *J. Inher. Metab. Dis.* **6**, 59–62.

Horn, N., Heydorn, K., Damsgaard, E., Tygstrup, I., and Vestermark, S. (1978). Is Menkes syndrome a copper storage disorder? *Clin. Genet* **14**, 186–7.

Hornabrook, R.W. and Nagurney, J.T. (1976). Essential tremor in Papua New Guinea. *Brain* **99**, 659–72.

Horoupian, D.S., Zucker, D.K., Moshe, S., and Peterson, H. De C. (1979). Behr syndrome: a clinicopathologic report. *Neurology, Minneap.* **29**, 323–7.

Horowitz, G. and Greenberg, J. (1975). Pallido-pyramidal syndrome treated with levodopa. *J. Neurol. Neurosurg. Psychiat.* **38**, 238–40.

Horton, W.A., and Schimke, R.N. (1970). A new mucopolysaccharidosis. *J. Pediat.* **77**, 252–8.

Horton, W.A., Eldridge, R., and Brody, J.A. (1976a). Familial motor neuron disease. Evidence for at least three different types. *Neurol., Minnep.* **26**, 560–5.

Horton, W.A., Wong, V., and Eldridge, R. (1976b). Von Hippel–Lindau disease: clinical and pathological manifestations in nine families with 50 affected members. *Arch. intern. Med.* **136**, 769–77.

Hostetler, K.Y., Hoppel, C.L., Romine, J.S., Sipe, J.C., Gross, S.R., and Higginbottom, P.A. (1978). Partial deficiency of muscle carnitine palmitoyltransferase with normal ketone production. *N. Engl. J. Med.* **298**, 553–7.

Houdou, S., *et al.* (1986). Joubert syndrome associated with unilateral ptosis and Leber congenital amaurosis. *Pediat. Neurol.* **2**, 102–5.

Housepian, E.M. and Pool, J.L. (1958). A systematic analysis of intracranial aneurysms from the autopsy file of the Presbyterian Hospital. *J. Neuropath exp. Neurol.* **17**, 409.

Houston, A.B., Brodie, M.J., Moore, M.R., Thompson, G.G. and Stephenson, J.B. (1977). Hereditary coproporphyria and epilepsy. *Arch. Dis. Child.* **52**, 646–50.

Howard, F.M., Till, K., and Carter, C.O. (1981). A family study of hydrocephalus due to aqueduct stenosis. *J. med. Genet.* **18**, 252–5.

Howard, R.O. and Albert, D.M. (1972). Ocular manifestations of subacute necrotizing encephalomyelopathy. Leigh's disease. *Am. J. Ophthal.* **74**, 386–93.

Howe, J., Saunders, M., and Clarke P. (1973). Familial benign intracranial hypertension. *Acta neurochir.* **29**, 173–5.

Hoyt, C.S. and Aicardi, E. (1979). Acquired mono-ocular nystagmus in monozygotic twins. *J. pediat. Ophthal. Strabismus.* **16**, 115–18.

Hoyt, W.F. (1960). Charcot–Marie–Tooth disease with primary optic atrophy. *Arch. Ophthal.* **64**, 145–8.

Hrbek, A. (1957). Beitrag zur pathogenese der Infeltkrämfe. *Ann. Paediat.* **50**, 191.

Hsia, Y.E., Bratu, M., and Herbordt, A. (1971). Genetics of the Meckel syndrome (dysencephalia splanchnocystica). *Pediatrics, Springfield.* **48**, 237–47.

Huang, K., *et al.* (1983). Adult spinal muscular atrophy. A report of four cases. *J. Neurol. Sci.* **61**, 249–59.

Hudgins, R.L. and Corbin, K.B. (1966). An uncommon seizure disorder: familial paroxysmal choreo-athetosis. *Brain* **89**, 199–204.

Hudgson, P., Gardner-Medwin, D., Pennington, R.J., and Walton, J.N. (1967). Studies of the carrier state in the Duchenne type of muscular dystrophy. Part I. (Effect of exercise on serum creatine kinase activity.) *J. Neurol. Neurosurg. Psychiat.* **30**, 416–19.

Hudgson, P., Gardner-Medwin, D., Worsfold, M., Pennington, R.J.T., and Walton, J.N. (1968). Adult myopathy from glycogen storage disease due to acid maltase deficiency. *Brain* **91**, 435–62.

Hudgson, P., Bradley, W.G., and Jenkison, M. (1972). Familial 'mitochondrial' myopathy. A myopathy associated with disordered oxidative metabolism in muscle fibres. Part I. Clinical electro-physiological and pathological findings. *J. neurol. Sci.* **16**, 343–70.

Hughes, B.P. (1962). Serum enzymes in carriers of muscular dystrophy. *Br. med. J.* **ii**, 963.

Huizinga, J. (1957). Hereditary acromelalgia (or 'restless legs'). *Acta genet. Statist. med.* **7**, 121–3.

Hülsmann, W.C., Bethlem, J., Meijer, A.E.F.H., Fleury, P., and Schellens, J.P.M. (1967). Myopathy with abnormal structure and function of muscle mitochondria. *J. Neurol. Neurosurg. Psychiat.* **30**, 519–25.

Humberstone, P.M. (1972). Nerve conduction studies in Charcot–Marie–Tooth disease. *Acta neurol. scand.* **48**, 176–90.

Hundley, J.D. and Wilson, F.C. (1973). Progressive diaphyseal dysplasia. Review of the literature and report of seven cases in one family *J. Bone Jt Surg.* **55A**, 461–74.

Hunt, A. and Lindenbaum, P.H. (1984). Tuberous sclerosis: a new estimate of prevalence within the Oxford Region. *J. Med. Genet.* **21**, 272–7.

Hunt, A.D., Stokes, J., McCrory, W.W., and Stroud, H.H. (1954). Pyridoxine dependency: report of a case of intractable convulsions in an infant controlled by pyridoxine. *Pediatrics, Springfield* **13**, 140–5.

Hunt, J.R. (1917). Progressive atrophy of the globus pallidus (primary atrophy of the pallidal system). *Brain* **40**, 58–148.

Hunt, J.R. (1921). Dyssynergia cerebellaris myoclonica—primary atrophy of the dentate system: a contribution to the pathology and symptomatology of the cerebellum. *Brain* **44**, 490–538.

Hunt, J.R. (1922). On the occurence of static seizures in epilepsy. *J. nerv. ment. Dis.* **56**, 351–6.

Hunter, A. and Pinsky, L. (1975). An evaluation of the possible association of malignant hyperpyrexia with the Noonan syndrome using serum creatin phosphokinase levels. *J. Pediat.* **86**, 412–15.

Hunter, A.G.W. and Rudd, N.L. (1976). Craniosynostosis I. Saggital synostosis: its genetics and associated clinical findings in 214 patients who lacked involvement of the coronal suture(s). *Teratology* **14**, 185–93.

Hunter, A.G.W. and Rudd, N.L. (1977). Craniosynostosis. II Coronal synostosis. Its familial characteristics and associated clinical findings in 109 patients lacking bilateral polysyndactyly or syndactlyly. *Teratology* **15**, 301–9.

Hunter, A.G.W., Rudd, N.L. and Hoffmann, H.J. (1976). Trigonocephaly and associated minor anomolies in mother and son. *J. med. Genet.* **13**, 77–9.

Hunter, J.B. and Critz, J.B. (1971). Effect of training on plasma enzyme levels in man. *J. appl. Physiol.* **31**, 20–3.

Hunter, R., Dayan, A.D., and Wilson, J. (1972). Alzheimer's disease in one monozygotic twin. *J. Neurol. Neurosurg. Psychiat.* **35**, 707–10.

Hurst, J. and Baraitser, M. (1988). Hereditary neurocutaneous angiomatous malformations: autosomal dominant inheritance in two families. *Clin. Genet.* **33**, 44–8.

Hurwitz, L.J., Lyttle, J.A., and Neill, D.W. (1965). Muscular dystrophy with a familial aminoaciduria of unusual pattern. *Lancet* **ii**, 722–3.

Hurwitz, S. and Braverman, I.M. (1970). White spots in tuberous sclerosis. *J. Pediat.* **77**, 587–94.

Huson, S.M., *et al.* (1986). Cerebellar haemangioblastoma and Von Hippel–Lindau disease. *Brain* **109**, 1297–310.

Huson, S.M., Harper, P.S. and Compston, D.A.S. (1988). Von Recklinghausen neurofibromatosis: a clinical and population study in South East Wales. *Brain*.

Husquinet, H. and Franck, G. (1980). Hereditary amyotrophic lateral sclerosis transmitted for five generations. *Clin. Genet.* **18**, 109–15.

Hutchison, J.H. and Hamilton, W. (1962). Familial dysautonomia in two siblings. *Lancet* **i**, 1216–18.

Huttenlocher, P.R. and Gilles, F.H. (1967). Infantile dystrophy. Clinical, pathologic and histochemical findings in a family with three affected siblings. *Neurol., Minneap.* **17**, 1174–84.

Huttenlocher, P.R., Landwirth, J., Hanson, V., Gallagher, B.B., and Bensch, K. (1969*a*). Osteochondro-muscular dystrophy. A disorder manifested by multiple skeletal deformities myotonia and dystrophic changes in muscle. *Pediatrics, Springfield* **44**, 945–58.

Huttenlocher, P.R., Schwartz, A.D., Klatskin, G. (1969*b*). Reye's syndrome: ammonia intoxication as a possible factor in encephalopathy. *Pediatrics, Springfield* **43**, 443.

Huttenlocher, P.R., Solitare, G.B., and Adams, G. (1976). Infantile diffuse cerebral degeneration with hepatic cirrhosis. *Arch. Neurol., Chicago* **33**, 186–92.

Hyllested, K. (1956). *Disseminated sclerosis in Denmark*. Jorgensen, Copenhagen.

Iannaccone, T., Griggs, R.C., Markesbery, W.R., and Joynt, R.J., (1974). Familial progressive external ophtalmoplegia and ragged red fibres. *Neurol., Minneap.* **24**, 1033–8.

Iancu, T.C., Almagor, G., and Savir, H. (1980). Ocular abnormalities in chronic familial hyperphosphatasemia. *J. pediat. Ophthal. Strabimus* **17**, 220–3.

Iancu, T., Komlos, L., and Shabtay, F. (1975). Incontinentia pigmenti. *Clin. Genet.* **7**, 103–10.

IIzuka, R.V. Hirayama, K., and Maekawa, K. (1984). Dentato-rubro-pallido-luysian atrophy: a clinic-pathological study. *J. Neurol. Neurosurg. Psychiat.* **47**, 1288–98.

IIkkos, D.G., Fraser, G.R., Matsouki-Gavra, E., and Petrochilos, M. (1970). Association of juvenile diabetes mellitus, primary optic atrophy and perceptive hearing loss in three sibs, with additional idiopathic diabetes mellitus insipidus in one case. *Acta endocr., Copenh.* **65**. 95–102.

Ilingworth, R.S. (1971). Myositis ossificans progressiva (Munchmeyer's disease). *Arch Dis. Childh.* **46**, 264–8.

Imahori, S., Bannerman, R.M., Grat, C.J., and Brennan, J.C. (1969).

Ehlers–Danlos syndrome with multiple arterial lessions. *Am. J. Med.* **47**, 96–77.

Indravasu, S. and Dexter, R. (1968). Infantile neuroaxonal dystrophy and its relationship to Hallervorden–Spatz disease. *Neurol., Minneap.* **18**, 693–9.

Ingram, T.T.S. (1964). *Paediatric aspects of cerebral palsy.* Livingstone, Edinburgh.

Ingwersen, O.S. (1967). Congenital indifference to pain. *J. Bone Jt Surg.* **49B**, 740–9.

Inokuchi, T., Umezaki, H., and Sana, T. (1975). A case of type I muscle fibre hypotrophy and internal nuclei. *J. Neurol. Neurosurg. Psychiat.* **38**, 475–82.

Ionasesco, V. and Drinca-Ionesco, M. (1964). Considerations clinico-satistiques sur la sclerose laterale amyotrohique. *Rev. roum. Neurol.* **1**, 279–94.

Ionasesco, V. and Zellweger, H. (1974). Duchenne muscular dystrophy in young girls? *Acta neurol. scand.* **50**, 619–30.

Ionasescu, V., *et al.* (1983). Giant axonal neuropathy: normal protein composition of neurofilaments. *J. Neurol. Neurosurg. Psychiat.* **46**, 551–4.

Ireland, D.C.R., Takayama, N., and Flatt, A.E., (1976). Poland's syndrome: a review of forty-three cases. *J. Bone Jt Surg.* **58A**, 52–8.

Isaacs, H.(1959). The treatment of myotonia congenita. *S. Afr. Med. J.* **33**, 984–6.

Isaacs, H. and Barlow M.B. (1973). Malignant hyperpyrexia. *J. Neurol. Neurosurg. Psychiat.* **36**, 228–43.

Isaacs, H., Heffron, J.J.A., and Badenhorst, M. (1975). Central core disease—a correlated genetic histochemical ultramicroscopic and biochemical study. *J. Neurol. Neurosurg. Psychiat.* **38**, 1177–86.

Isaacs, H., Heffron, J.J.A., Badenhorst, M. and Pickering, A. (1976). Weakness associated with the pathological presence of lipid in skeletal muscle: a detailed study of a patient with carnitine deficiency. *J. Neurol. Neurosurg. Psychiat.* **39**, 1114–23.

Isamat, F., Miranda, A.M., Bartumeus, F., and Prat, J. (1974). Genetic implications of familial brain tumours. *J. Neurosurg.* **41**, 573–5.

Iseki, E., *et al.* (1987). A case of adult neuronal ceroid-lipofuscinosis with the appearance of membranous cytoplasmic bodies localized in the spinal anterior horn. *Acta Neuropathol.* **72**, 362–8.

Itard, J.M.G. (1825). Memoire sur quelques fonctions involontaires des appareils de la locomation de la prehension et de la voix. *Arch. Gen. Med.* **8**, 385–407.

Ito, M. (1952). Studies of melanin X. I. Incontinentia pigmenti achromians, a singular case of nevus depigmentosus systematicus bilateralis. *Tohuku J. exp. Med.* **55**, Suppl. 57.

Itoh, H., Ohsato, K., Yao, T., Iada, M., and Watanabe, H. (1979). Turcot's syndrome and its mode of inheritance. *Gut* **20**, 414–19.

Itoh Y., *et al.* (1986). Congenital insensitivity to pain with anhidrosis: morphological and morphometrical studies on the skin and peripheral nerves. *Neuropediatrics*. **17**, 103–10.

Iwabuchi, S., Yoshino, Y., Goto, H., Chujo, T., Hagihara, T., and Kawano, K. (1976). Analysis of serum immunoglobulins in hereditary sensory radicular neuropathy. *J. Neurol. Sci.* **30**, 29–32.

Iwashita, H., Inoue, N., Araki, S., and Kuroiwa, Y. (1970). Optic atrophy, neural deafness and distal neurogenic amyotrophy. Report of a family with two affected siblings. *Archs Neurol., Chicago.* **22**, 35–64.

Izawa, K., Suzuki, M. Homma, Y., and Tsubaki, T. (1969). Familial primary amyloid neuropathy. *Clin. Neurol., Tokyo* **9**, 292.

Jackson, A.H. (1934). Familial spastic paralysis. *Arch. Neurol. Psychiat.* **31**, 1266–70.

Jackson, C.E. and Carey, J.H. (1961). Progressive muscular dystrophy: autosomal recessive type. *Pediatrics, Springfield* **28**, 77–84.

Jackson, C.E. and Strehler, D.A. (1968). Limb girdle muscular dystrophy. Clinical manifestations and detection of preclinical disease. *Pediatrics, Springfield* **41**, 495–502.

Jackson, C.E., Weiss, L., Reynolds, W.A., Forman, T.F., and Petersen, J.A. (1976). Craniosynostosis, mid facial hypoplasia and foot abnormalities: an autosomal dominant phenotype in a large Amish kindred. *J. Pediat.* **88**, 963–8.

Jackson, J.F., Currier, R.D., Terasaki, P.I., and Morton, N. (1977). Spino-cerebellar ataxia and HLA linkage. *N. Engl. J. Med.* **296**, 1138–41.

Jackson, M.A. (1949). Familial lumbo-sacral syringomyelia and significance of developmental errors of the spinal cord and column. *Med. J. Aust.* **1**, 433.

Jackson, R.C., *et al.* (1974). Muscular dystrophy: Duchenne type and Becker type within a kindred. *Am. J. Hum. Genet.* **26**, 44A.

Jackson, W.P.U. (1951). Osteo-dental dysplasia (cleido- cranial dysostosis). The 'Arnold' head. *Acta med. scand.* **139**, 292–307.

Jacob, A., Clark, E.R., and Emery, A.E.H. (1968). Genetic study of a sample of 70 patients with myasthenia gravis. *J. med. Genet.* **5**, 257–61.

Jacob, H. (1970). Muscular twitchings in Alzheimer's disease. In *Alzheimer's disease* (ed G.E.W. Wolstenholme and M. Connor) Ciba Foundation Symposium, pp. 75–93. Churchill, Edingburgh.

Jacob, H., Pyrkosch, W., and Strube H. (1950). Hereditary form of Creutzfeldt–Jakob disease (Backer family). *Arch. Psychiat.* **184**, 653–74.

Jacob, J.C., Andermann, F., and Robb, J.P. (1961). Heredofamilial neuritis with brachial predeliction. *Neurol., Minneap.* **11**, 1025–33.

Jacobi, M. (1947). Über leukodystropie und Pelizaus–Merzbachersche Krankheit. *Virch. Arch. Path. Anat. Physiol.* **314**, 460–80.

Jacobs, H. (1965) Myoclonus and ataxia occuring in a family. *J. Neurol. Neurosurg. Psychiat.* **28**, 272–5.

Jaffe, M., Shapira, J., and Borochowitz, Z. (1988). Familial congenital fiber type disproportion (CFTD) with an autosomal recessive inheritance. *Clin. Genet.* **33**, 33–7.

Jager-Roman, E., *et al.* (1986). Fetal growth, major malformations, and minor anomalies in infants born to women receiving valproic acid. *J.Pediatr.* **108**, 997–1004.

Jakob, H., and Kolkmann, F.W. (1973). Zur pigmentvariante der adulten form der amaurostische idiotie. *Arch. Neuropath.* **26**, 225–36.

Jalbert, P., Mouriquand, D., Beaudoing, A. and Jaillard, M. (1966). Myopathie progressive de type Duchenne et mounique XO/XX/XXX. Considerations sur la génese de la fibre muscular striée. *Ann. Génét.* **9**, 104–8.

James, E., and Van Leeuwen, G. (1970). Familial cebocephaly. Case description and survey of the anomaly. *Clin. Pediat.* **9**, 491–3.

Jammes, J., Mirhosseini, S.A., and Holmes, L.B., (1973). Syndrome of facial abnormalities, kyphoscoliosis and severe mental retardation. *Clin. Genet.* **4**, 203–9.

Jampel, R.S., Okazaki, H., and Bernstein, H. (1961). Ophthalmoplegia and retinal degeneration associated with spinocerebellar ataxia. *Arch. Ophthal., N.Y.* **66**, 247–59.

Janeway, R., Ravens, J.R., Pearce, L.A., Odor, D.L., Winston-Salem, N.C., and Suzuki, K. (1967). Progressive myoclonus epilepsy with Lafora inclusion bodies. I. Clinical genetic, histopathologic and biochemical aspects. *Arch. Neurol., Chicago* **16**, 565–82.

Jankovic, J. and Rivera, V.M. (1979). Hereditary myoclonus and progressive distal muscular atrophy. *Ann. Neurol.* **6**, 227–31.

Jankovic, J., Kirkpatric, J.B., and Blomquist, K.A. (1985). Late-onset Hallervorden–Spatz disease presenting as familial parkinsonism. *Neurology* **35**, 227–34.

Jankowicz, E., Berger, H., Kurasz, S., Winogrodzka, W., and Eljasz, L. (1977). Familial progressive external ophthalmoplegia with abnormal muscle mitochondria. *Eur. Neurol.* **15**, 318–24.

Jankowska, H. (1935). Heredity in torsion dystonia. (Abst.) *Zbl. ges. Neurol. Psychiat.* **74**, 359.

Janota, I. (1979). Widespread intranuclear neuronal corpuscles (Marinesco bodies) associated with a familial spinal degeneration with cranial and peripheral nerve involvement. *Neuropath. appl. Neurol.* **5**, 311–7.

Jansen, J. (1975). Sex-linked hydrocephalus. *Dev. med. Child. Neurol.* **17**, 633–40.

Jarmas, A.L., *et al.* (1981). Microcephaly, microphthalmia, falciform retina folds, and blindness. *Am. J. Dis. Child.* **135**, 930–3.

Jarvis, J.L. and Keats, T.E. (1974). Cleidocranial dysostosis: a review of 40 new cases. *Am. J. Roentg.* **121**, 5–16.

Jay, B. (1955). Malignant melanoma of the orbit in a case of oculodermal melanosis. *Br. J. Ophthal.* **49**, 359–63.

Jeavons, P.M. and Bower, B.D. (1964). *Infantile spasms.* Clinics in Developmental Medicine, No. 15. Heinnemann. London.

Jedrzejowska, H. and Milczarek, H. (1976). Recessive hereditary sensory neuropathy. *J. neurol. Sci.* **29**, 371–87.

Jellinger, K. (1968). Progressive pallidum atrophie. *J. neurol. Sci.* **6**, 19–44.

Jellinger, K. and Jirasek, A. (1971). Neuro-axonal dystrophy in man: character and natural history. *Acta neuropath.* Suppl. 5, 3–16.

Jellinger, K. and Seitelberger, F. (1969). Juvenile form of spongy degeneration of the central nervous system. *Acta. Neuropath.* **13**, 276–81.

Jellinger, K., Seitelberger, F. (1969). Pelizaeus-Merzbacher's disease. Transitional form between classical and connatal (Seitelberger) type. *Acta neuropath.* **14**, 108–17.

Jellinger, K., Seitelberger, F. (1970). Spongy glioneuronal dystrophy in infancy and childhood. *Acta neuropath.* **16**, 125–40.

Jenkins, A.C. (1966). Epidemiology of Parkinsonism in Victoria. *Med. J. Aus.* **2**, 496–502.

Jenkins, D.H. and Gill, W. (1972). A case of carcinoma of the colon in association with neurofibromatosis. *Br. J. Surg.* **59**, 322–3.

Jennekens, F.G.I., Busch, H.F.M., Van Hemel, N.M., and Hoogland, R.A. (1975). Inflammatory myopathy in scapulo-ilio-peroneal atrophy with cardiopathy—a study of two families. *Brain* **98**, 709–22.

Jensen, I. (1975). Genetic factors in temporal lobe epilepsy. *Acta neurol. scand.* **52**, 381–94.

Jéquier, M. and Déonna, T. (1973). A propos des degenerescences neuro-sensorielles. *Schweiz. Arch. Neurol. Psychiat.* **112**, 219–27.

Jéquier, M. and Streiff, E.B. (1947). Paraplegia dystrophie squelettique et degenerescence tapeto-retinienne familiales. *Arch. Julius Klaus-Stift. Vererbforsch.* **22**, 129–67.

Jéquier, M., Michail, J., and Streiff, E.B. (1945). Paraplégie familiale et dégénérescence tapeto-retinienne. *Confinia neurol.* **6**, 277–80.

Jersild, C., Fog, T., Hansen, G.S., Thomsen, M., Svejgaard, A., and Dupont, B. (1973). Histocompatibility determinants in multiple sclerosis with special reference to clinical course. *Lancet* **ii**, 1221–5.

Jersild, C., Dupont, B., Fog, T., Platz, P.J., and Svejgaard, A. (1975). *Transplant. Rev.* **22**, 148–63.

Jerusalem, F., Angelini, C., Engel, A.G., and Groover, R.V. (1973*a*). Mitochondria-lipid-glycogen disease of muscle. *Arch. Neurol., Chicago* **29**, 162–9.

Jerusalem, F., Engel, A.G., and Gomez, M.R. (1973*b*). Sarcotubular myopathy. A newly recognised, benign, congenital familial muscle disease. *Neurol., Minneap.* **23**, 897–906.

Jerusalem, F., Spiess, H., and Baumgartner, G. (1975). Lipid storage myopathy with normal carnitine levels. *J. neurol. Sci.* **24**, 273–82.

Jerusalem, F., Ludin, H., Bischoff, A., and Hartmann, G. (1979). Cytoplasmic body neuromyopathy presenting as respiratory failure and weight loss. *J. neurol. Sci.* **41**, 1–9.

Jervell, A. and Lange-Nielsen, F. (1957). Congenital deaf-mutism, functional heart disease with prolongation of the Q-T interval, and sudden death. *Am. Heart J.* **54**, 59–68.

Jervis, G.A. (1950). Early familial cerebellar degeneration. *J. nerv. ment. Dis.* **111**, 398–405.

Jervis, G.A. (1952). Hallervorden-Spatz disease associated with atypical amaurotic idiocy. *J. Neuropath. exp. Neurol.* **11**, 4–18.

Jervis, G.A. (1954). Concordant primary atrophy of the cerebellar granules in monozygotic twins. *Acta Genet. med. Gemell.* **3**, 153–62.

Jervis, G.A. (1963). Huntington's chorea in childhood. *Arch. Neurol., Chicago.* **9**, 244–57.

Jervis, G.A. and Donahue S. (1975). An unusual type of infantile lipofuscinosis. *Acta neuropath.* **31**, 109–16.

Jervis, G.A. and Pullarkat, R.K. (1978). Pigment variety of lipofuscinosis. *Neurol., Minneap.* **28**, 500–3.

Jessen, R.T., Thompson, S., and Smith, E.B. (1977). Cobb syndrome. *Arch. Derm., N.Y.* **113**, 1587–90.

Jestico, J.V., Urry, F.A., and Efphimiou, J. (1985). An hereditary, sensory and autonomic neuropathy transmitted as a X-linked recessive trait. *J. Neurol. Neurosurg. Psychiat.* **48**, 1259–64.

Jeune, M., Tommasi, M., Freycon, F., and Nivelon, J. (1963). Syndrome familial associate ataxie surdite et oligophrenie. Sclerose myocardique d'evolution fatale chez l'un des enfants. *Pediatrics, Springfield* **18**, 984–7.

Jimbow, K., Fitzpatrick, T.B., Szabo, G., and Hori, Y. (1975). Congenital circumscribed hypomelanosis: a characterisation based on electron microscopic study of tuberous sclerosis, nevus depigmentosus and piebaldism. *J. invest. Derm.* **64**, 50–62.

Johns, T.R., Campa, J., and Adelman, L. (1973). Familial myasthenia with 'tubular aggregates' treated with prednisolone. (Abstract.) *Neurol., Minneap.* **23**, 426.

Johnson, R.H. and Spalding, J.M.K. (1964). Progressive sensory neuropathy in children. *J. Neurol. Neurosurg. Psychiat.* **27**, 125–30.

Johnson, W., Schwartz, G., and Barbeau, A. (1962). Studies on dystonia musculorum deformans. *Arch. Neurol., Chicago* **7**, 301–13.

Johnson, W.G. and Chutorian, A.M. (1978). Inheritance of the enzyme defect in a new hexosaminidase deficiency disease. *Ann. Neurol.* **4**, 399–403.

Johnson, W.G., Chutorian, A., and Miranda A. (1977). A new juvenile hexosaminidase deficiency disease presenting as cerebellar ataxia. Clinical and biochemical studies. *Neurology*, **27**, 1012–18.

Johnson, W.G., Lovelace, R.E., Rubin, S.P., and Martin, J.R. (1978). A new X-linked disorder resembling classic Pelizaeus–Merzbacher disease. (Abstract.) *Ann. Neurol.* **4**, 194–8.

Johnson, W.G., *et al.* (1980). Congenital sialidosis: Biochemical studies: clinical spectrum in four sibs; two successful prenatal diagnoses. *Am. J. Hum. Genet.* **32**, 43A.

Johnston, A.W. and Mckay, E. (1986). X linked muscular dystrophy with contractures. *J. Med. Genet.* **23**, 591–5.

Johnston, A.W. and McKusick, V.A. (1962). A sex linked recessive form of spastic paraplegia. *Am J. hum. Genet.* **14**, 83–94.

Johnston, C.C., Lavy, N., Lord, I., Vellios, F., Merritt, A.D., and Deiss, W.P., Jr (1968). Osteopetrosis. A clinical, genetic, metabolic and morphologic study of the dominantly inherited, benign form. *Medicine, Baltimore* **47**, 149–67.

Johnston, H.A. (1964). Severe muscular dystrophy in girls. *J. med. Genet.*, **1**, 79–81.

Jokelainen, M. (1977). Amyotrophic lateral sclerosis in Finland. II. Clinical characteristics. *Acta neurol. scand.* **56**, 194–204.

Jokelainen, M., Palo, J., and Lokki, J. (1978). Monozygous twins discordant for amyotrophic lateral sclerosis. *Eur. Neurol.* **17**, 295–9.

Jokl, E. and Wolffe, J.B. (1954). Sudden non-traumatic death associated with physical exertion in identical twins. *Arch. genet. Med.* **3**, 245–6.

Jones, M.B. (1973). Fertility and age of onset in Huntington's chorea. In *Advances in neurology*, Vol. 1 (ed. A. Barbeau, T.N. Chase and G.W. Paulson), pp. 171–7. Raven Press, New York.

Joosten, E.M.G., Krijgsman, T.B., Gabreëls-Festen, A., Gabreëls, F., and Baars, P. (1974). Infantile globoid cell leucodystrophy: some remarks on clinical biochemical and sural nerve biopsy findings. *Neuropaediatrics* **5**, 191–209.

Jorda, A., *et al.* (1986). A new case of arginase deficiency in a Spanish male. *J. Inherit. Met. Dis.* **9**, 393–7.

Joseph, R., Lefebvre, J., Guy, E., and Job, J.C. (1958). Dysplasie craniodiaphysaire progressive. Ses relations avec la dysplasie diaphysaire progressive de CamuratiEngelmann. *Annl. Radiol., Paris.* **1**, 477–90.

Joubert, M., Eisenring, J.-J., Robb, J.P., and Andermann, F. (1969). Familial agenesis of the cerebellar vermis. *Neurol., Minneap.* **19**, 813–25.

Joynt, R.J. and Perret, G.E. (1961). Meningiomas in a mother and daughter. Cases without evidence of neurofibromatosis. *Neurol., Minneap.* **11**, 164–5.

Juberg, R.C. and Gershanik, J.J. (1976). Cervical vertebral fusion

(Klippel–Feil) syndrome with consanguineous parents. *J. med. Genet.* **13**, 246–9.

Juberg, R.C. and Hirsch, R. (1971). Expressivity of heritable telecanthus in five generations of a kindred. *Am. J. hum. Genet.* **23**, 547–54.

Juliao, O.F., Queiroz, L.S., and De Faria, Lopes, J. (1974). Portugese type of familial amyloid polyneuropathy. Anatomo-clinical study of a Brazilian family. *Eur. Neurol.* **11**, 180–95.

Julien, J., Vital, Cl., Wallat, J.M., Vallat, M., and le Blanc, M. (1974). Oculopharyngeal muscular dystrophy. A case with abnormal mitochondria and 'fingerprint' inclusions. *J. neurol. Sci.* **21**, 165–9.

Jung, S.-S., Chen, K.-M., and Brody, J.A. (1973). Proxysmal choreoathetosis: report of Chinese cases. *Neurol., Minneap.* **23**, 749–55.

Jusic, A., Dogan, S., and Stojanovic, V. (1972). Hereditary persistant distal cramps. *J. Neurol. Neurosurg. Psychiat.* **35**, 379–84.

Jusic, A., Radošević, Z., Grečevic, N., Hlavka, V., Petričević-Migić, R., and Hartl-Prpic, V. (1973). 'L'acropathie ulcéro-multilante familiale' with involvement of the distal mixed nerves and long bones fractures. *J. Neurol. Neurosurg. Psychiat.* **36**, 585–91.

Justice, R.M., Wenger, D.A., Naidu, S., and Rosenthal, I.M. (1977). Enzymatic studies in a new variant of GM_1 gangliosidosis in an older child. *Pediat. Res.* **11**, 407.

Kaback, M.M., Nathan, T.J., and Greenwood, S. (1977). Tay–Sachs disease screening and prenatal diagnosis. *Prog. clin. biol. Res.* **18**, 13–36.

Kaeser, H.E. (1964). Die familiare scapuloperoneale Muskelatrophie. *Dt. Z. Nervenheilk.* **186**, 379–94.

Kahana, E., and Feldman, S. (1976). Amyotrophic lateral sclerosis. A population study. *J. Neurol.* **212**, 205–13.

Kahana, E., Alter, M., Braham, J., and Sofer, D. (1974). Creutzfeldt–Jakob's disease: focus among Libyan Jews in Israel. *Science, N.Y.* **183**, 90–1.

Kahler, S.G., Burns, J.A., Aylsworth, A.S. (1984). A mild autosomal recessive form of osteopetrosis. *Am. J. Med. Genet.* **17**, 451–64.

Kajii, T., Matsuda, K., Ohsawa, T., Katsunuma, H., Ichida, T., and Arashema, S. (1974). Hurler/Scheie genetic compound (mucopolysaccharidosis IH/IS) in Japanese brothers. *Clin. Genet.* **6**, 394–400.

Kak, V.K., Gleadhill, C.A., and Bailey, I.C. (1970). The familial incidence of intracranial aneurysm. *J. Neurol. Neurosurg. Psychiat.* **33**, 29–33.

Kalimo, H., Lundberg, P.O., and Olsson, Y. (1979). Familial subacute necrotizing encephalomyelopathy of the adult form (adult Leigh's syndrome). *Ann. Neurol.* **6**, 200–6.

Kalinowsky, L. (1927), Familiare Erkrankung mit besonderer Beteiligung der Stammganglien. *Mschr. Psychiat. Neurol.* **66**, 168–90.

Kallmann, F.J., Schoenfeld, W.A., and Barrera, S.E. (1944). The genetic aspects of primary eunuchoidism. *Am. J. ment. Defic.* **48**, 203–36.

Kalyanaraman, K., Cancilla, P.A., Munsat, T., and Pearson, C.M. (1970). Hereditary hypertrophic neuropathy: report of two cases of an autosomal recessive variant. *Bull. Los Ang. neurol. Soc.* **35**, 58–68.

Kalyanaraman, K., Smith, B.H., and Schlagenhauff, R.E. (1974). Hereditary hypertrophic neuropathy with facial and trigeminal involvement. *Arch. Neurol., Chicago.* **31**, 15–17.

Kamoshita, S., Konishi, Y., Segawa, M., and Fukuyama, Y. (1976). Congenital muscular dystrophy as a disease of the central nervous system. *Arch Neurol., Chicago.* **33**, 513–16.

Kanter, W.R., Eldridge, R., Fabricant, R., Allen, J.C., and Koerber, T. (1980). Central neurofibromatosis with bilateral acoustic neuroma. *Neurol., Minneap.* **30**. 851–9.

Kaplan, A.M., Itabashi, H.H., Hanelin, L.G., and Lee, A.T. (1975*a*). Neurocutaneous melanosis with malignant leptomeningeal melanoma. A case with metastases outside the nervous system. *Arch. Neurol., Chicago* **32**, 669–71.

Kaplan, A.M., Bergeson, P.S., Gregg, S.A., and Curless, R.G. (1977). Malignant hyperthermia associated with myopathy and normal muscle enzymes. *J. Pediat.* **91**, 431–3.

Kaplan, P. (1983). X-linked recessive inheritance of agenesis of the corpus callosum. *J.Med Genet.* **20**, 122–24.

Kaplan, P., Hollenberg, R.D., and Fraser, F.C. (1975*b*). A spinal arteriovenous malformation with hereditary cutaneous hemangiomas. *Am. J. Dis. Childh.* **130**, 1329–31.

Kaplan, R.E. and Lacey, D.J. (1983). Benign familial neonatal infantile seizures. *Am. J. med. Genet.* **16**, 595–9.

Karch, S.B., and Urich H. (1975). Infantile polyneuropathy with defective myelination: an autopsy study. *Dev. Med. Child Neurol.* **17**, 504–11.

Karpati, G., Carpenter, S., and Nelson, R.F. (1970). Type I muscle fibre atrophy and central nuclei: a rare familial neuromuscular disease. *J. neurol. Sci.* **10**, 489–500.

Karpati, G., Carpenter, S., and Andermann, F. (1971). A new concept of childhood nemaline myopathy. *Arch. Neurol., Chicago.* **24**, 291–304.

Karpati, G., Eisen, A.E., Wolfe, L.S., and Feindel, L.W. (1974). Multiple peripheral nerve entrapments. An unusual phenotype variant of the Hunter syndrome (mucopolysaccharidosis II) in a family. *Arch. Neurol., Chicago* **31**, 418–22.

Karpati, G., *et al.* (1975). The syndrome of systemic carnitine deficiency—clinical, morphological, biochemical and pathophysiologic features. *Neurol. Minneap.* **25**, 16–24.

Karpati, G., Wolfe, L.S., and Andermann, F. (1977). Juvenile dystonic lipidosis: an unusual form of neurovisceral storage disease. *Neurol. Minneap.* **27**, 32–42.

Kato, M. and Araki, S. (1969). Paroxysmal kinesigenic choreoathetosis:

report of a case relieved by carbamazepine. *Arch. Neurol., Chicago* **20**, 508–13.

Katz, D.A., *et al.* (1984). Familial multisystem atrophy with possible thalamic dementia. *Neurology* **34**, 1213–17.

Katzman, R. (1976). The prevalence and malignancy of Alzheimer's disease. *Arch. Neurol., Chicago* **33**, 217–18.

Kaufman, H.H. and Brisman, R. (1972). Familial gliomas. Report of four cases. *J. Neurosurg.* **37**, 110–12.

Kaufman, S., Holtzman, N.A., Milstein, S., Butler, L.J., and Krumholz, A. (1975). Phenylketonuria due to a deficiency of dihydropteridine reductase. *N. Engl. J. Med.* **293**, 785.

Kaufman, S., *et al.* (1978). Hyperphenylalaninemia due to a deficiency of biopterin. *New Engl. J. Med.* **299**, 673.

Kay, D.W.K. (1986). The genetics of Alzheimer's disease. *Br. Med. Bull.* **42**, 19–23.

Kayden, H.J. (1972). Abetalipoproteinemia. *A. Rev. Med.* **23**, 285–96.

Kazakov, V.M., Bogorodinsky, D.K., Znoyko, Z.V., and Skorometz, A.A. (1974). The facio-scapulo-limb (or the facio-scapulo-humeral) type of muscular dystrophy. *Eur. Neurol.* **11**, 236–60.

Kazakov, V.M., Bogorodinsky, D.K., and Skorometz, A.A. (1976). The myogenic scapulo-peroneal syndrome. Muscular dystrophy in the K kindred: clinical study and genetics. *Clin. Genet.* **10**, 41–50.

Keddie, K.M.G. (1967). Presenile dementia, clinically of the Pick's disease variety occurring in a mother and daughter. *Int. J. Neurol. Psychiat.* **3**, 182–7.

Keiller, W. (1926). Four cases of olivo-ponto-cerebellar atrophy giving a history of heredity with three autopsies. *Sth. med. J., Nashville* **19**, 518.

Keleman, J., Rice, D.R., Bradley W.G., Munsat, T.L., D. Mauro, S., and Hogan, E.L. (1982). Familial myoadenylate deminase deficiency and exertional myalgia. *Neurol., Minneap.* **32**, 857–63.

Kelley, R.I. and Moser, H.W. (1984). Hyperpipecolic acidmeia in neonatal adrenoleukodystrophy. *Am. J. Med. Genet.* **19**, 791–5.

Kelley, R.I., *et al.* (1986). Neonatal adrenoleukodystrophy: new cases, biochemical studies and differentiation from Zellweger and related peroxisomal polydystrophy syndromes. *Am. J. Med. Genet.,* **23**, 869–901.

Kelly, T.E. (1973). Globoid cell leucodystrophy. Letter. *Arch. Dis. Child.* **48**, 165.

Kelly, T.E. (1984). Teratogenicity of anticonvulsant drugs I: Review of the literature. *Am. J. Med. Genet.* **19**, 413–34.

Kelly, T.E. and Graetz, G. (1977). Isolated acid neuraminidase deficiency: a distinct lysosomal storage disease. *Am. J. med. Genet.* **1**, 31–46.

Kelly, T.E. *et al.* (1975) Mucolipidosis III (pseudo-Hurler polydystrophy). Clinical and laboratory studies in a series of 12 patients. *Johns Hopkins med. J.* **137**, 156–75.

Kelly, T.E., Reynolds, L.W., and O'Brien, J.S. (1976). Segregation within a family of two mutant alleles for hexosammidase A. *Clin. Genet.* **9**, 540-3.

Kennedy, W.R., Alter, M., and Sung, J.H. (1968). Progressive proximal spinal and bulbar muscular atrophy of late onset. *Neurol. Minneap.* **18**, 671-80.

Kennedy, W.R., Sung, J.H., and Berry, J.F. (1977). A case of congential hypomyelination neuropathy. Clinical, morphological, and chemical studies. *Arch. Neurol., Chicago* **34**, 337-45.

Kenwrick, S., *et al.* (1986). Linkage studies of X-linked recessive spastic paraplegia using DNA probes. *Hum. Genet.* **73**, 264-6.

Kertesz, A. (1967). Paroxysmal kinesigenic choreo-athetosis: an entity within the paroxysmal choreoathetosis syndrome: description of 10 cases including one autopsied. *Neurol. Minneap.* **17**, 680-90.

Kerzin-Storrar, L., *et al.* (1988). Genetic factors in myasthenia gravis: A family study. *Neurology.* **38**, 38-42.

Kessler, G.B. (1968). Non-progressive proximal and generalised spinal muscular atrophy in siblings. *Bull. Los. Ang. neurol. Soc.* **33**, 21-5.

Kessler, C.H., *et al.* (1984). Hallervorden–Spatz syndrome restricted to the pallidal nuclei. *J. Neurol.* **231**, 112-16.

Kessler, S., Field, T., Worth, L., Mosbarger, H. (1987). Attitudes of persons at risk for Huntington Disease towards predictive counselling. *Am. J. Med. Genet.* **26**, 259-70.

Keyes, M.J. (1973). Voluntary nystagmus in two generations. *Arch. Neurol., Chicago* **29**, 63-4.

Khalifeh, R.R. and Zellweger, H. (1963). Hereditary sensory neuropathy with spinal cord disease. *Neurol. Minneap.* **13**, 405-11.

Khan, S.A. and Peterkin, G.A.G. (1970). Congenital indifference to pain. *Trans. Rep. St. Johns Hosp. derm. Soc.* **56**, 122-30.

Khodadad, G. (1971). Familial cirsoid aneurysm of the scalp. *J. Neurol. Neurosurg. Psychiat.* **34**, 664-7.

Kidd, H.A. and Cumings, J.N. (1947). Cerebral angiomata in an Icelandic family. *Lancet* **i**, 747-8.

Kihara, H., *et al.* (1980). Prenatal diagnosis of metachromatic leukodystrophy in a family with pseudo arylsulfatsase A deficiency by the cerebroside sulfate loading test. *Pediat. Res.* **14**, 224-7.

Kihara, S., and Nonaka, I. (1985). Congenital muscular dystrophy. A histochemical study with morphometric analysis on biopsied muscles. *J. Neurol. Sci.* **70**, 139-49.

Killian, J.M. and Kloepfer, H.W. (1979). Homozygous expression of a dominant gene for Charcot–Marie–Tooth neuropathy. *Ann. Neurol.* **5**, 515-22.

Kim, R.C., *et al.* (1981). Familial dementia of adult onset with pathological findings of a nonspecific nature. *Brain* **104**, 61-78.

King, C.R., Lovrein. E.W., and Reiss, J. (1977) Central nervous system arteriovenous malformations in multiple generations of a family with hereditary hemorrhagic telangiectasia. *Clin. Genet.* **12**, 372–81.

King, J.O. (1975). Progressive myoclonic epilepsy due to Gaucher's disease in an adult. *J. Neurol. Neurosurg. Psychiat.* **38**, 849–54.

King, J.O., Denborough, M.A., and Zapf, P.W. (1972). Inheritance of malignant hyperthermia. *Lancet* **i**, 365–70.

King, M.D., Dudgeon, J., and Stephenson, J.B.F. (1984). Joubert's syndrome with retinal dysplasia: neonatal tachypnoea as the clue to a genetic brain-eye malformation. *Arch. Dis. Childh.* **59**, 709–18.

Kingston, H.M., *et al.* (1984). Localisation of the Becker muscular dystrophy gene on the short arm of the X-chromosome by linkage to cloned DNA sequences. *Hum. Genet.* **67**, 6–17.

Kinney, H., Burger, P.C., Vogel, F.S.Y. (1980). Subacute diencephalic angioencephalopathy: report of an additional case. *J. Neurol. Sci.* **45**, 73–81.

Kinnunen, E., Koshenvuo, M., and Saprio, J. (1987). Multiple sclerosis in a nation wide series of twins. *Neurol.* **37**, 1627–9.

Kinoshita, M., Satoyoshi, E., and Kumagai, M. (1975*a*). Familial type I fibre atrophy. *J. neurol. Sci.* **25**, 11–17.

Kinoshita, M., Satoyoshi, E., and Matsuo, N. (1975*b*). 'Myotubular myopathy' and type I fibre atrophy in a family. *J. Neurol. Sci.* **26**, 575–82.

Kinsbourne, M. (1962). Myoclonic encephalopathy of infants. *J. Neurol. Neurosurg. Psychiat.* **25**, 271–6.

Kirkham, T.H. (1969). Familial Marcus-Gunn phenomenon. *Br. J. Ophtahal.* **53**, 282–3.

Kirkham, T.H. (1970). Inheritance of Duane's syndrome. *Br. J. Ophthal.* **54**, 323–9.

Kirschbaum, W.R. (1924). Zwei eigenartige Erkrankungen des Zentralnervensystems nach Art der spastichen Pseudoscklerose (Jakob). *Z. ges. Neurol. Psychiat.* **92**, 175–202.

Kishimoto, K., Nakamura, M., and Sotokawa, Y, (1959). On population genetics of Huntington's chorea in Japan. In *First Int. Congr. Neurol. Sci.,* Vol. 4 (ed. L. Von Bogaert and J. Rademecker), pp. 217–26. Pergamon Press, Oxford.

Kissel, P. and André J.M. (1976),. Maladie de Parkinson et anosmia in monozygotic twin sisters. *J. Génét. hum.* **24**, 113–17.

Kissel, P., André J.M., and André, M. (1973). Coexistence dans la méme famille d'un syndrome de Sjögren–Larsson et d'un syndrome de Rudd. *J. Génét. hum.* **21**, 15–22.

Kissel, P. and Arnould, G. (1954). Neuro-ectodermose familiale a type de maladie de Recklinghausen chez la mére, de maladie de Sturge–Weber chez la fille. *Rev. neurol.* **91**, 381–4.

Kissel, P., Arnould, G., and André, J.M. (1972). Incidence des accidents vascúlaires cerebraux au cours des conjonctivo dysplasies hereditaires. *J. Génét. hum.* **29**, 151–67.

Kistler, J.P., *et al.* (1977). Mannosidosis. *Arch. Neurol., Chicago* **34**, 45–51.

Kitahara, T., Ariga, N., Yamaura, A., Makino, H., and Maki, Y. (1979). Familial occurrence of moya-moya disease: report of three Japanese families. *J. Neurol. Neurosurg. Psychiat.* **42**, 208–14.

Kito, S., Itoga, E., Hiroshige, Y., Matsumoto, N., and Miwa, S. (1980). A pedigree of amyotrophic chorea with acanthocytosis. *Arch. Neurol., Chicago* **37**, 514–17.

Kjellin, K. (1959). Familial spastic paraplegia with amyotrophy, oligophrenia and central retina degeneration. *Arch. Neurol., Chicago* **1**, 133–40.

Kjellin, K. Müller, R., and Aström, K.E. (1960). The occurrence of brain tumours in several members of a family. *J. Neuropath. exp. Neurol.* **19**, 528–37.

Klasen, E.L. (1968). *Legasthenia*. Huber, Bern.

Klebanaoff, M.A. and Neff, J.M. (1980). Familial dysautonomia associated with recurrent osteomyelitis in a non-Jewish girl. *J. Pediat.* **96**, 75.

Klein D. (1958). La dystrophic myotonique (Steiner et la myotonie congenitale (Thomsen) en Suisse. *J. Génét. hum.* Suppl. 7, 1–328.

Klein, D. and Ammann, F. (1969). The syndrome of Laurence–Moon–Bardet–Biedl and allied diseases in Switzerland. *J. Neurol. Scr.* **9**, 479.

Klein, D., Mumenthaler, M., Kraus-Ruppert, R., and Rallo, E. (1968). Une grande famille valaisanne attenter d'epilepsic myoclonique progressive et de rctinitc pigmcntairc: ctudc cliniquc, genetique et anatomopatholique. *Humangenetik* **6**, 237–52.

Klein, H. and Dichgans J. (1969). Familiare juvenile glionerurable Dystrophie. *Arch. Psychiat. Nervenkheit.* **212**, 400–22.

Klein, R., Haddow, J.E., and De Luca, C. (1972). Familial congenital disorder resembling stiff-man syndrome. *Am. J. Dis. Child.* **214**, 730–1.

Klenke, E. and Kahlke, W. (1963). Uber das Vorkommen von 3, 7, 11, 15. Tetramethyl-Hexadecansaure (Phytansaure) in den Chotesterinestern un anderen Lipoidfraktionen der Organe bei einem Krankheitsfall unbekannter Genese (Verdacht auf Heredapathia atactica polyneuritiformis (Refsum syndrom). *Hoppe-Seyler's Z. Physiol. Chem.* **333**, 133–9.

Kline, L.B. and Giaser, J.S. (1979). Dominant optic atrophy. *Arch. Opthal.* **97**, 1680–6.

Klippel, M. and Durante, G. (1892). Contribution à l'Étude des affections nerveuses familiales et héréditaires. *Revue med.* **xii**, 745–85.

Kloepfer, H.W., Platou, R.V., and Hansche, W.J. (1964). Manifestations of a recessive gene for microcephaly in a population isolate. *J. Génét. hum.* **13**, 52–9.

610 *Bibliography*

Kopfstock, A. (1921). Familiares Vorkomman con Cyklopie und Arthinencephalie. *Mschr. Geburtsh. Gynaekol.* **56**, 59.

Knies, P.T. and Le Fever, H.E. (1941). Metabolic craniopathy: hypertostosis frontalis interna. *Ann. intern. Med.* **14**, 1858–92.

Knight, W.A., Murphy, W.K., and Gottlieb, J.A. (1973). Neuro-fibromatosis associated with malignant neurofibromas. *Arch. Derm., N.Y.* **107**, 747–50.

Kobayashi, A.L.P. (1975). Familial periodic paralysis. A report and review. *Nebraska med. J.* **60**, 110–13.

Kobayashi, Y., *et al.* (1982). Ultrastructural study of the childhood mitochondrial myopathic syndrome associated with lactic acidosis. *Eur. J. Pediat.* **143**, 25–30.

Kocen, R.S. and Thomas P.K. (1970). Peripheral nerve involment in Fabry's disease. *Arch. Neurol., Chicago* **22**, 81–8.

Kocen, R.S., Lloyd, J.K., Lascelles, P.T., Fosbrook, A.S., and Williams, D. (1967). Familial α-lipoprotein deficiency (Tangier disease) with neurological abnormalities. *Lancet* **i**, 1341–5.

Kocen, R.S., King, R.H.M., Thomas, P.K., and Haas, L.F. (1973). Nerve biopsy findings in two cases of Tangier disease. *Acta neuropath.* **26**, 317–27.

Koch, G. (1940). Beitrag zur Erblichkeit der Sturge–Weberschen Krankheid. *Z. ges. Neurol. Psychiat.* **169**, 614–23.

Koch, G. (1957), Ergebnisse aus der Nachuatersuchung der Berliner Zwillingsserie nach 20–25 Jahren (Vorlaufige Ergebnisse). *Arch. Genet.* **7**, 43–52.

Koch, G. (1966). Phakomatosen In *Humangenetik, V/1 Krankheiten des Nervensystems* (ed. P E. Becker), pp. 67–74. G. Thieme, Stuttgart.

Koch, G. (1972). Genetic aspects of the phakomatoses. In *Handbook of clinical neurology*, Vol. 14 (ed. P J. Vinken and G. W. Bruyn), pp. 448–561. Elsevier.

Koenig, R.H. and Spiro, A.J. (1970). Hereditary spastic paraparesis with sensory neuropathy. *Dev. med. Child Neurol.* **12**, 576–81.

Koeppen, A.H. and Hans, M.B. (1976). Supranuclear ophthalmoplegia in olivopontocerebellar degeneration. *Neurol., Minneap.* **26**, 764–8.

Koeppen, A.H., Barron, K.D., and Cox J.F. (1971). Striatonigral degeneration. *Acta neuropath.* **19**, 10–19.

Koeppen, A.H., Goedde, H.W., Hirth, L., Benkmann, H.-G., and Hiller, C. (1980). Genetic linkage in hereditary ataxia. *Lancet* **i**, 92–3.

Kohlschütter, A. Chappuis, D,, Meier, C., Tönz, O., Vassella, F., and Herschkowittz, N. (1974). Familial epilepsy and yellow teeth—a disease of the CNS associated with enamel hypoplasia. *Helv. pediat. Acta* **29**, 283–94.

Kohlschütter, A., *et al.* (1987). Infantile glycerol kinase deficiency—A condition requiring prompt identification. Clinical, biochemical, and morphological findings in two cases. *Eur. J. Pediat.* **146**, 575–81.

Kohn, G., *et al.* (1977). Prenatal diagnosis of mucolipidosis IV by electron microscopy. *J. Pediat.* **90**, 62–6.

Kok, O. and Bruyn, G.W. (1962). An unidentified hereditary disease. (Letter.) *Lancet* **i**, 1359.

Kolodny, E.H., Wald, I., Moser, H.W., Cogan, D.C., and Kuwabara, T. (1973). GM$_2$ gangliosidosis without deficiency in the artifical substrate cleaving activity of hexosaminidase A and B. *Neurol. Minneap.* **23**, 427.

Komai, T., Kishimoto, K., and Ozaki, Y. (1955). Genetic study of microcephaly based on Japanese material. *Am. J. hum. Genet.* **7**, 51–65.

Kondo, K., Tsubaki, T., and Sakamuto, F. (1970). The Ryukyuan muscular dystrophy—an obscure heritable neuromuscular disease found in the islands of southern Japan. *J. neurol. Sci.* **11**, 359–82.

Kondo, K., Kurland, L.T., and Schull, W.J. (1973). Parkinson's disease: genetic analysis and evidence of a multifactorial etiology. *Proc. Staff Meet. Mayo Clinic.* **48**, 465–75.

Konigsmark, B.W. and Gorlin, R.J. (1976). *Genetic and metabolic deafness.* Saunders, New York.

Konigsmark, B.W. and Lipton, H.L. (1971). Dominant olivopontocerebellar atrophy with dementia and extrapyramidal signs: report of a family through three generations. *Birth Defects*: Original Article Series, **VII**, No. 1, 178–91.

Konigsmark, B.W., and Weiner, L.P. (1970). The olivopontocerebellar atrophies: a review. *Medicine, Baltimore* **49**, 227–41.

Konigsmark, B.W., Knox, D.L., Hussels, I.E., and Moses, H. (1974). Dominant congential deafness and progressive optic atrophy. *Arch. Ophthal.* **91**, 99–103.

Konishi, Y. and Kamoshita, S. (1975). An autopsy case of classical Pelizaeus–Merzbacher's disease. *Acta neuropath.* **31**, 267–70.

Korinthenberg, F., Palm, D., Schlake, W., and Klein, J. (1984). Congenital muscular dystrophy, brain malformation and ocular problems (muscle, eye and brain disease) in two German families. *Eur. J. Pediat.* **142**, 64–8.

Kornfeld, M. (1972). Generalized lipofuscinosis (generalized Kuf's disease). *J. Neuropath, exp. Neurol.* **31**, 668–82.

Kornzweig, A.L. (1970). Bassen-Kornzweig syndrome. Present status. *J. med. Genet.* **7**, 271–6.

Korten, J.J., Notermans, S.L.H., Frenken, C.W.G.M., Gabreels, F.J.M., and Joosten, E.M.G. (1974). Familial essential myoclonus. *Brain* **97**, 131–8.

Korula, J., Namasivayam, R.K., and Shadangi, T.N. (1976). A case of familial spinocerebellar degeneration with hypobetalipoproteinemia. *Neurol. Madras* **24**, 41–5.

Koskenoja, M. and Suvanto, E. (1959). Gargoylism: report of an adult form with glaucoma in two sisters. *Acta opththal.* **37**, 234–40.

Koskiniemi, M., Donner, M. Majuri, H., Haltia, M., and Norio, R. (1974).

Progressive myoclonus epilepsy. A clinical and histopathological study. *Acta neurol. scand.* **50**, 307–32.

Koster, J.F., Niermeijer, M.F., and Loonen, M.C. (1976). Beta-galactosidase deficiency in an adult: a biochemical and somatic cell genetic study on a variant of GM_1-gangliosidosis. *Clin. Genet.* **9**, 427–32.

Kotagel, S., *et al.* (1986). AB variant GM2 gangliosidosis: cerebrospinal fluid and neuropathologic characteristics. *Neurology* **36**, 438–40.

Koto, A., *et al.*. (1978). Sensory neuropathy with onion-bulb formation. *Am. J. Dis. Childh.* **132**, 379–81.

Kouseff, BG., *et al.* (1976). Fucosidosis type 2. *Paediatrics, Springfield* **57**, 205–13.

Krabbe, K. (1916). A new familial, infantile form of diffuse brain sclerosis. *Brain* **39**, 74–114.

Kraft, M. (1970). Adult cerebral lipidosis. A case of Kuf's disease. *Acta Neurol. Scand.* **46**, Suppl. 43, 86.

Krangenbühl, H. and Yasargil, M.G. (1958). *Das Hiraneurysma.* Geigy, Basel.

Kramer, F. (1914). Drei Fälle von familiärer spastischer Erkrankung. *Allg. Z. Psychiat.* **71**, 531–2.

Kramer, R.E., *et al.* (1987). HLA-DR2 and narcolepsy. *Arch. Neurol.* **44**, 853–5.

Kramer, W. (1953). Poliodysplasia cerebri. *Acta Psychiat Scand.* **28**, 413–27.

Kramer, W. (1963). Syndromes of Klippel–Trenaunay and Sturge–Weber in the same patient. *Psychiat., Neurol. Neurochir.* **66**, 362–70.

Kraus-Ruppert, R., Ostertag, B., and Häfner, H. (1970). A study of the late form (type Lundborg) of progressive myoclonic epilepsy. *J. neurol. Sci.* **11**, 1–15.

Kreiborg S., Pruzansky, S., and Pashayan, H. (1972). The Saethre–Chotzen syndrome. *Teratology* **6**, 287–94.

Kreindler, A., Crighel, E., and Poilici, J. (1959). Clinical and electroence-phalographic investigations in myoclonic cerebellar degeneration. *J. Neurol. Neurosurg. Psychiat.* **22**, 232–7.

Kresse, H., Von Figura, K., and Bartsocas, C. (1976). Clinical and biochemical findings in a family with Sanfillipo disease type C. (Abstract.) *Clin. Genet.* **10**, 364.

Krill, A.E., Smith, V.C., and Pokorny, J. (1971). Further studies supporting the identity of congenital tritanopia and hereditary dominant optic atrophy. *Invest. Ophthal.* **10**, 457–65.

Kufs, A. (1928). Uber heredofamiliäre angiomatose des gehirns und der Retina, ihre Beziehungen zueinader und Angiomatose der Hant. *Zbl. ges. Neurol. Psychiat.* **112**, 651–86.

Kufs, H. (1925). Uber eine spatform der amaurostischen Idiotie und ihre heredofamiliaren Grandlagen. *Z. ges. Neurol. Psychiat.* **95**, 169–88.

Kugelberg, E,. and Welander, L. (1954). Familial neurogenic (spinal?) muscular atrophy simulating proximal dystrophy. *Acta psychiat. scand.* **29**, 42–3.

Kugelberg, E. and Welander, L. (1956). Heredofamilial juvenile muscular atrophy simulating muscular dystrophy. *Arch. Neurol. Psychiat., Chicago.* **75**, 500–9.

Kuhn, E., and Schroder, J.M. (1981). A new type of distal myopathy in two brothers. *J. Neurol.* **226**, 181–5.

Kuhn, E., Fiehn, W., Schroder, J.M., Assmus, H., and Wagner, A. (1979). Early myocardial disease and cramping myalgia in Becker-type muscular dystrophy: a kindred. *Neurol. Minneap.* **29**, 1144–9.

Kuhlendahl, H.D., Grob-Selbeck, G., Doose, H., and Jensen, H.-P. (1977). Cranial computer tomography in children with tuberous sclerosis. *Neuropediatrie* **8**, 325–32.

Kuhr, M.D. (1975). Doubtful benefits of Tay–Sachs screening. (Letter.) *N. Engl. J. Med.* **292**, 371.

Kunkel, L.M. (1986). Analysis of deletions in DNA from patients with Becker and Duchenne muscular dystrophy. *Nature* **322**. 73–77.

Kunstandter, R.H. (1965). Melkersson's syndrome. A case report of multiple recurrences of Bell's palsy and episodic facial edema. *Am J. Dis. Childh.* **110**, 559–61.

Kuo, P.T. and Bassett, D.R. (1962). Blood and tissue lipids in a family with hypobetalipoproteinaemia. *Circulation* **26**, 660.

Kurczynski, T.W. (1983). Hyperexplexia. *Arch. Neuro.* **40**, 246–8.

Kurent, J.E., Hirano, A., and Foley, J.M. (1975). Familial amyotrophic lateral sclerosis with spinocerebellar degeneration and peripheral neuropathy. (Abst.) *J. Neuropath. exp. Neurol.* **24**, 110.

Kuriyama, M., Ishu, K., Umezaki, H., and Tanaka, Y. (1975). A case of mucolipidosis. *Clin. Neurol., Tokyo* **15**, 580–4.

Kuriyama, M., *et al.* (1984). Mitochondrial encephalopathy with lactate-pyruvate elevation and brain infarctions. *Neurology.* **34**, 72–7.

Kurlan, R., Behr, J., and Shoulson, I. (1987). Hereditary myoclonus and chorea: the spectrum of hereditary nonprogressive hyperkinetic movement disorders. *Movement Disorders.* **2**, 301–6.

Kurland, L.T. (1957). Epidemiological investigations of amyotrophic lateral sclerosis. III. A genetic interpretation of incidence and geographic distrubution. *Proc. Staff Meet. Mayo Clin.* **32**, 449–62.

Kurland, L.T. (1958). Epidemiology, incidence, geographic distribution, genetic considerations (of Parkinsonism), In *Pathogenesis and treatment of Parkinsonism* (ed. W.S. Fields), pp. 5–49. Thomas, Springfield.

Kurland, L.T. (1959). The incidence and prevalence of convulsive disorders in a small urban community. *Epilepsia* **1**, 143–61.

Kurland, L.T. (1977). Epidemiology of amyotrophic lateral sclerosis with emphasis on antecedent events from case-control comparisons. In *Motor neuron disease* (ed. F.C. Rose). Pitman, London.

Kurland, L.T. and Alter, M. (1961). Current status of the epidemiology and genetics of myasthenia gravis. In *Myasthenia gravis: 2nd Int. Symp. Proc.* (ed. H.R. Viets), pp. 307–37. Thomas, Springfield, Ill.

Kurland, L.T. and Mulder, D.W. (1955). Epidemiologic investigations of amyotrophic lateral sclerosis. *Neurol. Minneap.* **5**, 249–68.

Kuroki, Y, Matsui, I., and Yamamoto Y. (1982). The 'happy puppet' syndrome in two siblings. *Hum. Genet.* **56**, 227–9.

Kuroiwa, Y. and Murai, Y. (1964). Hereditary sensory radicular neuropathy with special refecence to conduction velocity study. *Neurol., Minneap.* **14**, 574–7.

Kuroiwa, Y., *et al.* (1982). Computed tomographic visualization of extensive calcinosis in a patient with idiopathic familial basal ganglia calcification. *Arch. Neurol.* **39**, 603.

Kurokawa, T., *et al.* (1987). Late infantile Krabbe leukodystrophy: MRI and evoked potentials in a Japanese girl. *Neuropediatrics* **18**, 182–3.

Kurtzke, J.F. (1975). A reassessment of the distribution of multiple sclerosis: Parts one and two. *Acta neurol. scand.* **51**, 110–36, 173–57.

Kurtzke, J.F. and Bui, Q.H. (1977). Multiple sclerosis in a migrant population: II. Half-orientals immigrating in childhood. *Trans. Am. Neurol. Assoc.* **102**, 54–6.

Kurtzke, J.F. and Hyllested, K. (1986). Multiple sclerosis in the Faroe Island. II Clinical update, transmission and nature of M.S. *Neurol.* **36**, 307–28.

Kuwert, E.K. (1977). Genetical aspects of multiple sclerosis with special regard to histocompatibility determinates. *Acta neurol. scand.* **55**, Suppl. 63. 23–42.

Kwittken, J. and Barest, H.D. (1958). The neuropathology of hereditary optic atrophy (Leber's disease). *Am. J. Path.* **34**, 185–207.

Kytilä J. and Miettinen, P. (1961). On bilateral aplasia of the optic nerve. *Acta ophthal.* **39**, 416.

Lacy, J.R. and Pendry, J.K. (1976). *Infantile Spasms.* Raven Press, New York.

Ladda, R.L., Stoltzfus, E., Gordon, S.L., and Graham, W.P. (1978). Craniosynostosis associated with limb reduction malformation and cleft lip/palate: a distinct syndrome. *Pediatrics, Springfield.* **61**, 12–15.

La France, R., Griggs, R., Moxley, R., and McQuillen, J. (1977). Hereditary paroxysmal ataxia responsive to acetazolmide. (Astract.) *Neurol. Minneap.* **27**, 370.

Lagos, J.C. and Gomez, M.R. (1967). Tuberous sclerosis: reappraisal of a clinical entity. *Proc. Staff Meet. Mayo Clin.* **42**, 26–49.

Laing, J.W. and Smith, R.R. (1974). Intracranial arteriovenous malformations in sisters: a case report. *J. Mo. St. med. Ass.* **15**, 203–6.

Lake, B.D. (1968). Segmental demyelination of peripheral nerves in Krabbe's disease. *Nature, Lond.* **217**, 171-2.

Lake, B.D. and Cavanagh, N.P. (1978). Early-juvenile Batten's disease—a recognizable sub-group distinct from other forms of Batten's disease. *J. Neurol. Sci.* **36**, 265-71.

Lake, B.D., Cavanagh, N.P., and Wilson, J. (1977) Myopathy with minicore in siblings. *Neuropathol appl. Neorobiol.* **3**, 159-67.

Lambert, C.D. and Fairfax, A.J. (1976). Neurological association of chronic heartblock. *J. Neurol. Neurosurg. Psychiat.* **39**, 571-5.

Lamy, M. and de Grouchy, J. (1954). L' heredité de la myopathie: formes lasses. *J. Génét. hum.* **3**, 219-61.

Lance, J. W. (1963). Sporadic and familial varieties of tonic seizures. *J. Neurol. Neurosurg. Psychiat.* **26**, 51-9.

Lance, J.W. (1977). Familial paroxysmal dystonic choreoathelosis and its differentiation from related syndromes. *Ann. Neurol.* **2**, 285-93.

Lance, *et al.* (1979). Hyperexcitability of motor and sensory neurons in neuromyotonia. *Ann. Neurol.* **5**, 523-32.

Land, J.M., Morgan-Hughes, J.A., and Clark, J.B. (1981). Mitochondrial myopathy. Biochemical studies revealing a deficiency of Nadh-cytochrome B reductase activity. *J. Neurol. Sci.* **50**, 1-13.

Landau, W.M. and Gitt, J.J. (1951). Hereditary spastic paraplegia and hereditary ataxia. *Arch. Neurol. Psychiat.* **66**, 346-54.

Lander, C.M., Eadie, M.J., and Tyrer, J.H. (1976). Hereditary motor peripheral neuropathy predominantly affecting the arms. *J. neurol. Sci.* **28**, 389-94.

Landis, D., Rosenberg, R.N., Landis, S.C., Schut, L., and Nyhan, W.L. (1974). Olivopontocerebellar degeneration. Clinical and ultrastructural abnormalities. *Arch. Neurol., Chicago* **31**, 295-307.

Landouzy, L. and Dejerine, J. (1885). De la myopathic atrophique progressive. *Revue méd.* **5**, 81-117, 253-366: **6**, 977-1027.

Landrigan, P.J., Berenberg, W., and Bresnan, M. (1973). Behr's syndrome: familial optic atrophy, spastic diplegia and ataxia. *Dev med. Child. Neurol.* **15**. 41-7.

Landy, P.J. and Bain, B.J. (1970). Alzheimer's disease in siblings. *Med. J. Aust.* **2**, 832-4.

Lang, A.E., Quinn, N.P., and Marsden C.D. (1984). Hereditary generalized dystonia with symmetrical striatal lucencies: a probable form of Leigh disease. *Neurology* Suppl., **34**, 236.

Langdon, N., *et al.* (1984). Genetic markers in narcolepsy. *Lancet* ü, 1170-80.

Lange, E., Poppe, W. (1963). Klinischer Beitrag zum Krantheitsbild der progressiven Palliamatrophy (van Bogaert). *Psychiatria Neurol.* **146**, 176-92.

Lange, E. and Scholtze, P. (1970). Familial progressive pallidum atrophy. *Eur. J. Neurol.* **3**, 265-7.

Lanzkowsky, P., Erlandson, M.E., and Bezan, A.I. (1970). Congenital malabsorption of folate. *Am. J. Med.* **48**, 580–3.

Lapkin, M.L. and Golden, G.S. (1978). Basilar artery migraine. *Am. J. Dis. Childh.* **132**, 278–81.

Lapresle, J. (1956). Contribution à l'etude de la dystasia aréflexique hereditaire: état actuel de quatre cas princeps de Roussy et Mlle Lévy trente and aprés la premiere publication de cas auteurs. *Sem. Hôp. Paris.* **32**, 2473–82.

Lapresle, J., Fardeau, M., and Godet-Guillain, J. (1972). Myopathie distale et congenitale, avec hypertrophie des mollets—prescence d'anomalies mitochondriales a' la biopsie musculaire. *J. neurol. Sci.* **17**, 87–102.

Lapresle, J. and Salisachs, P. (1973). Onion bulbs in a nerve specimen from an original case of Roussy–Lévy disease. *Archs. Neurol., Chicago* **29**, 346–8.

Larsen, R., Ashwal, S., and Peckham, N. (1987). Incontinentia pigmenti: Association with anterior cell disease. *Neurology* **37**, 446–50.

Larsen, T.A., Dunn, H.G., and Jan J.E. (1985). Dystonia and calcification of the basal ganglia. *Neurology.* **35**, 533–7.

Larsson, T. (1966). Dystonia musculorum deformans. *Acta neurol. scand.* **42** Suppl. 17.

Larsson, T. and Sjögren, T. (1960). Essential tremor: a clinical and genetic population study. *Acta psychiat, sand.* **36** Suppl. 144, 1–176.

Larsson, T., Sjögren, T. and Jacobson, G. (1963). Senile dementia: a clinical sociomedical and genetic study. *Acta neurol. scand.* Suppl. 167.

La Russo, N.F., Summerskill, W.H.J., and McCall, J.T. (1976). Abnormalities of chemical tests for copper metabolism in chronic active liver disease: differentiation from Wilson's disease. *Gastroenterology* **70**, 653–5.

Lascelles, R.G., Baker, I.A., and Thomas, P.K. (1970). Hereditary polyneuropathy of Roussy–Levy type with associated cardiomyopathy. *Guy's Hosp. Rep.* **119**, 253–62.

Lassater, G.M. (1962). Reading epilepsy. *Arch. Neurol.,Chicago* **6**, 492–5.

Lassonde, M., Trudeau, J.G., and Girard, C. (1970). Generalized lentigines associated with multiple congenital defects (Leopard syndrome). *Can. med. Ass. J.* **103**, 293–4.

Latham, A.D. and Munro, T.A. (1937–38). Familial myoclonus epilepsy associated with deaf mutism in a family showing other psychobiological abnormalities. *Ann Eugen.* **8**, 166–75.

Laurance, B.M., Matthews, W.B., and Diggle, J.H. (1968). Hereditary quivering of the chin. *Arch. Dis. Child.* **43**, 249–51.

Laurence, J.Z. and Moon, R.C. (1866). Four cases of 'retinitis pigmentosa' occurring in the same family and accompanied by general imperfections of development. *Ophthal. Rev.* **2**, 32–41.

Laurence, K. M. and Cavanagh, J. B. (1968). Progressive degeneration of the cerebral cortex in infancy. *Brain.* **91**, 261-80.

Lauter, H. (1961). Genealogische Erhebungen in einer familie mit Alzheimerscher Krankheir. *Arch. Psychiat. Nervkrankh.* **202**, 126-39.

Laver, J., *et al.* (1983). Infantile lethal neuraminidase deficiency (sialidosis). *Clin. Genet.* **23**, 97-101.

Lavy, N. W., Palmer, C. G., and Merritt, A. D. (1966). A syndrome of bizarre vertebral anomalies. *J. Pediat.* **69**, 1121-5.

Layzer, R. B., Rowland, L. P., Ranney, H. M. (1967). Muscle phosphofructokinase deficiency. *Arch. Neurol.* **17**, 512-23.

Lazjuk, G. I., Lurie, I. W., and Nedzved, M. K. (1976). Further studies on the genetic heterogeneity of cebocephaly. *J. med. Genet.* **13**, 314-18.

Leak, D. (1961), Paroxysmal atrial flutter in peroneal muscular atrophy. *Br. Heart J.* **23**, 326-8.

Leavitt, F. H. (1928). Cerebellar tumors occuring in identical twins. *Arch. Neurrol. Psychiat., Chicago* **19**, 617-22.

Lebenthal, E., *et al.* (1970). Arthrogryposis multiplex congenita—23 cases in Arab Kindred. *Pediatrics, Springfield* **46**, 891-9.

Leblhuber, F., *et al.* (1986). Heterogeneity of hereditary motor and sensory neuropathy Type II (HMSN I): electroneurographical findings visual evoked potentials and blood group markers in a family with Charcot-Marie-Tooth (CMT). *Acta Neurol. Scand.* **74**, 145-9.

Lechtenberg, R. and Ferretti, C. (1981). Ataxia with aniridia of Gillespie: a case report. *Neurology* **31**, 95-7.

Ledic, P. and Van Bogaert, L. (1960). Heredodegenerescence cerebelleuse et spastique avec degenerescence maculaire. *J. Génét. hum.* **9**, 140-57.

Lee, B. I., *et al.* (1985). Familial paroxysmal hypnogenic dystonia. *Neurology* **35**, 1357-60.

Lee, K. R., Kishore, P. R. S., Wulfsberg, E., and Kepes, J. J. (1978). Supratentarial leptomeningeal hemangioblastoma. *Neurol., Minneap.* **28**, 727-30.

Lee, L. V., Pascasio, F. M., Fuentes, F. D., and Viterbo, G. H. (1976). Torsion dystonia in Panay, Philippines. *Adv. Neurol.* **14**, 137-51

Lees, F., Macdonald, A.-M. and Turner, J. W. A. (1964). Leber's diisease with symptoms resembling disseminated sclerosis. *J. Neurol. Neurosurg. Psychiat.* **27**, 415-21.

Legras, A. (1934). Multiple sclerosebij Tweelingen. *Ned Tijdschr. Geneesk.* **78**, 174.

Lehman, R. A. W., Stears, J. C., Wesenberg, R. L., and Nusbaum, E. D. (1977). Familial osteosclerosis with abnormalities of the nervous system and meninges. *J. Pediat.* **90**, 49-54.

Lehmann, E. C. H. (1957). Familial osteodystrophy of the skull and face. *J. Bone Jt Surg.* **39B**, 313-15.

Lehoczky, T., Halasy, M., Simon, G., and Harmos, G. (1965). Glycogenic myopathy. A case of skeletal muscle-glycogenosis in twins. *J. neurol. Sci.* **2**, 366–84.

Leibel, F. L., *et al.* (1980). Glutaric acidemia: a metabolic disorder causing progressive choreoathetosis. *Neurology*. **30**, 1163–8.

Leiber, B. (1985). Rett syndrome. A nosological entity. *Brain Dev.* **7**, 275–6.

Leii, D. A., Furlow, T. N., and Falgout, J. C. (1984). Benign familial chorea an association with intellectual impairment. *J. Neurol. Neurosurg. Psychiat.* **27**, 471–4.

Lennox, M. A. (1949). Febrile convulsions in childhood. *Am. J. Dis. Childh.* **78**, 868–82.

Lennox, W. G. (1947). The genetics of epilepsy. *Am. J. Psychiat.* **103**, 457–62.

Lennox, W. G. (1953). Significance of febrile seizures. *Pediatrics, Springfield* **11**, 341–57.

Lennox, W. G. and Davis, J. P. (1950). Clinical correlates of the fast and slow spike-wave electroencephalogram. *Pediatrics, Springfield* **5**, 626–44.

Leri, A. and Weill, J. (1929). Phenomene de Marcus Gunn (synergie palpebromaxillaire) congenital et heriditaire. *Bull. Soc. méd., Paris* **53**, 875–80.

Leroy, J. G. and O'Brien, J. S. (1976). Mucolipidosis II and III: different residual activity of beta-galactosidase in cultured fibroblasts. *Clin. Genet.* **9**, 533–9.

Leroy, J. G., Spranger, J., Feingold, M., Opitz, J. M., and Crocker, A. C. (1971). I-cell disease: a clinical picture. *J. Pediat.* **79**, 360–5.

Lessell, S., Torres, J. M., and Kurland, L. T. (1962). Seizure disorders in a Guamanian village. *Arch. Neurol., Chicago* **7**, 37–44.

Levin, L. S., *et al.* (1977). A heritable syndrome of craniosynostosis, short, thin hair, dental abnormalities and short limbs: Cranioectodermal dysplasia. *J. Pediat.* **90**, 55–61.

Levin, P. M. (1936). Multiple hereditary hemangioblastomas of the nervous system. *Arch. Neurol. Psychiat., Chicago* **3**, 384–91.

Levine, I. M., Estes, J. W., and Looney, J. M. (1968). Hereditary neurological disease with acanthocytosis. *Arch. Neurol., Chicago* **19**, 403–9.

Levine, M. D., Rimoin, D. L., and Lachman, R. (1975). Familial frontal dysplasia. *Birth Defects* **11**, 313–14.

Levine, S. L., Manniello, R. L., and Farrell, P. M. (1977). Familial dysautonomia: unusual presentation in an infant of non-Jewish ancestry. *J. Peidat.* **90**, 79–81.

Levy, H. L., Erickson, A. M., Lott, I. T., and Kurtz, D. J. (1973). Isovaleric acidemia: results of family study and dietary treatment. *Pediatrics, Springfield* **52**, 83–94.

Levy, J. A. and Wittig, E. O. (1962). Familial proximal muscular atrophy. *Arch. Neuropschiat.* **20**, 233–7.

Levy, M. De L. (1951). Au sujet de deux cas de maladie heredo- degenerative du systeme nerveux. *Revta port. Pediat. Pueric.* **14**, 313–18.

Lewis, P. (1964). Familial orthostatic hypotension. *Brain* **87**, 719–28.

Lewitt, P.A. *et al.* (1983). Episodic hyperhidrosis, hypothermia, and agenesis of corpus callosum. *Neurol.* **33**, 1122–9.

Lewkonia, R.M. and Buxton, P.H. (1973). Myositis in father and daughter. *J. Neurol. Neurosurg. Psychiat.* **36**, 820–5.

Lewkonia, R.M. and Lowry, R.B. (1983). Progressive hemifacial atrophy (Parry–Romberg syndrome). Report with review of genetics of nosology. *Am. J. med. Genet.* **14**, 385–90.

Lhermitte, F., Gautier, J.C., and Rosa, A. (1973). Neuropathie recurente familiale. *Revue neurol.* **128**, 419–24.

Liano, H., Ricoy, J.R., Diaz-Flores, L., and Gimeno, A. (1974), A sporadic case of presumed Pelizaeus-Merzbacher disease. *Eur. Neurol.* **11**, 304–34.

Lichtenstein, H. and Knorr, A. (1930). Uber einige Falle von fotschreitender Schwerhörogkeit bei hereditärer Ataxie. *Dt. Z. Nervenheilk.* **114**, 1–28.

Lichenstein, J.R., Bilbrey, G.L., and Mckusick, V.A. (1972). Probable genetic heterogenity with mucopolysaccharidosis II. Report of a family with the mild form. *Johns Hopkins med, J.* **131**, 425–35.

Lieberman J.S., *et al.* (1980). Peripheral neuropathy as an early manifestation of Krabbe's disease. *Arch. Neurol.* **37**, 446–7.

Liechti-Gallati, S., *et al.* (1987). Familial deletion in Becker type muscular dystrophy within the pXJ region. *Hum. Genet.* **77**, 267–8.

Lima, B., Neves, B., and Nora, M. (1987). Juvenile parkinsonism: clinical and metabolic characteristics. *J. Neurol. Neurosurg. Psychiat.* **50**, 327–32.

Lindau, A. (1926). Studien über kleinhirncystenßau, Pathogenese and Beziechungen zur Angiomatosis Retinae. *Acta path. microbiol. scand.* **1**, 1–128.

Linde, L.M. (1955). Dysautonomia, case report of a variant. *J. Pediat.* **46**, 453.

Lindenauer, S.M. (1965). The Klippel-Trenaunay syndrome. Varicosity, hypertrophy, and hemangioma with an arteriovenous fistula. *Ann. Surg.* **162**, 303–14.

Lindenberg, R. (1960). Discussion remark. *J. Neuropath, exp. Neurol.* **19**, 160.

Lindenov, H. (1945). *The etiology of deaf-mutism with special reference to heredity.* Munksgaarl, Copenhagen.

Lindhout, D., *et al.* (1980). The Joubert syndrome associated with bilateral choreoretinal coloboma. *Eur. J. Pediat.* **134**, 175–6.

Lindsay, J.M.J. (1971). Genetics of epilepssy: a model from critical path analysis. *Epilepsia* **12**, 47–54.

Lingam, S., Wilson, J., Naser, H., and Mowat, A.P. (1987). Neurological abnormalities in Wilsons disease are reversible. *Neuropediat.* **18**, 11–12.

Lipinski, C.H. and Stenzel, K. (1974). Foramina parietalia permagna. *Neurol. Pediat.* **5**, 376–82.

Lipton, E.L. and Morgenstern. S.H. (1955). Arthrogryposis multiplex congenita in identical twins. *Am. J. Dis. Childh.* **89**, 233–6.

Lison, M., Kombrut, B., Feinstein, A., Miss, Y., Boichis, H., and Goodman, R.M. (1981). Progressive spastic paraparesis, vitiligo, premature graying and distinct facial appearance: a new genetic syndrome in three sibs. *Am. J. med. Genet.* **9**, 361–7.

Liss, L., Paulson, G., and Sommer, A. (1973). Rigid form Huntington's chorea: a clinicopathological study of three cases. *Adv. Neurol.* **1**, 405–24.

Little, B.W., *et al.* (1986). Familial myoclonic dementia masquerading as Creutzfeldt–Jakob disease. *Ann. Neurol.* **20**, 231–9.

Littler, W.A. (1970). Heart block and peroneal muscular atrophy. *Q. J. Med.* **39**, 431–40.

Liu, H. (1970). Ultrastructure of globoid leucodystrophy (Krabbe's disease) with reference to the origin of globoid cell. *J. Neuropath. exp. Neurol.* **29**, 441–62.

Liu, M.C. and Sylvester, P.E. (1960). Familial diffuse progressive encephalopathy. *Arch. Dis. Childh.* **35**, 345–51.

Livingstone, I.R., Gardner-Medwin, D. Penning T. (1984). Familial intermittent ataxia with possible X-linked recessive inheritance. Two patients with abnormal pyruvate metabolism and a response to acetazolamide. *J. Neurol, Sci.* **64**, 89–97.

Lockman, L.A., Kennedy, W.R., and White, J.G. (1967). The Chediak-Higashi syndrome. *J. Pediat.* **70**, 942–51.

Loeb, H., Tondeur, M., Toppet, M., and Cremer, N. (1969). Clinical, biochemical and ultrastructural studies of atypical form of muco-polysaccharidosis. *Acta pediat. scand.* **58**, 220–8.

Logigian, E.L., *et al.* (1986). Myoclonus epilepsy in two brothers: clinical features and neuropathology of a unique syndrome. *Brain* **109**, 411–29.

Logue, V., Durward, M., Pratt, R.T.C., Piercy, M., and Nixon, W.L.B. (1968). The quality of survival after rupture of an anterior cerebral aneurysm. *Br. J. Psychiat.* **114**, 137–60.

Londe, P. (1984). Paralysie bulbaire progressive infantile et familiale. *Revue méd.* **14**, 212–54.

Longhi, R., *et al.*. (1985). Phenylketonuria due to dihydropterine reductase deficiency: presentation of two cases. *J. Inherit. Met. Dis.* **8**, Suppl. 97–8.

Lonsdale, D., Faulkner, W.R., Price, J.W., and Semby, R.R. (1969). Intermittent cerebellar ataxia associated with hyperpyruvic acidemia, hyperphenylalaninaemia and hyperalaninuria. *Pediatrics, Springfield.* **43**, 1025–34.

Loonen, M.C.B., *et al.* (1984). Combined sialidase (neuraminidase) and beta-galactosidase deficiency. Clinical, morphological and enzymological observations in a patient. *Clin. Genet.* **26**, 139–49.

Loonen, M.C.B., *et al.* (1985). Late-onset globoid cell leucodystrophy (Krabbe's disease). Clinical and genetic delineation of two forms and their relation to the early- infantile form. *Neuropediatrics* **16**, 137–42.

Lope, E.S., Junquera, S.R.Y.C., and Berenguel, A.B. (1974). Progressive myoclonic epilepsy with Lafora's bodies. *Acta neurol. scand.* **50**, 537–52.

Lopez, F., Velez, H., and Toro, G. (1969). Hartnup diesease in two Colombian siblings. *Neurol., Minneap.* **19**, 71–6.

Lorber, J. (1971). What are the chances for the next generation? *Link* **20**, 10.

Lordero, J.F. and Holmes, L.B. (1978). Phenotypic overlap of the BBB and G syndrome. *Am. J. med. Genet* **2**, 145–52.

Loria-Cortes, R., Quesada-Calvo, E., and Cordero-Chruerr, I. (1977). Osteopetrosis in children. A report of 26 cases. *J. Pediat.* **91**, 43–7.

Lother, K. (1959). Familiares Vorkommen von Foramina parietalia permagna. *Arch. Kindesheilk.* **160**, 156.

Louis-Bar, D. (1941). Sur un syndrome progressif comprenant des telangiectasies capillaires cutanées et conjonctivales symetriques, a disposition naevoide et des troubles cerebelleux. *Confina Neurol.* **4**, 32–42.

Louis-Bar, O. and Pirot, G. (1945). Sur une paraplegie spasmodique avec degenerescence maculaire chez dues freres. *Ophthalmoplogia* **109**, 32–43.

Love, S., Duchen, L.W. (1982). Familial cerebellar ataxia with cerebrovascular amyloid. *J.N.N.P.* **45**, 271–3.

Lovell, H.W. (1932). Familial progressive bulbar paralysis. *Arch. Neurol. Psychiat., Chicago* **28**, 394–8.

Low, P.A., Burke, W.J., and McLeod, J.G. (1978). Congenital sensory neuropathy with selective loss of small myelinated fibres. *Ann.Neurol.* **3**, 179–82.

Lowden, B.A. and Harris, G.S. (1976). Pheochromocytoma and Von Hippel–Lindau's disease. *Can. J. Ophtahal.* **11**, 282–9.

Lowden, J.A., Callahan, J.W., Norman, M.G., Thain, M., and Prichard, J.S. (1974). Juvenile GM_1 gangliosidossis. *Arch. Neurol., Chicago* **31**, 200–3.

Lowden, J.A., Cutz, E., Conen, P.E., Rudd, N., and Doran, T.A. (1973). Prenatal diagnosis of GM_1 gangliosidosis. *N. Engl. J. Med.* **288**, 225–8.

Lowden, J.A., Ives, E.J., Keene, D.L., Burton, A.L., Skomorowski, M.A., and Howard, F.(1978). Carrier detection in Sandhoff disease. *Am. J. hum. Genet.* **30**, 38–45.

Lowden, J.A., Laramge, M.A., and Wentworth, P. (1967). The subacute form of Niemann–Pick disease. *Arch. Neurol., Chicago* **20**, 227–38.

Lowden, J.A. and O'Brien, J.S. (1979). Sialidosis: a review of human neuraminidase deficiency. *Am. J. hum. Genet.* **31**, 1–18.

Lowe, C.R. (1973). Congenital malformations among infants born to epileptic women. *Lancet* **i**, 9–10.

Löwenberg, K. and Hill, T.S. (1933). Diffuse sclerosis with preserved myelin islands. *Archs Neurol. Psychat., Chicago* **29**, 1232–45.

Lowenberg, V. and Waggoner, R.W. (1934). Familial organic psychosis (Alzheimers type). *Arch. Neurol. Psychiat., Chicago* **31**, 737–54.

Löwenthal, A. (1954). Un groupe heredodegeneratit nouveau: les myoscleroses heredofamiales. *Acta neurol. belg.* **54**, 155–65.

Lowenthal, A., Bekaert J., Van Dessel, F., and Van Hauwaert, J. (1979). Familial cerebellar ataxia with hypogonadism. *J. Neurol.* **222**, 75–80.

Lowry, R.B. (1972). Congenital absence of the fibula and craniosynostosis in sibs. *J. med. Genet.* **9**, 227–9.

Lowry, R.B. (1974). Holoprosencephaly. *Am. J. Dis. Child.* **128**, 887.

Lowry R.B. (1982). Invited editorial comment: early onset Cockayne syndrome. *Am. J. Med. Genet.* **13**, 209–10.

Lowry, R.B. and Renwick, S.H.G. (1971). The relative frequency of the Hurler and Hunter syndromes. *N. Engl. J. Med.* **284**, 221.

Lowry, R.B. and Wood, B.J. (1975). Syndrome of epiphyseal dysplasia, short stature, microcephaly and nystagmus. *Clin. Genet.* **8**, 269–74.

Lucas, R.N. (1977). Migraine in twins. *J. Psychosom. Res.* **21**, 147–56.

Lüers, T. (1947). Über die familiáre juvenile form der Alzheimerschen Krankheiit mit neurologischen Hederscheinungen. *Arch. Psychiat. NervKrankh.* **179**, 132–45.

Lugaresi, E., Gambetti, P., and Giovannardi-Ross, R. (1966). Chronic neurogenic muscle atrophies of infancy – their nosological relationship with Werdnig–Hoffman disease. *J. neurol. Sci.* **3**, 399–409.

Lugaresi, E., Cirignotta, F., and Montagna, P., (1986). Nocturnal paroxysmal dystonia. *J. Neurol Neurosurg. Psychiat.* **49**, 375–80.

Lund, M. (1949). On epilepsy in Sturge–Weber's disease. *Acta psychiat. neurol. scand.* **24**, 569–86.

Lunberg, P.O., Stalberg E., and Thiele, B, (1974). Paralysis periodica paramyotonica. *J. neurol. Sci.* **21**, 309–21.

Lunberg, P.O. and Westerberg, C.C. (1969). A hereditary neurological disease with facial spasm. *J. neurol. Sci.* **8**, 85–100.

Lundborg, H.B. (1903). *Die progressive Myoklonys-epilepsie (Unverricht's Myoklonie).* Almqvist, Uppsala.

Lundborg, H.B. (1912). Der Ebgabg der progressiven Myoklonus-epilepsie (Myoklonusepilepsie s. Unverricht's familiare Myoklonie). *Z. ges. Neurol. Psychiat.* **9**, 353–8.

Lundemo, G. and Persson, H.E. (1985). Hereditary essential myoclonus. *Acta Neurol. Scand.* **72**, 176–9.

Lundsgaard, R. (1944). Leber's disease: a genealogic, genetic and clinical study of 101 cases of retrobulbar optic neuritis in 20 Danish families. *Acta ophthal.* Suppl. 21, 300–8.

Luo, Y. and Huang, K. (1984). Spongy degeneration of the CNS in infancy. *Arch. Neurol.* **41**, 164–70.

Lurie, I. W., Cherstvoy, E. D., Lazjuk, G. I., Nedzved, M. K., and Usdev, S. S. (1976). Further evidence for the autosomal recessive inheritance of the COFS syndrome. *Clin. Genet.* **10**, 343–6.

Lüthy, F. and Bischoff, A. (1961). Die Pelizaeus–Merzachersche Krankeheit. *Acta neuropath.* **1**, 113–14.

Lutschg, J., *et al.* (1985). Heterogeneity of congenitial motor and sensory neuropathies. *Neuropediatrics.* **16**, 33–8.

Lygidakis, C., Tsakanikas, C., Ilias, A., and Vassilopoulos, D. (1979). Melkersson–Rosenthal's syndrome in four generations. *Clin. Genet.* **15**, 189–92.

Lynn, R. B., *et al.* (1980). Agenesis of the corpus callosum. *Arch. Neurol.* **37**, 444–5.

Lyon, B. B. (1975). Peripheral nerve involvement in Batten–Spielmeyer–Vogt's disease. *J. Neurol. Neurosurg. Psychiat.* **38**, 175–9.

Maas, O. (1904). Ueber ein selten beschriebenes familiäres Nervenleiden. *Berl. klin. Wschr.* **41**, 832–3.

Mabry, C. C., Roeckel. I. E., Munich, R. L., and Robertson, D. (1965). X-linked pseudohypertrophicc muscular dystrophy with a late onset and slow progression. *N. Engl. J. Med.* **273**, 1061–70.

McArdle, B. (1962). Adynamia episidica hereditaria and its treatment. *Brain* **85**, 121–48.

McArthur, R. G., Hayles, A. B., Gomez, M. R., and Bianco, A. J. Jr. (1969). Carpal tunnel syndrome and trigger finger in childhood. *Am. J. Dis. Childh.* **117**, 463–9.

McAuley, D. L., Isenberg, D. A., and Gooddy, W. (1978). Neurological involvement in the epidermal naevus sundrome. *J. Neurol. Neurosurg. Psychiat.* **411**, 466–9.

Maccario, M., *et al.* (1983). A sibship with neuroaxonal dystrophy and renal tubular acidosis. *Ann. Neurol.* **13**, 608–15.

McColl, K. E. L., Moore, M. R., Thompson, G. G., and Goldberg, A. (1982). Screening for latent acute intermittent porphyria the value of measuring both leucocyte delta-aminolaevulinic acid synthetase and erythrocyte uroporphyrinogen—synthase activities. *J. med. genet.* **19**, 271–6.

McComas, C. E., *et al.* (1983). The constellation of adult acid maltase deficiency—clinical, electrophysiologic, and morphologic features. *Clin. Neuropath.* **2**, 182–7.

McCormick, K., *et al.* (1985). Partial pyruvate decarboxylase deficiency with profound lactic acidosis and hyperammonemia: response to dichloroacetate acid benzoate. *Am. Med. Genet.* **22**, 291–9.

McDonald, W.I. (1987). The pathogenesis of multiple sclerosis. *J. Roy. Coll. Physic.* 21, 287–94.

Mace, J.W., Sponaugle, H.D., Mitsunaga, R.Y., and Schanberger, J.E. (1971). Congenital hereditary nonprogressive external ophthalmoplegia. *Am. J. Dis. Childh.* 122, 261–3.

MacFarlin, D.E., Strober, W., and Waldmann, T.A. (1972). Ataxia-telangiectasia. *Medicine, Baltomore* 51, 281–314.

McGarry, J., Garg, B., and Silbert, S. (1983). Death in childhood due to facio-scapulo-humeral dystrophy. *Acta Neurol. Scand.* 68, 61–3.

MacIlroy, J.H. (1930). Hereditary ptosis with epicanthus: a case with pedigree extending over four generations. *Proc. Roy. Soc. Med.* 23, 285.

McIntyre, C.A. and Brown, H.W. (1965). Twins with cachectic dwafism. *J. Pediat.* 67, 1204–6.

MacKay, R.P. (1963). Course and prognosis in amyotrophic lateral sclerosis. *Arch. Neurol., Chicago* 8, 117–27.

MacKay, R.P. and Myrianthopoulos, N.C. (1958). Multiple sclerosis in twins and their relatives. Prelimary report on a genetic and clinical study. *Arch. Neurol. Psychiat., Chicago* 8, 117–27.

MacKay, R.P. and Myrianthopoulos (1966). Multiple sclerosis in twins and their relative. *Arch. Neurol., Chicago* 15, 449–62.

MacKeith, R.C., MacKenzie, I.C.K., and Polani, P.E. (1959). Definition of cerebral palsy. *Cerebr. Palsy Bull.* 5, 23.

McKeran, R.O., *et al.* (1985). Neurological involvement in type I (adult) Gaucher's disease. *J. Neurol. Neurosurg. Psychiat.* 48, 172–5.

McKinney, J.M. and Frocht, M. (1940). Adie's syndrome: a non-luetic disease simulating tabes dorsalis. *Am. J. med. Sci.* 199, 546–55.

McKusick, V.A. (1960). *Heritable disorders of connective tissue*, 2nd edn. Mosby, St. Louis.

McKusick V.A. (1971). X-linked muscular dystrophy; benign form with contractures. *Birth Defects* 7, 113–15.

McKusick, V.A. (1988). Mendelian inheritance in man. In *Catalogues of autosomal dominant, autosomal recessive and X-linked phenotypes*. 8th edn. Johns Hopkins University Press, Baltimore.

McKusick, V.A., Howell, R.R., Hussels, I.E., Neufeld, E.F., and Stevenson, R.E. (1972). Allelism, non-allelism and genetic compounds among the mucopolysaccharidoses: hypothesis. *Lancet* i, 993–6.

Maclaren, N.K., *et al.* (1976). GM_3-gangliosidosis: a novel human sphingolipodystrophy. *Pediatrics, Springfield* 57, 106–10.

McLaughlin, T.G., Krovetz, L.J., and Schiebler, G.L. (1964). Heart disease in the Laurence–Moon–Biedl syndrome. *J. Pediat.* 65, 388–99.

McLean, W.T. and McKone, R.C. (1973). Congenital myasthenia gravis in twins. *Arch. Neurol., Chicago* 29, 223–6.

McLeod, J.G., Baker, W. de C., Lethlean, A.K., and Shorey, C.D. (1972).

Centronuclear myopathy with autosomal dominant inheritance. *J. neurol. Sci.* **15**, 375–7.

McLeod, J.G., Low, P.A., and Morgan, J.A. (1978). Charcot–Marie–Tooth disease with Leber optic atrophy. *Neurol., Minneap.* **28**, 179–84.

McLeod J.G., Morgan, J.A., and Reye, C. (1977). Electrophysiological studies in familial spastic paraplegia. *J. Neurol. Neurosurg. Psychiat.* **40**, 611–15.

McLeod, J.G. and Prineas, J.W. (1971). Distal type of chronic spinal muscular atrophy. *Brain* **94**, 703–14.

McLeod, J.G. and Williams, I.M. (1971). Spinal muscular atrophy *Minn. Med.* **54**, 457–61.

Macleod, P.M., Wood, S., Jan, J.E., Applegarth, D.A., and Dolman, C. (1977). Progressive cerebellar ataxia, spasticity, psychomotor retardation, and hexosaminidase deficiency in a 10 year old child: juvenile Sandhoff disease. *Neurol. Minneap.* **27**, 571–3.

Macleod F.M., *et al.* (1985). Prenatal diagnosis of neuronal ceroid lipofuscinoses. *Am. J. Med. Genet.* **22**, 781–9.

McMenemey, W.H., Worster-Drought, C., Flind, J., and Williams, H.G. (1939). Familial presenile dementia. *J. Neurol. Psychiat.* **2**, 293–302.

McNamara, J.O., Curran, J.R., and Itabashi, H.H. (1975). Congenittal ichthyosis with spastic paraplegia of adult onset. *Arch. Neurol., Chicago* **32**, 699–701.

McNicholl, B., Egan-Mitchell, B., Murray, J.P., Doyle, J.F., Kennedy, J.D., and Crome. L. (1970). Cerebro-costo-mandibular syndrome. A new familial developmental disorder. *Arch. Dis. Childh.* **45**, 421–4.

McPherson, E. and Tayler, C.A. (1982). The genetics of malignant hyperthermia: evidence for heterogeneity. *Am. J. Med. Genet.* **11**, 273–85.

McPherson, E., Robertson, C., Cammarano, A., and Hall, J.G. (1976). Dominantly inherited ptosis, strabismus and ectopic pupils. *Clin. Genet.* **10**, 21–6.

MacPherson, R.I. (1974). Craniodiaphyseal dysplasia a disease or group of diseases? *J. Can. Ass. Radiol.* **25**, 22.

McQuillen, M. (1966). Familial limb-girdle myasthenia. *Brain* **89**, 121–32.

Macrae, H.M. and Newbigin, B. (1968). Von Hippel-Lindau disease. A family history. *Can. J. Opthal.* **3**, 28–34.

Macrae, W., Stieffel, J., and Todorov, A.A. (1974). Recessive familial spastic paraplegia with retinal degeneration. *Acta Genet. med. Gemell.* **23**, 249–52.

McWilliam, R.C. and Stephenson, J.B.P. (1978). Depigmented hair. The earliest signs of tuberous sclerosis. *Arch. Dis. Childh.* **53**, 961–3.

Madrid, R. and Bradley, W.G. (1975). The pathlogy of neuropathies with focal thickening of the myelin sheath (tomaculous neuropathy) : studies on the formation of the abnormal myelin sheath. *J. neurol. Sci.* **25**, 415–48.

Maere, M. and Muyle, G. (1938). Un syndrome d'ataxie cerebelleuse progressive avec oligophrenie chez deux jeunes israelities polonais *J. belge Neurol. Psychiat.* **38**, 96–107.

Maeshir O. H. *et al.* (1980). An unclassiful case of degenerative disease of the central nervous system – with reference to hereditary pallidal and dentate system atrophy. *Psychiatrica et Neurologica Japanica,* **20**, 863.

Magee, K. R. (1960). Familial progressive bulbar-spinal muscular atrophy. *Neuro., Minneap.* **10**, 295.

Magee, K. R. (1963). A study of paramyotonia congenita. *Arch. Neurol., Chicago* **8**, 461–70.

Magee, K. R. (1966). Paramytonia congenita. Association with cutaneous cold sensitivity and description of peculiar sustained posture after muscle contraction. *Arch. Neurol., Chicago* **14**, 590–4.

Magee, K. R. and de Jong, R. N. (1960). Neurogenic muscular atrophy simulating muscular dystrophy. *Arch. Neurol., Chicago* **2**, 677–82.

Magee, K. R. and de Jong R. N. (1965). Hereditary distal myopathy with onset in infancy. *Arch. Neurol., Chicago* **13**, 387–90.

Mahloudji, M. (1963). Hereditary spastic ataxia disseminated sclerosis. *J. Neurol. Neurosurg. Psychiat,* **26**, 511–13.

Mahloudji, M. (1969). A recessively inherited mixed polyneuropathy of early onset. *J. med. Genet.* **6**, 411–12.

Mahloudji, M., Daneshbod, K. and Karjoo, M. (1970). Familial spongy degeneration of the brain. *Archs Neurol., Chicago* **22**, 294–8.

Mahloudji, M., Amirhakimi, G. H., Haghighi, P., and Khodadoust, A. A. (1972). Marinesco–Sjögren syndrome. *Brain* **95**, 675–80.

Mahloudji, M. and Chuke, P. O. (1968). Familial spastic paraplegia with retinal degeneration. *Johns Hopkins med. J.* **123**, 142–4.

Mahloudji, M. and Pikielny, R. T. (1967). Hereditary essential myoclonus. *Brain* **90**, 669–74.

Mahloudji, M., Teasdall, R. D., Adamkiewics, J. J., Hartmann, W. H., Lambird, P. A., and McKusick, Y. A. (1969). The genetic amyloidoses, with particular reference to hereditary neuropathic amyloidosis, type II (Indiana or Rukavina type). *Medicine, Baltimore* **48**, 1–137.

Maia, M. (1974). Sjögren–Larsson syndrome in two sibs with peripheral nerve involvement and bisalbuminaemia. *J. Neurol. Neurosurg. Psychiat.* **37**, 1306–15.

Maia, M., Pires, M. M., and Guimaraes, A. (1988). Giant axonal disease: Report of three cases and review of the literature. *Neuropediatrics* **19**, 10–15.

Mailander, J. C. (1967). Hereditary gustatory sweating. *J. Am. med. Ass.* **202**, 203–4.

Makita, K. (1968). The rarity of reading disability in Japanese children. *Am. J. Orthopsychiat.* **38**, 599–614.

Malamud, N. and Cohen, P. (1958). Unusual form of cerebellar ataxia with sex-linked inheritance. *Neurol., Minneap.* **8**, 261–6.

Malamud, N. and Waggoner, R.W. (1943). Genelogic and clinico pathologic study of Pick's disease. *Arch. Neurol. Psyschiat. Chicago* **50**, 288–303.

Malek, R.S. and Greene, L.F. (1971). Urologic aspects of Hippel–Lindau syndrome. *J. Urol.* **106**, 800–1.

Malin. J.-P. (1979). Familial meralgia paresthetica with an autosomal dominant trait. *J. Neurol.* **221**, 133–6.

Mandell, A.J. and Smith, C.K. (1960). Hereditary sensory radicular neuropathy. *Neurol., Minneap.* **10**, 627–30.

Mandeville, F.B. and Sahyoun, P.F. (1949). Benign and malignant pheochromocytomas with necropsies: benign case with multiple neurofibromatosis and cavernous hemangioma of fourth ventricle: malignant case with widespread metastasis and bronchogenic carcinoma. *J. Urol.* **62**, 93–103.

Mankowsky, B.N. and Czerny, L.I. (1929). Sur Erage über die Heredität der Torsionsdystonie. *Mschr. Psychiat. Neurol.* **72**, 165–79.

Mann, J.B., Alterman, S., and Hills, A.G. (1962). Albright's hereditary osteodystrophy, comprising pseudohypoparathyroidism and pseudo-pseudohypoparathyroidism with a report of two cases representing the complete syndrome occuring in successive generations. *Ann. intern. Med.* **56**, 315–42.

Marie, P. (1893). Sur l'herédo-ataxie cérébelleuse. *Sem. med.* **13**, 444–7.

Marie, P. (1906). Forme spéciale de névrite interstitielle hypertophiquc progressive de l'enfance. *Revue neurol.* **14**, 557–8.

Marinesco, G. (1915). Sur duex cas de paralysie bulbaire progressive infantile et familiale. *C.r. Séanc. Soc. Biol.* **78**, 481–3.

Marinesco, G., Draganesco, S.T., and Vasiliu, D. (1931). Nouvelle maladie familiale caracterise′ par une cataracte congenitale et un arrét du development somatoneurophsique. *Encephalé* **26**, 97–109.

Markand, O.N., North, R.R., and D'Agostino, A.N., and Daly, D.D. (1969). Benign sex-linked muscular dystrophy. Clinical and pathological features. *Neurology, Minneap.* **19**, 617–33.

Markesbery, W.R., Griggs, R.C., Leach, R.P., and Lapham, L.W. (1974). Late onset hereditary distal myopathy. *Neurol., Minneap.* **24**, 127–34.

Markham, C.H. (1969). Huntington's chorea in childhood. In *Progress in neurogenetics* (ed. A. Barneau and J.R. Brunette), pp. 651–60. Excerpta Medica, Amsterdam.

Markham, C.H. and Knox J.W. (1965). Observations on Huntington's chorea in childhood. *J. Pediat.* **67**, 46–57.

Marmor, M.F. (1973). Hereditary vertical nystagmus. *Arch. Ophthal. N.Y.* **90**, 107–11.

Maroteaux, P., Fontaine, G., Scharfman, W., and Farriaux, J.P. (1971). L'hyperostose corticale generalisee a transmission dominante. *Arch. fr. Pédiat.* **28**, 685–91.

Marquardt, J.L. and Loriaux, L. (1974). Diabetes mellitus and optic atrophy. *Archs. intern. Med.* **134**, 32–7.

Mars, H., Lewis, L.A., Robertson, A.L., Butkus, A., and Williams, C.H. (1969). Familial hypo-ß-lipoproteinemia. *Am. J. Med.* **46**, 886–900.

Marsden, C.D. and Parkes, J.D. (1973). Abnormal movement disorders. *Br. J. hosp. Med.* **10**, 428–50.

Marsden, C.D., Obeso, J.A., and Lang, J.E. (1982). Adrenoleukomyelo-neuropathy presenting as a spinocerebellar degeneration. *Neurol., Minneap.* **32**, 1031.

Marsden, C.D., *et al.* (1986). Familial dystonia and visual failure with striatal CT lucencies. *J. Neurol. Neurosurg. Psychiat.* **49**, 500–7.

Marshall, D. (1954). Glioma of the optic nerve as manifestation of von Recklinghausen's disease. *Am. J. Ophthal.* **37**, 15–36.

Marshall, D., Saul, G.B., and Sachs, E. Jr (1959). Tuberous sclerosis: a report of 16 cases in two family trees revealing genetic dominance. *N. Engl. J. Med.* **261**, 1102–5.

Marshall, J. (1971). Familial incidence of cerebrovascular disease. *J. med. Genet.* **8**, 84–9.

Marshall, J. (1973) Familial incidence of cerebral hemorrhage. *Stroke* 4, 38–41.

Marshall, R.E. and Smith, D.W. (1970). Fronto-digital syndrome: a dominantly inherited disorder with normal intelligence. *J. Pediat.* **77**, 129–33.

Marstein, S., Jellum, E., Halpern, B., Eldjarn, L., and Perry, T.L. (1976). Biochemical studies of erythrocytes in a patient with pyroglutamic acidemia (5-oxo prolinemia). *N. Engl. J. Med.* **295**, 406–12.

Martin, J.J. and Martin, L. (1972). Infantile neuro-axonal dystrophy. *Eur. Neurol.* **8**, 239–50.

Martin, A.O., Perrin, J.C.S., Muir, W.A., Ruch, E., and Schafer, I.A. (1977). An autosomal dominant midline cleft syndrome resembling familial holoprosencephaly. *Clin. Genet.* **12**, 65–72.

Martin, G.I., Kaiserman, D., Liegler, D., Amorosi, E.D., and Nadel, H. (1976). Computer-assisted cranial tomography in early diagnosis of tuberous sclerosis. *J. Am. med. Ass.* **235**, 2323–4.

Martin, J.J. and Martin, L. (1974). Infantile form of Hallervorden–Spatz disease. An ultrastructural examination of motor end plates as a contribution to the differentiation between Hallervorden–Spatz disease and infantile neuro-axonal dystrophy. *Clin Neurol. Neurosurg.* **77**, 26–37.

Martin, J.J., Navarro, C., Roussel, J.M., and Michielsse, P. (1973*a*).

Familial capillarovenous leptomeningeal angiomatosis. *Eur. Neurol.* **9**, 202–15.

Martin, J.J., DeBarsy, Th., and Den Tandt, W.R. (1976). Acid maltase deficiency in non-identical twins-A morphological and biochemical study. *J. Neurol.* **213**, 105–18.

Martin, J.J., *et al.* (1983). Selective thalamic degeneration—report of a case with memory and mental disturbances. *Clinical Neuropath* **2**, 156–62.

Martin, W.E., Resch, J.A., and Baker, H.B. (1971). Juvenile Parkinsonism. *Arch. Neurol., Chicago,* **25**, 494–500.

Martin, W.E., Young, W.I., and Anderson, V.E. (1973*b*). Parkinson's disease—a genetic study. *Brain* **96**, 495–506.

Martinez-Martin, F. and Pareja, F.B. (1985). Familial writer's cramp. *J. Neurol. Neurosurg. Psychiat.* **48**, 487.

Martin-Sneessens, L. (1962). Formes a evolution tres prolongee de l'amytrohie spinale de Werdnig–Hoffman. *J. Génét. hum.* **11**, 251–69.

Martin-Sneesens. L. and Radermecker, J. (1965). Amyotrophie neurogene classique (Werdnig–Hoffman) et pseudo-myopathique infantile dans une fratrie. *J. Génét. hum.* **14**, 341–50.

Martsolf, J.T., Cracco, J.B., Carpenter, G.G., and O'Hara, A.E. (1971). Pfeiffer syndrome: an unusual type of acrocephalosyndactyly with broad thumbs and great toes. *Am. J. Dis. Child.* **121**, 257–62.

Marttila, R.J. and Rinne, U..K. (1976). Epidemiology of Parkinson's disease in Finland. *Acta neurol. scand.* **53**, 81–102.

Martyn, L.J. and Knox, D.L. (1972). Glial hamartoma of the retina in generalised neurofibromatosis: van Recklinghausen's disease. *Br. J. Ophthal.* **56**, 487–91.

Marxmiller, J., Trenkle, I., and Ashwal, S. (1985). Rud syndrome revisited: ichthyosis, mental retardation, epilepsy and hypogonadism. *Dev. Med. Child Neurol.* **27**, 335–43.

Massey, E.W. (1978). Familial occurrence of meralgia paresthetica. (Letter.) *Arch. Neurol., Chicago* **35**, 182.

Massey, E.W., Brannon, W.L., and Moreland, M. (1979). Nevus of Ota and intracranial arteriovenous malformation. *Neurol., Minneap.* **29**, 1625–7.

Massion-Verniory, L., Dumont, E., and Potvin, A.A. (1946). Retinite pigmentaire familiale compliquee d'une amyotrophie neurale. *Rev. neurol.* **78**, 561–71.

Masters, C.L., Gajdusek, D.C., and Gibbs, C.J. (1981). Creutzfeldt-Jakob disease virus isolations from the Gerstmann-Sträussler syndrome. *Brain.* **104**, 559.

Mastromauro, C., Myers, R.H., and Berkman, B. (1987). Attitudes toward presymptomatic testing in Huntington disease. *Am. J. Med. Genet.* **26**, 271–82.

Mata, M., *et al..* (1983). New form of familial Parkinson-dementia syndrome. Clinical and pathological findings. *Neurology* 33, 1439–43.

Matalon, R., Arbogast, B., Dorfman, A. (1974). Morquio's syndrome: a deficiency of chondroitin sulfate N-acetylhexosamine sulfate sulfatase. (Abstract.) *Pediat. Res.* 8, 431.

Matalon, R., Michals, K., Sebasta. D., *et al.* (1988). Aspartoacylase deficiency and *N*-acetylaspartic aciduria in patients with Canavan disease. *Am. J. Med. Genet.* 29, 463–71.

Mathai, K.V., Dunn, D.P., Kurland, L.T., and Reeder, F.A. (1968). Convulsive disorders in the Mariana Islands. *Epilepsie.* 9, 77–85.

Matsuda, I., *et al.* (1975). Prenatal diagnosis of I-cell disease. *Hum. Genet.* 30, 69–73.

Matsui, S.M., Mahoney, M.J., and Rosenberg, L.E. (1983). The natural history of the inherited methylmalonic acidemias. *N. Engl. J. Med.* 308, 857.

Matsumoto, A., *et al.* (1981). Infantile spasms: etiological factors, clinical aspects, and longterm prognosis in 200 cases. *Eur. J. Pediat.* 135, 239–44.

Matsunaga, M., Inoicuchi, T., Ohnishi, A., and Kuroiwa, Y. (1973). Oculopharyngeal involvement in familial neurogenic muscular atrophy. *J. Neurol. Neurosurg. Psychiat.* 36, 104–11.

Matsuo, M., *et al.* (1981). Congenital insensitivity to pain with anhidrosis in a 2-month-old boy. *Neurology* 31, 1190–2.

Matsuyama, H., Mattes, A. and Weber, H. (1968). Klinische und elektroencephalographische. Familienuntersuchungen bei Pyknolepsien. *Dt. med. Wschr.* 93, 429–35.

Matsuyama, H., Watanabe, I., Mihm, M.C., and Richardson, E.P. (1978). Dermatoleukodystrophy with neuroaxonal spheroids. *Arch. Neurol., Chicago* 35, 329–36.

Matthes, A. and Weber, H. (1968). Clinical and electroencephalographic family studies in pyknolepsies. *Dtsch. Med. Wochenschr.* 93, 429–35.

Matthew, N.T., Jacob, J.C., and Chandry, J. (1970). Familial ocular myopathy with curare sensitivity. *Arch. Neurol., Chicago* 22, 68–74.

Matthews, N.L. (1968). Lentigo and electrocardiographic changes. *N. Engl. Med. J.* 278, 780–1.

Matthews, W.B. (1950). Familial ataxia, deaf-mutism, and muscular wasting. *J. Neurol. Neurosurg. Psychiat.* 13, 307–11.

Matthews, W.B. (1957). Familial calcification the basal ganglia with response to parathormone. *J. Neurol. Neurosurg. Psychiat.* 20. 172–7.

Matthews, W.B. (1975). Epidemiology of Creutzfeldt-Jakob disease in England and Wales. *J. Neurol. Neurosurg. Psychiat.* 38, 210–13.

Matthews, W.B., Howell, D.A., and Stevens, D.L. (1969). Progressive myoclonus epilepsy without Lafora bodies. *J. Neurol. Neurosurg. Psychiat.* 32, 116–22.

Matthews, W. B. and Rundle, A. T. (1964). Familial cerebellar ataxia and hypogondism. *Brain* **87**, 463–8.

Matthews, W. B. and Wright, F. K. (1967). Hereditary primary reading epilepsy. *Neurol., Minneap.* **17**, 919–21.

Mattis, S., French, J. H., and Apin, I. (1975). Dyslexia in children and young adults: three independent neuropsychologocal syndromes. *Dev. med. Child Neurol.* **17**, 150–63.

Mattsson, B. (1974). Huntington's chorea in Sweden. I. Prevalence and genetic data. *Acta psychiat, scand.* Suppl. 225, 211–19.

Mawatari, S., Iwashita, H., and Kuroiwa, Y. (1972). Familial hypo-ß lipoproteinemia. *J. neurol Sci.* **16**, 93.

Mawatari, S. and Katayama, K. (1973). Scapuloperoneal muscular atrophy with cardiomyopathy. An X-linked recessive trait. *Arch. Neurol., Chicago* **28**, 55–9.

Max, S. R., (1974). GM3 (Hematoside) sphingolipodystrophy. *N. Engl. J. Med.* **291**, 929–31.

Max, S. R., Maclaren, N. K., Brady, R. O., Bradley, R. M., Rennels, M. B., Tanaka, J., Garcia, J. H., and Cornblath, M. (1974). Gm(3) hematoside sphingolipodystrophy. *New Eng. J. Med.* **291**, 929–31.

May, D. L. and White, H. H. (1968). Familial myoclonus cerebellar ataxia and deafness: specific genetically-determined disease. *Arch. Neurol., Chicago* **19**, 331–8.

May, W. W., Itabashi, H. H., and De Jong, R. N. (1968). Creutzfeldt–Jakob disease II. Clinical, pathologic and genetic study of a family. *Arch. Neurol., Chicago* **19**, 137–49.

Mayer, J. M., *et al.* (1986). Familial juvenile Parkinsonism with multiple system degeneration. A clinicopathological study. *J. Neurol. Sci.* **72**, 91–101.

Mayeux. F., and Fahn, S. (1982). Paroxysmal dystonic choreoathetosis in a patient with familial ataxia. *Neurology* **32**, 1184–6.

Meadow, S. R. (1968). Anticonvulsant drugs and congenital abnormalities. *Lancet* **ii**, 1296.

Meadows, J. C., and Marsden, C. D., (1969). A distal form of chronic spinal muscular atrophy. *Neurol., Minneap.* **19**, 53–8.

Meadows, J. C., Marsden, C. D., and Harriman, D. G. F. (1969). Chronic spinal muscular atrophy in adults. Part I. The Kugelberg–Welander syndrome. *J. neurol. Sci.* **9**, 527–50.

Meek, D., Wolfe, L. S., Andermann, E., and Andermann, F. (1984). Juvenile progressive dystonia. A new phenotype of GM$_2$ gangliosidosis. *Ann. Neurol.* **15**, 348–52.

Mehler, M. and Di Mauro, S. (1977). Residual acid maltase activity in late-onset acid maltase deficiency. *Neurol., Minneap* **27**, 178–84.

Mehta, L., Trounce, J. Q., Moore, J. R., and Young I. D. (1986). Familial calcification of the basal ganglia with cerebrospinal fluid pleocytosis. *J. Med. Genet.* **23**, 157–60.

Melchior, J.C., Benda, C.E., and Yakovlev, P.I. (1960). Familial idiopathic cerebral calcifications in childhood. *Am. J. Dis. Childh.* **99**, 787–803.

Melmon, K.L. and Rosen, S.W. (1964). Lindau's disease. Review of literature and study of a large kindred. *Am. J. Med.* **36**, 595–615.

Melnick, J.C. and Needles, C.F. (1966). An undiagnosed bone dysplasia. A two family study of four generations and three generations. *Am. J. Roentg.* **97**, 39–48.

Members of the Executive Committee of COMBAT (1976). Genetic counselling in Huntington's chorea. (Letter.) *Br. med. J.* **ii**, 420.

Menkes, JH. (1976). Dystonia musculorum deformans—a status report. *Bull. Los Ang. neurol. Soc.* **41**, 184–8.

Menkes, J.H., Alter, M., Steigleder, G.K., Weakley, D.R., and Sung. J.H. (1962). A sex-linked recessive disorder with retardation of growth, with peculiar hair, and focal cerebral and cerebeller degeneration. *Pediatrics, Springfield* **29**, 764–79.

Menkes, J.H., Philippart, M., and Clark, D.D., (1964). Hereditary partial agenesis of the corpus callosum. *Arch. Neurol., Chicago* **11**, 198–208.

Menzel, P. (1891). Beitrag zur Kenntnis der hereditären Ataxie and Kleinhirnatrophie. *Arch. Psychiat. NervKrankh.* **22**, 160.

Mercellis, R., Demeester, J., and Martin, J.J. (1980). Neurogenic scapuloperoneal syndrome in childhood. *J. Neurol. Neurosurg. Psychiat.* **43**, 888–96.

Meredith, J.M. and Hennigar, G.R. (1954). Cerebellar hemangiomas: A clinicopathologic study of fourteen cases. *Ann. Surg.* **20**, 410–23.

Meretoja, J. (1969). Familial systemic paramyloidosis with lattice dystrophy of the cornea, progressive cranial neuropathy, skin changes and various internal symptoms. *Ann. clin. Res.* **1**, 314–24.

Meretoja, J. (1973). Genetic aspects of familial amyloidosis with corneal lattice dystrophy and cranial neuropathy. *Clin. Genet.* **4**, 173–85.

Meretoja, J. and Teppo, L. (1971). Histopathological findings of familial amyloidosis with cranial neuropathy as principal manifestation. *Acta path. microbiol, scand.* **79A**, 432–40.

Mereu, T., Porter, I.H., and Hug, G. (1969). Myotonia, shortness of stature, and hip dysplasia. Schwartz-Jampel syndrome. *Am. J. Dis. Childh.* **117**, 470–8.

Merlini, L., *et al.* (1986). Emery–Dreifuss muscular dystrophy: Report of five cases in a family and review of the literature. *Muscle nerve* **6**, 481–5.

Merriam, G.R., Beitins, I.Z., and Bode, H.H. (1977). Father to son transmission of hypogonadism with anosmia. *Am. J. Dis. Childh.* **131**, 1216–19.

Merritt, A.D., Conneally, R.M., Rahman, N.F., and Drew, A.L. (1969). Juvenile Huntington's chorea. In *Progress in neurogenetics*, (ed. A. Barbeau and J.R. Brunette) pp. 645–50. Excerpta Medica, Amsterdam.

Merten, D.F., Gooding, C.A., Newton, T.H., and Malamud, N. (1974). Meningiomas of childhood and adolescence. *J. Pediat.* **84**, 696–700.

Merzbacher, L. (1910). Eine eigenartige familiär-hereditare Erkrankungsform (Aplasia axialis extra-corticalis congenita). *Z. ges. Neurol. Psychiat.* **3**, 1–138.

Metcalf, C.W., and Hirano, A. (1971). Amyotrophic lateral sclerosis. Clinicopathological studies of a family. *Arch. Neurol., Chicago.* **24**, 581–23.

Metrakos, J.D. and Metrakos, K. (1960). Genetics of convulsive disorders I. Introduction, problems, methods and base lines. *Neurol, Minneap.* **10**, 228–40.

Metrakos, K. and Metrakos, J.D. (1961). Genetics of convulsive disorders II. Genetic and electroencephalographic studies in centrencephalic epilepsy. *Neurol., Miinneap.* **11**, 470–83.

Metzger, J., Messing, R., and Simon, J. (1962). Cas familial d'anomalie de la charniere occipiloatloidienne: syndrome syringomyelique chez la mere, troubles bulbaires, frustes chez la fille. *Rev. neurol.* **106**, 767–8.

Meyer, E. (1979). Neurocutaneous syndrome with excessive macrohydrocephalus (Sturge–Weber–Kippel–Trenaunay syndrome). *Neuropaediology* **10**, 67–75.

Meyer, J.E. (1949). Uber eine Kombinerte Systomkrankung in Klein–Mittel und Endhirn. *Arch. Psychiat. NervKrankh.* **182**, 732–58.

Meyers, K.R., Golomb, H.M., Hansen, J.L., and McKusick, V.A. (1974). Familial neuromuscular disease with 'myotubes'. *Clin. Genet.* **5**, 327–37.

Michael, J.C. and Levin, P.M. (1936). Multiple telangiectases of the brain. A discussion of hereditary factors in their development. *Arch. Neurol. Psychiat., Chicago* **36**, 514–29.

Michal, S., Ptasinska-Urbanska, M., and Mitkiewicz-Bochenek, W. (1968). Atrophie optique heredofamiliale dominante associee a la surde-mutite. *Annl. Oculist.* **201**, 431–5.

Michielsen, F., *et al.* (1984). Cardiac involvement in juvenile ceroid lipofuscinosis of the Spielmeyer–Vogt–Sjögren type. Prospective noninvasive findings in two siblings. *Eur. Neurol.* **23**, 166–72.

Migeon, B.R., Spenkle, J.A., Liebaers, I., Scott, J.F., and Neufeld, E.F. (1977). X-linked Hunter's syndrome: the heterozygous phenotype in cell culture. *Am. J. hum. Genet.* **29**, 448–54.

Migeon, B.R., *et al.* (1981). Adrenoleukodystrophy: evidence for X-linkage inactivation and selection favouring the mutant allele in heterozygous cells. *Proc. Nat. Acad Sci.* **78**, 5066–70.

Miles, J.H., *et al.* (1984). Macrocephaly with hamartomas: Bannayan–Zonana syndrome. *Am. J. Med. Genet.* **19**, 225–34.

Milhorat, A.T. (1943). Studies in disease of muscle. XIV Progressive muscular atrophy of peroneal type associated with atrophy of the optic nerves; report on a family. *Archs Neurol. Psychiat. Chicago* **50**, 279–87.

Milla, P.J. (1978). I-cell disease. *Arch. Dis. Childh.* **53**, 513–15.

Millar, J.H. and Allison, R.S. (1954). Familial incidence of disseminated sclerosis in Northern Ireland. *Ulster med. J.* **23**, Suppl. 2, 29–92.

Millard, R.D., Maisels, D.O., Batstone, J.H.F., and Yates B.W. (1967). Craniofacial surgery in craniometaphyseal dysplasia. *Am. J. Surg.* **113**, 615–21.

Miller, G.M., *et al.* (1984). Schizencephaly: a clinical and CT study. *Neurology* **34**, 997–1001.

Miller, J.D., McCluer, R., and Kanfer, J.N. (1973), Gaucher's disease: neurologic disorder in adult siblings. *Ann. intern. Med.* **78**, 883–8.

Miller, J.Q. (1963). Lissencephaly in two siblings. *Neurol. Minneap.* **13**, 841–50.

Miller, M. and Hall, J.G. (1978). Possible maternal effect on severity of neurofibromatosis. *Lancet* **ii**, 1071–3.

Miller, R.G., Nielsen, S.L., and Sumner, A.J. (1976). Hereditary sensory neuropathy and tonic pupils. *Neurol. Minneap.* **26**, 931–5.

Miller, R.G., *et al.* (1985). Emery–Dreifuss muscular dystrophy with autosomal dominant transmission. *Neurology* **25**, 1230–3.

Miller, R.M. and Sparkes, R.S. (1977). Segmental neurofibromatosis. *Arch. Derm.* **113**, 837–8.

Millichap, J.G. (1968). *Febrile convulsions*. Macmillan, New York.

Millman, C.G. and Whittick, J.W. (1952). A sex linked variant of gargoylism. *J. Neurol. Neurosurg. Psychiat.* **15**, 253–9.

Minchom. P.E., Dormer, R.L., Hughes, I.A., *et al.* (1983). Fatal infantile mitochondrial myopathy due to cytochrome C oxidase deficiency. *J. Neurol. Sci.* **60**, 453–63.

Mito, T., *et al.* (1986). Infantile bilateral striatal necrosis. *Arch. Neurol.* **43**, 667–80.

Mitsumoto, H., Adelmann, L.S., and Liu H.C. (1982). A case of congenital Werdnig Hoffmann disease with glial bundles in spinal roots. *Ann. Neurol.* **11**, 214–16.

Mitsomoto, H., Sliman, R.J., Schafer, I.A., *et al.* (1985). Motor neuron disease and adult hexosaminidase, a deficiency in two families: evidence for multisystem degeneration. *Ann. Neurol.* **17**, 378–85.

Miyabayashi, S., Ito, T., Narisawa, K., *et al.* (1985). Biochemical study in 28 children with lactic acidosis in relation to Leigh's encephalomyelopathy. *Eur. J. Pediatr.* **143**, 278–83.

Miyatake, T., *et al.* (1979). Adult type neuronal storage disease with neuraminidase deficiency. *Ann. Neurol.* **6**, 232–44.

Miyoshi, K., Matsuoka, T., and Mizushima, S. (1969). Familial holotopostic striatal necrosis. *Acta neuropath.* **13**, 240–9.

Miyoshi, K., *et al.* (1986). Autosomal recessive distal muscular dystophy as a new type of progressive muscular dystrophy: Seventeen cases in eight families including an autopsied case. *Brain.* **109**, 31-54.

Mjönes, H. (1949). Paralysis agitans. *Acta psychiat. neurol.,* Suppl. 54, 1-195.

Möbius, P.J. (1888). Über angeborenen doppelseitige Abducens-Facialis-Lähmung. *Mschr. med. Woenschr.* **35**, 91-4.

Mochizuki, Y., Ohkubo, H., and Notomura, T. (1981). Familial bilateral carpal tunnel syndrome. *J.N.N.F.* **44**, 367-9.

Moerman, P. and Fryns, J.P. (1988). Holoprosencephaly and postaxial polydactyly: another observation. *J. Med. Genet.* **25**, 501-2.

Moffie, D. (1961). Familial occurence of neural muscular atrophy (Tooth-Marie-Charcot) combined with cerebral atrophy and Parkinsonism. *Psychiat. Neurol. Neurochir.* **64**, 381-91.

Moldofsky, H., Tullis, C., and Lamon, R. (1974). Multiple tic syndrome (Gilles de la Tourette syndrome). *J. nerv. ments. Dis.* **159**, 282-92.

Mollaret, P. (1929). *La maladie de Friedreich.* Paris.

Möller, E., Link, H., Matell, G., Olhagen, B., and Stendahl, C. (1975). In *Histocompatibility testing* (1975) (Report of the Sixth International Histocompatibility Workshop and Conference, Arhus, 29 June-5 July (ed. F. Kissmeyer-Bielsen). Munksgaard, Copenhagen.

Moller, P.M. (1952). Another family with von Hippel-Lindau's disease. *Acta ophthal.* **30**, 155-65.

Mollica, F., Pavone, G., Nuciforo, G., and Sorge, G. (1979). A case of cyclopia. Role of enviromental factors. *Clin. Genet.* **16**, 69-71.

Monaco, F., Pirisi, A., Sechi, G.P., and Mutani, R. (1979). Compilicated optic atrophy (Behr's disease) associated with epilepsy and amino acid imbalance. *Eur. Neurol.* **18**, 101-5.

Monaco, P. (1964). Sulle form di passagio tra Friedreich e atrofia muscolare di Charcot-Marie-Tooth. *Acta neurol.* **19**, 275-86.

Monnens, L. and Heymans, H. (1987). Peroxisomal disorders: clinical characterization. *J. Inherit.Met. Dis.* **10**, Suppl. 23-32.

Monnens, L., Gabreëls, F., and Willems, J. (1975). A metabolic myopathy associated with chronic lactic acidemia, growth failure and nerve deafness. *J. Pediat.* **86**, 983.

Montgomery, M.A., Clayton, P.J., and Friedhoff, A.J. (1982). Psychiatric illness in Tourette syndrome patients and first-degree relatives. In: Chase TN, Friedhoff AJ, eds. *Gilles de la Tourette syndrome*, Advances in neurology, Vol. 35 (ed. T.N. Chase and A.J. Friedhoff), pp. 335-9. Raven Press, New York.

Montpetit, V.J.A., Andermann, F., Carpenter, S., Fawcett, J.S., Zborowska-Sluis, D., and Giberson, H.R. (1971). Subacute necrotizing encephalomyelopathy: a review and a study of two families. *Brain* **94**, 1-30.

Montplaisir, J., *et al.* (1985). Familial restless legs with periodic movements in sleep. Electrophysiologic, biochemical, and pharmacologic study. *Neurology* **35**, 130–4.

Moore, W. B., Matthews, T. J., and Rabinowitz, R. (1975). Genitourinary anomalies associated with Kippel–Feil syndrome. *J. Bone Jt Surg.* **57**, 355.

Moosa, A., Brown, B. H., and Dubowitz, V. (1972). Quantitative electromyography: carrier detection in Duchenne type muscular dystrophy using a new automatic technique. *J. Neurol. Neurosurg. Psychiat.* **35**, 841–4.

Moosa, A. and Dawood, A. A. (1987). Centronuclear myopathy in Black African children—Report of 4 cases. *Neuropediatrics* **18**, 213–17.

Moosa, A. and Dubowitz, V. (1970). Peripheral neuropathy in Cockayne's syndrome. *Arch. Dis. Childh.* **45**, 674–7.

Morariu, M. A. (1977). A new classification of amyotrophic lateral sclerosis (ALS) and familial amyotrophic lateral sclerosis (FALS) *Dis. nerv. Sys.* **8**, 468–9.

Morariu, M. A. and Taranu (1968). Clinico-genetic consideration of basilar impression. *Studii Cerc. Neurol.* **13**, 441.

Moreno-Fuenmayor, H. (1980). The spectrum of frontonasal dysplasia in an inbred pedigree. *Clin. Genet.* **17**, 137–42.

Morgan, J. D. (1971). Incontinentia pigmenti (Bloch–Sulzberger syndrome). A report of four additional cases. *Am. J. Dis. Childh.* **122**, 294–300.

Morgan-Hughes, J. A. and Mair, W. G. P. (1973). Atypical muscle mitochondria in oculoskeletal myopathy. *Brain* **96**, 215–24.

Morgan-Hughes, J. A., Brett, E. M., Lake, B. D., and Tome, F. M. S., (1973). Central core disease or not? Observations on a family with a non-progressive myopathy. *Brain* **96**, 527–36.

Morgan-Hughes, J. A., Hayes, D. J., Cooper, M., and Clark, J. B. (1985). Mitochondrial myopathies: deficiencies localised to complex I and complex III of the mitochondrial respiratory chain. *Biochem. Soc. Trans.* **13**, 648–50.

Morgan-Hughes, J. A., *et al.* (1982). Mitochondrial encephalomyopathies: biochemical studies in two cases revealing defects in the respiratory chain. *Brain* **105**, 553–82.

Morley, K. J., Weaver, D. D., Garg, B. P., *et al.* (1982). Hyperexplexia: an inherited disorder of the startle response. *Clin. Genet.* **21**, 388–96.

Morquio, L. (1929). Sur une forme de dystrophie osseuse familiale. *Bull Soc. Pediat. Paris* **27**, 145.

Morishima, A. and Aranoff, G. (1986). Syndrome of septo-optic-pituitary dysplasia – the clinical spectrum. *Brain develop.* **8**, 233–9.

Morris, J. C., *et al.* (1984). Hereditary dysphasic dementia and the Pick–Alzheimer spectrum. *Ann. Neurol.* **16**, 455–66.

Morris, S. J., Kaplan, S. R., Ballan, K., and Tedesco, F. J. (1978). Blue rubber bleb nevus syndrome, *J. Am. med. Ass.* **239**, 1887.

Morse, W.I., Cochrane, W.A., and Landrigan, P.L. (1961). Familial hypoparathyroidism with pernicious anemia, steatorrhea and adrenocrtical inefficiency. *N. Engl. J. Med.* **264**, 1021–6.

Morton, N.E. and Chung, C.S. (1959). Formal genetics of muscular dystrophy. *Am. J. hum. Genet.* **11**, 360–79.

Moschel, R. (1954). Amaurostische idiotie mit einer besonderen form von pigmentablagerung. *Dt. Z. Nervenheilk.* **172**, 102–10.

Moser, H. and Emery, A.E.H. (1974). The manifesting carrier in Duchenne muscular dustrophy. *Clin. Genet.* **5**, 271–84.

Moser, H. and Vogt, J. (1974). Follow-up of serum creatine kinase in carriers of Duchenne muscular dystrophy. *Lancet* **ii**, 661–2.

Moser, H., Sugita, M., Harbison, M.D., and Williams, S.R. (1972). Liver glycolipids, steroid sulfates and steroid sulfatases in a form of metachromatic leucodystrophy associated with multiple sulfatase deficiences. In *Sphingolipids, sphingolipidoses and applied disorders* (ed. B. Volk and S. Aronson) p. 429. Plenum Press, New York.

Moser, H.W., *et al.* (1981). Adrenoleukodystrophy: increased plasma content of saturated very long chain fatty acids. *Neurol. Minneap.* **31**, 1241–9.

Moser, H.W., Moser, A.E., Trojak, J.E., and Supplee, S.W. (1983). Identification of female carriers of adrenoleukodystrophy. *J. Pediat* **103**, 54–9.

Moser, H.W., Moser, A.E., Singh, I., and O'Neill, B.P. (1984). Aderenoleukodystrophy: survey of 303 cases. Biochemistry, diagnosis, and therapy. *Ann. Neurol.* **16**, 628–41.

Moskowitz, M.A., Winickoff, R.N., and Heinz, E.R. (1971). Familial calcification of the basal ganglions, a metabolic and genetic study. *N. Engl. J. Med.* **285**, 72–7.

Mossakowski, M., Mathieson, G., and Cumings, J.N. (1961). On the relationship of metachromatic leucodystrophy and amaurotic idocy. *Brain* **84**, 385–604.

Mossman, H., *et al.* (1983). Hunter's disease in a girl: association with X: 5 chromosomal translocation disrupting the Hunter gene. *Arch. Dis. Childh.* **58**, 911–5.

Mostacciuolo, M.L., *et al.* (1987). Population data on benign and severe forms of X-linked muscular dystrophy. *Hum. Genet.* **75**, 217–20.

Mount, L.A. and Reback, S. (1940). Familial paroxysmal choreoatheosis: preliminary report on a hitherto undescribed clinical syndrome. *Arch. Neurol. Psychiat., Chicago* **44**, 841–7.

Mousa, A.R.A-A., *et al.* (1986). Autosomally inherited recessive spastic ataxia, macular corneal dystrophy, congenital cataracts, myopia and vertically oval temporally tilted discs. A report of a Beduin family-a new syndrome. *J. Neurol. Sci.* **76**, 105–21.

Moyes, P.D. (1968). Case reports and technical note. Familial bilateral acoustic neuroma affecting 14 members from four generations. *J. Neurosurg.* **29**, 78–82.

Moynahan, E.J. (1962). Familial congenital alopecia, epilepsy, mental retardation with unusual electroencephalograms. *Proc. Ray. Soc. Med.* **55**, 411–12.

Mudd, S.H. Uhlendorf, B.W., Freeman, J.M., Finkelstein, J.D., and Shih, V.E. (1972). Homocystinuria associated with decreased methylenetetrahydrofolate reductase activity. *Biochem. Biophys. Res. Commun.* **46**, 905–12.

Mudd, S.H., *et al.* (1985). The natural history of homocystinuria due to cystathionine beta-synthase deficiency. *Am. J. Hum. Genet.* **37**. 1–31.

Mueller, O.T., *et al.* (1984). Identification and chromosome location of genes involved in glycoprotein neuraminidase deficiency disorders. (Abstr.) *Am. J. Med. Genet.* **36**, 205S.

Muir, C.S. (1959). Hydranencephaly and allied disorders. *Arch. Dis. Childh.* **34**, 231–46.

Mulder, D.W., *et al.* (1986). Familial adult motor neuron disease: Amyotrohic lateral sclerosis. *Neurology.* **36**, 511–7.

Müller, D., Pilz, H., and Muelen, T. (1969). Studies on adult metachromatic leukodystrophy Part I. Clinical morphological and histochemical observations in two cases. *J. neurol. Sci.* **9**, 567.

Muller, J. and Zeman, W. (1965). Degenerescence systematisee opticocochleodentelle. *Acta neuropath.* **5**, 26–39.

Müller, R. (1952). Progressive motor neurone disease in adults. *Acta psychiat. scand.* **27**, 137–56.

Munro, J.M. and Loizou, L.A. (1982). Transient global amnesia – familial incidence. *J. Neurol. Neurosurg. Psychiat.* **45**, 1070.

Munsat, T.L. and Poussaint, A.F. (1962). Clinical mnanifestations and diagnosis of amyloid polyneuropathy. Report of three cases. *Neurol, Minneap.* **12**, 413–22.

Munsat, T.L., Thompson, L.R., and Coleman, R.F. (1969*a*). Centronuclear ('myotubular') myopathy. *Arch. Neurol., Chicago* **20**, 120–31.

Munsat, T.L., Woods, R., Fowler, W., and Pearson, C.M. (1969*b*). Neurogenetic muscular atrophy of infancy with prolonged survival. *Brain* **92**, 9–24.

Munsat, T.L., Piper, D., Cancilla, P., and Mednick, J. (1972). Inflammatory myopathy with facioscapulohumeral distribution. *Neurol. Minneap.* **22**, 335–347.

Munslow, R.A., and Hill, A.H. (1955). Multiple occurences of gliomas in a family. *J. Neurosurg.* **12**, 646–50.

Murdoch, J.L. and Nissim, J. (1971). Olivopontocerebellar degeneration with macular dystrophy. *Birth Defects*: Original Article Series **VII**, 246.

Murphy, E.A. (1968). The rationale of genetic counseling. *J. Pediat.* **72**, 121–30.

Murphy, E.A. and Mutalik, G.S. (1969). The application of Bayesian methods in genetic counselling. *Hum. Hered.* **19**, 126–51.

Murphy, J. and Murphy, S.F., (1986). Myasthenia gravis in identical twins. *Neurology* **36**, 78–80.

Murphy, J.V. (1973). Subacute necrotizing encephalomyelopathy (Leigh's disease): detection of the heterozygous carrier state. *Pediatrics, Springfield* **51**, 710–15.

Murray, J.C., Johnson, J.A. and Bird, T.D. (1985). Dandy—Walker malformation: etilogic heterogeneity and empiric recurrence risks. *Clin. Genet.* **28**, 272–83.

Murray, T.J. (1973). Congenital sensory neuropathy. *Brain* **96**, 387–94.

Mustajoki, P. (1978). Variegate porphyria. *Ann. intern. Med.* **89**, 238–44.

Mustajoki, P. (1980). Variegate porphyria. Twelve year's experience in Finland. *Q. J. Med.* **49**, 191–203.

Mustajoki, P. (1981). Normal erythrocyte uroporphyrinogen I synthase in a kindred with acute intermittent porphyria. *Ann. Int. Med.* **95**, 162–6.

Mustajoki, P. and Koskelo, P. (1976). Hereditary hepatic prophyrias in Finland. *Acta med. scand.* **200**, 171–8.

Mutoh, T., *et al.* (1986). A family with beta-galactosidase deficiency: Three adults with atypical clinical picture. *Neurology* **36**, 54–9.

Myle, G. and Fadilogu, S. (1967). Une fratrie de maladie de Hallervorden et Spatz (contribution a l'etude des formes dites pures). *Encephale* **56**, 343–59.

Myle, G. and Van Bogaert, L. (1940). Etudes anatomo-cliniques de syndromes hypercinetiques complexes. 1. Sur le tremblement par une lesion capsulaire tres limitee. *Msch. Psychiat. Neurol.* **103**, 28–43.

Myrianthopoulos, N. and Brown I.A. (1954). A genetic study of progressive spinal muscular atrophy. *Am. J. hum. Genet.* **6**, 387–411.

Myrianthopoulos, N., Waldrop, F.N., and Vincent, B.L. (1969). A repeat study of hereditary predisposition in drug-induced Parkinsonism. In *Progress in neurogenetics* (ed. A. Barbeau and J.R. Brunette), pp.486–91. Excerpta Medica, Amsterdam.

Nadjmi, B., Flanagan, M.J., and Christian, J.R. (1969). The Laurence–Moon–Biedl syndrome. *Am. J. Dis. Childh.* **117**, 352–6.

Nadler, H.L. and Gerbie, A.B. (1970). Role of amniocentesis in the intrauterine detection of genetic disorders. *N. Engl. J. Med.* **282**, 596–9.

Nagae, K., Goto, I., Ueda, K., and Morotomi, Y. (1972). Familial occurrence of multiple intracranial aneurysms. Case report. *J. Neurosurg.* **37**, 364–7.

Nagashima K., *et al.* (1985). Infantile neuroaxonal dystrophy: perinatal onset with symptoms of diencephalic syndrome. *Neurology* **35**, 735–8.

Naidoo, S. (1972). *Specific dyslexia.* Pitman, London.

Naidu, S., Moser, A.E., and Moser, H.W. (1988). Phenotypic variability of generalised peroxisomal disorders. *Pediatr. Neurol.* **4**, 5–12.

Naiman, J.L. and Fraser, F.C. (1955). Agenesis of the corpus callosum. A report of two .cases in siblings. *Arch. Neurol. Psychiat., Chicago* **74**, 182–5.

Nair, K. R. (1976). Acrodystrophic neuropathy. *Neurol., Madras* **24**, 94–9.

Naito, H. and Oyanagi, S. (1982). Familial myoclonus epilepsy and choreoathetosis. Hereditary dentatorubral–pallidoluysian atrophy. *Neurology. Minneap.* **32**, 798–807.

Naito, S., Nameron, N., Mickey, M. R., and Terasaki, P. I. (1972). Multiple sclerosis associated with HLA-A$_3$. *Tiss. Antigens.* **2**, 1–4.

Najjar, S. S. (1974). Muscular hypertrophy in hypothyroid children: the Kocher–Debre–Semelaigne syndrome. *J. Pediat.* **85**, 236–9.

Najjar, S. S. and Mahmud, J. (1968). Diabetes insipidus and diabetes mellitus in a six-year-old girl. *J. Pediat.* **73**, 251–3.

Nakano, K. K., Dawson, D. M., and Soence, A. (1972). Machado disease. A hereditary ataxia in Portugese emigrants to Massauchusetts. *Neurol. Minneap.* **22**, 49–55.

Nakano, K. K., Sakai, H., Kinoshita, J., Yagishita, S., and Itoh, Y. (1979). The fine structure of blood cells in ceroid-lipofuscinosis (Spielmeyer-Vogt's disease). *Neuropaediatrics* **10**, 56–66.

Nakano, T., *et al.* (1985). Adult GM$_1$-gangliosidosis: clinical patterns and rectal biopsy. *Neurology* **35**, 875–80.

Nakao, K., Togi, H., Furukawa, T., Mozai, T., and Toyokura, Y. (1966). A pedigree of familial amyloid neuropathy. *Clin. Neurol., Tolyo* **6**, 369–70.

Namba, T., Aberfeld, D. C., and Grob, D. (1970). Chronic proximal spinal muscular atrophy. *J. neurol. Sci.* **11**, 401–23.

Namba, T., Brunner, N. G., Brown, S. B., Mugurama, M., and Grob, D. (1971). Familial myasthenia gravis: report of 27 patients in 12 families and review of 164 patients in 73 families. *Arch. Neurol., Chicago* **25**, 49–60.

Narumi, S., Nishimura, K., Fuchizakwa, K., and Hidaka, T. (1976). Three cases of moya moya disease found in an inbred family. *Brain Nerve, Tokyo* **28**, 201–5.

Naveh, Y. and Friedman, A. (1976). Pfeiffer syndrome. Report of a family and review of the literature. *J. med. Genet.* **13**, 277–80.

Naveh, Y., Ludatscher, R., Alon, U., and Sharf, B. (1985). Muscle involvement in progressive diaphyseal dysplasia. *Pediatrics.* **76** 944–9.

Navon, R., Padeh, B., and Adam, A. (1973). Apparent deficiency of hexosaminidase A in healthy members of a family with Tay-Sachs disease. *Am. J. hum. Genet.* **25**, 287–93.

Navon, R., Geiger, B., Ben-Yoseph, Y., and Rattazzi, M. (1976). Low levels of hexosaminidase A in healthy individuals with apparent difficiency of the enzyme. *Am. J. hum. Genet.* **28**, 339–49.

Navon, R., Brand, N., and Sandbank, U. (1980). Adult GM$_2$ gangliosidosis:

neurological and biochemical findings in an apparently new type. *Neurology, Minneap.* **30**, 449–50.

Navon, R., Argov, Z., and Frisch, A. (1986). Hexosaminidase A deficiency in adults. *Am. J. Med. Genet.* **24**, 179–96.

Nazer, H., *et al.* *(1983). Wilson's disease, a diagnostic dilemma. Br. Med. J.* **287**, 313.

Nee, L.E., Caine, E.D., Polinsky, R.J., Eldridge, R., and Ebert, M.H. (1980). Gilles de la Tourette syndrome: clinical and family study of 50 cases. *Ann. Neurol.* **7**, 41–9.

Nee, L.E., et al. (1987). Dementia of the Alzheimer type: clinical and family study of 22 twin pairs. *Neurology* **37**, 359–63.

Needleman, H.L. and Root, A.W. (1963). Sex linked hydrocephalus. Report of two families with chromosomal study of two cases. *Pediatrics. Springfield* **31**, 396.

Neil, J.F., Glew, R.H., and Peters, S.P. (1979). Familial psychosis and diverse neurologic abnormalities in adult-onset Gaucher disease. *Arch. Neurol., Chicago* **36**, 95–9.

Neill, C.A. and Dingwall, M.M. (1950). A syndrome resembling progeria: a review of two cases. *Arch. Dis. Child.* **25**, 213–21.

Nelaton, A. (1952). Affection singuliere des os du pied. *Gaz. Hôp. civ. milit., Paris* **25**, 13–20.

Neligan, P., Harriman, D.G.F., and Pearce, J. (1977). Respiratory arrest in familial hemiplegic migraine: a clinical and neuropathological study. *Br. med. J.* **ii**, 732–4.

Nelson, E., Aurebeck, G., Osterberg, K., Berry, J., Jabbour, J.T., and Bornhofen, J.H. (1963). Ultrastructural and chemical studies on Krabbe's disease. *J. Neuropath exp. Neurol.* **22**, 414–34.

Nelson, J.W. and Amick, L.D. (1966). Heredofamilial progressive spinal muscular atrophy: a clinical and electromyographic study of a kinship. *Neurol., Minneap.* **16**, 306.

Nelson, K.B., and Ellenberg, J.H. (1978). Prognosis in children with febrile seizures. *Pediatrics, Springfield* **61**, 720–7.

Nelson, K.B. and Eng, G.D., (1972). Congenital hypoplasia of the depressor anguli oris muscle: differentiation from congenital facial palsy. *J. Pediat.* **81**, 16–20.

Nelson, T.E. and Flewellen, E.H. (1983). The malignant hyperthermia syndrome. *N. Engl. J. Med.* **309**, 416–18.

Neubürger, K. (1922). Zur histopathologie der multiplen sklerose im Kindesalter. *Z. Neurol. Psychiat.* **76**, 384–414.

Neufeld, E.F., Liebaers, I., Epstein, C.J., Yatziv, S., Milunsky, A., and Migeon, B.R. (1977). The Hunter syndrome in females: is there an autosomal recessive form of iduronate sulphatase deficiency? *Am J. hum. Genet.* **29**, 455–61.

Neugut, R.H., Neugut, A.I., Kahana, E., Stein, Z., and Alter, M. (1979).

Creutzfeldt–Jakob disease: familial clustering among Libyan-born Israelis. *Neurol., Minneap.* **29**, 225–31.

Neuhäuser, G., Kaveggia, E.G., and Opitz, J.M. (1976*b*). A craniosynostosis–craniofacial dysostosis syndrome with mental retardation and other malformations: – 'craniofacial dyssynostosis'. *Eur. J. Pediat.* **123**, 15–28.

Neuhäuser, G., and Opitz, J.M. (1975). Autosomal recessive syndrome of cerebellar ataxia and hypogonadotropic hypogonadism. *Clin. Genet.* **7**, 426–34.

Neuhäuser, G., Daly, F.F., Magnelli, N.C., Barreras, R.F., Donaldson, R.M., and Opitz, J.M. (1976*a*). Essential tremor, nystagmus and duodenal ulceration. *Clin. Genet.* **9**, 81–91.

Neuhuauser, G., Zu Rhein, G.M., Kaveggia E.G., and Opitz, J.M. (1977). Fatal CNS dysgenesis with severe microencephaly, mental retardation seizures, and paucity of myelin, autosomal recessive trait? *Eur. J. Ped.* **124**, 185–98.

Neumann, MA. (1959). Combined degeneration of globus pallidus and dentate nucleus and their projection. *Neurol., Minneap.* **9**, 430–8.

Neumann, M.A. and Cohn R. (1953). Incidence of Alzheimer's disease in a large mental hospital. *Arch. Neurol. Psychiat., Chicago* **69**, 615.

Neustein, H.B. (1973). Nemaline myopathy. A family study of three autopsied cases. *Arch. Path.* **96**, 192–5.

Neustein, H.B., *et al.* (1979). An X-linked recessive cardiomyopathy with abnormal mitochondria. *Pediatrics* **64**, 24–9.

Neville, B.G.R., Lake, B.D., Stephens, R., and Sanders, M.D. (1973). A neurovisceral storage disease with vertical supranuclear ophthalmplegia and its relationship to Niemann–Pick disease. *Brain* **96**, 97–120.

Neville, H.E., Brooke, M.H., and Austin, J.H. (1974). Studies in myoclonus epilepsy (Lafora body form). IV. Skeletal muscle abnormalities. *Arch. Neurol., Chicago* **30**, 466–74.

Nevin, N.C. and Johnston, W.P. (1980). Risk of recurrence after two children with central nervous system malformations in an area of high incidence. *J. med. Genet.* **17**, 87–92.

Nevin, N.C. and Pearce, W.G. (1968). Diagnostic and genetical aspects of tuberous sclerosis. *M. med. Genet.* **5**, 273–80.

New, G.B. and Kirch, W.A. (1933). Permanent enlargement of lips and face secondary to recurring swelling and associated with facial paralysis: clinical entity. *J. Am. med. Ass.* **100**, 1230–3.

Newell, F.W., Matalon, R., and Meyer, S. (1975). A new mucolipidosis with psychomotor retardation, corneal clouding, and retinal degeneration. *Am. J. Ophthal.* **80**, 440–9.

Niakan, E., Belluomini, J., Lemmi, H., Summitt, R.L., and Ch'ien, L. (1979). Disturbance of rapid-eye movement sleep in three brothers with Pelizaeus–Merzbacher disease. *Ann. Neurol.,* **631**, 253–7.

Niakan, E., Harati, Y. and Danon, M.J. (1985). Tubular aggregates: their association with myalgia. *J. Neurol. Neurosurg. Psychiat.* **48**, 882-6.

Nibbelink, D.W., Peters, B.H., and McCormick, W.F. (1969). On the association of pheochromocytoma and cerebellar hemangioblastoma. *Neurol., Minneap.* **19**, 445-60.

Nichols, FL., Holdsworth, D.E., and Reinfrank, R.F. (1961). Familial hypocalcemia latent tetany and calcification of the basal ganglia: report of a kindred. *Am. J. Med.* **30**, 518-28.

Nicholson, G.A., Gardner-Medwin, D., Pennington, R.J.T., and Walton, J.N. (1979). Carrier detection in Duchenne muscular dystrophy: assessment of the effect of age on detection-rate with serum-creatine-kinase-activity. *Lancet* **i**, 692-4.

Nicol, A.A. McI. (1957). Lindau's disease in five generations. *Ann. hum. Genet.* **22**, 7-15.

Niederwieser, A. *et al.* (1986). Prenatal diagnosis of 'dihydrobiopterin synthetase' deficiency, a variant form of phenylketonuria. *Eur. J. Pediat.* **145**, 176-8.

Nielsen, S.L. (1977). Striatonigral degeneration disputed in familial disorder (Letter.) *Neurol., Minneap.* **27**, 306.

Nienhuis, A.W., Coleman, R.F., Brown, W.J., Munsat, T.L., and Pearson, C.M. (1967). Nemaline myopathy: a histopathologic and histochemical study. *Am. J. clin. Path.* **48**, 1-13.

Nikoskelainen, E., Hoyt, W.F., and Nemmelin, K. (1982). Ophthalmological findings in Lebers hereditary optic neuropathy. *Arch. Opthal.* **100**, 1597.

Nikoskelainen, B.K., *et al.* (1987). Leber's hereditary optic neuro-retinopathy, a maternally inherited disease. A genealogic study in four pedigrees. *Arch. Ophthal.* **105**, 665-71.

Nino, H.E., Noreen, H.J., Dubey, D.P., Resch, J.A., Namboodiri, K., Elston, R.C., and Yunis, E.Y. (1980). A family with hereditary ataxia. HLA typing. *Neurology, Minneap.* **30**, 12-20.

Nisenbaum, C., Sandbank, U., and Kohn, R. (1965). Pelizaeus-Merzbacher's disease 'infantile acute type'. Report of a family. *Ann. Pediat.* **204**, 365-76.

Nishigaki, S., Ando, K., Takeda, At. *et al.* (1966). A pedigree with atypical hereditary neurogenic amyotrophy. *Clin. Neurol., Tokyo* **6**, 189.

Nishimura, H. (1970). Incidence of malformation in abortions. In. *Congenital malformations.* International Congress Series, Vol. 204 (ed. F.C. Fraser and V.A. McKussick) pp. 275-83. Excerpta Medical, Amsterdam.

Nishimura, R.N. and Barranger, J.A. (1980). Neurologic complications of Gaucher's disease, Type 3. *Arch. Neurol., Chicago* **37**, 92-3.

Nishizawa, M., *et al.* (1987). A mitochondrial encephalomyopathy with cardiomyopathy. A case revealing a defect of complex 1 in the respiratory chain. *J. Neurol. Sci.* **78**, 189–201.

Nivelon-Chevallier, A. and Nivelon, J.L. (1975). Forme familiale d' holoprosencephalie. *J. Génét. hum.* **23**, 215–23.

Nja, A. (1946) A sex linked type of gargoylism. *Acta pediat., Stockh.* **33**, 267–86.

Noack, M. (1959). Ein Beitrag zum Krankheitsbild der Akrocephalosyndaktylie (Apert). *Arch. Kinderheilk.* **106**, 168–71.

Noad, K.B. and Lance, J.W. (1960). Familial myoclonic epilepsy and its association with cerebellar disturbance. *Brain* **83**, 618–30.

Noetzel, M.J., Landau, W.M., and Moser, H.W. (1987). Adrenoleukodystrophy carrier state presenting as a chronic nonprogressive spinal cord disorder. *Arch. Neurol.* **44**, 566–7.

Nogen, A.G. (1980). Congenital muscular disease and abnormal findings on computerized tomography. *Dev. Med. Child Neurol.* **22**, 658–63.

Nonaka, I. *et al.* (1983). Congenital myopathy without specific features (minimum change myopathy). *Neuropediatrics* **14**, 273–41.

Nordborg, C., *et al.* (1981). A new type of non-progressive sensory neuropathy in children with atypical dysautonomia. *Acta Neuropath., Berl.* **55**, 135–41.

Norden, N.E., Öckerman, P.A. and Szaro, L. (1973). Urinary mannose in mannosidosis. *J. Pediat.* **82**, 686–8.

Norio, R, and Koskiniemi, M. (1979). Progressive myoclonus epilepsy. Genetic and nosological aspects with special reference to 107 Finnish patients. *Clin. Genet.* **15**, 382–98.

Norman, M.E. (1972). Neurofibromatosis in a family. *Am. J. Dis. Child.* **123**, 159–60.

Norman, M.G. (1968). Encephalopathy and fatty degeneration of the viscera in childhood I: Review of cases at the Hospital for Sick Children, Toronto (1954–1966). *Can. med. Ass. J.* **99**, 522.

Norman, M.G., Roberts, M., Sirois, J., and Trcmblay, L.J., (1976). Lissencephaly. *J. Can. Sci. Neurol.* **3**, 39–46.

Norman, R.M. (1940). Primary degeneration of the granular layer of the cerebellum: an unusual form of familial cerebellar atrophy occurring in early life. *Brain* **63**, 365–79.

Norman, R.M. (1964). The neuropathology of status epilepticus. *Med. Sci. Law* **4**, 46–51.

Norman, R.M. and Tingey, A.H. (1966). Syndrome of microcephaly, striocerebellar calcifications and leucodystrophy. *J. Neurol. Neurosurg. Psychiat.* **29**, 157–63.

Norman, R.M. and Urich, H. (1958). Cerebellar hypoplasia associated with systemic degeneration in early life. *J. Neurol. Neurosurg. Psychiat.* **21**, 159–66.

Norris, E. (1939). *Om Ordblindhet.* Copenhagen.

Norton. J.C. (1975). Patterns of neuropsychological test performance in Huntington's disease. *J. nerv. ment. Dis.* **161**, 276-9.

Norton, P., Ellison, P., Sulaiman, A.R., and Harb, J. (1983). Nemaline myopathy in the neonate. *Neurol. Minneap.* **33**, 351-6.

Novotny, E.J., *et al.* (1986). Leber's disease and dystonia: A mitochondrial disease. *Neurology* **36**, 1053-60.

Nuttall, F.Q. and Jones, B. (1968). Creatine kinase and glutamic oxaloacetic transaminase activity in serum: kinetics of change with exercise, and effect of physical conditioning. *J. Lab. clin. Med.* **71**, 847-54.

Nutting, P., Cole, B., and Schimke, R.N. (1969). Benign recessively inherited choreo-athetosis of early onset. *J. med. Genet.* **6**, 408-10.

Nuutila, A. (1970). Dystrophia retinae pigmentosa-dysacusis syndrome (DPD): a study of the Usher or Hallgren syndrome. *J. Génét. hum.* **18**, 57-88.

Nyberg-Hansen, R., Grönvik, O., and Refsum, S. (1972). Hereditary cerebellar ataxia associated with congenital cataracts. Four cases of the Marinesco-Sjögren syndrome with some unusual features. Proc. 20th Cong. Scand. Neurologists, Oslo 1972. *Acta neurol. sand.* Suppl. 51, 257-60.

Nyberg-Hansen and Refsum, S. (1972). Spastic paraparesis associated with optic atrophy in monozygotic twins. Proc. 20th Cong. Scand. Neurologists, Oslo 1972. *Acta neurol. scand.* **48**, Suppl. 51 261-2.

Nygaard, T.G. and Duvoisin, R.C. (1986). Hereditary dystonia-parkinsonism syndrome of juvenile onset. *Neurology* **36**, 1424-8.

Nyland, H. and Skre, H. (1977). Cerebral calcinosis with late onset encephalopathy: an unusual type of pseudo-pseudohypoparathyroidism. *Acta neurol. scand.* **56**, 309-25.

Nyssen, R. and Van Bogaert, L. (1934). La degenerescence systematisee opticocochleo-dentelee. *Rev. neurol.* **2**, 321-45.

Obesco, J.A., Rothwell, J.C. and Lang, A.E. (1983). Myoclonic dystonia. *Neurology* **33**, 825-30.

O'Brien, J.G. (1942). Subarachnoid haemorrhage in identical twins. *Br. med. J.* **i**, 607-9.

O'Brien, J.G., *et al.* (1972). Juvenile GM_1 gangliosidosis. Clinical, pathlogical, chemical and enzymatic studies. *Clin. Genet.* **3**, 411-34.

O'Brien, J.G., Tennant, L., Veath, M.L., Scott, C.R., and Bucknall, W.E. (1978). Charaterization of unusual hexosaminidase A (HEX A) deficient human mutants. *Am. J. hum. Genet.* **30**, 602-8.

O'Brien, J.S. (1978). The cherry-red spot myoclonus syndrome: a newly recognized inherited lysosomal storage disease due to acid neuraminidase deficiency. *Clin. Genet.* **14**, 55-60.

Oda, M. (1976). Thalamus degeneration in Japan. A review from clinical and pathological viewpoints. *Appl. Neurophysiol.* **39**, 178-98.

O'Donnell, P. P., Leshner, R. T. and Campbell, W. W. (1986). Hypertrophia musculorum vera in familial ataxia. *Arch. Neurol.* **43**, 146–7.

Oelschlager, R., White, H. H., Schimke, R. W. (1971). Roussy–Levy syndrome: report of a kindred and discussion of the nosology. *Acta neurol. scand.* **47**, 80–90.

Oepen, H. (1973). Discordant features of monozygotic twin sisters with Huntington's chorea. *Ad. Neurol.* **1**, 199–201.

Ogasawara, N. (1965), Multiple sclerose mit Rosenthalschen fasern. *Acta neuropath.* **5**, 61–8.

Ogata, A., Ishida, S., and Wada, T. (1987). A survey of 37 cases with basal ganglia calacification (BGC): CT-scan findings of BGC and its relationship to underlying diseases and epilepsy. *Acta Neurol. Scand.* **75**, 117–24.

Ogata, J., Okayama, M., Goto, I., Inomata, H., Yoshida, I., and Omae, T. (1978). Primary familial amyloidosis with vitreous opacities. Report of an autopsy case. *Acta neuropath.* **42**, 67–70.

Ogden, T. E., Robert, F., and Carmichael, E. A. (1959). Some sensory syndromes in children: indifference to pain and sensory neuropathy. *J. Neurol. Neurosurg. Psychiat.* **22**, 267–76.

Ohlson, A., Stark, G., and Sakati, A. (1980). Marble brain disease: recessive osteopetrosis, renal tubular acidosis and cerebral calcification in three Saudi Arabian families. *Dev. med. Child. Neurol.* **22**, 72–84.

Ohta, M., Akari, S., and Kuroiwa, Y. (1967). Familial occurrence of migraine with a hemiplegic syndrome and cerebellar manifestations. *Neurol., Minneap.* **17**, 813–17.

Ohta, M., Ellefson, R. D., Lambert, E. H., and Dyck, P. J. (1973). Hereditary sensory neuropathy Type II. *Arch. Neurol., Chicago* **29**, 23–37.

Ohtahara, S., Ishida, T., Oka, E., *et al.* (1976). On the age-dependent epileptic syndromes: the early infantile encephalopathy with suppression-burst. *Brain Dev.* **8**, 270–88.

Oizumi, J., *et al.* (1983). Neonatal pyruvate carboxylase deficiency with renal tubular acidosis and cystinuria. *J.Inherit. Met. Dis.* **6**, 89–94.

Okada, S. and O'Brien, J. S. (1968). Generalized gangliosidosis. Beta-galactosidase deficiency. *Science, N. Y.* **160**, 1002–4.

Okada, S., McCrea, M., and O'Brien, J. S. (1972). Sandhoff's disease (GM$_2$ gangliosidosis type 2): clinical, chemical and enzyme studies in five patients. *Pediat. Res.* **6**, 606–15.

Okada, S., *et al.* (1978). Hypersialyloligosacchariduria in mucolipidoses: a method for diagnosis. *Clinica chem. Acta* **86**, 159–67.

Okada, S., *et al.* (1983). A severe infantile sialidosis (beta-galcactosidase-alpha-neuraminidase deficiency) mimicking GM1-gangliosidosis type 1. *Eur. J. Pediatr.,* **140**, 295–8.

Okamura, K., Santa, T., Nagae, K., and Omae, T. (1976). Congenital oculoskeletal myopathy with abnormal muscle and liver mitochondria. *J. neurol. Sci.* **27**, 79-91.

Okihiro, M.M., Tasaki, T., Nakano, K.K., and Bennet, B.K. (1977). Duane syndrome and congenital upper-limb anomalies. *Arch. Neurol.* **34**, 174-9.

Oley, C. and Baraitser, M. (1988). Blepharophimosis, ptosis, epicanthus inversus syndrome (BPES). *J. Med. Genet.* **25**, 47-51.

Olivecrona, H. (1952). The cerebellar angioreticulomas. *J. Neurosurg.* **9**, 317-30.

Oliver, J. and Dewhurst, K. (1969). Childhood and adolescent forms of Huntington's disease. *J. Neurol. Neurosurg. Psychiat.* **32**, 455-9.

Olsen, W., Engel, W.K., Walsh, G.O., and Einaugler, R. (1972). Oculocraniosomatic neuromuscular disease with ragged red fibres. *Arch. Neurol., Chicago* **26**, 193-211.

Olson, L.C., Bourgeois, C.H, Cotton, R.B., Harikul, S., Grossman, R.A., and Smith, T.J. (1971). Encephalopathy and fatty degeneration of the viscera in North Eastern Thailand. Clinical syndrome and epidemiology. *Pediatrics, Springfield* **47**, 707-16.

Olsson, J.-E., Möller, E., and Link, H. (1976). HLA haplotypes in families with high frequency of multiple sclerosis. *Arch. Neurol., Chicago.* **33**, 808-12.

O'Neill, B., Butler, A.B., Young, E., Falk, P.M., and Bass, N.H. (1978). Adult onset GM_2 gangliosidosis. *Neurol., Minneap.* **28**, 1117-23.

Ono, J., *et al.* (1982). A case report of congenital hypomyelination. *Eur. J. Pediat.* **138**, 265-70.

Oohira, Y., Nagara, N., Akaboshi I, *et al.* (1985). The infantile form of sialidosis type II associated with congenital adrenal hyperplasia: possible linkage between HLA and the neuraminidase deficiency gene. *Hum. Genet.* **70**, 341-3.

Oonk, J.G., Van der Helm, H.J., and Martin, J.J. (1979). Spinocerebellar degeneration: hexosaminidase A and B deficiency in two adult sisters. *Neurol., Minneap.* **29**, 380-4.

Opitz, J.M. (ed.) (1986). The Rett syndrome. *Am. J. Med. Genet.* **24**, Suppl.1.

Opitz, J.M. and Kaveggia, E.G. (1974). The FG syndrome. An X-linked recssive syndrome of multiple congenital anomilies and mental retardation. *S. Kindrheilk* **117**, 1-18.

Opitz, J.M., Smith, D.W., and Summit, R.L. (1965). Hypertelorism and hypospadias. A newly recognized hereditary malformation syndrome. (Abst.) *J. Pediat.* **67**, 968.

Opitz, J.M. Frias, J.L., Gutenberger, J.E., and Pellett, J.R. (1969). The G syndrome of multiple congential anomalies. The clinical delineation of birth defects II Malformation syndromes. *Birth Defects:* Original Article Series **V**, 95-101.

Opitz, J.M., Kaveggia, E.G., Durkin-Stamm, M.V., and Pendleton, E. (1978). Diagnostic/genetic studies of severe mental retardation. *Birth Defect*: Original Article Series **XIV**, 1–38.

Opjordsmoen, S. and Nyberg-Hansen, R. (1980). Hereditary spastic paraplegia with neurogenic bladder disturbances and syndactylia. *Acta neurol. scand.* **61**, 35–41.

Oppenheimer, D.R. (1976). Diseases of the basal ganglia, cerebellum and motor neurons. In *Greenfield's neuropathlogy,* 3rd edn. (ed. W. Blackwood and J AN. Corsellis) P. 615. Arnold, London.

Origuchi. Y, *et al.* (1984). Citrullinemia presenting as uncontrollable epilepsy. *Brain Dev.* **6**, 328–31.

Orii, T., *et al.* (1972). A new type of mucolipidosis with β-galactosidase deficiency and glycopeptiduria. *Tahoku J. exp. Med.* **107**, 303–15.

Ormerod, J.A. (1904). An unusual form of family paralysis. *Lancet* **i**, 17–18.

Ortiz de Zarate, J.C. and Maruffo, A. (1970). The descending ocular myopathy of early childhood myotubular or centronuclear myopathy. *Eur. Neurol.* **3**, 1–12.

Ortiz de Zarate, J.C. and Rodriguez, J.A. (1958). Herencia en la epilepsia. *Rev. neurol. B. Aires* **16**, 16.

Osborn, S.B. and Walshe, J.M. (1969). The influence of genetic and acquired liver defects on radiocopper turnover in Wilson's disease. *Lancet* **ii**, 17–20.

Osler, W. (1880). On heredity in progressive muscular atrophy as illustrated in Farr family of Vermont. *Arch. Med., N.Y.* **4**, 316–20.

Osuntokun, B.O., Odeku, E.L., and Luzzato, L. (1968). Congenital pain asymbolia and auditory imperception. *J. Neurol. Neurosurg. Psychiat.* **31**, 291–6.

Ota, Y., Miyoshi, S., Ueda, O., Mukai, T., and Maeda, A. (1968), Familial paralysis agitans juvenile. A clinical anatomical and genetic study. *Folia psychiat. neurol. jap.* **12**, 112–21.

O'Tauma, L.A., Farmer, T.W., Kirkman, H.N., Radcliffe, W.B., and Guinto F.C. (1972). Familial cerebral vascular disease in association with midline nevus flammeus—a new neurocutaneous syndrome. *Trans. Am. neurol. Ass.* **97**, 317–19.

Otenasek, F.J. and Silver, M.L. (1961). Spinal hemangioma (hemangioblastoma) in Lindau's disease. Report of six cases in a single family. *J. Neurosurg.* **18**, 295–300.

Otsuki, S., Sato, M., and Shirahige, I. (1965). Familial myoclonus and ataxia. *Clin. Neurol., Tokyo* **5**, 221–4.

Ounsted, C. (1951). The significance of convulsion in children with purulent meningitis. *Lancet* **i**, 1245–8.

Ounsted, C. (1955). Genetic and social aspects of the epilepsies of Childhood. *Eugen. Rev.* **47**, 33–49.

Ounsted, C., Lindsay, J., and Norman, R. (1966). Biological factors in temporal lobe epilepsy. *Clinics in development medicine* No, 22. Spastics Society, London.

Ouvrier, R. A. (1978). Progressive dystonia with marked diurnal fluctuation. *Ann. Neurol.* **4**, 412–17.

Ouvrier, R. A., McLeod, J. G., and Conchin, T. E. (1987). The hypertrophic forms of hereditary motor and sensory neuropathy. A study of hypertrophic Charcot–Marie–Tooth disease (HMSN type I) and Dejerine-Sottas disease (HMSN type III) in childhood. *Brain* **110**, 121–48.

Owen, F., Poulter, M., Lofthouse, R., Collinge, J., Crow, T. J., Risby, D., *et al.* (1989). Insertion in prion protein gene in familial Creutzfeldt–Jakob disease. *Lancet* **i**, 51–2.

Owens, N., Hadley, R. C. and Kloepfer, H. W. (1960). Hereditary blepharophimosis, ptosis and epicanthus inversus. *J. Int. Coll. Surg.* **33**, 558–74.

Pachter, B. R., Pearson, J., Davidowitz, J., *et al.* (1976). Congenital total external ophthaloplegia infantile spinal muscular atrophy. *Invest. Ophthal.* **15**, 320–4.

Page, M., Asmat, A. C., and Edwards, C. R. W. (1976). Recessive inheritance of diabetes: the syndrome of diabetes insipidus, diabetes mellitus, optic atrophy and deafness. *Q. Jl Med.* **45**, 505–20.

Pages, M., *et al.* (1985). Multicore disease and Marfan's syndrome. *Eur. Neurol.* **24**, 170–5.

Pagon, R. A., *et al.* (1978). Hydrocephalus, agyria, retinal dysplasia, encephalocele (HARD ± E) syndrome: an autosomal recessive condition. *Birth Defects*: Original Article Series **XIV**, 233–41.

Paillas, J. F., Bonnal, J., Gastaut, M., and Naguet, R., (1951). Angiomatose encephalotrigeminee associee d'un syndrome de Klippel Trenaunay. *Acta neurol. belg.* **51**, 487–90.

Paine, R. S. (1960). Evaluation of familial biochemically determined mental retardation in children, with special reference to aminoaciduria. *N. Engl. J. Med.* **262**, 658–65.

Palamucci, L., Schiffer, D., Monga, G., Mollo, F., and de Marchi, M. (1978). Central core disease: histochemical and ultrastructural study of muscle biopsies of father and daughter. *J. Neurol.* **218**, 55–62.

Palant, D. J. and Carter, B. L. (1972). Klippel–Feil syndrome and deafness. *Am. J. Dis. Childh.* **123**, 218.

Pallis, C. A., and Schneeweiss, J. (1962). Hereditary sensory radicular neuropathy. *Am. J. Med.* **32**, 110–18.

Pallis, C. A., Duckett, S., and Pearse, A. G. E. (1967). Diffuse lipofuscinosis of the central nervous system. *Neurol., Minneap.* **17**, 381–94.

Pallister, P. D. and Opitz, J. M. (1985). Brief Clinical report: disequilibrium syndrome in Montana Hutterites. *Am. Hum. Genet.* **22**, 567–9.

Pampiglione, G. and Harden, A. (1974). An infantile form of neuronal 'storage' disease with characteristic evolution of neurophysiological features. *Brain* **97**, 355–60.

Pampiglione, G. and Moynahan, E. J. (1976). The tuberous sclerosis syndrome: clinical and E. E. G. studies in 100 children. *J. Neurol. Neurosurg. Psychiat.* **39**, 666–73.

Panse, F. (1942). *Die Erbchorrea: Eine Klinisch-Gebetische Studie.* Thieme, Leipzig.

Papadatos, C., Alexiou, D., Nicolopoulos, D., Mikropoulos, H., and Hadzigeorgiou, E. (1974). Congenital hypoplasia of depressor anguli oris muscle. A genetically determined condition? *Arch. Dis. Childh.* **49**, 927–31.

Papadia, F., *et al.* (1987). Progressive form of multiple pterygium syndrome in association with nemaline-myopathy: Report of a female followed for twelve years. *Am. J. Med. Genet.* **26**, 73–83.

Pape, K. E. and Pickering, D. (1972). Asymmetric crying facies: an index of other congenital anomolies. *J. Pediat.* **81**, 21–30.

Paradis, R. W. and Sax, D. S. (1972). Familial basilar impression. *Neurol., Minneap.* **22**, 554–60.

Parke, J. T., *et. al.* (1984). A syndrome of microcephaly and retinal pigmentary abnormalities without mental retardation in a family with coincidental autosomal dominant hyperreflexia. *Am. J. Med. Genet.* **17**, 585–94.

Parker, H. L. (1946). Periodic ataxia. In *Collected papers of the Mayo Clinic and Mayo Foundation 1946.* Vol. 38, p. 642. Saunders, Philadelphia.

Parker, N. (1985). Hereditary whispering dysphonia. *J. Neurol. Neurosurg. Psychiat.* **48**, 218–24.

Parkes, J. D., Langdon, N., and Lock, C. (1986). Narcolepsy and immunity. *Br. Med. J.* **292**, 359–60.

Parkinson, D. and Hall, C. W. (1962). Oligodrendrogliomas. Simultaneous appearance in frontal lobes of siblings. *J. Neurosurg.* **19**, 424–6.

Parrish, M. L., Roessman, U., and Levinsohn, M. W. (1979). Agenesis of the corpus callosum. A study of the frequency of associated malformations. *Ann. Neurol.* **6**, 349–54.

Parrish, R. M. and Stevens, H. (1977). Familial hemiplegic migraine. *Minn. Med.* **60**, 709–15.

Parsonage, M. J. and Turner, J. W. A. (1948). Neuralgic amyotrophy. The shoulder-girdle syndrome. *Lancet* **i**, 973–8.

Paschke, E., *et al.* (1986). Infantile type of sialic acid storage disease with sialuria. *Clin. Genet.* **29**, 417–24.

Pascual-Castroviejo, I., Casas-Fernandez, C., Lopez-Martin, V., and Martinez-Bermejo, A. (1977). X-linked dysosteosclerosis. Four familial cases. *Eur. J. Pediat.* **126**, 127–38.

Pascuzzi, R. M. (1986). Familial autoimmune myasthenia gravis and thymoma: occurence in two brothers. *Neurology* **26**, 423.

Pashayan, H.M., Singer, W., Bove, C., *et al.* (1982). The Angelman syndrome in two brothers. *Am. J. Med. Genet.* **13**, 295–8.

Paskind, H.A. and Stone, T.T. (1933). Familial spastic paralysis. *Arch. Neurol. Psychiat., Chicago* **30**, 481–500.

Passarge, E. and Lenz, W. (1966). Syndrome of caudal regression in infants of diabetic mothers: observations of further cases. *Pediatrics, Springfield* **37**, 67–5.

Passwell, J., Adam, A., Garfinkel, D., Streiffler, M., and Cohen, B.E. (1977). Heterogeneity of Wilson's disease in Israel. *Israel J. med. Sci.* **13**, 15–19.

Patel, A.M. and Richardson, A.E. (1971). Ruptured intracranial aneurysms in the first two decades of life. A study of 38 patients. *J. Neurosurg.* **35**, 571–6.

Patel, H., Dolman, C.L., and Byrne, M.A. (1972*a*). Holoprosencephaly with median cleft lip. *Am. J. Dis. Childh.* **124**, 217–21.

Patel, V., Watanabe, I., and Zeman, W. (1972*b*). Deficiency of alpha-*l*-fucosidase. *Science, N.Y.* **176.**, 426–7.

Patel, H., Tze, W.J., Crichton, J.U., McCormick, A.Q., Robinson, G.C., and Dolman, C.L. (1975). Optic nerve hypoplasia with hypopituitarism: septo-optic dysplasia with hypopituitarism. *Am. J. Dis. Childh.* **129**, 175–80.

Patel, H., *et al.* (1985). Multisystem atrophy with neuronal intranuclear hyaline inclusions. Report of a case and review of the literature. *J. Neurol. Sci.* **67**, 57–65.

Paterson, D. and Carmichael, E.A. (1924). Form of familial cerebral degeneration chiefly affecting the lenticular nucleus. *Brain,* **47**, 207–31.

Patrick, A.D., Young, E., Kleijer, W.J., and Niermeijer, M.F., (1977). Prenatal diagnosis of Niemann-Pick disease type A using chromogenic substrate. *Lancet* **ii**, 144.

Patterson, V.H., Hill, T.R.G., Fletcher, P.J.H., and Heron, J.R. (1979). Central core disease. Clinical and pathological evidence of progression within a family. *Brain* **102**, 581–94.

Patton, M.A., Baraitser, M., and Brett, E.M. (1986). A family with congenital suprabulbar paresis (Worster–Drought syndrome). *Clin. Genet.* **29**, 147–50.

Patton, M.A. *et al.* (1988). Intellectual development in Apert's syndrome: a long term follow up of 29 patients. *J. Med. Genet.* **25**, 164–7.

Paty, D.W., *et al.* (1977). HLA antigens and mitogen responsiveness in multiple scleroris. *Transplant. Proc.* **9**, 1845–8.

Pauli, R.M. (1984). Letter to the editors: sensorineural deafness and peripheral neuropathy. *Clin. Genet.* **26**, 383–4.

Paulian, E.D., (1922). Contributions cliniques a l'etude de la paralysie bulbaire infantile familiale. *Rev. neurol.* **38**, 275–8.

Pauls, D.L. and Leckman, J.F. (1986). The inheritance of Gilles de la

Tourette's syndrome and associated behaviours: evidence for autosomal dominant transmission. *New England J. of Med.* **315:16**, 993–6.

Paunier, L., Pagnamenta, F., Monnard, E., Felgenhauer, W. R., Behar, A., and Moody, J. (1973). Spinal muscular atrophy with various clinical manifestations in a family. *Helv. paediat. Acta* **28**, 19–25.

Pavlakis S. G., *et al.* (1984). Mitochondrial myopathy encepalopathy lactic acidosis, and stroke like episodes a distinctive clinical syndrome. *Ann. Neurol.* **16**, 481–8.

Pavone, L., Mollica, F., Grasso, A., Cao, A., and Gullotta, F. (1978). Schwartz-Jampel syndrome in two daughters of first cousins. *J. Neurol. Neurosurg. Psychiat.* **41**, 161–9.

Pavone, L., Mollica, F., Grasso, A., and Pero, G. (1980). Familial centronuclear myopathy. *Acta neurol. scand.* **62**, 33–40.

Pearce, W. G. (1978). Congenital nystagmus – genetic and environmental causes *Can. J. Ophthal.* **13**, 1–9.

Pearn, J. H., (1974). The spinal muscular atrophies of childhood. A genetic and clinical study. Ph.D. thesis, University of London.

Pearn, J. H. (1978). Autosomal dominant spinal muscular atrophy. A clinical and genetic study. *J. neurol. Sci.* **38**, 263–75.

Pearn, J. H. (1979a). Genetic studies of acute infantile spinal muscular atrophy (SMA type I.) An analysis of sex ratios, segregation ratios and sex influence. *J. med Genet.* **15**, 414–14, 418–23.

Pearn, J. H. (1979b). Distal spinal muscular atrophy. A clinical and genetic study of 8 kindred. *J. neurol. Sci.* **43**, 182–91.

Pearn, J. H. and Hudgson, P. (1978). A new syndrome – spinal muscular atrophy with adolescent onset and hypertrophied calves simulating Becker dystrophy. *Lancet* **i**, 1059–6.

Pearn. J. H. and Wilson, J. (1973). Chronic generalized spinal muscular atrophy of infancy and childhood. Arrrested Werdnig-Hoffmann disease. *Arch. Dis. Childh.* **48**, 768–74.

Pearn, J. H., Carter, C. O., and Wilson, J. (1973). The genetic identity of acute infantile spinal muscular atrophy. *Brain* **96**, 463–70.

Pearn, J. H., Bundey, S., Carter, C. O., Wilson, J., Gardner-Medwin, D., and Walton, J. N. (1978a). A genetic study of subacute and chronic spinal muscular atrophy in childhood. *J. neurol. Sci.* **37**, 227–48.

Pearn, J. H., Gardner-Medwin, D., and Wilson, J. (1978b). A clinical study of chronic childhood spinal muscular atrophy. *J. neurol. Sci.* **38**, 23–37.

Pearn, J. H. and Walton, J. N. (1978c). A clinical and genetic study of adult-onset spinal muscular atrophy. The autosomal recessive form as a discrete disease entity. *Brain*, **101**, 591–66.

Pearson, C. M. (1964). The periodic paralysia. Differential feature and pathological observations in permanent myopathic weakness. *Brain* **87**, 341–53.

Pearson, C. M. and Fowler, W. G. (1963). Hereditary non-progressive muscular dystrophy including arthrogryposis syndrome. *Brain* **86**, 75–88.

Pearson, K. (1933). Two new pedigrees of muscular dystrophy. *Anr. Eugen.* **5**, 179–91.

Pearson, J.S., Petersen, M.C., Lazarte, J.A., Blodgett, H.E., and Kley, I.B. (1955). An educational approach to the social problem of Huntington's chorea. *Proc. Staff Meet. Mayo Clin.* **30**, 349–57.

Pebenito, R., Feretti, C., Chaudary, R.R., *et al.* (1984). Idiopathic torsion dystonia associated with lesions of the basal ganglia. *Clin. Ped.* **23**, 232.

Pedersen, I.L. Mikkelsen, M., and Oster, J. (1976). The G syndrome. A four generation family study. *Hum. Hered.* **26**, 66–71.

Peiffer, J. and Zerbin-Rüdin, E. (1963). Sur Variationsbreite der Pelizaeus-Merzbbacherschen Krankheit. *Acta neuropath.* **3**, 87–107.

Pelias, M.Z. and Thurmon, T.F. (1979). Congenital universal muscular hypoplasia: evidence for autosomal recessive inheritance. *Am. J. hum. Genet.* **31**, 548–54.

Pelizaeus, F. (1885). Uber eine eigenthümliche Form Spastischer Lähmung mit Cerebralerschinungen auf hereditärer Grundlage (Multiple Sklerose). *Arch. Psychiat. NevKrankh.* **16**, 698–710.

Pembrey, M.E., *et al.* (1985). Prenatal diagnosis of ornithine carbamoyl transferase deficiency using a gene specific probe. *J. Med. Genet.* **22**, 462–5.

Pembrey. M. *et al.* (1988). The association of Angelman syndrome and deletions within 15q11–13. *J. Med. Genet.* **25**, 274.

Pena, S.D.J. and Shokeir, M.H.K. (1974). Autosomal recessive cerebro-oculo-facioskeletal (COFS) syndrome. *Clin. Genet.* **5**, 285–93.

Penchaszadek, V.B., Gutierrez, E.R., and Figueroa, E.P. (1980). Autosomal recessive craniometaphyseal dysplasia. *Ann. J. med. Genet.* **5**, 43–55.

Penders, C.A. and Delwaide, P.J. (1972). Formes clinique de la myotonie congenitale none dystrophique. *Acta neurol. belge.* **72**, 366–72.

Pendl, G. and Zimprich, H. (1971). Ein Beitrag zum Syndrom des Hypertelorismus. *Helv. paediat. Acta* **26**, 319–25.

Penn, A.S., Lisak, R.R., and Rowland, L.P. (1970). Muscular dystrophy in young girls. *Neurol. Minneap.* **20**, 147–59.

Penrose, L.S. (1938). *A clinical and genetic study of 1,280 cases of mental defect*. MRC Special Report Series, No. 229. HMSO, London.

Penrose, L.S. (1948). The problem of anticipation in pedigrees of dystrophia myotonica. *Ann. Eugen.* **14**, 125–36.

Penry, J.K., Porter, R.J., and Dreifuss, F.E. (1975). Simultaneous recording of absence seizures with video tape and electroencephalography. *Brain* **98**, 427–40.

Pepin, B., Mikol, J., Goldstein, B., Haguenau, M., and Godlewski, S.

(1976). Forme familiale de myopathie centronucleaire de l'adulté. *Rev. neurol.* **132**, 845–57.

Pepin, B., Mikol, J. Goldstein, B., Aron, J.J., and Lebuisoson, D.A. (1980). Familial mitochondrial myopathy with cataract. *J. Neurol. Sci.* **45**, 191–203.

Pepper, P. and Pendergrass, E.P. (1936). Hereditary occurrence of enlarged parietal foramina. *Am. J. Roentg.* **35**, 1.

Percy, A.K., Kaback, M.M., and Herndon, R.M. (1977). Metachromatic leukodystrophy: comparison of early and late onset forms. *Neurl. Minneap.* **27**, 933–41.

Pérez-Comas, A. and Garcia-Castro, J.M. (1974). Occipito-facial-cervico-thoracic-abdomino-digital dysplasia; Jarco-Levin syndrome of vertebral anomolies. *J. Pediat.* **85**, 388.

Perkinson, N.G. (1957). Melanoma arising in a *cafe-au-lait* spot of neurofibromatosis. *Am. J. Surg.* **73**, 1018–20.

Perniola, T., Krajewska, G., Carnevale, F., and Lospalutti, M. (1980). Congenital alopecia, psychomotor retardation convulsions in two sibs of a consanguineous marriage. *J. inher. metab. Dis.* **3**, 49–53.

Perrine, G. and Goodman, R. (1966). A family study of Huntington's chorea with unusual manifestations. *Ann. intern. Med.* **64**, 570–4.

Perry, T.B. and Fraser, F.C. (1973). Variability of serum creatine phosphokinase activity in normal women and carriers of the gene for Duchenne muscular dystrophy. *Neurol., Minneap.* **23**, 1316–23.

Perry, T.L., Hansen, S., Tischler, R., Bunting, R., and Berry, K. (1967). Carnosinemia: a new metabolic disorder associated with neurologic disease and mental defect. *N. Engl. J. Med.* **277**, 1219–27.

Perry, T.L., Bratty, P.J.A., Hansen, S. Kennedy, J., Urquhart, N., and Dolman, C.L. (1975). Hereditary mental depression and Parkinsonism with taurine deficiency. *Arch. Neurol., Chicago* **32**, 108–13.

Perry, T.L., Currier, R.D., Hansen, S., and MacLean, J. (1977). Aspartate-taurine imbalance in dominantly inherited olivopontocerebellar atrophy. *Neurol., Minneap.* **27**, 257–61.

Perry, T.L., Norman, M.G., Young, V.W., *et al.* (1985). Hallervorden-Spatz disease: cysteine accumulation and cysteine dioxygenase deficiency in the globus pellidus. *Ann. Neurol.* **18**, 482–9.

Pescia, G., Nguyenthe, H., and Deonna, T. (1983). Prenatal diagnosis of genetic microcephaly. *Prenatal diagnosis* **3**, 363–5.

Peterman, A.F., Hayles, A.B., Dockerty, M.B., and Love, J.G. (1958). Encephalotrigeminal angiomatosis (Sturge–Weber disease)—clinical study of thirty-five cases. *J. Am. med. Ass.* **167**, 2169–76.

Peterman, A.F., Lillington, G.A., and Jamplis, R.W. (1964). Progressive muscular dystrophy with ptosis and dysphagia. *Am. med. Ass. Arch. Neurol* **10**, 38–41.

Peters, A.C.E., *et al.* (1984). Fukuyama type congenital muscular dystrophy—two Dutch sibliings. *Brain Dev.* **6**, 406-20.

Peters, H.D., Opitz, J.M., Goto, I., and Reese, H. (1968). The benign and proximal progressive muscular atrophies. *Acta neurol. scand.* **44**, 542-60.

Petersen, G.M., *et al.* (1983). The Tay–Sachs disease gene in North American Jewish populations: geographic variations and origin. *Am. J. Hum. Genet.* **35**, 1258-69.

Petty, R.K.H., Harding, A.E., and Morgan-Hughes, J.A. (1986). The clinical features of mitochondrial myopathy. *Brain* **109**, 915-38.

Pfeiffer, S., Harzer, K., and Schlote, W. (1984). Diffuse disseminated sclerosis combined with partial arylsulfatase A (ASA) deficiency. Mixed heterozygosity of ASA and pseudo-ASA deficiency. *Neuropediatrics* **15**, 59-62.

Pfeiffer, R.A., Palm, D., Jünemann, G., Mandl-Kramer, S., and Heimann, E. (1974). Nosology of congenital non-progressive cerebellar ataxia. *Neuropaediatrie* **5**, 91-102.

Pfeitzer, P. and Müntefering, H. (1968). Cyclopism as a hereditary malformation. *Nature, Lond.* **217**, 1071-2.

Philcox, D.V., Sellars, S.L., Pamplett, R., and Beighton, P. (1975). Vestibular dysfunction in hereditary ataxia. *Brain* **98**, 309-16.

Philipp, E. (1949). Hereditary (familial) spastic paraplegia. Report of six cases in one family. *NZ. med. J.* **48**, 22-5.

Philippart, M., Martin, L., Martin, J.J., and Menkes, J.H. (1969). Niemann–Pick disease. *Arch. Neurol., Chicago.* **20**, 227-38.

Phillips, L.H., *et al.* (1985). Hereditary motor-sensory neuropathy (HMSN): probable X-linked dominant inheritance. *Neurology* **35**, 498-502.

Phillips, R.I. (1963). Familial cerebral aneurysms. Case report *J. Neurosurg.* **20**, 701-3.

Pickering, D., Laski, B., MacMillan, D.C., and Rose, V. (1971). 'Little Leopard' syndrome: description of three cases and review of 24. *Arch. Dis. Childh.* **46**, 85-90.

Pierides, A.M., *et al.* (1976). Study on a family with Anderson–Fabry's disease and associated familial spastic paraplegia. *J. med. Genet.* **13**, 455-61.

Pilley, S.F.J. and Thompson, H.S. (1976). Familial syndrome of diabetes insipidus diabetes mellitus optic atrophy and deafness (DIDMOAD) in childhood. *Br. J. Ophthal.* **60**, 294-8.

Pilz, H. (1972). Late adult metachromatic leukodystrophy. Arylsulfatase A activity of leukocytes in two families. *Arch. Neurol., Chicago* **27**, 87-90.

Pilz, H. and Hopf, H.C. (1972). A preclinical case of late adult metachromatic leukodystrophy: manifestation only with lipid abnormalities in urine, enzyme deficiency, and decrease of nerve conduction velocity. *J. Neurol. Neurosurg. Psychiat.* **35**, 360–4.

Pilz, H. and Schiener, P. (1973). Kombination von Morbus Addison und Morbus Schilder bei einer 43 jahrigen Frau. *Acta neuropath.* **26**, 357–60.

Pilz, H., *et al.* (1977). Adult metachromatic leukodystrophy. I. clinical manifestation in a female aged 44 years, previously diagnosed in the preclinical state. *Eur. Neurol.* **15**, 301–7.

Pilz, H., Von Figura, K., and Goebel, H.H. (1979). Deficiency of arylsulfatase B in two brothers aged 40 and 38 years. (Maroteaux–Lamy syndrome type B.) *Ann. Neurol.* **6**, 315–25.

Pincus, J.H. (1972). Subacute necrotizing encephalomyelopathy (Leigh's disease) a consideration of clinical features and etiology. *Dev. med. Child. Neurol.* **14**, 87–101.

Pincus, J.H. and Chutorian, A. (1967). Familial benign chorea with intention tremor: a clinical entity. *J. Pediat.* **70**, 724–9.

Pineda, M., *et al.* (1984). Familial agenesis of corpus callosum with hypothermia and apnoeic spells. *Neuropediatrics* **15**, 63–7.

Pinsky, L. and DiGeorge, A.M. (1966). Congenital familial sensory neuropathy with anhidrosis. *J. Pediat.* **68**, 1–13.

Pinsky, L., Miller, J., Shanfield, B., Watters, G., and Wolfe, L.S. (1974). GM_1 gangliosidosis in skin fibroblast culture: enzygmatic differences between type 1 and 2 and observations on a third variant. *Ann. J. hum. Genet.* **26**, 563–77.

Pinsky. L., Finlayson, M.H., Libman, I., and Scott, B.H. (1975). Familial amyotrophic lateral sclerosis with dementia: a second Canadian family. *Clin. Genet.* **7**, 186–91.

Pirskanen, R. (1976). Genetic associations between myasthenia gravis and the HL-A system *J. Neurol. Neurosurg. Psychiat.* **39**, 23–33.

Pirskanen, R. (1977). Genetic aspects in myasthenia gravis: a family study of 264 Finnish patients. *Acta neurol. scand.* **51**, 365–88.

Piton, J. and Tiffeneau, R. (1940). Maladie familiale du type de l'heredo ataxic. *Rev. neurol.* **72**, 774–7.

Plant, G. (1983). Focal paroxysmal kinesigenic choreothhetosis. *J. Neurol. Neurosurg. Psychiat.* **46**, 345–8.

Plant, G.T., William, A.C., Earl, C.J., and Marsden, C.D. (1984). Familial paroxysmal dystonia induced by exercise. *J.N.N.P.* **47**, 275–9.

Plant, E.T., William, A.C., Earl, C.J., and Marsden, C.D. (1984). Familial paroxysmal dystonia induced by exercise. *J. Neurol. Neurosurg. Pyschiat.* **47**, 275–9.

Plato, C.C., Cruz, M.T., and Kurland, L.T. (1967). Amyotrophic lateral sclerosis/Parkinsonism dementia complex of Guam: further genetic investigation. *Am. J. hum. Genet.* **19**, 133–41.

Platz, P., Ryder, L.P., Staub-Nielsen, L., Svejgaard, A., Thomsen, M., and Wolheim, M.S. (1975). HLA and idiopathic optic neuritis. *Lancet* **i**, 520–1.

Plewes, J.L. and Jacobson, I. (1971). Familial frontonasal dermoid cysts. Report of four cases. *J. Neurosurg.* **34**, 683–6.

Plott, D. (1964). Congenital laryngeal-abductor paralysis due to nucleus ambiguus dysgenesis in three brothers. *N. Engl. J. Med.* **271**, 593–7.

Plowright, O. (1928). Familial claw-foot with absent tendon jerks and with cerebellar disease. *Guy's Hosp. Rep.* **78**, 314–19.

Poch, G.F., Sica, E.P., Taratuto, A., and Weinstein, I.H. (1971). Hypertrophia musculorum vera. Study of a family. *J. neurol. Sci.* **12**, 53–61.

Poenaru, L., Dreyfus, J.-C., Boue, J., Nicolesco, H., Ravise, N., and Bamberger, J. (1976). Prenatal diagnosis of fucosidosis. *Clin. Genet.* **10**, 260–4.

Poewe, W., *et al.* (1985). The rigid spine syndrome – a myopathy of uncertain nosological position. *J. Neurol. Neurosurg. Psychiat.* **48**, 887–93.

Poffenbarger, A.C. (1968). Heredofamilial neuritis with brachial predeliction. *W. Va. med. J.* **64**, 425–9.

Pogacar, S., Ambler, M., Conklin, W.J., O'Neil, W.A., and Lee, H.Y. (1978). Dominant spinopontine atrophy. Report of two additional members of family W. *Arch. Neurol., Chicago* **35**, 156–62.

Polani, P.E. and Moynahan, E.J. (1972). Progressive cardiomyopathic lentiginosis. *Q. J. Med.* **41**, 205–25.

Polgar, J.G., *et al.*. (1972). The early detection of dystrophia myotonica. *Brain* **95**, 761–76.

Pollock, M. *et al.* (1983). Peripheral neuropathy in Tangier Disease. *Brain*, **106**, 911–28.

Pollock, M. and Hornabrook, R.W. (1966). The prevalence, natural history and dementia of Parkinson's disease. *Brain* **89**, 429–48.

Pomeroy, J., Efron, M.L., Dayman, J., and Hoefnagel, D, (1968). Hartnup disorder in a New England family. *N. Engl. J. Med.* **278**, 1214–16.

Pompe, J.C. (1932). Over idiopathische hypertrophy van het hart. *Ned. Tijdschr. Geneesk.* **76**, 304–11.

Pond, D.A., Bidwell, B.H., and Stein, L. (1960). A survey of epilepsy in fourteen general practices. I. Demographic and medical data. *Psychiat. Neurol. Neurochir.* **63**, 217–36.

Ponsot, G. and Lyon, G. (1977). La sclerose tubereuse de Bourneville. *Arch. fr. Pédiat.* **34**, 9–22.

Poser, C., Dewulf, A., and Van Bogaert, L. (1957). Atypical cerebellar degeneration associated with leucodystrophy. *J. Neuropath. exp. Neurol.* **16**, 209–37.

Poser, C.M., Johnson, M., and Bunch, L.D. (1965). Familial amyotrophic lateral sclerosis. *Dis. nerv. syst.* **26**, 607–702.

Poskanzer, D.C. and Kerr, D.N.S. (1961). A third type of periodic paralysis with normokalaemia and favourable response to sodium chloride. *Am. J. Med.* **31**, 328–42.

Potasman, *et al.* (1985). The Groll-Hirschowitz Syndrome. *Clin. Genet.*, **28**, 76–9.

Potter, E. L. (1961). *Pathology of the fetus and the infant*, 2nd edn. Year Book Medical, Chicago.

Powell, F.C., *et al.* (1983). Letter to the Editors: keratoderma and spastic paraplegia. *Clin. Genet.* **24**, 462.

Powell, H., Tindall, R., Schultz, P., Paa, D., O'Brien, J., and Lampert, P. (1975). Adrenoleukodystrophy. *Arch. Neurol., Chicago* **32**, 250–60.

Powers, J.M. and Schaumburg, H.H. (1973). The adrenal cortex in adrenoleukodystrophy, *Arch. Patch.* **93**, 305–10.

Pratt, R.T.C. (1967). *The genetics of neurological disorders.* Oxford University Press, London.

Pratt, R.T.C. (1970). The genetics of Alzheimer's disease. In *Alzheimerr's disease and related conditions* (ed. G.E.W. Wolstenholme and M. O'Connor), pp. 137–43. Churchill, London.

Pratt, R.T.C., Compston, N.D., and McAlpine, D. (1951). The familial incidence of disseminated sclerosis and its significance. *Brain* **74**, 191–232.

Praud, E., Labrune, B., Lyon, G., and Mallet, R, (1972). Une observation familiale d'hypoplasie progressive et bilaterale des branches de l'hexagone de Willis avec examen anatomique. *Arch. fr. Pediat.* **1**, 397–409.

Preiser, S.A. and Davenport, C.B. (1918). Multiple neurofibromatisis (von Recklinghausen's disease) and its inheritance in man. *Am. J. med. Sci.* **156**, 507–41.

Prensky, A.L. (1976). Migraine and migrainous variants in pediatric patients. *Pediat. Clins. N. Am.* **23**, 461–71.

Preston, F.W., Walsh, W.S., and Clarke, T.H. (1952). Cutaneous neurofibromatosis (von Recklinghausen's disease): clinical manifestations and incidence of sarcoma in 61 male patients. *Arch. Surg.* **64**, 813–27.

Price, E.B. (1971). Papillary cystadenoma of the epididymis. *Arch. Path.* **91**, 456–70.

Prick, M.J.J., *et al.* (1981). Progressive infantile poliodystrophy. Association with disturbed pyruvate oxidation in muscle and liver. *Arch. Neurol.*, **38**, 767–72.

Primrose, D.A. (1975). Epiloia in twins: a problem in diagnosis and counselling. *J. ment. Def. Res.* **19**, 195–203.

Probst, A. and Ohnacker, H. (1977). Sclérose tubereuse de bourneville chez un prématuré. Ultrastructure des cellules atypiques: présence de microvillosités. *Acta neuropath.* **40**, 157–61.

Probst, A., *et al.* (1981). Sensory ganglioneuropathy in infantile spinal muscular atrophy. *Neuropediatrics* **12**, 215–31.

Probst, F.P. (1979). *The prosencephalies*, Springer, Berlin.

Prusiner, S.B. (1987). Prions and neurodegenerative disease. *N. Engl. J. Med.* **317**, 1571–81.

Pruzanski, W. (1965). Congenital malformations in myotonic dystrophy. *Acta neurol. scand.* **41**, 34–8.

Przuntek, H. and Monninger, P. (1983). Therapeutic aspects of kinesigenic paraxysmal choreoathetosis of the Mount and Reback type. *J. Neurol.* **230**, 163–9.

Przyrembel, H., *et al.* (1976). Glutaricaciduria type II. *Clin. chim. Acta* **66**, 227.

Purcell, J.J. and Krachmer, J.H. (1979). Familial corneal hypesthesia. *Arch. Ophthal.* **97**, 872–4.

Qazi, Q.H. and Reed, T.E. (1973). A problem in diagnosis of primary versus secondary microcephaly. *Clin. Genet.* **4**, 46–52.

Quarrell, O.W.J., *et al.* (1987). Exclusion testing for Huntington's disease in pregnancy with a closely linked DNA marker. *Lancet* **i**, 1281–3.

Quattlebaum, T.G. (1979). Benign familial convulsions in the neonatal period and early infancy. *J. Pediat.* **95**, 257–9.

Quinlan, C.D. and Martin, E.A. (1970). Refsum's syndrome: report of three cases. *J. Neurol. Neurosurg. Pychiat.* **33**, 817–23.

Quinn, N.P. and Marsden, C.D. (1984). Dominantly inherited myoclonic dystonia with dramatic response to alcohol.

Rabending, G. and Klepel, H. (1970). Fotokonvulsivreaktion und Fotomyoklonus: Altersbhangige genetisch determinierte Varianten der gesteigerten Fotosensibilitat. *Neuropaediatrie* **2**, 164–72.

Radermecker, J. and Martin, J.J. (1972). Dystrophie neuro-axonale et maladie de Hallcrvorden—Spatz infantiles. Aspects electro-cliniques et anatomo-pathologiques diagnostic differentiel. *Bull. Acad. R. med. belg.* **12**, 459–502.

Radu, H. and Ionescu, V. (1972). Nemaline (neuro) myopathy—rod like bodies and Type I fibre atrophy in a case of congenital hypotonia with denervation. *J. neurol. Sci.* **17**, 53–60.

Radu, H., Seceleanu, A., Migeh, S., Torok, Z., Bordeianu, L., and Seceleanu, S. (1966). La Pseudomyopathic neurogene de Kugelberg–Welander. *Acta neurol. psychiat. belge.* **66**, 409–27.

Raggio, J.F., Thurmon, T.F., and Anderson, E.E. (1973). X-linked hereditary spastic paraplegia. *J. Los. Ang. St. med. Soc.* **125**, 4–6.

Rajic, D.S. and De Veber, L.L. (1966). Hereditary oculo-dento-osseous dysplasia. *Ann. Radiol.* **9**, 224–9.

Raju, T.N.K., *e tal.* (1977). Centronuclear myopathy in the newborn period causing severe respiratory distress *Pediatrics.* **59**, 29–36.

Ramaeckers, Th.V., *et al.* (1987). Diagnostic difficulties in infantile neuroaxonal dystrophy—a clinicopathological study of eight cases. *Neuropediatrics.* **18**, 170–5.

Ramos. A.O. and Schmidt, B.J. (1964). Neurokinin and pain-producing substance in congenital generalized analgesia. *Arch. Neurol., Chicago* **10**, 42–6.

Rampini, S., Isler, W., Baerlocher, K., Bischoff, A., Ulrich, R., and Pluss, H. J. (1970). Du kombination von metachromatischer leukodystrophie und Mukopolysaccharodise als selbestandiges Krankheitsbild (Mukosulfatidoses). *Helv. paediat. Acta* **25**, 436–61.

Rank, B. K. and Thomson, J. A. (1959). The genetic approach to hereditary congenital ptosis. *Aust. N. Z. J. Surg.* **28**, 274–9.

Rapin, I., Goldfischer, S., Katzman, R., Engel, J., and O'Brien, J. S. (1978). The cherry-red spot-myoclonus syndrome. *Ann. Neurol.* **3**, 234–42.

Rapin, I., Suzuki, K., and Valsamis, M. P. (1976). Adult (chronic) GM_2 gangliosidosis. *Arch. Neurol., Chicago* **33**, 120–30.

Rasmussen, P. and Waldenström E. (1978). Hereditary mirror movements – a case report. *Neuropaediatrie* **9**, 189–94.

Rautakorpi, I. and Rinne, U. K. (1978). Epidemiology and natural course of essential tremor. Proc. 22nd Scandinavian Congress of Neurology. *Acta neurol. scand.* **57** Suppl. 67, 222–3.

Ravitch, M. M. (1977). Poland's syndrome – a study of an eponym. *Plastic reconst. Surg.* **59**, 508–12.

Raymond F. (1900–01). *Lecons sur les maladies du systême nerveux*, 1st edn, Vol. 3, pp. 346–65; Vol. 6, pp. 203–27. Gustave Doin, Paris.

Raynaud, E.-J., *et al.* (1975). Metachromatic leukodystrophy. Ultrastructural and enzymatic study of a case of variant O form. *Arch. Neurol., Chicago* **32**, 834–6.

Rebollo, M., Val, J. F., Garijo, F., *et al.* (1983). Livedo reticularis and cerebrovascular lesions (Sneddon's syndrome). *Brain* **106**, 965–79.

Redding, F. K. (1970). Familial congenital ocular motor apraxia. (Abstract.) *Neurol., Minneap.* **20**, 405.

Reed, D., Plato, C., Elizan, T., and Kurland, L. T. (1966). The amyotrophic lateral sclerosis/Parkinsonism dementia complex: a ten-year follow-up on Guam. I. Epidemiologic studies. *Am. J. Epidemiol.* **83**, 54–73.

Reed, M. H., Shokier, M. H. K., and MacPherson, R. I. (1975). The hypertelorism – hypospadias syndrome. *J. Can. Ass. Radiol.* **26**, 240–8.

Reed, T. E. and Chandler, J. H. (1958). Huntington's chorea in Michigan. 1. Demography and genetics. *Am. J. hum. Genet.* **10**, 201–25.

Reed, T. E. and Neel, J. V. (1959). Huntington's chorea in Michigan. 2. Selection and mutation. *Am. J. hum. Genet.* **11**, 107–36.

Reed, W. B. and Sugarman, G. I. (1974). Unilateral nevus of Ota with sensorineural deafness. *Arch. Derm.* **109**, 881–3.

Reed, W. B., Sugarman, G. I. and Mathis, R. A. (1977). De Sanctis – Cacchione syndrome. *Arch. Derm.* **113**, 1561–3.

Refsum, S. (1945). Heredoataxia hemeralopica polyneuritiformis et tidligere ikke beskrevet familiaert syndrom? En foreløbig meddelelse. *Nord. Med.* **28**, 2682.

Refsum, S. and Skillicorn, S.A. (1954). Amyotrophic familial spastic paraplegia. *Neurol., Minneap.* **4**, 40–7.

Regensburg. I. (1930). Zur klinik des hereditären torsionsdystonischen Symptom-komplexes. *Meschr. Psychiat. Neurol.* **75**, 323–45.

Regli, F., Filippa, G., and Wiesendanger, M. (1967). Hereditary mirror movement. *Arch. Neurol., Chicago* **16**, 620–3.

Reich, S.D. and Wiernik, P.H. (1976). Von Recklinghausen neurofibromatosis and acute leukemia. *Am. J. Dis. Child.* **130**, 888–9.

Reimann, H.A., McKechnie, W.G., and Stanisavljevic, S. (1958). Hereditary sensory radicular neuropathy and other defects in a large family. *Am. J. med.* **25**, 573–9.

Renier, W.O., *et al.* (1981). Connatal Pelizaeus–Merzbacher with congenital stridor in two maternal cousins. *Acta Neuropath.* **54**, 11–17.

Renlund, M. (1984). Clinical and laboratory diagnosis of Salla disease in infancy and childhood. *J. Pediat.* **104**, 232–6.

Renlund, M. *et al.* (1983). Salla disease: a new lysosomal storage disorder with disturbed sialic acid metabolism. *Neurology* **33**, 57–66.

Reske-Nielsen, E., *et al.* (1987). Familial centronuclear myopathy: a clinical and pathological study.*Acta Neurol. Scand.* **76**, 115–22.

Resnick J.S. and Engel, W.K. (1967). Myotonic lid lag in hypokalaemic periodic paralyis. *J. Neurol. Neurosurg. Psychiat.* **30**, 47–51.

Rett, A. and Teubel, R. (1964). Neugeborenenkraempk in Rahmen einer epileptisch belasteten Familie. *Wien, klin. Wschr.* **76**, 609–13.

Reza, M.J., Kar, N.C., Pearson, C.M., and Kark, R.A. (1978). Recurrent myoglobinuria due to muscle cartitine palmityl transferase deficiency. *Ann. intern. Med.* **88**, 610–15.

Reznik, M. and Alberca-Serrano, R. (1964). Forme familiale d'hypertelorisme avec lissencephalie se presentant cliniquement sous form d'une Arrieration mentale avec epilepsie et paraplegie spasmodique. *J. neurol. Sci.* **1**, 40–58.

Rhein, J.W. (1916). Family spastic paralysis. *J. nerv. ment. Dis.* **44**, 115–44; 224–42.

Rho, Y.M. (1969) Von Hippel – Lindau's disease. A report of five cases. *Can. med. Ass. J.* **101**, 135–42.

Riccardi, V.M. Hässler, E., and Lubinsky, M.S. (1977). The FG syndrome. *Am. J. med. Genet.* **1**, 47–58.

Riccardi, V.M. and Marcus, E.S. (1978). Congenital hydrocephalus and cerebellar agenesis. *Clin. Genet.* **13**, 443–7.

Rice, G.P.A., Boughner, D.R., Stiller, C., and Ebers, G.C. (1980). Familial stroke syndrome associated with mitral valve prolapse. *Ann. Neurol.* **7**, 130–4.

Richard, J., De Ajuriaguerra, J., and Constantinidis, J. (1965). L'incidence familale de l'angiopathie dyshorique du cortex cerebral. *Int. J. Neuropsychiat.* **1**, 118–24.

662 *Bibliography*

Richards, B.W. (1960). Congenital ichthyosis, spastic diplegia and mental deficiency. *Br. med. J.* **ii** 714–15.

Richards, B.W. and Rundle, A.T. (1959). A familial hormonal disorder associated with mental deficiency, deaf mutism and ataxia. *J. ment. Defic. Res.* **3**, 33–55.

Richards, B.W., Rundle A.T. and Wilding, A.S. (1957). Congenital ichthyosis, spastic diplegia and mental deficiency. *J. ment. Defic. Res.* **1**, 118–29.

Richards, F., Cooper, M.R., Pearce, L.A., Cowan, R.J., and Spurr, C.L. (1974). Familial spinocerebellar degneration, hemolytic anemia and glutathione deficiency. *Arch. intern. Med.* **134**, 534–7.

Richards, R.D., Mebust, W.K., and Schimke, R.N. (1973). A prospective study on Von Hippel-Lindau disease. *J. Urol.* **110**, 27–30.

Richards, R.N. and Barnett, H.J.M. (1968). Paroxysmal choreoathetosis of Mount and Reback. *Neurology* **18**, 461–9.

Richardson, M.E. and Bornhofen, J.H. (1968). Early childhood cerebral lipidosis with prominent mycolonus. *Arch. Neurol., Chicago* **18**, 34–43.

Richieri-Costa, A., Rogatko, O., Levinsky, R. Finkel, N., and Frota-Pessoa, O. (1981). Autosomal dominant late spinal muscular atrophy, type Finkel. *Am. J. med. Genet..* **9**, 119–28.

Richter, R. (1940). A clinico-pathologic study of parenchymatous cortical cerebellar atrophy. *J. nerv. ment. Dis.* **91**, 37–46.

Richter, R.B. (1950). Late cortical cerebellar atrophy. *Am. J. hum. Genet.* **2**, 1–29.

Ricker, K. and Mertens, H.-G. (1968). The differential diagnosis of the myogenic (facio)-scapulo-peroneal syndrome. *Eur. Neurol.* **1**, 275–307.

Ricker, K., Mertens, H.-G., and Schimrigk, K. (1968). The neurogenic scapuloperoneal syndrome. *Eur. Neurol.* **1**, 257–74.

Riedel, K.G. *et al.* (1985). Ocular abnormalities in mucolipidosis IV. *Am. J. Ophthalmol.* **99**, 125–36.

Riggs, J.E. *et al.* (1984). Mitochondrial encephalomyopathy with decreased succinate cytochrome c reductase activity. *Neurology* **34**, 48–53.

Riley, C.M. and Moore, R.H. (1966). Familial dysautonomia differentiated from related disorders. Case reports and discussions of current concepts. *Pediatrics, Springfield* **37**, 435–46.

Riley, C.M. (1952). Familial autonomic dysfunction. *J. Am. med. Ass.* **149**, 153–5.

Riley, C.M, Day, R.L., Greely, D.M., and Langford, W.S. (1949). Central autonomic dysfunction with defective lacrimation. Report of 5 cases. *Pediatrics, Springfield* **3**, 468–78.

Riley, E. and Swift, M. (1979). Congenital horizontal gaze palsy and kyphoscoliosis in two sisters. *J. med. Genet.* **16**, 314–27.

Riley, H.A. (1930). Syringomyelia or myelodysplasia. *J. nerv. ment. Dis.* **72**, 1–27.

Riley, H. D. Jr and Smith, W. R. (1960). Macrocephaly, pseudopapilledema and multiple hemangiomata. A previously undescribed heredofamilial syndrome. *Pediatrics, Springfield* **26**, 293–300.

Rimoin, D. L., Fletcher, B. D., and McKusick, V. A. (1968). Spondylocostal dysplasia. A dominantly inherited form of short-trunk dwarfism. *Am. J. Med.* **45**,948.

Rimoin, D. L., Woodruff, S. L., and Holman, B. L. (1969). Craniometaphyseal dysplasia (Pyles disease) Autosomal dominant inheritance in a large kindred. *Birth Defects* **V**, 96–104.

Rimoldi, M., *et al.* (1982). Cytochrome-C-oxidase deficiency in muscles of a floppy infant without mitochondrial myopathy. *J. Neurol.* **227**, 201–7.

Ringel, S. P., Carroll, J. E., and Schold, S. C. (1977). The spectrum of mild X-linked recessive muscular dystrophy. *Arch. Neurol., Chicago* **34**, 408–16.

Rischbieth, R. H. (1973). Familial spastic paraplegia. *Proc. Aust. Ass. Neurol.* **10**, 31–3.

Ritscher, D., *et al.* (1987). Dandy—Walker (like) malformation, atrioventricular septal defect and a similar pattern of minor anomalies in 2 sisters: A new syndrome? *Am. J. Med. Genet.* **26**, 481–91.

Roach, E., De Meyer, W., Palmer, K., Conneally, M., and Merritt, A. (1975). Holoprosencephaly: birth data, genetic and demographic analysis of 30 families. *Birth Defects:* Original Artical Series **XI**, 294–313.

Roach, E. S. Williams, D. P. and Laster, D. W. (1987). Magnetic resonance imaging in tuberous sclerosis. *Arch. Neurol.* **44**, 301–3.

Robb, S. A. Pohl, K. R. E., Baraitser, M., Wilson, J., and Brett, E. M. (1989). The 'happy puppet' syndrome of Angelman: review of the clinical features. *Arch. Dis. Childhood* **64**, 83–6.

Robert, J. M., Pernod, J., and Bonnet, R. (1974). L'agenesie sacrococcygienne familiale. *J. Génét. hum.* **22**, 45–60.

Roberts, A. H. (1967). Association of a pheochromocytoma and cerebral gliosarcoma with neurofibromatosis. *Br. J. Surg.* **54**, 78–9.

Roberts P. D. (1959). Familial calcification of the cerebral basal ganglia and its relation to hypoparathyroidism. *Brain* **82**, 599–609.

Robertson, E. E. (1953). Progressive bulbar paralysis showing heredofamilial incidence and intellectual impairment. *Arch. Neurol. Psychiat., Chicago* **69**, 197–207.

Robertson, W. C. *et al.* (1979). Basal ganglia calcification in Kearns-Sayre syndrome. *Arch. Neurol.* **36**, 711–13.

Robinow, M. (1974). Mucolipidosis III. *Birth Defects* **X**, 267–73.

Robinow, M. and Sorauf, T. J. (1975). Acrocephalopolysyndactyly type Noack in a large kindred. New Chromosomal and malformation syndromes. *Birth Defects:* Original Article Series **XI**, 99–106.

Robins, M.M. (1966). Pyridoxine dependency convulsions in a newborn. *J. Am. med. Ass.* **195** 491–3.

Robinson, B.H. Sherwood, W.G., Taylor, J., Balfe, J. W., and Mamer, O.A. (1979). Acetoacetyl CoA thiolase deficiency: a cause of severe ketoacidosis in infancy simulating salcylism. *J. Pediat.* **95**, 228–33.

Robinson, B.H. *et al.* (1981). Lactic acidemia. neurologic deterioration and carbohydrate dependence in a girl with dihydrolipoyl dehyrogenase deficiency. *Eur. J. Pediat.* **136**, 35–9.

Robinson, G.C., Jan, J.E., and Dunn H.G. (1977*a*). Infantile polymyoclonus: its occurrence in second cousins. *Clin. Genet.* **11**, 53–6.

Robinson, G.C., Jan, J.E., and Miller, J.R. (1977*b*). A new variety of hereditary sensory neuropathy. *Hum. Genet.* **35**, 153–61.

Robinson, L.K. *et al.* (1985). Carpenter syndrome; natural history and clinical spectrum. *Am. J. Med. Genet.* **20**, 461–9.

Robinson, R.O. and Thornett, C.E.E. (1985). Benign hereditary chorea-response to steroids. *Dev. Med. Child Neurol.* **27**, 814–16.

Robles, J. (1966). Congenital ocular motor apraxia in identical twins. *Br. J. Ophthal.* **40**, 444.

Rodeck, C.H. *et al.* (1982). Fetal liver biopsy for prenatal diagnosis of ornithine carbamyl transferase deficiency. *Lancet* **II**, 297–300.

Rodin, F.H. and Barkan, H. (1935). Hereditary congenital ptosis: report of a pedigree and review of the literature. *Am. J. Ophthal.* **18**, 213–25.

Rodriguez, H.A. and Berthrong, M. (1966). Multiple primary intracranial tumors in Von Recklinghausen's neurofibromatosis. *Arch. Neurol., Chicago* **14**, 467–75.

Rodzilsky, B., Bolton, C.F., and Takeda, M. (1971). Neuroaxonal dystrophy—a case of delayed onset and protracted course. *Acta neuropath.* **17**, 331–40.

Roe, P.F. (1963). Hereditary spastic paraplegia. *J. Neurol. Neurosurg. Psychiat.* **26**, 516–19.

Roe, P.F. (1964). Familial motor neuron disease. *J. Neurol, Neurosurg. Psychiat.* **27**, 140–3.

Roelvink, N.C.A. *et al.* (1986). Concordant cerebral oligodendroglioma in identical twins. *J. Neurol. Neurosurg. Psychiat.* **49**, 706–8.

Roessman, U. and Schwartz. J.F. (1973). Familial striatal degeneration. *Arch. Neurol., Chicago* **29**, 314–17.

Rohkamm, R., Boxler, K., Ricker, K., and Jerusalem, F. (1983). A dominantly inherited myopathy with excessive tubular aggregates *Neurology* **33**, 331–6.

Romanul, F.C.A., Fowler, H.L., Radvany, J., Feldman, R.G., and Feingold,M. (1977). Azorean disease of the nervous system. *N. Engl. J. Med.* **296**, 1505–8.

Romeo, G., Borrone C., Gutti, R., and Durand, P. (1977). Fucosidosis in

Calabria: founder effect or high gene frequency. (Letter.) *Lancet* **i**, 368-9.

Ron, M.A. and Pearce, J. (1971*a*). Marinesco-Sjögren-Garland syndrome with unusual features. *J. neurol. Sci* **13**, 175-9.

Ron, M.A. and Pearce, J. (1971*b*). Refsum's syndrome with normal phytanic metabolism. *Acta neurol. scand.* **47**, 646-9.

Ron, M.A. and Pearce, J. (1972). Myotonia congenita in dizygotic twins. *Eur. Neurol.* **7**, 196-200.

Rondot, P., de Recondo, J., *et al.* (1983). Menzel's hereditary ataxia with slow eye movements and myoclonus. A clinico-pathological study. *J. Neurol. Sci.* **61**, 65-80.

Roos, D. and Thygesen, P. (1972). Familial recurrent polyneuropathy. *Brain* **95**, 235-48.

Roos, R., Gajdusek, D.C., and Gibbs, C.J. (1973). The clinical characteristics of transmissible Creutzfeldt-Jakob disease. *Brain* **96**, 1-20.

Ropers, H.H., Burmeister, P., Von Petrykowski, W., and Schindera, F. (1975). Leukodystrophy, skin hyperpigmentation and adrenal atrophy: Siemerling–Creutzfeldt disease. Transmission through several generations in two families. *Am. J. hum. Genet.* **27**, 547-53.

Rosatti P. (1972). Une famille atteinte d'hyperostose frontale interne (syndrome de Morgagni–Morel). A travers quatre generations successives. *J. Génét. hum.* **20**, 207-52.

Rosenbaum, H.E. (1960). Familial hemiplegic migraine *Neurol., Minneap.* **10**, 164-70.

Rosenberg, A.L., Bergstrom, L., Troost, T., and Bartholomew, B.A. (1970). Hyperuricaemia and neurological deficits. *N. Engl. J. Med.* **282**, 992-7.

Rosenberg, M.L. (1984). Congenital trigeminal anaesthesia, a review and classification. *Brain* **107**, 1073-82.

Rosenberg, R.N. and Chutorian, A. (1967). Familial opticoacoustic nerve degeneration and polyneuropathy. *Neurol., Minneap.* **17**, 827-32.

Rosenberg, R.N., Nyhan, W.L., Bay, C., and Shore, P. (1976). Autosomal dominant striatonigral degenerative. *Neurol., Minneap.* **26**, 703-14.

Rosenberg, R.N., Nyhan, W.L., Coutinho, P., and Bay, C. (1978). Joseph's disease: an autosomal dominant neurological disease in the Portuguese of the United states and the Azores Islands. *Adv. Neurol.* **21**, 33-57.

Rosenhagen, H. (1943). Die primäre atrophie des Brückenfusses und der unteren Oliven. *Arch. Psychiat. NervKrankh.* **116**, 163.

Rosenmann, A. and Arad, I. (1974). Arthrogryposis multiplex congenita: neurogenic type with autosomal recessive inheritance. *J. med, Genet.* **11**, 91-4.

Rosenthal, C. (1931). Klinisch–erbbiologischer beitrag zur Konstitutions-pathologie. *Z. Neurol, Psychol.* **131**, 475–501.

Rosenthal, N.P., Keesey, J., Crandall, B., and Brown, J. (1976). Familial neurological disease associated with spongiform encepthalopathy. *Arch. Neurol., Chicago* **33**, 252–9.

Rosing, H.S., Hopkins, L.C., and Wallace, D.C. (1985). Maternally inherited mitochondrial myopathy and myoclonic epilepsy. *Ann. Neurol.* **17**, 228–37.

Rosman, N.P. and Pearce, J. (1967). The brain in multiple neurofibromatosis (Von Recklinghausen's disease): a suggested neuropathological basis for the associated mental defect. *Brain* **90**, 829–38.

Ross, A.T. and Dickerson, W.W. (1943). Tuberous sclerosis. *Arch. Neurol. Psychiat., Chicago* **50**, 233–57.

Ross, E.M., Peckham, C.S., West, P.B., and Butler, N.R. (1980). Epilepsy in childhood: findings from the National Child development study. *Br. med J.* **i**, 207–10.

Ross, J. and Bury, J.S. (1893). *On peripheral neuritis.* London.

Ross, N. (1932). Congenital epicanthus and ptosis transmitted through 4 generations. *Br. med, J.* **i**, 378.

Ross, R.T. (1959). Multiple and familial intracranial vascular lesions. *J. Can. med. Ass.* **81**, 477–9.

Ross, R.T. Simpson, C.A., and Styles, S. (1974). Wohlfart Kugelberg Welander syndrome. *Can. J. neurol. Sci.* **1**, 130–8.

Rossi, L.N. *et al.* (1983). Hereditary motor and sensory neuropathies in childhood. *Dev. Med. Child Neurol.* **25**, 19–31.

Roth, M. (1948). On a possible relationship between hereditary ataxia and peroneal muscular atrophy: with a critical review of the problems of 'intermediate form's in the degenerative disorder of the central nervous system. *Brain* **71**, 416–33.

Rothner, A. and Brust, J. (1974). Pseudotumour cerebri: report of a familial occurrence. *Arch. Neurol., Chicago* **30**, 110–11.

Rothner, A.D., Yahr, F., and Yahr, M.D. (1976). Familial spastic paraparesis, optic atrophy and dementia. Clinical observations of affected kindred. *N. Y. St. J. Med.* **76**, 756–8.

Rothschild, H., Happel, L., Rampp, D., and Hackett, E. (1979). Autosomal recessive spastic paraplegia: evidence for demyelination. *Clin. Genet.* **15**, 356–60.

Rotter, R. (1932). Zum Problem des Vorkommens' progressiver Versteifung' bei dur Huntingtonschen Krankheit. *Z. ges. Neurol. Psychiat.* **138**, 376–413.

Rotthauwe, W.W., Mortier, W., and Beyer, H. (1972). Neuer Typ einer recessiv X-chromosomal vererbten Muskeldystrophie. *Humangenetik* **16**, 181–200.

Roussel, B. Leroux, B., Gaillard, D., Fandre, M. (1985). Chronic, diffuse tubulo-interstitial nephritis and hepatic involvement in the

Laurence – Moon – Bardet – Biedl syndrome. (In French.) *Helv. Paediat. Acta* **40**, 405.

Roussy, G. and Levy, G. (1926). Sept. cas d'une maladie familiale particuliere. *Rev. neurol.* **33**, 427–50.

Routsonis, K. and Georgiadis, G. (1984). Peroneal muscular atrophy and epilepsy with cerebellar ataxia and choreoathetosis in the same family. *J. Neurol. Sci.* **64**, 161–73.

Rowan, A.J., Heathfield, K.W.G., and Scott, D.F. (1970). Is reading epilepsy inherited? *J.Neurol. Neurosurg. Psychiat.* **33**, 476–8.

Rowland, L.P. and Eskenazi, A. (1956). Myasthenia gravis with features resembling muscular dystrophy. *Neurol., Minneap.* **6**, 667–71.

Rowland, L.P., Fahn, S., and Schotland, D.L.(1963). McArdle's disease, *Arch. Neurol., Chicago* **9**, 325–42.

Rowland, L.P., Lovelace, R.E., Schotland, D.L., Araki, S.,and Carmel, P. (1966). The clinical diagnosis of McArdle's disease. identification of another family with deficiency of muscle phosphorylase. *Neurol., Minneap.* **16**, 93–100.

Rowland, L.P., Fetell, M., Olarte, M., Hays, A., Singh, N., and Wanat, F.E. (1979). Emery-Dreifuss muscular dystrophy. *Ann. Neurol.* **5**, 111–17.

Rowland, L.P., Hays, A.P., Di Mauro, S., DeVivo, D.C., and Behrens, M. (1983). Diverse clinical disorders associated with morphological abnormalities of mitochondria. In *Mitochondrial Palthology in Muscle Diseases* (ed. G. Scarlato al C. Cerri), Piccin, Padova, Pr. 142–58.

Rowlatt, U. (1969). Cockayne's syndrome. Report of case with necropsy findings. *Acta neuropath*, **14**, 52–61.

Roy, P.R., Emanuel, R., Ismail, S.A., and Tayib, M.H.E. (1976). Hereditary prolongation of the Q-T interval: genetic observations and management of three families with 12 affected members. *Am. J. Cardiol.* **37**, 237–43.

Rozear, M.P., *et al.* (1987). Hereditary motor and sensory neuropathy: a half century follow-up. Neurol. **37**, 1460–5.

Rozen, R., *et al.* (1985). Gene deletion and restriction fragment length polymorphisms at the human ornithine transcarbamylase locus. *Nature* **313**, 815–17.

Rubenstein, A.E. and Yahr, M.D. (1977). Adult onset autonomic dysfunction coexistant with a familial dysautonomia in a consanguineous family. *Neurol. Minneap.* **27**, 168–70.

Rubin, M.B. (1972). Incontinentia pigmenti achromians. Multiple cases within a family. *Arch. Derm.* **105**, 424–5.

Rukavina. J.C., Block, W.D., Jackson, C.E., Falls, H.F., Carey, J.H., and Curtis, A.C. (1956). Primary systemic amyloidosis: a review and an experimental genetic, and clinical study of 29 cases with particular emphasis on the familial form. *Medicine, Baltimore* **35**, 239–334.

Rushton, A.R. and Shaywitz, B.A. (1979). Tuberous sclerosis: possible modification of phenotypic expression by an unlinked dominant gene. *J. med. Genet.* **16**, 32–5.

Russo, L. S., Aron, A., and Anderson, P. J. (1976). Alexanders disease. A report and reappraisal. *Neurol., Minneap.* **26**, 607–14.

Rustam, H., Hamdi, T., and Witri, S. (1975). Progressive familial myoclonic epilepsy. *J. Neurol. Neurosurg. Psychiat.* **38**, 845–8.

Saadat, M., Mokfi, H., Vakil, H., and Ziai, M. (1972). Schwartz syndrome: myotonia with blepharophimosis and limitation of joints. *J. Pediat.* **81**, 348–50.

Sachdeva, K. K., Singh, N., and Krishnamoorthy, M. S. (1977). Juvenile Parkinsonism treated with levodopa. *Arch. Neurol. Chicago* **34**, 244–5.

Sachs, R. (1967). Apraxie oculo-motrice congenitale de Cogan. A propos de trois nouveaux cas dont deux dans la meme fratrie. *Annl. Oculist.* **200**, 266–74.

Sacks, O. W., Aguilar, M. J., and Brown, W. J., (1966). Hallervorden-Spatz disease. Its pathogenesis and place among axonal dystrophies. *Acta neuropath.* **6**, 164–74.

Sadjadpour, K. and Amato, R. S. (1973). Hereditary non-progressive chorea of early onset—a new entity? In *Advances in neurology,* Vol. 1 (ed. A. Babeau, T. N. Chese, and G. W. Paulsen), pp. 79–91. Raven Press, New York.

Sadovnick, A. D. and Macleod, P. M. J. (1981). The familial nature of multiple scelerosis. *Neurol., Minneap.* **31**, 1039–41.

Såebo, J. A. (1952). Von Hippel—Lindau's disease. *Acta ophthal.* **30**, 129–93.

Saethre, M. (1931). Ein Beithfag zum Turmchädelproblems. (Pathogenese, Erblichkeit und Symptomatologie.) *Dt. Z. Nervenheilk.* **119**, 533.

Sahar, A. (1965). Familial occurrence of meningiomas. *J. Neurosurg.* **23**, 444–5.

St George-Hyslop, P. H., *et al.* (1987). The genetic defect causing familial Alzheimeir's disease maps on chromosome 21. *Science* **235**, 885–90.

Saint-Hilaire, M-H., *et al.* (1986). Jumping Frenchmen of Maine. *Neurology* **36**, 1269–71.

Sajid, M. H. and Copple, P. J. (1968). Familial aqueduct stenosis and basilar impression. *Neurol., Minneap.* **18**, 260–2.

Sak, H. G., Smith, A. A., and Dancis, J. (1967). Psychometric evaluations of children with familial dysautonomia. *Am. J. Psychiat.* **124**, 682.

Sakai, T., Ohta, M., and Ishino, H. (1983). Joseph disease in a non-Portugese family. *Neurol., Minneap.* **33**, 74–80.

Sakati, N., Nyhan, W. L., and Tisdale, W. K. (1971). A new syndrome with acrocephalopolysyndactyly, cardiac disease, and distinctive defects of the ear, skin and lower limbs. *J. Pediat.* **79**, 104–9.

Sakoda, S., *et al.* (1983). Genetic studies of familial amyloid polyneuropathy in the Arao district of Japan: the genealogical survey. *Clin. Genet.* **24**, 334–8.

Sakuraba, H., *et al.* (1983). β-Galactosidase-neuraminidase deficiency (Galactosialidosis): clinical, pathological and enzymatic studies in a postmortem case. *Ann. Neurol.* **13**, 497–503.

Sakurane, H.F., Sugai, T., and Saito, T. (1967). The association of blue rubber bleb nevus and Maffucci syndrome. *Arch. Derm.* **95**, 28–36.

Sala, E. and Savoldi, F. (1959). Contributo alla conoscenza delle forme ereditarie della mallattia di Fahr. *Sist. nerv.* **11**, 46–73.

Saldino, R.M., Steinbach, H.L., and Epstein, C.J. (1972). Familial acrocephalosyndactyly (Pfeiffer syndrome). *Am. J. Roentg.* **116**, 609–22.

Salet, J., *et al.* (1970). Demonstration de la nature familiale de l'hypomagnesemie congenitale chronique. *Arch. fr, Pediat.* **27**, 550–1.

Salisachs, P. (1974). Wide spectrum of motor conduction velocity in Charcot–Marie–Tooth disease. An anatomico-physiological interpretation. *J. neurol. Sci.* **23**, 25–31.

Salisachs, P. (1975). Unusual motor conduction velocity values in Charcot–Marie–Tooth disease associated with essential tremor. Report of a kinship. *Eur. Neurol.* **13**, 377–82.

Salisachs, P. (1976). Charcot–Marie–Tooth disease associated with 'essential tremor'. Report of seven cases and a review of the literature. *J. neurol. Sci.* **28**, 17–40.

Salisachs, P., Codina, A., Giminez-Roldan, S., and Zarranz, J.J. (1979). Charcot–Marie–Tooth disease associated with 'essential tremor' and normal and/or slightly diminished motor conduction velocity. *Eur. Neurol.* **18**, 49–58.

Salonen, R., *et al.* (1981). The hydrolethalus syndrome—A new hereditary malformation syndrome in Finland. *Duodecim.* **97**, 1312–9.

Salt, H.B., Wolff, O.H., Loyd, J.K., Fosbrooke, A.S., Cameron, A.H., and Hubble, D.V. (1960). On having no beta-lipoprotein: a syndrome comprising a-betalipoproteinemia, acanthocytosis and steatorrhea. *Lancet* **ii**, 325–9.

Samuel, V.N. (1966). Rigid and akinetic form of Huntington's chorea. Two cases occurring in the same family. *J. Indian med. Ass.* **47**, 336–8.

Sand, T. and Hestnes, A. (1985). Sensory signs and symptoms in scapuloperoneal atrophy: a report of a family. *Eur. Neurol.* **24**, 405–13.

Sandbank, U. and Lerman, P. (1972). Progressive cerebral poliodystrophy—Alpers disease. *J. Neurol. Neurosurg. Psychiat.* **35**, 749–55.

Sander, J.E., Layzer, R.B., and Goldsobel, A.B. (1980). Congenital stiffman syndrome. *Ann. Neurol.* **8**, 195–7.

Sanders, D.B. (1976). Myotonia congenita with painful muscle contractions. *Arch. Neurol., Chicago.* **33**, 580–2.

Sanders, D.G. (1973). Familial occurrence of Gilles de la Tourette syndrome: report of the syndrome occurring in a father and son. *Arch. gen. Psychiat.* **28**, 326–8.

Sanders, J., Schent, V. W. D., and van Veen, P. (1939). A family with Pick's diseases. *Verh. Koninkl. Ned. Akad. Wescn.* Section 2. Part 38, No. 3.

Sandhoff, K., Harzer, K., Wassle, W., and Jatzkewitz, H. (1971). Enzyme alteration and lipid storage in three variants of Tay-Sachs disease. *J. Neurochem.* **18**, 2469.

Sandyk, P. (1982). An atypical form of familial myoclonus epilepsy. A case report. *S Afr. med. J.* **61**, 366–7.

Sanner, G. (1973). The dysequilibrium syndrome. *Neuropaediatrie* **4**, 403–13.

Santangelo, T. G. (1934). Contributo clinico alla conoscenza della forme familiare della dysbasia lordotica progressiva. *G. Psichiat. Neuropatol.* **62**, 52–77.

Santavuori, P. (1973). EEG in the infantile type of so-called neuronal ceroid-lipofuscinosis. *Neuropaediatrie.* **4**, 375–87.

Santavuori, P., Haltia, M., and Rapola, J. (1974). Infantile type of so-called neuronal ceroid-lipofuscinosis. *Dev. med. Child. Neurol.* **16**, 64–53.

Santavuori, P., Leisti, J., Kruus, J., and Raitta, C. (1978). Muscle, eye and brain disease: a new syndrome. *Doc. Ophthalm.* **17**, 393–6.

Santen, R. J,. and Paulsen, C. A. (1973). Hypogonadotrophic eunuchoidism. I. Clinical study of the mode of inheritance. *J. clin. Endocr.* **36**, 47.

Santiago-Borrero, P. J., Santini, R., Jr., Perez-Santiago, E., and Maldonado, N., (1973). Congenital isolated defect of folic acid absorption. *J. Pediatr.,* **82**, 450–5.

Sarfarazi, M., Quarrel, O. W. J., Wolak, G., and Harper, P. S. (1987). An integrated microcomputer system to maintain a genetic register for Huntington's chorea. *Am. J. Med. Genet:* **28**, 999–1006.

Sargent, C., *et al.* (1985). Trigonocephaly and the Opitz C syndrome. *J.Med. Genet.* **22**, 39–45.

Sarsfield, J. K. (1971). The syndrome of congenital cerebellar ataxia, aniridia and mental retardation. *Dev. med. Child. Neurol.* **13**, 508–11.

Sasaki, H., *et al.* (1983). Myoclonus, cerebellar disorder, neuropathy, mitochondrial myopathy and ACTH deficiency. *Neurology* **33**, 1288–93.

Sass-Kortsak, A. (1975). Wilson's disease. A treatable liver disease in children. *Pediat. Clins. N. Am.* **22**, 963–84.

Satapathy, R. K. and Skinner, R. (1979). Serum creatine kinase levels in normal females. *J. med. Genet.* **16**, 49–51.

Sato, N., *et al.* (1987). Aicardi syndrome with holoprosencephaly and cleft lip and palate. *Pediat. Neurol.* **3**, 114–16.

Satoyoshi, E. and Kinoshita, M. (1977). Oculopharyngodistal myopathy—report of four families. *Arch. Neurol., Chicago* **34**, 89–92.

Saudubray, J. M., Marsac, C., Charpentier, C., Cathelineau, L., Besson-

Leaud, M., and Leroux, J. P. (1976). Neonatal congenital lactic acidosis with pyruvate carboxylase deficiency in two siblings. *Acta pediat. scand.* **65**, 717–24.

Saunders, E. S., Shortland, D., and Dunn, P. M. (1984). What is the incidence of holoprosencephaly? *J.Med. Genet.* **21**, 21–6.

Saunders, M., Ashworth, B., Emery, A. E. H., and Benedikz, J. E. G (1968). Familial myotonic periodic paralysis with muscle wasting. *Brain* **91**, 295–304.

Savage, D. C. L., Forbes, M., and Pearce, G. W. (1971). Idiopathic rhabdomyolysis. *Arch. Dis. Childh.* **46**, 594–607.

Say, B. and Coldwell, J. G. (1975). Hereditary defect of the sacrum. *Hum. Genet.* **27**, 231–4.

Sayli, B. S., Erüreten, I., and Topuz, U. (1969). Sjögren–Larsson syndrome in a Turkish family. *J.med. Genet.* **6**, 352–3.

Scarlato, G., Valli, G., and Bracch, I. M. (1978). Quantitative, histological and neurophysiological studies in scapuloperoneal syndrome. *Acta neurol.* **33**, 327–41.

Scarpelezos, S. (1948). Sur la notion d'heredite similaire dans la maladie de Parkinson. Données statisques sur 626 cas de'etats Parkinsoniens. *Rev. neurol.* **80**, 184–203.

Scelsi, R., Mazella, G. L., and Lombardi, M. (1976). Myoclonus epilepsy with cerebellar Lafora bodies. *J. Neurol. Neurosurg. Psychiat.* **39**, 357–61.

Schafer, I. A., Powell, D. W., and Sullivan, J. C. (1971). Lysosomal bone disease. (Abstract.) *Pediat. Res.* **5**, 391–2.

Schapira, K., Poskanzer, D. C., and Miller, H. (1963). Familial and conjugal multiple sclerosis. *Brain* **86**, 315–32.

Schaub, J., *et al.* (1978). Tetrahydrobiopterin therapy of atypical phenylketonuria due to defective dihydrobiopterin biosynthesis. *Arch. Dis. Childh.* **53**, 674–83.

Schaumburg, H. H. and Suzuki, K. (1968). Non-specific familial presenile dementia. *J. Neurol. Neurosurg. Psychiat.* **31**, 479–86.

Schaumburg, H. H., Powers, J. M. Raine, C. S., Suzuki, K., and Richardson, E. P. (1975). Adrenoleukodystrophy. A clinical and pathological study of 17 cases. *Arch. Neurol., Chicago* **32**, 577–91.

Schaumburg, H. H., *et al.* (1977). Adrenomyeloneuropathy II. *Neurol., Minneap.* **27**, 1114–17.

Scheie, H. G., Hambrick, G. W., and Barness, L. A. (1962). A newly-recognized forme fruste of Hurler's disease (gargoylism). *Am. J. Ophthal.* **53**, 753–69.

Schinzel, A. (1988). The acrocallosal syndrome in first cousins: widening of the spectrum of clinical features and further support for autosomal recessive inheritance. *J. Med. Genet.* **25**, 332–6.

Schinzel, A., and Litschgi, M. (1984). Autosomal recessive severe congenital microcephaly: antenatal ultrasonographic diagnosis and head growth from 15 to 24 weeks of gestation. *J. Med. Genet.* **21**, 355–8.

Schittz-Christensen, E. (1972). Genetic factors in febrile convulsions. *Acta neural. scand.* **48**, 538–46.

Schimke, R.N. (1974). Adult onset hereditary cerebellar ataxia and neurosensory deafness. *Clin. Genet.* **6**, 416–21.

Schimschock, J.R. Alvord, E.C., and Swanson, P.D. (1968). Cerebrotendinous xanthomatosis. Clinical and pathological studies. *Arch. Neurol., Chicago* **18**, 688–98.

Schmid, R. and Hammaker, L. (1961). Hereditary absence of muscle phosphorylase (McArdle's syndrome). *N. Engl. J. Med.* **264**, 223–5.

Schmidt B., et al. (1987). McArdle's disease in two generations: Autosomal recessive transmission with manifesting heterozygote. *Neurology*, **37**, 1558–61.

Schneider, D.E. and Abeles, M.M. (1937). Charcot-Marie-Tooth disease with primary optic atrophy, *J. nerv. ment. Dis.* **85**, 540–7.

Schneider, H., Vasella, F., and Karbowski, K. (1970). The Lennox syndrome. *Eur. Neurol.* **4**, 289.

Schneiderman, L.J., Sampson, W.I., Schoene, W.C., and Haydon, G.B. (1969). Genetic studies of a family with two unusual autosomal dominant conditions: muscular dystrophy and Pelger-Huet anomaly. Clinical, pathologic and linkage consideration. *Am. J. Med.* **46**, 380–93.

Schnitzler, E.R. and Robertson, W.C. Jr (1979). Familial Kearns-Sayre syndrome. *Neurol., Minneap.* **29**, 1172–4.

Schockett, S.S. Jr, Zellweger, H. Ionasescu, U., and McCormick, W.K. (1972). Centronuclear myopathy: disease entity or a syndrome? Light and electron microscopic study of two cases and review of the literature. *J. neurol. Sci.* **16**, 215–28.

Schoene, W.C., Asbury, A.K., Aström, K.E., and Masters, P. (1970). Hereditary sensory neuropathy. A clinical and ultrastructural study. *J.neurol. Sci.* **11**, 463–87.

Scholte, H.R., Jennekens, F.G.I., and Bouvy, J.J.B.J. (1979). Carnitine palmitoyltransferase II deficiency with normal Carnitine palmitoyltransferase I in skeletal muscle and leucocytes. *J. neurol. Sci.* **40**, 39–51.

Scholtz, C.L. and Swash, M. (1985). Cerebellar degneration in dominantly inherited spastic paraplegia. *J. Neurol. Neurosurg. Psychiat.* **48**, 145–9.

Schoolman, A. and Kepes, J.J. (1967). Bilateral spontaneous carotid-cavernous fistulae in Ehlers-Danlos syndrome. *J. Neurosurg.* **26**, 82–6.

Schoolman, A. and Rowland, L.P. (1964). Muscular dystrophy—features of ocular myopathy, distal myopathy and myotonic dystrophy. *Arch. Neurol., Chicago* **10**, 433–45.

Schotland, D.L., DiMauro, S., Bonilla, E., Scarpa, A., and Lee, C.-P. (1976). Neuromuscular disorder associated with a defect in mitochondrial energy supply. *Arch. Neurol., Chicago* 33, 475-9.

Schott, G.D. (1980). Familial cerebellar ataxia presenting with down beat nystagmus. *J. med. Genet.* 17, 115-18.

Schott, G.D. and Wyke, M.A. (1977). Obligatory bimanual associated movements. *J. neurol. Sci.* 33, 301-12.

Schreier, H., Rapin, I., and Davis, J. (1974). Familial megalencephaly or hydrocephalus? *Neurol., Minneap.* 24, 232-6.

Schreiner, A., Hopen, G., and Skrede, S. (1975). Cerebrotendinous xanthomatosis (cholestanolosis). *Acta neurol. scand.* 51, 405-16.

Schröder, J.M. and Bohl, J. (1978). Altered ratio between axon caliber and myelia thickness in sural nerves of children. In Peripheral Neuropathics (ed. N. Canal and G. Pozza.), pp. 49-62. Elsevier, Amsterdam, New York and Oxford.

Schroffner, W.G. and Furth, E.D. (1970). Hypogonadotropic hypogonadism with anosmia (Kallmann's syndrome) unresponsive to clomiphene citrate. *J. clin. Endocr.* 31, 267-70.

Schuchmann, L. (1970). Spinal muscular atrophy of the scapulo-peroneal type. *Z. Kindeheilk.* 109, 118-23.

Schuffler, M.D., Bird, T.D., Sum, S.M., and Cook, A. (1978). A familial neuronal disease presenting as intestinal pseudo-obstruction *Gastroenterology* 75,889-98.

Schulman, J.D., et al. (1970). A new variant of maple syrup urine disease (branched chain ketoaciduria). Clinical and biochemical evaluation. *Am. J. med.* 49, 118-24.

Schumann, S.H. and Miller, L.J. (1966). Febrile convulsions in families: findings in an epidemiologic survey. *Clin. Pediat.* 5, 604-8.

Schut, J.W. (1950). Hereditary ataxia: clinical study through six generations. *Arch. Neurol. Psychiat., Chicago* 63, 535-68.

Schut, J.W. and Haymaker, W. (1951). Hereditary ataxia: a pathologic study of five cases of common ancestry. *J. Neuropath. clin. Neurol.* 1, 183-213.

Schutgens, R.B.H. et al. (1982). Beta-ketolase deficiency in a family confirmed by *in vitro* enzymatic assays in fibroblasts. *Eur. J. Pediat.* 139, 39.

Schutt, W. (1963). Congenital cerebellar ataxia: a review of 32 cases. In Walsh, G. (ed.) Cerebellum, posture and cerebral palsy. *Little Club Clinics in Developmental Medicine* 8, 83-90.

Schutta, H.S., Pratt, R.T.C., Metz, H., Evans, K., K., and Carter, C.O. (1966). A family study of the late infantile and juvenile forms of metachromatic leukodystrophy. *J. med. Genet.* 3, 86-90.

Schwab, R.S. and England, A.C. (1958). Parkinson's disease. *J. chron. Dis.* 8, 448-509.

Schwartz, J.F. (1962). Photosensitivity in a family. *Am. J. Dis. Child.* **103**, 786–93.

Schwarz, G.A. and Liu, C. (1956). Hereditary (familial) spastic paraplegia. *Arch. Neurol. Psychiat., Chicago* **75**, 144–62.

Schwarz, G.A. and Yanoff, M. (1965). Lafora's disease. Distinct clinico-pathologic form of Unverricht's syndrome. *Arch. Neurol., Chicago* **12**, 172–88.

Schwartz, J.F., O'Brien, M.S., and Hoffman, J.C. (1979). Hereditary spinal arachnoid cysts, distichiasis and lymphedema. *Ann. Neurol.* **7**, 340–3.

Schwartz, M.F., Esterly, N.B., Fretzin, D.F., Pergament, E., and Rozenfeld, I.H. (1977). Hypomelanosis of Ito (incontinentia pigmenti achromians): a neurocutaneous syndrome. *J. Pediat.* **90**, 236–40.

Schwartz, M.S. and Swash, M. (1975). Scapuloperoneal atrophy with sensory involvement. Davidendow's syndrome. *J. Neurol. Neurosurg. Psychiat.* **38**, 1063–7.

Schwartz, O. and Jampel, R.S. (1962). Congenital blepharophimosis associated with unique generalized myopathy. *Arch. Ophthal., N.Y.* **68**, 52–7.

Scott, G.A. (1964). Melkersson's syndrome. *Br. J. clin. Pract.* **18**, 415–18.

Scott, J., Gollan, J.L., Samourian, S., and Sherlock, S. (1978). Wilson's disease, presenting as chronic active hepatitis. *Gastroenterology* **74**, 645–51.

Scotto, J.M. *et al.* (1982). Infantile phytanic acid storage disease, a possible variant of Refsum's disease: Three cases including ultrastructural studies of the liver. *J. Inherit. Met. Dis.* **5**, 83–90.

Scribanu, N. and Kennedy, C. (1976). Familial syndrome with dystonia, neural deafness and possible intellectual impairment. *Adv. Neurol.* **14**, 235–43.

Scrimgeour, E.M. and Mastaglia, F.L. (1984). Oculopharyngeal and distal myopathy. A case study from Papua New Guinea. *Am. J. Med. Genet.* **17**, 763–71.

Scriver, C.R., Pueschel, S., and Davies, E. (1966). Hyper-beta-alaninemia associated with beta-aminoaciduria and gamma- amino butyricaciduria, somnolence and seizures. *N. Engl. J. Med.* **274**, 635–43.

Scriver, C.R. *et al.* (1971). Thiamine-responsive maple syrup-urine disease. *Lancet* **i**, 310–12.

Seay, A.R., Ziter, F.A., and Petajan, J.H. (1977). Rigid spine syndrome. *Arch. Neurol.* **34**, 119–22.

Sedano, H., Cohen, M., Jirasek, J., and Gorlin, R. (1970). Frontonasal dysplasia. *J. Pediat.* **76**, 906–13.

Sedano, H.D., Gorlin, R.J., and Anderson, V.E. (1968). Pycnodysostosis. Clinical and genetic considerations. *Am. J. Dis. Child.* **116**, 70–7.

Sedzimir, C.B., Frazer, A.K., and Roberts, J.R. (1973). Cranial and spinal meningiomas in a pair of identical twin boys. *J. Neurol. Neurosurg. Psychiat.* **36**, 368–76.

Seedorff, T. (1970). Leber's disease. V. *Acta ophthal.* **48**, 186–213.

Seemanova, E., Lesny, I., Hyanek, J., Brachfeld, K., Rossler, M., and Proskova, M. (1973). X-chromosome recessive microcephaly with epilepsy, spastic tetraplegia and absent abdominal reflex. New variety of Paine syndrome? *Humangenetik* **20**, 113–17.

Seemanova, E., *et al.* (1985). Familial microcephaly with normal intelligence. immunodeficiency. and risk for lymphoreticular malignancies: a new autosomal recessive disorder. *Am. J. Med. Genet.* **20**, 639–48.

Segawa, M., Hosaka, A., Miyagawa, F., Nomura, Y., and Imai, H. (1976). Hereditary progressive dystonia with marked diurnal fluctuations. *Adv. Neurol.* **14**, 215–33.

Seigal, R.S. *et al.* (1979). Computerised tomography in oculocraniosomatic disease (Kearns-Sayre syndrome). *Radiology* **130**, 159–64.

Seitelberger, E. (1970). Pelizaeus–Merzbacher disease. In *Handbook of clinical neurology.* Vol. 10 (ed. P.J. Vinken and G.W. Bruyn). North Holland, Amsterdam.

Seitelberger, E. and Simma, A. (1962). On the pigment variant of amaurotic idiocy. In *Cerebral sphingolipidosis* (ed. S.M. Aronson and B.M. Volk), pp. 29–47. Academic Press, New York.

Seitelberger, E., Gootz, E., and Gross, H. (1963). Beitrag zur spätinfantilen Hallervorden–Spatzschen Krankheit. *Acta neuropath.* **3**, 16–28.

Seitelberger, E. and Gross, H. (1957). Über eine spätinfantile form der Hallervorden–Spatzschen Krankeit. *D.Z. Nervenheilk.* **176**, 104–25.

Seitz, D. (1957). Zur nosologischen Stellung des sogenannten scapuloperonealen Syndroms. *D. Z. Nervenheilk.* **175**, 547–52,

Seitz, R.J. *et al.* (1984).Congenital Leigh's disease: panencephalomyelopathy and peripheral neuropathy. *Acta Neuropath,* **64**, 167–71.

Seitz, R.J. *et al.* (1986). Hypomyelination neuropathy in a female newborn presenting as arthrogryposis multiplex congenita. *Neuropediatrics* **17**, 132–6.

Seizinger, B.R., *et al.* (1987). Genetic linkage of Von Recklinghausen neurofibromatosis to the nerve growth receptor gene. *Cell* **49**, 589–94.

Seizinger, B.R., *et al.* (1988). Von Hippel–Lindau disease maps to the region of chromosome 3 associated with renal cell carcinoma. *Nature,* **332**, 268–9.

Sells, C.J., Hanson, J.W., and Hal, J.G. (1979). The Summitt syndrome. Observations on a third case. *Am. J. med. Genet.* **3**, 27–33.

Selmanowitz, V.J. (1968). Nevus flammeus of the forehead. *J. Pediat.* **73**, 755–7.

Selmanowitz, V.J. and Porter, M.J. (1967). The Sjögren–Larsson syndrome. *Am. J. Med..* **42**, 412–22.

Selmanowitz, V.J., Orentreich, N. and Felsenstein, J.M. (1971). Lentiginosis profusa syndrome (multiple lentigines syndrome). *Arch Derm.* **104**, 393–401.

Selmar, P., Skov, T., and Skov E.G. (1986). Familial hypoplasia of the thenar eminence. Report of three cases. *J. Neurol. Neurosurg. Psychiat.* **49**, 105–6.

Sengers, R.C.A., Lamers, K.J.B., Bakkeren, J., Schretlen, E., and Trijbels, J. (1975*a*). Infantile Gaucher's disease: glucocerebrosidase deficiency in peripheral blood leucocytes and cultured fibroblasts. *Neuropediatrics* **6**, 377–82.

Sengers, R.C.A., Ten Haar, J.M.F., Trijbels, J.M.F., Wilkens, J.L., Daniels, O., and Stadhounders, A.M. (1975*b*). Congenital cataract and mitochondrial myopathy of skeletal and heart muscle associated with lactic acidosis after exercise. *J. Pediat.* **86**, 873–80.

Sengers, R.C.A., *et al.* (1984). Deficiency of cytochromes b and aa3 in muscle from a floppy infant with cytochrome oxidase deficiency. *Eur. J. Pediat.* **141**, 178–80.

Serratrice, G., Roux, H., Aquaron, R., Gambarelli, D., and Baret, J. (1969). Myopathies scapuloperonieres. (A propos de 14 observations dont 8 avec atteinte faciale.) *Sem. Hôp. Paris* **45**, 2678–83.

Serratrice, G., Gastaut, J.L., and Dubois-Gambarelli, D. (1973). Amyotrophie neurogene peripherique au cours du syndrome de Marinesco–Sjögren. *Rev.neurol.* **128**, 431–41.

Serratrice, G., Gastaut, J.L., Pellissier, J.F., and Pouget, J. (1976). Amyotrophies scapulo-peronieres chroniques de type Stark-Kaeser (a propos de 10 observations). *Rev. neurol.* **132**, 823–32.

Serratrice, G., *et al.* (1979). Scapuloperoneal myopathies and neuropathics. In *Peroneal atrophies and related disorder* (ed. G. Serratrice and H. Roux), pp. 233–52. Masson, New York.

Servidei, S., *et al.* (1986). Fatal infantile form of muscle phosphofructokinase deficiency. *Neurology* **36**, 1465–70.

Seuanez, H., Mane-Garzon, F., and Kolski, R. (1976). Cardio-cutaneous syndrome (the 'Leopard' syndrome). *Clin. Genet.* **9**, 266–76.

Shafiq, S.A., Dubowitz, V., Peterson, H. de C., and Milhorat, A.T. (1967). Nemaline myopathy: report of a fatal case with histochemical and electron microscopic studies. *Brain* **90**. 817–28.

Shafiq, S.A., Sande, M.A., Carruthers, R.R., Killip, T., and Milhorat, A.T. (1972). Skeletal muscle in idiopathic cardiomyopathy. *J. neurol. Sci.* **15**, 303–20.

Shahriaree, H. and Harkess, J.W. (1970). A family with spondylolisthesis. *Radiology* **94**, 631–3.

Shannon, M.W. and Nadler, H.L. (1968). X-linked hydrocephalus *J.med. Genet.* **5**, 326–7.

Shapira, Y. and Cohen, T. (1973). Agenesis of the corpus callosum in two sisters. *J. med. Genet.* **10**, 266–9.

Shapira, Y., Cederbaum, S.D., Cancilla, P.A., Nielsen, D., and Lippe,

B.M. (1975). Familial poliodystrophy, mitochondrial myopathy and lactate acidemia. *Neurol., Minneap.* **25**, 614–21.

Shapira, Y., Harel, S., and Russell, A. (1977). Mitochondrial encephalomyopathies: a group of neuromuscular disorders with defects in oxidative metabolism. *Isr. J. Med. Sci.* **13**, 161–4.

Shapiro, L.R., Raab, E.L., Leopold, I.H., and Hirschorn, K. (1969). Hereditary optic atrophy. An autosomal dominant with incomplete penetrance. *Arch. Ophthal.*, N.Y. **81**, 359–62.

Shapiro, L.J., *et al.* (1979). Metachromatic leukodystrophy without arylsulfatase A deficiency *Pediat. Res.* **13**, 1179–81.

Sharpe, J.A., Rewcastle, N.B., Llyod, K.G., Hornykiewicz, O., Hill, M., and Tasker, R.R. (1972). Striatonigral degeneration. Responses to levodopa—therapy with pathological and neurochemical correlation. *J. neurol. Sci.* **19**, 275–86.

Sharpe, J.A., Sulversides, J.L., and Blair, R.D.G. (1975). Familial paralysis of horizontal gaze: associated with pendular nystagmus, progressive scoliosis, facial contractions with myokymia. *Neurol. Minneap.* **25**, 1035–40.

Sharr, M., (1975). Adrenocortical insufficiency and diffuse cerebral sclerosis. Problems of presentation and diagnosis. *J. neurol. Sci.* **24**, 305–12.

Shaw, R.F. and Dreifuss, F.E. (1969). Mild and severe forms of X-linked muscular dystrophy. *Arch. Neurol., Chicago* **20**, 451–60.

Shelley, W.B. and Livingood, C.S. (1949). Familial multiple nevi flammei. *Arch. Derm. Syph.* **59**, 343–5.

Shepherd, M. (1955). Report of a family suffering from Friedreich's disease, peroneal muscular atrophy and schizophrenia. *J. Neurol. Neurosurg. Psychiat.* **18**, 297–304.

Sher, J.H., Rimalovski, A.B., Athanassiades, T.J., and Aronson, S.M. (1967). Familial centronuclear myopathy: a clinical and pathological study. *Neurol. Minneap.* **17**, 727–42.

Sheramata, W., Kott, H.S., and Cyr, D.P. (1971). The Chediak–Higashi–Steinbrinck syndrome. Presentation of three cases with features resembling spinocerebellar degeneration. *Arch. Neurol., Chicago* **25**, 289–94.

Shibasaki, H., *et al.* (1979). Pigmentary degeneration of the retina in heredodegenerative neurological disease. *Acta neurol. scand.* **59**, 331–42.

Shields, J. (1975). Clinical genetics of the presenile and senile dementias. Read before MRC Conference on the Senile and Presenile Dementias.

Shields, W.D. *et al.* (1977). Fibromuscular dysplasia as a cause of stroke in infancy and childhood. *Pediatrics* **59**, 899–901.

Shih, V.E., *et al.* (1971a). Sulfite oxidase deficiency. *N. Engl. J. Med.* **297**, 1022–8.

Shih, V.E., Bixby, E.M., Alpers, D.H., Bartsocas, C.S., and Thier, S.O.

(1971*b*). Studies of intestinal transport defect in Hartnup disease. *Gastroenterology* **61**, 445–53.

Shih, V.E., Salam, M.Z., Mudd, S.H., Uhlendorf, B.W., and Adams, R.D. (1972). A new form of homocystinuria due to $N^{5.10}$-methylenetetrahydrofolate reductase deficiency. *Pediat. Res.* **6**, 135.

Shillito, J. and Matson, D.D. (1968). Craniosynostosis: a review of 519 surgical patients. *Pediatrics, Springfield* **41**, 829–53.

Shimono, M., Ohta, M., Asada, M., and Kuroiwa, Y. (1976). Infantile neuroaxonal dystrophy. Ultrastructural study of peripheral nerve. *Acta neuropath.* **36**, 71–9.

Shimozana, N., Ohno, K., Takashima, S., *et al.* (1986). The syndrome of the absence of a septum pellucidum with porencephaly. *Brain Develop.* **8**, 632–6.

Shiraki, Y. and Yase, Y. (1975). Amyotrophic lateral sclerosis in Japan. In *Handbook of clinical neurology*, Vol. 22, (ed. P. Vinkin and G. Bruyn), pp. 353–419. North Holland, Amsterdam.

Shokeir, M.H.K. (1970*a*). Von Hippel–Lindau syndrome. A report on three kindreds. *J. med. Genet.* **7**, 155–7.

Shokeir, M.H.K. (1970*b*). X-linked cerebellar ataxia. *Clin. Genet.* **1**, 225–31.

Shokeir, M.H.K. (1977). Universal permanent alopecia, psychomotor epilepsy, pyorrhea and mental subnormality. *Clin. Genet.* **11**, 13–17.

Shokeir, M.H.K. and Kobrinsky, N.L. (1976). Autosomal recessive muscular dystrophy in Manitoba Hutterites. *Clin. Genet.* **9**, 197–202.

Shokeir, M.H.K. and Rozdilsky, B. (1985). Muscular dystrophy in Saskatchewan Hutterites. *Am. J. Med. Genet.* **22**, 487–93.

Short, E.M., Conn, H.O., Snodgrass, P.H., Campbell, A.G.M., and Rosenberg, L.E. (1973). Evidence for X-linked dominant inheritance of ornithine transcarbamylase deficiency. *N. Engl. J. Med.* **288**, 7–12.

Shoulson, I. and Chase, T.N. (1975). The Huntington's disease. *A. Rev. Med.* **26**, 419–26.

Shuman, R.M., Leech, R.W., and Scott, C.R. (1978). The neuropathology of the nonketotic and ketotic hyperglycinemias. Three cases. *Neurol., Minneap.* **28**, 139–46.

Shy, G.M. and Drager, G.A., (1960). A neurological syndrome associated with orthostatic hypotension: a clinical-pathologic study. *Arch. Neurol., Chicago* **2**, 511–27.

Shy, G.M. and Magee, K.R. (1956). A new congenital non-progressive myopathy. *Brain* **79**, 610–21.

Shy. G.M., Engel, W.K., Somers, J.E., and Wanko, T. (1963). Nemaline myopathy: a new congenital myopathy. *Brain* **86**, 793–810.

Shy, G.M., Gonatas, N.K., and Perez, M. (1966). Two childhood myopathies with abnormal mitochondria. I. Megaconial myopathy. II. Pleoconial myopathy. *Brain* **89**, 133–8.

Signorato, U. (1963). Alcuni rilievi sul ruolo del fattore ereditario nell genesi dell-epilessia. Aspetti metodologicie contributo personale. *Riv. sper. Freniat. Med. leg. Alien. ment.* **87**, 155–82.

Silengo, M.C. Cavallaro, S. and Franceschini. P. (1978). Recessive spondylocostal dysostosis: two new cases. *Clin. Genet.* **13**, 289–94.

Sillapää, M. (1976). Prevalence of migraine and other headaches in Finnish children starting school. *Headache* **15**, 288–90.

Silver, J.R. (1966). Familial spastic paraplegia with amyotrophy of the hands. *J. Neurol. Neurosurg. Psychiat.* **29**, 135–44.

Silverman, F.N. and Gilden, J.J. (1959). Congenital insensitivity to pain: a neurologic syndrome with bizzare skeletal lesions. *Radiology* **72**, 176–89.

Silverman, F.N., Strefling, A.M., Stevenson, D.K., and Lazarus, J. (1980). Cerebro-costomandibular syndrome. *J. Pediat.* **97**, 406–16.

Sim, M. and Bale, R.N. (1973). Familial pre-senile dementia: the relevance of a histological diagnosis of Pick's disease. *Br. J. Psychiat.* **122**, 671–3.

Simell, O. and Takki, K. (1973). Raised plasma ornithine and gyrate atrophy of the choroid and retina. *Lancet* **i**, 1031–3.

Similä, S. and Visakorp, J.K. (1967). Hyperprolinemia without renal disease. *Acta paediat. scand.* Suppl. 177, 122.

Simma, K. (1957). Zum verlauf der Hallervorden–Spatzschen Krankheit. *Psychiatria Neurol.* **133**, 39–46.

Simpson, J., Zellweger, H., Burmeister, L.F., Christee, R., and Nielsen, M.K. (1974). Effect of oral contraceptive pills on the level of creatine phosphokinase with regard to carrier detection in Duchenne muscular dystrophy. *Clinica chim. Acta* **52**, 219–23.

Singer, H., *et al.* (1980). Interrelationship among serum folate, CSf folatc, neurotransmitters and neuropsychiatric symptoms *Neurology* **30**, 419.

Singer, H.S. and Schafer, I.A. (1972). Clinical and enzymatic variations in GM_1 generalised gangliosidosis. *Am. J. hum. Genet.* **24**, 454–63.

Singer, P.A., Crampton, R.C., and Bass, N.H. (1974). Familial Q-T prolongation syndrome: convulsive seizures and paroxysmal ventricular fibrillation. *Arch. Neurol. Chicago* **31**, 64–6.

Singh, B.M., Ivamoto, H., and Strobos, R.J. (1973*a*). Slow eye movements in spinocerebellar degeneration. *Am. J. Ophthal.* **76**, 237–40.

Singh, H. and Shain, R. (1964). Heredofamilial ataxia with muscle fasciculations (a report of two cases in brothers). *Br. J. clin. Pract.* **18**, 91–2.

Singh, N., Grewal, M.S. and Austin, J.H. (1970). Familial anosmia. *Archs Neurol., Chicago* **22**, 40–4.

Singh, N., Mehta, M., and Roy, S. (1973*b*). Familial posterior column ataxia (Biemond's) with scoliosis. *Eur. Neurol.* **10**, 160–7.

Sjaastad, O., Berstad, J., Gesdahl, P., and Gjessing, L. (1976). Homocarnosinosis. 2. A familial metabolic disorder associated with spastic paraplegia, progressive mental deficiency and retinal pigmentation. *Acta neurol. scand.* **53**, 275-90.

Sjögren, T. (1943). Klinische und erbbiologische Untersuchengen uber die Heredoataxien. *Acta psychiat. neurol. scand.* Supply. 27, 1-197.

Sjögren, T. (1950). Hereditary congenital spinocerebellar ataxia accompanied by congenital cataract and oligophrenia. *Confinia neurol.* **10**, 293-308.

Sjögren, T. (1956). Oligophrenia combined with congenital ichthyosiform erthrodermia, spastic syndrome, and macula-retinal degeneration. *Acta genet., Basel* **9**, 80-91.

Sjögren, T. and Larsson, T. (1957). Oligophrenia in combination with congenital ichthyosis and spastic disorders. A clinical and genetic study. *Acta psychiat. scand.* **113**, 1-112.

Sjögren, T., Sjögren, H., and Lindgren, A.G.H. (1952). Morbus Alzheimer and morbus Pick. A genetic clinical and patho-anatomical study. *Acta psychiat. scand.* Suppl. 882, 1-152.

Skinner, R., Emery, A.E.H, Anderson, A.J.B., and Foxall, C. (1975). The detection of carriers of benign (Becker-type X-linked) muscular dystrophy. *J. med. Genet.* **12**, 131-4.

Skomer, C., Stears, J.U., and Austin, J. (1983). Metachromatic leukodystrophy (MLD) XV. Adult MLD with focal lesions by computed tomography. *Arch. Neurol.* **40**, 354-5.

Skre, H. (1974*a*). Hereditary spastic paraplegia in Western Norway. *Clin. Genet.* **6**, 165-83.

Skre, H. (1974*b*). Spino-cerebellar ataxia in Western Norway. *Clin Gent.* **6**, 265-88.

Skre, H. (1975). Friedreich's ataxia in western Norway. *Clin. Genet.* **7**, 287-98.

Skre, H. and Berg, K. (1974). Cerebellar ataxia and total albinism: a kindred suggesting pleiotropism or linkage. *Clin. Genet.* **5**, 196-204.

Skre, H. and Löken, A.C. (1970). Myoclonus epilepsy and subacute presenile dementia in heredo-ataxia. *Acta neurol. scand.* **46**, 18-42.

Skre, H. Bassöe, H.H., Berg, K., and Frövig, A.G. (1976). Cerebellar ataxia and hypergonadotropic hypogonadism in two kindreds. Chance concurrence, pleiotropism or linkage? *Clin. Genet.* **9**, 234-44.

Skre, H., Mellgren, S.I., Bergsholm, P., and Slagsvold, J.E. (1978). Unusual type of neurol muscular atrophy with a possible X-chromosomal inheritance pattern. *Acta neurol. scand.* **58**, 249-60.

Sladky, J.T. and Brown, M.J. (1984). Infantile axonal polyneuropathy with X-linked inheritance. *Ann. Neurol.* **16**, 402.

Slager, U.T., Kelly, A.B., and Wagner, J.A. (1957). Congenital absence of the corpus callosum. *N. Engl. J. Med.* **256**, 1171-6.

Slover, R. and Sujansky, E. (1979). Frontonasal dysplasia with coronal craniosynostosis in three sibs: penetrance and variability in malformation syndromes. *Birth Defects:* Original Articles Series **XV**, 75–83.

Sly, W.S., Lang, R., Avioli, J., Haddad, J., Lubowitz, H., and McAlister, W. (1972). Recessive osteopetrosis: new clinical phenotype. *Am. J. hum. Genet.* **24**, 34A.

Sly, W.S., Quinton, B.A., McAlister, W.H., and Rimoin, D.L. (1973). Beta-glucuronidase deficiency: report of clinical, radiologic and biochemical features of a new mucopolysaccharidosis. *J. Pediat.* **82**, 249–57.

Small, R.G. (1968). Coat's disease and muscular dystrophy. *Trans. Am. Ophthal. Otol.* **72**, 225–31.

Smit, L.M.E. *et al.* (1984). Familial porencephalic white matter disease in two generations. *Brain Dev.* **6**, 54–8.

Smith, A.J., and Strang. L.B. (1958). An inborn error of metabolism with the urinary excretion of alpha-hydroxy-butyric acid and phenylpyruvic acid. *Arch. Dis. Childh.* **33**, 109–13.

Smith, D.M., Zeman, W., Johnston, C.C., and Deiss, W.P. (1976). Myositis ossificans progressiva. Case report with metabolic and histochemical studies. *Metabolism* **15**, 521–8.

Smith, D.W. (1976*a*). *Recognizable patterns of human malformation.* 2nd edn. Saunders, Philadelphia.

Smith, D.W. (1976*b*). Blepharophimosis syndrome. In *Recognizable patterns of human malformation*, 2nd edn, p. 122. Saunders, Philadelphia.

Smith, I., Clayton, B.E., and Wolff, O.H. (1975). New variant of phenylketonuria with progressive neurological illness unresponsive to phenylalanine restriction. *Lancet* **i**, 1108.

Smith, J.K., Gonda, V.E., and Malamud, N. (1958). Unusual form of cerebellar ataxia. Combined dentato rubral and pallido-Luysian degeneration. *Neurology,* **8**, 205–9.

Smith, J.L., Hoyt, W.F., and Susal, J.D. (1973). Ocular fundus in acute Leber optic neuropathy. *Arch. Ophthal., N.Y.* **90**, 349–54.

Smith, L.A. and Heersema, P.H. (1941). Periodic dystonia. *Proc. Staff Meet. Mayo. Clin.* **16**, 842–6.

Smith, N.J., Espir, M.L.E., and Matthews, W.B. (1978). Familial myoclonic epilepsy with ataxia and neuropathy with additional features of Friedreich's ataxia and peroneal muscular atrophy. *Brain* **101**, 461–72.

Smith, R.F., Pulicicchio, L.U., and Holmes, A.V. (1970). Generalized lentigo electrocardiographic abnormalities conduction disorders and arrhythmias in three cases. *Am. J. Cardiol.* **25**, 501–6.

Smith, S.D., *et al.* (1983). A genetic analysis of specific reading disability. Genet. Asp. Speech Lang. pp. 169–78. Academic Press, New York.

Snead O.C.I.I.I., Acker, J.D. and Morawetz, R. (1979). Familial arteriovenous malformation. *Ann. Neurol.* **5**, 585–7.

Snyder, L.M., Necheles, T.F., and Reddy, J.W. (1970). G.6.P.D. Worcester: a new variant associated with X-linked optic atrophy. *Am. J. Med.* **49**, 125–32.

Snyderman, S.E. (1975). Maple syrup urine disease. In *the treatment of inherited metabolic disease* (ed D.N. Raine), p. 71. MTP, Lancaster.

Snyderman, S.E., *et al.* (1977). Argininemia. *J. Ped.*, **90**, 563.

Sobue, I., Saito, M., Iida, M., and Ando, K. (1978). Juvenile type of distal or segmental muscular atrophy of upper extremities. *Ann. Neurol.* **3**, 429–32.

Sobue, G., *et al.* (1986). Peripheral nerve involvement in familial choreo-acanthocytosis. *J. Neurol. Sci.* **76**, 347–56.

Soffer, D., Grotsky, H.W., Rapin, I., and Suzuki, K. (1979). Cockayne syndrome: unusual neuropathological findings and review of the literature. *Ann. Neurol.* **6**, 340–8.

Søgaard, I., and Jorgensen, J. (1975). Familial occurence of bilateral intracranial occlusion of the internal carotid arteries (moya moya). *Acta neurochir.* **31**, 245–52.

Sogawa, H., *et al* (1978). Chronic Niemann-Pick disease with sphingomyelinase deficiency in two brothers with mental retardation. *Eur. J. Pediat.* **128**, 235–40.

Sogg, R.L. and Hoyt, W.F. (1962). Intermittent vertical nystagmus in a father and son. *Arch. Ophthal., N. Y.* **68**, 515–17.

Solis-Cohen, S. and Weiss, E. (1925). Dystrophia adiposogenitalis with atypical retinitis pigmentosa and mental deficiency (Laurence–Biedl syndrome). *Am. J. med. Sci.* **169**, 489–505.

Solomon, G.E. and Chutorian, A.M. (1968). Opsoclonus and occult neuroblastoma. *N. Engl. J. Med.* **279**, 475–7.

Solomon, L.M. and Bolat, I.H. (1972). Adenoma sebaceum in encephalofacial angiomatosis (Sturge-Weber syndrome). *Acta derm.-vener., Stockh.* **52**, 386–8.

Solomon, L.M. and Esterly, N.B. (1975). Epidermal and other congenital organoid nevi. *Curr. Prob. Paediat.* **VI**, 1–56.

Solomon, L.M., Eng, A.M., Bené M., and Loeffel, E.D. (1980). Giant congenital neuroid melanocytic nevus. *Arch. Derm.* **116**, 318–20.

Somer, H., *et al.* (1985). Duchenne-like muscular dystrophy in two sisters with normal karyotypes: evidence for autosomal recessive inheritance. *Clin. Genet.* **28**, 151–6.

Somers, A.B., Levin, H.S., and Hannay, H.J. (1976). A neuropsychological study of a family with hereditary mirror movements. *Dev. med. Child Neurol.* **18**, 791–8.

Sommer, A-M. and Liu, P.H. (1984). *Am. J. Med. Genet.* **17**, 655–9.

Sonninen, V. and Savontaus, M-L. (1987). Hereditary multi-infarct dementia. *Eur. J. Neurol.* **27**, 209–15.

Soodan, V.M. and Goel, R.K. (1976). Scheie's syndrome: report in two siblings. *India Pediat.* **13**, 949–52.

Sourander, P. and Svennerholm, L. (1962). Sulphatide lipidosis in the adult with the clinical picture of progressive organic dementia with epileptic seizures. *Acta neuropath.* **1**, 384–96.

Sourander, P. and Walinder, J. (1977). Hereditary multi-infarct dementia. *Lancet* **i**, 1015.

Søvik, O., Van der Hagen, C.B., and Løken, A.C. (1977). X-linked aqueductal stenosis. *Clin. Genet.* **11**, 416–20.

Spaans, F., *et al.* (1986). Myotonic dystrophy associated with hereditary motor and sensory neuropathy. *Brain* **109**, 1149–68.

Sparkes, R.S., *et al.* (1987). Assignment of the myelin basic protein gene to human chromosome 18q22-qter. *Hum. Genet.* **75**, 147–50.

Sparrow, G.P., Samman, P.D., and Wells R.S. (1976). Hyperpigmentation and hypohydrosis (the Naegel–Franceschetti–Jadasshon syndrome): report of a family and review of the literature. *Clin. exp. Derm.* **1**, 127–40.

Spence, M.W., Ripley, B.A., Embil, J.A., and Tibbles, J.A.R. (1974). A new variant of Sandhoff's disease. *Pediat Res.* **8**, 628–37.

Spencer, F.R. (1917). Congenital ptosis. *Ophthal. Rec.* **26**, 254.

Sperling, M. and Herrmann, C. (1985). Syndrome of palatal myoclonus and progressive ataxia two cases with magnetic resonance imaging. *Neurology* **35**, 1212–14.

Spillane, J.D. (1940). Familial pes cavus and absent ankle jerks. *Brain* **63**, 275–90.

Spillane, J.D. and Wells, C.E.C. (1969). *Acrodystrophic neuropathy.* Oxford University Press, London.

Spillane J.D., Pallis, C., and Jones, A.M. (1957). Developmental abnormalities in the region of the foramina magnum. *Brain* **80**, 11–48.

Spira, P.J., McLeod, J.G., and Evans, W.E. (1979). A spinocerebellar degeneration with X-linked inheritance. *Brain* **102**, 27–41.

Spira, R. (1963). Neurogenic, familial, girdle type muscular atrophy (clinical electromyographic and pathologic studies). *Confinia neurol.* **23**, 245.

Spiro, A.J. (1970). Minipolymyoclonus. *Neurol. Minneap.* **20**, 1124–6.

Spiro, A.J. and Kennedy, C. (1965). Hereditary occurrence of nemaline myopathy, *Arch. Neurol., Chicago* **13**, 155–9.

Spiro, A.J. Fogelson, M.H., and Goldberg, A.C. (1967). Microcephaly and mental subnormality in chronic progressive spinal muscular atrophy of childhood. *Dev. med. Child. Neurol.* **9**, 594–601.

Spiro, A.J. Moore, C.L., Prineas, J.W., Strasberg, P.M., and Rapin, I. (1970). A cytochrome related inherited disorder of the nervous system and muscle. *Arch. Neurol. Chicago.* **23**, 103–12.

Spitz, M., Jakovic, J., and Killian, J.M.C. (1985). Familial tic disorder, Parkinson, motor neuron disease, and acanthocytosis. *Neurol.* **35**, 366–70.

Spranger, J., Wiedemann, H.R., Tolksdorf, M., Graucob, E., and Caesar, R. (1968). Lipomucopolysaccharidose. *Z. Kinderheilk*. **103**, 285–306.

Spranger, J., Gehler, J., and Cantz, M. (1977). Mucolipidosis 1—A. sialidosis *Am. J. med. Genet*. **1**, 21–9.

Spranger, J.W. Albrecht, C., Rohwedder, H.J., and Wiedemann, H.R. (1968). Die dysosteosklerose: eine sonderform der generalisierten Osteoklerose. *Fortschr. Roantgenstr*. **109**, 504–12.

Spritz, R.A., *et al*. (1978). Neonatal presentation of I-cell disease. *J. Pediat*. **93**, 954–8.

Staal, A. and Bots, G.T.A. (1969). A case of hereditary juvenile amyotrophic lateral sclerosis complicated with dementia. *Pschiat. Neurol. Neurochir*. **72**, 129–35.

Staal, A. and Went, L.W. (1968). Juvenile amyotrophic lateral sclerosis–dementia complex in a Dutch family. *Neurol., Minneap*. **18**, 800–6.

Staal, A., de Weerdt, C.J., and Went, L.N. (1965). Hereditary compression syndrome of peripheral nerves. *Neurol., Minneap*. **15**, 1008–17.

Staal, A., Stefanko, S.Z., Jennekens, F.G.I., *et al*. (1983). Autosomal recessive spino-olivo-cerebellar degeneration without ataxia. *J.N.N.P*. **46**, 468–652.

Stadlin, W. and Van Bogaert, L. (1949). Alterations chorioretiniennes chez deux soeures atteintes d'heredoataxie du type Friedreich. *Acta neurol. belg*. **49**, 705–30.

Stapletone, F.B. and Johnson, D. (1980). The cystic renal lesions in Tuberous Sclerosis. *J. Paediat*. **97**, 574–9.

Stark, G. (1972). Pelizaeus–Merzbacher disease. *Dev. med. Child Neurol*. **14**, 806–8.

Stark, P. (1958). Etude clinique et genetique d'une famille atteinte d'atrophie musculaire progressive neuronale (amyotrophie de Charcot–Marie). *J. Génét. hum*. **7**, 1–32.

Starkman, S., Kaul, S., Fried, J., and Behrens, M. (1972). Unusual abnormal eye movements in a family with hereditary spinoccrebellar degeneration. (Abstract.) *Neurol., Minneap*. **22**, 402.

Starrenveld, E. and Ashenhurst, E.M. (1975). Bilateral carpal tunnel syndrome in childhood—a report of two sisters with mucolipidosis III (pseudo Hurler polydystrophy). *Neurol., Minneap*. **25**, 234–8.

Staunton, H., *et al*. (1987). Hereditary amyloid polyneuropathy in North West Ireland. *Brain* **110**, 1231–45.

Steele, M.W. (1980). Lessons from the American Tay–Sachs programme. *Lancet* **ii**, 914.

Stefanis, C., Papapeteopoulos, Th., Scarpalezos, S., S., Lygidakis, G., and Panayiotopoulos, C.P. (1975). X-linked spinal and bulbar muscular atrophy of late onset. *J. neurol. Sci*. **24**, 493–503.

Steiman, G.S., Yudkoff, M., Berman, P.H., Blazer-Yost, B., and Segal, S.

(1979). Late onset non-ketotic hyperglycinemia and spinocerebellar degeneration. *J. Pediat*. **94**, 907–11.

Stendahl-Brodin, L., Möller, E., and Link, H. (1978). Hereditary optic atrophy with probable association with a specific HLA haplotype. *J. neurol. Sci*. **38**, 11–21.

Sternlieb, I. (1978). Diagnosis of Wilson's disease. (Editorial.) *Gastroenterology* **74**, 787–9.

Sternlieb, I. and Scheinberg, I.H. (1972). Chronic hepatitis as a first manifestation of Wilson's disease. *Ann. intern. Med*. **76**, 59–64.

Stevens, D.L. (1971). Tests for Huntington's chorea. *New Engl. Med. J.* **285**, 413–4.

Stevens, D.L. (1973). The classification of variants of Huntington's chorea. *Adv. Neurol*. **1**, 57–64.

Stevens, D.L., Hewlett, R.H., and Brownell, B. (1977). Chronic familial vascular encephalopathy. *Lancet* i, 1364–5.

Stevens, D.L. and Parsonage, M. (1969). Mutation in Huntington's chorea. *J. Neurol. Neurosurg. Psychiat*. **32**, 140–3.

Stevens, H. (1966). Paroxysmal choreo-athetosis: a form of reflex epilepsy. *Arch. Neurol., Chicago* **14**, 415–20.

Stevens, J.R. (1954). Familial periodic paralysis, myotonia, progressive amyotrophy, and pes cavus, in members of a single family. *Arch. Neurol. Psychiat., Chicago* **72**, 726–47.

Stevens, R.L. *et al*. (1981). Cerebroside sulfatase activator deficiency induced metachromatic leukodystrophy *Am. J. Hum. Genet*. **33**, 900–6.

Stevenson, A.C. (1953/4). Muscular dystrophy in Northern Ireland. I. An account of the condition in fifty-one families. *Ann. Eug*. **18**, 50–93.

Stevenson, A.C. and Fisher, O.D. (1956). Frequency of epiloia in Northern Ireland. *Br. J. prevt. soc. Med*. **10**, 134–5.

Stevenson, R.E., Taylor, H.A. Jr, and Parks, S.E. (1978). β-Galactosidase deficiency: prolonged survival in three patients following early central nervous system deterioration. *Clin. Genet*. **13**, 305–13.

Stevenson, R.E., Taylor, H.A., Schroer, R.J. (1983). Sialuria—clinical and laboratory features of a severe infantile form. *Proc. Gr. Genet. Center*. **1**, 73–8.

Stevenson, R.E. *et al*. (1983). Sialic acid storage disease with sialuria: clinical and bichemical features in the severe infantile type. *Pediatrics*, **72**, 441–9.

Stewart, G.J., Basten, A., Guinan, J., Bashir, H.V., Cameron, J., and McLeod, J.G. (1977). HLA-DW$_2$, viral immunity and family studies in multiple scelrosis. *J. neurol. Sci*. **32**, 157–67.

Stewart, J.M. and Stoll, S. (1979). Familial caudal regression anomalad and maternal diabetes. *J. med. Genet*. **16**, 17–20.

Stewart, R.E. (1978). Craniofacial malformations. Clinical and genetic considerations. *Pediat. Clins. N. Am*. **25**, 485–55.

Stewart, R.M. (1937). Amentia, familial cerebellar diplegia and retinitis pigmentosa. *Proc. Roy Soc. Med.* **30**, 849–50.

Stocks, P. (1923). Facial spasm inherited through four generations. *Biometrika* **14**, 311–15.

Stöhr, M., Schlote, W., Bundschu, H.D., and Reichenmiller, H.E. (1975). Myopathia myotonica. *J. Neurol.* **210**, 41–6.

Stoll, C., Levy, J.M., Bigel, P., and Francfort, J.J. (1974). Etude genetique dubleopharophimosis familial (maladie autosomique dominante) *J. Génét. hum.* **22**, 353–63.

Stone, T.T. (1950). Peripheral facial palsy. Multiple attacks in three brothers. *J. Am. med. Ass.* **143**, 1154–5.

Streeten, D.H.P., Kerr, L.P., Kerr, C.B., Prior, J.C. and Dalakos, T.G. (1972). Hyperbradykinism a new orthostatic syndrome. *Lancet* **ii**, 1048–53.

Strehl, E., Vanasse, M., and Brochu, P. (1985). EMG and needle muscle biopsy studies in arthrogryposis multiplex congenita. *Neuropediatrics* **16**, 225–7.

Streiff, E.B. (1947). Association de phakomatoses. A propos d'un cas de neurofibromatose de Recklinghausen avec hemangiome cutané naso-frontal. *Rev. Oto-Neuro-Ophthal.* **19**, 36–40.

Strobos, R.R.J., De la Toree, E., and Martin, J.F. (1957). Symmetrical calcification of the basal ganglia with familial ataxia and pigmentary macular degeneration. *Brain* **80**, 313–18.

Strömgren, E., Dalby, A., Dalby, M.A., and Ranheim, B. (1970). Cataracts, deafness, cerebellar ataxia, psychosis and dementia—a new syndrome. *Acta neurol. scand.* Suppl. **43**, 261–2.

Stromme, J.H., Nesbakken, R., Norman, F.T., Skjørten, F., Skyberg, D., and Johannesson, B. (1969). Familial hypomagnesemia. *Acta paediat. scand.* **58**, 433–44.

Strümpell, A. (1880). Beiträge sur Pathologie des Ruckenmarks. *Arch. Psychiat. NervKrankh.* **10**, 676–717.

Strunk, P. and Struck, G. (1965). Beitrag zur klinik und Pathomorphologie orthochromatischer leukodystrophien. *D. Z. Nervenheilk.* **186**, 496–510.

Stumpf, D.A., Austin, J.H., Crocker, A.C., and La France, M. (1973). Mucopolysaccharidosis type VI (Maroteaux–Lamy syndrome) I. Sulfatase B deficiency in tissue. *Am. J. Dis. Childh.* **126**, 747–55.

Stumpf, D.A, *et al.* (1987). Friedreich disease: V. Variant form with vitamin E deficiency and normal fat absorption. *Neurology* **37**, 68–74.

Stumpf, D.A., Parks, J.K., and Eguren, L.A. (1982). Friedreich ataxia: III, mitochondrial malic enzyme deficiency. *Neurol.* **32**, 221–7.

Stutchfield, P., *et al.* (1985). Glutaric aciduria type I, misdiagnosed as Leigh's encephalopathy and cerebral palsy. *Dev. Med. Child Neurol.* **27**, 514–21.

Sugarman, G.I., Landing, B.H., and Reel, W.B. (1977). Cockayne syndrome: clinical study of two patients and neuropathologic findings in one. *Clin. Paediat.* **16**, 225–32.

Sugita, K., *et al.* (1985). Tuberous sclerosis: Report of two cases studied by computer-assisted cranial tomography within one week after birth. *Brain devel.* **7**, 438–43.

Sugita, T.M. *et al.* (1982). Prenatal diagnosis of Cockayne syndrome using assay of colony-forming ability in ultraviolet light irradiated cells. *Clin. Genet.* **22**, 137–42.

Suhren, O., Bruyn, G.W., and Tuynman, J.A. (1966). Hyperexplexia. A hereditary startle syndrome. *J. neurol. Sci.* **3**, 577–605.

Sulaiman, A.R, Swick, H.M., and Kinder, D.S. (1983). Congenital fibre type disproportion with unusual clinico-pathological manifestation. *J. Neurol. Neurosurg. Psychiat.* **46**, 175–82.

Summitt, R.L. (1969*a*). Recessive acrocephalosyndactyly with normal intelligence. *Birth Defects.* **V**, 35–8.

Sumner, D., Crawfurd, M., and Harriman, D.G.F. (1971). Distal muscular dystrophy in an English family. *Brain* **94**, 51–60.

Sunder, J.H., *et al.* (1972). Pedigree with diabetes insipidus, diabetes mellitus and optic atrophy. *J. med. Genet.* **9**, 408–12.

Superneau, D. Wertelecki, W., and Zellweger, H. (1985). Letter to the Editor: the Marinesco–Sjögren syndrome described a quarter of a century before Marinesco. *Am. J. Med. Genet.* **22**, 647–8.

Surbek, B. (1961). L'angropathie dyshlorique (Morel) de l'ecorce cerebrale. *Acta neuropath.* **1**, 168–97.

Sutherland, J.M. (1956). Observation on the prevelance of multiple sclerosis in Northern Scotland. *Brain* **79**, 635–54.

Sutherland, J.M. (1957). Familial spastic paraplegia: its relationship to mental and cardiac abnormalities. *Lancet* ii, 169–70.

Sutherland, J.M., Tyrer, J.H., and Eadie, M.J. (1963). Atrophie spino-cérébelleuse familiale avec mydriase fixe. *Rev. neurol.* **108**, 439–42.

Sutherland, J.M., Edwards, V.E., and Eadie, M.J. (1975). Essential (hereditary or senile) tremor. *Med. J. Aust.* **2**, 44–7.

Suzuki, S., Kamoshita, S., and Ninomura, S. (1985). Ramsay Hunt syndrome in dentatorubralpallidoluysian atrophy. *Pediat. Neurol.* **1**, 298–301.

Suzuki, Y. and Mizuno, Y. (1974). Juvenile metachromatic leuko-dystrophy—deficiency of an arylsulfatase A component. *J. Pediat.* **85**, 823–5.

Suzuki, Y. and Suzuki, K. (1970). Partial deficiency of hexosaminidase component A in juvenile GM_2 gangliosidosis. *Neurol., Minneap.* **20**, 848–51.

Suzuki, Y. and Suzuki, K. (1971). Krabbe's glodoid cell leukodystrophy: deficiency of galactocerebrosidase in serum. laucocytes and fibroblasts. *Science, N.Y.* **171**, 73–4.

Suzuki, Y., Nakamura, N., Fukuoka, K., Shinada, Y., and Uono, M. (1977).

β-Galactosidase deficiency in juvenile and adult patients. Report of six Japanese cases and review of literature. *Hum Genet.* **36**, 219–29.

Suzuki, Y., Nakamura, N., Jimbo, T. Horiguchi S., and Fujii, T. (1978). Prenatal diagnosis in twin pregnancy. *J. Pediat.* **93**, 293–4.

Swaiman, K. F., Garg, B. P., and Lockman, L. A. (1975). Seablue histiocyte and posterior column dysfunction: a familial disorder: *Neurol., Minneap.* **25**, 1084–7.

Swaiman, K. F., *et al.* (1983). Sea-blue histiocytes, lymphocytic cytosomes, movement disorder and 59 Fe uptake in basal ganglia: Hallervorden–Spatz disease or ceroid storage disease with abnormal isotope scan. *Neurology* **33**, 301–5.

Swanson, A. G. (1963). Congenital insensitivity to pain with anhydrosis. *Arch. Neurol., Chicago* **8**, 299–306.

Swash, M. and Schwartz, M. S. (1981). Familial multicore disease with focal loss of cross-striations and ophthalmoplegia *J. Neurol. Sci.* **52**, 1–10.

Swash, M., Van den Noort, S., and Craig, J. W. (1970). Late-onset proximal myopathy with diabetes mellitus in four sisters. *Neurol., Minneap.* **20**, 694–9.

Sweeney, V. P., Sadovnick, A. D., and Brandejs, V. (1986). Prevalence of multiple sclerosis in British Columbia. *Can. J. Neurol. Sci.* **13**, 47–51.

Sweetman, L., W., Shafai, T., Young, P., and Nyhan, W. L. (1979). Prenatal diagnosis of propionic acidemia. *J. Am. med. Ass.* **242**, 1048–52.

Sweetman, L., Nyhan, W. L., Trauner, D. A., Merritt, A., and Singh, M. (1980). Glutaric aciduria Type II. *J. Pediat.* **96**, 1020–6.

Swerts, L. and Van den Bergh, R. (1976). Sclerose laterale amyotrophique familiale. Etude d'une famille atteiente sur trois generations. *J. Génét. hum.* **24**, 247–55.

Swift, M. R. and Horowitz, S. L. (1969). Familial jaw cysts in Charcot–Marie–Tooth disease. *J. med. Genet.* **6**, 193–5.

Swingler, R. J. and Compston, A. (1986). The distribution of multiple sclerosis in the United Kingdom. *J. Neurol. Neurosurg. Psychiat.* **49**, 1115–24.

Swisher, C. N., Menkes. J. H., Cancilla, P. A., and Dodge, P. R. (1972). Co-existence of Hallervorden–Spatz disease with acanthocytosis. *Trans. Am. neurol. Ass.* **97**, 212–16.

Sylvester, P. E. (1958). Some unusual findings in a family with Friedreich's ataxia. *Arch. Dis. Childh.* **33**, 217–21.

Sylvester, P. E. (1972). Spino-cerebellar degeneration, hypogonadism, deaf-mutism and mental deficiency. *J. ment. Defic. Res.* **16**, 203–14.

Symonds, C. P. (1951). Migrainous variants. *Trans. med. Soc. Lond.* **67**, 237.

Symonds, C. P. and Shaw, M. E. (1926). Familial claw foot with absent tendon jerks: a 'forme fruste' of the Charcot–Marie–Tooth disease. *Brain* **49**, 387–403.

Tagatz, G., Fialkow, P.J., Smith, D., and Spadoni, L. (1970). Hypogonadotropic hypogonadism associated with anosmia in the female. *N. Engl. J. Med.* **283**, 1326–9.

Takahashi, H., Nakamura, H., and Nakashima, R. (1974). Scapuloperoneal dystrophy associated with neurogenic changes. *J. neurol. Sci.* **23**, 575–83.

Takahashi, H., Nakamura, H., and Okada, E. (1972). Hereditary amyotrophic lateral sclerosis, histochemical and electron microscopic study of hyaline inclusions in motor neurones *Arch. Neurol., Chicago* **27**, 292–9.

Takahata, N., Ito, K., Yoshimura, Y., Nishihori, K., and Suzuki, H. (1978). Familial chorea and myoclonus epilepsy. *Neurol. Minneap.* **28**, 913–19.

Takamoto, K., *et al.* (1984). A genetic variant of Emery–Dreifuss disease muscular dystrophy with humeropelvic distributions, early joint contracture, and permanent atrial paralysis. *Arch. Neurol.* **41**, 1292–3.

Takebe, Y., Koide, N., and Takahashi, G. (1981). Giant axonal neuropathy. Report of two siblings with endocrinological and histological studies. *Neuropediatrics* **12**, 392–404.

Takei, Y. and Mirra, S.S. (1973). Striatonigral degeneration. A form of multiple system atrophy with clinical Parkinsonism. *Prog. Neuropath.* **2**. 217–51.

Takei, Y. and Solitare, G.B. (1972). Infantile spongy degneration of the central nervous system associated with glycogen storage and markedly fatty liver. *J. Neurol. Neurosurg. Psychiat.* **5**, 11–21.

Tal, Y. *et al.* (1980). Dandy–Walker syndrome: analysis of 21 cases. *Dev. Med. Child Neurol.* **22**, 189–201.

Tamura, K., Santa, T., and Kuroiwa, Y. (1974). Familial oculocranioskeletal neuromuscular disease with abnormal muscle mitochondria. *Brain* **97**, 655–72.

Tan, K.-L. (1972). The metopic fontanelle. *Am. J. Dis. Child,* **124**, 211–14.

Taniguchi, R. and Konigsmark, B.W. (1971). Dominant spino-pontine atrophy. *Brain.* **94**, 349–58.

Tanzi, R.E., *et al.* (1987). The genetic defect in familial Alzheimer's disease is not tightly linked to the amyloid beta-protein gene. *Nature* **32**, 156–7.

Tarlow, M.J., Lake, B.D., and Lloyd, J.K. (1973). Chronic lactic acidosis in association with myopathy. *Arch. Dis. Childh.* **48**, 489–92.

Tasker, W. and Chutorian, A.M. (1969). Chronic polyneuritis of childhood. *J.Pediat.* **74**, 699–708.

Tavill, A.S., Evanson, J.M., Baker, S.B. de C., and Hewitt, V. (1974). Idiopathic paroxysmal myoglobinuria with acute renal failure and hypercalcaemia. *Engl. J. Med.* **271**, 283–7.

Tawara, S. *et al.* (1983). Identification of amyloid prealbumin variant in familial amyloidotic polneuropathy (Japanese type). *Biochem. Biophys Res. Comm.* **116**, 880–8.

Taylor, A.M.R., *et al.* (1987). Variant forms of ataxia telangiectasia. *J. Med. Genet.* **24**, 669–77.

Taylor, D.C. and Ounsted, C. (1971). Biological mechanisms influencing the outcome of seizures in response to fever. *Epilepsia* **12**, 33–45.

Taylor, E.W., (1915). Progressive vagus-glossopharyngeal paralysis with ptosis. *J. nerv. ment. Dis.* **42**, 129–39.

Taylor, H.A., Thomas, G.H., Miller, C.S., Kelly, T.E., and Siggers, D. (1973). Mucolipidois III (Pseudo-Hurler polydystrophy). Cytological and ultrastructural observations in cultural fibroblast cells. *Clin. Genet.* **4**, 388–97.

Taylor, J. (1912). Peroneal atrophy. *Proc. Roy. Soc. Med.* **6**, 50.

Teasdall, R.D., Schuster, M.M., and Walsh, F.B. (1964). Sphincter involvement in ocular myopathy. *Arch. Neurol., Chicago* **10**, 446–8.

Teebi, A.S., Al-Awardi, S.A., and White, A.G. (1987). Autosomal recessive nonsyndromal microcephaly with normal intelligence. *Am. J. Med. Genet.* **26**, 355–9.

Telerman-Toppet, N., Gerard, J.M., and Coers, C. (1973). Central core disease. A study of clinically unaffected muscle. *J. neurol. Sci.* **19**, 207–23.

Telfer, M.A., Sugar, M., Jaeger, E.A., and Mulcahy, J. (1971). Dominant piebald trait (white forelock and leukoderma) with neurological impairment. *Am. J. hum. Genet.* **23**, 383–9.

Teller, H., Lindner, B., and Götze. W. (1953). Doppelseitiger Trigeminusnaevus bei eineiigen Zwilligen mit gleichartigen enzephalographischen Befunden. *Derm. Wschr.* **127**, 488–93.

Temtamy, S.A. and McKusick V.A. (1969). Synoposis of hand malformations with particular emphasis on genetic factors. *Birth Defects* **V**, 125–84.

Temtamy, S.A., El-Meligy, M., Badrawy, H.S., Meguid, S.A., and Safwat, H.M. (1974*a*). Metaphyseal dysplasis anectoderma and optic atrophy: an autosomal recessive syndrome. In *Skeletal dysplasia* (ed. D. Bergsma), pp. 61–71. Amsterdam.

Temtamy, S.A., El-Meligy, Salem, S., and Osman, N. (1974*b*). Hyperphosphatasia in an Egyptian child. *Birth Defects*: Original Article Series **10/12**, 196.

Temtamy, S.A. (1978). The genetics of hand malformations. *Birth Defects*: Original Article Series **XIV** (3).

Tenconi, R., *et al.* (1981). Chorio-retinal dysplasia, microcephaly, and mental retardation. An autosomal dominant syndrome. *Clin. Genet.* **20**, 347–51.

Terheggen, H.G., Lowenthal, A., Lavinha, F., and Colombo, J.P. (1975). Familial hyperargininemia. *Arch. Dis. Chill.* **50**, 57.

Terplan, K.L. and Cares, H.L.: (1972). Histopathology of the nervous system in carnosinase enzyme deficiency with mental retardation. *Neurology* **22**, 644–54.

Terry, R.D. (1976). Dementia. A brief and selective review. *Arch. Neurol., Chicago* **33**, 1–3.

Thevenard, A. (1953). L'acropathie ulcero-mutilante familiale. *Acta neurol. belg.* **53**, 1–23.

Thibault, J.H. and Manuelidis, E.E. (1970). Tuberous sclerosis in a premature infant. Report of a case and review of the literature. *Neurol., Minneap.* **20**, 139–46.

Thieffrey, S., Lyon, G., and Maroteaux, P. (1966). Leucodystrophie metachromatique (Sulfatidose) et mucopolysaccharidose associees chez un meme maladie. *Rev. neurol.* **114**, 193–200.

Thomas, C., Schweitzer, R., Isch, F., Collin, H., and Fardeau, M. (1978). Données nouvelles sur la genealogie la chinique et l'histologie d'une famille atteinte de paralysie periodique hyperkaliemique. *Revue neurol.* **134**, 45–58.

Thomas, G.H., Taylor, H.A., Reynolds, L.W., and Miller, C.S. (1973). Mucolipidosis III. (pseudo-Hurler polydystrophy). Multiple lysosomal enzyme abnormalities in serum and cultured fibroblast cells. *Pediant. Res.* **7**, 751–6.

Thomas, G.H., Tipton, R.E., Chien, L.T., Reynolds, L.W., and Miller, C.S. (1978*a*). Sialidase (α-N-acetylneuramimidase) deficiency: the enzyme defect in an adult with macular cherry-red spots and myoclonus without dementia. *Clin. Genet.* **13**, 369–79.

Thomas, J.V., Schwartz, P.L., and Gragoudas, E.S. (1978*b*). Von Hippel's disease in association with van Recklinghausen's neurofibromatosis. *Br. J. Ophthal.* **62**, 604–8.

Thomas, M., Adams, J.H., and Doyle, D. (1977). Neuroectodermal tumours in the cerebellum in two sisters. *J.Neurol, Neurosurg. Psychiat.* **40**, 886–9.

Thomas. P.K. (1975). Genetic factors in amyloidosis. *J. med. Genet.* **12**, 317–26.

Thomas, P.K., Abrams, J.D., Swallow, D., and Stewart, G. (1979). Sialidosis type I—cherry-red spot, myoclonus syndrome with sialidase deficiency and altered electrophoretic mobilities of some enzymes known to be glycoproteins. *J. Neurol. Neurosurg. Psychiat.* **42**, 873–80.

Thomas, P.K., Calne, D.B., and Elliot, C.-F. (1972). X-linked scapuloperoneal syndrome. *J. Neurol. Neurosurg. Psyschiat.* **35**, 208–15.

Thomas, P.K., Calne, D.B., and Stewart, G. (1974). Hereditary motor and sensory polyneuropathy (peroneal muscular atrophy). *Ann. hum.Genett.* **38**, 111–53.

Thomas, P.K., Schott, G.D., and Morgan-Hughes, J.A. (1975). Adult-onset scapuloperoneal myopathy. *J. Neurol. Neurosurg. Psychiat.* **38**, 1008–15.

Thomas, P.K., Workman, J.M., and Thage, O. (1984). Behr's Syndrome A family exhibiting pseudodominant inheritance. *J. neurol. Sci.* **64**, 137–48.

Thomasen, E. (1948). Mytonia. Thomasens disease (myotonia congenita) paramyotonica and dystrophia myotonica. *Op. ex Domo biol. hered. hum. Univ. Hafniensis* **17**, 1–251.

Thompson, A. F. and Alvarez, F. A. (1969). Hereditary amyotrophic lateral sclerosis. *J. neurol. Sci.* **8**, 101–10.

Thompson, C. E. (1978). Reproduction in Duchenne dystrophy. *Neurol. Minneap.* **28**, 1045–7.

Thompson, E. and Baraitser, M. (1987). Syndrome of the month: FG syndrome. *J. Med. Genet.* **24**, 139–43.

Thompson, E. M., Baraitser, M., and Hayward, R. D. (1984). Parietal foramina in Saethre–Chotzen syndrome. *J. Med. Genet.* **21**, 369–72.

Thompson, J. H. (1896). A wry-necked family. *Lancet* **ii**, 24.

Thompson, M. W. Murphy, E. G., and McAlpine, P. J. (1967). Assessment of creatine kinase test in the detection of carriers of Duchenne muscular dystrophy. *J. Pediat.* **71**, 82–93.

Thompson, M. W., *et al.* (1986). Linkage analysis of polymorphisms within the DNA fragment XJ cloned from the breakpoint of an X:21 trans location associated with X-linked muscular dystrophy. *J. Med. Genet.* **23**, 548–55.

Thomsen, J. (1876). Tonische Krämpe in willkurlich beweglichen Muskeln in Folge von erebter psychisher Disposition (ataxin muscularis?) *Arch. Psychiat. NervKrankh.* **6**, 702–18.

Thomson, J. (1949). Osteopetrosis in successive generations. *Archs Dis. Childh.* **24**, 143–8.

Thoren, C. (1962). Diabetes mellitus in Friedreichs ataxia. *Acta paediat.* **51**, Suppl. 135, 239–47.

Thorsby, E., Helgesen, A., Solheim, B. G., and Vandvik, B. (1977). HLA antigens in multiple sclerosis. *J. neurol. Sci.* **32**, 187–93.

Thrush, D. C. (1973*a*). Congenital insentivity to pain—a clinical, genetic and neurophysiological study of four children from the same family. *Brain* **96**, 369–86.

Thrush, D. C. (1973*b*). Autonomic dysfunction in four patients with congenital insentivity to pain. *Brain* **96**, 591–600.

Thrush, D. C., Morris, C. J., and Salmon, M. V. (1972). Paramytonia congenita: a clinical histochemical and pathological study. *Brain* **95**, 537–52.

Thrush D. C., Holti, G., Bradley, W. G., Campbell, M. J., and Walton, J. N. (1974). Neurological manifestations of xeroderma pigmentosum in two siblings. *J. neurol. Sci.* **22**, 91–104.

Thulin, B., McTaggart, D., and Neubuerger, K. T. (1968). Demyelinatising leukodystrophy with total cortical cerebellar atrophy. *Arch. Neurol., Chicago* **18**, 113–22.

Thurmon, T.F. and Walker, B.A. (1971). Two distinct types of autosomal dominant spastic paraplegia. The clinical delineation of birth defects in nervous system. *Birth Defects*: Original Article Series **VII**, 216–18.

Thurmon, T.F., Walker, B.A., Scott, C.I., and Abbott, M.H. (1971). Two kindreds with a sex-linked recessive form of spastic paraplegia. The clinical delineation of birth defects. VI. Nervous system. *Birth Defects*: Original Article Series **VII**, 219–21.

Thuwe, I., Lundström, B., and Walinder, J. (1979). Familial brain tumour. (Letter.) *Lancet* **i**. 504.

Thuy, L.F. *et al.* (1986). Multiple carboxylase deficiency due to deficiency of biotinidase. *J. Neurogenet.* **3**, 357–63.

Tibbetts, R.W. (1971). Spasmodic torticollis. *J. psychosom. Res.* **15**, 461–9.

Tieber, E. (1972). Anfalllmuster bei augenschluss—Bericht uber drei Fälle in einer Familie. *Neuropaediatric* **3**, 305–12.

Tiedemann, G. (1951). Zur Frage der Erblichkeit von Gefäßmälern. *Arch. Derm. Syph.* **192**, 327–57.

Till, K. (1975). *Paediatric neurosurgery for paediatricians and neurosurgeons*. Blackwell, Oxford.

Tinel, J. (1939). Le systéme nerveux vegetatif. *Prog. med., Paris* **18**, 826.

Ting-Ming, L.I., Alberman, E., and Swash, M. (1988). Comparison of sporadic and familial disease amongst 580 cases of motor neuron disease. *J.N.N.P.* **51**, 778–84.

Tishler, P.V. (1975). A family with coexistent Von Recklinghausen neurofibromatosis and Von Hippel–Lindau disease. *Neurol., Minncap.* **25**, 840–4.

Tishler, P.V., Knighton, D.J., and Schumaker, H.M. (1976). Screening test for intermittent acute porphyria. *Lancet* **i**, 303.

Titica, J. and Van Bogaert, L. (1946). Heredo-degenerative hemiballismus. A contribution to the question of primary atrophy of the corpus Luysii. *Brain* **69**, 251–63.

Tittarelli, R., Giagheddu, M., and Spadetta, V. (1966). Typical opthalmoscopic pictures of cherry-red spot in an adult with the myoclonic syndrome. *Br. J. Ophthal* **50**, 414–20.

Tobo, M., *et al.* (1984). Familial occurrence of adult type neuronal ceroid lipofuscinosis *Arch. Neurol.* **41**, 1091–4.

Tocantis, L.M. and Reimann, H.(1939). Perforating ulcers of the feet and osseous atrophy in a family with other evidence of dysgenesis (harelip, cleft palate) an instance of probable myelodysplasia. *J. Am. med. Ass.* **122**, 2251–5.

Todd, D.W., Christferson, L.A., and Leech, R.W. (1981). A family affected with intestinal polyposis and glioma. *Ann. Neurol.* **10**, 390–2.

694   Bibliography

Todorov, A. (1965). Le syndrome de Marinesco-Sjögren: premiere etude anatomclinique. *J. Genet. hum.* 14, 197–233.

Todorov, A., Jequier, M., Klein. D., and Morton, N.E. (1970). Analyse de la segregation dans la dystrophie myotonique. *J. Genet. hum.* 18, 387–406.

Tolmie, J.L. *et al.* (1987). Microcephaly: Genetic counselling and antenatal diagnosis after the birth of an affected child. *Am. J. Med. Genet.* 27, 583–94.

Tome, F.M.S., and Fardeau, M. (1973). 'Fingerprint inclusions' in muscle fibres in dystrophia myotonica. *Acta neuropath.* 24, 62–7.

Tomiwa, K., Baraitser, M., and Wilson, J. (1987). Dominantly inherited congenital cerebellar ataxia. *Pediat. Neurol.* 3, 360.

Tommerup, N. and Nielsen, F. (1983). A familial reciprocal translocation ε (3:7) (P21.1:P13) associated with the Greig polysyndactyly-craniofacial anomalies syndrome. *Am. J. Med. Genet.* 16, 313–21.

Tomsick, T.A., Lukin, R.R., Chambers A.A., and Benton, C. (1976). Neurofibromatosis and intracranial arterial occlusive disease. *Neuroradiology.* 11, 229–34.

Tondeur, M., Vamos-Hurwitz, E., Mockel-Pohl, S., Dereume, J.P., Cremer, N., and Loeb, H. (1971). Clinical, biochemical and ultra-structural studies in a case of chondrodystrophy presenting the I-cell phenotype in tissue culture. *J. Pediat.* 79, 366–78.

Tondeur, M. *et al.* (1982). Infantile form of sialic acid storage disorder: clinical, ultrastructural, and biochemical studies in two siblings. *Eur. J. Pediat.* 139, 142–7.

Tonnesen, T., *et al.* (1987). Experience with first trimester prenatal diagnosis of Menkes disease. *Prenatal Diag.* 7, 497–509.

Tonning, H.O., Warren, R.F., and Barrie, H.J. (1952). Familial hae-mangiomata of the cerebellum. Report of three cases in a family of four. *J. Neurosurg.* 9, 124–32.

Tonnis, W. and Lange-Cosack, H. (1953). Klinik, operative Behandlung und Prognose der arterio-venösen angiome des Gehirns und seine. *Dt. Z. Nervenheilk.* 170, 460–85.

Tooth, H.H. (1886). *Two peroneal types of progressive muscular atrophy.* Lewis, London.

Torbergsen, T. (1975). A family with dominant hereditary myotonia, muscular hypertrophy and increased muscular irritability, distinct from myotonia congenita. Thomsen. *Acta neurol. scand.* 51, 225–32.

Tourette G., del a (1885). Etude sur une affection nerveuse, characterisee par l' incoordination motrice accompagnee de l' echolalie et de coprolalic. *Arch. Neurol.* 9, 158–200.

Towfighi, J. and Gonatas, N.K. (1976). Hexachlorophine and the nervous system. In *Progress in neuropathology*, Vol. III (ed. H. Zimmerman), pp. 297–317. Grune & Stratton, New York.

Towfighi, J., Sassani, J. W., Suzuki, K., and Ladda, F. M. (1984). Cerebro-ocular dysplasia—muscular dystrophy (COD-MD) syndrome. *Acta Neuropathol* **65**, 110–23.

Tramer, M. (1951). Naevus flammeus hereditarius (mil Caer) bei einen Knaben. *Z. Kinderpsychiat.* **18**, 143–8.

Trauner, D. A., *et al.* (1981). Progressive neurodegenerative disorder in a patient with non-ketotic hyperglycinemia. *J. Pediat.* **98**, 272–7.

Trawiesa, D. C., Schwartzman, R. J., Glaser, J. S., and Savino, P. (1976). Familial benign intracranial hypertension. *J. Neurol. Neurosurg. Psychiat.* **39**, 420–3.

Trend, F. J. *et al.* (1985). Acid maltase deficiency in adults. Diagnosis and management in five cases. *Brain* **108**, 845–60.

Tridon, P., Renard, M., Picard, J., Weber, M., and Andre, J. M. (1969). Malformation vasculaire cerebrale et syndrome d'Ehlers–Danlos. *Rev. neurol.* **121**, 615–21.

Tripp, J. H., Lake, B. D., Young, E., Ngu, J., and Brett, E. M. (1977). Juvenile Gaucher's disease with horizontal gaze palsy in three siblings. *J. Neurol. Neurosurg. Psychiat.* **40**, 470–8.

Trockel, U., *et al.* (1983). Multiple exercise-related mononeuropathy with abdominal colic. *J. Neurol. Sci.* **60**, 431–42.

Troost, B. T., Savino, P. J., and Lozita, J. C. (1975). Tuberous sclerosis and Klippel–Trenaunay–Weber syndrome. *J. Neurol. Neurosurg. Psychiat.* **38**, 500–4.

Trotter, J. L. (1973). Striatonigral degeneration Alzheimer's disease and inflammatory changes. *Neurol., Minneap.* **23**, 1211–16.

Tsairis, P., Engel, W. K., and Kark, P. (1973). Familial myoclonic epilepsy syndrome associated with skeletal muscle mitochondrial abnormalities. (Abstract). *Neurol., Minneap.* **23**, 408.

Tsuboi, T. (1986). Seizures of childhood. A population-based and clinic-based study. *Acta. Neurol. Scand.* **110**, Suppl. 1–237.

Tsuboi, T. and Endo, S. (1977). Febrile convulsions followed by non-febrile convulsions. A clinical electroencephalographic and follow up study. *Neuropaediatrics* **8**, 209–23.

Tsuchiyama, A., *et al.* (1983). A case of pyruvate carboxylase deficiency with later prenatal diagnosis of an unaffected sibling. *J. Inherit. Met. Dis.* **6**, 885–8.

Tsuda, H., Fukushima, S., Takahashi, M., Hikisaka, Y., and Hayashi, K. (1976). Familial bilateral papillary cystadenoma of the epididymis. *Cancer, N. Y.* **37**, 183–9.

Tsujahata, M., Shimomura, C., Toshiro, Y., *et al.* (1983). Fatal meomatal nemaline myopathy: a case report. *J. N. N. P.* **46**, 856–9.

Tsukagoshi, H., Nakanishi, T., Kondo K., and Tsubaki, T. (1965). Hereditary proximal neurogenic muscular atrophy in adult. *Arch. Neurol., Chicago* **12**, 597–603.

Tsukagoshi, H., Shoji, H., and Furukawa, T. (1970). Proximal neurogenic muscular atrophy in adolescence and adulthood with X-linked recessive inheritance. *Neurol., Minneap.* **20**, 1188–93.

Tsukagoshi, H., Sugita, H., Furukawa, T., Tsubaki, T., and Ono, E. (1966). Kugelberg–Welander syndrome with dominant inheritance. *Arch., Neurol., Chicago* **14**, 378–81.

Tuck R.R. and Mcleod, J.G. (1983). Retinitis pigmentosa, ataxia, and peripheral neuropathy. *J. Neurol. Neurosurg. Psychiat.* **46**, 206–13.

Tunte, W. (1971). Fortpflanzungsfahigkeit, Heiratshaufigkeit und Zahl und Beschafenheit der Nachkommen bei Patienten mit Spina bifida aperta. *Humangenetik* **13**, 43–8.

Tuomaala, P. and Haapanen. E., (1968). Three siblings with similar anomalies in the eyes, bones and skin. *Acta ophthal.* **46**, 365–71.

Turcott, J., Despres, J.P., and St. Pierre, F. (1959). Malignant tumors of the central nervous system associated with familial poliposis of the colon. *Dis. Colon Rect.* **2**, 465–8.

Turnbull, D.M., *et al.* (1984). Short chain acyl-CoA dehydrogenase deficiency associated a lipid storage myopathy and secondary carnitine deficiency. *N. Engl. J. Med.* **311**, 1232–6.

Turner, E.V. and Roberts, E. (1938). A family with sex-linked hereditary ataxia. *J. nerv. ment. Dis.* **87**, 74–80.

Turner, J.W.A. and Heathfield, K.W.G. (1961). Quadriceps myopathy occurring in middle age. *J. Neurol. Neurosurg. Psychiat.* **24**, 18–21.

Turner, J.W.A. and Lees, F. (1962). Congenital myopathy—a fifty year follow-up. *Brain* **85**, 733–40.

Tyler, F.H. and Stevens, F.E. (1950). Studies in disorders of muscle. II. Clinical manifestations and inheritance of facioscapulohumeral dystrophy in a large family. *Ann. intern. Med.* **32**, 640–60.

Tyler, F.H., Stephens, F.E., Gunn, F.D., and Perkoff, G.T. (1951). Studies in disorders of muscle VII. Clinical manifestations and inheritance of a type of periodic paralysis without hypopotassemia. *J. Clin. Invest.* **30**, 492–502.

Uchida, I.A., McRae, K.N., Wang, H.C., and Ray, M. (1965). Familial short arm deficiency of chromosome 18 concomitant with arhinencephaly and alopecia congenita. *Am. J. hum. Genett.* **17**, 410–19.

Ullrich, D.P. and Sugar O. (1960). Familial cerebral aneurysms including one extracranial internal carotid aneurysm. *Neurol., Minneap.* **10**, 288–94.

Ullrich, O. (1930). Kongenitale, atonisch-sklerotische Muskeldystrophie. *Mschr. Kinderheilk.* **47**, 502–10.

Ulrich, J. and Cunz, A. (1966) Dis Alpersche. Krankheil Scheia. *Schweiz. Arch. Neurol. Neurochir. Psychiat.* **97**, 297–303.

Ulrich, J. and Herschkowitz, N. (1977). Seitelberger's connatal form of

Pelizaeus–Merzbacher disease. Case report, clinical pathological and biochemical findings. *Acta neuropath.* **40**, 129–36.

Ungley, C.C. (1933). Recurrent polyneuritis in pregnancy and the puerperium affecting three members of a family. *J. Neurol. Psychopath.* **14**, 15–26.

Unverricht, H. (1891). *Die Myoclonie.* Deuticke, Berlin.

Unverricht, H. (1895). Uber familiare myoclonie. *Dt. Z. Nervenheilk* **7**, 32–67.

Upadhyaya, M., *et al.* (1985). Hunter's syndrome: evidence supporting a location on the distal part of the X chromosome long arm. (Abst.). *J. Med. Genet.* **22**, 394–5.

Urechia, C.I., Retezeano, A. and Maller, O. (1950). La maladie de Hallervorden–Spatz. Deux cas de rigidite progressive familiale avec un examen anatomique. *Encephale* **39**, 197–219.

Usher, C.H. (1935). Bowman's lecture on a few hereditary eye affections. *Trans. ophthal. Soc. U.K.* **55**, 164–245.

Ushiyama, M. *et al.* (1985). Type III (chronic) GM_1-gangliosidosis. Histochemical and ultrastructural studies of histochemical and ultrastructural studies of rectal biopsy. *J. Neurol. Sci.* **71**, 209–23.

Vaizey, M.J., Sanders, M.D., Wybar, K.C. and Wilson, J. (1977). Neurological abnormalities in congenital amaurosis of Leber. Review of 30 cases. *Arch. Dis. Childh.* **52**, 339–402.

Van Allen, M.W., Frohlich, J.A. and Davis, J.R. (1969). Inherited predisposition to generalized amyloidosis. Clinical and pathological study of a family with neuropathy, nephropathy and peptic ulcer. *Neurol., Minneap.* **19**, 10–25.

Vanasse, M. and Dubowitz, V. (1981). Dominantly inherited peroneal muscular atrophy (hereditory motor and sensory neuropathy type I) in infancy and childhood. *Muscle Nerv.* **4**, 26–30.

Vanasse, M., Bedard, P., and Andermann, F. (1976). Shuddering attacks in children: an early clinical manifestation of essential tremor. *Neurol., Minneap.* **26**, 1027–30.

Van Biervliet, J.P., Bruinvis, L., and Ketting, D. (1977). Hereditary mitochondrial myopathy with lactic acidemia: a De Toni–Fanconi–Debre syndrome and a defective respiratory chain in voluntary striated muscle. *Pediat. Res.* **11**, 1088–93.

Van Biervliet, J.P., and Van Hemel, J.O. (1975). Familial occurrence of the G syndrome. *Clin. Genet.* **7**, 238–44.

Van Bogaert, L. (1929). Cyste cerebelleux associe a la syringo-myelio-bulbie chez une malade dont la soeur presente une syringomyelie cervicale typique. *J. Neurol. Psychiat.* **29**, 146–52.

Van Bogaert, L. (1946). Aspects cliniques et pathalogiques des atrophies pallidales et pallidoluysiennes progressives. *J.Neurol. Neurosurg. Psychiat.* **9**, 125–57.

Van Bogaert, L. (1948). Maladie nerveuses systematisees et problems de l'heredite. *Acta neurol. belg.* **48**, 308–29; 339–79.

Van Bogaert, L. (1950). Pathologie des angiomatoses. *Acta neurol. Belg.* **50**, 525–610.

Van Bogaert, L. (1957). Familial ulcers, mutilating lesions of the extremities and acro-osteolysis. *Br. med. J.* **ii**, 367–71.

Van Bogaert, L. (1961). Sur une idiotie amaurotique tardive a evolution tres prolongee sans troubles visuels et se Presentant, dans deux fratries, comme une heredoataxie de type spastique avec oligophrenie. *J. Génét. hum.* **10**, 1–19.

Van Bogaert, L. and Bertrand, I. (1949). Sur une idiotie familiale avec degenerescene spongieuse du nevraxe-Note preliminaire. *Acta neurol. belg.* **49**. 572–87.

Van Bogaert, L. and Klein, D. (1955). Observations sur l'heredite des idioties amaurotiques et de la spleno-hepatomegalie lipidenne (11 famillies). *J. Génét. hum.* **4**, 23–78.

Van Bogaert, L., Maere, M., and Desmedt, E. (1940). Sur les formes famillales precocos de la maladie d'Alzheimer. *Mschr. Psychiat. Neurol.* **102**, 294.

Van Bogaert, L., Scherer, H., Froehlich, A., and Epstein, E. (1937). Une deuxieme observation de cholesterinose tendineuse symetrique avec symptomes cerebraux. *Ann. Med.* **42**, 69–101.

Van Bogaert, L. and Martin, L. (1974). Optic and cochleovestibular degenerations in the hereditary ataxias. *Brain* **97**, 15–40; 41–48.

Van Bogaert, L. and Moreau, M. (1939). Combinaison de l'amyotrophie de Charcot–Marie–Tooth et de la maladie de Friedreich. *Encephalhé* **34**, 312–20.

Van Bogaert, L. and Radermecker, M. (1954). Scleroses laterales amyotrophiques et paralysies agitantes hereditairs dans une même famille, avec une forme de passage possible entre les deux affections. *Mschr. Psychiat. Neurol.* **127**, 185–203.

Van Broeckhoven, L. *et al.* (1987). Failure of familial Alzheimer's disease to segregate with the A4—amyloid gene in several European families. *Nature* **329**, 153–5.

Van Buchem, F.S.P., Hadders, H.N. and Ubbens, R. (1955). An uncommon familial systemic disease of the skeleton: hyperostosis corticalis generalisata familiaries. *Acta radiol.* **44**, 109–20.

Van Caillie, M., Morin, C.L., Roy, C.C., Geoffroy, G., and McLaughlin, B. (1977). Reyes's syndrome: Relapses and neurological sequelae *Pediatrics, Springfield* **59**, 245–9.

Vance, J.M. *et al.* (1987). Chorea-acanthocytosis: A report of three new families and implications for genetic counselling. *Am. J. Med. Genet.* **28**, 403–10.

Van der Berg, B.J. (1974). Studies on convulsive disorders in young children

IV. Incidence of convulsions among siblings. *Dev. med. Child. Neurol.* **16**, 457-64.

Van der Berg, B.J. and Yerushalmy, J. (1969). Studies on convulsive disorders in young children. I. Incidence of febrile and non-febrile convulsions by age and other factors. *Paediat. Res.* **3**, 298-304.

Van der Bergh, P., Bulcke, J.A., and Dom, R. (1980). Familial muscular cramps with autosomal dominant transmission. *Eur. Neurol.* **19**, 207-12.

Van der Bosch, J. (1959). Microcephaly in the Netherlands: a clinical and genetical study. *Ann. hum. Genet.* **23**, 91-116.

Van der Does de Willebois, A.E.M., Bethlem, J., Meyer, A.E.F.H., and Simons, A.J.R. (1968). Distal myopathy with onset in early infancy. *Neurol. Minneap.* **18**, 383-90.

Van der Kamp, J.J.P., Van Pelt, J.F., Liem, K.O., Giesberts, MA.H., Niepoth, L.T.M., and Staalman, C.R. (1976). Clinical variability in Sanfillipo B disease: a report on six patients in two related sibships. *Clin. Genet.* **10**, 279-84.

Van der Wiel, H.J. (1957). Hereditary congenital facial paralysis. *Acta genet. Statist. med.* **7**, 348. Amsterdam.

Van der Wiel, H.J. (1960). *Inheritance of Glioma. The genetic aspects of cerebral glioma and its relation to status dysraphicus.* Amsterdam.

Van der Zee, S.P.M., *et al.* (1971). Citrullinaemia with a rapidly fatal neonatal outcome. *Arch. Dis. Childh.* **48**, 847.

Van Dyke, D.H., Griggs, R.C., Markesbery, W., and DiMauro, S. (1975*a*). Hereditary carnitine deficiency of muscle. *Neurol. Minneap.* **25**, 154-9.

Van Dyke, D.H., Griggs, R.C., Murphy, M.J., and Goldstein, M.N. (1975*b*). Hereditary myokymia and periodic ataxia. *J. neurol. Sci.* **25**, 109-18.

Van Epps, C. and Kerr, H.D. (1940). Familial lumbosacral syringomyelia. *Radiology,* **35**, 160-73.

Van Erven, P.M.M. *et al.* (1987). Leigh syndrome, a mitochondrial encephalo(myo)pathy. *Clin. Neurol, Neurosurg.* **89-4**, 217-30.

Van Erven, P.M.M., Ruitenbeek, W., and Gabreels, F.J. (1986). Disturbed oxidative metabolism in subacute necrotizing encephalomyelopathy (Leigh's syndrome). *Neurodepiat.* **17**, 28-32.

Van Gehuchten P. (1956). Sur l'origine des cellules globoides dans un cas de sclerose diffuse. *Rev. neurol.* **94**, 253-8.

Van Gent, E.M., Hoogland, R.A., and Jennekens, F.G.I. (1985). Distal amyotrophy of predominantly the upper limbs with pyramidal features in a large kinship. *J. Neurol. Neurosurg. Psychiat.* **48**, 266-9.

Van Heycop ten Ham, M.W., and de Jager, H. (1963). Progressive myoclonus epilepsy with Lafora bodies. Clinical-pathological features. *Epilepsia* **4**, 95-119.

Vanier, T.M. (1960). Dystrophic myotonica in childhood. *Br. med. J.* **ii**, 1284–8.

Van Laere, J. (1966). Paralysie bulbo-pontine chronique progressive familiale avec surdite. Un cas de syndrome de Klippel–Trenaunay dans le meme fratrie. Problemes diagnostiques et genetiques. *Rev. neurol.* **115**, 289–95.

Van Leeuwen, M.A., and Lauwers, H. (1947). A. Partial form of familial myoclonus (the family G). *Brain* **70**, 479–855.

Van Leeuwen, M.A. and Van Bogaert, L. (1942). Sur l'atrophie optique hérédo-familiale compliqueé (Behr), forme de passage de l'atrophie de Leber aux hérédo-ataxies. *Mschr. Psychiat. Neurol.* **105**, 314–50.

Van Leeuwen, M.A. and Van Bogaert, L. (1949). Hereditary ataxia with optic atrophy of the retrobulbar neuritis type and latent pallido-luysian degeneration. *Brain* **72**, 340–63.

Van Munster, E.T.L. *et al.* (1986). The rigid spine syndrome. *J. Neurol. Neurosurg. Psychiat.* **49**, 1292–7.

Van Wijngaarden, G.K., and Bethlem, J. (1973). Benign infantile spinal muscular atrophy. *Brain* **96**, 163–670.

Van Wijngaarden, G.K., Hagen, C.J., Bethelm, J., and Meijer, A.E.F.H. (1968). Myopathy of the quadriceps muscles. *J. neurol. Sci.* **7**, 201–6.

Van Wijngaarden, G.K., Fluery, P., Bethlem, J., and Meijer, A.E.F.H. (1969). Familial 'myotubular' myopathy. *Neurol. Minneap.* **19**, 901–8.

Van Wijngaarden, G.K., Bethlem, J., Dingemans, K.P., Coërs, C., Telerman-Toppet, N., and Gerard, J.M. (1977). Familial loss of cross striations. *J.Neurol.* **216**, 163–72.

Van Wulfften-Patthe, P.M. (1957), The marks on the skin in some phacomatoses. *Folia psychiat, neurol, neurochir. neerl.* **60**, 173–86.

Vaquero, J., Cabezudo, J.M., and Areitio, E. (1981). Nevus planus of the skin associated to jugulo-tympanicum paraganglioma. A new neuro-cutaneous syndrome? *J. Neurol. Neurosurg. Psychiat.* **44**, 740.

Varadi, V., *et al.*. (1987). Prenatal diagnosis of X linked hydrocephalus without aqueductal stenosis. *J. Med. Genet.* **24**, 207–9.

Vasilescu, C., Alexianu, M., and Dan, A. (1984). Neuronal type of Charcot–Marie–Tooth disease with a syndrome of continuous motor unit activity. *J. neurol. Sci.* **63**, 11–25.

Vassella, F., Richterich, R., and Rossi, E. (1965). The diagnostic value of serum creatine kinase in neuromuscular and muscular disease. *Pediatrics, Springfield* **35**, 322–30.

Vassella, F., Emrich, H.M., Kraus-Ruppert, R., Aufdermaur, F., and Tönz, O. (1968). Congenital sensory neuropathy with anhidrosis. *Arch. Dis. Childh.* **43**, 124–30.

Vassella, F., Lütschg, J., and Mumenthaler, M. (1972). Cogan's congenital ocular motor apraxia in two successive generations. *Dev. med. Child. Neurol.* **14**, 788–803.

Vegter Van der Vlis, M., Volkers, W.S., and Went, L.N. (1976). Ages if death of children with Huntington's chorea and of their affected parents. *Am. hum. Genet.* **39**, 329–34.

Ventruto, V., Di Girolamo, R., Festa, B., Romano. A., Sebastio, G., and Sebastio, L. (1976). Family study of inherited syndrome with multiple congenital deformities: symphalangism, carpal and tarsal fusion, brachydactyly, craniosynostosis, strabismus, hip osteochondritis. *J. med. Genet.* **13**,394–8.

Vercryssen, A., Martin, J.J., Ceuterick, C., Jacobs, K., and Swerts, L. (1982). Adult ceroid-lipofuscinosis: diagnostic value of biopsies and neurophysiological investigations. *J. Neurol. Neurosurg. Psychiat.* **45**, 1056–9.

Verhaart, W.J.C. (1930). Over de Ziekte von Rck. *Ned. Tedsch. Geneesk.* **74**, 5586–98.

Verhaart, W.J.C. (1958). Degeneration of the brain stem reticular formation, other parts of the brain stem and the cerebellum. An example of heterogeneous systemic degeneration of the central nervous system. *J. Neuropathol Exp. Neurol.*, **17**, 382–91.

Verhaart, W.J.C. (1938). Symmetrical degeneration of the neostriatum in Chinese infants. *Arch. Dis. Childh.*, **13**, 225–34.

Verhiest, W., Brucher, J.M., Goddeeris, P., Lauweryns, J., and De Geest, H. (1976). Familial centronuclear myopathy associated with 'cardiomyopathy'. *Br. Heart J.* **38**, 504–9.

Verity, M.A. and Montasir, M. (1977). Infantile Gaucher's disease: neuropathology, acid hydrolase activities and negative staining observations. *Neuropaediatrie* **8**, 89–100.

Vernea, J. and Symington, G.R. (1977). The late form of pure familial spastic paraplegia. *Proc. Aust. Ass. Neurol.* **14**, 37–41.

Vianey–Liaud, C., Divry, P., Grecersen, N., and Mathieu, M. (1987). The inborn errors of mitochondrial fatty acid oxidation. *J. Inh. Mehrs Dis.*, **10**, Suppl. 1, 159–198.

Victor, M., Hayes, R., and Adams, R.D. (1962). Oculopharyngeal muscular dystrophy: a familial disease of late life characterized by dysphagia and progressive ptosis of the eyelids. *N. Engl. J. Med.* **267**, 1267–72.

Vidgoff, J., Lovrien, E.W., Beals, R.K., and Buist, N.R.M. (1977). Mannosidoses in three brothers. *Medicine, Baltimore* **56**, 335–48.

Vierling, J.M., Shrager, R., Rumble, W.F., Aamodt, R., Berman, M.D. and Jones, E.A. (1978). Incorporation of radiocopper into ceruloplasmin in normal subjects and patients with primary biliary cirrhosis and Wilson's disease. *Gastroenterology* **74**, 652–60.

Vignaendra, V. and Loh, T.G. (1978). Myoclonus epilepsy in two families—clinical and electrographic studies. *Aust. N.Z. J. Med.* **8**, 52–60.

Vinken, P. J. and Bruyn, G. W. (1972). *Handbook of neurology,* Vol. 14. North Holland, Amsterdam.

Visscher, B., *et al.* (1979). Genetic susceptibility to multiple sclerosis. *Neurol. Mineap.* **29**, 1354–60.

Visser, de M., *et al.* (1986). Riboflavin responsive lipid-storage myopathy and glutaric aciduria type II of early adult onset. *Neurology* **36**, 367–72.

Vogel, F., Haefner, M., and Diebold, K. (1965). Zur Genetik der progressiven Myoklonusepilepsien (Unverricht–Lundborg). *Humangenetik* **I**, 437–75.

Voigtländer, V. and Jung, E. G. (1974). Giant pigmented hairy nevus in two siblings. *Humangenetik* **24**, 79–84.

Voit, T., *et al.* (1986). Hearing loss in facio scapulohumeral dystrophy. *Eur. J. Pediat.* **145**, 280–55.

Volpe, J. J. and Adams, R. D. (1972). Cerebro-hepato-renal syndrome of Zellweger. An inherited disorder of neuronal migration. *Acta neuropath.* **20**, 175–98.

Von Hippel, E. (1904). Uber eine sehr seltene Erkrankung der Netzhant von Graefes. *Arch. Opthal., N. Y.* **59**, 83.

Von Motz, I. P., Bots, G., and Endtz, L. J, (1977). Astrocytoma in three sisters. *Neurol, Minneap.* **27**, 1038–41.

Von Ziegler, E. (1958). Bösartige, familiare frühinfantile Krampfkrankheit teilweise verbunden mit familiärer Balkenaplasie. *Helv. pediat, Acta* **13**, 169–84.

Vos, A. J. M. Joosten, E. M. G. and Gabreels-Festen, A. A. W. M. (1983). An atypical case of infantile globoid cell leukodystrophy. *Neuropediatrie* **14**, 110–12.

Vuia, O. (1977). Congenital spongy degeneration of the brain (Van Bogaert–Bertrand) associated with microencephaly and pontocerebellar atrophy. *Neuropaediatrie* **8**, 73–88.

Vuia, O. (1978). Congenital Pelizaeus–Merzbacher disease (Seitelberger type). Malformation and cystic degeneration of the central nervous system. *Neuropaediatrie* **9**, 172–84.

Vuia, O. and Gutermuth, M. (1973). Encephalopathie spongieuse du type Van Bogaert–Bertrand. *J. Génét hum.* **21**, 287–96.

Waardenberg, P. J. (1934). Ein merkwardige Kombination von angeborenen Missbildungen: Doppelseitiger Hydrophtalmus verbunden mit Akro-kephalasyndaktylie Herzfehler, Pseudo-hermaphroditismus und anderen Abweichungen. *Klin. Mbl. Augenheilk.* **92**, 29.

Wadia, N. H. and Swami, R. K. (1971). A new form of heredo-familial spinocerebellar degeneration with slow eye movements. *Brain* **94**, 358–74.

Wadia, R. S., Wadgaonkar, S. U., Amin, R. B., and Sardesai, H. V. (1976). An unusual family of benign 'X' linked muscular dystrophy with cardiac involvement. *J. med. Genet.* **13**, 352–6.

Wadlington, W. B. (1958). Familial trembling of the chin. *J. Pediat.* **53**, 316–21.

Waggoner, R. W., Löwenberg, K., and Speicher, K. G. (1938). Hereditary cerebellar ataxia. *Arch. Psychiat., Chicago.* **39**, 570–86.

Waggoner, R. W., Lowenberg-Scharenberg, K., and Schilling, M. E. (1942). Agenesis of the white mattter with idiocy. *Am. J. ment. Defic.* **47**, 20–4.

Wagman, A. D., Weiss, E. K., and Riggs, H. E. (1960). Hyperplasia of the skull associated with intra-osseous meningioma in the absence of gross tumor. *J. Neuropath. exp. Neurol.* **19**, 111–15.

Wald, I., Loesch, D., and Wolchnik, D. (1962). Concerning the relationship between dystrophia myotonica and peroneal muscular atrophy. *Psychiat. Neurol., Basel* **143** 392–7.

Waldenström, J. (1956). Studies on the incidence heredity of acute porphyria in Sweden. *Acta genet. Statist. med.* **61**, 122–31.

Waldinger, C. (1964). Pyridoxine deficiency and pyridoxine dependency in infants and children. *Post-grad. med. J.* **35**, 415–22.

Waldmann, T. A. and McIntire, K. R. (1972). Serum alpha-fetoprotein levels in patients with ataxia-telengiectasia. *Lancet* **ii**, 1112–15.

Wallace, B. J., Kaplan, D., Adachi, M., Schneck, L., and Volk, B. W. (1966). Mucopolysaccharidosis type III. Morphlogic and biochemical studies of two siblings with Sanfillipo syndrome. *Arch. Path.* **82**, 462–73.

Wallace, D. C. (1970). A new manifestation of Leber's disease and a new explanation for the agency responsible for its unusual pattern of inheritance. *Brain* **93**, 121–32.

Wallace, D. C. and Parker, N. (1973). Huntington's chorea in Queensland: the most recent story. In *Advances in Neurology*, Vol. 1, pp. 223–36. Raven Press, New York.

Wallace, D. C., Singh, G., Lott, M. T., Hodge, J. A., Schair, T. G., Lezza, A. M. S., Elsas II, L. J., and Nikoskelainen, E. (1988). Mitochondrial DNA mutation associated with Leber's Hereditary Optic Atrophy. *Am. J. Hum. Genet.* **43**, A392.

Wallace, S. J. (1985). Deficiencies with pyruvate dehydrogenase complex: clinical and pathological correlates. *Dev. Med. Child Neurol.* **27**, 257–60.

Waller, J. D., Greenberg, J. H., and Lewis, C. W. (1976). Hereditary haemorrhagic telangiectasia with cerebrovascular malformations. *Arch. Derm.* **112**, 49–52.

Walsh, F. B, (1957). *Clinical neuro-opthalmology.* Williams & Wilkins, Baltimore.

Walsh, F. B. and Hoyt. W. F. (1959). External ophthalmoplegia. *Am. J. Ophthal.* **47**, 28–34.

Walsh, R. J. and Morris, P. (1973). Developmental dyslexia: a genetic survey. *Med. J. Aust.* **2**, 766.

Walshe, J.M. (1967). The physiology of copper in man and its relation to Wilson's disease. *Brain,* **90**, 149–76.

Walshe, M.M., Evans, C.D., and Warin, R.P. (1966). Blue rubber bleb naevus. *Br. med. J.* **ii**, 931–2.

Walton, D.S., Robb, R.M. and Crocker, C.A. (1978). Ocular manifestations of group A. Niemann–Pick disease. *Am. J. Ophthal.* **85**, 174–80.

Walton, J.N. (1955/56). On the inheritance of muscular dystrophy. *Ann. hum. Genet.* **20**, 1–13.

Walton, J.N. (1956). *Subarachnoid haemorrhage.* Livingstone, Edinburgh.

Walton, J.N. and Gardner-Medwin. F. (1974). *Disorders of voluntary muscle,* 2nd edn (ed. J.N. Walton) p. 479. Churchill, London.

Walton, J.N. and Gardner-Medwim, D. (1981). Progressive muscular dystrophy and the myotonic disorders. In *Disorders of Voluntary Muscle,* 4th edition (ed. J.N. Walton), pp. 482–524. Churchill Livingstone.

Walton, J.N., Geschwind, N., and Simpson, J.A. (1956). Benign congenital myopathy with masthenic features. *J. Neurol. Neurosurg, Psychiat.* **19**, 224–31.

Ward, C.D., *et al.* (1983). Parkinson's disease in 65 pairs of twins and in a set of quadruplets. *Neurology* **33**, 815–24.

Ward, F. and Bower, B.D. (1978). A study of certain social aspects of epilepsy in childhood. *Dev. med. Child. Neurol.* **20** Suppl. 39, 1–63.

Warkany, J., Bofinger, M., and Benton, C. (1973). Median cleft face syndrome in half sisters. *Teratology* **8**, 273–85.

Warkany, J. and Weaver, T.S. (1940). Heredofamilial deviations II. Enlarged parietal foramens combined with obesity, hypogenitalism microphthalamus and mental retardation. *Am. J. Dis. Child* **60**, 1147–54.

Warner, J.O. (1978). Juvenile onset metachromatic leucodystrophy: failure of response on a low vitamin A diet. *Arch. Dis. Child.* **50**, 735–7.

Warner, T.G., Mock, A.K. Nyhan, W.L., and O'Brien, J.S. (1984). Alpha-mannosidosis:analysis of urinary oligosaccharides with high performance liquid chromatography and diagnosis of a case with unusually mild presentation. *Clin. Genet.* **25**, 248–55.

Warren, M.C., Lu, A.T., and Ziering, W.H. (1963). Sex-linked hydrocephalus with aqueductal stenosis. *J. Pediat.* **63**, 1104.

Warrie, C.B. and Pillai, T.D. (1967). Familial mysathenia gravis. *Br. med. J.* **iii**, 839–40.

Wassman, E.R., Eldridge, R., Abuzzahab, F.S., and Nee, L. (1978). Gilles de Tourette syndrome: clinical and genetic studies in a midwestern city. *Neurol. Minneap.* **28**, 304–7.

Watanabe, I., McCaman, P., Dyken, R., and Zeman, W. (1969). Absence of cerebral myelin sheaths in a case of presumed Pelizaeus–Merzbacher's disease. *J. Neuropath. Exp. Neurol.* **28**, 243–56.

Watanabe, I., *et al.* (1973). Early lesion of Pelizaeus–Merzbacher's disease Electron microscopic and biochemical study. *J. Neuropath. exp. Neurol.* **32**, 313–33.

Waters, D.D., Nutter, D.O., Hopkins, L.C., and Dorney, E.R. (1975). Cardiac features of an unusual X-linked humeroperoneal neuromuscular disease. *N. Engl. J. Med.* **293**, 101–22.

Waters, W.E. (1974). *The epidemiology of migraine.* Boehringer Ingelheim, Bracknell, Herts.

Waters, W.E. and O'Connor, P.J. (1970). The clinical validation of a headache questionarre. In *Background to migraine,* 3rd Migraine Symposium (ed. A.J. Cochrane) pp. 1–10. Heinemann, London.

Waters, W.E. (1975). Prevelance of migraine. *J. Neurol. Neurosurg. Psychiat.* **38**, 613–16.

Watkins, P.A., Naidu, S., and Moser, H.W. (1987). Adrenoleukodystrophy: Biochemical procedures in diagnosis, prevention and treatment. *J. Inhert. Metab. Dis.* **10**, Suppl. 1, 46–53.

Watson, C. and Davidson, S. (1975). The pattern of inheritance of cerebral light sensitivity. *Electroenceph. clin. Neurophysiol.* **9**, 378–9.

Watson, G.H. (1976). Pulmonary stenosis, cafe-au-lait spots and dull intelligence. *Arch. Dis. Childh.* **42**, 303–7.

Watters, G.V. and Fitch, N. (1973). Familial laryngeal abductor paralysis and psychomotor retardation. *Clin. Genet.* **4**, 429–33.

Watts, R.W.E., Purkiss, P., and Chalmers, R.A. (1979). A new variant form of phenylketonuria. *Q. Jl Med.* **191**, 1403–17.

Watts, R.W.E., Spellacy, E. and Adams, J.H. (1986). Neuropathological and clinical correlation in Hurler disease. *J. Inher. Metal. Dis.* **9**, 261–72.

Waziri, M., Zellweger, H., and Seibert, J. (1976). Parental consanguinity in pycnodysostosis. *Lancet* i, 257.

Weary, P.E. and Bender, A.S. (1967). Chediak–Higashi syndrome with severe cutaneous involvement: occurence in two brothers 14 and 15 years of age. *Arch. intern. Med.* **119**, 381–6.

Weber, F.P. and Greenfield, J.G. (1942). Cerebello-olivary degeneration: an example of heredofamilial incidence. *Brain* **65**, 220–31.

Weber, M., Barroche, G., Vespignani, H., Werner, J.E., and Tridon, P. (1980). Adrenomyeloneuropathise de adulte. *Rev. neurol.* **136**, 131–46.

Wechsler, I.S., and Sapirstein, M.R., and Stein, A. (1944). Primary and symptomatic amyotrophic lateral sclerosis: a clinical study of 81 cases. *Am. J. med. Sci.* **208**, 70–81.

Wee, A.S., Subramony, S.H., and Currier, R.D. (1986). 'Orthostatic tremor' is familial—essential tremor. *Neurology* **36**, 1241–5..

Weech, A.A. (1927). Combined acrocephaly and syndactylism occuring in mother and daughter. *Johns Hopkins Hosp. Bull.* **40**, 73.

Wefring, K.W. and Lamvik, J.O. (1967). Familial progressive poliodystrophy with cirrhosis of the liver. *Acta pediat. scand.* **56**, 295–300.

Wegenke, J.D., *et al.* (1975). Familial Kallmann syndrome with unilateral renal aplasia. *Clin. Genet.* **71**, 368–81.

Wegmuller, E., Ludin, H.P., and Mumenthaler, M. (1979). Paramyotonia congenita. A clinical, electrophysiological and histological study of 12 patients. *J. Neurol.* **220**, 251–7.

Weiner, L.P., Stoll, J., and MaGladery, J.W. (1967). Hereditary olivopontocerebellar atropy with retinal degeneration. *Arch. Neurol., Chicago* **16**, 364–76.

Weinschenk, C. (1962). Die erbliche Lese-Rechtschreibschwäche und ihre socialpsychiatrischen Auswirkungen. *Beih. schweiz. Z. Psychol.* **44**, 1–90.

Weiss, L., Reynolds, W.A., and Szymanowski, R.T. (1976). Frontometaphyseal dysplasia: evidence for dominant inheritance. *Am. J. Dis. Child.* **130**, 259–64.

Weissleder, R., *et al.* (1987). Limb girdle type muscular dystrophy associated with Wolff–Parkinson–White syndrome. *J. Neurol. Neurosurg. Psychiat.* **50**, 500–1.

Weissman, S.L., Khermosh, C., and Adam, A. (1963). Arthrogryposis in an Arab family. In *The genetic of migrant and isolate populations* (ed. Goldschmidt), p. 319. Williams & Wilkins, Baltimore.

Welander L. (1951). Myopathia distalis tarda hereditaria. *Acta med. scand.* Suppl. 265, 1–124.

Weleber, R.G. and Beals, R.K. (1976). Hajdu–Cheney syndrome—report of two cases and review of literature. *J. Pediat.* **88**, 243–9.

Welch, L.K., Appenzeller, O., and Bicknell, J.M. (1972). Peripheral neuropathy with myotonia, sustained muscle contraction and continuous motor unit activity. *Neurology, minneap* **22**, 161–90.

Welford, N.T. (1959). Facial paralysis associated with osteopetrosis (marble bones). *J. Pediat.* **55**, 67–72.

Weller, R.O. (1967). An electron microscopic study of hypertrophic neuropathy of Dejerine and Sottas. *J. Neurol. Neurosurg. Psychiat.* **30**, 111–25.

Wells, C.R. and Jankovic, J. (1986). Familial spastic paraparesis and deafness. A new X-linked neuro-degenerative disorder. *Arch. Neurol* **43**, 943–6.

Welsh, O. (1975). Study of a family with a new progeroid syndrome. *Birth Defects:* Original Article Series **XI**, 25.

Wendt, G.C. and Drohm, D. (1972). Die Huntingtonische Chorea. Eine populations-genetische Studie. *Fortschr. allgem. klin Humangenet.* **IV**.

Wendt. L.V., Hirvasniemi, A., and Similä, S. (1979). Nonketotic hyperglycinemia. A genetic study of 13 Finnish families. *Clin. Genet.* **15**, 411–17.

Wenger, D.A. Barth, G., and Githens J.H. (1977). Nine cases of sphingomyelin lipidosis, a new variant in Spanish American children. *Am. J. Dis. Childh.* **131**, 955–61.

Wenger, D. A., Sattler, M., Mueller, O. T., Myers, G. G., Schneiman, R. S., and Nixon, G. W. (1980). Adult GM_2-gangliosidosis: clinical and biochemical studies on two patients and comparison to other patients called variant or adult GM_1 gangliosidosis. *Clin. Genet.* **17**, 323–34.

Went, L. N., De Vries-De Mol, E. C., and Völker-Deiben, H. J. (1975). A family with apparently sex-linked optic atrophy. *J. med. Genet.* **12**, 94–8.

Werner, W. and Benedikt, O. (1971). Juvenile optic atrophy with dominant inheritance . *Klin. Mbl. Augenheilk.* **159**, 798–803.

Westphal, C. (1877). Eigenthumliche mit einschlafen verbundene anfalle. *Arch. Psychiat. NervKrankh.* **7**, 631–5.

Wetterberg, L. (1967). *A neuropsychiatric and genetical investigation of acute intermittent porphyria.* Scandinavian University Books.

Wheby, M. S. and Miller, H. S. (1960). Idiopathic paroxysmal myoglobinuria. Report of two cases occuring in sisters: review of literature. *Am. J. Med.* **29**, 599–610.

Wheelan, L. (1959). Familial Alzheimer's disease. *Ann. hum. Genet.* **23**, 300–10.

Whelan, D. T., Chang, F L P., and Cocksholt, P. W. (1983). Mucolipidosis II. The clinical. radiological and biochemical features in three cases. *Clin. Genet.* **24**, 90–6.

Wherret, J. R. and Rewcastle, N. B. (1969). Adult neurovisceral lipidosis. *Clin. Res.* **17**, 665.

Whitaker, J. N., Falchuck, Z. M., Engel, W. K., Blaese, R. M., and Strober, W. (1974). Hereditary sensory neuropathy. Association with increased synthesis of IGA. *Arch. Neurol., Chicago* **30**, 359–71.

White, J. C. (1969). Familial periodic nystagmus, vertigo and ataxia. *Arch. Neurol., Chicago* **20**, 276–80.

White, N. R. and Blaw, M. E. (1971). An unusual inheritance pattern for spinal muscular atrophy. *Dev. med. Child. Neurol.* **13**, 621–4.

Whitehouse, D. (1966). Diagnostic value of the cafe-au-lait spot in children. *Arch. Dis. Childh.* **41**, 316–19.

Whitely, A. M., Schwartz, M. S., Sachs, J. A., and Swash, M. (1976). Congenital myasthenia gravis: clinical and HLA studies in two brothers. *J. Neurol. Neurosurg. Psychiat.* **39**, 1145–50.

Whitley, C. B., et al., (1983). Warburg syndrome: Lethal neurodysplasia with autosomal recessive inheritance. *J. Pediat.* **102**, 547–51.

Whitty, C. W. M. (1953). Familial hemiplegic migraine. *J. Neurol. Neurosurg. Psychiat.* **16**, 172–7.

Whitty, C. W. M., Lishman, W. A., and Fitzgibbon, J. P. (1964). Seizure induced by movement: a form of reflex epilepsy. *Lancet*, **i**, 1403–6.

Whyte, M. P. and Debakan, A. S. (1976). Familial cerebellar degeneration with slow eye-movements, mental deterioration and incidental nevus of Ota (oculo-dermal melanocytosis). *Dev. med. Child. Neurol.* **18**, 373–80.

Wichman, A., Buchthal, F., and Pezeshkpour, G. H. (1985). Peripheral neuropathy in abetalipoproteinemia. *Neurol.* **35**, 1279–89.

Wichman, A., Frank, L. M., and Kelly T. E., (1985). Autosomal recessive congenital cerebellar hypoplasia. *Clin. Genet.* **27**, 372–82.

Wieacker, P., *et al.* (1983). Menkes kinky hair disease: a search for closely linked restriction fragment—length polymorphism. *Hum. Genet.* **64**, 139–42.

Wiederholt, W. (1973). Familial brachial neuropathy. Report on two kinships. (Abst.) Presented at the Xth International Congress on Neurology, Barcelona. Excerpta Medica ICS No. 296, p. 255. Amsterdam.

Wienker, T. F., Von Reutern, G. M., and Ropers, H. H. (1979). Progressive myoclonus epilepsy. A variant with probable X-linked inheritance. *Hum. Genet.* **49**, 83–9

Wiesmann, U., Vassela, F., and Herschkowitz, N. (1971). 'I-cell' disease: leakage of lysosomal enzymes into extracellular fluids. *N. Engl. J. Med.* **285**, 1090–1.

Wikstrom, J., *et al.* (1982). Classic amyotrophic lateral sclerosis with dementia. *Arch. Neurol.* **39**, 681–3.

Wild, H. and Behnert, J. (1964). Konkordante syringomyelie mit okzipito-zervikaler Dysplaise bei einerigem Zwillingspaar. *Munck. med. Wschr.* **106**, 1421–8.

Wilde, J., Moss, T., and Thrush, D. (1987). X-linked bulbo-spinal neuronopathy: a family study of three patients. *J. Neurol. Neurosurg. Psychiat* **50**, 279–84.

Will, R. G. and Matthews, W. B. (1984). A retrospective study of Creutzfeldt–Jakob disease. I England and Wales. *J. N. N. P.* **47**, 134–40.

Willems, P. J., *et al.* (1987). X-linked hydrocephalus. *Am. J. Med. Genet.* **27**, 921–8.

Willemse, J. (1986). Benign idiopathic dystonia with onset in the first year of life. *Dev. Med. Child Neurol.* **28**, 355–60.

Williams, A., Eldridge, R., McFarland, H., Houff, S., Krebs, H., and McFarlin, D. (1980). Multiple sclerosis in twins. *Neurol., Minneap.* **30**, 1139–47.

Williams, C. A. and Frias, J. L. (1982). The Angelman ('happy puppet') syndrome. *Am. J. Med. Genet.* **11**, 453–60.

Willner, J., DiMauro, S., Eastwood, A., Hays, A., Roohi, F., and Lovelace, R. (1979). Muscle carnitine deficiency: genetic heterogeneity. *J. neurol. Sci.* **41**, 235–46.

Willvonseder, R., Goldstein, N. P., McCall, J. T., Yoss, R. E., and Tauxe, W. N. (1973). A hereditary disorder with dementia, spastic dysarthria, vertical eye movement paresis, gait disturbance, splenomegaly and abnormal copper metabolism. *Neurol. Minneap.* **23**, 1039–49.

Wilson, J. (1963). Leber's hereditary optic atrophy—some clinical and aetiological considerations. *Brain* **86**, 347–62.

Wilson, J. and Carter, C. O. (1978). Genetics of tuberose sclerosis. *Lancet* i, 340.

Wilson, K. M., Evans, K. A., and Carter, C. O. (1965). Creatine kinase levels in women who carry genes for three types of muscular dystrophy. *Br. med. J.* i, 750-3.

Wilson, S. A. K. (1911-12). Progressive lenticular degeneration: a familial nervous disease associated with cirrhosis of the liver. *Brain* 34, 295.

Wilson, W. G., *et al.* (1983). Agenesis of the corpus callosum in two brothers. *J. Med. Genet.* 20, 416-18.

Winchester, R. J., Ebers, G., Fu, S. M., Espinosa, L., Zabiskie, J., and Kunkel, H. G. (1975). B-cell alloantigen Ag7a in multiple sclerosis. (Letter.) *Lancet* ii, 814.

Winkelman, N. W. (1932). Progressive pallidal degeneration. A new clinicopathologic syndrome. *Arch. Neurol. psychiat., Chicago* 27, 1-21.

Winkler, C. (1923). A case of olivo-pontine cerebellar atrophy and our conceptions of neo- and paleo-cerebellum. *Schweiz. Arch. Neurol. Psychiat.* 13, 684-702.

Winship, I. M. (1985). Sotos syndrome—autosomal dominant inheritance substantiated. *Clin. Genet.* 28, 243-6.

Winsor, E. J. T., Murphy, E. G., Thompson, M. W., and Reed, T. E. (1971). Genetics of childhood spinal muscular atrophy. *J. med. Genet.* 8, 143-8.

Winsor, E. J. T. and Welch, J. P. (1978). Genetic and demographic aspects of Nova Scotia Niemann–Pick disease (type D). *Am. J. hum. Genet.* 30, 530-8.

Winter, R. M., Swallow, D. M., Baraitser, M., and Purkiss, P. (1980). Sialidosis type 2 (acid neuraminidase deficiency) clinical and biochemical features of a further case. *Clin. Genet.* 18, 203-10.

Winters, J. L. and McLaughlin, L. A. (1970). Myotonia congenita. A review of four cases. *J. Bone Jt Surg.* 52A, 1345-50.

Winters, P. R., Harrod, M. J., Molenich-Heetred, S. A., Kirkpatrick, J., and Rosentberg, R. N. (1976). Alpha-l-iduronidase deficiency and possible Hurler–Scheie genetic compound: clinical, pathologic and biochemical findings. *Neurol., Minneap.* 26, 1003-7.

Wisniewski, K., *et al.* (1985). Sanfilippo disease, type A with some features of ceroid lipofuscinosis. *Neuropediatrics* 16, 98-105.

Wit, J. M., *et al.* (1985). Cerebral gigantism (Sotos syndrome). Compiled data of 22 cases. Analysis of clinical features, growth and plasma somatomedin. *Eur. J. Pediat.* 144, 131-40.

Witkop, C. J. and Henry, F. V. (1963). Sjögren–Larsson syndrome and histidinemia: hereditary biochemical disease with defects of speech and oral functions. *J. Speech Heart Disorders.* 28, 109-23.

Wiznia, R. A., Freedman, J. K., Mancini, A. D., and Shields, J. A. (1978). Malignant melanoma of the choroid in neurofibromatosis. *Am. J. Ophthal.* 86, 684-7.

Wohlfart, G. (1942). Zwei Falle von Dystrophia muscularum progressiva mit fibrillären Zuckungen und atypischen Muskelbefund. *D. Z. Nervenheilk.* **153**, 189–204.

Wohlfart, G. and Höök, O. (1951). A chemical analysis of myoclonus epilepsy (Univerricht–Lundborg) myoclonic cerebellar dyssynergy (Hunt) and hepato-lenticular degeneration (Wilson). *Acta psychiat. scand.* **26**, 219–45.

Wohlfart, G., Fex, J. and Eliasson, S. (1955). Hereditary proximal spinal muscular atrophy—a clinical entity simulating progressive muscular dystrophy. *Acta psychiat. scand.* **30**, 395–406.

Wohlwill, F. J., Berstein, J., and Yakovlev, P. I. (1959). Dysmyelinogenic leukodystrophy. Report of a case of a new presumably, familial type of leukodystrophy with megalobarencephaly. *J. Neuropath. exp. Neurol.* **18**, 359–83.

Wolberg–Buchholz, K., *et al.* (1985). Familial lysosomal storage disease with generalized vacuolization and sialic aciduria. Sporadic Salla disease. *Neuropediatrics* **16**, 67–75.

Wolf, H. and Feldman, G. L. (1982). The biotin-dependent carboxylase deficiencies. *Am. J. Hum. Genet.* **34**, 699–716.

Wolf, B., Paulsen, E. P., and Hsia, Y. E. (1979). Asymptomatic propionyl CoA carboxylase deficiency in a 13-year-old girl. *J. Pediat.* **95**, 563–5.

Wolf, B., *et al.* (1985). Clinical findings in four children with biotinidase deficiency detected through a statewide neonatal screening program. *N. Engl. J. Med.* **313**, 16–19.

Wolfe, L. S., Callahan, J., Fawcett, J. S., Andemann, F., and Scriver, C. R. (1970). GM_1 gangliosidosis without chondrodystrophy or visceromegaly. *Neurol. Minneap.* **20**, 23–44.

Wolfe, S. M. and Henkin, R. I. (1970). Absence of taste in type II familial dysautonomia. Unresponsiveness to methacholin despite the presence of taste buds. *J. Pediat.* **77**, 103–8.

Wolfenden W. H., Calvert, A. F., Hirst, E., Evans, W., and McLeod J. G. (1973). Familial amyotrophic lateral sclerosis. *Proc. Aust. neurol. Ass.* **9**, 51–5.

Wolfram, D. J. (1938). Diabetes mellitus and simple optic atrophy among siblings: report of four cases. *Proc. Staff Meet. Mayo Clin.* **13**, 715–18.

Wolpert, J. (1916). Klin. beitrag z progressiven familiaren zere bralen Diplegie. *Z. ges. Neurol. Psychiat.* **34**.

Wong, P. W. K., Justice, P., and Berlaw, S. (1977*a*). Detection of homozygotes and heterozygotes with methlenetetrahydrofolate reductase dificiency. *J. Lab. clin. Med.* **90**, 283–8.

Wong, P. W. K., Justice, P., Hruby, M., Weiss, E. B., and Diamond, E. (1977*b*). Folic acid nonresponsive homocystinuria due to methylenetrahydrofolate reductase deficiency. *Pediatrics. Springfield* **59**, 749–56.

Woo, S. L. C., *et al.* (1984*a*). Prenatal diagnosis of classical phenylketonuria by gene mapping. *J. Am. Med. Ass.* **251**, 1998–2002.

Woo, S.L.C. Lidsky, A., Law, M., and Kao, F.T. (1984*b*). Regional mapping of the human phenylalanine hydroxylase gene and PKU locus to 12q21-qter. (Abst.) *Am. J. Med. Genet.* **36**, 210S.

Woods, B.T. and Schaumburg, H.H. (1972). Nigro-spino-dentatal degeneration with nuclear ophthalmoplegia. *J. neurol. Sci.* **17**, 149–66.

Woodworth, J.A., Becket, R.S., and Netsky, M.G. (1959). A composite of hereditary ataxias. A familial disorder with features of olivo-ponto-cerebellar atrophy, Leber's optic atrophy and Friedreich's ataxia. *Arch. intern. Med.* **104**, 594–606.

Woody, N.C., Hutzler, J., and Dancis, J. (1966). Further studies of hyperlysinemia. *Am. J. Dis. Child.* **112**, 577–80.

Woody, N.C. and Pupene, M.B. (1971). Derivation of pipecolic acid from l-lysine by familial hyperlysinemics. *Pediat. Res.* **5**, 511–13.

Woratz, G. (1964). *Neurak Muskelatrophie mit dominantem X-chromosomalen Erbgang*. Akademic–Verlag. Berlin.

Worsfold, M., Park, D.C., and Pennington, R.J. (1973). Familial 'mitochondrial' myopathy. A myopathy associated with disordered oxidative metabolism in muscle fibres. Part 2. Biochemical findings. *J. neurol. Sci.* **19**, 261–74.

Worster-Drought, C., Greenfield, J.C., and McMenemy, W.H. (1940). A form of familial presenile dementia with spastic paraplysis. *Brain* **63**, 237–54.

Worth, H.M. and Wollin, D.G. (1966). Hyperostosis corticalis generalisata congenita. *J. Can. Ass. Radiol.* **17**, 67–74.

Wulff, O.H. and Trojaborg, W. (1985). Adult metachromatic leukodystrophy: neurophysiologic findings. *Neurology* **35**, 1776–8.

Wyburn Mason, R. (1943). Artcriovcnous ancurysm of mid-brain and retina, facial naevi and mental changes. *Brain* **66**, 163–203.

Wynne-Davies, R. (1975). Congenital vertebral anomalies; etiology and relationship to spina bifida cystica. *J. med. Genet.* **12**, 280–8.

Wynne-Davies, R. and Lloyd-Roberts, G.C. (1976). Arthrogryposis multiplex congenita. Search for prenatal factors in 66 sporadic cases. *Arch. Dis. Childh.* **51**, 618–23.

Yaffe, M.G., *et al.* (1979). An amyotrophic lateral sclerosis-like syndrome with hexosaminidase-A deficiency: a new type of GM(2) gangliosidosis. (Abstract). *Neurology* **29**, 611 only.

Yagishita, S., Itoh, Y., Nakano, T., Oizumi, J., Okuyama, Y., and Aoki, K. (1978). Infantile neuroaxonal dystrophy. Schwann cell inclusions in the peripheral nerve. *Acta neuropath.* **41**, 257–9.

Yakovlev, P.I. and Guthrie, R.H. (1931). Congenital Ectodermoses (neurocutaneous syndromes) in epileptic patients. *Arch. Neurol. Psychiat., Chicago* **26**, 1145–94.

Yakura, H., Wakisaka, A., Fujiimoto, S., and Itakura, K. (1974) Hereditary ataxia and HL-A genotypes. (Letter). N. Engl. J. Med. **291**, 154–5.

Yamaguchi, K., Santa, T., Inoue, K., and Omae, T. (1978). Lipid storage myopathy in Von Gierke's disease. *J. neurol. Sci.* **38**, 195–205.

Yamamoto, K., Ito, K., and Yamaguchi, M. (1966). A family showing smell disturbances and tremor. *Jap. J. hum. Genet.* **11**, 36–8.

Yamamoto, T., *et al.* (1985). Familial Creutzfeldt–Jakob disease in Japan. Three cases in a family with white matter involvement. *J. Neurol. Sci.* **67**, 119–30.

Yamamura, Y., Sobue, I., Ando, K., Iida, M., Yanagi, T., and Kono, C. (1973). Paralysis agitans of early onset with marked diurnal fluctuation of symptoms. *Neurol., Minneap.* **23**, 239–44.

Yamano, T., *et al.* (1985). Ultrastructural study on a severe infatile sialidosis (Galactosidase–Neuraminidase deficiency). *Neuropediatrics* **16**, 109–12.

Yanagisawa, N., Goto, A., and Narabayashi, H. (1972). Familial dystonia musculorum deformans and tremor. *J. neurol. Sci.* **16**, 125–36.

Yates, J.R.W. and Emery, A.E.H. (1985). A population study of adult onset limb-girdle muscular dystrophy. *J. Med. Genet.* **22**, 250–7.

Yates, J.R.W., *et al.* (1986). Emery-Dreifuss muscular dystrophy: localization to Xq27.3-qter confirmed by linkage to the factor viii gene. *J. Med. Genet.* **23**, 587.

Yatziv, S. and Russel, A. (1981). An unusual form of metachromatic leukodystrophy in three siblings. *Clin. Genet.* **19**, 222–7.

Yatziv, S., Erickson, R.P., and Epstein, C.J. (1977). Mild and severe Hunter syndrome (MPS II) within the same sibships. *Clin. Genet.* **11**, 319–26.

Yawger, V.S. (1917). Familial head nystagmus in four generations associated with ocular nystagmus. *J.Am. med. Ass.* **69**, 773.

Yeatman, G.W. (1984). Mental retardation clasped thumb syndrome. *Am. J. Med. Genet.* **17**, 339–44.

Yee, R.D., Wong, E.K., Baloh, R.W., and Honrubia, V. (1976). A study of congenital nystagmus waveforms. *Neurol., Minneap.* **26**, 326–33.

Yerby, M.S., Shaw, C.-M., and Watson, J.M.D. (1986). Progressive dementia and epilepsy in a young adult: unusual intraneuronal inclusions. *Neurology* **36**, 68–71.

Yokochi, M., *et al.* (1984). Juvenile parkinsonism-some clinical, pharmacological and neuropathological aspects. *Adv. Neurol.* **40**, 407–13.

Yokoi, S., Austin, J., and Witmer, F. (1966). Isolation and characterisation of Lafora bodies in two cases of myoclonus epilepsy. *Trans. Am. neurol. Ass.* **91**, 116–19.

Yoshimura, M. (1983). Cortical changes in the parkinsonian brain: a contribution to the delinèation of "diffuse Lewy body disease". *J. Neurol.* **229**, 17–32.

Yoshioka, M., Okuno, T., Honda, Y., and Nakano, Y. (1980). Central nervous system involvement in progressive muscular dystrophy. *Arch. Dis. Childh.* **55**, 584–96.

Yoss, R.E. and Daly, D.D. (1957). Criteria for the diagnosis of the narcoleptic syndrome. *Proc. Staff Meet. Mayo Clin.* **32**, 320-8.

Young, D.F., Eldridge, R., and Gardner, W.I. (1970*a*). Bilateral acoustic neuroma in a large kindred. *J. Am. med. Ass.* **214**, 347-53.

Young, E., Wilson, J., Patrick, A.D., and Crome, L. (1972). Galactocerebrosidase deficiency in globoid cell leucodystrophy of late onset. *Arch. Dis. Child* **47**, 449-50.

Young, G.F., Leon-Barth, C.A., and Green, J.. (1970*b*). Familial hemiplegic migraine, retinal degeneration, deafness and nystagmus. *Arch. Neurol., Chicago* **23**, 201-9.

Young, I.D. (1987). Syndrome of the month: craniofrontonasal dysplasia. *J. Med. Genet.* **24**, 193-6.

Young, I.D. and Harper. P.S. (1980). Hereditary distal spinal muscular atrophy with vocal cord paralysis. *J. Neurol. Neurosurg. Psychiat.* **43**, 413-8.

Young, I.D., *et al.* (1985). Agenesis of the corpus callosum and macrocephaly in siblings. *Clin. Genet.* **28**, 225-30.

Young, I.D. and Madders, D.J. (1987). Unknown syndrome: holoprosencephaly, congenital heart defects and polydactyly. *J. Med. Genet.* **24**, 714-6.

Young, I.D., Moore, J.R., and Tripp, J.H. (1987). Sex-linked recessive congenital ataxia. *J. Neurol. Neurosurg. Psychiat.* **50**, 1230-2.

Young, L.W., Radebaugh, J.F., Rubin, P., Sensenbrenner, J., and Fiorelli, G. (1971). New syndrome manifested by mandibular hypoplasia, acroosteolysis, stiff joints and cutaneous atrophy (mandibuloacral dysplasia) in two unrelated boys. *Birth Defects:* Original Article Series **VIII**, 291-7.

Yuasa, T., *et al.* (1986). Joseph's disease: Clinical and pathological studies in a Japanese family. *Ann. Neurol.* **19**, 152-7.

Yudel, A., Dyck, P.J., and Lambert, E.H. (1965). A kinship with the Roussy–Levy syndrome: a clinical and electrophysiological study. *Arch. Neurol., Chicago* **13**, 432-40.

Yuill, G.M. and Lynch, P.G. (1974). Congenital non-progressive peripheral neuropathy with arthrogryposis multiplex. *J. Neurol. Neurosurg. Psychiat.* **37**, 316-23.

Zackai, E.H., Sly, W.S., and McAlister, W.H. (1972). Microcephaly, mild mental retardation, short stature and skeletal anomalies in siblings. *Am. J. Dis. Child* **124**, 111-19.

Zackai, E.H., Mellman, M.J., Neiderer, B., and Hanson, J.W. (1975). The fetal trimethadione syndrome. *J. Pediat.* **87**, 280-4.

Zahalkova, M., Vrzal., and Kloboukova, E. (1972). Genetical investigations in dyslexia. *J. med. Genet.* **9**, 48-52.

Zalin, A., Darby, A., Vaughan, S., and Raferty, E.B. (1974). Primary neuropathic amyloidosis in three brothers. *Br. med. J.* **i**, 65-6.

Zaremba. J. (1968). Tuberous sclerosis: a clinical and genetical investigation. *J. ment. Defic. Res.* 12, 63–80.

Zaremba. J. (1978). Jadassohn's naevus phakomatosis. 2. A study based on review of thirty-seven cases. *J. ment. Defic. Res.* 22, 103–23.

Zaremba, J., Stepien. M., Jelowicka, M., and Ostrowska, D. (1979). Hereditary neurocutaneous angioma: a new genetic entity? *J. med. Genet.* 16, 443–7.

Zatz, M., Frota-Pessoa, O., Levy, J. A., and Peres, C. A. (1976*a*). Creatine-phosphokinase (CPK) activity in relatives of patients with X-linked muscular dystrophies: a Brazilian study. *J. Genet. hum.* 24, 153–68.

Zatz, M., Penha-Serrano, C., and Otto, P. A. (1976*b*). X-linked recessive type of pure spastic paraplegia in a large pedigree: absence of detectable linkage with Xg. *J. med. Genet.* 13, 217–22.

Zawuski, G. (1960) Zur erblichkeit der Alzhemerschen krankheir. *Arch. Z. ges. Neurol.* 201, 123–32.

Zee, D. S., Lance, M., Cook, J. D., Robinson, D. A., Eng, D., and Engel, W. K. (1976). Slow saccades in spinocerebellar degeneration. *Arch. Neurol., Chicago* 33, 243–51.

Zellweger, H. and Hanhart, E. (1972). The infantile proximal spinal muscular atrophies in Switzerland. *Helv. paediat. Acta* 27, 355–60.

Zellweger, H., Brown, B. I., McCormick, W. F., and Jun-Bi, T. (1965). A mild form of muscular glycogenosis in two brothers with alpha-1,4-glucosidase deficiency. *Ann. Pediat* 205, 413–37.

Zellweger, H., Afifi, A., and McCormick, W. F. (1967). Severe congenital muscular dystrophy. *Am. J. Dis. Child* 114, 591–602.

Zellweger, H., Schneider, H., and Schuldt, D. R. (1969). A new genetic variant of spinal muscular atrophy. *Neurol., Minneap.* 19, 865–9.

Zellweger, H., Simpson, J., McCormick, W. F., and Ionasescu, V. (1972). Spinal muscular atrophy with autosomal dominant inheritance. *Neurol., Minneap.* 22, 957–63.

Zeman, W. (1975). Degenerescence systematiscc optico-cochlea-dentelee. In *Handbook of clinical neurology*, Vol. 21 (ed. P. J. Vinken and G. W. Bruyn) pp. 535–51. Elsevier, Amsterdam.

Zeman, W. (1976). Dystonia: an overview. *Adv. Neurol.* 14, 91–103.

Zeman, W. and Scarpelli, D. G. (1958). The non-specific lesions of Hallcrvorden–Spatz disease. *J. Neuropath. exp. Neurol.* 17, 622–30.

Zeman, W., Scarpelli, D. G., and Jenkins, J. T. (1959). Idiopathic dystonia musculorum deformans. 1. The hereditary pattern. *Am. J. hum. Genet.* 11, 188–202.

Zeman, W., Kaelbling, R., and Pasamanick, B. (1960). Idiopathic dystonia musculorum deformans. II. The formes frustes. *Neurol., Minneap.* 10, 1068–75.

Zeman, W., Demyer, W., and Falls, H. F. (1964). Pelizaeus–Merzbacher disease, a study in nosology. *J. Neuropath, exp. Neurol.* 23, 334–54.

Zeviani, M., Van Dyke, D. H., Servidei, S., *et al.* (1986). Myopathy and fatal cardiopathy due to cytochrome c oxidase deficiency. *Arch. Neural.* **43**, 1198–202.

Ziegler, D.K. and Rogoff, J. (1956). Rare variant of myotonia atrophica—clinical and electro-myographic study of a family. *Brain* **79**, 349–57.

Ziegler, D.K., Schimke, R.N., Kepes, J.J., Rose, D.L., and Klin Kerfuss, G. (1972). Late-onset ataxia, rigidity and peripheral neuropathy. *Arch. Neurol., Chicago* **27**, 52–66.

Ziegler, D.K., Van Speybroech, N.W., and Seitz, E.F. (1974). Myoclonic epilepsia partialis continua and Friedreich ataxia. *Arch. Neurol., Chicago* **31**, 308–11.

Ziegler, D.K., Hassanein, R.S., Harris, D., and Stewart, R. (1975). Headache in a non-clinic twin population. *Headache* **14**, 213–18.

Zietz, S. and Engel. A.G. (1987). Are there two forms of carnitine palmitoyltransferase in muscle? *Neurology* **37**, 1785–90.

Zifkin, B., Andermann, E., Andermann, F., and Kirkham, T. (1980). An autosomal dominant syndrome of hemiplegia, migraine, nystagmus and tremor. *Ann. Neurol.* **8**, 329.

Zilber, N., Korczyn, A.D., Kahana, E., Fried, K., and Alter, M. (1984). Inheritance of idiopathic torsion dystonia among Jews. *J. med. Genet.* **21**, 13–20.

Zintz, R. and Villiger, W. (1967). Elektronenmikroskoplsche Befunde bei 3 Fallen von chronisch progressives okulärer Meskeldystrophis. *Ophtalmologica, Basel* **153**, 439–59.

Ziter, F.A., Wiser, W.C., and Robinson, A. (1977). Three-generation pedigree of a Moebius syndrome variant with chromosome translocation. *Arch. Neurol.* **34**, 437–442.

Zlotogora, J. and Bach, G. (1983). Deficiency of lysosomal hydrolases in apparently healthy individuals. *Am. J. Med. Genet.* **14**, 73–80.

Zlotogora, J. and Bach, G. (1984). Heterozygote detection in Hunter syndrome. *Am. J. Med. Genet.* **17**, 661–5.

Zonana, J., Sotos. J.F., Romshe, C.A., Fisher, D.A., Elders, M.J., and Rimoin, D.L. (1977). Dominant inheritance of cerebral gigantism. *J. Pediat.* **91**, 251–6.

Index